TUBULAR STRUCTURES XII

PROCEEDINGS OF THE 12TH INTERNATIONAL SYMPOSIUM ON TUBULAR STRUCTURES, SHANGHAI, CHINA, 8–10 OCTOBER, 2008

Tubular Structures XII

Editors

Z.Y. Shen, Y.Y. Chen & X.Z. Zhao
Department of Structural Engineering, Tongji University, Shanghai, China

CRC Press
Taylor & Francis Group
Boca Raton London New York Leiden

CRC Press is an imprint of the
Taylor & Francis Group, an **informa** business

A BALKEMA BOOK

CRC Press/Balkema is an imprint of the Taylor & Francis Group, an informa business.

© 2009 Taylor & Francis Group, London, UK.
Except *Re-evaluation of fatigue curves for flush ground girth welds*, by Y.H. Zhang, S.J. Maddox & A. Stacey.

Typeset by Vikatan Publishing Solutions (P) Ltd., Chennai, India.
Printed and bound in Great Britain by Antony Rowe (A CPI-group Company), Chippenham, Wiltshire.

All rights reserved. No part of this publication or the information contained herein may be reproduced, stored in a retrieval system, or transmitted in any form or by any means, electronic, mechanical, by photocopying, recording or otherwise, without written prior permission from the publisher.

Although all care is taken to ensure integrity and the quality of this publication and the information herein, no responsibility is assumed by the publishers nor the author for any damage to the property or persons as a result of operation or use of this publication and/or the information contained herein.

Published by: CRC Press/Balkema
P.O. Box 447, 2300 AK Leiden, The Netherlands
e-mail: Pub.NL@taylorandfrancis.com
www.crcpress.com – www.taylorandfrancis.co.uk – www.balkema.nl

ISBN: 978-0-415-46853-4 (hbk)
ISBN: 978-0-203-88281-8 (ebook)

Table of contents

Preface	XI
Publications of previous symposia on tubular structures	XIII
Organisation	XV
Acknowledgements	XVII

ISTS Kurobane lecture

High strength steels and cast steel nodes for tubular structures-investigations, applications and research results 3
R. Puthli

Plenary session A: Applications & case studies

Design and research of welded thin-wall box components for the National Stadium 21
Z. Fan, X.W. Fan, Y. Peng, Z. Wang & H.L. Sun

Aquatic Center Maria Lenk – Tubular structure roof – Design, construction 37
F.C. D'Alambert

Structures and erection methods of concrete filled steel tubular arch bridges 43
B.C. Chen

Design of steel towers for Skylink Tramway 51
Q. Ye, M. Venuto & D. Murphy

Analysis and design of hollow steel section folded plate—the hall of Petroleum Science and Technology Museum 61
J.M. Ding, H.L. Wu & Z.J. He

Welded tubular structure for the Zadar dome 67
E. Hemerich, I. Šošić, D. Lazarević & M. Anđelić

Parallel session 1: Composite construction

The interaction of steel tube and concrete core in concrete-filled steel tube columns 75
U. Starossek & N. Falah

Design calculations on concrete-filled thin-walled steel tubes subjected to axially local compression 85
L.H. Han, W. Liu & Y.F. Yang

Examination of yield stress limitation for concrete filled steel tube (CFT) column using SM570TMC steel 93
M.J. Lee, Y.S. Oh & E.T. Lee

Behaviour of slender high strength concrete filled tubular columns 103
J.M. Portoles, D.Hernández-Figueirido, M.L. Romero, J.L. Bonet & F.C. Filippou

Stability of concrete filled CFRP-steel tube under axial compression 111
G.S. Sun, Y.H. Zhao & W. Gu

Parallel session 2: Fatigue & fracture

Fatigue modeling for partially overlapped CHS K-joints with surface crack 119
S.P. Chiew, C.K. Lee, S.T. Lie & T.B.N. Nguyen

Simplified micromechanics model to assess constraint effect on brittle fracture at weld defects 127
T. Iwashita, Y. Kurobane & K. Azuma

Advanced numerical modelling of fatigue size effects in welded CHS K-joints 135
L. Borges & A. Nussbaumer

Further experimental study into applicability of FAD approach to beam-column connections with weld defects 145
K. Azuma, Y. Kurobane & T. Iwashita

Approaches for fatigue design of tubular structures 153
F.R. Mashiri, X.L. Zhao & P. Dong

Tests of welded cross-beams under fatigue loading 163
F.R. Mashiri & X.L. Zhao

Parallel session 3: Static strength of joints

Parametric finite element study of branch plate-to-circular hollow section X-connections 173
A.P. Voth & J.A. Packer

Load-carrying capacity study on T- and K-shaped inner-stiffened CHS-RHS joints 183
G.F. Cao & N.L. Yao

Experimental study on overlapped CHS KK-joints with hidden seam unwelded 193
X.Z. Zhao, Y.Y. Chen & G.N. Wang

Behaviour of semi-rigid steelwork connections of I-section beams to tubular columns 201
A.H. Orton

Shear capacity of hollow flange channel beams in simple connections 209
Y. Cao & T. Wilkinson

Parallel session 4: Seismic

Comparative study of stainless steel and carbon steel tubular members subjected to cyclic loading 219
K.H. Nip, L. Gardner & A.Y. Elghazouli

Seismic response of circular hollow section braces with slotted end connections 227
G. Martinez-Saucedo, R. Tremblay & J.A. Packer

Seismic performance of encased CFT column base connections 235
W. Wang, Y.Y. Chen, Y. Wang, Y.J. Xu & X.D. Lv

Tubular steel water tank tower dynamic analysis 243
J. Benčat, M. Cibulka & M. Hrvol

Elasto-plastic behavior of circular column-to-H-shaped-beam connections employing 1000 N/mm²-class super-high-strength steel 249
N. Tanaka, T. Kaneko, H. Takenaka & S. Sasaki

Plenary session B: Specification & code development

New IIW (2008) static design recommendations for hollow section joints 261
X.L. Zhao, J. Wardenier, J.A. Packer & G.J. van der Vegte

Evaluation of the new IIW CHS strength formulae for thick-walled joints 271
X. Qian, Y.S. Choo, G.J. van der Vegte & J. Wardenier

Comparison of the new IIW (2008) CHS joint strength formulae with those of the previous IIW (1989) and the new API (2007) 281
J. Wardenier, G.J. van der Vegte, Y. Makino & P.W. Marshall

Extending existing design rules in EN1993-1-8 (2005) for gapped RHS K-joints for maximum chord slenderness (b_0/t_0) of 35 to 50 and gap size g to as low as $4t_0$ 293
O. Fleischer & R. Puthli

Plenary session C: Design rules & evaluation

Recent development and applications of tubular structures in China 305
Z.Y. Shen, W. Wang & Y.Y. Chen

Evaluation of new CHS strength formulae to design strengths 313
G.J. van der Vegte, J. Wardenier, X.-L. Zhao & J.A. Packer

Structural design rules for elliptical hollow sections 323
L. Gardner & T.M. Chan

Application of new *Eurocode* 3 formulae for beam-columns to Class 3 hollow section members 331
N. Boissonnade, K. Weynand & J.P. Jaspart

Re-evaluation of fatigue curves for flush ground girth welds 341
Y.H. Zhang, S.J. Maddox & A. Stacey

Parallel session 5: Fatigue & fracture

Experimental determination of stress intensity factors on large-scale tubular trusses 351
A. Nussbaumer & L. Borges

Stress intensity factors of surface cracks in welded T-joints between CHS brace and concrete-filled CHS chord 359
M. Gu, L.W. Tong, X.L. Zhao & X.G. Lin

Failure assessment of cracked Circular Hollow Section (CHS) welded joints using BS7910: 2005 367
S.T. Lie & B.F. Zhang

Effect of chord length ratio of tubular joints on stress concentration at welded region 375
Y.B. Shao, S.T. Lie & S.P. Chiew

Welded KK-joints of circular hollow sections in highway bridges 381
U. Kuhlmann & M. Euler

Experimental comparison in hot spot stress between CFCHS and CHS K-joints with gap 389
L.W. Tong, C.Q. Sun, Y.Y. Chen, X.L. Zhao, B. Shen & C.B. Liu

Parallel session 6: Composite construction

Behaviour and design of hollow and concrete filled stainless steel square sections subjected to combined actions *B. Uy & Z. Tao*	399
Test and analysis on double-skin concrete filled tubular columns *J.S. Fan, M.N. Baig & J.G. Nie*	407
Flexural limit load capacity test and analysis for steel and concrete composite beams with tubular up-flanges *C.S. Wang, X.L. Zhai, L. Duan & B.R. Li*	413
Parametric analysis of blind-bolted connections in a moment-resisting composite frame *H. Yao, H.M. Goldsworthy & E.F. Gad*	421
Steel-concrete composite full strength joints with concrete filled tubes: design and test results *O.S. Bursi, F. Ferrario & R. Pucinotti*	431
Development of an effective shearhead system to eliminate punching shear failure between flat slabs and tubular columns *P.Y. Yan, Y.C. Wang & A. Orton*	441

Special session: CIDECT President's Student Awards Research Award Competition

Cast steel connectors for seismic-resistant tubular bracing applications *J.C. de Oliveira*	451
High strength concrete-filled tubular steel columns in fire *O. Bahr*	461
Experimental study on SCFs of welded CHS-to-concrete filled CHS T-joints under axial loading and in-plane bending *W.Z. Shi*	469

Design Award Competition

Modular viewing tower *C. Joost & T.O. Mundle*	479
Design of a single large-span roof structure for Rotterdam Zoo 'Blijdorp' *S.J.C. Kieboom*	487
Design and development of a public leisure center made up of tropical garden and an indoor swimming pool in the City of Valencia [Spain] *E. Fernandez Lacruz*	491

Parallel session 7: Welding & cast steel

Selection of butt welding methods for joints between tubular steel and steel castings under fatigue loading *M. Veselcic, S. Herion & R. Puthli*	499
Welding recommendations for modern tubular steels *M. Liedtke, W. Scheller & J. Krampen*	507
Non-linear finite element simulation of non-local softening for high strength steel material *R.Y. Xiao, F.M. Tong, C.S. Chin & F. Wang*	515

Experimental research and design for regular and irregular cast steel joints in tubular structures *Y.Y. Chen, X.Z. Zhao & L.W. Tong*	521
Non-linear analysis of ultimate loading capacity of cast tubular Y-joints under axial loading *Y.Q. Wang, Y. Jiang, Y.J. Shi & P. Sun*	529

Parallel session 8: Static strength of members/frames

Structural performance of stainless steel oval hollow sections *T.M. Chan & L. Gardner*	535
Study on columns with rounded triangular section in Henan TV tower *M.J. He, F. Liang & R.L. Ma*	545
Numerical modeling and design approach of aluminum alloy tubular columns *J.H. Zhu & B. Young*	553
Experimental research on beams with tubular chords and corrugated steel webs *J. Gao & B.C. Chen*	563
Simulation of CFRP strengthened butt-welded Very High Strength (VHS) circular steel tubes in tension *H. Jiao & X.L. Zhao*	571

Parallel session 9: Static strength of joints

Reanalysis of the moment capacity of CHS joints *G.J. van der Vegte, J. Wardenier, X.D. Qian & Y.S. Choo*	579
Experimental research and parameter analysis on rigidity of unstiffened tubular X-joints *G.Z. Qiu & J.C. Zhao*	589
Numerical investigations on the static behaviour of CHS X-joints made of high strength steels *O. Fleischer, S. Herion & R. Puthli*	597
Local joint flexibility of tubular circular hollow section joints with complete overlap of braces *W.M. Gho*	607
Experimental research on large-scale flange joints of steel transmission poles *H.Z. Deng, Y. Huang & X.H. Jin*	615

Parallel session 10: Fire

Fire resistance design of externally protected concrete filled tubular columns *Y.C. Wang & A.H. Orton*	625
Interaction diagrams for concrete-filled tubular sections under fire *J.B.M. Sousa Jr., R.B. Caldas & R.H. Fakury*	635
Experimental study of the behaviour of joints between steel beams and concrete filled tubular (CFT) columns in fire *J. Ding, Y.C. Wang & A. Orton*	641
A numerical investigation into the temperature distribution of steel hollow dual tubes *M.B. Wong*	651
Thermoelastic behaviour of elastically restrained tubular steel arches *Y.L. Pi & M.A. Bradford*	659
Author index	669

Preface

This Proceedings book is a collection of the papers contributed to the *12th International Symposium on Tubular Structures*, held in Shanghai, China, from October 8 to 10, 2008.

This academic conference series, the International Symposium on Tubular Structures, or ISTS, has a 24 year history dating back to 1984 when the first one was organised in Boston. The 11 previous symposia are described in the "Publications of the previous symposia on tubular structures". The venues of the first 10 conferences moved from North America to Asia (twice), to Europe (on six occasions), and to Australia (once). The cities that hosted these ISTS were in turn: Boston (USA), Tokyo (Japan), Lappeenranta (Finland), Delft (The Netherlands), Nottingham (UK), Melbourne (Australia), Miskolc (Hungary), Singapore (Singapore), Düsseldorf (Germany), and Madrid (Spain). The 11th ISTS was held in Québec City (Canada), which seems to have started a new cycle from a city in Northern America, moving to Asia, then to other continents. It might be a mere coincidence that Shanghai was chosen to host the 12th ISTS after Québec City, but it is a fact that in the fast developing China—of which Shanghai is just a window—the usage of steel structures, including tubular structures in buildings and infrastructure, is increasing tremendously.

The organising work of the 12th ISTS received valuable guidance from the International Institute of Welding (IIW) Subcommission XV-E on Welded Tubular Structures. After the 11th ISTS, the committee gathered together twice to discuss affairs concerning the 12th ISTS, once in Scotland in September, 2007 and again in Singapore in December, 2007. As one of the supporting organizations, the Comité International pour le Développement et l'Étude de la Construction Tubulaire (CIDECT) gave solid support as they have done for previous symposia. CIDECT financially supported the publication of the Proceedings book and the student papers competition. Beside the continuous support of IIW and CIDECT, the symposium has been helped by other academic societies and industrial associations, including National Natural Science Foundation of China (NSFC), China Steel Construction Society (CSCS), the Department of Structural Engineering of Tongji University, and the Japan Iron and Steel Federation (JISF). Design institutions and industrial companies sponsoring this symposium are: Architectural Design & Research Institute of Tongji University, Jinggong Steel Construction Group, Qingdao East Steel Tower Group (ETG), Tianjin Pipe Company Ltd. (TPCO) and ArcelorMittal. It is their generous support that made it possible for this symposium to continue the tradition of allowing presenting authors of papers to attend the symposium without payment of registration fees.

These Proceedings contain 75 peer-reviewed papers. All papers were subjected to full peer technical review by members of the International Programme Committee and other experts in the field of tubular structures. Of these, 68 papers have resulted from an initial "Call for Papers". One paper is the invited Kurobane Lecture, by Professor R.S. Puthli of Germany, selected by the ISTS International Programme Committee. Six student papers are also included; these have been selected by a judging panel established by CIDECT as the finalists of a Research Award Competition and a Design Award Competition. The winners of these two competitions are announced at ISTS12, after the publication of these Proceedings. Sixteen students from 10 countries participated in the competitions, indicating the interest of young researchers and engineers in steel tubular structures.

The Editors would like to express their great gratitude to the reviewers for their hard and excellent reviewing work. These experts have undoubtedly helped the contributed papers to show more clearly and more fully the authors' innovative and meaningful achievements. The Editors also express their appreciation to colleagues at the University of Toronto, who shared with us their valuable experience acquired in organising the 11th ISTS. The Editors' appreciation also goes to the post-graduate students at Tongji University who helped to check the formats of the manuscripts.

Although the symposium series has been organised twice before in Asia, this is the first time in China. The year of 2008 is really special to China. For the Beijing Olympic Games, which is also held in China for the first time in the past 100 years, a number of landmark buildings and facilities—especially grand stadiums, airport terminals, railway stations and hotels—have been erected in the country and put into use this year, illustrating the great progress achieved in modern steel materials and construction techniques. On the other hand, the snow disaster that happened this spring in the south of China, and the Wen-chuan earthquake in May, have reminded people of the serious consequences of structural failure and warned us to pay much attention to the safety of buildings. Safety, economy, durability and sustainability, are the enduring key words for structural engineering, and tubular structures are no exception.

Zu-yan Shen, Yiyi Chen & Xianzhong Zhao
Editors
Tongji University, China
June 2008

Publications of the previous symposia on tubular structures

J.A. Packer & S. Willibald (Eds.) 2006. *Tubular Structures XI*, 11th International Symposium and IIW International Conference on Tubular Structures, Québec, Canada, 2006. London/Leiden/New York: Taylor & Francis (including A.A. Balkema Publishers).

M.A. Jaurrieta, A. Alonso & J.A. Chica (Eds.) 2003. *Tubular Structures X*, 10th International Symposium on Tubular Structures, Madrid, Spain, 2003. Rotterdam: A.A.Balkema Publishers.

R. Puthli & S. Herion (Eds.) 2001. *Tubular Structures IX*, 9th International Symposium on Tubular Structures, Düsseldorf, Germany, 2001. Rotterdam: A.A.Balkema Publishers.

Y.S. Choo & G.J. van der Vegte (Eds.) 1998. *Tubular Structures VIII*, 8th International Symposium on Tubular Structures, Singapore, 1998. Rotterdam: A.A.Balkema Publishers.

J. Farkas & K. Jármai (Eds.) 1996. *Tubular Structures VII*, 7th International Symposium on Tubular Structures, Miskolc, Hungary, 1996. Rotterdam: A.A.Balkema Publishers.

P. Grundy, A. Holgate & B. Wong (Eds.) 1994. *Tubular Structures VI*, 6th International Symposium on Tubular Structures, Melbourne, Australia, 1994. Rotterdam: A.A.Balkema Publishers.

M.G. Coutie & G. Davies (Eds.) 1993. *Tubular Structures V*, 5th International Symposium on Tubular Structures, Nottingham, United Kingdom, 1993. London/Glasgow/New York/Tokyo/Melbourne/Madras: E & FN Spon.

J. Wardenier & E. Panjeh Shahi (Eds.) 1991. *Tubular Structures*, 4th International Symposium on Tubular Structures, Delft,The Netherlands, 1991. Delft: Delft University Press.

E. Niemi & P. Mäkeläinen (Eds.) 1990. *Tubular Structures*, 3rd International Symposium on Tubular Structures, Lappeenranta,Finland, 1989. Essex: Elsevier Science Publishers Ltd.

Y. Kurobane & Y. Makino (Eds.) 1987. *Safety Criteria in Design of Tubular Structures*, 2nd International Symposium on Tubular Structures, Tokyo, Japan, 1986. Tokyo:Architectural Institute of Japan, IIW.

International Institute of Welding 1984. *Welding of Tubular Structures/Soudage des Structures Tubulaires*, 1st International Symposium on Tubular Structures, Boston, USA, 1984. Oxford/New York/Toronto/Sydney/Paris/Frankfurt: Pergamon Press.

Organisation

This volume contains the Proceedings of the **12th International Symposium on Tubular Structures – ISTS12** held in Shanghai, China, from October 8 to 10, 2008. ISTS12 has been organised by the International Institute of Welding (IIW) Subcommission XV-E, Comité International pour le Développement et l'Étude de la Construction Tubulaire (CIDECT) and Tongji University, China.

INTERNATIONAL PROGRAMME COMMITTEE

Prof. J.A. Packer, *Committee Chair*
University of Toronto, Canada

Prof. Y.Y. Chen
Tongji University, Shanghai, China

Prof. Y.S. Choo
National University of Singapore, Singapore

Prof. K. Jarmai
University of Miskolc, Hungary

Prof. Y. Kurobane
Kumamoto University, Japan

Prof. M.M.K. Lee
Southampton University, United Kingdom

Mr. M. Lefranc
Force Technology Norway, Sandvika, Norway

Prof. P.W. Marshall
MHP Systems Engineering, Houston, USA/ National University of Singapore, Singapore

Dr. A.C. Nussbaumer
Ecole Polytechnique Fédérale de Lausanne, Switzerland

Prof. R.S. Puthli
University of Karlsruhe, Germany

Prof. Z.Y. Shen
Tongji University, Shanghai, China

Dr. G.J. van der Vegte
Delft University of Technology, The Netherlands

Prof. J. Wardenier
Delft University of Technology, The Netherlands / National University of Singapore, Singapore

Prof. X.L. Zhao, *Chair of IIW Subcommission XV-E*
Monash University, Melbourne, Australia

LOCAL ORGANISING COMMITTEE

Prof. Z.Y. Shen, *Chair*
Tongji University, Shanghai, China

Prof. Y.Y. Chen, *Co-Chair*
Tongji University, Shanghai, China

Dr. X.Z. Zhao, *Secretary*
Tongji University, Shanghai, China

Prof. J.M. Ding
Tongji University, Shanghai, China

Prof. L.H. Han
Tsinghua University, Beijing, China

Prof. G.Q. Li
Tongji University, Shanghai, China

Prof. L.W. Tong
Tongji University, Shanghai, China

Dr. W. Wang
Tongji University, Shanghai, China

Prof. B. Young
The University of Hong Kong, Hong Kong, China

Prof. Q.L. Zhang
Tongji University, Shanghai, China

Prof. Y. C. Zhang
Harbin Institute of Technology, Harbin, China

Acknowledgements

The Organising Committee wish to express their sincere gratitude for the financial assistance from the following associations: IIW (International Institute of Welding), CIDECT (Comité International pour le Développement et l'Étude de la Construction Tubulaire), NSFC (National Natural Science Foundation of China), CSCS (China Steel Construction Society), JISF (Japan Iron and Steel Federation), as well as all the corporate sponsors: Architectural Design & Research Institute of Tongji University, Jinggong Steel Construction Group, Qingdao East Steel Tower Group (ETG), Tianjin Pipe Company Ltd. (TPCO) and ArcelorMittal.

The technical assistance of the IIW Subcommission XV-E is gratefully acknowledged. We are also thankful to the International Programme Committee as well as the members of the Local Organising Committee. The valuable assistance of ISTS11 Local Organising Committee is highly appreciated. Finally, the editors (who also served as reviewers) wish to acknowledge the kind assistance of the following reviewers:

Y.S. Choo
J.M. Ding
L. Gardner
L.H. Han
S. Herion
J.P. Jaspart
J. Krampen
M. Lefranc
G.Q. Li
P. Marshall
F.R. Mashiri
A. Nussbaumer

J.A. Packer
R. Puthli
P. Schaumann
L.W. Tong
R. Tremblay
B. Uy
G.J. van der Vegte
J. Wardenier
T. Wilkinson
B. Young
Q.L. Zhang
X.L. Zhao

Zu-yan Shen, Yiyi Chen & Xianzhong Zhao
Editors
Tongji University, China
June 2008

ISTS Kurobane lecture

High strength steels and cast steel nodes for tubular structures- investigations, applications and research results

R. Puthli
Research Centre for Steel, Timber and Masonry, University of Karlsruhe, Germany

ABSTRACT: The use of (very) high strength steels has been restricted due to lack of appropriate design standards, in spite of modern steels possessing sufficient toughness and weldability. Structural steels of up to 1100 N/mm^2 yield stress can be produced and the technical potential is not yet fully exhausted. Cast steel nodes have been found to be particularly attractive in Germany and Switzerland, as demonstrated by a number of recently erected imposing roof structures and bridges. The fatigue design of some of the nodes was supported by extensive tests at Karlsruhe University for the German Railways (Deutsche Bahn AG) in the early 1990s. Research is now directed towards guidelines for acceptable standards. This paper will present some results obtained from a number of recent research projects at Karlsruhe University.

1 INTRODUCTION

Rolled weldable fine grain structural steels in Europe are standardized as S275, S355, S420 and S460 and available as normalized with suffix N or NL (EN 10025-3 2005) or thermo-mechanically rolled with suffix M or ML (EN 10025-4 2005). The N and M suffices have an impact strength category at –20 °C and the NL and ML suffices at –50 °C. The normalized and thermo-mechanical steels up to 460 N/mm^2 yield strength are now permitted for use in all Eurocodes and may be considered as mild steel today. Structural hollow sections are covered by two codes, for products that are hot finished (EN 10210-1 2006, EN 10210-2 2006) and cold formed (EN 10219-1 2006, EN 10219-2 2006). They have an additional suffix H for hollow sections at the end of each steel grade. The hot finished hollow sections only cover normalized steels up to 460 N/mm^2, whereas the cold formed hollow sections include both normalized and thermo-mechanical steels, up to 460 N/mm^2.

High strength steels (HSS) and very high strength steels (VHS steels), such as obtained by quenching and tempering, have three grades of toughness with suffices Q, QL and QL1 added to the grade. The impact strength categories are –20 °C, –40 °C and –60 °C respectively. These steels are available as flat products (plate or strip) with steel grades S460, S500, S550 S620, S690, S890 and S960 according to EN 10025-6 (2005). The thermo-mechanically rolled high strength steels according to EN 10149-1 & -2 (1995) and EN 10149-3 (1996) are available in grades S460, S500, S550, S600, S650 and S700. These steels are only partially covered by the design codes and the availability of rolled sections is a problem. Also, the design and fabrication rules are in their infancy. Higher steel grades for plates up to 1100 N/mm^2 and even 1300 N/mm^2 are presently possible, although not yet covered by product standards. It is also not possible to find suitable weld materials above a yield stress of 890 N/mm^2, giving so-called under-matched welds. Kolstein & Dijkstra (2006) have however demonstrated that in some cases this can be compensated by larger weld reinforcement.

The obvious advantage of these HSS and VHS steels is to reduce weight and costs due to fabrication and transport for the same design resistance, although this may not be proportional to the yield strength. The greatest clear advantage is in members subjected to tensile loads. An added advantage, however, is in reduction of weld volume and residual stresses. A new European standard EN 1993-1-12 (2007) has recently being introduced to allow the design codes to include HSS up to 700 N/mm^2, in many cases with conservative provisions, in lieu of adequate research results.

On the basis of the state of the art at present, normal strength steels are considered to be up to 460 N/mm^2, HSS up to 700 N/mm^2 and VHS steels above 700 N/mm^2. Research is ongoing at several European establishments, particularly at the University of Karlsruhe, on HSS and VHS steels, on static and fatigue strength determination of joints and connections using structural hollow sections.

In Germany, S690 QL1 is presently allowed, through a general technical approval from the DIBt (Deutsches Institut für Bautechnik, German Organization for Technical Approvals).

The fatigue recommendations in IIW Document XIII-2151-07/XV-1254-07 (2007) includes all steels up to S960 without differentiating in the fatigue classes, and states that for high strength steels, a higher fatigue class may be used if verified by tests. For improvement methods, steels up to only S900 are considered but differentiates for mild steel (nominal yield stress below 350 N/mm²) and higher grade steels. In IIW Joint Working Group JWG-XIII-XV-195-07 (2007), it is stated that although the IIW fatigue recommendations go up to high strength QT steels, they show no significant differences in comparison to the lower strength steels, and therefore no special activity is envisaged, except that they will monitor ongoing research and standardization work.

Traditionally, offshore structures have been fabricated with normalized steels up to S350 grades, but increasingly in the past two decades, higher strength steels have been used by the desire to save weight and costs. The most important use has been on topsides of jacket structures which, in addition to weight savings, allowed crane barge installation of more complete modules. More recently, jacket members of smaller structures have involved such steels although not in nodal joints, because of potential problems with (corrosion) fatigue and particularly hydrogen embrittlement related to sulphate reducing bacteria. However, S550 to S700 quenched and tempered steel grades have been used in mobile jack-up drilling rigs, which can be dry docked, inspected and repaired at least every 5 years. Recently, three production jack-ups have been erected as permanent installations in the North Sea with these higher grades, but with S350 steels in the lower part of the structures.

The problems in structural design with HSS and VHS steels are acknowledged as the lack of information in welded joints on the deformation capacity and a more specific assessment of fatigue life, which are addressed in this paper.

Another related area presently being further developed and of particular interest for (multi-) planar hollow section tubular joints is individually designed and manufactured cast steel joints. These are appropriately pre-cast in a flowing form and a gradually changing wall thickness, and the sharp corners are carefully rounded to reduce or eliminate stress concentrations. Simultaneously, the welded joint between the cast steel and steel tubulars is removed away from critical areas, additionally allowing a simple welded joint between the two parts in one plane, normal to the member axis.

Steel castings have been used as primary or secondary elements in offshore structures since the 1980s. There were no specific standards for steel castings for primary offshore structural applications and it was customary to match, where practicable, the appropriate steel grade such as specified in EN 10025-6 (2005). Presently, a draft API Specification 2SC (2007) is being formulated for manufacture of structural steel castings.

The excellent through thickness properties of castings have allowed them to be used in lifting attachments, padeyes and spreader bars in the offshore industry. Broughton et al. (1997) describe the use of cast steel nodes for the offshore steel jacket in the Ekofisk field, with 14 cast nodes, weighing up to 78 tons, at the transition from vertical to battered legs. The nodes are also used as lifting points. The work on cast steel nodes for support structures of roofs and bridges in Germany has been inspired by the pioneering offshore work.

High strength steel castings with good combinations of strength and toughness are available, with yield strengths up to 1000 N/mm². Unlike rolled steel products, they cannot derive their strength from processing to finer grains, so are usually alloyed with nickel and chromium to suppress transformation temperatures for toughness and produce low carbon martensite or bainite microstructures.

A modern foundry can guarantee cast steel grades with carbon content below 0.5% with non-alloy, low alloy and high alloy compositions comparable to fine grain steels and are covered by European product standards such as EN 1559-1 (1997), EN 1559-2 (2000) and EN 10293 (2005). The disadvantage with cast steel is the possible formation of blow holes, so that special care is required in manufacturing cast steel products. Therefore, a technically sophisticated casting process with appropriate competence in metallurgy and materials research as well as knowledge of the correct moulding materials are prerequisites for an optimal end product.

In comparison to directly welded joints, the stress distribution for cast steel joints is significantly more favourable, so that improved static and fatigue strength, good access to the weld area and a simplified geometry and maintenance is possible. Research in this area is driven by the increasing demand from awarding authorities, planners and architects for light and aesthetic lattice girders for large building roof structures and bridges using circular hollow sections. In addition to the popularity of conventional directly welded braces on chord faces, the demand by architects and planners for aesthetically and individually fabricated cast steel joints is increasing. Such joints are appealing for structures subjected predominantly to fatigue loading such as bridges and structures where multi-planar joints with many braces are preferably jointed without eccentricity of system lines. Cast steel joints also lend themselves ideally to forming treelike support structures for large roofs and bridges. Bridges built in this way range from slender and gently curved pedestrian bridges and over-bridges for agricultural traffic to heavily loaded highway and motorway bridges and railway bridges.

An important point to consider in the execution of such construction works is in the geometry and design of joints, where fatigue plays a major role. It is assumed that with cast steel joints, full joint strength can be achieved and therefore no static joint design is required. Through advances in automatic cutting and welding of the braces and also improved welding techniques, conventional hollow section joints can be welded accurately and cost effectively. In spite of these significant improvements in joining techniques, the directly welded joints remain a weakness for predominantly dynamic loading for large, heavily loaded members. The notches and grooves inherent in the welded joints lead to high stress concentrations at weld toes and are consequently detrimental to fatigue strength.

Research is therefore ongoing at Karlsruhe University and elsewhere, pertaining to this problem. A number of ongoing projects financed by FOSTA (Forschungsvereinigung Stahlanwendung e.V.), CIDECT and the German Federal Ministry of Education and Research (BMBF), with no final published report, have only the start date of the project mentioned in the references.

2 STATIC STRENGTH OF HIGH STRENGTH TUBULAR JOINTS

2.1 *General*

In the last decade, very little work has been carried out on HSS joints, partly because of the attention being directed, particularly in Europe, to consolidating existing and ongoing research work on normally available grades of steel for the new standards. Noordhoek & Verheul (1998) carried out the only significant detailed experimental work on X-joints with circular hollow sections (CHS) with steel grades up to S690. The effect of yield to ultimate stress ratio on the static strength was investigated. They concluded that existing codes which limited yield stress to 2/3 of ultimate were conservative, and values between 0.7 and 0.8 were more appropriate. This however did not consider comparisons with lower grades when considering a deformation criterion, which has only been generally accepted since then. Other experiments were carried out at the Universities of Delft and Karlsruhe up to steel grades of S690 on rectangular hollow sections (RHS). Early experimental work at Karlsruhe University by Mang (1978) on RHS K-Joints using S690 showed a reduction factor of 0.67 compared to lower grades.

The HSE (2003) research report gives a comprehensive review of high strength steels for offshore use and states that since 1996, several panels have been drafting a new ISO standard for offshore structures, including one for static strength of tubular joints (ISO CD 19902, 2001). Improved design equations have been developed to include steel grades up to S500, with a limiting yield stress ratio of 0.8. For higher grades, the panel allows a limiting value of 0.8 for compression strength for grades between S500 and S800, provided adequate ductility can be demonstrated in the heat affected zone and parent metal. Based upon later examination of available static strength data, it was concluded that for tension loading, this factor should be lowered to 0.5, on account of initial cracking governing over ultimate strength. Jiao & Zhao (2004) have tested VHS thin steel tubes ($t < 3$ mm) of grade S1350, with butt welded end to end tubes and tube to flange-plate fillet welded tubes and have made the same conclusions on the basis of the weakness of the HAZ. The HSE (2003) report concludes that overall, the design of high strength welded joints for static strength is unclear, based upon the very limited data available, so that they are not recognised as offshore fixed platform structural grades. Cautionary statements are made for their use.

Numerical work has been carried out in Delft by Lu et al. (1994), Liu & Wardenier (2001) and Liu & Wardenier (2004), mainly on RHS joints with steel grade S235, S355 and S460, with the proposal for S460 to limit yield stress to 80% of the ultimate stress and additionally use a reduction factor of 0.9 on the static strength formula in EN 1993-1-8 (2005) for chord face plastification, chord punching shear and brace effective width. They found that related to the yield strength, this gave a total relative reduction of 15% in joint strength for S460 with respect to S235 steel grade.

To confirm these findings, independent experimental and numerical work is presently in progress at Munich and Karlsruhe Universities respectively, to include steels up to grade S690 for cold-formed and hot-rolled RHS X-joints. Only the numerical work using nominal dimensions has been carried out so far, as presented in this paper.

2.2 *Mitre joints in CHS up to steel grade S890*

Hitherto, the formulae for RHS mitre joints in EN 1993-1-8 (2005) based on work at Karlsruhe University has also been used for CHS. Work on the design of stiffened and unstiffened mitre joints (L-joints) on CHS has since been presented by Karcher & Puthli (2001a) and the results confirmed by Choo et al. (2001) for unstiffened joints (see Fig. 1). The Karlsruhe work includes grades up to S890, while at Singapore grades up to S460 were used. At Karlsruhe, 38 static tests were performed, 19 for unstiffened and 19 for stiffened joints. On the basis of a parametric study using 67 FE models, a design formula has been determined for $9 < d/t < 100$ for grades between S235 and S890.

For stiffened mitre joints, a reduction of the full plastic moment capacity of the tube is not found to be necessary. All stiffened joints with $d/t \leq 70\varepsilon^2$ reach full plastic moment capacity and have sufficient rotation capacity. The stiffened joints with $70\varepsilon^2 \leq d/t \leq 90\varepsilon^2$ even attain the full plastic moment capacity but do not have enough rotation capacity. The stiffened joints with $90\varepsilon^2 \leq d/t \leq 100$ ($d/t = 100$ was the maximum slenderness considered) that were within the range investigated, reach 80% to 100% of the full plastic moment capacity and have insufficient rotation capacity.

The un-stiffened joints always fail by local buckling of the tube. Particularly under closing moments, insufficient resistance is obtained. The yield stress plays less of a role, since the reduction factor is to be multiplied by ε. A reduction factor based on a deformation criterion of 0.015 radians according to EN 1993-1-1 (2005) has been deduced, based upon statistical evaluation. This criterion is more severe than the 3% ovalization criterion.

The reduction constant to be multiplied with the full plastic moment capacity of a tube is given by:

$$\kappa = \left(\frac{d/t}{20} + 0.77\right)^{-1.2} \cdot \varepsilon \quad (1)$$

where: $\varepsilon = \sqrt{235/f_y}$ (2)

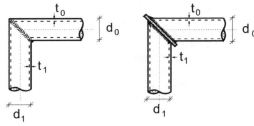

$\beta = d_1 / d_0 = 1.0$ in all cases.

Figure 1. Mitre joints.

2.3 X-joints in CHS up to steel grade S690

The new EN 1993-1-12 (2007) for HSS grades up to 700 N/mm² has introduced a reduction factor of 0.8 in the design load. EN 1993-1-8 (2005) already gives a reduction factor of 0.9 for S460.

In the past two years, two structural projects have been completed in Germany using CHS lattice girders with chords to grade S690. A bridge in Munich has chord members of S690 and an exhibition hall (Neue Messe Stuttgart) with 3000 tons of CHS has S690 steel grades at critical locations, with a general technical approval from the DIBt for the steel grade S690 and approval of the design given for individual cases by the building authorities.

In view of all this, a research programme has commenced at Munich (experimental work) and Karlsruhe (numerical work), see Puthli et al. (2007) for hot rolled and cold-formed CHS X-joints using steel grades up to S690, financed by FOSTA, CIDECT, V & M Tubes and Voest Alpine Krems. The projects are FOSTA Project P715 (2006) and CIDECT Project 5BT (2007).

Preliminary analogous considerations for S460 and S690, in comparison to S235 and S355, have confirmed that a single adaptation of existing formulae using stress related factors alone is not adequate. This is due to the complex structural behaviour based also upon the geometrical parameters of the joint. The final results are expected in 2009.

Another paper in these proceedings by Fleischer et al. (2008) will present progress on this project, so no further details are given here.

3 FATIGUE STRENGTH OF TUBULAR JOINTS IN HSS AND VHS STEELS

The current design codes and regulations implicitly assume that the yield stress of steel has no influence on fatigue strength for non-offshore applications. Some work, however, has been done at Karlsruhe University and elsewhere since 2001 to investigate whether this is so.

Work on offshore structures predates other research on fatigue due to the harsh environmental conditions and stringent requirements offshore. Fatigue design is one of the major requirements in offshore structures, so that it is well established for steels up to S355. The draft ISO standard (ISO CD 19902, 2001) for offshore structures states that the limited test data on plates (up to S540) and tubular joints (up to S700) suggests a similar fatigue performance to lower strength steels, even in seawater under free corrosion and cathodic protection, but that test data should be used to determine appropriate fatigue curves. However, for steel grades of S700 to S800, seawater is considered to decrease fatigue performance due to greater susceptibility to cracking due to hydrogen embrittlement. The DnV Offshore Standard (2000) includes fatigue curves and fracture mechanics constants for all steels up to only S500, although steel grades up to S690 are covered. In effect, most offshore codes limit steel grades up to S500 for fatigue design and propose tests for high strength steels to develop a database.

With modern steels, due to the high steel qualities and low degree of impurities of atoms, fatigue failure due to atomic bond imperfections is reduced. On the other hand, the small grain size leads to voids extending over several grains, with more susceptibility to crack propagations. Therefore, high strength steels can exhibit better fatigue strength, but are more sensitive to notches. Consequently, no significant improvement in fatigue resistance of as-welded high strength steel joints is observed, but the performance increases with weld toe improvements, as discussed later.

3.1 Mitre joints

Karcher & Puthli (2001b) presented design recommendations for fatigue strength of CHS and RHS stiffened and un-stiffened mitre joints based upon 81 fatigue tests, where the influence of welding procedure and of the steel grade was investigated. The welding procedures used were metal-arc and gas shield welding. The test data showed that, especially for RHS mitre joints, a differentiated evaluation for the steel grades is possible. The higher steel grades gave a higher fatigue resistance at lower nominal stress ranges. This tendency for higher steel grades to have a flatter S-N line is also confirmed by other researchers such as Kolstein & Dijkstra (2006).

However, the number of tests was not enough to differentiate the results. In accordance with EN 1993-1-9 (2005), a constant slope of $m = 5$ is used for the fatigue $S-N$ line and the data statistically evaluated for a 95% survival probability. The details and comparison of the latest European code of practice for crane structures, CEN/TS 13001-3-1 (2005) is given in Tables 1–2.

It may be noted that the European crane standard includes wall thicknesses from 2 mm to 8 mm, whereas the work in Karlsruhe was from 4 mm to 8 mm, so that these results cannot yet be directly compared.

3.2 Other joint details

The wide range of high strength steel grades that are available as flat products (plate and strip) is not yet produced as hollow sections. Also, grades above S355 sometimes have long delivery times. This is not surprising, considering the low demand because of the lack of adequate design provisions. Hot-rolled hollow sections can be produced up to steel grades

Table 1. Range of parameters in the fatigue tests.

Parameter	d or b mm	t mm	Grade
CHS	82.5 to 101.6	4.0 to 6.3	S235 to S890
RHS	100.0 to 120.0	4.0 to 8.0	S235 to S460

Table 2. Detail categories in N/mm² for mitre joints at 2×10^6 cycles.

	Karcher & Puthli (2001b) $t = 4$–8 mm	CEN/TS 13001-3-1 (2005) $t = 2$–8 mm
CHS stiffened	90	50
CHS un-stiffened	45	50
RHS stiffened	71	45
RHS un-stiffened	63	45

S890 (CHS) and S690 (RHS). Cold-formed hollow sections are available up to S690.

Research at Karlsruhe University has been concentrated on details and attachments. This is not only because of non-availability of higher structural steel grades for hollow sections, but also because some of the structures being considered are (offshore) wind energy converters, bridges and wide-span roofs, for which large scale testing is limited or impracticable.

Also, since this research is in its infancy, it is considered more sensible to commence with problems of small scale welded details. The work dedicated to HSS and VHS steel research is financed at Karlsruhe by four large research projects, two of which are now completed. These projects are: FOSTA Project P512 (2005), ECSC Project 4553 (2006), FOSTA Project P633 (2003), and the BMBF Project REFRESH (2006) that commenced in 2006 and also forms part of a European EUREKA project. The chemical composition of the steels used for these projects is given in Table 3. It may be noted that the compositions required for offshore steels are different.

The German work has been, or is being carried out, together with Aachen and Braunschweig Universities for the different projects, together with an advisory team and financial support from the crane (mobile and stationary), wind energy and bridge building industries. The European ECSC Project 4553 (2006) involved partners from Germany, Sweden and France. The results for welded details that could be used in further work with hollow sections or large tubulars for fatigue strength determination of HSS and VHS steels are presented below and discussed in relation to factors influencing fatigue design.

Table 3. Chemical composition (percentage by weight) of the high strength steels used.

	C	Si	Mn	P	S	Cr	Ni	Mo	V	Cu	Al	Nb	N	Ti	B
S690	0.14	0.31	1.0	0.008	0.003	0.37	0.05	0.006	0.049	0.01	0.047	0.023	0.004	0.015	0.002
S960	0.16	0.23	1.23	0.008	0.002	0.20	0.06	0.636	0.035	0.01	0.062	0.015	0.003	0.003	0.001
S1100	0.21	0.50	1.40	0.02	0.01	0.80	3.0	0.70	0.08	0.30	0.020	0.04	0.015	0.02	0.005

3.3 Influence of steel and weld quality

The French partners CETIM in the ECSC Project 4553 (2006) presented results of their work on convex double-V butt welded 8 mm plates with normal stress across the weld, where 21 tests on S690 and 21 tests on S960 plates showed an increase in the fatigue strength with increase in the parent metal steel grade. However, this increase is small in comparison to the increase in yield strength and not found in other joint types, particularly where weld quality or welding method is not appropriately controlled. They reported that detail categories were 90 and 110 respectively, for a slope $m = 3$, as opposed to 90 in EN 1993-1-9 (2005).

For obvious economic reasons, structural steel fabricators do not have the stringent quality control that has been specified for crane manufacturers in NPR-CEN/TS 13001-3-1 (2005), which give much higher detail categories by strictly controlling weld quality on the basis of EN ISO 5817 (2006). The background information on NPR-CEN/TS 13001-3-1 (2005) is however not available and various specialists are seeking clarifications on a number of clauses in it, which do not agree with other standards such as EN 1993-1-9 (2005).

In any case, it is possible that if high quality welds are produced, the scatter of results will be narrow enough to show the small improvement in detail categories with increase in yield stress. However, the most dramatic and promising increase in fatigue strength is with post weld treatment, and is worth considering for at least the critical joints in a structure. For instance, for post weld treated butt welded plates with normal stress across the weld, EN 1993-1-9 (2005) allows a detail category of 112 in comparison to 90 for the as welded plates described above.

3.4 Influence of post-weld treatment

Post-weld treatment is increasingly being used in improving fatigue life of new structures and extending that of existing structures, so that it is being researched at present in the BMBF REFRESH Project (2009). Some preliminary work was done in the ECSC Project 4553 (2006), where two methods were investigated, namely burr grinding and TIG dressing (Bergers et al. 2007). Figures 2–3 show the result of these treatments for 80 mm long (8 mm thick, 60 mm high) symmetrical attachments to both sides of a 100 mm wide, 8 mm thick plate.

In the first method, a smooth transition in the weld zone is obtained while removing the burrs. In the second method, the weld is re-melted with tungsten inert gas electrodes without using filler metal. The weld seam is thus toughened and smoothened. On the basis of tests on S690, S960 and S1100 steel grades and evaluating all tests together, detail categories of 63 (for as welded specimens), 91 (for burr ground specimens) and 91 (for TIG dressed specimens) were obtained, giving a 44% improvement in the detail category for both weld treatment methods and not 50% as specified in IIW Document XIII-2151-07/XV-1254-07 (2007).

Burr grinding is less expensive, making it a preferable method. Other tests in Karlsruhe with more diligent and professional grinding have given even better results, illustrating the importance of experienced professionals.

In the BMBF REFRESH Project (2009), high frequency peening methods are considered, such as the Ultrasonic Impact Treatment (UIT) and the HiFIT treatment developed by DYNATEC GmbH. The improvements are due to a combination of a smoother shape and a high compressive residual stress induced at the weld toe. The two methods only differ in the means of activating the pin, the UIT through an ultrasonic magnetostrictive transducer and an oscillating system and HiFIT pneumatically. Initial results on butt welded plates exhibit a twofold improvement in fatigue resistance. Further information is given by Ummenhofer & Weich (2007).

3.5 Size/wall-thickness influence

It is obvious that as long as stability (buckling) is not a governing factor, a smaller wall thickness for HSS and VHS steels results in lower weight and weld volumes, giving savings in production, handling and transport costs. Additionally, there could be a hidden bonus in the fact that lower wall thicknesses can give a higher detail category and thus increase fatigue life.

In the framework of the FOSTA Project P633 (2003), 8 specimens with 30 mm thick double-V (60°) symmetrical butt welded 30 mm plates of steel grade S460 were tested (Puthli et al. 2006a), as

Figure 2. Burr ground weld toes.

Figure 3. TIG dressed weld toes.

Materials	S 460 M, t=30mm
Welding Techniques	Submerged-arc welding
Weld Material	OK Autrod 12.24
Weld Preparation	butt joint with double V Weld
Opposite Side	grinding and dye penetration test
Heat Treatment before	50°C
Working temperature	50°C - 200°C
Powder	OP 121 TT

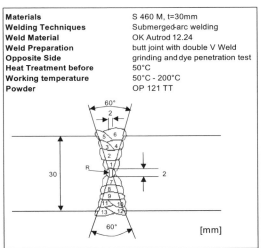

Figure 4. Welding details of 30 mm thick plates (S 460).

can be seen in Figure 4. This gave a detail category of 87 for the 30 mm welded plate. EN 1993-1-9 (2005) requires a thickness correction for plates over 25 mm, so that to obtain the detail category for plate thickness of 25 mm, the size effect exponent of 0.2 is to be used, giving a detail category of 90, which agrees with the detail category of 90 in EN 1993-1-9 (2005) for symmetrically butt welded plates with normal stress across the weld. The European work on 8 mm plates for S690 and S960 by the French partners in the ECSC Project 4553 (2006) also gave a detail category of 90, such that no improvement in fatigue strength below a wall thickness of 25 mm was observed.

The fatigue recommendations in IIW Document XIII-2151-07/XV-1254-07 (2007) specify a size effect exponent of 0.3 for transverse T-joints, plates with transverse attachments and longitudinal stiffeners; and 0.1 for base metal, butt welds ground flush and longitudinal welds or attachments. All other details have a size effect exponent of 0.2 as in EN 1993-1-9 (2005). Also, improvement in fatigue strength for wall thicknesses below 25 mm is allowed only when verified by component tests.

Schumacher et al. (2003) have carried out an investigation on the thickness effect on the basis of their work on the hot spot approach for bridges, where chord wall thicknesses ranged from 20 mm to 60 mm and the d/t ratios were less than 15. They included the existing database for wall thicknesses between 6 mm and 80 mm and concluded that for the hot spot stress method, a large degree of scatter exists due to the combination of many parameters and variables that contribute to this effect, such as the brace to chord thickness ratio. They conclude that none of their results based on statistical and analytical methods justify a higher size effect exponent than 0.25 as originally proposed by Gurney (1981).

Marshall & Wardenier (2005), as well as the work in the IIW Document XIII-2151-07/XV-1254-07 (2007), give a differentiation in the thickness effect for steel plates. A comprehensive survey of tubular versus non-tubular hot spot stress methods are summarized in Table 4 by Marshall & Wardenier (2005) for the proposed AWS hot spot S-N curves for tubular as well as non-tubular applications. On examining test data and notch stress analyses, they conclude that the size effect exponent is dependent on the severity of the notch and varies from 0.1 to 0.4. Curves X1, X, Y and Z are used to reflect differences in local notch severity, which are missed by finite element analyses at the txt level (shell element sizes of the order of plate thickness).

For CHS and RHS joints, stress concentration factors (SCFs) are given in recommendations such as in CIDECT (Zhao et al. 2001), and can for example be used with the X-curve in Table 4.

Table 4. Details of proposed AWS D1.1 hot spot S-N curves (Marshall & Wardenier 2005).

	X1	X	Y	Z
Fatigue class N/mm² at 2×10^6 for $t = 25$ mm	104	100	90	80
Fatigue class N/mm² at 2×10^6 for $t = 16$ mm	114	114	105	96
Knee at N	$N=5 \times 10^6$	$N=10^7$	$N=2 \times 10^7$	$N=5 \times 10^7$
Slope m before and after knee	3/5	3/5	3/5	3/5
Size exponent tubular joints for $t \geq 16$ mm	0.2	0.3	0.3	
Size exponent tubular joints for $t < 16$ mm	$0.06 \log N_f$	$0.06 \log N_f$	$0.07 \log N_f$	
Size exponent for plates with $t \geq 25$ mm		0.3	0.35	0.4
Size exponent for plates with $t < 25$ mm		0*	0*	0*

*The size effect for the plated details with $t < 25$ mm may require further analysis.

Table 5. Results of fatigue tests on attachments to hollow sections.

Length mm	Shape	Steel grade	No. of tests	S_C $P_s = 95\%$	Slope m	S_C for $m = 3$	EC3[1]
80	■	S460	4	75.3	3.1	83.4	71
80	■	S960	3	76.7	2.3	87.1	71
200	▬	S690	9	72.9	2.8	61.5	56
200	▬	S960	9	52.8	2.6	59.1	56
200	▬	S1100	8	73.7	3.6	69.7	56
300	45°	S460	3	71.6	3.1	70.5	71
300	45°	S690	6	54.0	3.0	61.2	71
300	45°	S960	6	54.7	2.5	69.3	71
300	45°	S1100	6	49.9	2.4	61.8	71
300	25°	S690	3	51.6	2.7	58.8	71
300	25°	S960	3			64.4	71
300	25°	S1100	3	59.9	2.2	74.0	71
300	R60	S460	3			59.5	80[2]
300	R60	S690	5	43.7	1.9	72.8	80[2]
300	R60	S960	6	53.0	2.8	67.9	80[2]
300	R60	S1100	6	66.8	2.7	80.9	80[2]
300	▭	S460	5	72.7	3.5	64.1[3]	
300	▭	S690	6	81.8	3.0	85.0[3]	
300	▭	S960	6	66.7	2.5	81.1[3]	
300	▭	S1100	6	59.6	2.6	111[3]	

1) EN 1993-1-9 (2005).
2) In EN 1993-1-9 Table 8.4, $R > 150$ mm; here $R = 60$ mm.
3) Rectangular attachments with welds running inward as in Figure 8. EN 1993-1-9 (2005) does not cover these cases.

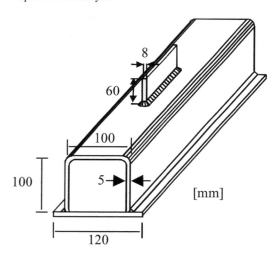

Figure 5. Typical test specimen with longitudinal attachment.

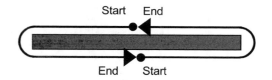

Figure 6. Correct welding sequence.

For plates, the IIW Document XIII-2151-07/XV-1254-07 (2007) specifies two types of welded details. Type (a) is for stress transverse to weld toe on plate surface involving shell bending and membrane stress, where SCF is affected by thickness. Type (b) is for stress transverse to weld toe on plate edge with geometric singularities in the membrane stress (danger of accelerated crack propagation with increase in crack size), where the SCF is not affected by thickness.

The size or thickness effects need to be further investigated, if the full potential of HSS and VHS steels is to be realised. Therefore, future work in Karlsruhe, for instance, will include further investigations into the fatigue curves and size effects that are most suitable for the applications mentioned here. This will not only include the hot spot stress method for tubular joints, but also non-tubular joints. These influences on the nominal stress (classification) method will also be studied for non-tubular and plate to tubular connections.

It seems that obtaining a universally applicable wall thickness correction and fatigue curve is still elusive, since the geometric parameters, extrapolation method used and the method of welding (notch effect) are shown to affect them. The following discussion on longitudinal attachments to a built-up hollow section chord will show how an improved welding detail affects the detail category and therefore implicitly also the wall thickness effect.

Figure 7. Wrong welding sequence.

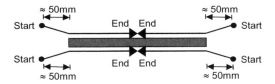

Figure 8. Inward running welds.

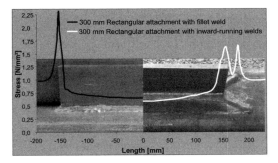

Figure 9. Comparison of stress peaks.

3.6 Influence of welding details

It has been known for some time that the welding sequence can have an influence on fatigue life, so that no starts and stops of welding should occur at fatigue sensitive positions. Also, any method of relieving the stress concentration, such as grinding for instance, increases fatigue life.

In the FOSTA Project P512 (2005), built-up hollow section chord specimens with longitudinal attachments as shown in the typical detail in Figure 5 were subjected to fatigue loading (Puthli et al. 2006b). The correct welding sequence for the attachments as agreed with the industry partners was as in Figure 6.

Different lengths and shapes of attachments were tested to determine their fatigue resistances and are listed in Table 5, which only includes correctly welded details according to Figure 6 for all but the last series, which were welded as in Figure 8. If the welding is performed wrongly, as in Figure 7, a loss in fatigue strengths (detail category) between 10% and 30% to those given in Table 5 is obtained.

The best method of improving the weld detail was the inward running welds starting some distance away from the corners of the attachment and running along the sides and up the middle of the attachment, as in Figure 8.

A finite element analysis on a 300 mm long attachment comparing the stress peaks for the details as in Figures 6, 8 shows that because of the two small peaks obtained (see Fig. 9), a better fatigue design is obtained at little extra cost. The improvement for the higher steel grades is also more evident than for other cases.

4 FRACTURE TOUGHNESS

The ability of weldments to resist brittle fracture is demonstrated by impact toughness (Charpy V-notch absorbed impact energy) tests and fracture toughness (critical stress intensity factor) based upon crack tip opening displacement (CTOD) or wide plate tests. The specimens are extracted from the welding procedure tests, where the CTOD tests are to be in a welding position which gives the highest heat input.

For offshore structures, the Charpy specimen locations are specified in HSE (2002) to be taken at subsurface (≤ 1 mm for $t \leq 20$ mm; ≤ 3 mm for $t > 20$ mm) for heat affected zones and weld metal. When the thickness exceeds 50 mm, or 40 mm in highly stressed regions, CTOD tests are additionally required for as welded joints. If exemption is sought from post weld heat treatment, CTOD tests are required, which should be carried out at a temperature not higher than the specified minimum design temperature. For the North Sea, this is -10 °C in air or the splash zone and 0 °C for submerged parts.

Modern high strength weldable steels make use of ultra-fine grain refinement to produce high yield strength and improved toughness. Grain growth controllers are also used to maintain the properties in the heat affected zones. VHS steels rely more heavily on alloying to give the added strength, but this raises the transition temperatures from ductile to brittle states. Weld metals also require comparable strength and toughness on solidifying, so that they also require addition of alloys. Toughness for VHS steels and comparable weld metals is therefore not as high as for lower grades of steel. Therefore, for steel grades above S690, it is prudent to restrict high weld heat input, and fracture toughness should be based upon adequate levels of toughness for the purpose instead of fixed values.

Most specifications give good guidance for steels below S460, but are limited for higher strengths. A minimum value of Charpy impact strength dependent upon yield strength is generally recommended. For high strength quenched and tempered steels, the

IACS (1994) recommends that the average impact energy should be 1/10 times the specified minimum yield strength (R_e) in the longitudinal direction and 2/3 of this value transversely, to allow for anisotropic influences in the microstructure. For S690 steels, this means an average impact energy of 69 Joules (longitudinally) and 46 Joules (transversely) at a test temperature of −60 °C with a minimum individual value of 70%, which is 48 and 32 Joules respectively.

However, toughness in modern high strength steels has generally been improved by the manufacturers over the years, so that the governing criteria in the future will be buckling and fatigue phenomena. Since moderate improvements in fatigue strength are expected with higher grades of steel, this could still give weight savings and reduced fabrication costs, particularly if fatigue performance can be proven to increase with reduced wall thickness.

The work in Karlsruhe on HSS and VHS steels cover thin walled structures used in the crane industry up to S1000 and the work on offshore wind converters with thicker walls is limited to steel grade S460. However, fracture toughness of HSS and VHS steels will also be considered in future work.

5 CAST STEEL NODES

It is important to consider the feasibility of the use of steel castings at the beginning of a project, with emphasis on satisfactory assurance of integrity and mechanical properties, including fracture toughness, fatigue and weldability. The detailed design of the castings should also be discussed with the manufacturer at an early stage.

Castings should be supplied in the fully heat treated condition. Any repair welding of the castings should undergo qualification tests and the location of weld repairs recorded. All repairs should be completed before final heat treatment. Fracture mechanics analyses could be used for defect acceptance criteria. Non-destructive testing (NDT) procedures depending on size and shape of the component should also consistently identify unacceptable defects during manufacture and after final heat treatment. A prototype casting of a typical detail could be used to demonstrate fitness for purpose. The properties should also comply with the Charpy impact and fracture toughness requirements stipulated for steel members of the same thickness.

The required quality and soundness of the steel in the castings, allowing for discontinuities such as blowholes, cracks and sand inclusions, are to be considered and appropriate requirements drawn up for service stresses and fabrication procedures.

In the pioneering work in the offshore industry, cast steel joints have been allowed since the 1980s for primary and secondary structures subjected to long term fatigue loading, provided accepted standards were used in the fabrication.

The fatigue design curve in HSE (2002) is based upon 173 N/mm² at 2×10^6 cycles and a slope (m) of 4 for a base thickness of 38 mm. The thickness influence is based on an exponent of 0.15. This curve is only allowed for castings that satisfy defect acceptance criteria compatible with current offshore practice on ultrasonic testing of ferritic steel castings including quality levels, such as EN 12680-1 (2003). This curve is at present based upon limited data and HSE (2002) recommends that consideration should be given to design using theoretical fatigue curves derived from fracture mechanics analysis methods. To verify the position of maximum stress range in the casting, a finite element analysis is recommended by HSE (2002) for fatigue sensitive joints.

For cast steel to tubular steel joints, it is noted that the brace to circumferential butt weld could become the critical location for potential fatigue cracking. The assessment of this weld, according to HSE (2002) is to be carried out with what is termed the P-curve and the appropriate SCF. The P-curve is normally to be used for plate to plate and tube to plate weld details, which has a slope m of 3 and ranges from 91 N/mm² (in air) to 63 N/mm² (in seawater with free corrosion) at 2×10^6 cycles.

In Germany, Karlsruhe University was involved in experimental investigations supporting the consultants Schleich, Bergermann & Partners and the Deutsche Bahn AG (German Railways) for the prestigious Humbolt Hafen Bridge in Berlin from 1989 until completion, under the overall supervision of Prof. Friedrich Mang. In the absence of any suitable design codes or adequate product standards, special approval for the structure from the building authorities could only be obtained from large scale testing.

In order to set up acceptable standards and fatigue design methods, the large project, FOSTA Project P591 (2003) was started in 2003 with support from steel mills, foundries, fabricators and railway and highway bridge authorities for use of cast steel nodes in bridges using steel hollow section girders to support the deck structure. The main project partners are EPFL Lausanne (Switzerland), Schlaich, Bergermann & Partners and Karlsruhe University (coordinating partner). The Karlsruhe work is additionally financed by CIDECT Project 7W (2002).

The loading situation on bridges with high shear forces at the nodes, particularly in the support regions, require the use of very stocky hollow sections. The tubular diameters range from 200 mm in the braces to 600 mm and more in the chords, with d/t ratios under 15. Consequently, the parameters lie outside the range of present design recommendations such as EN 1993-1-9 (2005) and CIDECT (Zhao et al. 2001). The work in the offshore industry is also not directly relevant for bridge applications and the fabrication techniques employed.

Figure 10 shows typical examples of cast steel nodes used on bridges erected in Germany.

Another paper in these proceedings by Veselcic et al. (2008) gives details of the main body of the research work in the FOSTA Project P591 (2003), where the influence of different details, lips and backing rings or plates on fatigue performance is studied. Therefore, this paper only addresses material selection and quality control aspects of cast steel for structural application in bridges, for use with the selected steel grade for the hollow sections to be welded to the cast steel nodes.

5.1 Material selection for the casting

A client has a wide choice in the material selection, comparable to rolled steel. The latest developments allow low alloy high strength steels with yield strengths up to 1000 N/mm² that can be welded without (or with very little) preheating. Wall thicknesses up to 300 mm can be cast without problems. EN 1993-1-1 (2005) does not make any specific mention of castings, but refers to the European product standards or the European Technical Approvals (ETAs). Depending on the percentage of alloying elements, a differentiation is made between non-alloyed, low-alloy or high-alloy castings. They all exhibit good weldability with structural steel, an isotropic fine grain and high yield stress in combination with high toughness. To manufacture a satisfactory casting and to avoid any blow holes, special care should be taken in the production.

For the Humbolt Hafen Bridge in Berlin, a superseded German product standard was used, equivalent to Grade G20Mn5 in EN 10293 (2005), which is available in the normalized (N) or quenched and tempered (QT) variety. However, due to small changes in chemical composition in comparison to the withdrawn German standard, the QT variety of this casting only has a yield strength of 300 N/mm² instead of the former 360 N/mm² used in Berlin. As an example, two grades of castings are given in Table 6, which are being used in the FOSTA Project P591 (2003). However, the yield and ultimate stress for the castings used in the project were 360 N/mm² and 500 N/mm² for G20Mn5, in accordance with the recently withdrawn German standard.

To improve the yield stress, one of the project partners, the foundry Friedrich Wilhelms-Hütte (Thyssen Umformtechnik & Guss GmbH) has introduced a technically highly developed casting method using secondary metallurgy in a so-called Vacuum-Argon-Refining-Process (VARP) for low alloyed cast steels. An electric arc induction furnace is used for pre-smelting, where the contents are melted, purified,

Figure 10. Some typical bridges using cast steel nodes.

electrically preheated, poured in a decanting vessel and taken to the converter. The yield and tensile stresses shown in Table 7 are appreciably higher than in EN 10293 (2005), where the G20Mn5 (+N) casting has values of 300 N/mm² and 480 N/mm² respectively for 30 mm wall thickness and the G10MnMoV6-3 has values of 400 N/mm² and 500 N/mm² for 100 mm wall thickness respectively. The higher grade casting will be used in future work in an extension of the present project. These castings have a general technical approval (AbZ No. Z-30.9-18, 2006) for use in Germany, based upon work in Karlsruhe.

5.2 Quality control in manufacture and fabrication of the casting and welded joint

Manufacturing castings is not normally a problem unless large members are being cast, such as for the Humbolt Hafen Bridge, where 13 ton nodes were cast for the arch springings. Careful control of the timing in decanting into the converter and the optimization of the cooling and heat treatment process was required. The solidification alone took 24 hours.

The quality level of steel castings is defined by the product standards EN1559-1 (1997) and EN 1559-2 (2000), where tolerances in chemical composition and allowable casting defects are specified, based upon the client's requirements. It is strongly advisable to employ only a well qualified foundry at the outset.

Methods for destructive tests on samples and non-destructive tests in the casting are given. Depending

Table 6. Chemical composition (percentage by weight) of the castings used in FOSTA Project P591 (2003).

Description	C	Si	Mn	P	S	Cr	Mo	Ni	V	T
G20Mn5	0.17–0.23	0.60	1.0–0.60	≤0.020	0.020 (≤0.030 for $t<28$)	–	–	≤0.80	–	–
G10MnMoV6-3	≤0.12	≤0.60	1.2–1.80	≤0.025	0.020	–	0.20–0.40	–	0.05–0.10	–

Table 7. Guaranteed material properties of castings with the VARP process.

Description	Wall thickness mm	0.2% strain limit N/mm^2	Tensile stress N/mm^2	Max. strain %	Impact strength at room temperature Joules	Impact strength at –40 °C Joules
G20Mn5 + N VARP (FSB 600)	≤100	360	500	22	70	27
G10MnMoV6-3 + QT2 VARP (FSB 900)	≤100	550	640	18	120	27

(*Source*: Friedrich Wilhelms-Hütte).

on the importance of the product and the client's requirements, visual/tactile (EN 1370, 1997), dye penetration (EN 1371-1, 1997) and magnetic powder (EN 1369, 1997) testing is carried out for surface roughness and quality determination. Ultrasonic (EN 12680-1 & -2, 2003) and radiographic testing (EN 12681, 2003) are carried out for the quality deeper within the product. Charpy V impact toughness and Brinell hardness testing are also specified. More care in the quality control is taken in the areas likely to be subjected to tensile stress or fatigue, as well as close to the proposed weld area. These tests are usually carried out by the foundry, but it is also always advisable to employ an independent organization to carry out spot checks.

Any detected faults are to be repaired by gouging and repair welding where possible, or otherwise strengthening the affected parts. In unacceptable cases, a new casting is to be made.

The welds between cast nodes and tubulars should also be ultrasonically tested and be of a quality similar to that for normal bridge construction as given in EN ISO 5817 (2006) and tested by ultrasonic testing according to EN 1712 (2002), which is being revised as a draft in 2007. The quality level of welds is always high and often higher than the casting to be welded. An optimal ultrasonic testing is ensured by grinding the weld seam and using a perpendicular transmitter-receiver probe. Alternatively, for as-welded joints, an angle transmitter-receiver probe could be used. The latter could only be reliably carried out by highly experienced weld inspectors. In any case, in view of the large number of regulations for ultrasonic inspection, the requirements should be set down and agreed at an early stage of the project to avoid confusion and disagreements.

6 CONCLUDING REMARKS

The use of high strength steel up to S700 has been shown to have found application in a number of areas in practice, in spite of inadequate standards. In offshore structures, they have found popularity in topsides, so that larger modules can be lifted by crane barges. Care has to be taken in jacket structures in material selection to avoid the danger of susceptibility to hydrogen embrittlement in seawater from sulphate reducing bacteria, so that HSS and VHS steels are used with caution. VHS steels (above steel grade S700) in particular have found great popularity for thin-walled slender structures in the mobile and fixed crane industry because of the substantial weight reductions offered, that save fabrication and preheating time and transportation costs, and improve performance. Research is therefore ongoing in this area in Karlsruhe and elsewhere on static as well as fatigue problems. Some of the research findings in Karlsruhe are presented in this paper and some solutions for improving joint details are given.

The development of new low alloy and weldable steel castings with high strength and good toughness (transition temperatures not above –40 °C) has allowed their use in offshore structures, roof supporting structures and bridges, without any apparent drawbacks.

There is sufficient experience in manufacture and use, with well documented standards for technical delivery conditions and inspection of the castings and welds. However, good experience and quality standards are required from the casting foundry. The requirements for the welds between castings and steel members are the same as for steel to steel joints. By correct joint preparation and employing a fully qualified welder, the joints can be fabricated safely and economically.

ACKNOWLEDGEMENTS

It is a great honour to have been asked by the International Institute of Welding, Subcommission XV-E to present this 5th "Kurobane Lecture" at the 12th International Symposium on Tubular Structures. The Karlsruhe work presented here reflects the dedicated research carried out by my team of researchers, Dr.-Ing. Stefan Herion, Dipl.-Ing. Oliver Fleischer, Dipl.-Ing. Jennifer Hrabowski/Bergers and Dipl.-Ing. Marian Veselcic, whose efforts are gratefully acknowledged.

Further, the author and his team are grateful for the financial and coordinating support of the Forschungsvereinigung Stahlanwendung e.V. (FOSTA) and Dipl.-Ing. Franz-Josef Heise and Dipl.-Ing. Gregor Nüsse in particular, for many years of support. The financial support of CIDECT and the German Federal Ministry of Education and Research (BMBF) is also much appreciated. Thanks are also due to the many industrial and project partners that unfortunately have to remain anonymous, as there are so many involved. However, special mention must be made to Dipl.-Ing. Jürgen Krampen of Vallourec & Mannesmann Tubes, who has constantly supported our work at Karlsruhe for very many years and also for being a loyal friend.

Lastly, I would like to express my sincere thanks to Prof. Dr. Ir. Jaap Wardenier for the close and fruitful cooperation in research on tubular structures during the 1980s and early 1990s in the Netherlands and for further cooperation and his lasting friendship during my tenure in Germany since then.

REFERENCES

ABZ No. Z-30.9-18. 2006. *Allgemeine bauaufsichtliche Zulassung: Bauteile aus hochfestem Stahlguss FSB 600*. Deutsches Institut für Bautechnik (DIBt).
API Specification 2SC 2007. *Specification for the manufacture of structural steel castings for primary offshore applications. Draft 3.1—5 November 2007*.
Bergers, J., Herion, S. & Puthli, R. 2007. Factors influencing the fatigue design of high-strength fine grained steels. *Proc. 3rd Intern. Conf. on Steel and Composite Structures*, Manchester, UK, 909–914.
BMBF Project REFRESH. 2006. Lebensdauerverlängerung bestehender und neuer geschweißter Stahlkonstruktionen. (Extension of fatigue life of existing and newly welded steel structures). *Final Report*: to be published in 2009.
Broughton, P., Hayes, R., Wood, A. & Komaromy, S. 1997. Cast steel nodes for the Ekofisk 2/4J jacket. *Proc. Institution of Civil Engineers, Structures and Buildings*, Vol. 122, No. 3, 266–280, Telford, London, UK.
CEN/TS 13001-3-1. 2005. *Cranes—General design—Part 3.1: Limit states and proof of competence of steel structures*.
Choo, Y.S., Ren, Y.B., Liew, J.Y.R. & Puthli, R. 2001. Static strength of unstiffened CHS knee joint under compression loading. *Tubular Structures IX, Proc. 9th Intern. Symp. on Tubular Structures*, Düsseldorf, Germany 213–219: Balkema.
CIDECT Project 5BT 2007. Adoption and extension of the valid design formulae for joints made of high strength steels up to S690 for cold-formed and hot-rolled sections. *Final Report*: to be presented in 2009.
CIDECT Project 7W 2002. Fatigue of end-to-end connections. *Final Report*: to be presented in 2008.
DnV Offshore Standard 2000. *Metallic materials*. OS-B101.
ECSC Project 4553. 2006. Efficient lifting equipment with extra high strength steel. *Final Report*: Brussels, Belgium.
EN 1369. 1997. *Founding—Magnetic particle inspection*.
EN 1370. 1997. *Founding—Surface roughness inspection by visual/tactile comparators*.
EN 1371-1. 1997. *Founding—Liquid penetration inspection—Part 1: Sand, gravity die and low pressure die castings*.
EN 1559-1. 1997. *Founding, technical conditions for delivery—Part 1: General*.
EN 1559-2. 2000. *Founding, technical conditions for delivery—Part 2: Additional requirements for steel casting*.
EN 1712. 2002. *Non-destructive testing of welds—Ultrasonic testing of welded joints—Acceptance levels*.
EN 1993-1-1. 2005. *Eurocode 3: Design of steel structures—Part 1.1: General rules and rules for buildings*.
EN 1993-1-8. 2005. *Eurocode 3: Design of steel structures—Part 1.8: Design of joints*.
EN 1993-1-9. 2005. *Eurocode 3: Design of steel structures—Part 1.9: Fatigue*.
EN 1993-1-12. 2007. *Eurocode 3—Design of steel structures—Part 1.12: Additional rules for the extension of EN 1993 up to steel grades S700*.
EN 10025-3. 2005. *Hot rolled products of structural steels—Part 3: Technical delivery conditions for normalized/normalized rolled weldable fine grain structural steels*.
EN 10025-4. 2005. *Hot rolled products of structural steels—Part 4: Technical delivery conditions for thermomechanical rolled weldable fine grain structural steels*.
EN 10025-6. 2005. *Hot rolled products of structural steels—Part 6: Technical delivery conditions for flat products of high yield strength structural steel in the quenched and tempered condition*.
EN 10149-1. 1995. *Specification for hot rolled flat products made of high yield strength steels for cold forming—Part 1: General delivery conditions*.
EN 10149-2. 1995. *Specification for hot rolled flat products made of high yield strength steels for cold forming—*

Part 2: Delivery conditions for thermomechanically rolled steels.

EN 10149-3. 1996. *Specification for hot rolled flat products made of high yield strength steels for cold forming—Part 3: Delivery conditions for normalized or normalized rolled steels.*

EN 10210-1. 2006. *Hot finished structural hollow sections on non-alloy and fine grain steels—Part 1: Technical delivery conditions.*

EN 10210-2. 2006. *Hot finished structural hollow sections on non-alloy and fine grain steels—Part 2: Tolerances dimensions and sectional properties.*

EN 10219-1. 2006. *Cold formed structural hollow sections of non-alloy and fine grain steels—Part 1: Technical delivery conditions.*

EN 10219-2. 2006. *Cold formed structural hollow sections of non-alloy and fine grain steels—Part 2: Tolerances, dimensions and sectional properties.*

EN 10293. 2005. *Steel castings for general engineering use.*

EN 12680-1. 2003. *Founding—Ultrasonic examination—Part 1: Steel castings for general purposes.*

EN 12680-2. 2003. *Founding—Ultrasonic examination—Part 2: Steel castings for highly stressed components.*

EN 12681. 2003. *Founding—Radiographic examination.*

EN ISO 5817. 2006. *Welding—Fusion-weld joints in steel, nickel, titanium and their alloys (beam welding excluded)—Quality levels for imperfections (ISO 5817:2003 + Cor. 1:2006).*

Fleischer, O., Herion, S. & Puthli, R. 2008. Numerical investigations on the static behaviour of CHS X-joints made of high strength steels. *Tubular Structures XII, Proc. 12th Intern. Symp. on Tubular Structures*, Shanghai, China: Balkema.

FOSTA Project P512. 2005. Beurteilung des Ermüdungsverhaltens von Krankonstruktionen bei Einsatz hoch- und ultrahochfester Stähle. (Evaluation of fatigue behaviour of crane structures using high strength and very high strength steels). *Forschungsvereinigung Stahlanwendung e.V., Final Report*: Düsseldorf, Germany (in German).

FOSTA Project P591. 2003. Wirtschaftliches Bauen von Straßen- und Eisenbahnbrücken aus Stahlhohlprofilen. (Economical construction of highway and railway bridges using structural hollow sections). *Forschungsvereinigung Stahlanwendung e.V., Final Report*: to be published in 2008, Düsseldorf, Germany (in German).

FOSTA Project P633. 2003. Detaillösungen bei Ermüdungsfragen und dem Einsatz hochfester Stähle bei Türmen von Offshore-Windenergieanlagen. (Solution of details under fatigue considerations and the use of high strength steel for towers of offshore wind energy converters). *Forschungsvereinigung Stahlanwendung e.V., Final Report*: to be published in 2008, Düsseldorf, Germany (in German).

FOSTA Project P715. 2006. Überprüfung und Korrektur der Abminderungsbeiwerte für Hohlprofilverbindungen aus Werkstoffen mit Streckgrenzen zwischen 355 N/mm^2 und 690 N/mm^2. (Control and correction of reduction factors for structural hollow sections with yield stresses between 355 N/mm^2 and 690 N/mm^2). *Forschungsvereinigung Stahlanwendung e.V., Final Report*: to be published in 2009, Düsseldorf, Germany (in German).

Gurney, T.R. 1981. Some comments on fatigue design rules for offshore structures. *Proc. 2nd Intern. Symp. on Integrity of Offshore Structures*: 219–234: Applied Sciences Publishers, Barking, Essex, UK.

HSE 2002. Steel. *Health and Safety Executive, Offshore Technology Report 2001/015 edited by BOMEL Ltd.,* HSE Books, HMSO, UK.

HSE 2003. Review of the performance of high strength steels used offshore. *Health and Safety Executive, Research Report 105 prepared by Cranfield University*, HSE Books, HMSO, UK.

IACS 1994. Unified requirements, Section W16, high strength quenched and tempered steels for welded structures. *Intern. Assoc. Classification Societies*, London, UK.

IIW Document XIII-2151-07/XV-1254-07. 2007. Recommendations for fatigue design of welded joints and components. *Intern. Institute of Welding Joint Working Group XIII–XV.*

IIW Joint Working Group JWG-XIII-XV-195-07. 2007. *Minutes of the meeting in Cavtat, Croatia, 2007-07-01.*

ISO CD 19902. 2001. *Petroleum and natural gas industries. Fixed steel offshore structures.* International Standards Organization, Geneva: as proposed by ISO TC67/SC7/WG3.

Jiao, H. & Zhao, X.-L. 2004. Tension capacity of very high strength (VHS) circular steel tubes after welding. *Advances in Structural Engineering.* Vol. 7, No. 4, 85–96.

Karcher, D. & Puthli, R. 2001a. The static design of stiffened and unstiffened CHS L-joints. *Tubular Structures IX, Proc. 9th Intern. Symp. on Tubular Structures*, Düsseldorf, Germany, 221–228: Balkema.

Karcher, D. & Puthli, R. 2001b. Design recommendations for stiffened and unstiffened L-joints made of CHS and RHS under fatigue loading. *Proc. 11th Intern. Offshore and Polar Engineering Conf., Vol. IV*: Stavanger, Norway, 51–58.

Kolstein, M.H. & Dijkstra, O.D. 2006. Tests on welded connections made of high strength steel up to 1100 MPa. *Tubular Structures XI, Proc.11th Intern. Symp. on Tubular Structures*, Quebec City, Canada, 567–574: Balkema.

Liu, D.K. & Wardenier, J. 2001. Multiplanar influence on the strength of RHS multiplanar gap KK-joints. *Tubular Structures IX, Proc. 9th Intern. Symp. on Tubular Structures*, Düsseldorf, Germany, 203–212: Balkema.

Liu, D.K. & Wardenier, J. 2004. Effect of the yield strength on the static strength of uniplanar K-joints in RHS (steel grades S460, S355 and S235*). IIW Document XV-E-04-293.*

Lu, L.H., Winkel, G.D. de, Yu, Y. & Wardenier, J. 1994. Deformation limit for the ultimate strength of hollow section joints. *Tubular Structures VI, Proc. 6th Intern. Symp. on Tubular Structures*, Melbourne, Australia, 341–347: Balkema.

Mang, F. 1978. Untersuchungen an Verbindungen von geschlossenen und offenen Profilen aus hochfesten Stählen. AIF—Nr. 3347: Karlsruhe University.

Marshall, P.W. & Wardenier, J. 2005. Tubular versus non-tubular hot spot stress methods. *Proc. 15th Intern. Offshore and Polar Engineering Conf., Vol. IV*: Seoul, Korea, 254–263.

Noordhoek, C. & Verheul, A. 1998. Static strength of high strength steel tubular joints, *CIDECT Report No. 5 BD-9/98.* Delft University of Technology, Delft, The Netherlands.

Puthli, R., Herion, S. & Bergers, J. 2006a. Influence of wall thickness and steel grade on the fatigue strength of towers of offshore wind energy converters. *Proc. 16th Intern. Offshore and Polar Engineering Conf., Vol. IV*: San Francisco, USA, 145–151.

Puthli, R., Herion, S. & Bergers, J. 2006b. Influence of longitudinal attachments on the fatigue behaviour of high strength steels. *Proc. 16th Intern. Offshore and Polar Engineering Conf., Vol. IV*: San Francisco, USA, 120–126.

Puthli, R., Fleischer, O. & Herion, S. 2007. Adaption end extension of the valid design formulae for joints made of high-strength steels up to S690 for cold-formed and hot-rolled sections. *CIDECT Report No. 5 BS-8/07*: Karlsruhe University, Karlsruhe, Germany.

Schumacher, A., Nussbaumer, A. & Hirt, M.A. 2003. Fatigue behaviour of welded CHS bridge joints: emphasis on the effect of size. *Tubular Structures X, Proc. 10th Intern. Symp. on Tubular Structures*: Madrid, Spain, 365–374: Balkema.

Ummenhofer, T. & Weich, I. 2007. Concepts for fatigue design of welds improved by high frequency peening methods. *Proc. 17th Intern. Offshore and Polar Engineering Conf., Vol. I*: Lisbon, Portugal, 356–362.

Veselcic, M., Herion, S. & Puthli, R. 2008. Technical and economical design considerations for end-to-end connections under fatigue loading. *Tubular Structures XII, Proc. 12th Intern. Symp. on Tubular Structures*: Shanghai, China: Balkema.

Zhao, X-L., Herion, S., Packer, J.A., Puthli, R., Sedlacek, G., Wardenier, J., Weynand, K., Wingerde, A.M. van & Yeomans, N.F. 2001. *Design guide for CHS and RHS welded joints under fatigue loading*. CIDECT (ed.), Verlag TÜV Rheinland GmbH, Cologne, Germany.

Plenary session A: Applications & case studies

Design and research of welded thin-wall box components for the National Stadium

Z. Fan, X.W. Fan, Y. Peng, Z. Wang & H.L. Sun
China Architecture Design and Research Group, Beijing, China

ABSTRACT: Since width-thickness ratio of steel plates for welded thin-wall box component is quite large, local buckling on the compressed plates occurs easily. The conception of effective width can be used in calculating the bearing capacity of thin-wall element. Referring to the design codes of steel structures in China and other countries, a new method to estimate effective width for welded thin-wall box section is presented in this paper based on the stress distribution of thin-wall box section. The favorable effect of low compression stress on the effective width of plates and the shift of centroid between effective section and box section are taken into account. The bearing capacity formulas for thin-wall box sections under tension, compression, bi-axial bending, biaxial shear and torsion are provided. The FEM is used to analyze the warp stresses of the thin-wall box section under free torque and restraint torque conditions respectively. Various wall-thicknesses, including the flange thickness not equal to the web, and different height to width ratio of the sections are taken into consideration. The transverse stiffeners with square hole are introduced within box sections and the effect on the warping stresses is studied. Due to the complicated behavior of the twisted member and many influencing factors, and based on which the method of placing the transverse stiffeners is proposed to improve the stiffness of the members and stress distribution, effectively restrain the out-of-plane deformations of the arc walls.

1 INTRODUCTION

Located in the south of Chengfu Road, Beijing and within the central district of the Olympic Green, the National Stadium is the main venue for the 29th 2008 Beijing Olympiad, and will host the Games' opening and closing ceremony as well as track and field events. The National Stadium has a saddle-shaped top surface, long axle of 332.3 m and short axle of 296.4 m. The height of the roof is 40.1~68.5 m. With a fixed seating capacity of 80,000 people and temporary seating capacity of 11,000 people, the National Stadium covers a total floor area of 2580,000 m². The design working life of the building is 100 years; the seismic fortification intensity is 8 and the Earthquake Protection Category is B.

The long-span roof of the National Stadium is supported by 24 truss columns which are spaced out 37.958 m apart. The length of roof opening is 186.7 m, and its width is 127.5 m. The reinforced concrete frame-shear wall structural system is adopted for the bowl structure, completely separated from the large span steel structure.

The geometric modeling of roof structure of the National Stadium is rather complicated. The architecture design and the structure system are in total conformity. The members on the facade and elbow area are twisted box members, warping and bending with the variance of the roof surface curvature so that smooth outlines of the twisted box members are guaranteed. The primary structure which is composed of the trusses-columns and the main truss together with the secondary structure of the façade and the roof constitutes the unique architecture design of the National Stadium—Bird's Nest. The steel roof of the national stadium consists of welded thin-walled box section members, which is unique in the world. Beijing National Stadium, upon its completion, will become a landmark building in Beijing.

2 RESEARCH ON THE DESIGN METHOD OF THIN-WALL BOX MEMBER UNDER COMPLICATED FORCES

2.1 *Effective section of thin-wall box members*

To reduce the weight on steel, smaller wall thickness plates should be adopted in the design of box section. If width-thickness ratio of steel plates for welded thin-wall box members is quite large, local buckling on the compressed plates occur easily.

In the designing the effective width can be firstly determined by its post-buckling strength, the bearing capacity of the member can be calculated. However, there is a great difference in the stipulation of effective width in different national codes. Referring to the design codes of steel structures in China and foreign countries, there are mainly two problems in the design of welded thin-wall box members. One is to provide the methods for solving the effective width of box sections when the member is under axial compression, pure bending, compression bending without considering the possible local buckling of the compressed flange of the member under tension bending. When the width-thickness ratio is very large, local buckling may happen to the compressed zone of the web of tension flexural members and pure bending members. However, method of calculating the corresponding effective width is absent in the Code for Design of Steel Structures (GB50017). The other problem is that members are often under bi-axial bending and the stress distribution of members in the actual structure is very complicated. Yet there is no specific stipulation on how to calculate the effective width of thin-wall member with box section under bi-axial bending in the design codes of steel structure in China and foreign countries.

In Eurocode-3, sections can be divided into four categories.

Plastic section: Cross-sections with plastic rotation capacity. A plastic hinge can be developed with sufficient rotation capacity to allow redistribution of moments in the structure.

Compact section: Cross sections with plastic moment capacity. The plastic moment capacity can be developed. But local buckling may prevent the development of plastic hinge with sufficient rotation capacity at the section.

Semi-compact section: Cross section in which the stress at the extreme compression fiber can reach the design strength, but the plastic moment capacity cannot be developed.

Slender section: Cross-sections in which it is necessary to make explicit allowance for the effects of local buckling which prevent the development of the elastic capacity in compression and/or bending. Local buckling is very obvious and affects the elastic bearing capacity of section under compression and bending. And the width-thickness ratio of steel plates for the welded thin-wall box member is $b_0/t_w > 40\sqrt{235/f_y}$.

Theoretical analysis and test study show that the effective section can be used to calculate the bearing capacity of the thin-wall member with box section for its post-buckling strength (Chen Shaofan. 2001, Chen Ji. 2003). The effective width of steel plates for the box section is mainly decided by the width-thickness ratio of the plates, the influence of the adjacent plates, steel strength, slenderness ratio of the member and the stress distribution. In the calculation of the strength and stability of the box member, if the width-thickness ratio of the plates is larger than certain value, only part of the section is effective when the member is under axial compression, compression bending or tension bending. Due to little influence of local buckling on the stiffness, cross section can be adopted in the deflection calculation of the structure.

2.2 Categories of the plates and effective width evaluation

Based on the stress distribution on the sections of welded thin-wall box members, the basic way of determining the effective width of steel plates is as follows.

1. Regard welded members with box section as the ones consisting of four independent plates and the plates are serving mutually as webs;
2. When using the cross section in calculation, stress distribution of the plates falls into four categories as shown in Table 1. σ_1 is the minimal compression stress at the extreme edge of the compression zone on plates. σ_2 is the maximum compressive stress of the other side of the edge. The values of σ_1 and σ_2 are negative when it is tension stress; positive when it is compression stress. Stability coefficient of members and plastic adaptation coefficient of sections should be left out of consideration in the calculation.
3. Evaluate the forces on each plate respectively, take the influence of the stress along the width of plates into consideration and determine the corresponding effective width.
4. Determine the effective section of the member based on the effective width of each plate.
5. No initial imperfection and welding residual stress are considered.

Plate Category I: the whole section of the plate is under tension ($\sigma_2 < 0$). In this case, the whole section of the plate is effective.

Plate Category II: Compression zone and tension zone coexist in the plate. Meanwhile, $|\sigma_1| \geq |\sigma_2|$.

In compliance with the stipulation on the effective web width of pure bending plates with slender section in Eurocode-3 and extending the code to the plates under tension bending,

2a) When $b_0/t_w \leq 120\sqrt{235/f_y}$, the whole section of the plates is effective;

2b) When $b_0/t_w > 120\sqrt{235/f_y}$, on the side close to the compressed flange, the effective web width is $b_{e1} = 24t_w\sqrt{235/f_y}$; on the side near the tension flange, the effective width of web is $36t_w\sqrt{235/f_y}$ the width of the tension zone. When $60t_w\sqrt{235/f_y}$

the width of the tension zone $\geq b_0$, the whole section of the plate is effective.

Plate Category III: Compression zone and tension area coexist in the plate. Meanwhile, $|\sigma_1| < |\sigma_2|$.

In compliance with the stipulation on effective web width of pure bending members with semi-compact section in Eurocode-3 and extending the code to the compression bending plates, according to the regulations on the webs of the compression members in 5.4.2, 5.4.3, 5.4.6 of Code for Design of Steel Structures (GB50017-2003),

3a) when b_0/t_w meets Condition of judgment (1) or (2), the whole plate is effective;

3b) when b_0/t_w cannot meet Condition of judgment (1) or (2), the effective width of web on the compressed zone is $b_{e1} = 20t_w\sqrt{235/f_y}$; on the side close to the tension flange, the effective width of web is $20t_w\sqrt{235/f_y}$ + the width of tension zone.

The conditions of judgment on the width-thickness ratio of plates are as follows:

When $0 \leq \alpha_0 \leq 1.6$,

$$\frac{h_0}{t_w} \leq 0.8(16\alpha_0 + 0.5\lambda + 25)\sqrt{\frac{235}{f_y}} \quad (1)$$

When $1.6 < \alpha_0 \leq 2.0$,

$$\frac{h_0}{t_w} \leq 0.8(48\alpha_0 + 0.5\lambda - 26.2)\sqrt{\frac{235}{f_y}} \quad (2)$$

In which $\alpha_0 = \dfrac{\sigma_2 - \sigma_1}{\sigma_2}$,

and λ represents the slenderness ratio. When $\lambda < 30$, $\lambda = 30$; when $\lambda > 100$, $\lambda = 100$.

Plate Category IV: the whole section of the plate is under compression ($\sigma_1 < 0$)

According to the regulations on the webs of compression flexural members in 5.4.2, 5.4.3, 5.4.6 of Code for Design of Steel Structures (GB50017–2003).

When b_0/t_w meets Condition of judgment (1) or (2), the whole section is effective;

When b_0/t_w fails to meet Condition of judgment (1) or (2), effective width and thickness of both ends for the plate are.

$$b_e = 20t_w\rho\sqrt{\frac{235}{f_y}} \quad (3)$$

In the above formula, ρ is correction coefficient of the plate's effective width with the main consideration being the favorable influence on the effective width when the maximum compression stress of the plate is relatively low.

$$\rho = \sqrt{\frac{\max|\sigma_A, \sigma_B, \sigma_C, \sigma_D|}{\sigma_2}} \quad (4)$$

In the above formula, $\sigma_{A,B,C,D}$ represent the stress values at the four angle points of the section when the cross section is adopted in calculation.

The way of evaluating the stress distribution and the corresponding effective width of the four categories of plates are shown in Table 1. See Table 2 and 3 for the characteristics of the neutral axis and effective section of thin-walled uni-axial and biaxial flexural members with box section.

2.3 Formula for calculating member strength

In space structure, thin-wall box members may subject to various internal forces such as tension, compression, bending, shearing and torsion. The warping stress generated by restraint torsion is relatively low, so it can be ignored when calculating the normal stress of thin-wall box members. After obtaining the characteristics of thin-wall members with box section through the above-mentioned method, calculation of the normal section stress, shear stress and compression stability can be realized respectively. At this time, the effect of shift between the centroid of effective section and of the cross section shall be considered.

(1) Normal section stress formula

When the whole section is under tension, the strength of the tension flexural member is calculated according to the following formula:

$$\frac{N}{A_n} \pm \frac{M_x}{\gamma_x W_{nx}} \pm \frac{M_y}{\gamma_y W_{ny}} \leq f \quad (5)$$

According to the regulations stipulated in Eurocode-3, when only part of the section is effective and taken into consideration, the effect of the additional

Table 1. Stress Distribution and Effective width.

$\varepsilon = \sqrt{235/f_y}$

Table 2. Stress and effective section of thin-walled box member under uni-axial bending.

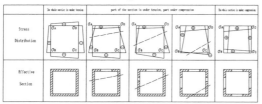

Table 3. Stress and effective section of thin-walled box member under biaxial bending.

bending moment generated by shift between the centroid of effective section and of the cross section shall be considered. If compressed zone exist on the section, the strength formula of tension flexural members or compression flexural members are as follows:

$$\frac{N}{A_{en}} \pm \frac{M_x + Ns_x}{\gamma_x W_{enx}} \pm \frac{M_y + Ns_y}{\gamma_y W_{eny}} \leq f \quad (6)$$

In which, N is the axial tension or axial compression of the member.;

M_x and M_y are bending moments about X-axis and Y-axis respectively;

γ_x and γ_y stand for the plastic adaptation coefficient that is corresponding to the section modulus. As for the box section, $\gamma_x = \gamma_y = 1.05$;

A_n, W_{nx} and W_{ny} represent net section area, net section modulus about X-axis and Y-axis respectively;

A_{en}, W_{enx} and W_{eny} represent effective net section area, effective net section modulus about X-axis and Y-axis respectively;

s_x and s_y stand for the shift between the centroid of effective section and cross section;

f is the design strength of the steel.

(2) Shear stress formula

When local buckling occurs on the welded thin-wall box member, there is little influence on the shear bearing capacity of the section. Therefore, cross section can be used in the calculation of shear stress.

Here the combined effect of shear and torsion should be considered. Referring to 4.1.2 of the Code for Design of Steel Structures (GB50017-2003) and also the formula of free torsion in elastic mechanics, shear stress should be calculated according to the following formula:

$$\frac{V_x S_y}{2 I_y t_f} + \frac{T}{2 A_s t_f} \leq f_v \quad (7)$$

$$\frac{V_y S_x}{2 I_x t_w} + \frac{T}{2 A_s t_w} \leq f_v \quad (8)$$

In which, V_x and V_y represent shear force on the flange and the web plate of section for calculation;

T is torsion exerted on the section;

S_x and S_y are the area moments of cross section about X-axis and Y-axis respectively;

I_x and I_y are the moment of inertia about X-axis and Y-axis respectively;

t_f and t_w are flange thickness and web thickness;

A_s is area of the shaded part formed by central lines for the box section which is illustrated in Figure 2;

f_y is the design value of shear strength of steels.

(3) Combined stress checking

For welded thin-wall box members where normal stress and shear stress are both relatively high, the following formula can be used to calculate the combined stress:

$$\sqrt{\sigma^2 + 3\tau^2} \leq \beta_1 f \quad (9)$$

In this formula, β_1 is the increasing coefficient of strength design value for combined stress.

(4) Stability calculation

As for thin-wall compressed flexural box members where moment is exerted on two axes, after the effect of shift between the centroid of effective section and

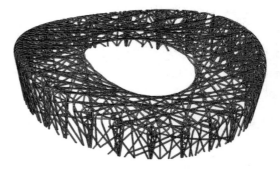

Figure 1. Model of the National Stadium.

of the cross section is considered, the stability of the member should be calculated according to the following formula:

$$\frac{N}{\varphi_x A_e} + \frac{\beta_{mx}(M_x + Ns_x)}{\gamma_x W_{ex}\left(1 - 0.8\dfrac{N}{N'_{Ex}}\right)} + \eta\frac{\beta_{ty}(M_y + s_y)}{\varphi_{by} W_{ey}} \leq f \quad (10)$$

$$\frac{N}{\varphi_y A_e} + \eta\frac{\beta_{tx}(M_x + Ns_x)}{\varphi_{bx} W_{ex}} + \frac{\beta_{my}(M_y + Ns_y)}{\gamma_y W_{ey}\left(1 - 0.8\dfrac{N}{N'_{Ey}}\right)} \leq f \quad (11)$$

In which, A_e, W_{ex} and W_{ey} are effective area, effective section modulus about X-axis and Y-axis respectively;

φ_x and φ_y are the stability coefficients about X-axis and Y-axis respectively;

φ_{bx} and φ_{by} are the global stability coefficients of flexural member in bending. For closed section, the value is 1.0;

η is the influence coefficients of section. For closed section, it equals 0.7;

β_{mx} and β_{my} are the equivalent bending moment coefficients about X-axis and Y-axis respectively for compression flexural member;

β_{tx} and β_{ty} are the equivalent bending moment coefficients about X-axis and Y-axis respectively for compression flexural member;

N'_{Ex} and N'_{Ey} are parameters,

$N'_{Ex} = \dfrac{\pi^2 EA}{1.1\lambda_x^2}$, $N'_{Ey} = \dfrac{\pi^2 EA}{1.1\lambda_y^2}$, where A is gross area;

λ_x and λ_y are the slenderness ratios of the member about X-axis and Y-axis respectively.

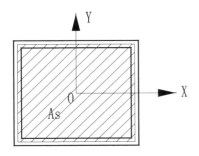

Figure 2. Method of calculating box section A_s.

2.4 Analysis of examples

Example 1: The section of the welded thin-wall box member $1200 \times 1200 \times 10 \times 10$, the steel is Q345, $f = 310$ MPa. Width-thickness ratio of plates is 11. The member is 20 m long and under uni-axial force. GB50017 and the method motioned in this paper are employed respectively to identify effective width of flange and web for the thin-wall box member, corresponding effective section, effective section moment, section shift and normal section stress, which are shown in Table 4.

From Table 4, it can be seen that when the width-thickness ratio of plates is relatively large, the laws of effective section area, effective section modulus, effective section shift and normal section stress for thin-wall box members under uni-axial stress and various internal forces are as follows:

1. For member under axial tension, the whole section is under tension and the method in this paper is identical with GB50017;
2. For tension flexural member, when the whole section is under tension, the method explored in this paper is the same with GB50017. When the flange of box member is under compression, according to the method in this paper, only part of the compressed flange is effective. With the increase of eccentricity, invalid area will appear in the web and the effective section area obtained according to the method herein is obviously smaller than that obtained by GB50017. After taking the effect of shift between the centroid of effective section and of the cross section into consideration and when pure bending state is almost achieved, the stress value obtained through the method in this paper is 55% higher than that obtained through GB50017;
3. For pure bending member, although the effective section area obtained through the method in this paper is larger than that obtained through GB50017, the stress value is relatively high due to consideration of the effect of effective section shift;
4. For compression flexural member, favorable effect on effective width and effective section shift when the stress value of flange under low compression is relatively small is taken into consideration according to the method in this paper. And stress value is a little higher than that in GB50017;
5. For member under axial compression, the whole section is under even compression and the method in this paper is identical to GB50017.

Example 2: The section of the welded thin-wall box member $1200 \times 1200 \times 20 \times 20$. The steel is Q345, $f = 295$ MPa. The width-thickness ratio of plates is 58. The member is 20 m long and under unidirectional

Table 4. Effective section and stress of box member □1200 × 1200 × 10 × 10.

Loading state	N (kN)	M (kN-m)	GB50017 A_e/A_{en} (mm²)	GB50017 W_e/W_{en} (mm³) ×10⁶	GB50017 σ_{max} (Mpa)	Method in this paper Flange category (upper/lower)	Method in this paper Web category	Method in this paper A_e/A_{en} (mm²)	Method in this paper W_e/W_{en} (mm³) ×10⁶	Method in this paper s_y (mm)	Method in this paper σ_{max}^* (Mpa)	$(\sigma_{max}^* - \sigma_{max})/\sigma_{max}$ (%)
Axial tension	14780	0	47600	18.73	310	1/1	1	47600	18.73	0	310	0
Tensile bending	13850	373	47600	18.73	310	1/1	1	47600	18.73	0	310	0
Tensile bending	1184	3764	47600	18.73	216	4b/1	2b	39102	12.31	127	310	43.5
Tensile bending	241	3831	47600	18.73	200	4b/1	2b	37580	11.89	148	310	55.0
Pure bending	0	3843	22100	11.76	282	4b/1	2b	37205	11.81	153	310	9.9
Compression bending	−278	3754	22100	11.76	305	4b/1	3b	33540	11.52	198	310	1.6
Compression bending	−3249	218	13600	6.991	310	4b/4b	4b	14225	7.067	26	307	−1.0
Axial compression	−3614	0	13600	6.991	310	4b/4b	4b	13600	6.991	0	310	0

Table 5. Effective section and normal stress of box member □1200 × 1200 × 20 × 20.

Loading state	N (kN)	M (kN-m)	GB50017 A_e/A_{en} (mm²)	GB50017 W_e/W_{en} (mm³) ×10⁶	GB50017 σ_{max} (Mpa)	Method in this paper Flange category (upper/lower)	Method in this paper Web category	Method in this paper A_e/A_{en} (mm²)	Method in this paper W_e/W_{en} (mm³) ×10⁶	Method in this paper s_y (mm)	Method in this paper σ_{max}^* (Mpa)	$(\sigma_{max}^* - \sigma_{max})/\sigma_{max}$ (%)
Axial tension	27850	0	94400	36.52	295	1/1	1	94400	36.52	0	295	0
Tensile bending	20863	2837	94400	36.52	295	1/1	1	94400	36.52	0	295	0
Tensile bending	1054	8553	94400	36.52	234	4b/1	2a	85050	28.60	65	295	26.1
Pure bending	0	8727	84400	30.72	271	4b/1	2a	84400	28.13	70	295	8.9
Compression bending	−979	8240	84400	30.72	271	4b/1	3a	84400	28.13	70	295	8.9
Compression bending	−13010	351	54420	24.23	295	4b/4b	4b	55377	24.38	10	294	−0.3
Axial compression	−13730	0	54420	24.23	295	4b/4b	4b	54420	24.38	0	295	0

force. GB50017 and the method motioned in this paper are employed respectively to identify effective width of flange and web for the thin-wall box member, corresponding effective section area, effective resist moment, section offset and normal section stress, which are shown in Table 5.

From Table 5, it can be seen that when the width-thickness ratio of plates is relatively small, the laws of effective section area, effective section modulus, effective section shift and normal section stress for thin-wall box member under uniaxial force are as follows:

1. For member under axial tension, the whole section is under even tension and the method in this paper is identical with GB50010;
2. For tension flexural member, pure bending member and compression flexural member, the tendency of difference in stress values obtained by the method in this paper and by GB50017 is similar to that in

Table 6. Effective Section and Stress of Box Member □1200 × 1200 × 10 × 10.

Loading state	N (kN)	M_x (kN·m)	M_y (kN·m)	GB50017				Method in this paper							
				A_e/A_{en} (mm²)	W_{ex}/W_{exn} (mm³) ×10⁶	W_{ey}/W_{eyn} (mm³) ×10⁶	σ_{max} (Mpa)	Flange category (upper/lower)	Web category (left/right)	A_e/A_{en} (mm²)	W_{ex}/W_{exn} (mm³) ×10⁶	W_{ey}/W_{eyn} (mm³) ×10⁶	S_x (mm)	S_y (mm)	σ_{max}^* (Mpa)	$(\sigma_{max}^* - \sigma_{max})/\sigma_{max}$ (%)
Axial tension	14670	0	0	47600	18.73	18.73	310	1/1	1/1	47600	18.73	18.73	0	0	310	0
Tensile bending	13050	352	352	47600	18.73	18.73	310	1/1	1/1	47600	18.73	18.73	0	0	310	0
Tensile bending	570	1813	1813	47600	18.73	18.73	196	3b/1	3b/1	32060	11.76	11.82	135	153	310	58.2
Tensile bending	115	1816	1816	47600	18.73	18.73	187	3b/1	3b/1	30893	11.29	11.29	163	163	310	65.8
Pure bending	0	1819	1819	30600	12.86	12.86	269	4b/1	4b/1	30600	11.17	11.17	165	165	310	15.2
Compression bending	−144	1792	1792	30600	12.86	12.86	270	4b/2a	4b/2a	30600	11.17	11.17	165	165	310	15.2
Compression bending	−1962	587	587	13600	6.991	6.991	310	4b/3b	4b/3b	21687	8.632	8.632	152	152	227	−26.8
Axial compression	−3614	0	0	13600	6.991	6.991	310	4b/4b	4b/4b	13600	6.991	6.991	0	0	310	0

Table 7. Effective section and normal stress of box member □1200 × 1200 × 10 × 10.

Loading state	N (kN)	M_x (kN-m)	M_y (kN-m)	GB50017				Method in this paper								
				A_e/A_{en} (mm²)	W_{ex}/W_{exn} (mm³) ×10⁶	W_{ey}/W_{eyn} (mm³) ×10⁶	σ_{max} (Mpa)	Flange category (upper/lower)	Web category (left/right)	A_e/A_{en} (mm²)	W_{ex}/W_{exn} (mm³) ×10⁶	W_{ey}/W_{eyn} (mm³) ×10⁶	S_x (mm)	S_y (mm)	σ_{max}^* (Mpa)	$(\sigma_{max}^* - \sigma_{max})/\sigma_{max}$ (%)
Axial tension	28750	0	0	94400	36.52	36.52	295	1/1	1/1	94400	36.52	36.52	0	0	295	0
Tensile bending	16700	2271	2271	94400	36.52	36.52	295	1/1	1/1	94400	36.52	36.52	0	0	295	0
Tensile bending	10800	3467	3467	94400	36.52	36.52	295	2a/1	2a/1	94400	36.52	36.52	0	0	295	0
Tensile bending	5404	3967	3967	94400	36.52	36.52	264	3b/1	3b/1	86629	32.48	32.48	33	33	295	11.7
Pure bending	0	4262	4262	74410	30.38	30.38	267	4b/1	4b/1	74410	27.52	27.52	79	79	295	10.5
Compression bending	−5144	3273	3273	74410	30.38	30.38	274	4b/2a	4b/2a	74410	27.52	27.52	79	79	295	7.7
Compression bending	−8016	1731	1731	54420	24.23	24.23	295	4b/3b	4b/3b	59245	24.60	24.60	32	32	277	−6.1
Axial compression	−13730	0	0	54420	24.23	24.23	295	4b/4b	4b/4b	54420	24.23	24.23	0	0	295	0

Example 1. However, due to the decrease of width-thickness ratio of plates, effective area of the box member's compressed flange increases and invalid area seldom occur in the web. Therefore, the difference between the two methods becomes less obvious;

3. For member under axial compression, the whole section is under compression and the method in this paper is identical to GB50017.

Example 3: The section of the welded thin-wall box member is □1200 × 1200 × 10 × 10. The steel is Q345, $f = 310$MPa. The width-thickness ratio of plates is 118. The member is 20 m long and under biaxial forces. The method motioned in this paper and the stipulation on compressed web in GB50017 are employed respectively to identify effective width of flange and web for the thin-wall box member, corresponding effective section area, effective resist moment, section shift and normal section stress, which are shown in Table 6. When the width-thickness ratio of plates is relatively large, the laws of effective section area, effective section modulus, effective section shift and normal section stress for thin-wall box member under biaxial force and various internal forces are as follows:

1. For biaxial eccentric tension member, when the tension or compression zone of the whole section is very small, the method in this paper is identical with GB50017. With the increase of eccentricity, the scope of compression zone expands and invalid area occurs for the tension plate. The effective section area obtained through the method in this paper is obviously smaller than that obtained through GB50017. After the effect of shift between the centroid of effective section and of cross section is considered and when pure bending state is almost achieved, the stress value of the method in this paper is 65% higher than that of GB50017;
2. For biaxial pure bending member, the method in this paper is the same with GB50017. Nevertheless, the consideration of the influence of effective section shift in the method mentioned in this paper leads to a higher stress value than that obtained through GB50017;
3. For biaxial eccentric compression member, the method in this paper takes the favorable influence on effective width when the stress value of flange under low compression is relatively small and also the effect of effective section shift into consideration. The stress obtained hereby can be as much as 26.8% smaller than that obtained by GB50017. When the whole section is under even compression, the method herein is the same with GB50017.

Example 4: The section of the welded thin-wall box member is □1200 × 1200 × 20 × 20. The steel is Q345, $f = 295$MPa. The width-thickness ratio of plates is 58. The member is 20 m long and under biaxial force. The method motioned in this paper and the stipulation on compressed web in GB50017 are employed respectively to identify effective width of flange and web for the thin-wall box section, corresponding effective section area, effective resist moment, section shift and normal section stress, which are shown in Table 7. From Table 7, it can be seen that when the width-thickness ratio of plates is relatively large, for biaxial eccentric tension member, biaxial flexural member and biaxial compression member, the tendency of difference in stress value obtained by the method in this paper and by GB50017 is similar to Case3. However, due to the decrease of width-thickness ratio of plates, the proportion of effective area of box members' compressed plates increase and invalid area seldom happen to plates, thus there is little difference in these two methods.

3 RESEARCH ON WARPING EFFECT OF WELDED THIN-WALL BOX MEMBER

Due to the large width-thickness ratio of thin-wall box member, warping deformation occurs easily when the member is under torque. So for, warping effect of welded thin-wall box members under restraint torque has been left out of consideration in codes for design of steel structures both in China and in foreign countries. How to correctly evaluate warping effect of welded thin-wall box members has become one of the key problems to guarantee the security of structures and to enhance the utilization of materials.

When determining the boundary conditions of restraint torque, considering that restraint on deformation at the joints of members is very strong, rigid zones are set at the two ends sections of thin-wall box member. One is fixed end, the other free end. And torque is exerted on the centroid of the rigid zones. The torque values of free torque and restraint torque are both $T = 5000$kN·m.

Various wall-thicknesses, including the flange thickness not equal to the web, and different height to width ratio of the sections are taken into consideration to evaluate the influence on warping effect.

3.1 The thin-wall box member with equal thickness for all plates

Shear stress and positive warping stress of thin-wall box members with different thicknesses under torque are shown in Table 8. From Table 8, it can be seen that for thin-wall box members, when the thickness of flange and that of web is the same, the analytical solution of shear stress under free torque is identical to that of shear stress under restraint torque and the positive warping stress in both conditions is zero. The result of shear stress through FEM is close to the analytical solution and the value of positive warping stress is very small. From this it can be seen that for thin-wall box members with equal thickness for flange and web, the effect of warping stress caused by torque can be ignored.

Stress distribution of thin-wall box member □ $1200 \times 1200 \times 10 \times 14$ under restraint torque is shown in Fig. 3, from which it can be seen that shear stress along the wall of the member is distributed relatively even and that positive warping stress varies along the length of the member with the maximum value on the four edge lines at the end of the member.

3.2 Thin-wall box member with unequal thickness for flange and web

Shear stress and positive warping stress of thin-wall box members with unequal thickness for flange and web under torque are shown in Table 9. When the thickness of flange and that of web is not the same, positive warping stress under free torque is zero and the shear stress calculated through finite element method is a little higher than the analytical solution. When it is under restraint torque, positive warping stress increase as the difference between web and flange become increasingly obvious; analytical solution is close to FEM solution and the latter is a little larger than the former with variance less than 5%. This is because there is part of assumption in the

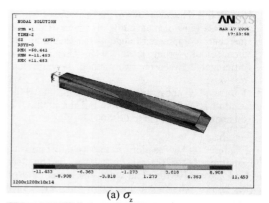

(a) σ_z

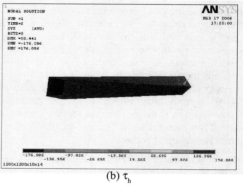

(b) τ_h

Figure 3. Stress Distribution of Thin-wall Box Member □ $1200 \times 1200 \times 10 \times 14$ under restraint torque.

Table 8. Warping stress (MPa) for the member under torque when thickness for flange and web is equal.

Size of section □$1200 \times 1200 \times t_w \times t_f$	Stress	Free torque		Restraint torque	
		Analytical solution	FEM solution	Analytical solution	FEM solution
$t_w = t_f = 10$ mm	τ_b	173.611	176.540	173.611	178.224
	τ_k	173.611	176.540	173.611	178.224
	σ	0.000	0.357	0.000	0.450
$t_w = t_f = 12$ mm	τ_b	144.676	147.579	144.676	150.494
	τ_k	144.676	147.579	144.676	150.494
	σ	0.000	0.290	0.000	0.217
$t_w = t_f = 14$ mm	τ_b	124.008	126.903	124.008	128.502
	τ_k	124.008	126.903	124.008	128.502
	σ	0.000	0.242	0.000	0.005
$t_w = t_f = 18$ mm	τ_b	96.451	99.334	96.451	100.813
	τ_k	96.451	99.334	96.451	100.813
	σ	0.000	0.179	0.000	0.033
$t_w = t_f = 20$ mm	τ_b	86.806	89.684	86.792	91.100
	τ_k	86.806	89.684	86.792	91.100
	σ	0.000	0.157	0.000	0.025

Table 9. Warping stress (MPa) of thin-wall box member with unequal thickness for flange and web ($h = b = 1200$).

Size of section $\Box h \times b \times t_w \times t_f$	Stress	Free torque Analytical solution	Free torque FEM solution	Restraint torque Analytical solution	Restraint torque FEM solution
$t_w = 10$	τ_b	144.676	148.063	144.418	149.082
$t_f = 12$	τ_k	173.611	171.037	173.920	177.335
	σ	0.000	1.722	5.432	7.639
$t_w = 10$	τ_b	124.008	127.489	123.723	129.048
$t_f = 14$	τ_k	173.611	176.102	174.010	177.897
	σ	0.000	1.716	11.281	9.578
$t_w = 10$	τ_b	96.451	100.593	96.060	102.339
$t_f = 18$	τ_k	173.611	171.46	174.310	178.057
	σ	0.000	1.710	17.083	11.666
$t_w = 10$	τ_b	86.806	91.212	86.253	93.158
$t_f = 20$	τ_k	171.611	171.532	174.716	178.255
	σ	0.000	1.709	21.7692	14.252

Table 10. Influence of height-width ratio on thin-wall box member.

Size of section $\Box h \times b \times t_w \times t_f$	Stress	Restraint torque Analytical solution	Restraint torque FEM linear solution	Restraint torque FEM nonlinear solution
$h/b = 1.0$	τ_b	173.611	178.224	176.908
$h = 1200$ mm	τ_k	173.611	178.224	176.908
$t_w = t_f = 10$ mm	σ	0.000	0.450	4.047
$h/b = 1.2$	τ_b	208.079	213.819	213.893
$h = 1200$ mm	τ_k	208.588	214.846	214.146
$t_w = t_f = 10$ mm	σ	8.627	9.705	8.391
$h/b = 1.5$	τ_b	259.614	269.718	267.484
$h = 1200$ mm	τ_k	261.219	269.718	268.253
$t_w = t_f = 10$ mm	σ	25.421	25.131	13.334
$h/b = 2.0$	τ_b	344.98	357.107	357.225
$h = 1200$ mm	τ_k	349.466	364.922	361.709
$t_w = t_f = 10$ mm	σ	61.16	57.200	26.551

analytical solution while the FEM numerical simulation is pretty close to the real situation.

When stiffeners are installed inside the member, the size of the stiffener is as follows:

1. stiffener thickness t_s: When $t_f = 10$ mm, $t_s = 8$ mm; when $t_f = 12$ mm, 14mm, $t_s = 10$ mm; when $t_f = 18$ mm, $t_s = 12$ mm; when $t_f = 20$ mm, $t_s = 14$ mm;
2. Flange stiffener height $h_s = 1/10b$, web stiffener thickness $b_s = 1/10h$.

After the member is stiffened, warping deformation of the member is restrained and the positive warping stress decreases greatly. Transverse stiffener works in the control of warping effect of thin-wall box member.

3.3 *Influence of height-width ratio on thin-wall box member*

The influence of height-width ratio on thin-wall box member under restraint torque is displayed in Table 10, from which it can be seen that for thin-wall box members with the same flange and web thickness, the positive warping stress rises rapidly with the increase of height-width ratio of the section. After transverse stiffener is placed, positive warping stress decreases obviously.

4 HYSTERETIC CURVE MODEL FOR WELDED THIN-WALL BOX MEMBER

A large number of welded thin-wall box members are used in the construction of the National Stadium. As the box sections consist of plates with large width to thickness ratio, they are likely to lose stability subjected to compression. In the design, post-buckling strength of the plates can be used to calculate the strength capacity of members based on the effective width of the plates. The effective width-thickness ratio of members with box section is mainly related to the width-thickness ratio of plates, the restraint influence of adjacent plates, steel strength, and member slenderness ratio and stress distribution. In the present seismic design of steel structures, the ductility of structures in case of earthquakes is mainly enhanced by limitation the slenderness ratio of compression members and the width-thickness ratio of compression plates. So far, the research on hysteretic curve of welded thin-wall box member is relatively rare.

Member slenderness ratio is ignored in FEMA356. For compression member, the ratio between axial compression and bearing capacity of the section is crucial to its ductility.

The main purpose of this paper is to identify the hysteretic property adopted in rare earthquake condition analysis by studying the important parameters such as plate width-thickness ratio, member slenderness ratio and axial compression ratio and combining with FEMA356.

The chords of primary trusses and truss columns are mainly to subject axial forces. The chord of primary trusses is about 14–20 m long; its slenderness ratio approximately 35–50 and typical section size 1000 × 1000 mm. Nonlinear FEM analysis method is adopted and the element SHELL181 is chosen. The effect of initial imperfection and residual stress on box section are considered.

4.1 Influence of width-thickness ratio of plates

For thin-wall box members with slenderness ratio of 30 and axial compression ratio of 0.3, the hysteretic property curves when the width-thickness ratio b/t is 30 and 50 are shown in Fig. 4, from which it can be found that width-thickness ratio of plates b/t exerts a great influence on the hysteretic property of the member.

With the increase of member width-thickness ratio, the maximum envelop value of the hysteretic curve decreases gradually; the ultimate bearing capacity decreases and the lateral stiffness also decreases, which is because with the increase of displacement, local buckling occurs to plates with large width-thickness ratio. When member slenderness ratio is relatively small, local buckling and corresponding buckling are the main failure factors.

And at this time, local stability of plates has a relatively great influence on the bearing capacity of the member.

Members with relatively small width-thickness ratio have good ductility and deformation. Its hysteretic curve is relatively full and has strong deformation ability.

When the axial compression ratio N/N_u is large, plastic buckling and failure occur to the member. Local buckling happens to the plates under the effect of small cyclic displacement. The envelope of hysteretic curve decreases rapidly after it hits the peak value and bearing capacity decreases quickly. The hysteretic curve of the member is flat and the energy dissipation performance is poor.

When local buckling and failure occurs to plates, the weakening of bearing capacity becomes extremely obvious.

4.2 Influence of slenderness ratio

For thin-wall box members with width-thickness ratio of 50 and axial compression ratio of 0, the hysteretic property curves are shown in Fig. 5 when the slenderness ratios λ are 30 and 60.

Slenderness ratios λ of members has an effect on the hysteretic property of P-M-M hinge. With the increase of the member slenderness ratio, enveloping area of the hysteretic curve decreases and the energy dissipation ability weakens. For members with a large slenderness ratio, lateral stiffness decreases gradually and the envelope of hysteretic curve plunges rapidly after it hits the peak value.

4.3 Influence of axial compression ratio

For thin-wall box members with width-thickness ratio of 30 and slenderness ratio of 30, the hysteretic property curves when the axial compression ratio μ is 0.0, 0.3 and 0.6 are shown in Figure 6, from which it can be found that axial compression ratio exerts a great influence on hysteretic curve.

The axial compression ratio N/N_u determines the $P\text{-}\Delta$ effect directly. Under relative great axial load, local buckling may happen to the plates or yield state

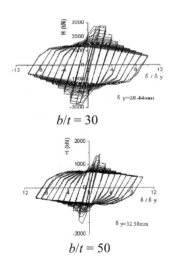

$b/t = 30$

$b/t = 50$

Figure 4. Influence of width-thickness ratio of plates.

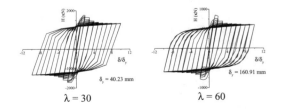

$\lambda = 30$ $\lambda = 60$

Figure 5. Influence of slenderness ratio.

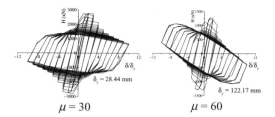

Figure 6. Influence of axial compression ratio.

may be achieved in advance. Therefore, axial load level greatly affects the bearing capacity and hysteretic curve of the member. As shown in the figure, with the increase of N/N_u, the maximum envelope of hysteretic curve decreases. Members under heavy axial load have relatively poor ductility. And bearing capacity is completely missing in the condition of small story drift.

When the axial compression ratio N/N_u is small, plastic buckling and failure happens to the member and local buckling will not happen before the plates enter into the plastic state. When the plate stress increases with the cyclic displacement and reaches the yield stress, buckling occurs to plates with the continuous increase of cyclic displacement and the bearing capacity of members decrease. Members with plastic buckling and failure have full hysteretic curves and good energy dissipation ability. The decrease of envelope of the hysteretic curve slows down.

When the axial compression ratio N/N_u is relatively large, plastic buckling and failure happens to the member; local buckling occurs to plates under the effect of relatively small cyclic displacement. The envelope of the hysteretic curve decreases rapidly after it hits the peak value and bearing capacity decreases quickly. Hysteretic curve of the member is flat and its energy dissipation ability is poor.

With the increase of N/N_u, local buckling occurs more easily to the plates and second-order effect tends to be serious. The fullness of hysteretic curve is poor; energy dissipation ability and lateral deformation decrease.

5 TWISTED THIN-WALL BOX MEMBERS

Members on the elbow of the roof are twisted box-sections approximately 1200 mm × 1200 mm. The twisted box-members bend and twist according to the curvature of the roof surface to ensure the smoothness of the members and continuity at intersections. Since the orientation of the section changes continuously along the box-member's axes and the wall of members is thin, the stress distribution on the plates of members is uneven, out-of-plane deflections as

Figure 7. Inner stiffeners for the box-sections at the top of the column.

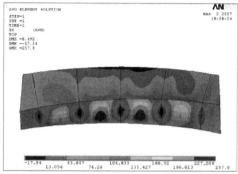

Mises stress (MPa) distribution

Scaled model test

Figure 8. Result of FEM and scaled model test.

well as local buckling of the plates occur easily. Inner stiffeners with square hole are introduced to effectively restrain the out-of-plane deflections on the arc flanges, so that the overall stiffness of the members and evenness of the stress distribution on the plates are improved significantly. The twisted box-members used in the project are shown in Figure 7.

Based on the actual situation of the twisted box members, a section of outer column between the adjacent diagonals at the top of the truss-column is chosen for analysis. The behavior of twisted box members is

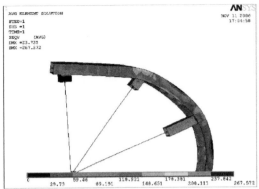
Von Mises stress (MPa) distribution

Top tension test

Figure 9. Local structure at the top of truss-column.

similar to that of arc members in tension. The rise of the arc member generates significant ad additional moment in mid-span. After the additional moment is superposed with the tension force, a big tension stress value is produced at the inner arc flange in mid-span. The flange plate shows obvious out-of-plane deflections and tends to be straightened.

Otherwise, the stress on the outer arc flange is smaller. The behavior of the webs of the twisted box members is similar to that of general box-sections. After the twisted box-members are stiffened inside, the stress distribution is improved, and the out-of-plane deflections of the plate are effectively restrained. In addition, the increasing thickness of the member flange can improve the elastic stiffness of the member, and decrease the stress concentration, so that the stress is more evenly distributed. The stress distribution of the twisted box members in tension and an experiment of 1:4 scaled specimen are shown in Figure 8.

In order to faithfully simulate the actual loading behavior and boundary conditions of the twisted box-members in the structure, a part of truss-columns where the twisted box-members are located is selected for detail analysis. Tangential tension force is applied along the top of the specimen, and the result of von Mises stress is shown in Figure 9. Furthermore, 1:4 scaled test on the part of truss-column is performed, as shown in Figure 9.

6 CONCLUSIONS

1. Based on the stress distribution of plates, a new method to estimate effective section for welded thin-wall box member is put forward. In which the favorable effect on effective width for the member with the low compressive stress for plates are considered. So is the shift between the centroid of effective section and cross section; for pure bending members and compression flexural members, though the effective web zone is rather large, the normal section stress value obtained according to the method in this paper is relatively high because of the consideration of effective section shift. With the decrease of plate width-thickness ratio, effective area of compression plates increases and the method in this paper becomes close to GB50017.

2. When thin-wall box members with unequal thickness for flange and web are under restraint torque, positive warping stress increases as the difference between web and flange become increasingly obvious. The positive warping stress rises rapidly with the increase of height-width ratio of the section. After stiffener is placed, warping deformation of box section is restrained and positive warping stress decreases obviously.

3. The research results of hysteretic curve for welded thin-wall box members reveal that width-thickness ratio make great difference in the hysteretic property of members. The member with relatively small width-thickness ratio has good ductility and deformation ability. Its hysteretic curve is relatively full. With the increase of width-thickness ratio, local buckling and failure occurs to section wall and the weakening of bearing capacity becomes extremely obvious. With the increase of the member slenderness ratio, enveloping area of the hysteretic curve decreases and the energy dissipation ability weakens. With the increase of axial compression ratio of the member, P-Δ effect becomes obvious; local buckling may happen to the plates or yield state may be achieved in advance. In addition, axial bearing capacity of members is completely missing in the condition of small story drift.

4. The behavior of the twisted member is rather complicated and bearing capacity and member

stiffness is influenced by factors such as wall thickness and stiffener interval. Stress distribution can be improved greatly and out-of-plane deformations of plates can be controlled by reducing the stiffener interval.

REFERENCES

ANSI/AISC 360–05. 2005. *An American National Standard, Specification for Structural Steel Buildings*.

Chen Shaofan. 2001. *Design Principles of Steel Structures (2nd Edition)*. Beijing: Science Press.

Chen Ji. 2003. *Stability Theory and Design of Steel Structures (2nd Edition)*. Beijing: Science Press.

Eurocode 3. 1995. *Design of Steel Structures, Part 1.1, General Rules and Rules for Buildings, EN 1993–1-1*, European Committee for Standardigation (EC Brussels, Belgium, 1995).

Fan Zhong, Fan Xuewei, Liu Xianming. 2004. Application of Stiffened Thin-wall Box Members in the Project of the National Stadium. *Papers for the 4th National Seminar on Modern Structural Projects*: 11–18. Ningbo.

Li Kaixi. 1990. *Warping of Elastic Thin-wall Bars*. Beijing: China Architecture and Building Press.

National Standards of P.R.China. 2002. *Technical Code for Cold-formed Thin-wall Steel Structures (GB50018–2002)*. Beijing: China Planning Press.

National Standards of P.R.China. 2003. *Codes for Design of Steel Structures (GB50017–2003)*. Beijing: China Planning Press.

Aquatic Center Maria Lenk – Tubular structure roof – Design, construction

F.C. D'Alambert
Projeto Alpha Engenharia de Estruturas, São Paulo, Brazil

ABSTRACT: This paper presents the concepts used for the development of a structural project, production and assembly of the tubular structures that compose the covering of the Aquatic Center Maria Lenk, where the swimming and diving competitions of the Pan American Games 2007, headquartered in the city of Rio de Janeiro in Brazil, were disputed.

1 INTRODUCTION

1.1 *Pan American Games—2007*

The city of Rio de Janeiro was the host of the Pan American Games in 2007, offering to the athletes and the tourists that participated in the events new, modern, comfortable and safe stadiums.

With a capacity for 15,000 people, the Aquatic Center Maria Lenk was especially built for PAN Rio 2007, and for the swimming and diving competitions.

1.2 *Sequence*

This paper will be presented with emphasis on the following topics:

– Structural conception
– Architectural and structural originality
– Construction methods
– Structural behavior—deflections and slenderness
– Respect to the environment
– Better use of materials and aesthetics

2 STRUCTURAL CONCEPTION

2.1 *General information*

The sports stadium, with a total constructed area of 42,000 m² and a covered area of roughly 7,500 m², is part of the "Sports City" compound (Racing circuit + Arena) in the Rio neighborhood of Jacarepaguá.

2.2 *Historical*

The architectural project was developed by Arquipan.

The idea in the development of the architectural project was one of providing a covering to integrate fully with the beautiful urban landscape of the place, formed by mountains, natural lakes and the sea, while at the same time keeping to its modern and heady roots.

2.3 *Structural system*

The covering of the Aquatic Center Maria Lenk is formed by 2 symmetrical blocks (Figure 1), each block having 7 principal trusses of 29.50 m in length and 16 m between them, with 2 trusses located in the extremities of the covering with 10 m cantilevered.

Among the main trusses we have 13 lines of secondary beams (joists), that besides supporting the steel cladding of the covering, they join and stabilize the main elements.

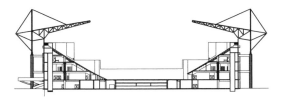

Figure 1. Architectural section.

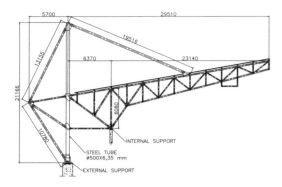

Figure 2. Truss type a.

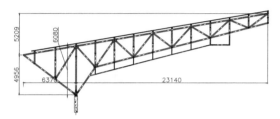

Figure 3. Truss type b.

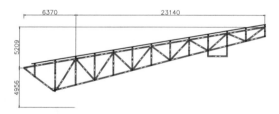

Figure 4. Extremity beams (Type c).

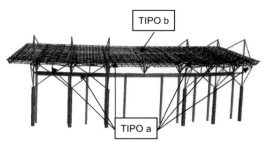

Figure 5. Geometric scheme.

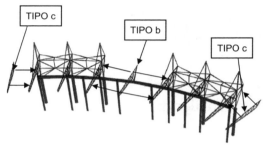

Figure 6. Static behavior.

Type c—With the same form as the extremity beams, they are also suspended by the secondary beams (joists), as shown in Figure 4 and Figure 6.

2.4 Loads

The roof's project for Aquatic Center Maria Lenk took into account the proper distribution for dead loads, live loads (0.25 kN/m^2), equipment, etc, thermal effect of the variation (+/− 15° Celsius) and 6 cases of wind effects.

There were obtained 50 combinations which allowed static and dynamic analysis of the structure for the 1st and 2nd order.

3 ARCHITECTURAL AND STRUCTURAL ORIGINALITY

The Aquatic Park Maria Lenk is the most modern and complete center for aquatic sports in Latin America, providing beauty, safety and infrastructure for international swimming and diving competitions.

The covering structure has an important role in an architectural context, providing protection for fans against sun and rain as well as being structurally unique and a point of reference in the urban landscape of Rio de Janeiro (Figure 7 and Figure 8).

Type a—Composed of 6 trapezoidal trusses, they are internally supported in a beam of concrete that composes the stands and are integrated into tubular masts 21 m high in steel tubes and 18 m in tubular concrete columns, forming the binary balance of the system (Figure 2 and Figure 5).

Type b—Similar to type **a**, however these have the internal support in the stands, without the external mast for stabilization. In this way the balance of the central part of the covering is made by the secondary beams (joists) that connect the main trusses (Figure 3).

Figure 7. Aerial view.

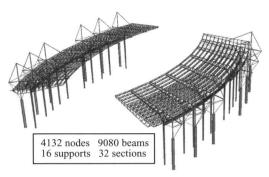

Figure 9. Structural model.

Figure 8. Lateral view.

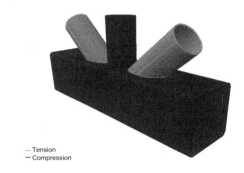

Figure 10. Typical welded connection.

4 STRUCTURAL BEHAVIOR—DETAILS-DISPLACEMENT

4.1 *Structural model*

The structural analysis calculation was made with STRAP 12.5 developed for calculation of flat and spatial structures and Finite Elements Method (FEM), as shown in Figure 9.

    4132 nodes   9080 beams
    16 supports   32 sections

4.2 *Constructive details*

The use of tubular profiles in the composition of the trusses was of fundamental importance to the accomplishment of the project. The period for the execution of the work was short, therefore it was necessary to standardize solutions, allowing serial production of the elements and reducing the necessary times for creation and for the assembly of the structures.

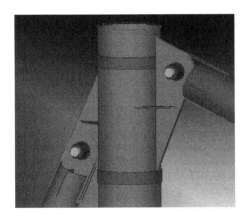

Figure 11. Principal column—Image of superior connection.

– Connections

Procedures were developed for analysis of the structure's connections (Figure 10), and for the more standardized connections the concepts of the book Hollow Structural Section-Connections and Trusses (Packer & Henderson, 1997) were used.

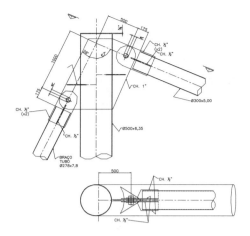

Figure 12. Principal column—steel design.

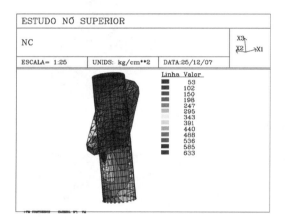

Figure 13. Principal column- FEM of superior connection.

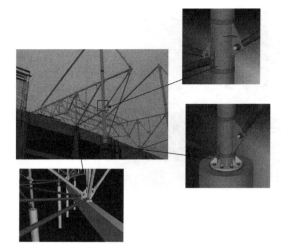

Figure 14. Principal connections—details.

Figure 15. Standardized connections.

With the results obtained through structural analysis, the main connections were studied, each one with the Finite Elements Method (FEM) allowing a better understanding of the distribution of the internal stresses and to make a correct design of plate's thickness, welds, etc. (Figure 11–15).

4.3 Displacement

The use of the tubular profiles allowed the principal trusses to have great rigidity, facilitating the process of movement of the pieces and assembly of the groups.

The chords were composed of profiles type square tubing ($200 \times 200 \times 8.00$ mm) and connections were totally screwed in, allowing speed and precision in the field.

The principal trusses were prepared on the ground and moved to the final position with 150-ton cranes, supported on the existent concrete beam and scaffolds located temporarily inside the stadium (Figure 16).

Once the assembly of the structure was concluded and the final checks made, all the scaffolding was released simultaneously.

A meticulous topographical accompaniment of the displacements of the structure was made, which was repeated several times, coinciding with the stages of shipment of the group.

Displacement structural analysis = 164 mm—
Measured displacement = 142 mm

Figure 16. Scaffolds system.

4.4 Fabrication and erection tolerances

An important factor in the control of the field works was in the attention to the tolerances demanded by Brazilian Code of Steel construction—NBR 8800—and all calculation parameters were based on these premises.

5 ADAPTATION TO THE ENVIRONMENT

Rio de Janeiro has tradition in the preservation culture to the environment (Eco Rio 92), and the location of the sporting compound is teaming with lakes, marshes and native vegetation. As such, the adoption of an industrialized system for both the stands, manufactured in pre-cast concrete, and the covering, with tubular profiles of steel, attended municipal requirements. Furthermore, the structure of the covering is 100% recyclable.

6 CONCLUSION

The covering of the Aquatic Park Maria Lenk is the proof that the techniques and available concepts in structural engineering can foster the development of light and sturdy structures.

The use of tubular profiles made it possible to meet the short deadlines imposed, facilitating the project, the structural production calculations and especially assembly of the structures.

The beauty of the work can be seen not only for its harmonious architectural design and its integration into the environment, but also for the appropriate use of tubular profiles, which provide a slender and elegant structure.

7 TECHNICAL TEAM

Construction- Consórcio Delta Sanerio
Architecture- Carlos Porto-Gilson Santos e Paulo Casé
Manufacturer of the steel structure- CPC

REFERENCES

NBR 8800. 1986. *Projeto e Execução de Estruturas de Aço de Edifícios*.
NBR 6355. 2003. *Perfis Estruturais de aço formados à frio-Padronização*.
Galambos, T.V. 2004. *Guide for stability design criteria for metal structures*. John Wiley & Sons.
Packer, J.A. & Henderson, J.E. 1997. *Hollow structural section connections and trusses – A design guide*. Canadian Institute of Steel Construction.

Structures and erection methods of concrete filled steel tubular arch bridges

B.C. Chen
College of civil engineering, Fuzhou University, Fuzhou, China

ABSTRACT: From 1990, more than 230 Concrete Filled Steel Tubular (CFST) arch bridges have been built in China. In this paper, five main types of CFST arch bridges are introduced with some typical bridges, i.e. deck (true) arch, half-through true arch, through deck-stiffened arch, through rigid-frame tied arch and fly-bird-type arch (half-through tied rigid-frame arch). Based on statistics data of built bridges, arch rib cross-section and materials, ratio of rise-to-span, and erection methods are analyzed.

1 INTRODUCTION

In Concrete Filled Steel Tubes (CFST) structures, the in-filled concrete delays local buckling of the steel tube, while the steel tube reinforces the concrete to resist tension stresses and improve its compression strength and ductility. Moreover, in construction, the tube also acts as a formwork for the concrete. The first two CFST arch bridges were built in the USSR in 1930s and 1940s. Since 1990, many CFST arch bridges have been building in China. Up to March 2005, more than 230 CFST arch bridges, with the span over 50 m, have been built or are under construction (Chen 2004, 2007a & 2007b). Among them, 131 bridges have a main span longer than 100 m and 33 significant bridges with a span greater than 200 m. In the following, all of the analyses are based on the data of CFST arch bridges with a span equal to or longer than 100 m.

2 TYPES OF CFST ARCH BRIDGES

In this paper, CFST arch bridges are classified into five main types (Chen et al. 2004). Among them, 131 bridges have a mains, i.e. deck (true) arch, half-through true arch, through deck-stiffened arch, through rigid-frame tied arch and fly-bird-type arch (half-through tied rigid-frame arch). It should be noted that for the deck and half-through arch with thrust, the span is clear span; while for no-thrust arch, the span is from the center line of pier to pier.

2.1 Deck arch bridge

In deck bridge, the arch ribs can be several vertical dumbbell (two tubes) shaped CFST ribs in medium span bridges or two vertical truss (four tubes) CFST ribs connected by lateral bracings of steel tubes. Generally, the decks are RC or Prestressed Concrete (PC) structures, and the spandrel columns are CFST or steel structures. The true arch bridge has a great crossing capacity. The deck arch has been built for spans over 150 m.

The Huangbaihe Bridge and the Xialaoxi Bridge are examples of the deck bridges, each with the main clear span of 160 m, width of 18.5 m, carrying traffic to the giant Three Gorges Dam site. In these two bridges, the four arch ribs are CFST dumbbell cross-sections connected with steel tube bracing members. The spandrel columns are also the CFST members (Fig.1) (Zheng & Chen, 2000).

Until now, the Fengjie Meixihe Bridge has the longest span of 288 m in CFST deck arch bridges, shown in Figure 2, but the Zhijinghe Bridge, under construction with a main span of 430 m, will break this span record. However, based on different concrete arch bridges, only 8% of investigated CFST arch bridges are deck bridges, most of them are half-through and through bridges.

2.2 Half-through true arch bridge

Half-through bridge is a good choice when the rise of the arch bridges is much higher than the road elevation for the long span. Half-through (true) arch bridges are

Figure 1. Huangbaihe Bridge.

Figure 2. Fengjie Meixihe Bridge.

counted for 62 (47%) of the investigated 131 CFST arch bridges. Moreover, it can reduce the height of the spandrel columns. Many long-span CFST half-through true arch bridges have been built.

The Sanan-Yongjiang Bridge with a span of 270 m has the longest span when it completed. The four-tube truss cross-section of arch rib is composed of ϕ1020∗92 mm steel tube filled with C50 concrete, as shown in Figure 3.

The span record of 460 m is kept by Wushan Yangtze River Bridge, shown in Figure 4, which is also the record of the longest CFST arch bridge in the world.

2.3 Through deck-stiffened arch bridge

CFST through deck-stiffened arch bridge is composed by CFST arch ribs and PC or steel tied girders. The hangers are high strength strands and the deck structure can be concrete or steel-composite structures, including cross beams and deck slabs. The construction difficulty of this type of bridges will increase with the span of the bridge because the horizontal reactions are not available until the tied girder is completed. Generally, such bridge type is a good option for mid-span bridge, saying from 50 m to 150 m.

The longest span of this type of bridges is found in Moon Island Bridge in Liaoning Province with a span of 202 m, while Second Yellow River Highway

Bridge in Zhengzhou, completed in 2004, has the largest scale of this bridge type. There are two separate bridges in the road section, and each bridge carries 4 lanes in each direction and has a net width of 19.484 m. The main bridge is composed of 8 spans of CFST tied arch bridges with each span of 100 m, as shown in Figure 5.

Double-deck-bridge also appeared in CFST arch bridge, e.g., the Qianjiang No. 4 Bridge in Zhejiang Province, completed in 2004, with span arrangement of 2×85 m + 190 m + 5×85 m + 190 m + 2×85 m, as shown in Figure 6.

Figure 3. Sanan-Yongjiang Bridge.

Figure 4. Wushan Yangtze River Bridge.

Figure 5. Second Yellow River Highway Bridge.

Figure 6. Qianjiang No. 4 Bridge.

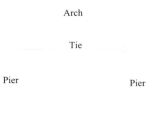

Figure 7. Rigid-frame tied through CFST arch.

2.4 Through rigid-frame tied arch bridge

In CFST rigid-frame tied arch bridges, arch ribs are fixed to the piers to form a rigid frame, so the arch rib can be erected similar to true arch using cantilever launching method. For small span bridge, the piers can stand small thrust forces caused by light self-weight of steel tubular arch rib and for large span bridge, temporary tied bars can be used. The construction of this type of bridges is easier than that of tied arch with deck girder stiffened bridges. The difficulty with the latter arises from the fact that the horizontal reactions are not available until the deck is completed.

High strength strands are employed as tied bars, which are pre-stressed to produce horizontal compression forces to balance the thrust of the arch ribs produced by dead loads. The tied bar should be tensioned step by step along with the increase of the dead load in the construction to balance the thrust, therefore, the tension sequence and tension forces of the tied bar must be determined in the design. In order to prevent crack which appears in reinforced concrete piers during the construction by in-span horizontal forces caused by prestressing of tied bars and out-span thrusts from the arches, a construction monitoring based on carefully calculation according to the construction stages is necessary (Yang & Chen 2007).

In through arch bridge, no side span is required like cable-stayed bridge or continuous girder bridge when a single main long span is needed to cross a railway or highway (Fig.7). The structure of the joint between the arch spring and arch seat on the top of the piers is very complicated, because arch rib, pier and end cross beams are joined together and tied bars are anchored there. The span of this type of bridges is generally 80 m 150 m.

Beizhan Bridge is a through rigid-frame tied arch bridge with a span of 150 m. The clear width of the bridge is 23.5 m. The four chord members in its truss rib are steel tubes of 750 mm diameter and 12 mm thickness, filled with C50 concrete (Figs 6, 8).

The longest span of such bridge type is 280 m in No. 3 Hanjiang Bridge (Fig. 9) in Wuhan. Most of the

Figure 8. Beizhan Bridge.

Figure 9. No. 3 Hanjiang Bridge.

CFST through rigid-frame tied arch has single span, while some of them have two or even more spans with separate tied bars for each span.

2.5 Fly-bird-type arch bridge

The most interesting structure type in CFST arch bridges is the so called fly-bird-type arch. This type of bridges generally consists of three spans (Fig. 10). The

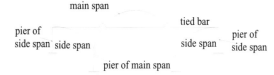

Figure 10. Fly-bird-type arch.

Figure 11. Maochaojie Bridge.

Figure 12. Ya-ji-sha Bridge.

central span is a half-through CFST arch and the two side spans are cantilevered half-arches. Both the main arch ribs and the side arch ribs are fixed to the piers and pre-stressed steel bars are anchored at the ends of the side spans to balance the arch thrusts. This bridge type has a large spanning capacity.

There are 9 of such bridges with a span above 200 m. The longest two bridges of this type are the Maochaojie Bridge, with a main span of 368 m and completed in 2006 (Fig.11), and the Yajisha Bridge, with a main span of 360 m and completed in 2000 (Fig.12). The latter was erected by combining vertical and horizontal swing method.

In CFST fly-bird-type arch bridge, it's necessary to have a good balance between the central span and side spans and minimize the bending moments in arch spring sections (especially in the side RC half arch).

The dead load should be taken into major consideration in design because it generally occupies a large part of the total load. Compared with the side spans, the central half-through arch has a longer span, so high rise-to-span ratio and light material (CFST) of arch rib should be used to decrease the thrust forces. In contrast, for the side half-arch with shorter span, low rise-to-span ratio and heavy material (generally RC) of arch rib should be used to increase the thrust. Sometimes, steel-concrete composite deck structure is applied in central span and RC deck structure is used in side spans. Furthermore, there are two end beams at each end of side spans, which are not only necessary for connecting the arch bridge and the approach, as well as anchoring the tied bar, but also helpful for balancing the horizontal thrust of main span. The key issue in design of such bridges is that under dead load the central span should act as a fixed arch, while the side spans act as half arches rather than cantilever girder (Chen et al. 2006).

3 GEOMETRIC PARAMETRIC ANALYSES

3.1 *Rise-span ratio and arch axis*

Rise-span ratio is an important parameter for an arch. The lower rise-span ratio means the greater thrust and axial force of the arch. The arch is suitable for bearing axial force, but it will lead to more redundant force for statically indeterminate structure. On the other hand, the greater thrust will result in the increasing of substructure cost. The span-rise ratio adopted depends on several factors as axial line shape, landform, etc. It can be seen from Figure 13 that there is little relationship between span and rise-span ratio of arches. For bridges with shorter span, the rise-span ratio can be adopted in a wider range.

But for longer span bridges, the rise-span ratio is adopted near 1/5. Figure 14 shows that most bridges have a rise-span ratio between 1/4~1/5 in which 1/5 is the most common value. In all investigated cases, most of the bridges with rise-span ratio greater than 1/4 or less than 1/5.5 are scenery bridges.

3.2 *Cross section of rib*

Based on the different cross sections of arch ribs, there are singular tube arch, dumbbell type arch, and multi-tube type arch (also called as truss type), as shown in Figure 15, as well as other shapes.

With regard to the dumbbell shaped cross-section, the two CFST tubes are connected by two steel web slabs. Both the single and dumbbell shaped section are solid rib. Generally speaking, single tube section is suitable for shorter bridges, the maximum span of such bridges is 80 m, while dumbbell shaped section

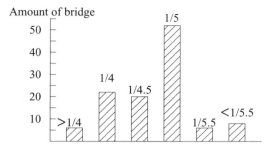

Figure 13. Statistics of rise-span ratio.

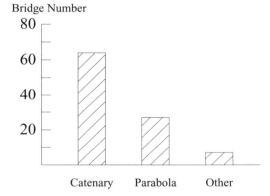

Figure 14. Arch axis statistics.

Figure 15. Cross-section type.

is widely used in the bridges with a span about 100 m, 160 m is the largest span in this kind of built bridge until now.

When more than two CFST tubes are used in a rib, they are connected by steel tubular web members to form a truss section. According to our investigation, three, four and six CFST tubes have been used in CFST arch bridges, but the three-tube section and six-tube section is only in some special cases.

As for the 119 bridges which cross-section is known, the arch rib of single tube, dumbbell, truss cross section are accounted for 12 (10%), 42 (35%) and 65 (55%) respectively. Figure 16 shows the relationship between the cross section of arch rib and span.

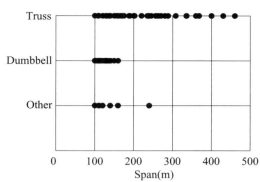

Figure 16. Cross-section type.

It indicates that truss section is the most widely used, and it should be preferential in large span bridges.

It is found that the height-span ratio (the height of cross-section) of CFST arch bridges is between 1/40~1/60 for constant cross-section and between 1/20~1/100 for varying cross-section.

3.3 *Material of arch rib*

The properties of steel and concrete are important for the mechanical behavior of the CFST structures. The Q345 steel is adopted in 80% of the investigated bridges and the Q235 steel in 20%. To achieve a better and more economical performance of CFST structures, the grade of concrete inside should be neither too low nor too high. It indicates that with the development of construction techniques, the concrete grade is adopted from C40 to C50 which is used most widely for the bridges built after 1995. C30 concrete is only used by 4 bridges built in early 1990's. C60 was only adopted in Wushan Yangtze River Bridge and Fengjie Meixi Bridge. Therefore, the match of steel and concrete strengths is important for the mechanical behavior of CFST arch bridges. In general, for CFST arch ribs, concrete C30 and C40 can be used to match the Q235 steel and C40 and C50 are appropriate for the Q345 steel.

4 ERECTION METHOD

Cantilever method, swing method, and scaffolding method are three major erection methods for the construction of CFST arch bridges (Chen 2005).

The relationship between construction methods and spans of arches is shown in Figure 17.

In construction of a CFST arch bridge, the steel tubular arch is erected at first and then concrete is

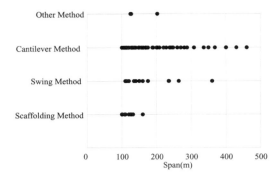

Figure 17. Relationship between erection method and span.

Figure 18. Erection of Wushan Yangtze River Bridge.

pumped into steel tubes to form CFST arch ribs. Though the thin-walled steel tubular arch has a lighter self-weight than concrete or shaped steel arch rib, it is still a key issue in construction when the span is longer. The popular erection methods used in CFST arch bridges are cantilever launching method and swing method as used in concrete arch bridges. However, these two construction methods have been improved with the development of CFST arch bridges.

In cantilever method, both main and auxiliary cables are used to maintain stability and balance during construction. These cables are stayed and controlled by jacks instead of windlass in the past. Therefore, the alignment of arch ring can be controlled by the adjustment of internal force of the fasten cables more easily than before. This method was adopted by 67% of the bridges. In these 8 bridges with a span no less than 300 m, 7 used this method, including the world's largest bridge, the Wushan Yangtze River Bridge with a 460 m span (Fig.18). It is evident that cantilever method has the most potential in construction of CFST arch bridges for its wide range of application, especially for long span bridges.

Another main erection method used in CFST arch bridges is swing method, which has been rapidly developed in recent years in China. This method includes vertical swing method and horizontal swing method. It is more suitable in some special conditions, such as gorge, or high requirements for clearance. Therefore, an appropriate landform and structural configuration are necessary for this method. Swing method is not used as prevalently as cantilever method. Statistics shows that there are about 15% adopted this construction procedure.

In vertical swing method, the semi-arch ribs are fabricated in low position and hoisted up into design level. This method is different with the one used in other countries where half-arches are built on the springs vertical and then rotated down on their lower end to close at the crown. This up-lift vertical swing method is mainly used in CFST arch bridge, for the tubular structure is much lighter than concrete arch ribs. Among those investigated bridges, Liantuo Bridge, Jing-hang Canal Bridge, and Wuzhou Guijiang Bridge are constructed by vertical swing method.

The first two CFST arch bridges constructed by horizontal swing method are the Huangbaihe Bridge and Xialaoxi Bridge near the Three Gorge Dam (Duan et al. 2001). The first CFST arch bridge in railway—Beipanjiang Railway Bridge opened in 2001 was also constructed by this method. Based on jakes pushing system as the drawing power, the rotated capacity in swing method is much greater than that by using windlass pulling system as mentioned before in concrete arch bridges. And the arch span can be much longer because the self-weight of steel tubular arch rib is much lighter than that of concrete arch.

A new method by combining vertical and horizontal swing method has been developed in CFST arch bridges, such as the fly-bird-type CFST arch bridge—Wenfenglu Bridge and Yajisha Bridge. For Yajisha Bridge, as shown in Figure 19, the half arch of the main span and cantilever half arch near it composed a rotation unit. First, the cantilever half arch was erected and main half arch was fabricated along the riverbank. Then, the main half arch was rotated vertically into right position. After that, the two half-arches were rotated horizontally, one 90° and the other about 117°. Finally, the arch rib was closed by a 1 m long rib-segment. The total weight of each horizontal rotating body is 136,850kN.

It can be seen from Figure 16 that the scaffolding method is mainly used for the bridges with shorter span, most of which are through deck-stiffened arch bridge, due to their short span and difficulty in applying other no-scaffolding method. One advantage of CFST arch bridge is the convenience construction procedure: the hollow steel arch rib could be erected

Figure 19. Erection of Ya-ji-sha Bridge.

first, and then the CFST arch ribs can be formed by pouring concrete into the tubes. As a result, it is easy to use no-scaffolding construction method for the erection of hollow steel tube arch rib with less self-weight. So for other type of CFST arch bridges, scaffolding method was seldom adopted and, if adopted, it is only for the bridges with short span.

For different construction conditions, various combined methods can be adopted. For example, the two side segments of Moon Island Bridge were erected by vertical swing method firstly. Then, the middle segment was vertically lifted up by hoist. For middle or shorter span, the construction procedure has more varieties. Jing-hang Canal Bridge in Xiyi Expressway with a span of 90 m was built by dividing the rib into two segments. The rib of the bridge was located by floating pontoon and then lifted. The rib of Kuokou Bridge in Fujian was erected by floating into position integrally.

5 CONCLUSION

The CFST structure has been applied prevalently and rapidly to arch bridges since 1990. Based on the rising demand of transportation in developing China, CFST arch bridges have a strong potential in bridge engineering. The analysis of design theory, construction techniques, and structure detail design specifications have become significant areas and been funded by government agencies. As a preliminary study, fundamental information of CFST arch bridges proposed in this paper should be a valuable reference for the practicing engineers, construction managers, and contractors in bridge engineering.

REFERENCES

Chen Baochun. 2004. Recent advances on design theory of CFST arch bridges. *Metropolitan Habitats and Infrastructure—IABSE Symposium*. IABSE Reports (88): 244–245. Shanghai.

Chen Bao-chun. 2007a. *Concrete Filled Steel Tubular Arch Bridges* (Second Edition). Beijing: China Communications Press, China.

Chen Bao-chun. 2007b. An overview of concrete and CFST arch bridges in China. *Proceedings of the Fifth International Conference on Arch Bridge*: 29–44. Madeira, Portugal.

Chen Bao-chun, Chen You-jie, Qin Ze-bao & Hiroshi Hikosaka. 2004. Application of concrete filled steel tubular arch bridges and study on ultimate load-carrying capacity. *Proceedings of the Fourth International Conference on Arch Bridge*: 38–52. Barcelona, Spain.

Chen Bao-chun, Gao Jing & Zheng Huai-ying. 2006. Studies on behaviors of CFST "Fly-bird-type" arch bridge. *Proceedings of the International Conference on Bridges*: 205–212. Dubronvnik, Croatia, SECON HDGK.

Yang Ya-Lin & Chen Bao-Chun. 2007. Rigid-frame tied through arch bridge with concrete-filled steel tubular ribs. *Proceedings of the Fifth International Conference on Arch Bridge*: 863–868 zheng. Madeira, Portugal.

Zheng zhenfei, Chen Baochun & Wu Qingxiong. 2000. Recent development of CFST arch bridge in China. *Composite and Hybrid structures, Proceedings of the 6th ASCCS International Conference:* 205–212. Los Angeles.

Design of steel towers for Skylink Tramway

Q. Ye
Weidlinger Associates, Inc., New York, USA

M. Venuto
P.L.S., Delaware River Port Authority, Camden, USA

D. Murphy
Delaware River Port Authority, Camden, USA

ABSTRACT: An aerial tramway has been designed to cross the Delaware River, linking the states of Pennsylvania (PA) and New Jersey (NJ). It consists of a suspended gondola system supported by one steel tower and one concrete anchorage on each side of the river. The steel towers will be about 93 meter high and are made of circular pipe sections with diameters varying from 2.1 m to 3.7 m. A few major challenges were overcome during the design, such as: improving the fatigue resistance of tubular connections; studying the vortex shedding behavior of the circular tower legs; fabricating the tower legs; and developing erection procedure.

1 INTRODUCTION

An aerial tramway, named Skylink Tramway, has been designed to cross the Delaware River, linking the states of Pennsylvania and New Jersey. Upon completion, it will be owned and operated by the Delaware River Port Authority (DRPA).

The aerial tramway consists of a suspended gondola system supported by one steel tower and one concrete anchorage on each side of the river, as shown in Figure 1. The gondola system designed by LEITNER POMA consists of two sets of cables, two haul ropes and one rescue rope.

The two concrete tower foundations were already built through a previous contract, and anchor rods were installed for connections to future tower legs. The steel towers will be about 93 m high, and the distance between them is approximately 579 m.

DRPA contracted Weidlinger Associates to conduct complete designs for the steel towers and concrete anchorages. This paper will address some special issues that came up and were overcome during the design of steel towers.

2 PIPE SECTIONS VS BOX SECTIONS

After completion, the steel towers will be subjected to insignificant axial loads, but very high wind loads due to their elevated height and proximity to the river. Therefore, bending behaviors of tower legs dominated their design.

In order to reduce wind loads and achieve an economical design, pipe sections have been selected for the tower legs and the middle strut. As shown in Table 1, the upper portions of tower legs, which are located between the middle and top struts and are curved for esthetic purposes, have a constant outer diameter of 2.134 m; the outer diameters of lower legs vary from 2.134 m to 3.658 m; the middle struts have a 2.134 m constant outer diameter. Due to the need of providing a walk platform on the top struts, box sections were chosen.

The reasons of choosing pipe sections are two-fold:

i. the pipe sections have smaller drag coefficient (about 1.0) than that of box sections (about 2.0), resulting less wind loads;
ii. compared with box sections, the pipe sections have much higher local buckling capacities, as shown in the example in Table 2.

Having benefits notwithstanding, using pipe sections has its own set of problems, and most notably are:

i. the greater potential of vortex-shedding induced resonant vibrations;
ii. and potential fatigue damage at various tubular connections.

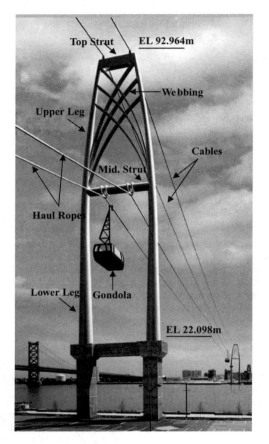

Figure 1. Steel tower on pennsylvania side.

Table 1. Pipe section diameter and thickness.

Member	Outer diameter (m)	t (mm)
Upper leg	2.134	19.0
Lower leg	2.134–3.658	25.4
Middle strut	2.134	25.4

To investigate and resolve the above-mentioned concerns, as well as to validate the design wind loads, Weidlinger proposed wind tunnel testing which was conducted by Rowan Williams Davies & Irwin Inc. (RWDI), Canada.

3 WIND TUNNEL TESTING

To study the wind effects on the steel towers, a 1:150 scale full aeroelastic model of the tramway (Figure 2) was designed, fabricated and tested in RWDI's boundary layer wind tunnel for 15 wind directions at various wind speeds. The structural damping of the individual tower model without the ropes was adjusted to 0.5% of critical. Accelerometers, strain gauges, laser displacement transducers and force balances were installed on the tower models for measuring tower responses.

The tested wind speeds cover the full scale wind speed from monthly speed to a 1000-year return period speed. During the testing, the entire range of wind speeds was first scanned to investigate if there were any signs of vortex-induced oscillations. The test results verified that the vortex-induced oscillations are not of concern and the wind-induced tower response were mainly caused by static wind loading and wind buffeting.

The tests also generated fatigue wind loading for the design of tubular connects at the middle strut and tower base locations.

In addition, the wind tunnel testing provided effective static design wind loads in both the longitudinal and transverse directions. One notable conclusion is that the measured static wind reactions at the bases of tower legs are much smaller than those calculated in accordance with relevant provisions in ASCE-07. In the transverse direction, the measured maximum moments are about 40% of that of ASCE-07; in the longitudinal direction, the ratio is about 27%. To be conservative ASCE-07 wind loads were still used for the design.

Table 2. Local buckling stress, pipe section vs square section.

Section	Pipe	Square
Area (m^2)	0.292	0.292
D or b (m)	3.658	2.865
t (mm)	25.4	25.4
Local buckling stress	$F_{xc} = \left[1.64 - 0.23 \cdot \left(\dfrac{D}{t} \right)^{1/4} \right] \cdot F_y$	$F_{xc} = k \cdot \dfrac{\pi^2 \cdot E}{12 \cdot (1 - \mu^2) \left(\dfrac{b}{t} \right)^2}$
F_{xc} (MPa)	291.0	64.8
Source	API D.2.2-4b	Steel Structures, 6.14.28; $k = 5.5$.

Figure 2. Wind tunnel testing in RWDI, Canada.

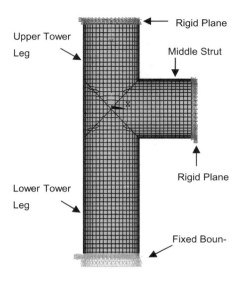

Figure 3. FE model for middle strut to tower leg connection.

4 MIDDLE STRUT TUBULAR CONNECTIONS

4.1 Strength design of connection

Connections between thin-walled tubular members have always been a design challenge, mainly due to the fatigue sensitive welding details and weak punching shear capacities. The critical connections in steel towers are the intersections of the middle strut and tower legs, since they will subject to the maximum moment under transverse winds.

A finite element model was created for studying the stress conditions at these connections and arriving at an optimal solution for stress reduction. As shown in Figure 3, the model comprises of portions of the tower legs and the middle strut. The bottom of the tower leg was fixed, and rigid planes were created at the top of the tower leg and the free end of the middle strut for applying loads, i.e. axial forces and moments. Under the factored ASCE design wind loads, which has a 60s-average wind speed of 120 mph at Elevation 54.9 m with Exposure Category D topography and a 100-year return period, the maximum vertical stresses in tower legs are shown in Figure 4, and the maximum compressive stress exceeds the yielding stress, which is 345MPa.

Various stiffening options were tested for reducing stresses in tower legs, including installing horizontal or sloped plate stiffeners inside the tower legs at different elevations adjacent to the middle strut, and the most effective solution is by installing curved internal stiffener plates, as shown in Figure 5.

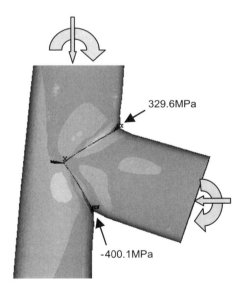

Figure 4. Vertical stress in legs under maximum wind loads.

The maximum vertical stresses in the column under the 100-year design wind loads are reduced to 238MPa (Figure 6). Also, with internal stiffeners, there is much less local deformation in tower leg shell plate, as evidenced by comparing deformed shapes in Figure 4 and Figure 6. Therefore, it was concluded that the stiffeners are effective and the stresses are acceptable at the strength limit state.

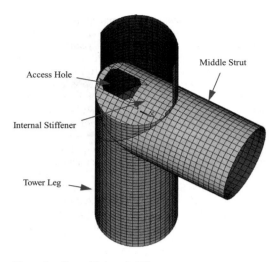

Figure 5. Curved internal stiffeners.

that over the design life, the DRPA Towers will be subjected to a vast number of cycles of different magnitudes. After processing the reading of strain gages, which were installed on the middle strut near tower legs, RWDI generated the spectra of moment ranges versus numbers of cycles, and Weidlinger further processed the data by Miner's rule to obtain the effective fatigue moment at any given number of cycle, as shown in Figure 7.

With the fatigue moments, analysis was first performed on the finite element middle strut model shown in Figure 3; then a much more refined local model was created at the most critical region, which in this case was the lowest point of the strut-to-leg connection, and analyzed for obtaining local stress concentrations. The maximum fatigue stress ranges were calculated and are presented in Figure 8, which shows that the calculated fatigue stress ranges are well below the AWS Hot Spot Stress Curve X (t = 25 mm), therefore there will be

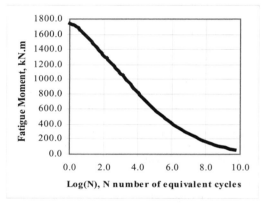

Figure 7. Effective fatigue moment at mid. Strut of NJ tower.

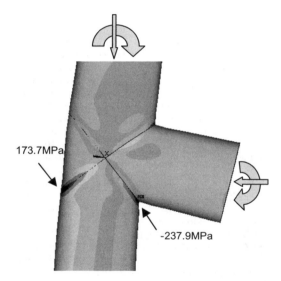

Figure 6. Vertical stress in tower legs with curved stiffeners under maximum wind loads.

4.2 Fatigue design of connection

Cyclic loading for these connections mainly come from the wind loads, and may cause fatigue damage. Since there is no provision currently available to determine the effective fatigue wind loads, we turned to wind tunnel testing for providing this information.

The steel towers will be subjected to wind speeds of various intensities during their lifetime. As the wind speed varies, the number and magnitude of cycles present in the loading will vary as well. This implies

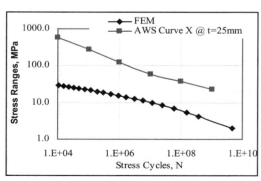

Figure 8. Hot-Spot fatigue stresses at mid. Strut of NJ tower.

essentially no fatigue damage during the design life of towers.

4.3 Detailing of middle strut connections

A portion of the middle strut and the tower leg, as well as internal stiffeners, will be shop welded to form a T connection or node (Figure 9), which will then be field bolted to the rest of the towers.

Complete Joint Penetration Groove welds are specified for these connections and their details conform to the prequalified joint details with improved concave profile for fatigue as specified in AWS D1.1 2004. At the most critical locations, which are around the highest and lowest points of the connections, the weld profiles are further improved for fatigue by back gouging and adding fillet weld reinforcement, as shown in Figure 10.

Due to the fact that the tower leg shell plates will subject to out-of-plane tensile stress at these locations, there was a risk of lamellar tearing damage. To counter this risk, a special steel material, which conforms to API Spec 2H with S4 (through thickness testing) and S5 (low sulfur steel) supplemental testing requirements, was specified.

5 TOWER LEG BASE CONNECTIONS

5.1 Detailing of tower leg base sections

Besides the middle strut connections, the design of tower leg base connections to the concrete piers was another challenge to overcome. As stated in INTRODUCTION, the tower foundations were already constructed and protruding anchor rods were installed for connecting tower legs, however the existing anchor rods followed an oval pattern which was the then assumed shape of tower legs at their base.

Therefore, special segments of tower legs are needed for the transition from circular sections to oval sections, and Figure 11 shows such a segment as designed. This segment consists of two half circular sections joined together by triangular connecting

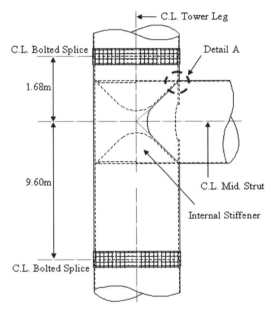

Figure 9. Detailing of middle strut to tower leg connections.

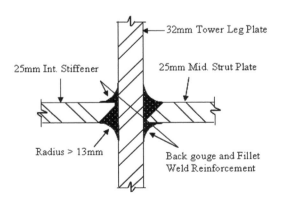

Figure 10. Detail A, reinforcement of CJP welds.

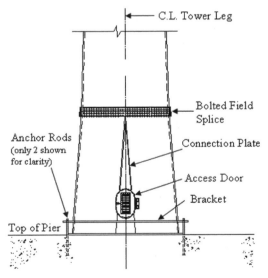

Figure 11. Tower leg base section.

plates in-between. These plates are stiffened internally by angles. Also, one access door per leg is needed. At the top, a bolted field splice was designed for connecting the circular tower leg; at its bottom, a steel bracket was designed for connection with anchor rods, as shown in Figure 12.

5.2 Strength and fatigue design of connections

A detailed finite element model was created for this segment, as shown in Figure 13. Shell elements were used for most of the structural elements, except the anchor rods, which were modeled by beam elements.

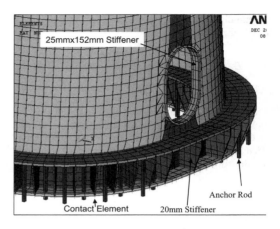

Figure 14. Modeling details of leg base.

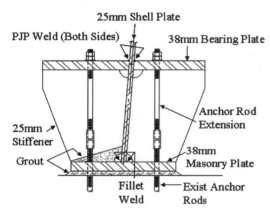

Figure 12. Tower leg base bracket details.

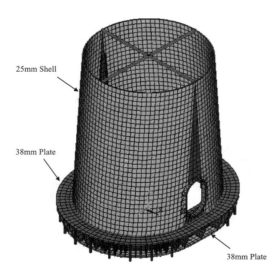

Figure 13. Finite element model of tower leg base.

A rigid plane was created on the top of the segment for applying loads transmitted from the tower leg. As shown in Figure 14, all the stiffeners and anchor rods were explicitly modeled; the access door and stiffener were also accurately modeled; contact elements were created underneath the masonry base plate for capturing possible uplifting behavior and calculating anchor rod tensions.

Factored design wind loads were applied on top of the segment and nonlinear analysis was performed. The maximum compressive vertical stress occurred next to the access door and is 219 MPa, which is less than the 345 MPa yield point, as well as the locale buckling stress, 291 MPa. The maximum anchor rod tension is 402 kN, which is OK since it's less than its proof load of 614 kN.

Fatigue evaluation of the welded bracket connections was also conducted in a manner similar to that for the middle strut connections. Multiple local models were created and effective fatigue moments were applied on the model. Proper weld details were developed for satisfying the fatigue requirements.

6 FABRICATION OF TOWER LEGS

6.1 Lower tower legs

The portions of tower legs below the middle struts are tapered circular sections. For fabrication, the maximum plate size commonly available is 25 mm × 3.658 m × 17.424 m. As shown in Figure 15, each of these plates can be easily rolled into a tapered section with a maximum length of 3.658 m; then form a closed circular section by longitudinal seam welds; afterwards, transverse girth welds are used to join two adjacent sections together.

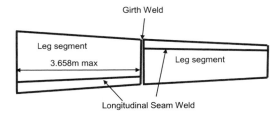

Figure 15. Fabrication of lower tower legs.

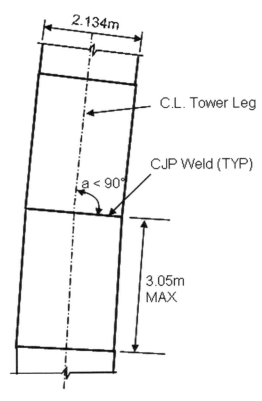

Figure 16. Miter joints in upper tower legs.

6.2 *Upper tower legs*

The portions of tower legs above the middle strut are of constant diameter, but are curved in shape. Two fabrication methods were evaluated, which are:

- **Induction Pipe Bending:** first fabricate straight legs by the same method as that for the lower legs; then slowly pull the legs through a bending machine, where the leg is locally heated by an electrical belt; afterwards the legs are gradually bent into the curved shape;

- **Miter Joints:** in this method, straight leg sections are first fabricated; then their two ends are cut into a sloped shape; afterwards, adjacent segments are joined together by complete joint penetration welds, as shown in Figure 16.

Between these two methods, the induction pipe bending method is more cost effective due to its usage of automatic welding process for all the seam and girth welds. However, the current maximum bending diameter in America is 1.68 m, which is smaller than the diameter of the upper leg, 2.13 m, which is governed by the access requirement inside the tower legs. Therefore, this method is not suitable.

The miter joints are feasible, but due to the need of large amount of manual welding at these joints, the unit cost of steel after fabrication is much higher than that of induction bending. Nonetheless, this method is the only workable option, so it is specified for the upper legs.

7 TOWER ERECTION METHOD

How will the towers be erected? This is an important question, and the answer needs to address both economical and safety issues. To be economical, the towers need to be erected in a short period so that the cost of labor and equipment will be minimized; to be safe, proper temporary supports need to be constructed and the construction needs to strictly follow specified steps.

Figure 17 shows the preferred erection stages:

1. erect tower base sections and lower tower legs; a maximum of 3 field splices per lower tower leg is allowed;
2. erect the middle strut section, which include the strut-to-column connections and middle strut itself;
3. lift and install portion of two upper tower legs, which are connected together by part of the webbing and temporary transverse struts;
4. install the remaining portions of upper tower legs and the top strut; remove temporary transverse struts;
5. install the remaining webbing elements;
6. tower erection complete and it is ready for cable erection.

At each stage, bolted field splices are designed for connecting different sections together. Field welded splices are not allowed due to the difficulty of quality control.

The weights of tower members at each stage are shown in Table 3.

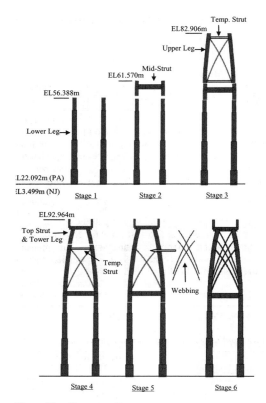

Figure 17. Tower erection sequence.

Table 3. Tower member weights at erection stages, metricton.

Stage	Member	PA tower	NJ tower
1	Lower Leg	87.1	126.1
2	Middle Strut	54.4	54.4
3	Upper Leg	79.8	79.8
4	Top Strut	56.2	56.2
5	Webbing	10.9	10.9

7.1 Temporary supports

Not shown in Figure 17 are the temporary supports needed to resist the wind loads and minimize wind vibrations. The selected option is to drive piles in the river bed on both sides of each tower, then erect wire ropes connecting the top of piles and the middle strut. More details on this will be given in Section 7.3.

7.2 Barge cranes

To implement the above tower erection stages, a method utilizing barge cranes was first envisioned, since the towers are located by the river. If feasible, this will be the fastest and most economical method. However, the largest barge crane in America has a lifting height of about 70 m (with 68 metric tons lifting capacity), but the towers will reach a total height of about 93 m above water.

To make this method feasible, it was suggested to extend the boom of the barge crane by about 25 m. The owner of this barge crane expressed a few concerns about this boom extension and lifting operation, such as there will be extra costs and time needed for the boom extension, and there is safety concern about lifting a 56.2 metric-ton top strut at 93 m above water using this particular barge. Therefore, this option is abandoned.

7.3 Tower cranes

Tower cranes have been commonly used for construction of tall structures, such as high rise buildings. The steps of tower erection using tower cranes are shown in Figure 18 and 19, and are described as follows:

1. build foundations for anchoring cranes;
2. erect the towers to the middle strut level;
3. brace the cranes against the towers;
4. the cranes self-climb to an elevation above the top strut;
5. erect the upper portions of towers;
6. disassemble the cranes;
7. after the main cables are erected, remove the temporary ropes and piles

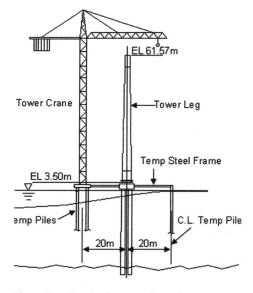

Figure 18. Erecting lower portions of towers.

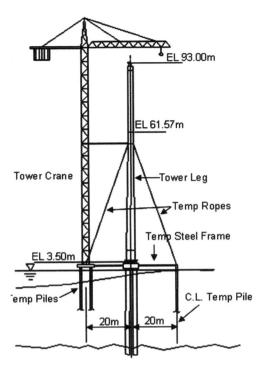

Figure 19. Erecting upper portions of towers.

To allow for the use of relatively smaller tower cranes, which are more economical, the tower legs can be further divided into smaller segments and be joined together in the field by bolted splices. However, due to esthetic reasons, the maximum number of bolted splices is generally limited to two or three.

8 SUMMARY

The following conclusions can be made after the completion of this project:

i. Due to higher local buckling stress of circular sections, high-rise steel towers made of circular sections are more efficient and economical than the traditional rectangular steel towers.
ii. Vortex shedding induced vibrations were not observed during the wind tunnel testing.
iii. The stiffened connections between middle struts and tower legs satisfy the fatigue requirements.
iv. Straight portions of tower legs can be cost-effectively fabricated, while the curved portions are more costly, unless their diameters are reduced to fit the largest pipe bending machines.
v. To safely erect the towers, temporary supports are needed and may have big impact on the final cost of the project.

ACKNOWLEDGEMENT

This project was funded by Delaware River Port Authority (DRPA). DMJM/AECOM performed project management for DRPA. LEITNER POMA was the designer of the cable and rope system. Peter Marshall served as a consultant on this project and provided valuable advice on the fabrication and erection of towers. RWDI performed wind tunnel testing. Mueser-Rutledge checked the capacity of existing tower foundations.

Analysis and design of hollow steel section folded plate—the hall of Petroleum Science and Technology Museum

J.M. Ding & H.L. Wu
Department of Building Engineering, Tongji University, Shanghai, China

Z.J. He
Architectural Design and Research Institute of Tongji University, Shanghai, China

ABSTRACT: This paper presents a design case on the structure of the hall of Petroleum Science and Technology Museum in Daqing, a lightweight and transparent enclosure structure which shows a shape of diamond. The project features a combination of tubular structure and folded plate structural system, which represents a classical yet pleasing steel structure. This innovative structure utilizes 10 v-shaped folded plates that cover a large volume public space and 4 v-shaped folded plates that support the facade glass curtains and a part of weight of the roof. The other part of roof weight is supported by the exhibition buildings which are behind the hall. In the view of mechanical character, the roof structure functions as a spatial shell. In order to release the shell thrust on the supporting structure, single-direction sliding bearing and tension ring beam have been used to form a "self-balance system". This paper introduces the components of structural system and the static performance. The linear and nonlinear buckling analysis is conducted to study the structural stability. The result demonstrates that this structure is good in mechanical properties.

1 INTRODUCTION

Folded plate structures have inspired and fascinated people for decades. In the last century, folded plate structure was one of the most popular structural systems that were utilized in many long-span structures. For example, the building for United Nations Educational, Scientific and Cultural Organization in Paris was a significant folded plate structure designed by P.L.Nevil. However, an oppressive and opaque interior space's effect is created by concrete folded plate structures, which often influence the feeling of people, so the concrete folded plate structures have not been utilized broadly in recent years.

Today, tubular structures are widely recognized as unique and innovative structural solutions, which create new and dramatic forms while enclosing large spaces and providing new design methods for people to enjoy natural light in interior space. So, is it possible for structural engineers to integrate the features of tubular structure and folded plate structural form into a classical yet pleasing structure design? The answer is positive.

This paper presents structural system, structural analysis and detail design of the hollow steel section folded plate used in the structure of the hall of Petroleum Science and Technology Museum in Daqing that are helpful to structural engineers.

2 PROJECT OVERVIEW

The Petroleum Science and Technology Museum is designed to be an advanced and professional museum (Figure 1), which functions as a large capacity and multi-use public project. The owners of this new and innovative museum hope to broaden the influence of petroleum science and technology culture, and make people get to know the history of petroleum exploitation in China.

The hall, located in the center of the Petroleum Science and Technology Museum, will be used to perform some great celebrations. At the same time, it functions as a connection of other parts of the exhibition buildings (Figure 2). The hall occupies an area of 1654 m². Together with the adjacent 3 parts of exhibition buildings, they will buildup a large complex museum.

The hall is particularly unique because it shows a shape of diamond (Figure 3). Architect called for a structural design of the "transparent diamond", which is based on the following considerations.

Figure 1. Overall of the museum.

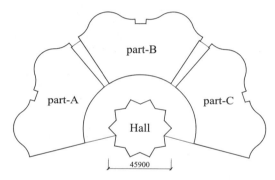

Figure 2. Site location of hall.

- Architectural and structural morphology should be concordant and unified;
- Structure morphology should be regular and elegant;
- The roof structure should be transparent to release oppression of the interior space.

Therefore, one of the primary goals of the engineering team was to develop the most efficient structural system that met architectural requirement. By collaborated with the architect, structural engineers integrated the features of tubular structure and folded plate structural form into this project, and the hollow steel section folded plate is applied to the roof structure and the curtain wall structure.

3 STRUCTURAL SYSTEM OVERVIEW

This innovative hall structure utilizes 10 v-shaped folded plates that cover a large volume public space and 4 v-shaped folded plates that support the facade glass curtains and a part of weight of the roof. At the end of v-shaped folded plate columns, there are 4 fixed bearings that restrict 3-dimensional translations. And at the back of the hall, 5 single-direction sliding bearings that can only deform along the radial direction connect the hall with in-situ reinforced concrete exhibition buildings (Figure 4).

3.1 Structural arrangement of roof

The roof structure (Figure 5 and Figure 6) is composed of 10 v-shaped folded plates, which are laid out circularly. Every V-shaped element consists of two in-plane

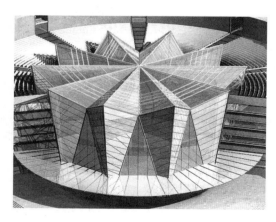

Figure 3. Perspective view of hall.

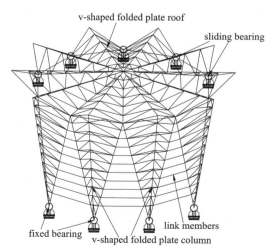

Figure 4. Perspective view of structural model.

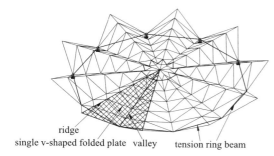

Figure 5. Perspective view of roof model.

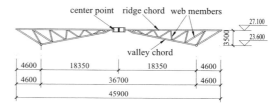

Figure 7. Section plane of roof model.

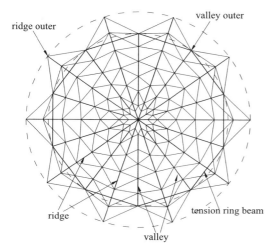

Figure 6. Plan form of roof model.

truss. The declining intersection lines of adjacent v-shaped elements are the folded plate valleys. The outer ends of the valleys locate at a circle of 36.7 m diameter and the height of the outer valleys is 23.6 m. The horizontal intersection lines of adjacent v-shaped elements are the folded plate ridges. The outer ends of the ridges locate at a circle of 45.9 m diameter and the height of the ridges is 27.1 m. Within every in-plane truss, web members are utilized to provide enough shear stiffness for the v-shaped folded plate (Figure 7).

The inner end of valleys and ridges of the roof converges at the center point, which can't bear bending moment as a truss, therefore the folded plate roof does not function as a folded plate at all, but rather as a spatial shell that is consisted of 5 arches. It results that the spatial shell has a much stronger vertical stiffness than the folded plate and also improves the structural efficient. However, shell action can create a big thrust on structural support. In traditional solution, strong support which constraints all translations of shell bases would be used to balance structural horizontal thrust force. The solution can cause a serious problem to the supports, especially under extreme temperature variations. The great thrust force would happen because of the different thermal expansions of concrete and steel. The great thrust leads to a great impact on supporting structure and structural detail design, and increase much more cost on the project.

Engineers deal with this challenge using a method called "self-balance system", an innovative solution that combines single-direction sliding bearing with the tension ring beam (Figure 5 and Figure 6). The single-direction sliding bearing releases the thrust force from the hall roof, so the in-situ reinforced concrete exhibition buildings would only take a vertical force of the hall roof structure. At the same time, the tension ring beams linking the adjacent outer ends of the plate valleys can prevent the outspread of v-shaped folded plates and balance the thrust force of the plate valleys. Therefore, this solution utilizes pressure property of concrete and tension property of steel sufficiently.

3.2 Structural arrangement of curtain wall

The curtain wall structure (Figure 4) consists of 4 v-shaped folded plate columns and a number of link members. The link members are applied between every two columns. They will provide lateral support to the columns, and increase the spatial stiffness of columns. They also transfer the curtain wall's gravity load and the lateral wind load to the v-shaped folded plate columns.

3.3 Structural member choice

Due to the aesthetical requirements by the architect, all structural steel members are circular hollow sections. Yield strength of the steel is 345 Mpa. Considering temperature of the site is so low in winter, impact toughness of the steel is up to classical C, which assures the steel's impact toughness around −20 °C. The main member sections are laid out in Table 1.

Table 1. Main member sections of the hall.

Structural member	Circular hollow section	
	Diameter(mm)	Thickness(mm)
Rigid chord	245	10
Valley chord	245	10
Web member	203	8
Tension ring	273	14
Curtain link	203	8

4 STRUCTURAL ANALYSIS

4.1 Structural analysis considerations

It is important that spatial shell structural overall stability should be ensured under all loading conditions. As to this hall, structural deformation should also be controlled strictly because glass curtain wall is used. Considering the site, the hall was analyzed under numerous loading conditions including snow loading, snow drifting, and wind uplift. Additionally, thermal impacts throughout the full range of climate temperature variations on the structure were also analyzed. SAP2000 and ANSYS structural analysis software were used to establish three-dimensional structural models and account for static performances and overall stability of the structure.

4.2 Structural loading

According to Chinese Load Code for the design of building structures, basic wind loading in Daqing is 0.65 KN/m². Considering the structural shape, the eventual wind loading is calculated to be 1.07 KN/m². The roof surface of diamond shaped is inclined to snow amassment. So it is important to have an accurate measure of the snow loading that make the building design to be further optimized and safety. The maximum snow loading is considered as 1.75 KN/m². The full range of climate temperature variations during a year is estimated around ±40 °C.

4.3 Structural static performances

Under the load combination of dead loading and live loading, the force distributions of valley chords are shown in Figure 8. From figure 8, it is found that the force of valley chords is increasing from the center to the outer side, which is just as the mechanical characteristics of spatial shell. It is interesting that though the geometry of the roof structure is symmetry, the forces of valley chords are uneven. The maximum force value of valley chords connected with curtain wall columns is 481 kN, which is bigger than the value of valley chords supported on single-direction

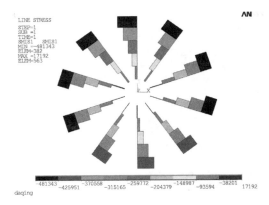

Figure 8. Force distributions of valley chords.

sliding bearings. The reason of this phenomenon is that the supporting stiffness of sliding bearing is less than the curtain wall columns, which leads to different supporting boundary condition. But the roof forms a self-balance system due to the design of tension ring beams, which reduces the difference of the lateral stiffness of roof supporting. Therefore, the difference of all valley chords' forces is controlled in a small range.

Under the action of typical load cases, forces of ridge chords, valley chords and tension ring beams, displacements of roof crown joint and the sliding bearing are shown in Table 2. The analysis demonstrates the maximum structural deformation under snow loading is 73 mm, which is 1/540 of the span of roof structure. This result satisfies the standard of Chinese Code for design of steel structures, which limitation is 1/400 of the span of roof structure. Temperature variation only causes a little influence on the structure because of "self-balance system". Under the range of ±40°C variation, the maximum change of the force value of the valley chord is only 11 kN.

4.4 Overall stability analysis of the structure

It is important for structure engineering to consider the stability of steel structure, especially spatial shell structural system. In the design process of this hall structure, the critical buckling factor and ultimate bearing capability have been got by structural overall stability analysis.

Nonlinear analysis is carried out by using the finite element program ANSYS. Different beam elements are used to meet the needs of different mechanical characteristics. Beam4 element is used for the analysis of buckling and geometrically nonlinear (large displacement) overall stability and beam188 element is used for the analysis of double nonlinear (large displacement, elastic-plastic) overall stability. Three

Table 2. Main result of structural static performance.

Analysis case	Force of rigid chord (KN)	Force of valley chord (KN)	Force of tension ring beam (KN)	Vertical displacement of roof crown joint (mm)	Displacement of sliding bearing (mm)
Dead + Snow	−532	−850	1125	−68	7.8
Dead + Snow −40 °C	−539	−839	1127	−73	2.4
Dead + Wind +40 °C	5	−288	111	−0.67	6.5
Dead + Wind −40 °C	−7	−266	115	−10	−4.2

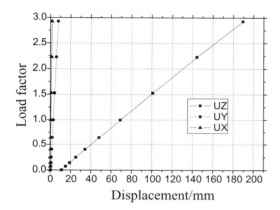

Figure 9. Load-deflection curve of roof crown joint.

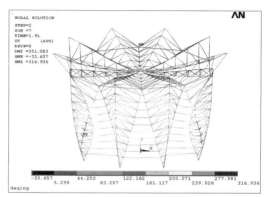

Figure 11. Deformation of structural ultimate bearing capacity.

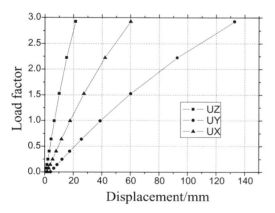

Figure 10. Load-deflection curve of critical column joint.

load combinations were considered to the structure including full-span snow load, half-span snow load and wind load. The most unfavorable load combination is applied to the analysis of structural ultimate bearing capacity.

The linear buckling analysis indicates that the v-shaped folded plate column will first buckle under 3.65 times of load combination of dead loading and full-span snow loading. Under the same load combination, typical load-deflection curves of the roof crown joint and curtain wall column's joint are shown in Figure 9 and Figure 10. From the two figures, we can see when the load factor is up to 2.93, the hall structure can not bear more load. Figure 10 shows that the slope of load-deflection curve is deceasing with the increasing of load factor, which demonstrates that the structural vertical stiffness is reducing. When ultimate bearing capacity of the hall structure reaches, torsional buckling happens on the v-shaped folded plate columns (Figure 11).

5 STRUCTURAL DETAIL DESIGN

Structural detail design should be true of mechanical characteristics of analysis model. At the same time, it should be satisfied with safety, enduring and aesthetical requirement.

The valley chords of roof structure are supported at the ends by the in-situ reinforced concrete column single-direction sliding bearings (Figure 12). On the perpendicular direction to the sliding direction, the sliding bearings are restrained. The displacement of

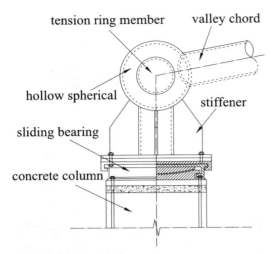

Figure 12. End supports of roof by sliding bearing.

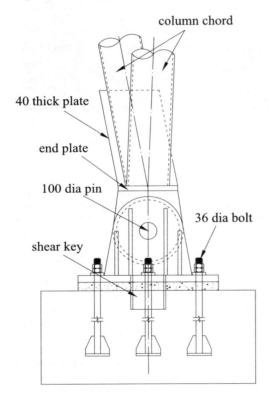

Figure 13. End supports of column by pin.

the sliding bearing is limited to 50 mm to prevent the roof structure from collapsing under the action of seldom occurred earthquake. Under the actions of frequently occurred earthquake, snow load, wind load and temperature, the sliding bearings can move unrestrainedly. At the convergence of members, a hollow spherical is designed to transmit the force from all of the members, and is convenient to connect all of the members.

The v-shaped folded plate columns are supported at the ends through pin connections (Figure 13). Because three chords of the column converge at one point and the angle among the chords is small, the connection is complicated. In detail design, a connecting plate which is 40 mm thick is adopted to transmit force from column to the pin which has a diameter of 100 mm. Bolts and shear keys transmit the forces into basements.

6 CONCLUSION

The hall of Petroleum Science and Technology Museum in Daqing is a unique and innovative hollow steel section folded plate structure. This structural solution creates a thin structure that covers a large volume public space, and allows natural light to penetrate in interior spaces through roof and curtain wall, and forms a dramatic architectural effect from both interior and exterior of the hall. The hollow steel section folded plate structural system also serves as good mechanical characters, and can take advantage of structural material appropriately, and improve structural efficiency.

This fine project demonstrates engineers can create the concord and unification of architectural and structural morphology eventually by integrating tubular structure with other classical structural system and using innovative solutions.

REFERENCES

Benjamin, B.S. 1984. Structures for Architects, Van Nostrand Reinhold Company, Inc.
Chen, W.F. 2000. Structural stability: from theory to practice. Engineering structures, 22:116–122.
Nie, G.H., Cheung Y.K. 1995. A nonlinear model for stability analysis of reticulated shallow shells with imperfections. Journal of Space Structures, 8(4):215–230.
Siegel, C. 1981. Structure and Form in Modern Architecture, Crosby Lockwood & Son Ltd.

Welded tubular structure for the Zadar dome

E. Hemerich & I. Šošić
Civil Engineering Institute of Croatia, Zagreb, Croatia

D. Lazarević & M. Anđelić
University of Zagreb, Faculty of Civil Engineering, Zagreb, Croatia

ABSTRACT: Described are construction details of the steel welded tubular dome structure with span of 89.5 m built in Zadar, Croatia. Discussed are primary structure—a lamella monolayer space dome composed of the three-directional trimmed steel tube beam system mesh, as well as the role of the secondary load-bearing structure of the external envelope integrated into the load bearing system and made to act compositely.

1 DESCRIPTION OF THE OVERALL STRUCTURAL SYSTEM

The dome is a part of the regularly shaped spherical body. The sphere is approximately 194 m in diameter. The dome itself measures 140 m in span, it is about 30 m high, and its central angle is at about 92°. From foundations to 20 m in height the dome is designed as a reinforced concrete structure. From that height and up to the top of the dome, the steel monolayer space dome about 90 m in diameter is planned. The top ends with an opening measuring about 12 m in diameter. The audacity of the entire dome is about 30/140 ≈ 1/4.7, while the audacity of its steel part is 10/90 = 1/9. The dome disposition, with its main structural elements, is presented in Figures 1 and 2.

Although the spherical envelope is geometrically quite simple, it is a complex structure made of two materials, with a great number of big openings, and with a big latticed part (see Figures 3 and 4). The problem of local and global stability, and hence of the overall reliability of the structure, dominantly determines the entire system's susceptibility to imperfections during construction work.

As this conference mainly focuses on steel tubular structures, the following text will specifically deal with problems relating to construction of the steel part of the dome.

Figure 1. Section view.

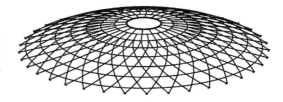

Figure 2. 3d view.

2 PRIMARY LOAD-BEARING TUBULAR STEEL STRUCTURE OF THE DOME

2.1 Concept

According to its concept, the primary load-bearing tubular steel structure of the dome can be described as a lamella monolayer space dome composed of the three-directional trimmed beam system mesh.

The steel part of the dome is 89.50 m in axial diameter, and its height in apex is 11.0 m. All tubes of the primary load bearing structure are hot-process longitudinally-welded round tubes (procedure: high frequency electric resistance induction welding), the quality is S355 J2H and the dimensions are Ø323.9 × 7.1 mm (the only exception being the tube at the top ring which measures Ø508 × 12.5 mm). The tubes are compliant with EN 10210–1 and 2. The total of 217 tons of tubes was installed.

The steel structure of the dome is supported by the relatively rigid ring situated at the top part of the reinforced-concrete structure.

The entire steel tube latticework of the dome is realized by butt welded and fillet welded joints (connections between ring-shaped tubes and fillet welds of nodes). In workshop, the ring tubes are cold processed into precast circular segments, and tube ends of the fill are workshop-shaped and prepared for welding. The realization of joints by welding ensures that there will be no reduction in shell stiffness at joint location, and that no significant unexpected deformation will occur in case of imperfections at the realization stage.

Figure 3. View from the air.

Figure 4. View from the inside.

In order to simplify the realization, and also to facilitate node shaping, the rings are staggered so that no more than four elements are used per one node.

2.2 Local and global stability

The numerical testing was conducted to test the local and global stability of the primary load-bearing steel structure of the spherical shell. The objective was to determine the way the plastic failure and geometrical imperfections during realization affect reduction of the ideal critical buckling load. The geometrical non-linear analysis was made using finite elements that enable considerable displacements and rotations. The arc length method was used to overcome singular points (bifurcation and boundary points), as this method ensures stability of the analysis (or more precisely of the tangent stiffness matrix) during analysis based on the falling unstable branch of the P-Δ diagram (arc length method is a version of the Newton-Raphson method). At that, we used the FEAP software, version 7.5 which is (partly through our efforts as well) currently developed at the University of Berkeley (USA) under the guidance of Professor R. L. Taylor. We have the entire code in the FORTRAN programming language (complied with g77) so that we can intervene as necessary.

Calculations have shown that the descending (unstable) branches are close to the climbing (stable) branch, which additionally complicates the convergence analysis. These branches are separated by the multiple (fivefold) bifurcation point which defines the computed coefficient of safety $\gamma = 5.7$ for the load consisting of the weighted snow and live load (Figure 5). More precisely, we are dealing here with two double bifurcation points and one boundary point that are located close to one another (within one 1% increment of the total load), so that we can assume that all points are affected by the same ideal critical buckling load. Many closely positioned forms of buckling, affected by the same or almost the same critical load—this is a typical property of rotational symmetric problems in the theory of stability. This property is responsible for the great susceptibility to imperfection.

If we look at the diagrams we see that, in case of loads close to bifurcation points, only several millimeters would suffice for the structure to "leap" to the unstable branch and achieve stability position at a much lower factor of safety—of about 2. It is clear the errors during realization can be much greater, so that the bifurcation point can not be achieved. As five buckling forms are quite close to one another, all five will exert influence on dome behavior after the loss of stability (we are referring here to post-critical behavior). It would therefore be logical to assume in bifurcation point an infinitesimally close

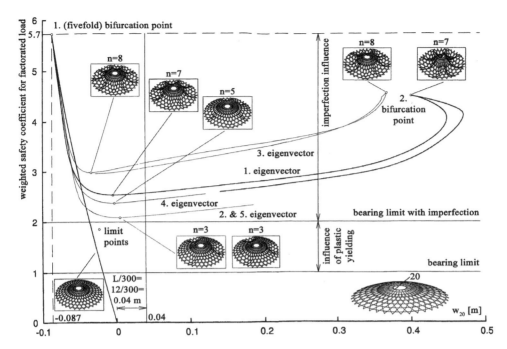

Figure 5. Safety coefficient for factorated load.

deformed form of the dome (that would assume the same force as an undeformed form) in form of each of these five eigenvectors. On top of that, any combination of these five forms can constitute a buckling form. As we move away from the bifurcation point, these combinations become more and more pronounced.

It can be seen from curves given in Figure 5 that the bottom limit of the dome resistance to load can be estimated at the safety factor of 2. Consequently, because of the influence of assumed geometrical irregularities the critical buckling force falls to:

$$\eta_{imp} = \frac{2}{5.7} = 0.35$$

of the load value p_{cr}^{lin} at the buckling of the ideally realized dome.

According to buckling forms from Figure 5, it can be concluded that plastic failure of members is inevitable at the top of waves, as here we have small radii in deformed state (n denotes the number of half-waves) and the influence of residual stresses. According to the following formula can be applied for the reduction of critical dome force due to plastic failure:

$$\eta = \frac{\sqrt{E_s \cdot E_t}}{E}$$

where E_s and E_t are secant elastic modulus and tangent elastic modulus, respectively. The usual estimate of the bottom limit for the secant modulus is $E_s = 0.7 \cdot E$, while the tangent modulus value of $E_t = 0.35 \cdot E$ can be used for the ratio of $f_{kr}/f_y \approx 1$. Then, based on the previous formula we have:

$$\eta_{pl} = \sqrt{0.7 \cdot 0.35} \approx 0.5$$

or $\gamma_{pl} = 2$. The factor obtained in this way is marked on the diagrams (Figure 5). It can be concluded that the total factor of safety:

$$\gamma = \lambda_{imp} \cdot \lambda_{pl} = 2.85 \cdot 2.0 = 5.70$$

is sufficient to ensure the dome resistance to weighted load.

According to these analyses, the critical load amounts to:

$$\eta = \eta_{imp} \cdot \eta_{pl} = 0.35 \cdot 0.5 \approx 0.18$$

of the ideal value p_{cr}^{lin}.

Detailed analyses, effective local buckling length of critical elements, comparison with a solid dome alternative, qualitative comparison with test results given in relevant literature, all this is contained in the document.

2.3 Load-bearing capacity of nodes (joint strength)

The entire network of the dome's primary load-bearing steel structure consists of equal-sized tubes all measuring Ø323.9 × 7.1 mm (the only exception is the top ring tube measuring Ø508 × 12.5 mm; quality S355J2H) and is composed of welded conventional tubular T and X joints, cf. Figure 6. As in this situation we mostly deal with compressive members in connection, we made use of the high D/t ratio (cost-effectiveness, low slenderness ratio) while at the same time we took into account the fact that the load bearing capacity of such joints is much lower in case of a high D/t ratio.

2.4 Assembly

Due to the fact that the steel dome assembly is an exceptionally demanding task, as the entire load-bearing system is highly susceptible to possible errors in the final geometry ("audacity" of structure: $f/D \approx 0.1$), the following requirements were set for the workshop fabrication and assembly:

– all geometrically complex member joints at the ring tube connection points are checked on a auxiliary structure during trial assembly in workshop,
– all assembly joints are welded so as to avoid uncontrolled displacements at joint positions,
– the entire structure is assembled on the scaffold which enables an optimum inspection of the position of elements during assembly and welding operations,
– rings at the sphere surface are precambered with the calculated deflection for the state of the unweighted total load.

The assembly of the dome structure is presented in Figure 6.

A minimum deviation from design values has been obtained thanks to systematic geodetic inspections of structural positions on the scaffold, and because of strict respect of requirements set for this project.

3 SECONDARY LOAD-BEARING STRUCTURE OF THE EXTERNAL ENVELOPE

3.1 Structural system

In order to increase the load-bearing capacity of the primary system, the secondary load-bearing structure of the external envelope was integrated into the load bearing system and made to act compositely with the primary structure, and hence their joint action greatly increases the stiffness of the dome structure and its resistance to additional loads due to soffit structure, installations and walkways, and external loads due to wind, snow, earthquake and temperature. However, as the structure is highly complex, it would be almost impossible to exactly numerically express this increase in stiffness and bearing capacity.

Figures 7 and 8 show a typical detail of the secondary load-bearing external envelope consisting of meridian omega sections measuring 160 × 100 × 65 × 5 mm and rigidly connected with the primary load-bearing structure and structural trapezoidal-shaped steel sheets.

3.2 Physical properties of the roof sandwich

The roof sandwich, composed as shown in Figures 7 and 8, must meet, not only the traditional requirements applied to roof coverings (local bearing capacity, thermal protection and condensate protection), but also the requirements relating to noise protection.

Figure 6. Steel structure on a scaffold during construction, with welded node detail visible.

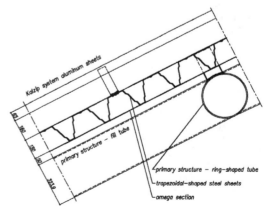

Figure 7. Secondary load bearing envelope—radial section.

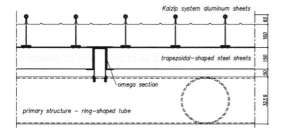

Figure 8. Secondary load bearing envelope—tangential section.

For this purpose, perforated sheet steel will be used in soffit, in addition to perforated webs made of trapezoidal sheet metal. The Kalzip® system aluminum sheet with its typical connections to the underlying structure was selected for an external exposed portion of the entire spherical shell. An expansion joint was installed between the top and the bottom parts. It separates sheet metal covers in the longitudinal direction. Each of these expansion fields has one solid point. On the steel part of the dome, the design called for installation of aluminum sheets Kalzip 65/xxx/1.0 (variable sheet width, conical sheets), Klipp E160 on the steel trapezoidal sheet with perforated webs TR 150/280/0.88 and 0.75 mm.

4 ADDITIONAL ANALYSES AND THEORETICAL MODELLING OF JOINT ACTION OF THE PRIMARY AND SECONDARY LOAD-BEARING STRUCTURE

The finishing work on the structure is still under way at the time this paper is written. The entire soffit, light well and installations (air conditioning ducts with walkways, main and accessory lights, and lighting installations) still need to be installed at the steel dome. The bench marks appropriately installed all over the structure will enable detailed inspection, based on geodetic monitoring, of the overall load-bearing system. It will be possible to estimate the real bearing capacity of the structure, and the effect of the joint composite action of the primary and secondary load-bearing system, by comparison with data from the original analysis and through additional theoretical modeling of the structure.

5 CONCLUSION

As most international design recommendations suggest that the relationship between the design load of the structure, and the elastic critical buckling load for global stability based on initial elastic stiffness, should be more than 10, it may seem that the safety factor of $\gamma = 5.7$, as analyzed and defined for this structure, is too low. However, many authors agree that the use of excessively high safety factors, as a means to counter for geometrical and structural irregularities, is not justified in case of civil engineering structures. This is especially true for structures for which correct structural measures have been taken to make the secondary structure act compositely with the primary structure so as to increase stiffness of the overall system, with a realistic increase in critical load to a higher level when compared to ideal values.

REFERENCES

Allen, H.G.; Bulson, P.S.: Background to Buckling, McGraw-Hill Book Company, London, 1980.
MAPA 3—PROJEKT KONSTRUKCIJE, Dio 12. Naknadni proračuni i uštede na čeličnome dijelu kupole, "D & Z", d.o.o., 23000 Zadar, Jerolima Vidulića 7, prosinac 2004.
Taylor, R. L.: FEAP-A Finite Element Analysis Program, Version 7, 5; University of California at Berkely, 2004.
Technische Information 2, Bemessung vorwiegend ruhend beanspruchter MSH-Konstruktionen, VALLOUREC & MANNESMANN Tubes, 2005.

Parallel session 1: Composite construction
(reviewed by L. Gardner & X.Z. Zhao)

The interaction of steel tube and concrete core in concrete-filled steel tube columns

U. Starossek & N. Falah
Hamburg University of Technology (TUHH), Germany

ABSTRACT: The load transfer by natural bond or by mechanical shear connectors and the interaction between the steel tube and the concrete core in concrete-filled steel tube columns are investigated experimentally and analytically. A total of 71 specimens are tested under axial compression loading applied to the steel tube alone, to the concrete core alone or to the steel tube and concrete core simultaneously at the top of the specimen, whereas the steel tube and concrete core are supported simultaneously at its bottom. The influence of a number of parameters is studied including the bond conditions, the cross section, the strength and type of concrete and the length of the specimen. For comparative study, nonlinear finite element models are developed using the general-purpose finite element software ABAQUS.

1 INTRODUCTION

When concrete-filled steel tubes (CFT) are used as continuous columns of frame structures with simple beam-column connections, the shear forces at the beam ends are not directly transferred to the concrete core but to the steel tube. The load transfer from the steel tube to the concrete core depends on the bond strength between steel and concrete. Although a large number of studies on the bond strength by means of push-out tests have been reported in the literature, they provide limited information on the real bond behavior in CFT columns. The main reason is that push-out tests do not represent the loading condition in an actual CFT column. In a real column the relative slip between steel and concrete is much less than in a push-out test, and the force that can be applied to a push-out specimen is much smaller than that which can be resisted by a real column. In addition, the Poisson's ratio of steel is greater than that of concrete, at least in the elastic range.

Concerning the design of CFTs, current design codes and standards assume that there is complete interaction (perfect bond) between steel and concrete. The ultimate value for bond stress is given as 0.4 MPa in Eurocode 4; the assumed transfer length of shear force should not exceed twice the smaller one of the two overall cross-sectional dimensions. For the case where the bond strength is exceeded, mechanical shear connectors have to be inserted. However, there are no rational rules for determining how and where shear connectors are installed at the interface between concrete core and steel tube. There is still a shortage of information on the performance of shear connectors. In continuous CFT columns with small cross-sectional dimensions, it is difficult to arrange mechanical shear connectors inside the tube in the regions where the highest transfer stresses occur.

2 TEST PROGRAM

2.1 Test specimens

The program comprised the test of 71 specimens and was organized in four phases as described in Table 1 below. The ends of the steel tubes were accurately cut and machined to the required length in a lathe at the steel factory, ensuring that the two ends were parallel to each other and normal to the sides. The inside of the tubes was wire brushed to remove any rust and loose scale, and it was carefully cleaned from deposits of grease and oil. The bottom end of each tube was welded to a square steel end plate of 15 mm thickness. Prior to casting the concrete, holes were drilled from the outside of the tubes, and then threads were made to install bolts via these holes. Bolts M16 grade 5.6 black bolts of 65 mm length were used as shear connectors in those specimens that contain the letter "B" in their designations see Table 1. Two, four, or six bolts were installed in the test specimens (2B, 4B, 6B). The inner surface of the tubes for specimens designated with "UD" was polished and oiled

Table 1. Properties of test specimens and test results.

No	Designation	Dimension (mm)	Length (mm)	Steel f_y (MPa)	Concrete f_c (MPa)	Bond stress (N/mm^2)	Shear connector F_t (kN)	Ultimate load F_t (kN)	F_{cal} (kN)	F_t/F_{cal}
1	C1-H2	168.3 × 5	750	511	–	–	–	1405	1308	1.074
2	C1-S1	168.3 × 5	750	511	46	1.00	–	–	–	–
3	C1-S2	168.3 × 5	750	511	46	0.96	–	–	–	–
4	C1-RS1	168.3 × 5	750	511	46	0.99	–	–	–	–
5	C1-RS2	168.3 × 5	750	511	46	0.63	–	–	–	–
6	C1-UD-S1	168.3 × 5	750	511	46	0.00	–	–	–	–
7	C1-UD-S2	168.3 × 5	750	511	46	0.00	–	–	–	–
8	C1-2B-Sc1	168.3 × 5	750	511	46	0.80	140	–	132	1.061
9	C1-2B-Sc2	168.3 × 5	750	511	46	1.10	143	–	132	1.083
10	C1-4B-Sc1	168.3 × 5	750	511	46	0.70	260	–	264	0.985
11	C1-4B-Sc2	168.3 × 5	750	511	46	0.68	254	–	264	0.962
12	C1-6B-Sc1	168.3 × 5	750	511	46	0.71	405	–	396	1.023
13	C1-6B-Sc2	168.3 × 5	750	511	46	0.78	380	–	396	0.96
14	C1-UD-2B-Sc	168.3 × 5	750	511	46	0.00	145	–	132	1.098
15	C1-UD-4B-Sc	168.3 × 5	750	511	46	0.00	263	–	264	0.996
16	C1-A1	168.3 × 5	750	511	46	–	–	2553	2560	0.997
17	C1-A2	168.3 × 5	750	511	46	–	–	2488	2560	0.972
18	C1-90S1	168.3 × 5	750	511	46	0.89	–	–	–	–
19	C1-90S2	168.3 × 5	750	511	46	1.27	–	–	–	–
20	C1-120S1	168.3 × 5	750	511	46	0.84	–	–	–	–
21	C1-120S2	168.3 × 5	750	511	46	0.75	–	–	–	–
22	C2-H2	244.5 × 7.1	750	411	–	–	–	1973	2176	0.907
23	C2-S1	244.5 × 7.1	750	411	32	0.95	–	–	–	–
24	C2-S2	244.5 × 7.1	750	411	32	1.04	–	–	–	–
25	C2-A1	244.5 × 7.1	750	411	32	–	–	3505	3507	0.999
26	C2-A2	244.5 × 7.1	750	411	32	–	–	3366	3507	0.960
27	C3-H2	114.3 × 5	750	475	–	–	–	789	815	0.968
28	C3-S1	114.3 × 5	750	475	32	1.08	–	–	–	–
29	C3-S2	114.3 × 5	750	475	32	1.20	–	–	–	–
30	C3-A1	114.3 × 5	750	475	32	–	–	1153	1170	0.985
31	C3-A2	114.3 × 5	750	475	32	–	–	1179	1170	1.008
32	S1-H2	150 × 150 × 6.3	750	411	–	–	–	1365	1488	0.917
33	S1-S1	150 × 150 × 6.3	750	411	32	0.68	–	–	–	–
34	S1-S2	150 × 150 × 6.3	750	411	32	0.69	–	–	–	–
35	S1-A1	150 × 150 × 6.3	750	411	32	–	–	1903	2092	0.910
36	S1-A2	150 × 150 × 6.3	750	411	32	–	–	1951	2092	0.933
37	C1-S1*	168.3 × 5	750	511	35	1.05	–	–	–	–
38	C1-S2*	168.3 × 5	750	511	35	1.00	–	–	–	–
39	C1-S3*	168.3 × 5	750	511	35	0.86	–	–	–	–
40	C1-2B-Sc1*	168.3 × 5	750	511	35	1.03	106	–	100	1.060
41	C1-2B-Sc2*	168.3 × 5	750	511	35	0.70	112	–	100	1.120
42	C1-4B-Sc1*	168.3 × 5	750	487	35	0.75	211	–	200	1.055
43	C1-4B-Sc2*	168.3 × 5	750	487	35	0.80	220	–	200	1.100
44	C1-UD-2B-Sc*	168.3 × 5	750	487	35	0.00	110	–	100	1.100
45	C1-UD-4B-Sc*	168.3 × 5	750	487	35	0.00	213	–	200	1.065
46	C1-A1*	168.3 × 5	750	487	35	–	–	2110	2270	0.930
47	C1-A2*	168.3 × 5	750	487	35	–	–	2260	2270	0.996
48	C1-Sy1	168.3 × 5	750	487	35	0.81	–	–	–	–
49	C1-Sy2	168.3 × 5	750	487	35	0.80	–	–	–	–
50	C1-2B-Sy	168.3 × 5	750	487	35	–	114	–	100	1.140
51	C1-4B-Sy	168.3 × 5	750	487	35	–	226	–	200	1.130
52	C1-LS1	168.3 × 5	750	511	46	0.86	–	–	–	–
53	C1-LS2	168.3 × 5	750	511	46	0.55	–	–	–	–
54	C1-75H-S1	168.3 × 5	750	511	75	0.63	–	–	–	–
55	C1-60H-S1	168.3 × 5	750	511	60	1.48	–	–	–	–
56	C1-60H-S2	168.3 × 5	750	511	60	1.10	–	–	–	–
57	C1-75H-S2	168.3 × 5	750	511	75	0.66	–	–	–	–

(*continued*)

Table 1. (*continued*)

No	Designation	Dimension (mm)	Length (mm)	Steel f_y (MPa)	Concrete f_c (MPa)	Bond stress (N/mm²)	Shear connector F_t (kN)	Ultimate load F_t (kN)	F_{cal} (kN)	F_t/F_{cal}
58	C1-SP1	168.3 × 5	750	487	44	3.00	–	2057	2167	0.949
59	C1-SP2	168.3 × 5	750	487	44	3.20	–	2203	2167	1.017
60	C1-2B-SP1	168.3 × 5	750	487	44	–	–	2075	2167	0.958
61	C1-2B-SP2	168.3 × 5	750	487	44	–	–	2200	2167	1.015
62	C1-4B-SP1	168.3 × 5	750	487	44	–	–	2150	2167	0.992
63	C1-4B-SP2	168.3 × 5	750	487	44	–	–	2107	2167	0.972
64	C1-M-S1	168.3 × 5	1200	487	44	0.87	–	–	–	–
65	C1-M-S2	168.3 × 5	1200	487	44	0.81	–	–	–	–
66	C1-M-S3	168.3 × 5	1200	487	44	0.94	–	–	–	–
67	C1-M-A1	168.3 × 5	1200	487	44	–	–	2256	2449	0.921
68	C1-M-A2	168.3 × 5	1200	487	44	–	–	2201	2449	0.899
69	C11-S	168.3 × 5	750	487	32	1.00	–	–	–	–
70	C11-2B-S	168.3 × 5	750	487	32	0.80	117	–	100	1.170
71	C11-A	168.3 × 5	750	487	32	–	–	2345	2272	1.032

* denotes tests with end plates thicker than the end plates in the first phase of tests.

thoroughly before casting the concrete to eliminate the bond between steel tube and concrete core. The tubes were filled with plain concrete in layers in the vertical position. The concrete core was compacted using a high-frequency vibrating poker introduced into the steel tube to eliminate air pockets in the concrete and to give a homogeneous mix. All test specimens were tested at the age of 28 days, except those designated with "C1-90S" or "C1-120S" that were tested at the age of 90 days or 120 days, respectively, after casting the concrete.

2.2 *Material*

Four types of ordinary concrete, one type of expansive concrete "L", and two types of high-strength concrete were used to fill the steel tubes. The ordinary concrete and expansive concrete were manufactured in a ready mix concrete factory, while the high-strength concrete was mixed in the laboratory. To find the properties of concrete three concrete cubes and six concrete cylinders were cast in standard steel moulds (150 × 150 × 150) and (300 × 150) for each concrete mix at the same time as the casting of the concrete in the test specimens. The cubes and cylinders were stored in the curing room in a curing tank for 28 days. The steel tube S355 J2H has been supplied from the steel factory to fabricate the test specimens for all cross-sections. A 750 mm long steel tube from each charge was used for preparation of tension specimens. Three tension coupons were cut, and machined from the side of steel tube, and were tested in uniaxial tension in order to determine the mechanical properties of steel tube.

2.3 *Instrumentation*

Electrical strain gauges were used to measure strains in the steel tube. Eighteen strain gauges were installed in the longitudinal direction in two columns (9 on each column), four strain gauges in the transverse direction in two rows (two on each raw). For measuring interior strain in concrete core under a loading test, electrical embedded strain gauges were used. Four linear variable differential transducers (LVDTs) with an accuracy of 0.001 mm (measuring range 0 ... 50 mm) were installed on each side of top end plate to measure the axial deformation of specimens. Axial deformation was obtained from the average of the LVDTs at each side. Test specimens instrumentations are shown in Figure 1.

2.4 *Experimental procedure*

The tests were carried out under axial compression in a 5000 kN capacity compression machine in the steel and concrete laboratory at Hamburg University of Technology. The axial load was applied to the steel tube alone (Type of loading S,Sc,RS, and Sy) or to the concrete core alone (Type of loading SP) or to the steel tube and concrete core simultaneously (Type of loading A) at the top of test specimen, whereas the steel tube and concrete core are supported simultaneously at the bottom of the test specimen. The loading process was preformed under displacement-controlled type of loading with rate 0.005 mm/sec. this rate of loading was adopted until 5 mm axial shortening was reached, and the rate was then changed to 0.01 mm/sec. However, a load-controlled type of loading was

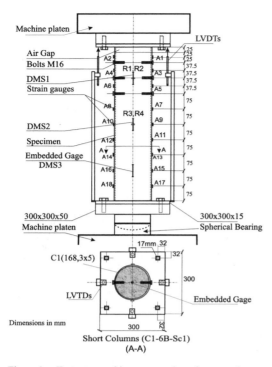

Figure 1. Test setup and instrumentation of test specimen.

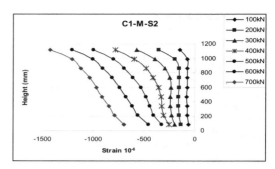

Figure 2. Distributions of longitudinal steel strain (C1-M-S2).

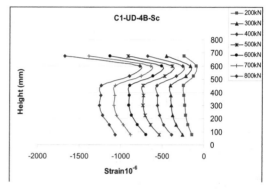

Figure 3. Longitudinal steel strain (C1-UD-4B-Sc*).

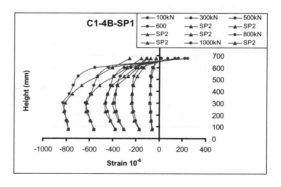

Figure 4. Longitudinal steel strain (C1–4B-SP1) & (C1-SP2).

used with rate of loading 2.6 kN/sec, was subjected to cyclic loading and then a displacement-controlled loading was applied with the same rate like the other test specimens. The tests were performed under monotonic loading, and followed until the end of test for most of the test specimens (Type A, S, Sc, and SP). In the second type of loading, the specimens were subjected to repeated loading at the calculated service load of 170 kN for specimens (C1-Sy1, 2), and at two levels of calculated service load of 170 kN and 570 kN for specimens (C1-2B-Sy, and C1-4B-Sy). The load was cycled ten times, and the load range ΔP was 100 kN. Upon completion of the cyclic load regime, the load was increased continually until the end of test. Strain gauges, electrical transducers (LVDTs) and load cell were connected to a data acquisition system. The specimens were removed from the test machine and carefully examined after the test, and then some test specimens were opened to investigate the failure of concrete core and shear connectors.

2.5 Experimental results and their evaluation

In Figure 2, longitudinal strain of steel tubes, are plotted against the positions of strain gauges on the steel tube at different levels of loading. It can be seen from this figure, that when the stain at the top of the steel tube increased, the strain distribution became nonparallel to the specimen axis. This is due to the transfer of load from the steel tube to the concrete core and the formation of bond stresses between the steel and concrete. Whereas, the strain compatibility occurs at the portions, where the strain distributions are parallel to specimens' axis, this indicates that the axial load transfer from the steel to the concrete core is halted at these portions. The axial load, which is applied only to the top of the steel tube, is gradually transferred to the concrete core. The average of longitudinal stains measurements at different locations along the whole length of the steel tube was used to infer the load transfer from the steel to the concrete (bond stress). The bond stress (τ_p) is calculated using test data as follow:

$$\tau_p = \frac{[(A_1 - A_3) + (A_3 - A_5) + \cdots + (A_{15} - A_{17})]}{\pi(D - 2t)L} \quad (1)$$

where: A_1, A_3, A_5, …… = average reading of the strain gauges on steel tube. L = length of the specimens. D = steel tube diameter. t = wall thickness of the steel tube. To calculate the relative slip between the steel tube and concrete core using the test data, the length of column is divided to three equal regions. Then the relative slip is equal to the summation of the difference between the strain on steel tube and the strain on concrete core for each region. The relative slip was calculated using the following formula

$$S_p = \left(\frac{A_1 + A_3 + A_5}{3} - DMS1\right)l$$
$$+ \left(\frac{A_7 + A_9 + A_{11}}{3} - DMS2\right)l \quad (2)$$
$$+ \left(\frac{A_{13} + A_{15} + A_{17}}{3} - DMS3\right)l$$

where: A_1, A_3, A_5, …… = average reading of the strain gauges on steel tube. $DMS1$, $DMS2$, $DMS3$ = reading of the embedded gauges on concrete core. l = length of the region ($L/3$). As the applied load at the top of steel tube increase the transferred load to the concrete core increase. Thus, the relative slip at steel-concrete interface become the highest at the top of specimen and decreased gradually toward the bottom. So the required shear connectors must be installed at steel-concrete interface from the top to the bottom. Figure 5 shows bond stress-relative slip relationship for the whole length of specimen. The maximum bond stress was achieved at relative slip of approximately 0.25–0.3 mm and at applied load of 500 kN, 700 kN for short specimens, and for long specimens respectively. The specimens with ordinary concrete give average bond strength of 0.97 N/mm², which is more than twice value of 0.4 N/mm² given in EC4. As can be seen from the Table 1 the repeated loading of ten cycles at one load level had no noticeable adverse effect on the bond strength. The average bond strength of specimens with high strength concrete, and with

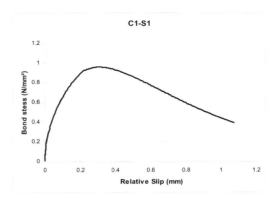

Figure 5. Bond stress-relative slip (C1-S1).

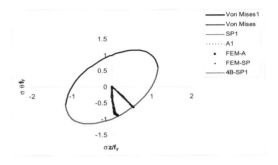

Figure 6. Von Mises yield criterion.

expansive concrete, as compared to those with ordinary concrete, had nearly smaller bond strength. The type of concrete was observed to have no effect on the bond strength. The average bond strength for square ship specimens was 0.685 N/mm², which was 70% of the bond strength values for circular ship specimens. The strain distribution in the (C1-UD-S1) specimens, in which the steel-concrete interface was oiled prior to casting the concrete, clearly indicates that no load transfer from the steel tube to concrete core occurred. In push-out tests, however, the bond strength of oiled specimens was about half the value of dry specimens (Shakir-Khalil 1993a, b). This discrepancy can be ascribed to the Poisson's effect. The test results for specimens (C1-90S) and (C1-120S) were compared with results of the (C1-S) specimens, which were tested at an age of 28 days. It was observed that the influence of the concrete age on the bond strength is small. In other words, the reduction of the bond strength due to the shrinkage of concrete was small. The relationship between the longitudinal strain of concrete core and the applied load for (C1-S1) is presented in Figure 7. The strain distribution on concrete core shows that the bond strength starts to

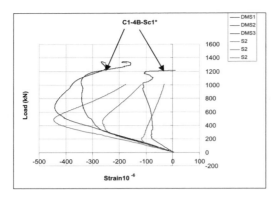

Figure 7. Longitudinal concrete strain (C1–4B-Sc1*) & (C1-S2).

fail at the top region, while the further load transfer from the steel tube to the concrete core continues in the remaining regions. This means that the failure does not happen progressively. However, the failure in the top region is more ductile than in the remaining regions. As can be seen from the concrete strain distribution, the slope of strain degradation at the bottom region is bigger than at the top region. The tests data shows that the transferred load to concrete core increase with increasing the length of column. The average transferred load of long specimens (C1-M-S) is 400 kN, which is three times large than value of 130 kN calculated using EC4 recommendation. For the specimens (SP), which the load is applied on concrete core at the top, the concrete core expands in the lateral direction (Possion's effect). Therefore high bond strength is created due to high interlocking, and high coefficient of friction at the steel-concrete interface. The two surfaces do not separate to each other until the failure of specimen. The average bond stress for this type of loading at the yield strength of steel tube is 3.0 N/mm². The following formula is used to calculate the bond strength in this type of loading until yield strength of steel

$$\tau_b = (\varepsilon_s E_s)\frac{t_s}{L} \qquad (3)$$

Where ε_s = reading of the strain gauge at the bottom of steel tube, t_s = wall thickness of steel tub, L = column length, E_s = modulus of elasticity for steel.

As the applied load is increased, the shear bond stresses at the steel-concrete interface increase, so more load can be transferred to the steel tube. The contribution of the steel tube to the total axial resistance must be taken into the account "SP". The passive confining pressure provided by steel tube in the top region is extremely activated. The passive confining pressure is decreased gradually toward the bottom of specimen. This is clearly due to load transfer from the concrete to the steel tube as shows in Figure 4. However, the passive confining pressure is nearly uniformly distribution along the length of specimens on the concrete core for specimens provide with shear connectors. The failure of concrete is accompanied by cracks, which is distributed in the radial direction. This type of failure is crushing of concrete, so the angle of internal friction is equal to zero, and the maximum shear stress is equal to the cohesion. It was found in tests results and in FEM analyses that the maximum longitudinal stress of steel at the ultimate load was approximately 60% of the yield strength see Von Mises yield criterion in Figure 6. To calculate the ultimate load for CFTs under this type of loading (SP) a design proposal for load introduction using loading plate was derived based on tests by Roik et al (1988) was used with addition of the contribution of steel tube as found in test results.

$$F_{ul,Rd} = (f_{ck} + 35)\frac{1}{\gamma_c}\sqrt{\frac{A_c}{A_1}}A_1 + 0.6 f_y A_s \qquad (4)$$

where A_c = total concrete area, A_1 = area below the loading plate, f_{ck} = characteristic concrete compressive strength, f_y = yield steel strength, A_s = steel area, γ_c = material safety factor for concrete. The safety factor is taken as 1.0 for the evaluation. The comparison of the results of the test with Equation 4, shows good agreement as given in Table 1.

The transferred load by shear connectors F_t as given in Table 1, is calculated directly using the reading of strain gauges on steel tube at the end of elastic range as follow

$$F_t = N_s - \varepsilon_s E_s A_s \qquad (5)$$

where N_s = the applied load at the top of steel tube, ε_s = longitudinal strain on the steel tube below the shear connector, A_s = area of steel tube, E_s = modulus of elasticity for steel. Figure 3 shows the strain distribution on steel tube for specimen with shear connectors. The test data and FEM analyses (Bussler 2007) show that the concrete stress below the shear connectors reaches very high stress values, because of the confinement that is provided by steel tube, see Figure 10. The comparison between the strain distribution on concrete core for the specimens which were provided with shear connectors and the strain distribution on concrete core for the other specimens shows that the shear connectors enhance the bond resistance in the elastic range and the load transfer becomes more ductile see Figure 8. The tests results showed that the bond stress and shear connectors worked together in transferring forces between the steel and

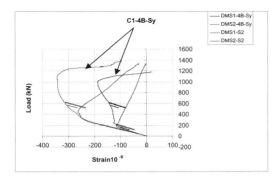

Figure 8. Longitudinal concrete strain (C1–4B-Sy) & (C1-S2).

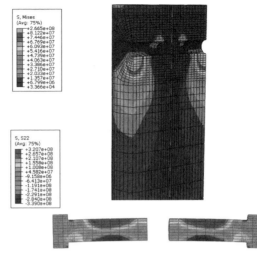

Figure 10. Stresses in concrete and shear connectors.

The predicted ultimate load of shear connectors is based on the bearing capacity of the concrete at the yield strength of steel tube.

$$F_{u1,Rd} = A_{co} f_{ck} \sqrt{\frac{A_{c1}}{A_{co}}} \leq \frac{\pi d^2}{4} \frac{0.8 f_u}{\gamma_m} \quad (6)$$

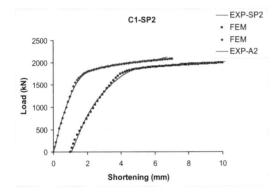

Figure 9. Comparison between experimental and FEM results.

$$A_{co} = b \times d \quad (7)$$

$$A_{c1} = 1.5b\,(1.5 - 2)d \quad (8)$$

where b = length of bolt, d = diameter of the shank of bolt, f_{ck} = characteristic concrete compressive strength, f_u = ultimate strength of bolt, A_{co} = bearing area below the bolt, A_{c1} = computational area. γ_m = a partial safety factor of 1.25, see figure11. The safety factors are taken as 1.0 for the evaluation. A good agreement was found between the results of Equation 6 and the test results as given in Table 1.

From the difference in the lateral deformation of two materials and the strain compatibility in Z-direction the following formula was derived to calculate the ultimate strength of axially loaded short CFTs columns (A type of loading).

$$N_u = 1.26 A_s f_y + A_c f_{cu} \quad (9)$$

where f_y = yield steel strength, f_{cu} = ultimate concrete compressive strength, A_c, A_s = areas of concrete and steel respectively. It can be found that, generally good

concrete because of installation the shear connectors at the top of specimen where the relative slip is larger. As can be seen from the figure and the ultimate load values given in Table 1, the repeated load of ten cycles at two levels had no effect on the ultimate load of shear connectors. Before the steel tube reaches the yield strength, the load is transferred by direct bearing pressure to the concrete. However, after reaching the yield strength, the bolts start to rotate under the effect of bending and high frictional shear stresses are created to the effect that the load transfer is enhanced by pinching and contraction effects. In the plastic range, the bearing stresses begin to decline gradually with the rotation of the shear connectors. The bearing stresses on the concrete and the bending stresses in the shear connector are presented in Figure10. It was found from the test data and FEM analyses that the transferred loads by shear connector increase with increasing the compressive strength of concrete core. The failure was deformation the steel tube in the bolts region and crushing the concrete below the bolts as shows in Figure 14.

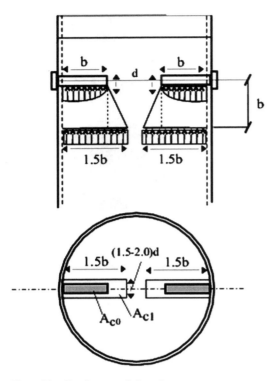

Figure 11. Bearing stress below shear connector.

agreement between the Equation 9 and test results see Table 1. For full derivation see Starossek & Falah (2008).

3 FINITE ELEMENT MODEL

Two and three dimensional nonlinear (material and geometry) finite element models were developed using the commercial finite element program ABAQUS/Standard 6.7. Two-dimensional axisymmetric models were used to model the concrete filled steel tube column with circular cross section, the concrete core, steel tube, and end plates were modeled using CAX4I axisymmetric solid element. The axiaymmetric element defines by four nodes having two degree of freedom at each node. Because of the symmetry of the square column, and columns with shear connectors only a quarter of the square column was modeled using three-dimensional solid element.

In the modeling of the concrete core, steel tube, and end plates a three-dimensional eight-node element C3D8 with three degree of freedom at each node was used. For shear connectors model see Bussler (2007). The damaged plasticity model in ABAQUS was used to simulate the concrete material behavior in the concrete filled steel tube columns. The uniaxial compressive stress-strain curve of concrete obtained from the standard cylinder tests was used in concrete material model and Poisson's ratio was $\upsilon = 0.2$. An elastic-plastic model with the von Mises yield criterion was used to describe the constitutive behavior of the steel. The complete stress-strain relation obtained from uniaxial tension-coupon tests has been used in steel material model and the Poisson's ratio was taken as $\upsilon = 0.3$. The interaction between the steel tube and the concrete core interface was modeled by a surface-based interaction with a contact pressure-overclosure model in the normal direction, and a nonlinear spring elements in the tangential directions to the surface. A comparison of the computational results from finite element analyses and the experimental results was made to verify the finite element models. Finite element analysis and experimental data are in good agreement see Figure 9. The present finite element model allows a comprehensive parametric study considering a wide rang and combination of the parameters of interest to be investigated. To investigate the effect of diameter and thickness ratio and uniaxial compressive strength of concrete on the behavior of confined concrete core, a combination of these two parameters Figure 11 Bearing stress below shear connector was performed. Three different type of unconfined concrete compressive strength f_c (34, 44. and 68MPa) were used. For each uniaxial compressive strength of concrete a total of six different diameter and thickness ratios (D/t) (25, 34, 45, 60, 85, and 100) were considered. Figure 12 presents the relationship between the confinement ratio (f_{cc}/f_c) and diameter and thickness ratio (D/t) for a circular CFT column with concrete uniaxial compressive strength of 34, 44, and 68 MPa. By using regression analysis, empirical equations was developed to determine the maximum compressive strength of concrete core (f_{cc}) under confinement pressure, and lateral confining pressure based on the results obtained relating (f_{cc}/f_c), (f_3/f_c) and (D/t).

$$f_{cc} = 5.596 \left(\frac{D}{t} \right)^{-0.172} (f_c)^{0.792} \quad (10)$$

$$f_3 = 154.6 \left(\frac{D}{t} \right)^{-0.855} (f_c)^{-0.164} \quad (11)$$

The maximum confined compressive strength of concrete (f_{cc}) can be accurately calculated with knowledge of uniaxial compressive strength of concrete and the diameter and thickness of steel tube. The stress-strain relationship for confined concrete core by steel

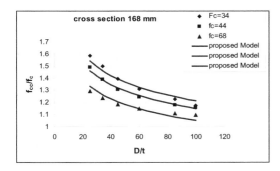

Figure 12. Relationship between (f_{cc}/f_c) and (D/t).

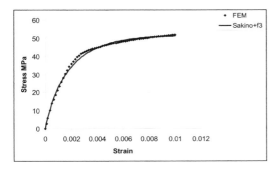

Figure 13. Stress-strain relationship for confined concrete.

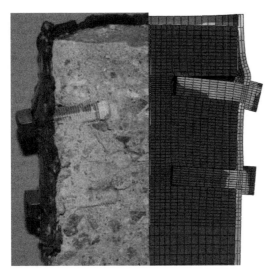

Figure 14. Failure of specimen with shear connectors.

tube can be predicted using Equation 11, in Sakino and Sun (1994) stress-strain model as follow

$$K = \frac{f_{cc}}{f_c} = 1 + 4.1\left(159.6\left(\frac{D}{t}\right)^{-0.855}(f_c)^{-1.164}\right) \quad (12)$$

Figure 13 and 14 shows a good agreement between the finite element results, and proposed model for confined concrete.

4 CONCLUSIONS

Based on the experimental and analytical results of this study the following conclusion can be drawn. It was found that the bond strength between steel tube and concrete core is about 0.8–1.0 N/mm², which is larger than the value of 0.4 N/mm² given in EC4. The age of concrete and the type of concrete were found to have no significant effect on the bond strength. The transferred load increases with increasing the introduction length. The load transfer capacity is further enhanced by using shear connectors. In that case, load transfer is provided by direct bearing stress in the elastic range and, at a higher loading level, by frictional shear stresses produced as the bolts start to rotate under the effect of bending and the tube deforms or buckles locally in the bolts region in the plastic range. For practical use, a minimum number of four bolts is suggested, with two bolts being placed on each side. The bolts can be installed from the outside. In the specimens where the load was applied on the concrete only, the shear connectors provided for a nearly uniform confining pressure on the concrete core. For this type of loading, the contribution of steel to the total axial strength must be added. The failure of bond did not occur progressively, when the load was applied on the steel tube only at the top of the column.

REFERENCES

Bussler, T. 2007. Numerische Untersuchungen zum Tragverhalten von Schubbolzen in Verbundstützen. *Diplomarbeit*, Hamburg University of Technology, Germany.

Johansson, M. 2003. Composite action in connection regions of concrete–filled steel tube columns. *Steel and Composite Structures*, 3(1): 47–64.

Morishita, Y., Tomii, M. & Yoshimura, K. 1979a. Experimental studies on bond strength in concrete filled circular steel tubular columns subjected to axial loads. Trans. Japan Concrete Inst, Tokyo, 1, 351–358.

Roeder, C.W, Cameron, B. & Brown, C.B. 1999. Composite action in concrete filled tubes. *Journal of Structural Engineering*, 125(5), May, 477–484.

Roik, K. & Schwalbenhofer, K. 1988. Experimentelle Untersuchung zum Plastischen Verhalten von Verbundstützen. Bericht zu P125, Düsseldorf, Germany.

Sakion, K. 2004. Confined concrete in concrete-filled steel tubular columns. *Int. Symposium on Confined Concrete*, Hunan University. Changsha, China, 267–278.

Shakir-Khalil, H. 1993a. Push-out strength of concrete-filled steel hollow sections. *The Structural Engineer*, 71(13), 230–233 and 243.

Starossek U. & Falah N. 2008. Force transfer in concrete filled steel tube columns. *Proc. 5th European Conference on Steel and Composite Structures*, Graz-Austria.

Design calculations on concrete-filled thin-walled steel tubes subjected to axially local compression

L.H. Han
Tsinghua University, Beijing, China

W. Liu
Wuzhou Engineering Design and Research Institute, Beijing, China

Y.F. Yang
Fuzhou University, Fuzhou, China

ABSTRACT: Based on a Finite Element Analysis (FEA) modeling for Concrete-filled Steel Tube (CFST) subjected to axially local compression, this paper investigates the effects of sectional type, local compression area ratio, steel ratio, strength of steel and concrete, and endplate thickness on sectional capacities. The parametric studies provide information for the development of formulae for the calculation of the sectional capacity of CFST columns subjected to axially local compression.

1 INTRODUCTION

It is well known that concrete-filled steel tube (CFST) constructions are currently being increasingly used in the construction of buildings, due to their excellent static and earthquake-resistant properties.

In practice, CFST columns are often subject to axially local compression, for example, the pier of girder bridge, the underneath bearing members of rigid frame, reticulate frame and arch structures. Figure 1 illustrates a schematic view of the CFST columns under axially local compression, where A_c is the cross-sectional area of concrete; and A_L is the local compression area.

In recent years, the authors have studied the behaviors of steel tube confined concrete (STCC) and CFST columns under axially local compression both theoretically and experimentally (Han et al. 2008a, b). However, such a problem has not been addressed satisfactorily to provide further information for drafting the design codes on the composite columns.

The present paper is thus an attempt to study the performance of CFST stub columns under axially local compression further. Detailed parametric studies were carried out to investigate the influences of: (1) sectional type; (2) local compression area ratio; (3) steel ratio; (4) strength of steel and concrete; and (5) endplate thickness on the sectional capacities of CFST columns subjected to axially local compression. The parametric studies provide information for the development of formulae for the calculation of the sectional capacity of the composite columns subjected to axially local compression. Comparisons are made with predicted sectional capacities using the simplified model with those of the tests, and generally good agreement has been achieved.

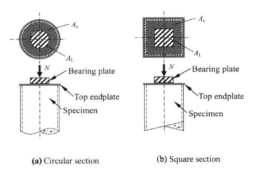

Figure 1. A schematic view of the CFST columns subjected to axially local compression.

2 LOAD-DEFORMATION CURVE

2.1 General characteristics

ABAQUS (2003) software was used in (Han et al. 2007) for the finite element analysis (FEA) on CFSTs under pure torsion, and recently was extended to the FEA on CFST stub column subjected to axially local compression, the predicted load versus deformation curves of the composite columns under axially local compression agreed well with the measured results (Han et al. 2008b).

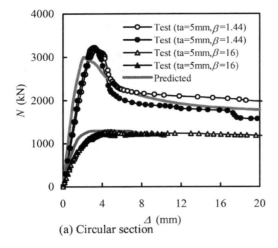

(a) Circular section

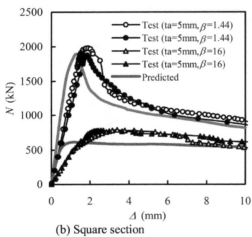

(b) Square section

Figure 2. Comparisons of load versus deformation relationships.

Figure 2 shows the comparisons of typical load versus deformation curves of CFST under axially local compression between predicted results by FEA and the test results, where, the outside diameter and dimension of circular and square member (D and B), equal to 206 mm and 177 mm respectively; t_a is the endplate thickness, and β is the local compression area ratio.

Figure 3 shows the comparisons of the sectional capacity of CFST columns under axially local compression between the predicted results (N_{uc}) by FEA and the tested results (N_{ue}). It can be seen that, generally, the predicted results show generally good agreement with the test results, and tend to safety.

Figure 4 schematically shows the typical curves of axial load (N) versus axial deformation (Δ) for CFSTs subjected to axially local compression.

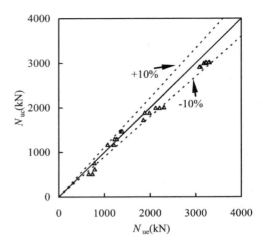

Figure 3. Comparisons of predicted sectional capacity and tested results.

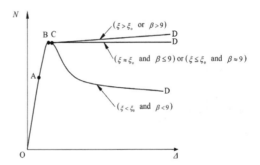

Figure 4. Typical N-Δ relationships.

It was found that the curves are a function of the confinement factor (ξ) and the local compression area ratio (β). The confinement factor (ξ) is defined as follows:

$$\xi = \frac{A_s \cdot f_y}{A_c \cdot f_{ck}} = \alpha \cdot \frac{f_y}{f_{ck}} \qquad (1)$$

where A_s is the cross-sectional area of steel; f_y is the yielding strength of steel; f_{ck} is the characteristic compressive strength of concrete and equals to $0.67 f_{cu}$ for normal strength concrete, f_{cu} is the cube compressive strength of concrete; and α is defined as the steel ratio.

The local compression area ratio (β) is defined as following:

$$\beta = \frac{A_c}{A_L} \qquad (2)$$

It can be seen that, in general, with the increase of ξ and β, the descending branch of the N-Δ curve becomes gentle, and there is no descending branch when ξ is greater than ξ_o or β is greater than 9. It was found that the values of ξ_o for concrete-filled CHS (Circular Hollow Section) and SHS (Square Hollow Section) can be given by 1.1 and 4.5 respectively (Han et al. 2005). Furthermore, the descending branch of CFSTs with square sections is steeper than that of CFSTs with circular sections, and appears earlier.

There are three or four stages in N-Δ curves, shown as in Figure 4:

1. Stage 1: Elastic stage (from point O to point A). During this stage, steel and concrete are in the elastic state, and there is no confined stress between them. The proportional stress (about 75 percent of f_y) of steel occurs at point A.
2. Stage 2: Elastic-plastic stage (from point A to point B). During this stage, concrete is confined by the steel tube because the Poisson's ratio of the concrete is larger than that of steel. The confinement enhances as the longitudinal deformation increases. At point B, the steel tube near the top of the composite column enters into plastic state, and the maximum longitudinal stress of concrete is attained.
3. Stage 3: Strain hardening stage (from point B to point C or from point B to point D). During this stage, N-Δ curves tend to go upwards. The shape of the curve depends on the value of confinement factor (ξ) and local compression area ratio (β). When ξ is larger than ξ_o or β is greater than 9, the curve goes up steadily to point C and D. When ξ and β are relatively small, the curve starts to go down after a short increase to point C. The smaller the confinement factor and local compression area ratio, the earlier the curve starts to fall down.
4. Stage 4: Falling stage (from point C to point D). This occurs only when the confinement factor (ξ) is less than ξ_o, and the local compression area ratio (β) is less than 9. The smaller the confinement factor and the local compression area ratio, the steeper the falling stage.

2.2 Parametric analysis

The possible parameters that influence the N-Δ curve of CFST under axially local compression including:

- Sectional type: circular and square;
- Local compression area ratio, β;
- Steel ratio, α;
- Strength of steel and concrete; and
- Thickness of the top endplate, t_a.

2.2.1 Local compression area ratio

Figure 5 shows the influences of local compression area ratio (β) on N-Δ curves of CFST under axially local compression. The calculating conditions are: cross-sectional dimension, $D(B) = 400$ mm; $\alpha = 0.1$; length of the member, $L = 1200$ mm; $t_a = 0.1$ mm; $f_y = 345$ MPa; and $f_{cu} = 60$ MPa.

It can be seen from Figure 5 that, the lower the local compression area ratio, the higher the stiffness and the sectional capacity of the member, and the steeper the descending branch of N-Δ curve. Under the same local compression area ratio, the deformation corresponding to sectional capacity of circular member is bigger than that of square member, and the falling stage of N-Δ curve of circular member is gentler than that of the square member. Furthermore, the decrease of sectional capacity of circular member is slower than that of square member. These can be explained that, under axially local compression, the confinement of circular steel tube to core concrete is more effective than that of square steel tube (Han et al. 2008b).

2.2.2 Steel ratio

Figure 6 indicates the influences of steel ratio (α) on N-Δ curves of CFST under axially local compression. The calculating conditions are: $D(B) = 400$ mm; $\beta = 4$; $L = 1200$ mm; $t_a = 0.1$ mm; $f_y = 345$ MPa; and $f_{cu} = 60$ MPa.

It is shown that, the higher the steel ratio, the higher the sectional capacity of the member and the gentler

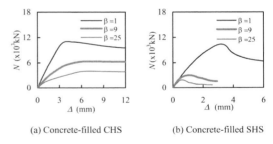

(a) Concrete-filled CHS   (b) Concrete-filled SHS

Figure 5. Influences of local compression area ratio on N-Δ curves.

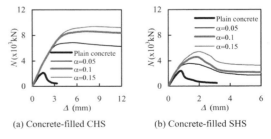

(a) Concrete-filled CHS   (b) Concrete-filled SHS

Figure 6. Influences of steel ratio on N-Δ curves.

the descending branch of the curve. However, the steel ratio has moderate influence on the stiffness. The stiffness and sectional capacity of plain concrete are significantly lower than those of composite member.

2.2.3 Steel yielding strength

Figure 7 indicates the influences of steel yielding strength (f_y) on N-Δ curves of CFST under axially local compression. The calculating conditions are: $D(B) = 400$ mm; $\beta = 4$; $L = 1200$ mm; $t_a = 0.1$ mm; $\alpha = 0.1$; and $f_{cu} = 60$ MPa.

It can be found from Figure 7 that, the steel yielding strength has little influence on the stiffness of the member, however, the sectional capacity of the member increase with the increase of the steel yielding strength. As the confinement of circular steel tube to its concrete core is better than that of square steel tube, the effects of f_y on the behaviours of circular CFST are more remarkable than those of square CFST.

2.2.4 Concrete strength

Figure 8 indicates the influences of concrete compressive cube strength (f_{cu}) on N-Δ curves of CFST under axially local compression. The calculating conditions are: $D(B) = 400$ mm; $\beta = 4$; $L = 1200$ mm; $t_a = 0.1$ mm; $\alpha = 0.1$; and $f_y = 345$ MPa.

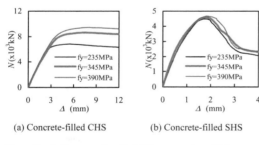

(a) Concrete-filled CHS (b) Concrete-filled SHS

Figure 7. Influences of steel yielding strength on N-Δ curves.

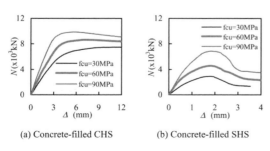

(a) Concrete-filled CHS (b) Concrete-filled SHS

Figure 8. Influence of concrete strength on N-Δ curves.

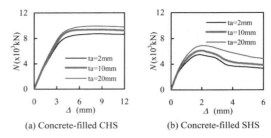

(a) Concrete-filled CHS (b) Concrete-filled SHS

Figure 9. Influence of endplate thickness on N-Δ curves.

It is shown that, the higher the concrete strength, the higher the stiffness and the sectional capacity of the member, and the steeper the descending branch of N-Δ curve.

2.2.5 Endplate rigidity

Figure 9 indicates the influences of endplate thickness (t_a) on N-Δ curves of CFST under axially local compression. As shown in the following description, the endplate rigidity can be expressed as a function of the endplate thickness. The calculating conditions are: $D(B) = 400$ mm; $\beta = 4$; $L = 1200$ mm; $\alpha = 0.1$; $f_y = 345$ MPa; and $f_{cu} = 60$ MPa.

It is shown that, the thicker the endplate, the higher the stiffness and the sectional capacity of the member, and the gentler the descending branch of N-Δ curves.

3 SECTIONAL CAPACITIES

3.1 Strength index

For convenience of analysis, strength index (K_{LC}) is defined to quantify the effects of above parameters on the sectional capacity of CFST columns subjected to axially local compression. It is expressed as:

$$K_{LC} = \frac{N_{uL}}{N_u} \quad (3)$$

where N_{uL} is the axial compressive capacity of locally loaded CFST, and N_u is the axial compressive capacity of fully loaded CFST, which can be calculated using the FEA method or simplified formulas listed in the related designing codes.

According to the concept of rigidity radius reflecting the influence range of the concentrated force in the theory of elastic basement plate (Terzaghi 1955), the endplate rigidity radius of CFST under axially local compression can be defined as follows:

$$r_0 = \left(\frac{D_w}{k}\right)^{0.25} \text{(mm)} \quad (4)$$

where D_w is the flexural rigidity of the endplate, and can be determined by the following formula based on the theory of elastic mechanics, i.e.

$$D_w = \frac{E_s \cdot t_a^3}{12 \cdot (1 - \mu_s^2)} \quad (5)$$

The factor k is defined as the rigidity coefficient of CFST according to WinKler hypothesis:

$$k = \frac{\bar{E}}{L_k} \quad (6)$$

where $\bar{E}$ is the elastic modulus of concrete-filled steel tubular stub columns, and $\bar{E} = (E_s A_s + E_c A_c)/A_{sc}$, $A_{sc} = (A_s + A_c)$ is the cross-sectional area of the composite section; L_k is the influencing height of the deflection, and is related to the dimension of the column, $L_k = D$ or B for composite sections with circular or square sections respectively.

From Equations 4–6, for concrete-filled CHS, it can be found that

$$r_0 = \left(\frac{D_w}{k}\right)^{\frac{1}{4}} = \left(\frac{E_s \cdot t_a^3 \cdot D}{12 \cdot (1 - \mu_s^2) \cdot \bar{E}}\right)^{\frac{1}{4}} \text{ (mm)} \quad (7)$$

In order to further clarify the influencing range of endplate rigidity, the relative rigidity radius of the endplate (n_r) is defined as:

$$n_r = \frac{r_0}{r} = \frac{\left(\frac{D_w}{k}\right)^{\frac{1}{4}}}{\left(\frac{D}{2}\right)} = \left(\frac{16}{12 \cdot (1 - \mu_s^2)}\right)^{\frac{1}{4}} \cdot \left(\frac{E_s \cdot t_a^3 \cdot D}{\bar{E} \cdot D^4}\right)^{\frac{1}{4}} \quad (8)$$

For CFSTs with square sections, D should be replaced by B in Equation 8.

Based on the theoretical model, the influence of the changing the local compression area ratio (β), the steel ratio (α), the strength of steel and concrete, and the relative rigidity radius of the endplate (n_r) on the strength index (K_{LC}) of CFST members subjected to axially local compression was discussed. The dimensions of the composite column are: $D(B) = 400$ mm, $L = 1200$ mm.

3.2 *Important parameters*

3.2.1 *Local compression area ratio and steel ratio*
Figure 10 shows the influences of the local compression area ratio (β) and steel ratio (α) on K_{LC}. The calculating conditions are: $f_y = 345$ MPa; $f_{cu} = 60$ MPa; $t_a = 0.1$ mm; $\beta = 1$ to 25; and $\alpha = 0.04$ to 0.2.

It can be seen that K_{LC} decreases with the increase of β. For concrete-filled CHS, K_{LC} decreases slowly with the increase of β, however, for concrete-filled SHS, K_{LC} decreases quickly when β is relatively small (i.e. $\beta \leq 9$), and then tend to stable. It also can be found from Fig. 10 that, for concrete-filled CHS, K_{LC} increases with the increase of α, however, the increment decreases gradually, and the steel ratio has little influence on K_{LC} when β is close to 1 or 25. For concrete-filled SHS, the steel ratio has little influence on K_{LC}, and generally K_{LC} decreases with the increase of α. These are due to the fact that, the confinement of circular steel tube to is core concrete is better than that of square steel tube.

3.2.2 *Steel yielding strength*
The influences of the steel yielding strength (f_y) on K_{LC} are indicated in Figure 11. The calculating conditions are: $α = 0.1$; $f_{cu} = 60$ MPa; $t_a = 0.1$ mm; $\beta = 1$ to 25; and $f_y = 235$ MPa to 420 MPa.

It can be seen that, for concrete-filled CHS, K_{LC} increases with the increase of the steel yielding strength. However, for concrete-filled SHS, the influence of the steel yielding strength is relatively small; in general, K_{LC} decreases with the increase of the steel yielding strength. This may be due to the fact that, on the one hand, the higher the steel yielding strength, the stronger the confinement of the steel tubes to its core concrete; on the other hand, the longitudinal force carried by steel tube increases due to the increase of

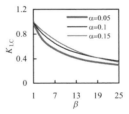

(a) Concrete-filled CHS

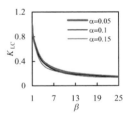

(b) Concrete-filled SHS

Figure 10. Influences of local compression area ratio and steel ratio on K_{LC}.

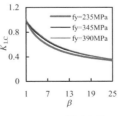

(a) Concrete-filled CHS

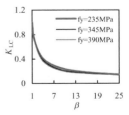

(b) Concrete-filled SHS

Figure 11. Influences of steel yielding strength on K_{LC}.

the steel yielding strength. The influences of above two aspects on circular and square CFSTs are different.

3.2.3 Concrete strength

Figure 12 illustrates the influences of the concrete compressive cube strength (f_{cu}) on K_{LC}. The calculating conditions are: $\alpha = 0.1$; $f_y = 345$ MPa; $t_a = 0.1$ mm; $\beta = 1$ to 25; and $f_{cu} = 30$ MPa to 90 MPa.

It can be seen that, for concrete-filled CHS, K_{LC} decreases with the increase of f_{cu}. However, for concrete-filled SHS, the influence of f_{cu} is relatively small, and in general K_{LC} decreases with the increase of f_{cu}. This may be explained that, on the one hand, the higher the concrete strength, the weaker the confinement of steel tube to its concrete core; on the other hand, the longitudinal force carried by steel tube decreases due to the increase of f_{cu}. Also, the influences of above two aspects on circular and square CFSTs are different.

3.2.4 Relative rigidity radius of the endplate

Figure 13 indicates the influences of the relative rigidity radius of the endplate (n_r) on K_{LC}. The calculating conditions are: $\alpha = 0.1$; $f_y = 345$ MPa; $f_{cu} = 60$ MPa; and $\beta = 1$ to 25.

It can be seen from Figure 13 that K_{LC} increases with the increase of n_r. K_{LC} increases quickly when n_r is relatively small and increases slowly when K_{LC} is close to unity. At the same value of n_r and β, K_{LC} of square CFST is lower than that of circular CFST. This means that the confinement of circular steel tube to its concrete core is better than that of square steel tube.

4 SIMPLIFIED MODEL

Using the relations between the strength index (K_{LC}) and the various parameters that determine it, the following formulae for the strength index (K_{LC}) can be obtained by using regression analysis method.

1. Concrete-filled CHS

$$K_{LC} = (A \cdot \beta + B \cdot \beta^{0.5} + C) \cdot (D \cdot n_r^2 + E \cdot n_r + 1) \leq 1 \quad (9)$$

where

$$A = (-0.17\xi^3 + 1.9\xi^2 - 6.84\xi + 7)/100$$

$$B = (1.35\xi^3 - 14\xi^2 + 46\xi - 60.8)/100$$

$$C = (-1.08\xi^3 + 10.95\xi^2 - 35.1\xi + 150.9)/100$$

$$D = (-0.53\beta - 54\beta^{0.5} + 46)/100$$

$$E = (6\beta + 62\beta^{0.5} - 67)/100.$$

2. Concrete-filled SHS

$$K_{LC} = (A \cdot \beta^{-1} + B \cdot \beta^{-0.5} + C) \cdot (D \cdot n_r + 1) \leq 1 \quad (10)$$

where

$$A = (35.45\xi + 26.29)/100$$

$$B = (-40.62\xi + 74.58)/100$$

$$C = (5.2\xi - 0.93)/100$$

$$D = (103.2\beta^{0.5} - 53.11)/100.$$

To verify the validity of the formulae, the strength index (K_{LC}) of CFST subjected to axially local compression predicted using the formulae is compared with those calculated using the FEA model, which, as shown in Figure 14, predict the strength index (K_{LC}) with reasonable accuracy.

The sectional capacity calculated with the formulae (N_{uc}) is compared with the experimental sectional capacity (N_{ue}) determined by the tests of (Han et al. 2008b), as shown in Figure 15. It can be found that the accuracy of using the formulae to predict the experimental sectional capacity is reasonable, and in general, the predictions are somewhat conservative.

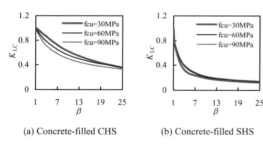

(a) Concrete-filled CHS   (b) Concrete-filled SHS

Figure 12. Influences of concrete strength on K_{LC}.

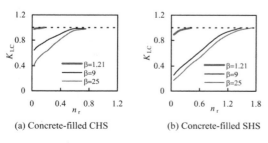

(a) Concrete-filled CHS   (b) Concrete-filled SHS

Figure 13. Influences of relative rigidity radius of the endplate on K_{LC}.

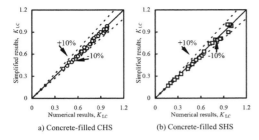

Figure 14. Comparisons of strength index of CFST with endplate.

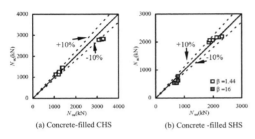

Figure 15. Comparisons of sectional capacity of CFST with endplate.

The validity limits of Equation 9–10 are: $\beta = 1$ to 25; $\alpha = 0.04$ to 0.2; $f_y = 235$ MPa to 420 MPa; $f_{cu} = 30$ MPa to 90 MPa; $t_a = 0$ mm to 25 mm; and $\xi = 0.2$ to 4.

5 CONCLUSIONS

Based on the results in this paper, the following conclusions can be drawn within the scope of the research:

1. The most important parameters that determine the strength index of CFST under axially local compression are the local compression area ratio, the steel ratio, the strength of steel and concrete, and the endplate rigidity.
2. The simplified formulae are developed for calculating the section capacity of the composite columns subjected to axially local compression.
3. Generally good agreement had been achieved between the comparisons of simplified model and the test results.

ACKNOWLEDGEMENTS

The research reported in the paper is part of Project 50425823 supported by National Natural Science Foundation of China. The financial support is highly appreciated.

REFERENCES

Han, L.H., Liu, W. & Yang, Y.F. 2008a. Behavior of thin walled steel tube confined concrete stub columns subjected to axial local compression. *Thin-Walled Structures* 46(2): 155–164.

Han, L.H., Liu, W. & Yang, Y.F. 2008b. Behavior of concrete-filled steel tubular stub columns subjected to axially local compression. *Journal of Constructional Steel Research* 64(4): 377–387.

Han, L.H., Yao, G.H. & Tao, Z. 2007. Performance of concrete-filled thin-walled steel tubes under pure torsion. *Thin-Walled Structures* 45(1): 24–36.

Han, L.H., Yao, G.H. & Zhao, X.L. 2005. Tests and calculations for hollow structural steel (HSS) stub columns filled with self-consolidating concrete (SCC). *Journal of Constructional Steel Research* 61(9): 1241–1269.

Hibbitt, Karlson & Sorenson. 2003. *ABAQUS version 6.4: theory manual, users' manual, verification manual and example problems manual.* Hibbitt: Karlson and Sorenson Inc.

Terzaghi, K. 1955. Evaluation of coefficients of subgrade reaction. *Geotechnique* (5): 297–326.

NOTATION

A_c	Concrete cross-sectional area
A_L	Area of local compression
A_s	Steel cross-sectional area
B	Outside width of square steel tube
CFST	Concrete-filled steel tube
CHS	Circular hollow section
D	Outside diameter of circular steel tube
E_c	Concrete modulus of elasticity
E_s	Steel modulus of elasticity
f_{ck}	Characteristic concrete strength ($f_{ck} = 0.67 f_{cu}$ for normal strength concrete)
f_{cu}	Concrete compressive cube strength
f_y	Yielding strength of steel
K_{LC}	Strength index
L	Length of the specimen
N	Axial load
n_r	Relative rigidity radius of endplate
N_u	Axial compressive capacity of fully loaded CFST
N_{uc}	Predicted bearing capacity
N_{ue}	Experimental bearing capacity
N_{uL}	Axial compressive capacity of locally loaded CFST
SHS	Square hollow section
t_a	Wall thickness of the top endplate
α	Steel ratio ($= A_s/A_c$)
β	Local compression area ratio ($= A_c/A_L$)
Δ	Axial deformation
ξ	Confinement factor $\left(\xi = \dfrac{A_s \cdot f_y}{A_c \cdot f_{ck}}\right)$

Examination of yield stress limitation for concrete filled steel tube (CFT) column using SM570TMC steel

Myung Jae Lee
School of Architecture & Building Science, Chung-Ang University, Korea

Young Suk Oh
Department of Architectural Engineering, Daejeon University, Korea

E.T. Lee
School of Architecture & Building Science, Chung-Ang University, Korea

ABSTRACT: Concrete-filled steel tube (CFT) has been used throughout the world in structures of varying heights and structural configuration. AISC-LRFD(2001) indicated that the specified minimum yield stress of structural steel and reinforcing bars used in calculating the strength of a composite column shall not exceed 415 MPa and AISC(2005) allowed a broad range to 525 MPa. The purpose of this paper is to examine the yield stress limitation in view of the performance characteristics of composite tubular columns using SM570 steel made by POSCO in Korea. 4 steel columns and 12 short and long composite columns using SM490TMC(F_y = 325 MPa) and SM570TMC structural steel (F_y = 440MPa) were tested and evaluated. Suitability of AISC(2005) and Korean Building Code(2005) design specification and the potential benefits for composite columns were investigated through the observation of the behaviors of these columns with various types.

1 INTRODUCTION

Composite action between different structural materials has been utilized in bridge and building construction for a number of years. Traditionally, constructed of steel and concrete, composite structures take advantage of the strength characteristics of each material. For example, a composite bridge girder utilizes the properties of steel in the tension zone and the compressive strength of the concrete in the deck. Similarly, a column made of a steel tube filled with concrete maximizes the steel strength by restraining the primary fialure mode, i.e., local buckling of the tube wall, while simultaneously improving the load sharing between steel and concrete. The tube also provides confinement for the concrete, preventing early spalling and effectively raising the available strength of the material. Two materials are thus mutually beneficial (Bridge & Roderik 1977, Galambos & Ravindra 1978, Gardner & Jacobson 1967, Janss 1974, Johnson & May 1978, Kim & Lee 2005, Kloppel & Goder 1957, Knowles & Park 1969, Loke 1968, Morino 2001, Ravindra & Galambos 1978, Salani & Sims 1964, Wang et al . 2004, Zong et al. 2002).

Modifying the existing strength formulas for steel columns to address the unique characteristics of composite compression members, design criteria for compostie columns were developed by Furlong and others (Furlong 1976, Task Group 20 1979). In essence, the procedures were based on the use of a modified yield stress, radius of gyration, and modulus of elasticity for the composite cross section, in combining the contributions of the steel and the concrete. In this fashion, composite columns were given design criteria that took advantage of the combined strength of steel and concrete.

AISC *Load and Resistance Factor Design(LRFD) Specification* for composite columns primarily based on a 1979 Structural Stability Research Colncil (SSRC) paper (Task Group 20 1979). It restricts the yield stress to 55 ksi (380 MPa). The limitation was introduced because of the perceived need for strain compatibility between the steel and unconfined concrete, as the latter appears in encased composite columns. Specifically, a limit strain of 0.0018 was imposed on the composite element, because this is the value at which unconfined concrete becomes unstable and begins to spall. With a modulus of elasticity of steel of 29,000 ksi(200 GPa), this strain corresponds to a yield stress of approximately 55 ksi(380 MPa).

In AISC (2001), the specified minimum yield stress of structural steel and reinforcing bars used in calculating the strength of composite column shall not exceed 60 ksi. The 60ksi(415 Mpa) limitation for the yield stress is very conservative for tubular composite columns, where the concrete confinement provided by the tubular walls is very significant. Kenny et al. have proposed raising the value of F_y for such columns to whatever the yield stress is for the steel grade used, but not higher than 80ksi(550Mpa).

In AISC Steel Manual (2005), the design yield strength is limited to 75 ksi(525 Mpa). This limitation was made from an assumption that the yield of steel tube was prior happened than compressive failure of concrete. However, because the time that the compressive failure of concrete is happened is related with the concrete strength, according to the combination of steel and concrete strengths, in the case of SM570TMC steel, 440 N/mm^2 yield strength may be applied. In KBC (2005), the specified minimum yield stress of structural steel and reinforcing bars used in calculating the strength of composite column shall not exceed 415 MPa.

It is purpose of this paper to examine the yield stress limitation in view of the performance characteristics of composite tubular columns and to evaluate whether or not 440 N/mm^2 yield strength formula can be used when SM570TMC steel is used for compoiste column. Concrete that is confined by the continuous shell provided by steel tubes are shown to be capable of sustaining substantially higher strains than 0.0018 before either material fails. The potential benefits for composite structures may be significant.

2 EXPERIMENTAL PROGRAM

With the parameters of CFT tube type and concrete strength, 4 short columns and 8 long columns were made and tested axially as follows:

– rectangular and circular steel tube short compressive test specimens
– rectangular and circular composite short compressive test specimens
– rectangular and circular composite long compressive test specimens

All tests were conducted at room temperature of RIST laboratory. For analysis purposes, the actual properties of the steel and grout materials were needed. These were obtained by the usual steel tension tests and chemical analyses, and concrete cylinder compression tests. In addition, deformations and strains were measured with linearly variable displacement transducers (LVDTs) and strain gages, respectively, on the full scale test specimens.

3 MATERIAL PROPERTIES

3.1 SM570TMC and SM490TMC steels

A primary objective of this work was to extend the existing design equations to include higher strength steel materials (TMC steel: Thermo Mechanical Controlled Steel). Tesile test specimens according to KSB0802(same as ASTM) were made for SM490TMC and SM570TMC steel plates (6T, 8T, 10T in thickness). Three specimens for each thickness were performed at Research Institute of Steel Technology (RIST) using 100 ft (980kN) Universal Testing Machine.

Tension tests of the material in SM570TMC and SM490TMC steels confirmed the specified minimum yield stress determined at 0.2 strain offset.

As shown in Figures 1–2, in case of SM490TMC steel, elastic region, yield point, pefectly plastic region, and strain hardening region can be apparently shown. On the other hand, in case of SM570TMC steel, as the general behavior of high strength steel, the yield point does not shown in stress-strain curves (continuous yielding).

3.2 Concrete

Two cylinders of the material used in filling the steel tubes were tested to be determined the 28-day

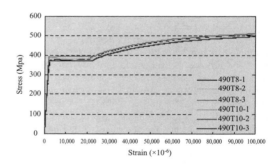

Figure 1. Stress-strain curves of SM490TMC steel.

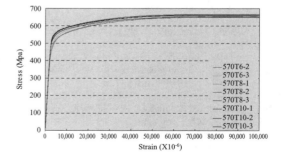

Figure 2. Stress-strain curves of SM570TMC steel.

unconfined compressive strength. Test results were consistent, with value of 40 Mpa as indicated in Table 5.

4 COMPRESSIVE SHORT COLUMN TESTS OF SM570TMC STEEL

Four SM570TMC steel (yield strength F_y = 440 N/mm², tensile strength F_u = 570 N/mm²) tube tests were performed with the test parameters; section types (□ or ○) and width-to-thickness ratios of steel plates as shown in Tables 1–2. Figures 3–4 show the specimens' dimensions of square and circular steel tube short columns respectively.

As shown in Tables 3–4, rectangular NCFT160R specimen is compact section and NCFT240R specimen noncompact. On the other hand, circular NCFT240 specimen is noncompact section and NCFT300C slender section. NCFT160R specimen had the low strength ratio even it was a compact section. On the other hand, NCFT300C specimen had the high strength ratio even though it is a slender section.

Stress-strain relationships of each specimen are shown in Figures 5–6. Test results showed the

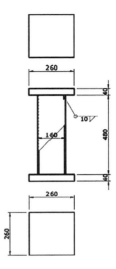

Figure 3. Short column of square steel tube (NCFT160R).

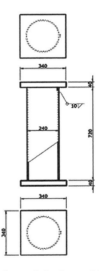

Figure 4. Short column of circular steel tube (NCFT240C).

Table 1. Layout of square steel tube specimen.

Name of specimen	Section mm	Length mm	Width thickness ratio
NCFT160R	□-160 × 160 × 8	480	20
NCFT240R	□-240 × 240 × 8	720	30

Table 2. Layout of circular steel tube specimen.

Name of specimen	Section mm	Length mm	Diameter thickness ratio
NCFT240C	Φ-240 × 6	720	40
NCFT300C	Φ-300 × 6	900	50

Table 3. Stress increasing ratio of each specimen.

Name of specimen	Section type	Yield load kN	Maximum load kN	Yield stress Mpa	Maximum stress Mpa	Increasing ratio of stress
NCFT 160R	b/t = 18 < λ_r = 21.7 compact section	2,623	3,055	539	628	1.164
NCFT 240R	b/t = 28 > λ_r = 27.4 thin walled	4,004	3,979	539	536	0.994
NCFT 240C	D/t = 40 < λ_r = 42.8 non-compact	2,333	3,029	529	687	1.298
NCFT 300C	D/t = 50 > λ_r = 42.8 thin walled	2,932	3,761	529	679	1.283

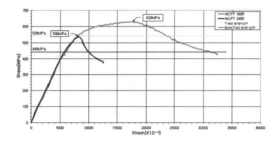

Figure 5. Stress-strain relationships of square steel tube short column.

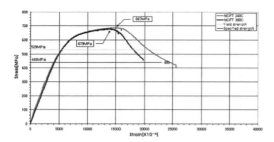

Figure 6. Stress-strain relationships of circular steel tube short column.

Table 4. Section type for nominal yield strength, stress increasing ratio and plastic deformation ability.

Name of specimen	Section type	Stress increasing ratio	Plastic deformation capacity	
			axial force ratio = 0.0	axial force ratio = 0.4
NCFT 160R	Compact section	1.427	8.84	9.80
NCFT 240R	non-compact section	1.218	2.82	3.39
NCFT 240C	non-compact section	1.552	12.74	13.63
NCFT 300C	non-compact section	1.545	12.48	13.38

maximum stresses different according to the width-to-thickness ratio of steel plates. Photos are in Figures 7–10.

Test results showed that the maximum yield strength of each specimen was larger than nominal yield strength $F_y = 440$ N/mm² with good plastic deformation capacity which was 7 to 9 times larger than the elastic range except the slender sections.

Test setup   After test

Figure 7. Photos of NCFT160R specimen.

Test setup   After test

Figure 8. Photos of NCFT240R specimen.

Test setup   After test

Figure 9. Photos of NCFT240C specimen.

Test setup After test

Figure 10. Photos of NCFT300C specimen.

Table 5. Layout of CFT short column specimens.

Name of specimen	Steel type	Concrete MPa	Length mm
CFT160R	SM570TMC	40	480
CFT240R			720
CFT240C			720
CFT300C			900

5 COMPRESSIVE COMPOSITE SHORT COLUMN OF SM570TMC STEEL (CFT)

As shown in Table 5, two types (□ or ○) of steel shapes were made using SM570TMC steel and the intended concrete stregnth was 40Mpa. In CFT short column tests as shown in Figures 11–12, according to the width-to-thickness or diameter-to-thickness ratios the strength increasing ratios were changed.

Figures 13–14 showed the typical load-strain diagrams for short columns. All of the specimens had exhibited a linear relationship until at least 75 percents of the maximum load had been reached. After this point, increasingly large strains were observed, as is common with stub column tests.

Cumulative strength can be calculated from the following equation and arranged in Table 6.

$$N_U = \sigma_{sY} \cdot A_s + f_{ck}' \cdot A_c \qquad (1)$$

where σ_{sY} = yield stress of steel; A_s = sectional area of steel; f_{ck}' = strength of concrete; A_c = sectional area of concrete; and N_U = cumulative strength.

In CFT specimens, the concrete was yielded first, and then the steel tube was yielded (Figures 15–18). Table 7 summarizes the material and geometric properties used to calculate the nominal strength P_n, for the test specimens under investigation. Cumulative strengths were used for this purpose. In order to establish a measure of the performance of the model used to formulate the equations, the test load is then divided by the nominal strength. These ratios range from 0.979 to 1.531, with an average value of 1.278. The previous tests resulted in an average test-to-design ratio of 1.38 (Kenny et al. 1994).

Therefore, when SM570TMC steel is used in composite column, the strength increasing rates are

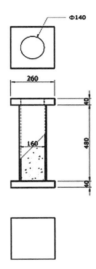

Figure 11. CFT short column of square steel tube (CFT160R).

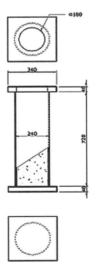

Figure 12. CFT short column of circular steel tube (CFT240C).

slightly different according to the combination of concrete. However, maximum strength is well over the cumulative strength by the nominal yield stress 440 N/mm² and lead to good application.

Table 7. Strength increasing ratio and yielding pattern.

Name of specimen	Cumulative strength, P_n kN	Maximum load, P_{test} kN	P_{test}/P_n	Yielding pattern
CFT160R	3,449	3,780	1.096	concrete
	2,968		1.274	→steel
CFT240R	6,009	5,882	0.979	concrete
	4,977		1.182	→steel
CFT240C	3,965	5,468	1.379	concrete
	3,572		1.531	→steel
CFT300C	5,535	7,334	1.325	concrete
	5,042		1.455	→steel

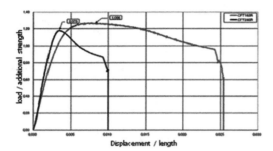

Figure 13. Load-strain relationships of square CFT short column.

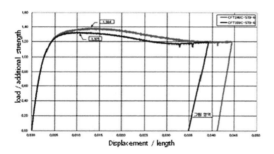

Figure 14. Load-strain relationships of circular CFT short column.

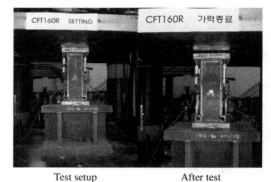

Test setup   After test

Figure 15. Photos of CFT160R specimen.

Table 6. Cumulative strength of each specimen.

Name of specimen	σ_{sY} MPa	A_s mm²	f'_{ck} MPa	A_c mm²	N_u kN
CFT160R	539	4860 (0.76)	40	20736 (0.24)	3,449
CFT240R	539	7424 (0.67)	40	50176 (0.33)	6,009
CFT240C	529	4409 (0.59)	40	40807 (0.41)	3,965
CFT300C	529	5539 (0.53)	40	65111 (0.47)	5,535

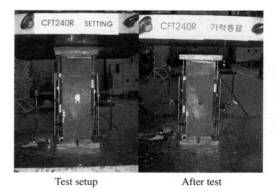

Test setup   After test

Figure 16. Photos of CFT240R specimen.

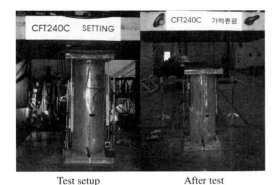

Figure 17. Photos of CFT240C specimen.

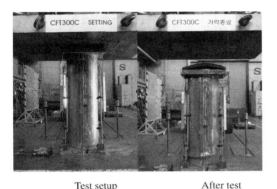

Figure 18. Photos of CFT300C specimen.

Table 8. Long CFT column specimens.

Name of specimen	Section	Length	Number
LCFT150-0.5	□-150 × 150 × 10	1,580 mm	2
LCFT150-0.8		2,740 mm	2
LCFT200-0.5	□-200 × 10	1,920 mm	2
LCFT200-0.8		3,280 mm	2

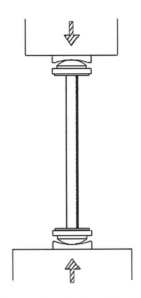

Figure 19. End conditions for long CFT column.

6 COMPRESSIVE LONG COLUMN TESTS OF CIRCULAR AND RECTANGULAR CFT

Rectangular tube(□-150 × 150 × 10) and circular tube(□-200 × 10) were made using SM570TMC steel. Total 8 long column specimens using the slenderness parameters 0.5 and 0.8 were made as shown in Table 8.

In order to minimize the end effects for the longer specimens, pinned-end fixtures were specified for these specimens as shown in Figure 19. These fixtures had been designed to allow free rotation of the column ends in all directions. Each tube was loaded continuously to failure, i.e., until the tube was no longer able to sustain its maximum load.

As shown in Figure 20, 4 LVDTs were installed in top and bottom ends and 1 LVDT was installed in middle in order to investigate the lateral displacement. Initial displacements were in good condition with maximum average 1/1,370 in each plane. Initiation of buckling was determined from visual and technical measurements.

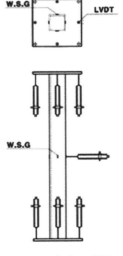

Figure 20. Measurements for long CFT column.

Experimental results of 4 round and 4 rectangular CFT long columns are in Table 9 with the cumulative strength values simply using the yield strength of steel and the maximum compressive strength of concrete ignoring the confinement effects. It was shown that the higher maximum strength than cumulative strength was observed.

Load-vertical and load–horizontal displacement and stress-strain relationship are shown in Figures 21–23 respectively. Photos of initial setting and after-tests are in Figures 24–25. As shown in Figures 26–27, the

Table 9. Result of long CFT column test.

Name of specimen	Concrete stiffness MPa	Maximum load kN	Additional strength kN
LCFT150-0.5A		4,272	
LCFT150-0.5B		4,241	
LCFT150-0.8A		3,969	2,981
LCFT150-0.8B		3,973	
LCFT200-0.5A	36	5,764	
LCFT200-0.5B		5,705	
LCFT200-0.8A		5,328	3,403
LCFT200-0.8B		5,439	

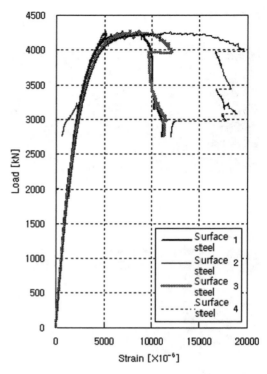

Figure 23. Load-strain relationship.

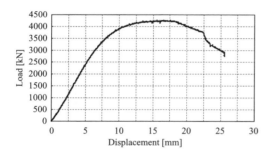

Figure 21. LCFT150-0.5 A load-verticality strain relation.

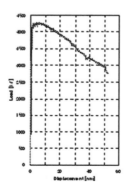

Figure 22. load-horizontal strain relationship.

Figure 24. Photos of square CFT long column (LCFT150-0.8A).

Test setup

After test

Figure 25. Photos of circular CFT long column (LCFT200-0.8B).

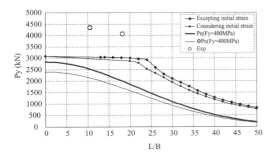

Figure 26. Maximum stresses of square CFT long column.

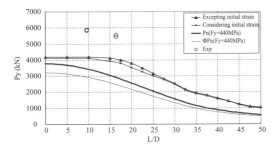

Figure 27. Maximum stresses of circular CFT long column.

maximum strength of SM570TMC CFT column is much higher than the design equation and the analysis values which 440 N/mm² was applied.

Figure 23 shows that the longer specimens behaved linearly unitl the maximum load was attained. At peak loads, the effects of eccentricity, due to the bent shape of the tube, greatly reduced any further load-carrying capacity of the columns. The maximum sustainable force therefore quickly dropped off as the deformations increased.

7 CONCLUSIONS

To evaluate the design yield strength when using the composite column of SM570TMC steel, existing design code were surveyed and material tensile test and CFT short and long column compressive tests of SM490TMC and SM570TMC steels were performed.

It is verified from steel tests that tensile values of both SM570TMC and SM490TMC were over code values. Maximum strength of short CFT column were much higher than the cumulative strength by the nominal yield strength 440 N/mm² as same as long CFT column tests. Therefore, it is recommended that 440 N/mm² for the composite column design criteria for tubular members can be used as the design yield strength SM570TMC steel.

REFERENCES

American Institute of Steel Construction. 1986. *Manual of Steel Construction-Load and Resistance Factor Design* (First Edition). Chicago.

American Institute of Steel Construction. 2001. *Manual of Steel Construction-Load and Resistance Factor Design* (Third Edition). Chicago.

American Institute of Steel Construction. 2005. *Manual of Steel Construction* (Thirteenth Edition). Chicago.

American Society of Testing and Materials. 1989. *Standard Test Methods and Definitions for Mechanical Testing of Steel Products*. ASTM Standard No. A370-89, ASTM, Philadelphia: PA.

Bridge, R.O. & Roderick, J.W. 1977. The Behavior of built-up composite columns. *Research Report No. 306*, School of Civil Engineering, University of Sydney. Sydney.

Furlong, Richard W. 1976. AISC Column logic makes sense for composite columns, Too. AISC. *Engineering Journal* 13(1), First Quarter: 1–7.

Galambos, T.V. & Ravindra, M.K. 1978. Properties of steel for use in LRFD. *Journal of the Structural Division*, ASCE 104(ST9): 1459–1468.

Gardner, N.J. & Jacobson, E.R. 1967. Structural behavior of concrete-filled steel tubes. *Journal of the American Concrete Institute* 64(7): 404–413.

Janss, J. 1974. Calculation of ultimate loads on metal columns encased in concrete. *Report No. MT89*, Industrial Center of Scientific and Technical Research for Fabricated Metal. Brussels.

Johnson, R.P. & May, I.M. 1978. Tests on restrained composite columns. *The Structural Engineer* 56B(2).

Kenny, J.R., Bruce, D.A., & Bjorhovde, R. 1994. Removal of yield stress limitation for composite tubular columns. *Engineering Journal*, AISC, First Quarter: 1–11.

Kim, Cheol-Hwan & Lee, E.T. 2005. Cyclic behavior of beam to-concrete filled steel tube column connection. *International Journal of Steel Structures* 5(4): 399–406.

Korean Building Code. 2005. *Manual of Structural Design*. Chapter 7. Korea.

Kloppel, K. Von & Goder, W. 1957. Investigations of the load carrying capacity of concrete-filled steel tubes and development of a design formula. *Der Stahlban* 26(1), January and February (in German).

Knowles, R.B. & Park, R. 1969. Strength of concrete-filled steel tubular columns. *Journal of the Structural Division* ASCE 95(ST12): 2565–2587.

Loke, Y.O. 1968. The behavior of composite steel-concrete columns. Ph.D. Dissertation. University of Sydney. Sydney.

Morino, S. 2001. Concrete-filled steel tube column system—it's advantages. *International Journal of Steel Structures* 1(1): 33–44.

Ravindra, M.K. & Galambos, T.V. 1978. Load and resistance factor design for steel. *Journal of the Structural Division*, ASCE 104(ST9): 1337–1353.

Salani, H.J. & Sims, J.R. 1964. Behavior of mortar-filled steel tubes in compression. *Journal of the American Concrete Institute* 61(11).

Task Group 20, Structural Stability Research Council. 1979. A Specification for the Design of Steel-Concrete Composite Columns. AISC *Engineering Journal* 16(4), 4th Quarter: 101–115.

Wang, In-Soo, Kang, S.K., Son, S.H., & Yeom, K.S. 2004. Construction of CFT column on the sports club of tower palace 3 project. *International Journal of Steel Structures* 4(1): 33–42.

Zong, S.T., Cheng, H.T. & Zhang, S.M. 2002. The continuity of behaviors for circular, square and octagonal forms for concrete filled steel tube(CFT) members under axial compression. *International Journal of Steel Structures* 2(2): 81–86.

Behaviour of slender high strength concrete filled tubular columns

J.M. Portoles & D. Hernández-Figueirido
Universitat Jaume I, Castellon, Spain

M.L. Romero* & J.L. Bonet
Universidad Politécnica de Valencia, Valencia, Spain

F.C. Filippou
University of California at Berkeley, California, USA

ABSTRACT: In recent years an increment in the utilization of concrete tubular columns was produced due to its high stiffness, ductility and fire resistance. On the other hand the use of High Strength Concrete (HSC) is more common due to the advances in the technology. The use of this material presents different advantages, mainly in elements subjected to high compressions as building supports or bridge columns.

However, there is a notably lack of knowledge in the behavior of high strength concrete filled tubular columns which produces that the existing simplified design models for normal strength concretes are not valid for them. This paper presents the results of a research project, where a numerical and an experimental study of high strength CFT's is *performed*.

1 INTRODUCTION

The utilization of hollow steel sections is well-known in multi-story buildings because a substantial reduction of the cross section is obtained. Moreover, the high strength concrete is spent more and more mainly for precast concrete structures, but the influence of the high strength on this type of columns (CFT's) is not well studied.

The actual simplified design methods are similar to the methods for reinforced concrete, assuming the steel section as an additional layer of reinforcement.

Each country has its different code of design for composite sections (Japan, Australia, Canada, United States, Europe, etc.). The design code of the Euro-Code 4 (2004) (allows only the utilization of concrete with strength lower than 50 MPa (cylinder strength); therefore for high strength concretes the method and the interaction diagrams are not valid. Furthermore, as the section is reduced, for an equal length of the element, the slenderness is increased and the buckling is more relevant.

1.1 Normal strength concrete

The utilization in Europe of normal strength concrete filled tubular columns is well-known some decades ago when appeared the first monograph from the CIDECT (1970) simplifying its applicability for practical engineers. Later research works gave rise to the monograph nº5, CIDECT (1979). All this documents were the base to make the Euro-Code 4 (2004), with a special section for CFT's. In Spain, the ICT (Instituto de la Construcción Tubular) published a practical monograph, CIDECT (1998), with the idea to make easier the design of this type of sections.

1.2 High strength concrete

However, there are not a lot of investigations regarding high performance materials for CFT's, focused in the buckling. The research on high strength concrete (HSC) has demonstrated that the tensile capacity does not increase in the same proportion as the compression capacity. For hollow sections filled with concrete the tension problem is not as much important because the concrete cannot split of. Therefore in this type of section is where more advantage is taken.

1.2.1 *Experimental tests*
The more important contributions are concentrated in the last 5–10 years:

Grauers [1993] performed experimental tests over 23 short columns and 23 slender columns, stating that the methods of the different codes were valid but should be extended in order to analyze the effect of other parameters. Bergman (1994) studied the confinement mainly for normal strength concrete and partially for high strength concrete, but applying only axial load.

Aboutaba et al. (1999) compared classical columns of HSC with concrete filled tubular columns with HSC. Those ones presented more lateral stiffness and ductility.

Certainly, the research of Rangan & Joyce (1992) and Kilpatrick & Rangan (1999) has advanced a lot in the field. They presented experimental results from 9 columns for uniaxial bending and 24 columns for double curvature. However they used small sections and proposed a simplified design method regarding the Australian code, which does not follow the same hypothesis than the Eurocode-4, as i.e. the buckling diagrams.

Liu et al. (2003) compared experimentally the capacity of 22 rectangular sections with the different codes (AISC, ACI, EC4) and they concluded that the Euro-Code 4 was on the unsafe side while other codes over-designed the sections.

Varma et al. (2002, 2004) studied initially the behavior of the square and rectangular tubular columns and recently they have investigated cyclic loads and the existence of plastic hinges.

Gourley et al. (2001) have published a research report about the state of the art of concrete filled tubular columns.

1.2.2 Numerical models

Concerning the numerical models, it can be stated that there are not a lot of specific studies that applied the finite element method or sectional analysis to this type of structure. Most of them as Hu et al. (2003), Lu et al. (2000), and Shams & Saadeghvaziri (1999) study normal strength concretes. Only, recently Varma et al. (2005) have implemented a fibber model applied to square tubular sections but for short columns, without taking into account the buckling.

There is also an important research work from Johansson and Gylltoft (2001) where the the effects of three ways to apply a load to a columns is investigated numerically and experimentally.

If a good sectional characterization (moment-curvature) was performed, it can be inferred that the actual simplified methods are valid as a first approach to study the strength of these supports.

Recently, Zeghiche and Chaoui (2005) have published a small study for circular sections following this procedure. They affirmed that more numerical and experimental tests should be performed to check the validity of the buckling design methods of the EC4 for high strength concrete and double curvature.

Due to that the authors are performing a research project to study the effect of high strength concrete in the buckling. It has three parts: experimental study, one-dimensional numerical model and three-dimensional model. In this paper the initial results of the experimental and one-dimensional (1-D) part is presented.

For the 1-D part a nonlinear finite element numerical model for circular concrete filled tubular sections will be presented. The method has to be computationally efficient and must represent the behaviour of such columns, taking into account the effect of high strength concrete and second order effects.

Also an experimental study of circular concrete filled tubular sections is presented. The experimental tests selected corresponds to circular tubular columns filled with concrete (CFT) with pinned supports at both ends subjected to axial load and uniaxial bending. In these tests the eccentricity of the load at the ends is fixed and the maximum axial load of the column is evaluated.

2 NUMERICAL MODEL

The numerical model based on the finite element method is deeply illustrated in Romero et al. (2006), but in order to clarify the presentation it is summarized below.

2.1 Formulation

The finite element selected is a classical one-dimensional 13 degrees of freedom (d.o.f.'s) element. It has 6 d.o.f.'s at each node (three displacement and three rotations), and a longitudinal degree of freedom in the mid-span to represent a non-constant strain distribution to represent the cracking. The Navier-Bernoulli hypothesis is accepted for the formulation. Also perfect bond between the concrete and structural steel is assumed. In the model the local buckling of the hollow steel section is neglected by stating at least the minimum thickness pointed out in the EC-4 (2004) (Art 4.8.2.4.).

$$\frac{D}{t} \leq 90 \cdot \varepsilon^2 \qquad (1)$$

where: D external diameter of the circular section
 t thickness of the circular hollow steel section
 ε $= \sqrt{235/f_y}$
 f_y yielding stress of the structural steel (MPa)

The model includes the second order effects by the formulation of large strains of the element (using the nonlinear deformation matrix $-B_L-$ and the geometric stiffness matrix $-K_g-$) and large displacements (by stating the force equilibrium in the deformed shape). The model automatically obtains the maximum load by using the total potential energy analysis "V" of the structure, Gutiérrez et al. (1983); detecting if the structure reaches an stable equilibrium position ($\delta^2 V > 0$), unstable equilibrium ($\delta^2 V < 0$) or instability ($\delta^2 V = 0$). Moreover, the Newton-Raphson method was selected to solve the nonlinear system of equations for a known load level. An arc-length method, Crisfield (1981), was used for displacement control.

2.2 Constitutive equations of the materials

A bilinear (elastic-plastic) stress-strain diagram for the structural steel was assumed, EC-4 (2004) (Art 3.3.4). The equation proposed by the CEB (1991) for the Model Code was used for the columns filled with normal strength concretes ($f_c \leq 50$ MPa); for the cases of high strength concrete ($f_c > 50$ MPa) was selected the equation from the CEB-FIP (1995). The tension-stiffening effect was considered by a gradual unload method, Bonet (2001). The permanent deformations due to cyclic loads were not included in the model. Therefore it is assumed that the maximum load is not dependent on the adopted path.

3 EXPERIMENTAL PROGRAM

Thirty-seven test specimens of NSC and HSC columns were performed in this experimental program; see Table 1. They are designed to investigate the effects of four main parameters on their behavior: slenderness (Length/Diameter), ratio D/t (Diameter/thickness), strength of concrete (Fck) and eccentricity. The column lengths, L, are 2135 and 3135 mm and the cross-sections are circular with a 100, 125, and 160 of outer diameter. The thickness of the steel tubes are 3, 4, 5 and 5.7 mm. The strength of concrete varies from 30, 70 to 110 MPa and the axial load is applied with two eccentricities: 20 and 50 mm. All of the tests were performed in the laboratory of the Department of Mechanical Engineering and Construction of the Universitat Jaume I in Castellon, Spain.

3.1 Materials

3.1.1 Concrete
All columns were cast using concrete batched in the laboratory. The concrete compressive strength fck was determined both from the cylinder and cubic compressive tests at 28 days. The specimens were cast in a inclined position.

3.1.2 Steel
The steel used was S275 JR. The yield strength f_y, the ultimate strength fu, the strain at hardening, the ultimate strain and modulus of elasticity E of the steel were obtained from the Eurocode 2.

3.2 Test Setup

The specimens are tested in a special 2000 kN capacity testing machine. The eccentricity of the applied compressive load was equal at both ends, so the columns are subjected to single curvature bending. It was necessary to built up special assemblages at the pinned ends to apply the load eccentrically. Figure 1 presents a general view of the test fo 2 meters long. Five LVDTs were used to measure symmetrically the deflection of the column at midlength (L/2), and also at four additional levels. The strains were measured at the mid-span section using strain gages, Figure 2.

Table 1 presents all the experiments.

Specimens 1 and 27 are not valid due to impact loads during the experiment. Also the specimen 2 is not valid due to problems in the contact between inner concrete and the end plate.

The values of steel strength have been taken theoretically using the steel grade. Steel elastic modulus has been taken 210000 MPa.

Figure 1. Test setup.

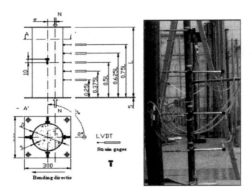

Figure 2. LVDT's for measurement the deformed shape.

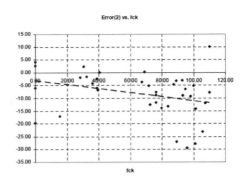

Figure 3. Error versus fck.

Table 1. Tests for CFT columns.

#	Name	Ltotal (mm)	e (mm)	Dext (mm)	t (mm)	fy theor. (Mpa)	fck (Mpa)
1	C100–3–2–00–20–1	2135	20	100	3	275	0.00
2	C100–3–2–30–20–1	2135	20	100	3	275	15.43
3	C100–3–2–30–20–2	2135	20	100	3	275	30.36
4	C100–3–2–30–50–1	2135	50	100	3	275	32.45
5	C100–3–2–60–20–1	2135	20	100	3	275	71.70
6	C100–3–2–60–50–1	2135	50	100	3	275	83.69
7	C100–3–2–90–20–1	2135	20	100	3	275	110.06
8	C100–3–2–90–50–1	2135	50	100	3	275	107.17
9	C100–3–3–30–20–1	3135	20	100	3	275	38.94
10	C100–3–3–30–50–1	3135	50	100	3	275	36.15
11	C100–3–3–60–20–1	3135	20	100	3	275	76.13
12	C100–3–3–60–50–1	3135	50	100	3	275	76.15
13	C100–3–3–90–20–1	3135	20	100	3	275	99.76
14	C100–3–3–90–50–1	3135	50	100	3	275	92.81
15	C100–5–2–00–50–1	2135	50	100	5	275	0.00
16	C100–5–2–30–20–1	2135	20	100	5	275	40.47
17	C100–5–2–30–50–1	2135	50	100	5	275	28.19
18	C100–5–2–60–20–1	2135	20	100	5	275	67.12
19	C100–5–2–60–50–1	2135	50	100	5	275	68.79
20	C101p6–5–2–00–20–1	2135	20	101.6	5	275	0.00
21	C101p6–5–2–00–20–2	2135	20	101.6	5	275	0.00
22	C101p6–5–2–90–20–1	2135	20	101.6	5	275	109.83
23	C101p6–5–2–90–50–1	2135	50	101.6	5	275	94.70
24	C101p6–5–3–30–20–1	3135	20	101.6	5	275	39.47
25	C101p6–5–3–30–50–1	3135	50	101.6	5	275	38.95
26	C101p6–5–3–60–20–1	3135	20	101.6	5	275	76.17
27	C101p6–5–3–60–50–1F	3135	50	101.6	5	275	97.85
28	C101p6–5–3–90–20–1	3135	20	101.6	5	275	87.46
29	C101p6–5–3–90–50–1	3135	50	101.6	5	275	88.86
30	C125–4–3–60–20–1	3135	20	125	4	275	89.14
31	C125–4–3–60–50–1	3135	50	125	4	275	95.76
32	C125–4–3–90–20–1	3135	20	125	4	275	105.69
33	C125–4–3–90–50–1	3135	50	125	4	275	100.86
34	C160–6–3–60–20–1	3135	20	160.1	5.7	275	79.77
35	C160–6–3–60–50–1	3135	50	160.1	5.7	275	72.28
36	C160–6–3–90–20–1	3135	20	160.1	5.7	275	92.98
37	C160–6–3–90–50–1	3135	50	160.1	5.7	275	101.22

Table 2. Comparison of experimental tests (N_exp) and the dsign load from the Eurocode (N_ed).

#	e/D	D/t	L/D	λ	fck(Mpa)	N_exp(KN)	N_ed(KN)	Error
1	0.20	33.33	21.35	2.10	0.00	180.80	145.35	−19.61
2	0.20	33.33	21.35	0.76	15.43	181.45	150.58	−17.01
3	0.20	33.33	21.35	0.85	30.36	181.56	186.08	2.49
4	0.50	33.33	21.35	0.86	32.45	117.49	115.79	−1.45
5	0.20	33.33	21.35	1.04	71.70	248.58	235.87	−5.11
6	0.50	33.33	21.35	1.09	83.69	151.59	131.72	−13.11
7	0.20	33.33	21.35	1.19	110.06	271.04	250.19	−7.69
8	0.50	33.33	21.35	1.18	107.17	154.24	136.04	−11.80
9	0.20	33.33	31.35	1.31	38.94	140.32	137.01	−2.36
10	0.50	33.33	31.35	1.29	36.15	93.75	89.82	−4.19
11	0.20	33.33	31.35	1.56	76.13	159.55	147.61	−7.48
12	0.50	33.33	31.35	1.56	76.15	106.80	97.51	−8.70
13	0.20	33.33	31.35	1.70	99.76	160.33	152.07	−5.15
14	0.50	33.33	31.35	1.66	92.81	102.75	99.71	−2.96
15	0.50	20	21.35	1.68	0.00	142.19	133.71	−5.96
16	0.20	20	21.35	0.86	40.47	270.02	270.48	0.17
17	0.50	20	21.35	0.81	28.19	161.26	158.41	−1.76
18	0.20	20	21.35	0.95	67.12	313.55	302.43	−3.55
19	0.50	20	21.35	0.95	68.79	183.81	184.44	0.34
20	0.20	20.32	21.01	1.66	0.00	223.37	232.95	4.29
21	0.20	20.32	21.01	1.66	0.00	226.61	232.95	2.80
22	0.20	20.32	21.01	1.06	109.83	330.40	363.54	10.03
23	0.49	20.32	21.01	1.02	94.70	213.46	199.89	−6.36
24	0.20	20.32	30.86	1.24	39.47	212.48	198.24	−6.70
25	0.49	20.32	30.86	1.23	38.95	144.83	136.34	−5.86
26	0.20	20.32	30.86	1.42	76.17	246.82	217.98	−11.68
27	0.49	20.32	30.86	1.51	97.85	164.95	149.67	−9.26
28	0.20	20.32	30.86	1.47	87.46	231.35	220.74	−4.59
29	0.49	20.32	30.86	1.47	88.86	153.16	148.27	−3.19
30	0.16	31.25	25.08	1.29	89.14	489.47	357.38	−26.99
31	0.40	31.25	25.08	1.32	95.76	322.97	227.98	−29.41
32	0.16	31.25	25.08	1.36	105.69	474.17	364.77	−23.07
33	0.40	31.25	25.08	1.34	100.86	317.90	229.43	−27.83
34	0.12	28.09	19.58	0.96	79.77	1012.50	873.69	−13.71
35	0.31	28.09	19.58	0.93	72.28	642.16	562.14	−12.46
36	0.12	28.09	19.58	1.00	92.98	1011.50	921.86	−8.86
37	0.31	28.09	19.58	1.03	101.22	686.21	589.01	−14.16

4 EXPERIMENTAL RESULTS

Table 2 presents a comparison between the experiments (Nexp) and the design load of the Euro Code 4 (N_ed), where:

$$\lambda = \sqrt{\frac{N_{plRd}}{N_{creff}}} \qquad (2)$$

is the relative slenderness.
and:

$$Error = \frac{N_{ed} - N_{exp}}{N_{exp}} \times 100 \qquad (3)$$

With the error defined like that, the errors grater than zero mean the design is unsafe.

Also the Figure 3 presents the tendency of the error in terms of the concrete strength.

Some of the errors are not admissible (20–30%). All of them correspond to the ones with cross section C125 × 4 mm. There should be a mistake in the experiments.

A first analysis of these results initially shows that the error increases when concrete strength or D/t ratio increases. This error tends to negative with fck or D/t, what means that the EC4 is conservative. The euro code 4 could be slightly improved for high strength concrete.

5 NUMERICAL RESULTS

The same cases are analysed using the numerical model. Figure 4 presents the global behaviour of one of the numerical model versus the experiments.

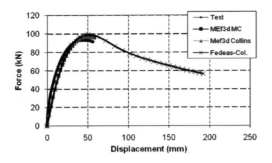

Figure 4. Load-displacement relationship of the test C100-3-3-30-50-1. (normal strength concrete).

The numerical model compares different models for concrete: Model Code (Mef3d MC), Collins et al. (1993), (Mef3d Collins) and the numerical model FEDEAS from the professor Filippou (2004) (Fedeas-Col) suing the Collins model for high strength concrete. All the numerical models presents similar results so in advance only the first one is used.

The errors between the experiments and the numerical models are not the same for the maximum forces than for the displacement corresponding to the maximum forces.

Table 3 presents the errors in "forces". The mean value and the standard deviation is close to 3% and 5% respectively for both models, which represents that the numerical model has a good accuracy. But some

Table 3. Errors of forces between the numerical models and experiments.

Test	Fmax	Mef3d Mc	Forces Mef3d Col	Error Mef3d Mc	Error Mef3dCol
1	180.8	136,183	136.194526	−24.68%	−24.67%
2	181.45	174,142	174.90083	−4.03%	−3.61%
3	181.56	205,215	205.130864	13.03%	12.98%
4	117.49	129,133	128.591334	9.91%	9.45%
5	248.58	259.758	262.439918	4.50%	5.58%
6	151.59	154.279	155.321858	1.77%	2.46%
7	271.04	299.103	291.69384	10.35%	7.62%
8	154.24	163.407941	161.008007	5.94%	4.39%
9	140.32	151.56	151.193717	8.01%	7.75%
10	93.75	97.155	97.082952	3.63%	3.56%
11	159.55	172.354	168.905686	8.03%	5.86%
12	106.8	101.593	106.924852	−4.88%	0.12%
13	160.33	180.176	175.123332	12.38%	9.23%
14	102.75	110.86	108.968067	7.89%	6.05%
15	142.19	139.644	139.647928	−1.79%	−1.79%
16	270.02	282.124	280.316256	4.48%	3.81%
17	161.26	167.979	167.2263	4.17%	3.70%
18	313.55	309.704	314.864234	−1.23%	0.42%
19	183.81	187.389	193.261551	1.95%	5.14%
20	223.37	218.886	218.899504	−2.01%	−2.00%
21	226.61	218.886	218.899504	−3.41%	−3.40%
22	330.4	387.376	381.935873	17.24%	15.60%
23	213.46	223.396	222.217669	4.65%	4.10%
24	212.48	215.941	215.394249	1.63%	1.37%
25	144.83	137.911	137.617121	−4.78%	−4.98%
26	246.82	239.165	235.406194	−3.10%	−4.62%
27	164.95	160.945	158.421475	−2.43%	−3.96%
28	231.35	244.043	239.194952	5.49%	3.39%
29	153.16	159.323	157.119374	4.02%	2.59%
30	489.47	420.547	411.152943	−14.08%	−16.00%
31	322.97	262.958	258.671054	−18.58%	−19.91%
32	474.17	437.002	424.243425	−7.84%	−10.53%
33	317.9	264.995	260.32025	−16.64%	−18.11%
34	1012.5	980.124	983.0469	−3.20%	−2.91%
35	642.16	618.181	623.667654	−3.73%	−2.88%
36	1011.5	1024.954	1031.09913	1.33%	1.94%
37	686.21	674.284	675,693,775	−1.74%	−1.53%
			MEAN	3.35%	3.10%
			STD DEV	5.78%	5.04%

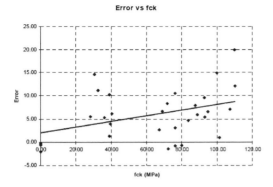

Figure 5. Error of the numerical model (using MC) in terms of fck.

of the cases have inadmissible errors (again the cases with 125 mm of diameter).

The same table of errors but for "displacements" is avoided for simplicity of the paper. In this case the mean value of the error is −6% and the standard deviation is 15%, which means that the numerical model has more stiffness than the experiment.

Figure 5 presents the tendency of the error in terms of the strength of concrete. If it increases, the error is higher which means that the numerical model has to be improved for high strength concrete.

6 CONCLUSIONS

The paper is focussed in the presentation of 37 experimental tests and a nonlinear numerical model for the analysis of concrete filled tubular columns using the finite element method. A comparative study with the design load obtained with the last version of the Eurocode 4 (2004) is presented. The code is in most cases conservative and it could be improved for high strength concrete (HSC) but further tests should be performed. The numerical model is more rigid and needs to be improved for the cases of higher eccentricities, where the displacement corresponding to the maximum load is not well predicted.

This issue is very important to create new buckling curves for high strength concrete.

ACKNOWLEDGEMENTS

The authors wish to express their sincere gratitude to the Spanish "Ministerio de Educacion" for help provided through project BIA2005–255 and the FEDER founds, and to the "ICT, Instituto de Construccion Tubular", partner of CIDECT, for their advice.

REFERENCES

Aboutaba R.S & Machado R.I. 1999, Seismic resistance of steel-tubed high-strength reinforced-concrete columns, J STRUCT ENG-ASCE 125 (5): 485–494.

Bergmann, R. 1994. Load introduction in composite columns filled with High strength concrete, Proceedings of the 6th Int Symposium on Tubular Structures, Monash University, Melbourne, Australia.

Bonet, J.L. 2001. Método simplificado de cálculo de soportes esbeltos de hormigón armado de sección rectangular sometidos a compresión y flexión biaxial, PhD Thesis, Civil Engineering Dept., Technical University of Valencia.

CIDECT. 1970. Monograph n° 1: Concrete filled hollow section steel columns design manual. British edition.

CIDECT. 1979. Monograph n° 5: Calcules Poteaux en Prolifes Creux remplis de Beton.

CIDECT. 1998. Guía de Diseño para columnas de perfiles tubulares rellenos de hormigón bajo cargas cíclicas estáticas y dinámicas. TUV-Verlag.

Collins M.O., Mitchell D. and McGregor J.P. 1993. Structural design considerations for high strength concrete, Concrete International, ACI, USA 27–34.

Comité Euro-internacional du beton. 1991. CEB-FIB Model Code 1990. C.E.B. Bulletin N° 203–204 and 205.

Comité Euro-internacional du beton. 1995. High Performance Concrete. Recommended extensions to the Model Code 90 research needs. C.E.B.. Bulletin N° 228.

Crisfield M.A. 1981. A fast incremental/iterative solution procedure that handles "snap-through". Computers & Structures 13 (1–3): 55–62.

Eurocode 4. 2004. Proyecto de Estructuras Mixtas de Hormigón y Acero, Parte 1–1: reglas generales y reglas para edificación. (in Spanish).

Filippou, F.C. & Constantinides, M. 2004. FEDEASLab. Getting Started Guide and Simulation Examples. Research Report. University of California at Berkeley, CA, USA.

Gourley B.C., Tort C., Hajjar J.F. & Schiller P.H. 2001. A Synopsis of Studies of the Monotonic and Cyclic Behavior of Concrete-Filled Steel Tube Beam-Columns. Structural Engineering Report No. ST-01-4. Department of Civil Engineering. University of Minnesota, Minneapolis, Minnesota.

Grauers M. 1993. Composite columns of hollow sections filled with high strength concrete. Research report. Chalmers University of Technology, Goteborg.

Gutiérrez G. & Sanmartin, A.: 1983. Influencia de las imperfecciones en la carga crítica de estructuras de entramados planos. Hormigón y Acero 147 : 85–100.

Hu H.T., Huang C.S., Wu M.H., et al. 2003. Nonlinear analysis of axially loaded concrete-filled tube columns with confinement effect, J STRUCT ENG-ASCE 129 (10): 1322–1329.

Johansson M. & Gylltoft K. 2001. Structural behavior of slender circular steel- concrete composite columns under various means of load application. Steel and Composite Structures 1(4): 393–410.

Kilpatrick A.E. & Rangan B.V. 1999. Tests on High-Strength Concrete-Filled Steel Tubular Columns. ACI STRUCT J 96 (2): 268–275.

Liu D.L., Gho W.M. & Yuan H. 2003. Ultimate capacity of high-strength rectangular concrete-filled steel hollow section stub columns. J CONSTR STEEL RES 59 (12): 1499–1515.

Lu X.L, Yu Y., Kiyoshi T., et al. 2000. Nonlinear analysis on concrete-filled rectangular tubular composite columns. STRUCT ENG MECH 10 (6): 577–587.

Neogi P.K., Sen H.K. & Chapman J.C. 1969. Concrete-Filled Tubular Steel Columns under Eccentric Loading. Structural Engineer 47(5): 187–195.

Rangan B. & Joyce M. 1992. Strength Of Eccentrically Loaded Slender Steel Tubular Columns Filled With High-Strength Concrete, ACI STRUCT J 89 (6): 676–681.

Romero M.L., Bonet J.M. Portoles J.L. and Ivorra S. 2006. A Numerical and experimental Study Of Concrete Filled Tubular Columns With High Strength Concrete, Proceedings of Tubular Structures XI, Packer & Willibald(Eds), Taylor and Francis Group, London,: 503–511.

Shams M. & Saadeghvaziri M.A. 1999. Nonlinear response of concrete-filled steel tubular columns under axial loading. ACI STRUCT J 96 (6): 1009–1017.

Srinivasan C.N. 2003. Discussion of "Experimental behavior of high strength square concrete-filled steel tube beam-columns" by Amit H. Varma, James M. Ricles, Richard Sause, and Le-Wu Lu, J STRUCT ENG-ASCE 129 (9): 1285–1286.

Varma A.H., Ricles J.M., Sause R., et al. 2002. Experimental behavior of high strength square concrete-filled steel tube beam-columns, J STRUCT ENG-ASCE 128 (3): 309–318.

Varma A.H., Sause R., Ricles J.M., et al. 2005. Development and validation of fiber model for high-strength square concrete-filled steel tube beam-columns. ACI STRUCTURAL JOURNAL 102 (1): 73–84.

Zeghiche J. & Chaoui K. 2005. An experimental behaviour of concrete-filled steel tubular columns. JOURNAL OF CONSTRUCTIONAL STEEL RESEARCH 61 (1): 53–66.

Stability of concrete filled CFRP-steel tube under axial compression

G.S. Sun, Y.H. Zhao & W. Gu
Institute of Road and Bridge Engineering, Dalian Maritime University, Dalian, China

ABSTRACT: An investigation on stability of concrete filled CFRP-steel tube is developed in two ways including experimental study and theoretical analysis. 24 long columns with different lengths and CFRP thicknesses but same diameters were tested under compressive loading. Test results indicate that loading-deflection behavior as well as the ultimate load is strongly depended on the lengths of the columns. And the longer specimens show obviously stability-based failure. For the purpose of establishing the stability-based criterion of this composed system, a set of equations for estimating critical loadings in terms of length/diameter ratio are derived by applying Euler's theory. This experimentally proved criterion is involves both the elastic and elastic-plastic stability of the specimens.

1 INTRODUCTION

The compressive strength of concrete under tri-axial compression is much higher than that in uni-axial compression. This advantage is utilized in infilling concrete into tubes made of different materials and served as the members subjected mainly to compression. These kinds of members make use of the composite properties of both the concrete core and outer tube in complex stress conditions, so that the advantages would be developed sufficiently and the load bearing capacity enhanced. The most common tube materials are steel and fiber reinforced polymer (FRP), into them the concrete is filled to construct the composed columns (Han 2004, Cai, 2003, Amir et al. 2000, Omar & Shahawy 2000, Amir & Rizkalla 2001). In recent years, a new system, concrete filled CFRP-steel tube (Wang & Zhao 2003, Gu 2007) has been proposed and investigated experimentally and theoretically.

Concrete filled CFRP-steel tube is a composed system made by wrapping around the steel tube with carbon fiber reinforced polymer (CFRP) sheet and then filling concrete into the tube (Fig. 1), which is evolved and developed on the basis of concrete filled steel tube structure and concrete filled CFRP tube structure. Because of the synergy of three materials, the mechanical properties and corrosion resistance of concrete filled CFRP-steel tube is even more superior to traditional concrete filled steel tube structure for the excellent performance in reinforcement of CFRP and the corrosion prevention of polymer. The system can be used as the column or other part sustaining mainly compressive loading in bridges and buildings. And it may also resolve the problems of lack of thicker-walled steel tubes and the relatively high cost induced by high-strength-steel in concrete filled CFRP tubular structures.

Some significant researches (Wang & Zhao 2003, Gu & Zhao 2006, Gu et al. 2004) were preformed for investigating the possibility of this new system. On the earliest work some short cylinders were tested for getting the ultimate loads (Gu et al. 2004). The columns were assumed to be the axial-symmetrical ones under simple compression. The analytical solution based on the method of equilibrium ultimate showed a nice agreement with the experimental data (Gu & Zhao, 2006). Then much more specimens including some longer ones were made for the

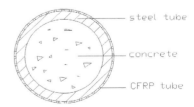

Figure 1. Cross-section of concrete filled CFRP-steel tubular column.

experimental study on the ability of carrying loads as well as the stability (Gu & Zhao 2007, Wang et al. 2006). Sufficient information from the preliminary work showed the evident improvement of the mechanical properties of the new system comparing with that without CFRP. However, before the system is accepted by engineering applications, exhaustive researches are still in need, especially those about rules or criteria for structure design.

In this paper an investigation on stability of concrete filled CFRP-steel tube is developed experimentally theoretically. 24 long columns were tested under uni-axial compressive loading. For the purpose of establishing the stability-based failure criteria for long columns, a set of equations for estimating critical loadings are suggested, in terms of length/section-area ratio. Numerical results from the theory are proven to be closed enough to experimental ones.

2 EXPERIMENTAL STUDY

A test is carried out with 500t Long Column Test Machine, on which 24 columns specimens including 16 concrete filled CFRP-steel tube and 8 normal concrete filled steel ones with different lengths, different thicknesses of CFRP, but same diameter and thickness of steel tubes, are tested under uni-axial monotonic proportional compressive loading.

2.1 Materials

Concrete is made of Portland cement, silica-based sand and limestone as the course aggregates with the largest size 20 mm, and 1.0% (in weight fraction) FDN was added. The proportions of constituents and the mechanical properties are summarized in table 1 & 2. The average compressive strength f_{cu} from 28-day cube tests is 67 MPa. Carbon fiber sheet used here is made in EPO Company, Germany, which is fabricated with TORAY T700 12K from Japan. The material properties are shown in Table 3. Two kinds of structural glues, JGN-C and JGN-P are the products of Liaoning Building Science Research Institute, China. Seamless steel tubes are made of Q345 steel.

2.2 Specimens

16 CFRP wrapped specimens and other 8 without CFRP are classified into 3 groups, all with the same thickness of steel tubes $t = 4.5$ mm and outer diameter D, but changed length L and CFRP layer. Specimens of group 1 are normal concrete filled steel tubes; group 2 with 1 layer of CFRP, and group 3 with 2 layers. The geometrical details of specimens are listed in Table 4, with t_c the thickness of

Table 1. Mix proportion of concrete.

Cement kg.m^{-3}	m (c): m (s): m (r)	Water-cement ratio
490	1:1.35:2.20	0.35

m (c): m (s): m (r)-mass ratio of cement, sand, crushed stone.

Table 2. Property of steel.

	f_y/MP	f_u/MP	E_s/MP
Mean value	360	525	206×10^3

Table 3. Mechanical properties of carbon fiber.

Tensile strength MPa	Elastic modulus MPa	Density g/cm3	1-layer thickness mm
4900	228,000	1.80	0.11

Table 4. Main parameters of specimens.

Specimen	L/mm	L/D	t_c/mm	F_u/kN
0–1	400	3	0	1858.9
0–2	530	4	0	1897.5
0–3	600	4.5	0	1856.1
0–4	800	6	0	1856.1
0–5	1200	9	0	1781.7
0–6	1800	13.5	0	1583.1
0–7	2400	18	0	1323.8
0–8	3000	22.5	0	1268.7
1–1	400	3	0.111	2068.5
1–2	530	4	0.111	2040.1
1–3	600	4.5	0.111	2009.6
1–4	800	6	0.111	1990.1
1–5	1200	9	0.111	1814.3
1–6	1800	13.5	0.111	1646.5
1–7	2400	18	0.111	1379.6
1–8	3000	22.5	0.111	1296.3
2–1	400	3	0.222	2264.3
2–2	530	4	0.222	2249.4
2–3	600	4.5	0.222	2247.8
2–4	800	6	0.222	2140.2
2–5	1200	9	0.222	2015.2
2–6	1800	13.5	0.222	1698.5
2–7	2400	18	0.222	1500.4
2–8	3000	22.5	0.222	1310

the carbon fiber sheet, L/D the slenderness ratio of specimens.

For a typical specimen, the steel tube is placed upright with the bottom end welded to a square steel base plate (150 × 150 × 20 mm) before the concrete casting. Then fiber sheet is bonded laterally outside the steel tube after surface cleansing and smoothing.

After concrete is filled in and ground carefully when hardening, another plate is welded to the free end, to ensure that load is applied evenly over the cross section of the column.

2.3 Testing set-up

Test equipment used in this study is a 500t Long Column Test Machine and data collected and analyzed with UCOM 70, a computer controlled data treatment equipment. Two displacement sensors are placed to measure the overall axial shortening of the specimens. Strain gauges are used inside and outside CFRP, along and normal to fiber direction, to measure the longitudinal and circumferential strain of the steel tube as well CFRP.

Uni-axial monotonic proportional compressive loading is applied in steps, at the beginning with a small increment by 1/10 of the estimated ultimate load for each step until the specimen approaches to yield limit, then loading slowly until failure. In order to obtain the values of deflection and strain, continually recording is carried during the loading process. Testing set-up and the distributions of strain gauges are shown in Figure 2.

2.4 Failure process

At the initial stage of loading process, the specimens deform linearly and show no obvious geometrical changes. The values of strain gauges inside and outside CFRP sheets are coincide with each other. For the shorter column, as the load increases continually, the load-displacement curve becomes nonlinear beyond the yield point and then middle part of the column starts to expand. When load reaches about 70–80% of the maximum value, specimen deforms significantly, and CFRP split happens locally. When the maximum load is reached, CFRP sheet breaks, a continual snap is heard. After the failure of CFRP, the specimen behaves as a normal concrete filled steel tube. For the longer column, with the increasing of loading, lateral displacement increases obviously along with a decline of the load. While deformation still accretes as the load reaches the critical value.

2.5 Experimental results

Figure 3 presents load-middle span deflection relations. A quick glance suggests that the strengthening effect is sensitive to the thickness of CFRP.

The compressive strength of the columns seen in these figures depends also on L/D.

The resulting curves of ultimate load P_u versus slenderness ratio L/D are shown in Figure 4. From this figure, it becomes clear that the increase of length has a detrimental effect on the ultimate strength. It is also observed that P_u is sensitive to the thickness

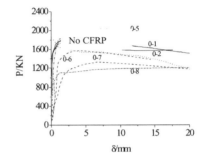

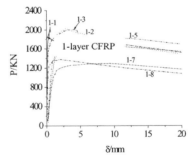

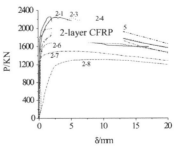

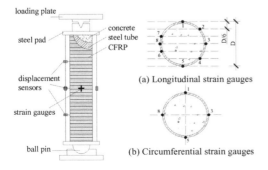

Figure 2. Location of the strain gages.

Figure 3. Load-middle span deflection relations.

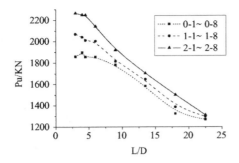

Figure 4. Ultimate load versus slenderness relations.

of CFRP only for short columns. In the case of very long ones, the confining of CFRP on the structure is neglectable.

3 STABILITY CRITERION

Experiment showed that longer columns became bending under a uni-axial load and finally buckled. The greater the ratio L/D, the lower the critical load is (Fig. 4). A general accepted method for judging the compressive stability of concrete filled steel columns is to use a modified Euler equation (Han & Zhong 2003, Pan & Zhong 1992). The accuracy of the formula results was reported to be closed enough to the experimental data. In this study Euler's theory is used to estimate the critical loads of compressed concrete filled CFRP-steel tubes with different L/D values.

3.1 Effective modulus of composed columns

The columns with large value of L/D would buckle in early stage of deformation (Fig. 4). The longitudinal elastic modulus of composed columns under axial compression is determined by:

$$E_{csc} = \frac{\sigma_{csc}^p}{\varepsilon_{csc}^p} \qquad (1)$$

where σ^p_{csc} is the proportional stress, ε^p_{csc} is the corresponding strain (Zhong 1994), with

$$\sigma_{csc}^p = \left(0.192\frac{f_y}{235} + 0.488\right)\sigma_{csc}^y$$
$$\varepsilon_{csc}^p = 0.67\frac{f_y}{E_s} \qquad (2)$$

in which f_y and E_s are yield stress and elastic modulus of steel, σ^y_{csc} the compressive strength of column defined by (Gu 2007):

$$\sigma_{csc}^y = f_{ck}(1 + 2\xi_s + 2\xi_f) \qquad (3)$$

where f_{ck} is the uni-axial compressive strength of concrete, ξ_s, ξ_y the confining factors of steel and CFRP tubes, with

$$\xi_s = \frac{A_s f_y}{(A_c f_{ck})}, \quad \xi_f = \frac{A_f f_f}{(A_c f_{ck})} \qquad (4)$$

where A_c, A_s and A_f denote the cross section areas of concrete core, steel tube, and CFRP tube, respectively; f_f the tensile strength of CFRP.

For columns not very long, buckling may occur beyond yield point. Then tangent modulus E^t_{csc} would be used in Euler's equation instead of the elastic one, as:

$$E_{csc}^t = \frac{(\sigma_{csc}^y - \bar{\sigma})\bar{\sigma}}{(\sigma_{csc}^y - \sigma_{csc}^p)\sigma_{csc}^p} E_{csc} \qquad (5)$$

here $\bar{\sigma}$ is the average stress of columns.

3.2 Critical Load

By applying Euler's equation, the critical load for the member which buckles within elastic state is given by

$$N_{cr} = \pi^2 E_{csc} \frac{I_{csc}}{L^2} \qquad (6)$$

For that fails in elastic-plastic state, the critical load is

$$N_{cr} = \pi^2 E_{csc}^t \frac{I_{csc}}{L^2} \qquad (7)$$

With $I = \pi D^4/64$. From equation 6 & 7, the critical slenderness ratio of elastic stability and elastic-plastic stability can be derived as

$$\frac{L}{D} = \frac{\pi}{4}\sqrt{\frac{E_{csc}}{f_{csc}^p}}, \quad \text{and} \quad \frac{L}{D} = \frac{\pi}{4}\sqrt{\frac{E_{csc}^t}{f_{csc}^p}}, \qquad (8)$$

3.3 Results and discussion

Numerical solutions from the above equations is computed and compared with experimental results, as seen in Figure 5. The diameter of all specimens is 133 mm.

The figures above show that the critical loads from equations and experiments are almost coincide with each other for shorter columns, although a little difference exists for longer ones. The suggested theory including Euler's equation and its application in inelastic specimens, as well as the assumption of confining factors for composed columns, is practicable and precise enough for estimating the buckling behavior of concrete filled CFRP-steel tubes without reference to their lengths.

4 CONCLUSIONS

This paper presents the research on a composed system, concrete filled CFRP-steel long columns. The primary conclusion from the experimental study on 24 specimens is the dependency of load carrying capacity on the slenderness of columns and the thickness of CFRP tubes as well. The greater the specimen the lower the ultimate load is. And thicker CFRP enhances the loading capacity. Based on test results, a modified Euler equation is suggested to estimate the critical loading for compressed column. Comparisons show no significant differences between the theoretical and experimental results. So the criterion is valid for the specimens regardless of the lengths.

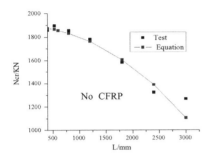

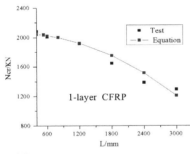

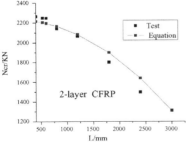

Figure 5. Critical loads versus lengths.

REFERENCES

Amir, M. Shahawy, M. Khoury, C.E.I. & Naguib, W. 2000. Large beam-column tests on concrete-filled composite tubes. *ACI Structure Journal* 97 (2): 268–276.

Amir, Z.F. & Rizkalla S.H. 2001. Confinement model for axially loaded concrete confined by circular FRP tubes. *ACI Structure Journal* 98 (4): 451–461.

Cai, S.H. 2003. Modern steel tube confined concrete structures, *China Communications Press*. Beijing, China (in Chinese).

Gu, W. et al. 2006. Ultimate loading of concrete filled CFRP-steel short columns. *Engineering Mechanics*. 23 (1): 149–153. (in Chinese)

Gu, W., Zhao, Y.H. & Sun, G.S. 2007. Strength calculation of concrete filled CFRP-steel tube. *Industrial Construction*. 37 (4): 42–44. (in Chinese)

Han, L.H. 2004. Concrete filled steel tubular structures. *Science Press*. Beijing, China (in Chinese).

Han, L.H. & Zhong, S.T. 2004. The Mechanic of Concrete filled Steel Tube. *Dalian University of technology Press*. Dalian, China. (in Chinese)

Omar, C. & Shahawy, M. 2000. Performance of fiber-reinforced polymer-wrapped reinforced concrete column under combined axial-flexural loading. *ACI Structure Journal* 97 (4): 659–668.

Pan, Y.G. & Zhong, S.T. 1992. The theoretical analysis of the stability bearing load capacity on concrete filled steel tube. *Journal of Building Structure* 13 (1): 43–52. (in Chinese)

Sun, G.S. 2007. Research on the Performance of Circular Concrete Filled CFRP-Steel Tubular Columns", *Shenyang Jianzhu University Mster Thesis*. Shenyang, China. (in Chinese)

Wang, Q.L. & Zhao, Y.H. 2003. A presumption on the concrete filled CFRP-steel composite tube structures. *Journal of Jilin University (Engineering and Technology Edition)*. 33 (Sup): 352–355. (in Chinese)

Wang, Q.L. Guan, C.W. & Zhao, Y.H. 2004. Theoretical Analysis about concentrically compressed concrete filled hollow CFRP-steel stub columns with circular cross-section. *Proceeding of the 2nd International Conference* on *Steel and Composite Structure*. 684–695. Techno-Press, Seoul, Korea.

Yu, Q. 2001. Several key issues of the studies and applications of FRP-confined concrete columns. *Industrial Construction* 31 (4): 1–4. (in Chinese)

Zhao, Y.H, Gu, W. & Wang, Q.L. 2004. Experimental study on concentrically compressed concrete filled hollow CFRP-steel stub columns with circular cross-section. *Proceeding of the 2nd International Conference on Steel and Composite Structures*. 712–721. Techno-Press, Seoul, Korea.

Zhong, S.T. 1999. Multi-story structures of concrete filled steel tubes. *Helongjiang Science Press*. Herbin, China.

*Parallel session 2: Fatigue & fracture
(reviewed by S. Herion & A. Nussbaumer)*

Fatigue modeling for partially overlapped CHS K-joints with surface crack

S.P. Chiew, C.K. Lee, S.T. Lie & T.B.N. Nguyen
School of Civil and Environmental Engineering, Nanyang Technological University, Singapore

ABSTRACT: During practical assessment of residual life of tubular joints, the Stress Concentration Factor (SCF) and the Stress Intensity Factor (SIF) are the two most important parameters and their estimated values are often obtained by numerical (finite element) modeling. However, to generate a proper mesh and model for complicated joint configurations like the partially overlapped circular hollow section (CHS) K-joints is not a trivial task. In this paper, a new automatic mesh generation procedure for the discretization and modeling of partially overlapped CHS K-joints is presented. In the suggested modeling and mesh generation procedures, a surface crack of any length can be created at any position along the joint intersection. With the development of the purposed mesh generator, the estimation of the SCF for an uncracked joint and the estimation of the SIF and residual fatigue life for a cracked joint could be carried out quickly and conveniently. The presented numerical results showed that the proposed method is both efficient and reliable.

1 INTRODUCTION

In heavily loaded offshore and bridge frames, partially overlapped CHS K-joints are often used due to their high residual capacity. However, the repetitive nature of the loading these structures subjected may cause fatigue damage to the joints. If the assessment of the fatigue damage and the fatigue life of a partially overlapped CHS K-joint are needed, investigations on the SCF and the SIF of the joint are required by either experimental study or by numerical modeling. For partially overlapped CHS K-joints, test result in the literature is largely limited to ultimate strength studies (Dexter & Lee 1999a). The only equations on the stress state around the intersection of partially overlapped CHS K-joints published by Efthymiou & Durkin (1985) were based merely on numerical parametric analyses and extrapolation from simple T/Y-joints. Recently, fatigue test on full scale specimens were reported by Lee et al. (2007) where distinctive features on the fatigue behavior of the joints tested were described. These test results provide much valuable information on the fatigue behavior of partially overlapped CHS K-joints. In addition, the results could also be used as reliable sources for the verification of numerical models developed for of this kind of joint.

In order to obtain reliable SCF and SIF predictions, the numerical modeling procedure developed for the analysis of partially overlapped CHS K-joint should involve a consistent mesh generation scheme which is capable of generating a good quality mesh that follows closely the geometrical characteristics of a real joint. Following the systematic method pioneered by Cao et al. (1997) and further pursued by Lie et al. (2003), as well as based on the geometrical model discussed in Chiew et al. (2006), an automatic mesh generation scheme is developed for partially overlapped CHS K-joints with cracks. Although the work follows the path paved by previous researchers, several new techniques for the mesh generation procedure are introduced in order to handle the complex geometry of this kind of joints. The numerical model created based on the purposed mesh generation scheme is then used in an example for the calculation of the SCF and the SIF values for the uncracked and cracked joints, respectively. In addition, an estimation of residual fatigue life of the numerical model with surface crack is also presented.

2 MESH GENERATION FOR PARTIALLY OVERLAPPED CHS K-JOINTS

2.1 *Method of mesh generation*

The existence of several intersection curves and intersection points makes the solid mesh generation procedure for a partially overlapped CHS K-joint not a trivial task. To reduce the complexity of the model/mesh generation procedure, the mesh is first created in the form of a surface mesh, which is then converted into a solid mesh using an extrusion algorithm (Xu 2006). Weld details and crack profiles can be added in subsequent steps. This method of mesh generation is flexible in such a way that at different modeling stages, meshes with different levels of details can be generated and employed for special applications. For example, a surface mesh without any welding details (SURF_0 W) could be extracted out for quick assessment of the fatigue strength of the joint while a surface mesh with welding details (SURF_1 W) could be formed with little addition effort for quick SCF estimation. For a more intensive and detailed study on the distribution of hot spot stress (HSS) around the joint intersection, full solid meshes (SOLID_1 W and SOLID_0 W) with or without welding details could be used. Eventually, in the case that a full numerical study for a crack joint is needed, a full solid mesh with welding and crack details ((SOLID_CR) can be generated. In these meshes, different kinds of element are used depending on the targeted purpose. For example, the entire mesh SURF_0 W is represented by second order triangular and quadrilateral shell elements, while mesh SURF_1 W is a mixed mesh consists of both shell elements and solid prisms and tetrahedrons for the welds and joint details. Second order brick elements, prisms, tetrahedrons and pyramid elements are used in the meshes SOLID_0 W and SOLID_1 W. In addition, the quarter point crack elements are used in the mesh SOLID_CR around the crack front for accurate SIF prediction. A flow chart for the mesh generation scheme is presented in Figure 1.

2.2 Surface mesh generation

As a basis for the development of the surface mesh, a geometrical model has been developed (Chiew et al. 2006), in which a weld model and a method for finding the intersection points among the braces and the chord are presented. As different element size densities are required in different parts of the joint to accurately model the geometry for the connection, the whole structure is divided into several zones. An adaptive surface mesh generator (Lee 1999) was applied to discretize these zones into surface meshes. The zones are shown in Figure 2 and concise descriptions for them are given below.

- Zone A1: This zone lies inside the intersection curve formed when the through brace intersects with the chord.
- Zone A2: This zone lies inside the intersection curve formed when the overlap brace intersects with the chord.
- Zone A3: This zone lies inside the intersection curve formed when the overlap brace intersects with the through brace.
- Zone ThruB: This zone belongs to the through brace surface, connects with zones A1, A2. It is extended to the total length of the through brace.
- Zone LapB: This zone resembles zone ThruB. In general, this zone is different from zone ThruB. It is because the joint configuration is not symmetric when the angles between the braces and the chord are different.

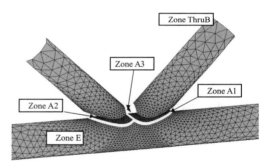

(a) Various Zones in mesh SURF_0W

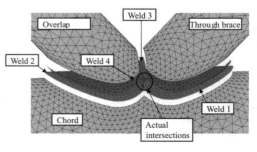

(b) SURF_0W with quadrilaterals in weld positions

Figure 2. Surface mesh of a partially overlapped CHS K-joint.

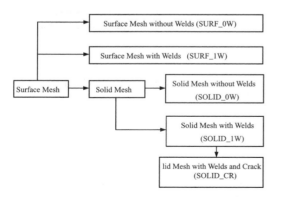

Figure 1. Mesh generation scheme.

- Zone E: This zone belongs to the outer side of the chord surface, and is extended to the full length of the chord.

During the discretization of the above zones, triangular elements are generated (Figure 2a). In order to connect the respective zones, quadrilateral elements are used at where the welding is applied the corresponding CHS (Figure 2b).

2.3 Solid mesh generation

The surface mesh can be converted into a solid mesh by connecting corresponding nodes on respective boundary surfaces using an algorithm to relate the nodal connectivity (Xu 2006). A typical extruded mesh without any welding details, named as mesh SOLID_0 W, is shown in Figure 3a. For the implementation of this algorithm, the entire joint is divided into four sub-spaces namely, the outer space (Space 1) and the inner space (Space 2) for the chord, the through brace (Space 3) and the overlap brace (Space 4) as shown in Figure 3b. The procedure is able to create a mesh with denser extrusion number of layers around the joint intersections.

In each sub-space, the boundary surface mesh should be generated. As described in Chiew et al. (2006), due to the cut off effect the weld makes to the full thickness, the intersection and weld curves are not always parallel. Therefore, the boundary surface for Space 1, which is the exterior surface of the three tips, is created first. Following that, the boundary surfaces for Spaces 3, 4 and 2 which are the interior surfaces of the through brace, overlap brace and chord respectively, are created subsequently. The mid-surface mesh which connects to all other surfaces together is generated afterward.

The weld profile can be conveniently added into the extruded solid mesh to form mesh SOLID_1 W as shown in Figure 3c. In the solid mesh for the welds, 3D prism elements are used. The common area of all the welds is filled up by tetrahedron elements as shown in Figure 3d.

2.4 Generation of surface crack details

It has been observed (Lee et al. 2007) that for partially overlapped CHS K-joints, the fatigue cracks occurred not only on the chord side of the weld toe as normally seen in other types of tubular joints. Depending on the configuration of the joint and the loadings applied, it is possible that a surface crack can be formed at the brace side of the weld toe. In practice, during fatigue assessment, a surface (SURF_1 W) or solid mesh (SOLID_1 W) with welding details would be created first to study the SCF along the joint to identify the location of the peak HSS which determines

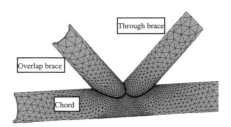

(a) Extruded mesh (SOLID_0W)

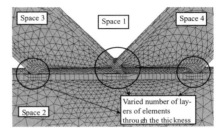

(b) Different number of layers of element along the thickness of the CHS

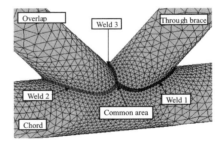

(c) Extruded mesh with welding details (SOLID_1W)

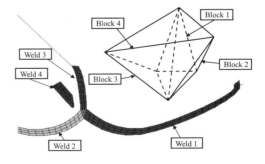

(d) Close up of the common area

Figure 3. Solid mesh of a partially overlapped CHS K-joint.

on which sides a surface crack shall be formed. Once the location of a surface crack is known, the extent of the crack, which is defined implicitly by the parameters l_{cr1} and l_{cr2}, could be determined by specifying their positions relative to the three reference points as shown in Figure 4.

The exact locations of the crack tips determine the numbers of element to be extracted from the solid mesh for modification to include the surface crack details. The solid mesh after extracting the crack zone at the weld toe on the through brace is shown in Figure 5a. The elements around the crack front are tailor-designed as a spider web consisting of concentric rings of elements (Figure 5b). The value of SIF can be extracted through the integral of energy release rate in the volume of these elements.

2.5 Ranges of application of the mesh generator

The automatic mesh generator can cater for a wide range of geometrical parameters (Figure 6a) of partially overlapped CHS K-joints. The recommended range of geometry in which the mesh generator can handle, is listed as below:

– Intersecting angle: $30° \leq \theta_1, \theta_2, \theta \leq 90°$,
– β ratio: $0.1 \leq \beta_1, \beta_2 \leq 1.0$,
– γ ratio: $0.03 \leq 1/\gamma \leq 0.25$,
– Percentage of overlap: $20\% \leq p \leq 80\%$,
– Position of crack: anywhere along the three welds at either the chord or brace sides.
– Angle of crack surface: $-20° \leq \omega \leq 20°$.

In particular, the mesh generator can handle special cases of identical chord and braces dimensions as well as large overlapped percentage of 80%, which are not commonly covered by other commercial software packages. Dexter & Lee (1999b) reported cases that numerical analysis cannot be performed due to

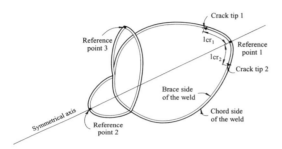

Figure 4. Definitions of locations of crack tips.

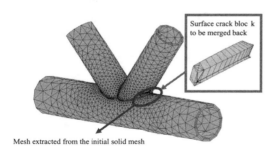

(a) Extraction and insertion of crack block in mesh SOLID_CR

(b) Assembly of elements enclosing the crack front

Figure 5. The mesh SOLID_CR.

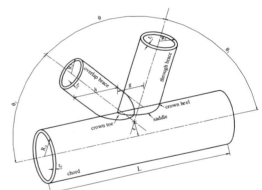

$$\alpha = \frac{L}{R_0}, \gamma = \frac{R_0}{t_0}, \zeta = \frac{g}{2R_0}, \beta_1 = \frac{R_1}{R_0}, \tau_1 = \frac{t_1}{t_0}, \beta_2 = \frac{R_2}{R_0}, \tau_2 = \frac{t_2}{t_0}$$

(a) Parameters defining the mesh SOLID_1W

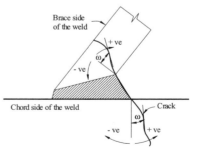

(b) Parameters defining the crack surface

Figure 6. Parameters for the mesh generator.

limited capacity of the ANSYS. It should also be mentioned that, the mesh generator has been developed in such a way that the practice range specified by the CIDECT Guide (1991) can be covered. Note that other than the application for partially overlapped K-joint, the current mesh generation approach could also be applied to other commonly used joint types such as T, Y and gap K joints. As reported in the literature, the mesh generator developed for them (Lie et al. 2003) are limited to a maximum value of $\beta = 0.8$.

3 NUMERICAL EXAMPLE

3.1 Numerical modelling of a full scale partially overlapped CHS K-joint

The partially overlapped CHS K-joint used in the numerical example has the identical geometrical and loading parameters as the one tested by Lee et al. (2007). The dimension and loading directions of the joint are given in Figure 7 and Table 1. The tested specimen was designed as a practical CHS joint in such a way that overlapping is inevitable. Experimental results (Lee et al. 2007) indicated some untypical fatigue behaviors of the joint when it was subjected to the combined loading 200 AX + 45IPB (kN). The crack eventually appeared at the *brace side* of the weld toe connecting the through brace and the chord.

Both static and fatigue test were carried out. Measurements were performed to record the detailed dimensions of the welding and the actual shape of the surface crack developed (Figure 6b). In order to verify the SCF values around the weld connecting the through brace and chord, the HSS were obtained by using the extrapolation method recommended by CIDECT in both actual measurements and the FE modelling using the software ABAQUS (2006). In order to verify the SIF values, the experimental results from full scale fatigue testing conducted by Lee et al. (2007) were converted to SIF values by using the Paris' law:

$$\frac{da'}{dN} = C(\Delta K)^m \quad (1)$$

In Equation 1, a' is the crack depth on an inclined crack surface. N is the number of accumulated cycles

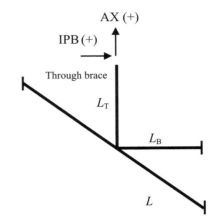

Figure 7. Specimen layout and loading scheme.

associating with a'. da'/dN is crack growth rate which could be obtained from experimental data while ΔK is the range of SIF. C and m are material parameters corresponding to API-5 L pipes tested in ambient temperature condition. The values of $C = 1.427 \times 10^{-12}$ $(m/\text{cycle})(\text{MPa} \times m^{1/2})^{-3.523}$ and $m = 3.523$ (Barsom & Novak 1977) are used in this paper.

3.2 Deformation shape and SCF estimations

The FE predictions of the deformation and the HSS distribution around the weld connecting the through brace and chord (Weld 1) are shown in Figure 8. In addition, the experimental and numerical SCF results from different models of the example are plotted in Figure 9.

In the numerical example, three different numerical models, namely, solid mesh without any welding details (SOLID_0 W), solid mesh with the welding details (SOLID_1 W) proposed by Chiew (2006) and solid mesh with welding details based on exact measurement were used. In general, higher SCF distributions are predicted along the brace side of the weld in all numerical models. In addition, it can be observed from Figure 9 that the SCF from the measured weld thickness is able to give the best prediction among all the models with a maximum relative error of 12%. Numerical models from the meshes SOLID_1 W and SOLID_0 W reproduced conservatively the values of SCF on the heel and saddle sides of the weld. However, when it comes close to the intersection point, the prediction of SCF is not conservative. It might be due to the complex notch effect of the actual weld that has not been catered for in the geometrical and the numerical models. However, as the SCF at that location is the lowest along the intersection, such

Table 1. Specimen dimensions.

Parameters (mm, degree)							
R_0	t_0	R_1, R_2	t_1, t_2	θ_1, θ_2	L	L_T	L_B
136.5	25.0	122.3	19.1	45	5840	2446	2110

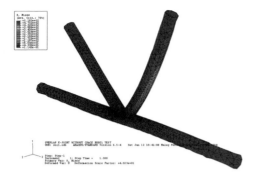

(a) Deformation

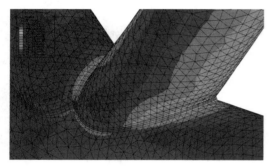

(b) Hot spot stress distribution around weld 1

Figure 8. Predictions of the stress distribution.

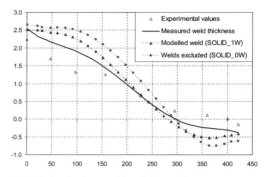

(a) SCF for the brace side of the weld

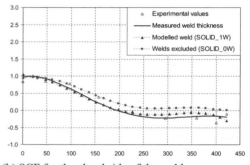

(b) SCF for the chord side of the weld

Figure 9. Comparison of SCF values between experimental data and results from different numerical models.

underestimation does not impair the practical value of the models. Hence, in case actual measurement of weld thickness is not available, the mesh SOLID_1 W is recommended for SCF prediction.

3.3 SIF estimation

The experimental and numerical results of SIF obtained from different models of the example are plotted in Figure 10. It can be seen that the numerical model using measured weld and crack profile represents the best match to the experimental data with a maximum relative error of about 10%. For comparison purpose, other numerical models using the proposed weld model (Chiew et al. 2006) and a constant angle for the crack surface ranging from $\omega = 0°$ to $-15°$ (Figure 6b) are created. The reason for the use of negative crack surface angle ω is based on the observation that the actual crack surface propagated outward from the weld. It is also noted that for this example, crack occurred at the brace side of the weld instead of the conventional chord side. Numerical results from these models show good correlation

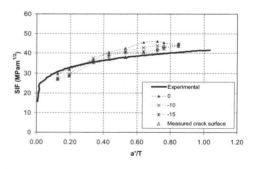

Figure 10. Comparison of SIF values between experimental data and results from different numerical models.

with test results with maximum error of 13% for the crack surface $\omega = 0°$.

From the figure, it can also be concluded that if the exact geometry of the crack surface angle is not available, a value of $\omega = 0°$ is recommended for the numerical analysis of the partially overlapped CHS K-joint.

3.4 Residual fatigue life estimation

In fracture mechanics, the concept of similitude applies. It is implied that the crack tip conditions, or under certain conditions the fatigue crack growth, are uniquely defined by a single loading parameter such as the SIF. For example, the crack growth rate da'/dN for a given combination of material and stress ratio can be expressed in the form of Paris' law (Equation 1) where any effects of environment, frequency are assumed to be included in the material constants C and m.

Based on the above assumption, if the estimated SIF values are known at some fixed intervals of the fatigue life, the number of cycles required for crack growth can be calculated by integrating Equation 2:

$$\int_{N_{initial}}^{N_{final}} dN = N_{final} - N_{initial} = \int_{a'_{initial}}^{a'_{final}} \frac{da'}{C(\Delta K)^m} \qquad (2)$$

The integral in Equation 2 gives the number of cycles required for the crack to grow from an initial size $a'_{initial}$ at cycle number $N_{initial}$ to a final size a'_{final} at cycle number N_{final}. Suppose that ΔK can be approximated by numerical analysis, then Equation 2 can be applied to estimate the residual life, $N_r(a') = N_{final} - N_{initial}$, of the crack. However, as ΔK can only be approximated at discrete values of a' in the range $[a'_{initial}, a'_{final}]$, numerical integration is required to compute Equation 2. As a result, the accuracy of $N_r(a')$ depends on the accuracy of ΔK estimated, and the value of a'_{final} set to be the failure criteria for the partially overlapped CHS K-joint.

Equation 2 is applied to integrate the SIF values from different numerical models. In Figure 11, the estimated residual life predictions obtained by integrating Equation 2 for different models are plotted against the actual life recorded during the test. As expected, the model based on measured geometry gave the most accurate prediction. The FE model using the angle $\omega = 0°$ on the crack surface (Figure 6b) gave prediction more conservative than the exact model. Hence, it can be concluded that the numerical model based on the developed mesh generator can lead to practical prediction of residual life of the partially overlapped K-joint.

Note that during the residual life prediction, many uncertainties are involved. For example, they are the uncertainties in geometry, material parameters and the estimated SIF values. In practice, an error of 10% in the SIF calculation is considered acceptable. Generally, the crack growth rate is roughly proportional to the fourth power of SIF values, the relative error in the crack growth prediction could be up to 30%. Taking into account all the errors that can enter the analysis, it is obvious that a substantial safety factor should be used in fatigue life prediction using SIF values.

4 CONCLUSIONS

In this paper, an automatic mesh generation procedure for fatigue analysis of partially overlapped CHS K-joints is proposed. The mesh generator is able to generate several kinds of mesh for different analysis purposes. Most importantly, it is able to generate a solid mesh with welding details and surface crack of any length and locates at either sides of the joint intersection. With the development of the purposed mesh generator, the calculation of the SIF and residual fatigue life for this type of joint could be carried out quickly and conveniently. The presented numerical results with measured weld and crack profiles showed that the proposed method is both efficient and reliable. The weld model as included in mesh SOLID_1 W and the crack surface angle of $\omega = 0°$ is proposed for numerical analysis of partially overlapped CHS K-joints in case actual measurement is not available.

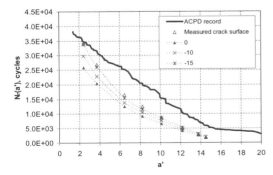

Figure 11. Life prediction using different numerical models.

REFERENCES

ABAQUS 2006. User Manual (Ver. 6.5), USA. Hibbit, Karlsson and Sorensen Inc.

Barsom, J.M., and Novak, S.R. 1977. Subcritical crack growth in steel bridge members, NCHRP Report 181. Transportation Research Board, National Research Council.

Cao, J.J., Yang, G.J., and Packer, J.A. 1997. FE mesh generation for circular tubular joints with and without cracks. Proceedings of the 7th International Offshore and Polar Engineering Conference, Honolulu, USA, Vol. IV, 98–105.

Chiew, S.P., Lee, C.K., Lie, S.T. and Nguyen, T.B.N. 2006. Geometrical Modelling of Partially Overlapped CHS

K-joints. The 8th International Conference on Steel, Space and Composite Structures, Malaysia, 425–435.

CIDECT 1991. Design Guide for Circular Hollow Section Joints under Predominantly Static Loading. Paris, France.

Dexter, E.M. and Lee, M.M.K. 1999a. Static strength of axially loaded tubular K joints I: Behaviour, Journal of Structural Engineering 125(2): 194–201.

Dexter, E.M. and Lee, M.M.K. 1999b. Static strength of axially loaded tubular K joints II: Ultimate capacity, Journal of Structural Engineering 125(2): 202–210.

Dowling, N.E. 1999. Mechanical Behaviour of Materials, 3rd ed., Prentice Hall, NJ

Efthymiou, M. and Durkin, S. 1985. Stress concentrations in T/Y and gap/overlap K-joints. Proceedings of the Conference of Behaviour of Offshore Structures, 429–440. Amsterdam: Elsevier Science Publishers.

Lee, C.K. 1999. Automatic adaptive mesh generation using metric advancing front approach Engineering Computations 16(2): 230–263.

Lee, C.K., Lie, S.T., Chiew, S.P., Sopha, T. and Nguyen, T.B.N. 2007. Experimental Studies on Fatigue Behaviour of Partially Overlapped CHS K-joints, International Maritime-Port Technology and Development Conference, Singapore, 280–285.

Lie, S.T., Lee, C.K., and Wong, S.M. 2003. Model and mesh generation of cracked tubular Y joints, Engineering Fracture Mechanics 70: 161–184.

Paris. P.C., Gomez, and Anderson, 1972. Extensive Study of Low Fatigue Crack Growth Rates in A533 and A508 Steels, Proceedings of the 1971 National Symposium on Fracture Mechanics, Part I, ASTM STP 513, 141–176.

Xu, Q.X. 2006. Analysis of thin-walled structural joints using 3D solid element, PhD Thesis, CEE, NTU, Singapore.

Simplified micromechanics model to assess constraint effect on brittle fracture at weld defects

T. Iwashita
Ariake National College of Technology, Fukuoka, Japan

Y. Kurobane
Kumamoto University, Kumamoto, Japan

K. Azuma
Sojo University, Kumamoto, Japan

ABSTRACT: In recent years, experimental and analytical investigations were conducted to predict the occurrence of brittle fracture from weld defects (Iwashita et al. 2003, 2006). These investigations took into account the effects of a loss of plastic constraint at the crack tips. These effects are important because many of the weld defects in beam to column connections are small cracks although the material toughness is measured using a specimen with a deep notch. This paper shows the ability of the simplified model to predict brittle fracture under low plastic constraint conditions. Fracture toughness tests of SENB specimens with deep and shallow notches were performed and the Weibull stress approach and TSM were used to consider the effects of a loss of plastic constraint on fracture toughness for shallow-notched specimens.

1 INTRODUCTION

Recent experimental and analytical investigations by the authors (Iwashita et al. 2003) indicated that occurrences of brittle fractures from weld defects can be predicted rather accurately. Based on a series of numerical analyses, Iwashita et al. (2006) presented matrices showing the critical loads as a function of the size and location of weld defects and the material toughness. These investigations took into account the effects of a loss of plastic constraint at the crack tips. These effects are important because many of the weld defects in beam to column connections are small cracks although the material toughness is measured using a specimen with a deep notch. Crack tips of the weld defects show much greater fracture toughness than the material toughness determined by conventional testing procedures such as a single edge notched bend (SENB) test in which the notch root is under high plastic constraint (See Fig. 1).

The Weibull stress σ_W is known to be a suitable crack near-tip parameter to predict the initiation of brittle fracture (e.g. Ruggieri & Dodds 1996a). However for calculating the Weibull stresses, the calibration of the micromechanics parameters for each material under study is required, which is time and money consuming. On the contrary, a toughness scaling model (TSM), such as the one proposed by Anderson & Dodds (1991) employs the J_c value measured by conventional fracture specimens with the only assumption that crack fronts attain equivalently stressed volumes. The latter TSM is much simpler and was found to give accurate predictions of test results for welded joints with various defects (Iwashita et al. 2003).

This paper shows the ability of the latter, simplified model to predict brittle fracture under low plastic constraint conditions. Fracture toughness tests of SENB specimens with deep or shallow notches were performed. Brittle fracture was found to occur at different toughness levels depending on the specimens because these specimens have different levels of plastic constraint at the crack tips. A finite element analysis was used to evaluate the corrected fracture toughness of each specimen type when crack tip regions attain predetermined stressed volumes. The evaluated corrected fracture toughness values were found to be close to the toughness values observed in the tests as well as predicted by the Weibull stress. This shows that the simplified model is effective to predict brittle fractures starting from various defects existing in beam-column joints.

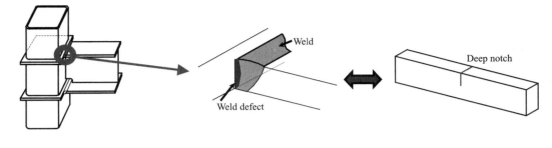

(a) Beam-to-column connection (b) Weld defect in connection (c) Notch in SENB specimen

Figure 1. Effect of plastic constraint.

2 WEIBULL STRESS AND TOUGHNESS SCALING MODEL

Because of their relevance in this research, this section introduces the Weibull stress approach and the toughness scaling model (TSM).

2.1 Weibull stress

Fracture toughness for cleavage fracture shows a large scatter, especially in the ductile-to-brittle transition region. The scatter in toughness results is caused by the scatter of the size of the weak spots. The local approach was proposed by Beremin (1983) for evaluating the statistical behavior of toughness results. This approach introduced a stress parameter σ_W, termed the Weibull stress, as a fracture resistance instead of the conventional fracture toughness parameter such as K_c, J_c and δ_c. The calibration of the micromechanics parameters for each material is required for calculating the Weibull stresses. Recent literature (e.g. Riesch-Oppermann et al. 2002) was used for calculating the Weibull stresses in this paper. The Weibull stress is defined as:

$$\sigma_w^m = \frac{1}{V_0} \int_{V_p} \sigma_1^m dV \qquad (1)$$

where σ_1 is the maximum principal stress, m is the Weibull slope and V_p denotes the fracture process zone. The Weibull slope, m, varies between 10 and 50 for structural steel. In general, the trend is that m has a low value in the case of low-level fracture toughness. In this research, the fracture process zone is taken as the region where the maximum principal stress at crack tips is $2\sigma_y$ or greater. It was found that the Weibull stress did not vary depending on the maximum principal stress between $2\sigma_y$ and $3\sigma_y$. The quantity V_0 is a unit volume and usually set to 1 mm³.

The Weibull stress is a random variable with a Weibull distribution. Its cumulative distribution function $F(\sigma_W)$ is given as:

$$F(\sigma_W) = 1 - \exp\left(-\left(\frac{\sigma_W}{\sigma_u}\right)^m\right) \qquad (2)$$

with the variables m and σ_u as parameters of the distribution. The Weibull parameters can be determined by a statistical method such as the maximum likelihood method. σ_u is the scale parameter which corresponds to the Weibull stress when the cumulative probability of failure equals 63.2%, $F = 0.632$.

2.2 Toughness Scaling Model (TSM)

Under small scale yielding (SSY) conditions, a single parameter, the J-integral, characterizes crack tip conditions and can be used as a geometry independent fracture criterion. Single parameter fracture mechanics breaks down in the presence of excessive plasticity and fracture toughness is dependent on the size and geometry of the test specimen. Based on the results of numerical analyses, Anderson & Dodds (1991) found that the apparent toughness of a material was highly dependent on the constraint conditions at the crack tip of the test specimens. In particular they examined the effects of crack depth on SENB specimens.

Figure 2 displays two SENB specimens with different notch depths (deep notch and shallow notch). This figure also shows the stress state at the crack tips when the same J value is applied to both specimens. The stress state shows a plane section of the volume surrounded by a contour of a certain value of the maximum principal stress, σ_1. The volume of the specimen with a shallow notch is smaller than that of the specimen with a deep notch, although the same J value is applied to each specimen. This example shows the dependency of the fracture toughness on the size and geometry of the test specimens. Anderson & Dodds (1991) proposed that the probability of brittle fracture is equal for the two specimens when the volumes are equal. As an example, Figure 3 illustrates J vs. volume curves. The vertical axis plots the applied J-integral while the horizontal axis depicts the volume surrounded by a contour of a certain value of σ_1 at the crack tip. The maximum principal stress is assumed as $\sigma_1 = 3\sigma_y$. From Figure 3, when the critical J value J_c, is obtained through SENB testing of a shallow notched specimen, the value of J_c can be corrected by determining the J value at which the crack tip volume of the deep notched specimen equals that (v_1) of the shallow notched specimen at fracture. Using the corrected J_c, it is possible to evaluate the occurrence of brittle fracture even if the level of plastic constraint is low, as for the defect in Figure 1(b).

Tagawa et al. (1999) showed that the TSM is a combination of the Weibull slope and level of strain hardening for the Weibull stress approach. Ruggieri & Dodds (1996a) showed that ductile crack extension generates near-tip stress fields sufficiently different in character from stationary near-tip fields to question the validity of TSM predictions. But the TSM employs the J_c value measured by conventional fracture specimens with the only assumption that crack fronts attain equivalently stressed volumes. In addition, the TSM requires no calibration to calculate the corrected J_c. Hence, the TSM is much simpler than the Weibull stress approach.

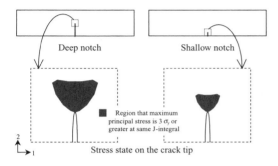

Figure 2. SENB specimens with a different notch depth.

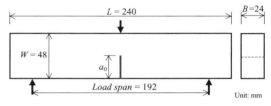

Figure 4. Specimen geometry.

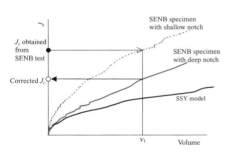

Figure 3. J-integral vs. volume curves.

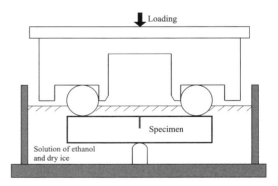

Figure 5. Loading scheme.

Table 1. Material properties.

Material	Steel grade (in JIS code)	Test temp. (°C)	Yield stress (MPa)	Tensile strength (MPa)	Maximum uniform strain (%)	Young's modulus (GPa)	Yield ratio (%)	$_vE_0$ (J)	$_vE$ at test temp. (J)
A	equivalent of SM490 A	−30	346	604	12.2	213	57.3	19.9	6.39
B	SN490B	−30	383	579	15.3	222	66.1	109	9.01
C	SN490B	−50	373	562	16.1	220	66.4	144	13.5

Table 2. SENB test results.

Specimen	a_0 (mm)	COD_f (mm)	P_{max} (kN)	J_c (N/mm)	J_c-FE (N/mm)
A-D1	24.3	0.282	31.3	25	28
A-D2	24.1	0.514	34.5	63	73
A-D3	23.8	0.287	31.8	28	29
A-D4	23.9	0.456	34.8	51	62
A-D5	24.2	0.577	34.8	74	86
A-D6	24.1	0.549	35.5	71	80
A-D7	23.8	0.346	34.0	36	40
A-D8	24.1	0.708	36.8	100	114
A-D9	24.0	0.353	33.8	40	41
A-D10	24.3	0.596	34.7	77	90
A-S1	5.3	0.166	97.7		53
A-S2	5.2	0.338	109		124
A-S3	4.9	0.445	114		165
A-S4	4.9	0.259	105		92
A-S5	5.1	0.289	107		104
A-S6	5.0	0.397	110		147
A-S7	4.8	0.416	110		154
A-S8	5.1	0.219	99.5		75
A-S9	5.0	0.602	116		223
A-S10	5.2	0.236	101		82
B-D1	26.3	0.360	27.7	25	30
B-D2	25.8	–	31.4	53	65
B-D3	25.5	0.900	33.1	83	88
B-D4	29.3	2.31	28.0	311	–
B-D5	25.7	0.630	31.6	74	76
B-D6	24.4	0.415	35.4	45	47
B-D7	24.3	0.639	39.1	94	90
B-D8	24.2	2.77	49.7	–	–
B-D9	24.2	1.10	41.8	174	181
B-D10	24.5	0.815	37.6	114	125
B-S1	4.9	0.400	112		134
B-S2	4.6	0.388	113		129
B-S3	4.0	0.841	130		276
B-S4	4.5	0.372	113		124
B-S5	3.8	0.790	130		261
B-S6	5.5	0.264	108		90
B-S7	5.1	0.240	109		80
B-S8	5.8	0.259	108		88
B-S9	5.3	1.01	135		345
B-S10	4.9	0.226	105		74
C-D1	25.1	1.87	40.8	293	344
C-D2	24.8	1.73	41.0	267	319
C-D3	24.6	1.05	37.8	146	185
C-D4	24.6	1.12	38.9	165	200
C-D5	24.7	0.529	35.3	61	74
C-D6	24.4	0.500	36.8	56	68
C-D7	24.6	1.08	38.4	163	191
C-D8	24.4	1.31	39.9	194	239
C-D9	24.4	1.33	40.1	205	242
C-D10	24.3	0.934	39.1	136	161
C-S1	6.0	1.39	129		485
C-S2	5.9	–	158		–
C-S3	6.1	0.500	106		197
C-S4	5.8	0.915	120		345
C-S5	5.9	1.75	138		578
C-S6	6.0	0.970	121		362
C-S7	5.8	0.935	122		352
C-S8	5.4	0.251	104		91
C-S9	5.7	1.38	132		482
C-S10	5.9	0.351	109		134

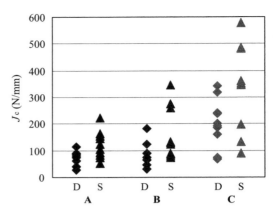

Figure 6. J_c plots of each specimen.

3 SENB TESTS

3.1 Test specimens

Specimens with deep notches or shallow notches were made based on BS 7448 (BSI 1997). The specimen geometry is shown in Figure 4. a_0 refers to the crack length. The crack length of the specimens is a_0/W = about 0.5 for a deep notch and a_0/W = about 0.1 for a shallow notch. Three types of material were prepared in this research. The materials are slightly different in the fracture toughness. The tensile coupon test and Charpy impact test results of the materials are summarized in Table 1. $_vE$ is the absorbed energy and $_vE_0$ is the absorbed energy at 0 °C. All materials were tested in low temperature and had a relatively low fracture toughness.

3.2 Test setup and procedure

The loading rig and specimen are shown in Figure 5. To reduce the temperature of the specimens to −30 degrees and −50 degrees, the specimens were cooled by using a solution of ethanol and dry ice. The temperature was measured using a thermocouple sensor. The load, the loading rig displacement and the crack opening displacement (COD) were measured during the tests.

3.3 Test results

Brittle fracture occurred in all of the specimens except two. The test results are summarized in Table 2. COD_f shows the crack opening displacement at fracture. J_c is the fracture toughness defined by BS 7448 (1997) for the deep notch specimens. The value of J_c for shallow notch specimens can not be calculated directly from test data. Hence, in this paper, the fracture toughness obtained from FE analyses is used for comparison, denoted by J_c-FE. The shaded rows in Table 2 indicate exceptions. The test temperature of A-D1 was too low and the crack depth of B-D4 was too large. B-D8 and C-S2 did not show brittle fracture.

J_c plots of each specimen are shown in Figure 6. The solid diamonds and triangles represent the J_c plots of the deep (D) and shallow (S) notch specimens for each material, respectively. Although the data spread of J_c is relatively large, it is obvious that there is a difference between deep notch and shallow notch with respect to the fracture toughness J_c. The results indicate that the difference of plastic constraint affects the fracture toughness. This will be evaluated further in a later section.

4 FE ANALYSES

4.1 Model of FE analyses

Figure 7 shows the 3-D FE meshes for a specimen with a deep notch. Symmetry about the x = 0 plane and z = 0 plane permits the use of a quarter-model with the number of nodes about 12,000. The FE models employ 8-noded brick elements with reduced integration technique. The crack front consists of 15 rings of element with 16 elements in each ring for J calculations. The minimum element dimension is 0.03 mm, in the area of 0.3 mm around the crack tip. The plasticity of the materials was defined by the von Mises yield criterion. The material stress-strain data at the relevant test temperature were obtained through standard tensile coupon tests of the materials. The test specimens showed small ductile cracks and several specimens showed a crack depth over 0.2 mm. If brittle fracture occurs with large ductile crack growth, another approach is needed to predict occurrence of brittle fracture. The effects of ductile crack growth are ignored in the current FE analyses.

4.2 FE analysis results

Figure 8 shows a comparison between the test and FE analysis results for load vs. COD curves of the deep notch specimens of material A. COD is crack opening displacement. For deep notch specimens, the analysis and test results are comparable. The results of the

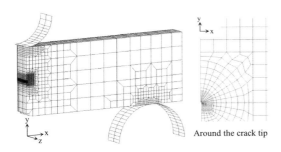

Figure 7. Finite element model for deep notch specimen.

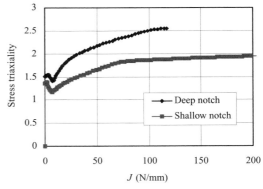

Figure 9. Stress triaxiality vs. J curves for material A.

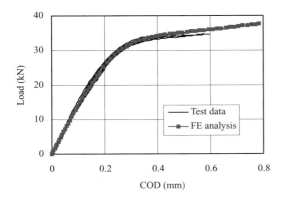

Figure 8. Load vs. COD curves for material A.

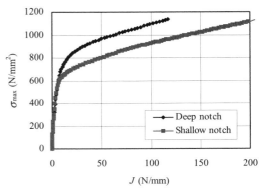

Figure 10. σ_{max} vs. J curves for material A.

analyses for materials B and C and shallow notched specimens of material A, also follow the test results. Figure 9 shows the stress triaxiality vs. J-integral curves for the SENB specimen models of material A. The stress triaxiality, T_s is calculated as follows:

$$T_s = \frac{\sigma_h}{\sigma_{eq}} \quad (3)$$

where σ_h is the hydrostatic pressure and σ_{eq} is the von Mises equivalent stress. Stresses were calculated from elements forming the crack tip, being 1 mm apart from the crack tip. The deep notch model is subjected to high stress triaxiality as compared with that of the shallow notch model. σ_{max} of the deep notch model is also higher than that of the shallow notch model at the same J value as shown in Figure 10. It is shown that the deep notch model is subjected to high plastic constraint around the crack tips. The FE results indicate that the risk of brittle fracture for the deep notch specimen is higher than that for the shallow notch specimen. Further, these results follow the test results that J_c of the deep notch is smaller than that of the shallow notch. It is therefore important to consider the effects of plastic constraint on fracture toughness for the prediction of brittle fracture.

5 RESULTS AND OBSERVATIONS

5.1 Weibull stress

Figure 11 shows cumulative probability F vs. Weibull stress σ_W plots, computed through Equation 1, with the plotted marks indicating the Weibull stress at brittle fracture. The solid diamonds and triangles represent the plots of the deep (D) and shallow (S) notch specimens, respectively. Cumulative probability F indicates median rank probabilities and is calculated as follows:

$$F = \frac{i - 0.3}{n + 0.4} \quad (4)$$

where i denotes the rank number and n defines the total number of specimens. In all cases, there is a significant difference between the experimental data for deep and shallow notches as shown in Figure 6, but σ_W plots are similar for the two geometries. Table 3 lists the Weibull slope m and scale parameter σ_u. The scale parameter indicates the Weibull stress σ_W when the cumulative probability $F = 63.2\%$. The value of σ_u for deep notches is also close in value to that for shallow notches for all of the materials. In contrast, the Weibull slope of materials B and C was significantly different for shallow and deep notches because the calibration by the maximum likelihood method did not work well although Table 3 shows the same m value for the deep and shallow notches. The Weibull slope of the shallow notch specimens marked by an asterisk in Table 3 are assumed to have the same m value as the deep notch specimens because the m values of the deep notch specimens were obtained smoothly by the maximum likelihood method and hence, are deemed to be accurate. One of the causes for the different m values for deep notch and shallow notch may be ductile crack growth (Ruggieri & Dodds 1996b). Ductile crack growth over 0.2 mm occurred in some specimens with a shallow notch for materials B and C.

These results indicate the Weibull stress is reasonable to consider the effects of plastic constraint on fracture toughness and can predict the occurrence of

Table 3. Weibull slope and scale parameter for Weibull stress plots.

Material specimen	A		B		C	
	D	S	D	S	D	S
m	23.5	23	15.5	15.5*	18	18*
σ_u	1443	1457	1522	1521	1546	1651

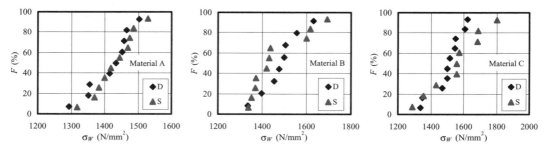

Figure 11. Cumulative probability F vs. σ_w plots.

brittle fracture, although the calibration of the Weibull slope for shallow notch specimens of materials B and C was not satisfactory.

5.2 Toughness scaling model

Figure 12 not only shows all J_c data (same as Fig. 6) but also the corrected J_c plots for the shallow notch specimens indicated by the open triangles. The shallow notch specimens have a higher fracture toughness J_c than the deep notch specimens but the corrected J_c values agree reasonably well.

Figure 13 shows cumulative probability F vs. J_c plots. F is also calculated by Equation 4. The symbols are similar to those in Figure 12. There is significant difference between the experimental data for deep and shallow notch specimens. The corrected J_c for shallow notch specimens, however, are close to the J_c for deep notch specimens and have the similar distribution for the specimens. Table 4 shows the Weibull slope and scale parameter for J_c plots. In the case of the Weibull slope, the m value of the corrected J_c is close to the value of the deep notch specimen for material A. However, for materials B and C, the values of the corrected J_c are different from the values for the deep notch. The Weibull slope has the value "2" ($m = 2$) from small scale yielding (SSY) for a pure weakest link model of fracture (e.g. Beremin 1983, Minami et al. 1992). The Weibull slope of all specimens is relatively close to the theoretical value "2", especially for material A. The scale parameter here indicates the fracture toughness J_c when the cumulative probability $F = 63.2\%$. The scale parameters of the corrected J_c are also close to the values of the deep notch specimens for all materials. These results indicate that the TSM is reasonable to describe the effects of plastic constraint on fracture toughness even though this approach requires no calibration to calculate the corrected J_c.

Figure 14 shows the stress triaxiality vs. J-integral curves for beam-to-column connections with weld defects. Each model has a 5 mm crack depth. The defect type of T, HS and HR is weld end defects and the defect type of R20, R10 and Ri10 is root defects. Detail of defect shape refers to those quoted in Iwashita et al. (2006). The highest value of the stress triaxiality in Figure 14 is about 1.7 and the value for shallow notch specimen is about 2.0. This result indicates that

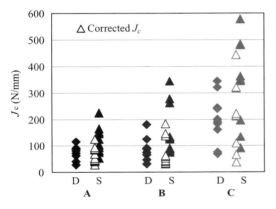

Figure 12. Corrected J_c plots of each specimen.

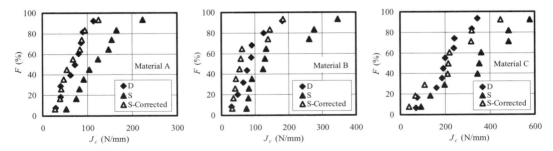

Figure 13. Cumulative probability F vs. J_c plots.

Table 4. Weibull slope and scale parameter for J_c plots.

Material specimen	A			B			C		
	D	S	S-corrected	D	S	S-corrected	D	S	S-corrected
Weibull slope m	2.37	2.56	2.32	1.95	1.78	1.58	1.97	1.70	1.34
Scale parameter	78	138	76	101	185	91	235	392	251

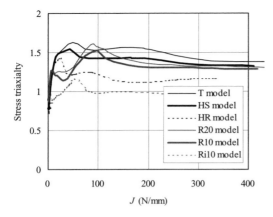

Figure 14. Stress triaxiality vs. *J*-integral curves.

the TSM is effective to predict brittle fracture starting from defects which have a medium level of plastic constraint (e.g. range of stress triaxiality is from 1.5 to 2.0). However, there are many weld defects in case of low level of plastic constraint (e.g. Ri10 model). Additional tests and further analyses for low levels of plastic constraint are required.

6 CONCLUSIONS

Fracture toughness tests of SENB specimens with a deep or shallow notch were performed for three types of material. The Weibull stress approach and TSM were used to consider the effects of a loss of plastic constraint on fracture toughness for shallow notch specimens. Both methods predicted the effects well. However the calibration of the micromechanics parameters for each material under study, which is time and money consuming, is required for calculating the Weibull stresses. On the contrary, TSM requires no calibration to calculate the corrected J_c and hence, is much simpler. Therefore it is reasonable to use the TSM method, which is a simplified model, to predict the occurrence of brittle fracture from weld defects. Further, the test specimens showed small ductile cracks. If brittle fracture occurs with large ductile crack growth, another approach is required to predict the occurrence of brittle fracture. In addition, the specimens with a shallow notch showed a low level of loss of plastic constraint. The effects of ductile crack growth and lower plastic constraint require further investigation.

ACKNOWLEDGEMENTS

This research was partly supported by the Japan Society for the Promotion of Science Grant-in Aid for Scientific Research. The authors would like to thank Dr. G.J. van der Vegte (Delft University of Technology) for his comments on this paper.

REFERENCES

Anderson, T.L. & Dodds, R.H. 1991. Specimen size requirements for fracture toughness testing in the ductile-brittle transition region. *Journal of Testing and Evaluation*, Vol. 19(2), pp. 123–134.

Beremin, F.M. 1983. A local criterion for cleavage fracture of a nuclear pressure vessel steel, *Metallurgical Transactions* 14 A, pp. 2277–2287.

British Standards Institution. 1997. BS 7448-4. *Fracture mechanics toughness tests—Part 4: Method for determination of fracture resistance curves and initiation values for stable crack extension in metallic structures.* London.

Iwashita, T., Kurobane, Y., Azuma, K. & Makino, Y. 2003. Prediction of brittle fracture initiating at ends of CJP groove welded joints with defects: study into applicability of failure assessment diagram approach, *Engineering Structures*, Vol. 25, Issue 14, pp. 1815–1826.

Iwashita, T., Kurobane, Y. & Azuma, K. 2006. Assessment of risk of brittle fracture for beam-to-column connections with weld defects, *Proceedings of the 11th International Symposium and IIW International Conference on Tubular Structures*, pp. 601–609.

Minami, F., Bruckner-Foit, A., Munz, D. & Trolldenier, B. 1992. Estimate procedure for the Weibull parameters used in the local approach. *International Journal of Fracture* 54, pp.197–210.

Riesch-Oppermann, H. & Diegele, E. 2002. *Elements of a fracture mechanics concept for the cleavage fracture behaviour of RAFM steels using local fracture criteria*, FZKA Report 6668, Forschungszentrum Karlsruhe.

Ruggieri, C. & Dodds, R.H. 1996a. A transferability model for brittle fracture including constraint and ductile effects: a probabilistic approach, *International Journal of Fracture* 79, pp. 309–340.

Ruggieri, C. & Dodds, R.H. 1996b. Probabilistic modeling of brittle fracture including 3-D effects on constraint loss and ductile tearing, *Journal de Physique IV*, pp. 353–362(C6).

Tagawa, T., Chaves, C.E., Yang, H., Yoshinari, H. & Miyata, T. 1999. Specimen size requirements for cleavage fracture toughness based on the Weibull stress criterion, *Journal of the Society of Naval Architects of Japan*, Vol. 189, pp. 485–497. (in Japanese)

Advanced numerical modelling of fatigue size effects in welded CHS K-joints

L. Borges & A. Nussbaumer
Swiss Federal Institute of Technology Lausanne, ICOM—Steel Structures Laboratory, Lausanne, Switzerland

ABSTRACT: The fatigue design of tubular space trusses for bridge applications requires additional knowledge with respect to their fatigue resistance. As an introduction, the paper reviews the fatigue assessment of tubular joints and exposes basic concepts on the size effects. In order to carry out a thorough study on the geometrical size effects in CHS K-joints, an advanced 3D crack propagation model was implemented using the Dual Boundary Elements Method (DBEM). The incremental crack propagation process allows for the calculation of the Stress Intensity Factors (SIF) along the doubly curved crack front at different crack depths. The fatigue life for three main basic load cases is then calculated using the Paris law. The model was validated against several fatigue tests on bridge-like tubular trusses conducted at ICOM/EPFL.

1 INTRODUCTION

1.1 Fatigue of tubular joints

The fatigue design stress, σ_{hs}, calculated for CHS bridge joints using empirical parametric equations found in design specifications (Zhao et al. 2000) is high, typically two to five times higher than the nominal stress, $\sigma_{nom,i}$, in the truss members. When the design stress is applied to corresponding $S_{R,hs}$–N design lines, a further "size effect" correction is made depending on the wall thickness, T, of the fatigue critical member (chord or brace), which can, in many cases, translate into a further penalty to the fatigue resistance of the joint (Fig. 1).

Schumacher (Schumacher et al. 2006) has carried out an experimental and numerical study to investigate two specific aspects of welded CHS bridge K-joint fatigue: the joint stresses (hot-spot stresses and stress concentration factors) at critical fatigue locations and the influence of the size effect on the fatigue resistance of these joints using a single site deterministic approach.

1.2 Size effects

Stated simply, the size effect for welded joints is the phenomenon whereby the fatigue strength of a larger or thicker joint is lower than a smaller or thinner joint of the same geometry, subjected to the same magnitude of stresses.

A significant push on research into the size effect in the 1980s provided more convincing evidence,

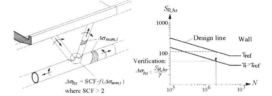

Figure 1. Fatigue design of welded CHS bridge joints: load effects and resistance.

both experimental and analytical, of a trend for lower fatigue strength in thicker specimens of, for the most part, the same geometry (Noordhoek et al. 1987). The research work included various types of joint geometries (plate joints, tubular joints), joint preparation (as-welded, post-weld heat treated), test conditions (in air, in sea water) and load types (tensile load, bending load). Results from this work led to a few relatively minor modifications and additions to the original Gurney rule (Gurney 1979). It is recalled that three factors are generally cited as contributing to the size effect (Marshall et al. 1992): the statistical size effect, the metallurgical or technological size effect and the geometrical size effect. The latter is often considered as the dominant effect in welded structures. It refers to the through-thickness stress gradient that arises at geometrical discontinuities (e.g. notches) and/or due to bending and torsional loads, which can be addressed by stress analysis. Due to the presence of a steeper stress gradient, a grain close to the surface of a small specimen will experience a lower strain than a grain close to the surface of a thick specimen, for the same stress at the surface.

2 NUMERICAL MODEL

2.1 Introduction

The model includes the calculation of the stress intensity factors (SIF) for different crack depths using a 3D boundary element model. The model presented in the current investigation uses Boundary Elements Method (BEM) commercial code (BEASY 2003). The crack growth rate is computed using the Paris law and then the number of cycles to failure is obtained by integration.

The standard model described herein is thereafter used to carry out a parametric study on the fatigue behaviour of different K-joint geometries; this is in progress and out of the scope of this paper.

2.2 Standard model for fatigue life computation

The schematic in Figure 2 describes the procedure used to carry out a crack propagation analysis. Informations such as the *joint geometry* (namely the elements dimensions and weld size), the *crack geometry* (namely the crack site, crack shape, crack angle and crack depths for the different increments) and the *load case* (basic load cases or complex load case combining different basic load cases in a single model) are input. Then, the mesh points coordinates are calculated in Excel for both the joint and the crack geometry corresponding to each crack depth.

The geometry of the joint is exported in Beasy data format. The crack model is also exported in Beasy syntax in a separate file. This is repeated sequentially for all the combinations of basic load cases (or just the complex case) and crack depths. The *Beasy crack adder* routine is then used to add the cracks to the joint models and automatically re-mesh the crack vicinity. For each load case and crack depth increment there is a results file containing the stress intensity factors K_I, K_{II}, K_{III} along the crack front mesh points. These data are imported and treated using *Wolfram Mathematica* (Wolfram 1988) software. The equivalent stress intensity factors are then calculated. The crack growth rates are computed using the Paris law and the number of cycles calculated using discrete linear integration algorithm.

2.3 Finite Element Method (FEM) and Boundary Element Method (BEM)

The estimation of a fatigue crack life using the theory of Linear Elastic Fracture Mechanics (LEFM) involves the calculation of stress intensity factors (SIF) at a number of different crack depths.

Different methods can be used to estimate SIFs. Most of them involve the use of expressions deduced from parametric studies on specific geometry ranges. A more complex way involves advanced modelling of the crack by finite element or boundary element codes. The finite element method has been widely used in fracture mechanics applications. Recent investigations applied the finite element method to simulate the crack behaviour in CHS joints (Cao et al. 1998, Shao 2005). An intrinsic feature of the finite element method is the need for continuous remeshing of the three-dimensional volume to follow the crack extension; this is a practical disadvantage of this method (Mellings et al. 2003). In the boundary element method, only the boundary of the domain of interest is discretised (Portela et al. 1993). This is done using elements which are interconnected at discrete points called nodes. One disadvantage of this method is that it can only be used for linear elastic problems. However this is not an issue in modelling fatigue life in long life region (not oligocyclic).

2.4 Boundary element model

In order to create a boundary element model simulating a cracked uniplanar K-joint, different aspects have to be considered. Firstly, the geometry of the boundaries that define the joint elements and respective intersections have to be parameterized so that different geometries in the parametric study range can be modelled. In the current study, symmetry was not taken advantage of. This choice was made because it makes the current model more versatile and valuable for an extension of the present investigation to study asymmetrical cracks.

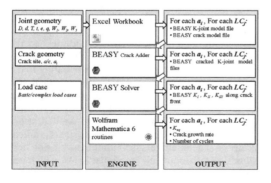

Figure 2. Standard model for fatigue life computation.

The crack path, or, in 3D, the surface defining crack faces, has also to be defined. A number of zones are created to confine regions of similar mesh density and material properties. The mesh discretising the boundaries is chosen and the external forces and boundary conditions applied to mesh points.

2.5 Geometry definition

In order to define the joint model, its boundaries were parametrically defined. The cylinder parametric equation is first used to define the boundaries of chord and diagonals. Then the weld and tubes intersections are defined using the intersection cylinder-cylinder system of equations. Overlapped joints are not considered in this study as they are usually less used in fatigue critical structures (except in mining equipment).

2.5.1 Chord and diagonals

The chord and diagonals boundaries consist of concentric cylinders of diameters D and $(D-2T)$ for the chord respectively d and $(d-2t)$ for the diagonals.

2.5.2 Weld geometry

The weld profile has an important influence on the stress concentration at the weld toe and thus on the stress intensity factors (SIF) for surface cracks. Therefore the welds should be modeled as close as possible to the reality. As it is a difficult task to simulate the weld profile realistically, many investigations did not consider it. This assumes, according to (Lee et al. 1995), an underestimation of the fatigue life up to 20%. In the present study, the weld is defined using three auxiliary curves for each diagonal-chord weld (see Fig. 3):

- The intersection of the inner boundary of the diagonal with the chord outer boundary;
- The intersection of the outer boundary of the diagonal with the chord, shifted by $(W_1\cos[\theta], 0, W_2\sin[\theta])$;
- The intersection of the chord with an imaginary cylinder with the same angle θ as the diagonal but diameter equal to $d^* = d + 2W_2$ and translated of $(W_3, 0, 0)$;

Figure 4 shows the weld dimensions in two sites around the diagonal-chord intersection: the weld crown toe and the weld crown heel. L_w is the weld footprint length, θ_w is the weld toe angle, ψ is the local dihedral angle. Subscripts "br" and "ch" are added to distinguish between brace-side and chord-side parameters.

These dimensions can be calculated using the parameters used to geometrically define the weld: W_1, W_2 and W_3 as follows:

$$L_{w'} = \frac{t}{\sin(\pi - \psi)} \quad (1)$$

For weld crown toe (hot-spot sites 1 and 11): $\psi = \theta$

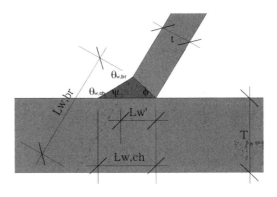

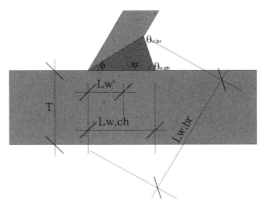

Figure 4. Dimensions needed to define weld geometry- left, crown toe and right, weld crown heel.

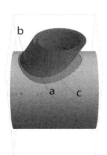
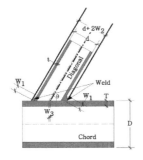

Figure 3. Weld geometry.

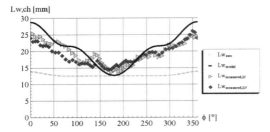

Figure 5. Weld footprint length comparison—AWS (AWS. 2000) recommended minimum, S5 BEM model and S5-2 j2 measured.

$$L_{w,ch,toe} - L_{w'} = \frac{W_2}{\sin(\theta)} - W_3 \quad (2)$$

For weld crown heel (hot-spot sites 31 and 3):
$\psi = \pi - \theta$

$$L_{w,ch,heel} - L_{w'} = W_3 + \frac{W_2}{\sin(\theta)} \quad (3)$$

$$L_{w,br} = W_1 + \frac{T}{\sin(\theta)} \quad (4)$$

Figure 5 compares the weld toe geometry in the model obtained following the procedure described to the weld profile of the fatigue test specimen S5-2 (joint2) (Nussbaumer et al. 2008).

The model of the weld profile closely represents the real weld profile. Both weld profiles respect the AWS requirements.

2.6 Material properties

The joint material is elastic linear with Young Modulus $E = 210 \cdot 10^3$ MPa, Poisson ratio, $\nu = 0.3$. Rigid rings are 100 times more rigid and have same Poisson ratio.

The Paris law constants $C = 2.0 \cdot 10^{-13}$ (mm/cycle)(N/mm$^{-3/2}$)^m and $m = 3$ were considered for deterministic fracture mechanics calculations.

2.7 Boundary conditions

The joint is fixed in the 3 directions at the chord right extremity. External forces are introduced using rigid rings of absolute length $\pi D^*/10$ (D^* diameter of the tube to which the rigid ring is connected) to preserve plane cross-sections (Schumacher 2003).

2.8 Crack location and geometry

Stress analysis of the uncracked joint clearly identifies hot-spot site 1 as the point where stress concentration is higher.

Therefore, and supported by previous and current experimental evidence (Schumacher 2003, van Wingerde et al. 1997), hot-spots 1 were primarily considered in the present study. Figure 6 shows the different hot-spots and stress concentration at the weld toes.

Also supported by experimental evidence, (Nussbaumer et al. 2008), the crack front corresponding to hot-spot site 1 is obtained by projecting a semi-ellipse over a conic surface. The conic surface directrix is the weld toe and the apex belongs to the xoz plane at a depth of $1.78 \cdot D$ (see Fig. 7). The crack angle determines the x coordinate of the apex.

2.9 Meshing

The mesh is a key aspect when performing parametric studies on the influence of changing the size of the

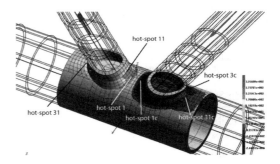

Figure 6. Stress concentration in the joint (σ_{xx})—Series S5 fatigue test loading conditions.

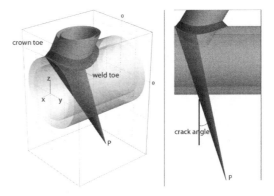

Figure 7. Crack surface geometry.

structural elements. In the present study, the proportions of the joint elements may change and it is therefore very important to assure that the results reflect the effect of size changes and not the effect of mesh being somewhat different. The following paragraphs describe the joint and crack meshing.

2.9.1 Joint mesh

The boundary elements model includes about 8100 mesh points and 2000 elements distributed in 8 zones as shown in Figure 8. Zones are groups of elements which can be considered as substructures of the component. Among these 8 zones, Zone 2 (see Fig. 8), where the crack is located and the stress is highly nonlinear, has a dense mesh; Zones 6, 7 and 8 are rigid rings for the external force introduction.

Even though BEASY has extensive automatic meshing capabilities, an *Excel Workbook* was prepared to calculate the mesh point coordinates, elements (including the weld profile and crack) and zones definitions. This allowed for the creation of user controlled, validated model files from the joint geometric parameters in a BEASY compatible format.

The entire model is meshed with reduced quadratic four-sided elements Q38 whenever possible and exceptionally with triangular quadratic elements (BEASY 2003).

2.9.2 Crack mesh

As described before, the crack faces belong to the conic surface (see Fig. 7) and the crack front is doubly-curved. This complex geometry makes modelling of the crack and of the propagation of such a crack in a tubular joint a complicated task. The mesh of the crack surface has to be carefully chosen. A good quality mesh depends on the shape of the elements defining the crack surface. Although Beasy software provides a crack growth tool allowing for automatic crack propagation from an initial crack, a manual, stepwise, crack modelling was preferred. The following reasons justify this option:

- due to the sharp weld toe geometry, automatic crack growth requires a very small step size so that the crack path remains at the weld toe.
- the big amount of time spent to automatically grow a crack from a_0 to $T/2$ and the model sizes made it impossible to carry out the parametric study in a reasonable amount of time;
- the need for identical crack path for the different basic load cases to make it possible to isolate/ superpose their influence;
- manual crack growth allows the control of the crack shape, a/c, and thus an indirect inclusion of the coalescence phenomenon;

The manual crack growth corresponds to the calculation of a set of models with built-in cracks of different given shapes and depths.

The crack follows the conic surface (Fig. 7). The mesh points are calculated to suit the curved shape. The number of elements and mesh points remains constant for the different crack depths. However Beasy automatically remeshes the area near the crack in order to optimally adapt the crack mesh in the existing joint mesh (see Fig. 9).

2.10 Basic load cases and complex load case

A main objective of the present study is to identify the effect of the different basic load cases in the fatigue behaviour of a tubular K-joint.

When we consider only the in plane actions, five basic load cases can be isolated (see Table 1). Several considerations justifying the selected method of decomposing the applied forces can be found in (Schumacher 2003).

The developed boundary element model allows for the application of the five basic cases independently or combined (complex load). A constant nominal stress range is applied. Complex load results of the

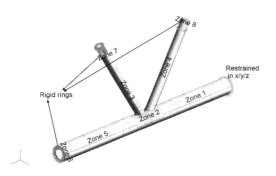

Figure 8. Zoning of the boundary element model.

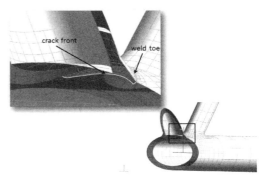

Figure 9. Detail of surface crack in weld toe and autoremeshed area.

Table 1. Individual basic load cases and boundary conditions (from (Schumacher 2003)).

No	Load case		Schematic	Nominal stress
LC1	Balanced axial brace	F_{ax_br}		σ_{ax_br}
LC2	Un-balanced in-plane bending brace	M_{ipb1_br}		σ_{ipb1_br}
LC3	Balance in-plane bending brace	M_{ipb2_br}		σ_{ipb2_br}
LC4	Axial chord	F_{ax_ch}		σ_{ax_ch}
LC5	In-plane bending chord	M_{ipb_ch}		σ_{ipb_ch}

simultaneous application of the selected basic cases, thus resulting in in-phase actions.

Out of the 5 basic load cases, 3 are considered predominant in this study: LC1, LC4 and LC5.

When applying actions to the model, the stress ratio, R, is equal to zero or $\sigma_{max} = \Delta\sigma$ is used for simplicity. Underlying assumption is that there are no crack closure effects.

2.11 Stress intensity factors

For mixed-mode problems, the equivalent stress intensity factor, K_{eq}, is calculated at every mesh-point along the crack front using equation

$$K_{eq} = \sqrt{(K_I + |K_{III}|)^2 + 2K_{II}} \quad (5)$$

where K_I, K_{II}, K_{III} are the stress intensity factors for the three crack opening modes calculated along the crack front (Gerstle 1986).

For load cases LC1, Figure 10 shows crack opening mode I is predominant over modes II and III. This is also the case for load cases LC4 and LC5. The equivalent stress intensity factor, K_{eq} is superposed with the stress intensity factor for opening mode I, K_I. Schumacher also showed a predominant mode I behaviour even neglecting crack angle. In these computations, crack angle is assumed constant and taken as the bisectrix between the weld profile and the chord wall at the crown weld toe.

2.12 Fatigue crack growth model

The crack growth process is simulated using an incremental crack-extension analysis. For each crack extension, a stress analysis is carried out and the stress intensity factors are computed. SIF values in-between are linearly interpolated. The Paris law (Paris 1960) was used to compute the crack growth rate between the different crack sizes:

$$\frac{da}{dN} = C \times \Delta K_{eq}^m \quad (6)$$

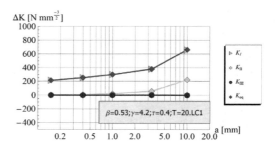

Figure 10. Comparison of the crack opening modes for the different basic load case LC1 ($\Delta\sigma_{nom} = 100$ MPa).

where C and m are the crack growth rate constants for the material and ΔK_{eq} is the equivalent stress intensity factor range.

Crack initiation is not considered in the current crack growth model. This simplification is justified by the fact that for welded joints without treatment the crack initiation phase represents only a small fraction of the fatigue crack through thickness propagation. Results of the fatigue tests carried out (Nussbaumer et al. 2008) support this hypothesis as the initiation phase was about 10% of the total fatigue life (N_3). It is also assumed that the fatigue load is clearly above the fatigue limit (stress intensity factor thresholds not considered), and constant amplitude cycles are applied.

2.13 Initial crack size, crack increments and failure criterion

As explained in paragraph 2–8 the crack grows in a predefined conic surface. The critical crack depth, a_c, is taken as equal to half of the wall thickness of the cracked element, $T/2$. Although a critical crack depth equal to the wall thickness is more common as failure criterion, it has been set to $a_c = T/2$ by others (Walbridge 2005). This option is justified by the fact that small variations in the critical crack depth have little influence in the final number of cycles as the majority of the crack life is spent at smaller crack depths (Moan et al. 2000). As the crack front gets close to the "bottom" boundary of the cracked member (the inner boundary of the chord) numerical problems may arise and a crack remeshing would be needed. Since through-thickness cracking must consider internal forces redistribution, the problem is no more limited to the joint study but must consider whole structure behaviour.

A constant initial crack size of $a_0 = 0.15$ mm is considered regardless of the geometry of the joint (Walbridge 2005). Crack growth increments correspond to crack depths of $a_i = \{0.15$ mm, $T/50, T/20, T/8, T/6, T/2\}$.

2.14 Model validation

The validation of the model was carried out:

- By comparing the numerical results to the behaviour of the tested specimen (see (Nussbaumer et al. 2008)). The comparison includes the nominal strains (in the chord and braces), the strain in the joint (hot-spot strain) and the stress intensity factors for different crack depths;
- Model accuracy evaluation;
- A mesh convergence test was carried out to investigate the sensitivity of the model with the mesh density.

These different comparisons and tests are carried out in following paragraphs.

2.15 Comparison to fatigue test results

Tables 2 and 3 summarize the geometry dimensions and non-dimensional parameters of the model used to simulate the fatigue tests. Weld dimensions were considered as: $W_1 = 15.0$ mm $W_2 = 10.0$ mm $W_3 = 5.0$ mm. The nominal stresses introduced correspond to the values calculated using a simplified bar-model of the truss beam. The eccentricity is simulated using model suggested in CIDECT recommendations (Zhao et al. 2000).

Figure 11 compares the stress intensity factor range ΔK computed using the model described and the values measured using the Alternating Current Potential Drop system at different locations at the weld toe crown (P1 to P8). A good match was found between both results.

2.16 BEASY—Accuracy evaluation

BEASY provides post-processing tools to evaluate the model accuracy such as accuracy reports or the stress error norm plot.

The stress error norm is higher (11.89 MPa) in zone 5, near zone 2, due to the element grading. For zone 2, where the surface crack is located and accurate results are needed, the stress error norm is less then 6 MPa.

The zone stress error norms are obtained considering the equilibrium convergence in each zone, by summing the surface forces in each direction. They are as low as 0.01% for zone 2 and as high as 5% for zone 6 (rigid ring in the chord).

2.17 Combining basic load cases

The possibility of combining the results of basic load cases is be very useful as it allows the modelling of various real loading conditions without having to rerun crack propagation simulations (to obtain SIF) each time.

Superposition of effects is valid because:

- a linear elastic fracture mechanics (LEFM) analysis is carried out on linear elastic material;

Table 3. Non-dimensional parameters of model used to compare with fatigue test results.

β	γ	τ	α	ζ	e/D
–	–	–	–	–	–
0.53	4.21	0.40	25.7	0.12	0.13

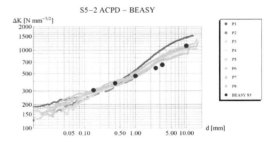

Figure 11. Comparison between "measured" and computed stress intensity factor range vs. crack depth.

Table 2. Geometric parameters taken to create model used to compare with fatigue test results.

D	T	d	t	e	g	L_{ch}	H
mm	mm	mm	mm	mm	mm	mm	mm
168.3	20.0	88.9	8.0	22.0	19.9	2166.0	1800.0

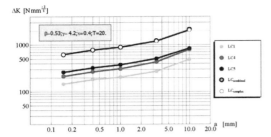

Figure 12. Stress intensity factor results obtained for the basic load cases acting isolated and combined and corresponding complex load case.

- the crack path is established *a priori* and remains constant for the different load cases;
- crack closure is not considered;
- proportional load cases are considered to be in phase;

The stress intensity factors for the different opening modes are combined, $K_{m,c}$ and then the combined equivalent stress intensity factor, $K_{eq,c}$ is calculated.

$$K_{m,c} = \sum_{LCi\,=\,1}^{lcn} K_{m,LCi}, \quad \text{where } m = \{I, II, I \quad (7)$$

$$K_{eq,c} = \sqrt{(K_{I,c} + |K_{III,c}|)^2 + 2K_{II,c}} \quad (8)$$

Figure 12 shows stress intensity factor range results obtained for the basic load cases acting separately and combined ($K_{eq,LC1}$, $K_{eq,LC4}$, $K_{eq,LC5}$, $K_{eq,c}$) and the result for the model with complex load case where all the load cases were introduced simultaneously ($K_{eq,complex}$).

3 CONCLUSIONS

This chapter presents the description and validation of a CHS K-joint crack propagation model. The 3-dimensional boundary element model is solved using BEASY software. The model was validated through comparisons between calculated stresses with measured stresses (strains) and between calculated stress intensity factors with experimentally measured stress intensity factors.

A standard model for fatigue life computation is developed to be used in the parametric study.

The following conclusions can be drawn:

- The boundary element method can be used reliably to calculate the stress intensity factors of surface cracks in CHS joints.
- Specific tools could be developed to model this complex crack shape in the doubly curved weld toe.
- Automatic crack growth calculation was not made in this particular case due to the complex geometry. The needed small increments make it too costly in time and computer resources to propagate a crack in the weld toe.
- With an identical crack path for the three load cases considered, crack opening mode I is predominant for the three load cases considered.
- Results from the numerical model compare well with the experimental determined SIF and thus it results in a good estimation of the fatigue life of the joints.

ACKNOWLEDGMENTS

The authors would like to acknowledge the funding assistance provided by the Swiss National Science Foundation (SNF). Part of the research carried out herein is included in the project P591, supervised by the Versuchsanstalt für Stahl, Holz und Steine at the Technische Universität Karlsruhe, which is supported financially and with academic advice by the Forschungsvereinigung Stahlanwendung e. V. (FOSTA), Düsseldorf, within the scope of the Stiftung Stahlanwendungsforschung, Essen.

REFERENCES

AWS Structural Welding Code 2000. American Welding Society, d 1.1-2000.

Beasy 2003, *Beasy User Guide*. Computational mechanics Beasy Ltd, Ashurst, Southhampton, UK.

Cao, J.J., Yang, G.J., Packer, J.A. & Burdekin, F.M. 1998. Crack modeling in FE analysis of circular tubular joints. *Engineering fracture mechanics* 61: 537–553.

Gerstle, W.H. 1986. *Finite and boundary element modelling of crack propagation in two- and three-dimensions using interactive computer graphics*. Ph.D. Thesis, Cornell University, Ithaca, NY.

Gurney, T.R. 1979. *Fatigue of welded structures*. Cambridge University Press, Cambridge.

Lee, M.M.K. & Wilmshurst, S.R. 1995. Numerical modelling of CHS joints with multiplanar double-k configuration. *Journal of Constructional Steel Research* 32(3): 281–301.

Marshall, P.W. 1992. *Design of welded tubular connections, basis and use of AWS provisions*. Elsevier Science Publishers. Amsterdam.

Mellings, S., Baynham, J., Adey, R.A. & Curtin, T. 2003. Durability prediction using automatic crack growth

simulation in stiffened panel structures. *Structures and Materials* 12: 193–202.
Moan, T., Song, R. 2000. Implications of inspection and repair on system fatigue reliability of offshore structures. *Journal of offshore mechanics and arctic engineering, transactions of the ASME* 122: 173–180.
Noordhoek, C., Van Delft, D.R.V. & Verhuel 1987. A. The influence of plate thicknesses on the fatigue behaviour of welded plates up to 160 mm with an attachment or butt weld. *In: SIMS* 81: 281–301.
Nussbaumer, A., Borges, L. 2008. Experimental determination of stress intensity factors on large-scale tubular trusses fatigue tests. *In: ISTS12*, Shanghai:
Örjasäter, O. 1995. *Effect of plate thickness on fatigue of welded components*. IIW-ywg XIII-XV-118–93: 1–19, IIW.
Paris, P.C. 1960. *The growth of cracks due to variations in load*. Ph.D. Dissertation, Lehigh University, Bethlehem, Pennsylvania, USA.
Portela, A., Aliabadi, M.H., Rooke, D.P. 1993. Dual boundary element incremental analysis of crack propagation. *Computers & Structures* 46: 237–247.
Rhodia Chimie Rhodorsil—sil 01 021 1 2002. Fiche technique, Rhodia Chimie—Silicones Europe, 55 av. Des frères Perret, bp 60, 69192, St-Fons-Cedex, France.
Schumacher, A. 2003. *Fatigue behaviour of welded circular hollow section joints in bridges*. Ph.D. Thesis EPFL n °2727, swiss federal institute of technology (EPFL), Lausanne.
Schumacher, A. & Nussbaumer, A. 2006. Experimental study on the fatigue behavior of welded tubular k-joints for bridges. *Engineering Structures* 28: 745–755.
Shao, Y. 2005. *Fatigue behaviour of uniplanar chs gap k-joints under axial and in-plane bending loads*. Ph.D. Thesis, pp. 283, Nanyang Technological University. School of Civil and Environmental Engineering, Singapore.
Van Wingerde, A.M., van Delft, D.R.V., Wardenier, J. & Packer, J.A. 1997. *Scale effects on the fatigue behaviour of tubular structures*. WRC Proceedings, IIW.
Walbridge, S. 2005. *A probabilistic fatigue analysis of post-weld treated tubular bridge structures*. Ph.D. Thesis EPFL n °3330, Ecole polytechnique fédérale de Lausanne (EPFL).
Wolfram, S. & Mathematica 1988, *A system for doing mathematics by computer*, USA. Addison-Wesley, Redwood City, USA.
Zhao, X.L., Herion, S., Packer, J.A. et al. *2000 Design guide for circular and rectangular hollow section joints under fatigue loading*. CIDECT, Comité international pour le développement et l'étude de la construction tubulaire, tüv-Verlag Rheinland, Köln.

Further experimental study into applicability of FAD approach to beam-column connections with weld defects

K. Azuma
Sojo University, Kumamoto, Japan

Y. Kurobane
Kumamoto University, Kumamoto, Japan

T. Iwashita
Ariake National College of Technology, Fukuoka, Japan

ABSTRACT: This paper concerns the applicability of modified FAD approach to the assessment of defects existing at the weld terminations. Welded plate bend models, which were designed to represent a connection of an I-section beam to an RHS column member with through diaphragms, were tested under cyclic loads. Through cracks, which were made by using machine cutting and fatigue cracks, were installed on both sides of weld terminations. The prediction of brittle fracture with a variety of defect sizes by using the FE analysis was conducted prior to the test, and defect sizes were decided so that the brittle fracture would not occur for one specimen and would occur for the others. Brittle fracture occurred from tips of defects for the specimens with through cracks, while ductile cracks grew stably and brittle fracture did not occur for the specimen with surface cracks. For the specimens with through cracks, however, the ductile cracks grew stably prior to final rupture and the connection showed deformation capacity.

1 INTRODUCTION

During the 1995 Kobe earthquake, brittle fractures occurred frequently caused by cracks growing from the corner of cope holes or weld tab regions in the beam bottom flanges. One of the post-earthquake proposals is to use improved profiles of cope holes (AIJ. 1996). These new details, however, revealed other weld defects. Previous testing of welded plate bend models with partial joint penetration (PJP) groove welded joints showed sufficient deformation capacity, although PJP groove welds formed internal defects because the roots of the welds were reinforced by additional fillet welds (Azuma et al. 2000, 2003, 2006). Fracture toughness properties of numerically modelled connections were evaluated by using a failure assessment diagram (FAD) approach (BS 7910, 1999), which was modified by considering the effect of enhanced apparent toughness of material due to the loss of crack tip constraint (Iwashita et al. 2003). Assessment of unfused regions predicted that brittle fracture would not occur.

This paper concerns the applicability of modified FAD approach to the assessment of defects existing at the weld terminations. Welded plate bend models, which were designed to represent a connection of an I-section beam to an RHS column member with through diaphragms, were tested under cyclic loads. Complete joint penetration groove welds were used for the connection between the beam flange and the diaphragm. The specimens were fixed to a strong reaction frame with high strength bolts through the diaphragm plate. The load was applied to the end of the cantilever by a hydraulic ram statically. Through cracks, which were made by using machine cutting and fatigue cracks, were installed on both sides of weld terminations. The plastic constraint at the crack tips of the specimen was close to that of the double edge notched tension (DENT) specimens.

The prediction of brittle fracture with a variety of defect sizes by using the FE analysis was conducted prior to the test, and defect sizes were decided so that the brittle fracture would not occur for one specimen and would occur for the others. Brittle fracture occurred from tips of defects for the specimens with defects of through cracks, while ductile cracks grew stably and brittle fracture did not occur for the specimen with defects of surface cracks. For the specimens with through cracks, however, the ductile cracks grew stably prior to final rupture and the connection performed sufficient deformation capacity. It was confirmed that current assessment method was too conservative and that the ductile tearing analysis had to be included into FAD approach.

2 CYCLIC TESTING OF WELDED PLATE BEND MODELS

2.1 Specimens and loading procedures

Two welded plate bend models, which were designed to represent a connection of an I-section beam to an RHS column member with through diaphragms, were tested under cyclic loads. The specimens consisted of 25 mm thick flange plate, 32 mm thick diaphragm and 25 mm thick rib plates. All plates were grade SN490B. Complete joint penetration (CJP) groove welds were used for the connection between the beam flange and the diaphragm. Weld metal was produced by electrodes designated as YGW-11. Stringer passes were used for all welding and heat inputs during welding were approximately 20 kJ/cm for each layer. The configuration of the specimen is shown in Figure 1. For the WRTC specimen, through cracks, which were made by using machine cutting and fatigue cracks, were installed on the weld toes at the edge of the diaphragm plate. The plastic constraint at the crack tips of the specimen was close to that of the double edge notched tension (DENT) specimens. For the GWSC specimens, surface cracks, which were made by using machine cutting, were installed on the weld toes at the edge of the flange plates. The locations of each defect are shown in Figure 2.

Cyclic loads in the horizontal direction were applied to the end of the cantilever by a hydraulic ram statically,

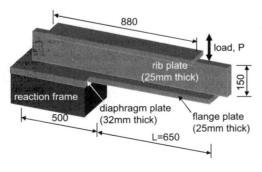

Figure 1. Specimen configuration.

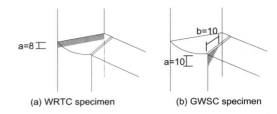

Figure 2. Locations of the defects.

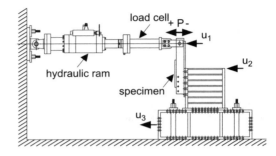

Figure 3. Positions of load application and displacement measurements.

while the specimens were fixed to a strong reaction frame with high strength bolts through the diaphragm plate. Figure 3 shows the position of load application and the displacement measurements. The bending moment M and the rotation angle θ of the cantilever were calculated by the following equations, respectively:

$$M = P \cdot L \quad (1)$$

$$\theta = \frac{u_1 - u_2 - u_3}{L} \quad (2)$$

Cyclic loading consisted of a few cycles in the elastic range and then cantilever rotations of θ_p, $2\theta_p$, $3\theta_p$,... with positive (tension) and negative (compression) displacement at each amplitude, until failure. The full plastic moment M_p was calculated using measured yield strengths of materials. The rotations at full plastic moment θ_p, namely M_p divided by the elastic stiffness of the cantilever, were calculated. The elastic stiffness was determined by using slopes at unloading portions of hysteresis loops.

2.2 Material properties

The material properties, in terms of engineering stress-strain, were obtained by tensile coupon tests for the flanges, diaphragms and rib plates, which are summarized in Table 1.

The fracture toughness was obtained by Charpy impact tests. Test pieces were taken from plates welded under the same welding conditions as those for the specimens. The positions of notch roots were base metal of flange plate, diaphragm plate, HAZ (heat affected zone) and DEPO (deposited weld metal). Test pieces were cooled to temperatures between −80 °C and 60 °C by using dry ice and alcohol or hot water. Three test pieces were tested at each temperature. The results of Charpy impact test are shown in Table 2. The energy transition curve was obtained by fitting test results into the follow equation:

$$vE(T) = \frac{_vE_{shelf}}{e^{-b(T-vTE)}+1} \quad (3)$$

2.3 Deformation capacity and failure modes

Figure 4 shows hysteresis loops for the two specimens. The moment is that at the edge of the diaphragm and is non-dimensionalized by dividing it by the full-plastic moment of the cantilever. The moment is herein defined as the positive moment when the flange plate is in tension. Figure 5 shows moment vs. rotation skeleton curves for both specimens. Table 3 summarized the maximum moments with the corresponding rotations and the cumulative plastic rotation factors η_s. η_s was obtained from the skeleton curves for each specimen (See Appendix for the definition of skeleton curves and η_s^+).

The WRTC specimen sustained brittle fracture from the tips of the defects. The specimen after

Table 1. Results of tensile coupon tests.

Location	σ_y MPa	σ_u MPa	E.L. %	E GPa
Flange plate	368.5	558.1	31.9	221.3
Rib plate Diaphragm plate	393.7	574.1	31.7	224.0

Table 2. Results of Charpy impact test.

	$_vE(0)$ J	$_vE_{shelf}$ J	$_vTE$ °C
Flange plate Rib plate	81	195	6
Diaphragm plate	170	237	−20
HAZ	136	198	−15
DEPO	114	171	−12

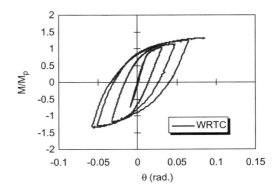

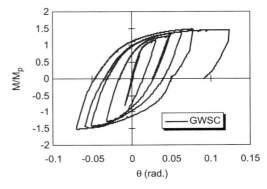

Figure 4. Hysteresis loops.

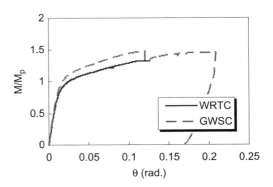

Figure 5. Moment vs. rotation skeleton curves.

Table 3. Cumulative plastic deformation factors.

Specimen	M_{max} kNm	θ_{max} 10^{-2} rad.	M_{max}/M_p	η^+	η_s^+
WRTC	300.9	8.55	1.32	17.57	7.98
GWSC	311.6	12.35	1.48	28.12	13.46

failure is shown in Figure 6. Ductile cracks initiated at tips of defects and extended for 3.4 mm. The FGSC specimen failed owing to combined local and lateral buckling of the cantilever. Ductile cracks initiated at tips of defects and extended stably, until the rotation of the cantilever reached 0.125 radians.

2.4 *FE analysis*

A finite element analysis of the specimens was carried out using the ABAQUS (2007) general-purpose finite element package. The models were constructed from 8-noded linear 3D elements. This element is nonconforming and isoparametric and employs the reduced integration technique with hourglass control. The plasticity of the material was defined by the von Mises yield criterion. The isoparametric hardening law was used for this analysis. The ABAQUS program requires the stress-strain data to be input in the form of true stress and logarithmic strain, and the stress-strain curves were transformed accordingly. The material data in the analysis were calculated from tensile coupon test results. Mesh models were generated for a half of the specimens because of symmetry in configuration.

The fatigue cracks initiated at the tips of machine cutting were generated by the nodes in the defect area on the contact surfaces between the flange or diaphragm and the weld metal. These surfaces were separated in each element as double nodes. A monotonic load was applied to the end of cantilever. The moment-rotation curves are compared with the skeleton curves that were obtained from experimental results for each specimen. The analysis results coincide well with the test results as seen in Figure 7.

Figure 8 shows the contour plot of equivalent stress around the defect when the deformation reached the final failure stage in WRTC specimen. The greatest

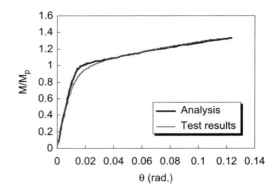

Figure 7. Moment vs. rotation skeleton curves obtained from FE analysis.

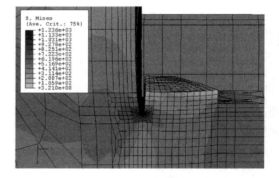

Figure 8. Contour plot of von Mises equivalent stress around the defects.

stress concentration was found at the tips of the defect.

Figure 9 shows the stress triaxiality, T_s, vs. J-integral curve. The stress triaxiality, which is related to plastic constraint, was defined using the following equation:

$$T_s = \frac{\sigma_h}{\sigma_{eq}} \quad (4)$$

T_s was taken as a peak value found below the blunted crack tips in FE analysis models. Plotted marks on each curve represent the fracture point. FEA results for SENB (simple edge notched bend) models are also plotted in this figure for reference. The fracture points show a tendency for J at fracture to be large, while the stress tiaxiality is low, which is in contradiction with SENB models showing fracture at a lower J value. For the WRTC specimen, the brittle fracture occurred at final stage of the test. The development of the J-integral is large as shown in Figure 9, so that ductile cracks extended before fracture. Crack growth increases the

Figure 6. WRTC specimen after failure.

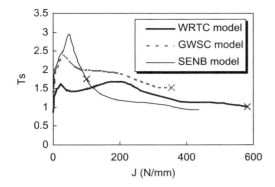

Figure 9. Stress triaxiality vs. J-integral curve.

Table 4. The seizes of defect (See Figure 2).

	a	b
	mm	mm
WRTC-1	12	–
WRTC-2	16	–
GWSC-1	12	12

chance of sampling hard particles and may lead to brittle fracture even if the stress triaxiality is low. It is therefore necessary to consider crack growth to predict stress triaxiality and J-integral more adequately.

2.5 Parameter study based on size of defects

Additional analyses were conducted as parameter study based on the defect size. Three modes, two with defects of WRTC type, designated as WRTC-1 and WRTC-2, and one with defects of GWSC type, designated as GWSC-1, were analyzed. The defect sizes of each specimen are specified in Table 4. The element types and material data in the analysis were assumed as the same as those for the models of tested specimen. The failure assessment of brittle fracture was conducted following the procedure described in Section 3.3.

3 ASSESSMENT OF WELD DEFECTS

3.1 Assessment procedure

Fracture toughness properties of the simplified beam-to-column connections specimens, which sustained brittle fractures, were assessed by using a modified FAD approach (Iwashita et al. 2003). but plastic constraint is taken into account. The same approach was applied to the numerically modeled defects to evaluate the fracture toughness properties of tested specimens and numerical models.

BS 7910 gives guidance on fracture mechanics based method for assessing the acceptability of defects in structures. Assessment is generally made by means of a FAD based on the principles of fracture mechanics. The defect is assessed by evaluating the fracture and plastic collapse parameters and plotting the corresponding point on the FAD. The vertical axis of the FAD is a ratio of the applied fracture toughness to the required fracture toughness, K_r. The horizontal axis is the ratio of the applied load to that required to cause plastic collapse, L_r. A limiting curve is plotted on the diagram. Calculations for a flaw provide the co-ordinates of an assessment point. The location of the point is compared with the limiting curve to determine the acceptability of the flaw.

The assessment procedure is given as follows:

1. Determination of assessment limiting curve
2. Calculation of plastic collapse parameters (L_r) using the ratio of moment
3. Calculation of fracture parameters (K_r) taking into account the effect of plastic constraint

3.2 Application of FAD

The assessment curve of Level 3C in the BS 7910 is applied to the FAD. L_r for the assessment curve is calculated from the following equation:

$$L_r = \frac{M}{M_p} \quad (L_r \leq L_{r,\max}) \quad (5)$$

$L_{r,\max}$ is determined using the following equation:

$$L_{r,\max} = \frac{\sigma_y + \sigma_u}{2\sigma_y} \quad (6)$$

K_r is calculated from the following equation:

$$K_r = \sqrt{\frac{J_e}{J}} \quad (7)$$

where J and J_e are the values at same applied load and obtain from FE analysis.

K_r and L_r at the fracture point of each specimen were determined by using the following equations:

$$L_r = \frac{M_{\max}}{M_p} \quad (8)$$

$$K_r = \sqrt{\frac{J_{ef}}{J_c}} \qquad (9)$$

3.3 Plastic constraint effects

The fracture toughness of a material is frequently measured by a three points bending test using SENB specimens. SENB specimens may be subjected to much greater plastic constraint at the crack tips as compared with tips of surface cracks in wide plate specimens. Therefore, critical fracture toughness for a wide plate under tensile loads is possibly under-estimated.

The stress state at crack tips resembles that of a notched bar with a volume of material surrounded by a contour of a certain value of the maximum principal stress, σ_1. The volume for the specimen with shallow notch is smaller than that for the specimen with deep notch, even if the same J value is applied to each specimen. This is an example of the fracture toughness depending on the size and geometry of the specimens. Anderson and Dodds (1991) proposed that the probability of brittle fracture may be equal for two specimens when the volumes are equal. The volume could be replaced with the area. The area in this case is the plane area at the crack tip surrounded by the contour of a certain value of σ_1. The maximum principal stress is assumed as $\sigma_1 = 3\sigma_y$.

Figure 10(a) and (b) show J-integral vs. area curves for WRTC models and for GWSC models, respectively. The vertical axis is the applied J-integral while the horizontal axis is the area surrounded by a contour of a certain value of σ_1 at the crack tip. When the critical J value, J_c, is obtained from SENB testing of a deeply notched specimen, the value of J_c can be corrected by determining the J value at which the crack tip area of the shallowly notched specimen equals that of the deeply notched specimen at J_c. The corrected J_c is called the apparent J_c. Notch toughness J_c of base metal of specimens is 88.99 N/mm, which was obtained from SENB testing. Apparent J_c were 327 J/mm for GWSC model and 307 J/mm for GWSC-1 model.

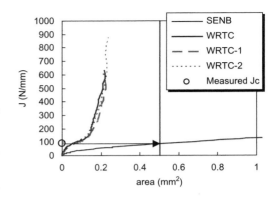

Figure 10(a). J-integral vs. area curves for WRTC model.

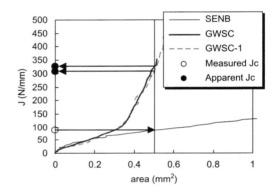

Figure 10(b). J-integral vs. area curves for GWSC model.

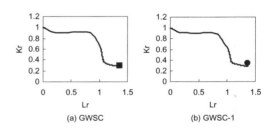

Figure 11. Failure assessment diagram for GWSC models.

3.4 Assessment

Following the procedures discussed in previous sections, the assessment of susceptibility to brittle fracture is made for numerically modeled joints with defects. Figure 11 shows the modified FAD plotting the failure point that is obtained by using the apparent J_c in the calculation of K_r.

Plastic constraint at tips of defects of WRTC specimens was much lower than that of SENB model. For the specimens and numerical models, apparent J_c could not be obtained because the area at crack tips surrounded by the contour of maximum principal stress did not reached that of the SENB model, as shown in Figure 10. The J-integral vs. area curve of each model is almost same, while crack depth is different from each other. Therefore, it is unlikely to have a brittle fracture starting from the tips of defects. These show the prediction is contrary to the test result, which the specimen sustained brittle fracture. Note that ductile crack growth is ignored in the calculations.

It was observed that ductile cracks grew slightly from the tips of defects of the GWSC specimens. The failure point for the GWSC specimen plotted on the assessment curves, while the failure point for the GWSC-1 model plotted over the assessment curve. Although this results show that brittle fractures would occur from the tips of the ductile cracks when the ductile cracks grew more, the ductile cracks grew only slowly from these defects in the testing.

4 CONCLUSIONS

A modified fracture mechanics approach was examined on two welded plate bend models and three numerical models using apparent J_c obtained by FE analyses. Deflection measurements as well as FE analysis results showed that stress sustained at the tips of the defects before cracks extended significantly were about equal in magnitude irrespectively of the defect size. Nevertheless, crack growth and failure behavior varied significantly with the location of defects. Some difficulties found in the proposed approach lies in how to evaluate the effect of ductile crack growth. To appraise this approach, the ductile tearing analysis had to be included into FAD approach. Connections and loading conditions had to be represented by reproducible numerical models. Further experimental verifications to evaluate the fracture toughness of various joints with weld defects are required to make the proposed assessment method more reliable.

APPENDIX A

The skeleton curve constructed from moment versus rotation hysteresis loops is defined here. To simplify the following description hysteresis loops are assumed to increase their rotation range incrementally with the cycle as shown in Figure A1. When the load during the 2nd cycle exceeds the peak load on the 1st cycle, the exceeding portion of the loop during the 2nd cycle is connected to the point at the peak load on the 1st cycle. The same process is repeated until the highest moment is reached. The piece-wise continuous curve thus constructed from hysteresis loops is called the skeleton curve. The skeleton curves are drawn on both positive and negative rotation sides.

The plastic component of rotation at the maximum moment, non-dimensionalized by dividing it by θ_p,

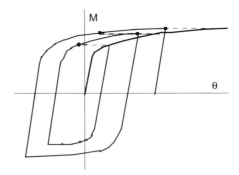

Figure A1. Definition of skeleton curve.

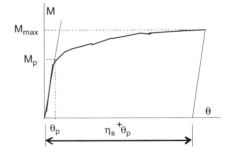

Figure A2. Definition of cumulative plastic rotation factor.

is called the cumulative plastic rotation factor and is denoted by η_s^+ (See Figure A2). The skeleton curve can be constructed also from stress versus strain hysteresis loops.

The skeleton curve for the elastic-plastic material or structure roughly coincides with the monotonic stress-strain or load-deflection curve.

ACKNOWLEDGMENT

This work was partly supported by the Japanese Society for the Promotion of Science Grant-in-Aid for Scientific Research under the number 17760464. The authors would like to thank the Japan Iron and Steel Federation for providing steel material for this test. Thanks are extended also to B.Sc. students for hard work in laboratories.

REFERENCES

AIJ Kinki. 1997. Full-scale test on plastic rotation capacity of steel wide-flange beams connected with square tube steel columns. Committee on Steel Building Structures,

Kinki Branch of Architectural Institute of Japan, Osaka, Japan. (in Japanese).

AIJ. 1996. Technical recommendations for steel construction for buildings. Part 1 guide to steel-rib fabrications. Architectural Institute of Japan, Tokyo, Japan. (in Japanese).

Azuma, K., Kurobane, Y. and Makino, Y. 2000. Cyclic testing of beam-to-column connections with weld defects and assessment of safety of numerically modeled connections from brittle fracture. Engineering Structures, Vol. 22, No. 12, Elsevier Science Ltd., pp. 1596–1608.

Azuma, K., Kurobane, Y. and Makino, Y. 2003. Full-scale testing of beam-to-column connections with partial joint penetration groove welded joints. Proceedings of the 10th International Symposium on Tubular Structures, pp. 419–427.

Azuma, K., Kurobane, Y. Iwashita, T. and Dale, K. 2006. Applicability of Partial Joint Penetration Groove Welded Joints to Beam-to-RHS Column Connections and Assessment of Safety from Brittle Fracture. Proceedings of the 11th International Symposium on Tubular Structures, pp. 611–619.

Azuma, K., Kurobane, Y. Iwashita, T. and Dale, K. 2006. Assessment of safety from brittle fracture initiating at tips of PJP groove welded joints. Proceedings of the 4th International Symposium on Steel Structures, Vol. 1, pp. 186–197.

BSI. 1997. Guidance on methods for assessing the acceptability of flaws in metallic structures. BS 7910.

ABAQUS 2007. ABAQUS v6.6 Manuals (User's Manuals I, II and III), Hibbitt, Karlsson and Sorensen, Inc.

Iwashita, T., Kurobane, Y., Azuma, K., and Makino, Y. 2003. Prediction of brittle fracture initiating at ends of CJP groove welded joints with defects: study into applicability of failure assessment diagram approach. Engineering Structures, Vol. 25, Issue 14, Elsevier Science Ltd., pp. 1815–1826.

Anderson, T.L. and Dodds, R.H., Jr. 1991. Specimen size requirements for fracture toughness testing in the ductile-brittle transition region. Journal of Testing and Evaluation,. Vol. 19(2), pp. 123–134.

NOMENCLATURE

b	material constant
E	Young's modulus
$E.L.$	elongation at failure
J	J-integral obtained by elastic-plastic analysis
J_c	critical J-integral of material obtained by SENB testing
J_e	J-integral obtained by elastic analysis
J_{ef}	J-integral at the fracture point obtained by elastic analysis
L	length between loading point and the edge of the diaphragm
M	bending moment at the edge of the diaphragm
M_{max}	maximum moment of the cantilever
M_p	full plastic moment of the cantilever
T_s	stress triaxiality
u_1, u_2, u_3	horizontal displacement
$_vE(0)$	Charpy absorbed energy at 0 (°C)
$_vE_{shelf}$	shelf energy obtained from Charpy tests
$_vTE$	energy transition temperature
η_s	cumulative total plastic rotation factor
θ	rotation of cantilever segment between loading point and the edge of the diaphragm
θ_p	beam rotation at M_p obtained from elastic stiffness
σ_h	hydrostatic stress
σ_{eq}	von Mises's equivalent stress
σ_u	tensile strength obtained from coupon tests
σ_y	yield stress of virgin steel
σ_{1-3}	principal stress

Approaches for fatigue design of tubular structures

F.R. Mashiri
School of Engineering, University of Tasmania, Australia

X.L. Zhao
Department of Civil Engineering, Monash University, Australia

P. Dong
Centre of Welded Structures Research, Battelle, USA

ABSTRACT: This paper summarizes the methods that are currently most used for the design of tubular structures subjected to fatigue loading. The methods that are mostly used for the fatigue design of tubular structures are the classification method, the hot spot stress method and the mesh-insensitive structural stress method. The classification method is considered to be the traditional method for evaluating the fatigue strength of structural details including nodal tubular connections. In the classification method, nominal stresses are used in the design. The hot spot stress method, on the other hand, uses the so called "hot spot stress". Both the hot spot stress method and the classification method have been incorporated into design standards for many years. The mesh-insensitive structural stress method on the other hand is relatively new and has been recently adopted by the 2007 ASME Div 2 and 2007 API 579/ASME FFS-1 Codes and Standards.

1 INTRODUCTION

There are various fatigue design guidelines that are used around the world for the design of tubular structures. The standards for fatigue design used in North America include the American Petroleum Institute's recommended practice (API 2005) and the American Welding Society's structural steel welding code (AWS 1996). These fatigue design standards recommend the use of both the classification method and the hot spot stress method. Eurocode 3 (EC3) Part 1.9 contains the recommendations for fatigue design of tubular connections from the European Committee for Standardization (CEN 2003). The design rules in EC3 also use both the classification and the hot spot stress method. British standards such as the Department of Energy guideline (Department of Energy 1990) and BSI 7608-1993 (BSI 1993), are also used for fatigue design of tubular structures. The Department of Energy guideline uses the hot spot stress method for the design of tubular nodal joints whereas BSI 7608-1993 uses the classification method. Fatigue design of tubular structures in Australia is covered in AS4100-1998 (SAA 1998) using the classification method. These recommendations are similar to those in EC3.

In recent years a new family of fatigue design curves has been developed for the hot spot stress method by the International Institute of Welding subcommission XV-E (IIW 2000). The design recommendations have been adopted by CIDECT (International Committee for the Development and Study of Tubular Construction) as Design Guide No.8 (Zhao et al. 2000).

The mesh-insensitive structural stress method, which has led to the development of the so called "Master S-N Curve", has also gained prominence in fatigue design. The Master S-N Curve (Dong 2005), developed by Pingsha Dong from the Centre for Welded Structures at Battelle in Ohio, USA, has received wide spread recognition. The method has been adopted in the new 2007 ASME Div 2 (ASME, 2007) and the joint 2007 ASME/API fitness for service Standards (API 2007).

This paper will summarize the different methods currently and mostly used in standards for the design of tubular structures. A discussion is given to highlight some of the advantages and disadvantages of using the classification, hot spot stress and mesh-insensitive structural stress methods. It is noted that both the classification and hot spot stress methods are heavily reliant on experimental investigation for their development and reliability. However, the fact that the hot spot stress S-N curves have been determined based on the thickness of the member undergoing failure means that fatigue design in tubular joints of any type can be carried by determining the stress concentration factors using finite element methods without the need for prototype testing.

The mesh-insensitive structural stress method on the other hand is also seen as having a great potential in the fatigue design of tubular connections because of its use of a master S-N curve for any type of detail provided a reliable finite element model can be developed for the connection (Dong, 2005 & 2008).

2 CLASSIFICATION METHOD

The classification method can be considered to be the traditional method for fatigue design. The classification method uses nominal stress in the fatigue design of tubular constructional details. The detail categories used in the classification method are based on a statistical analysis of the stress range-number of cycles (S-N) data of a given constructional detail to determine a lower bound curve (Zhao et al. 2000). The detail category also referred as the "class" corresponds to the stress range at 2 million cycles on the design curve. The design curve which is the lower bound curve of the S-N data of a given constructional detail can be defined as the mean-minus-two-standard-deviations S-N curve (Department of Energy 1990). Maximum and minimum nominal stresses are determined for a representative service stress cycle using the simple beam theory. The nominal stress for given axial force and bending moment:

$$\sigma_{nom} = P/A + M/Z \qquad (1)$$

where, P is the axial force, A is the cross sectional area, M is the bending moment, and Z is the elastic sectional modulus.

A review by Fricke (2003) on fatigue of welded joints showed that the nominal stress approach is based on the statistical evaluation of fatigue test data of different constructional details obtained in the 1970 s as reported by Gurney and Maddox (1973) and Olivier and Ritter (1979). Fricke (2003) also points out that through the International Institute of Welding commissions, international consensus was reached on a set of S-N curves and corresponding structural details were published by IIW (Hobbacher 1996).

2.1 Existing fatigue design guidelines

Fatigue design guidelines based on the classification method are given in the following standards, Australian Standard AS4100 (SAA 1998), Eurocode 3, Part 1.9 (CEN 2003), American Welding Society AWS D1.1 (AWS 1996), Japanese Society of Steel Construction (JSSC 1995), American Institute of Steel Construction (AISC 1993), International Institute of Welding (Hobbacher 1996) and the Canadian Standards Association (CSA 1989). The application of the classification method is limited for tubular connections to members, attachments and lattice girders. For lattice girders, detail categories are only available for uniplanar K- and N-joints, but parameters are very limited (Zhao et al. 2000). A large variation in fatigue behaviour may occur for joints within the same category, which may result in a considerable variation in fatigue life (van Wingerde et al. 1997). Typical tubular connection details that are covered in the classification method are shown in Figure 1. For lattice girders, curves in Figure 2 can be used for estimating fatigue life.

2.2 Limitations/advantages of the method

The classification method is limited in that it requires the determination of nominal stresses for a given joint geometry and loading mode. Both the nominal stress and loading mode can be difficult to determine in real life, where complex loading patterns sometimes exist (Dong 2005). For the design of tubular nodal joints, the use of the classification method is also limited by the specified validity range.

2.3 Procedures in existing fatigue design guidelines

Figure 1 summarizes the types of tubular connection details that are can be designed using fatigue design guidelines in existing standards using the classification method. The steps in designing tubular joints using the classification method are as follows: (a) Determine the representative service load cycle from rainflow counting analysis of stress-time histories measured from the field or determine stresses through structural analysis. If a simplified truss model with pinned connections is used, apply magnification factors to account for secondary stresses in lattice girders, (b) Use the connection under consideration to select a corresponding detail category, (c) Use the S-N curve corresponding to the appropriate detail category and constructional detail to estimate the fatigue life at the service stress range. Apply partial safety factors to the fatigue life predicted.

3 HOT SPOT STRESS METHOD

In the hot spot stress method, the so called "hot spot stress" is used in the design. The hot spot stress can be defined as the "structural stress" which is the stress at weld toe including only structural stress concentration but not including local stress concentration due

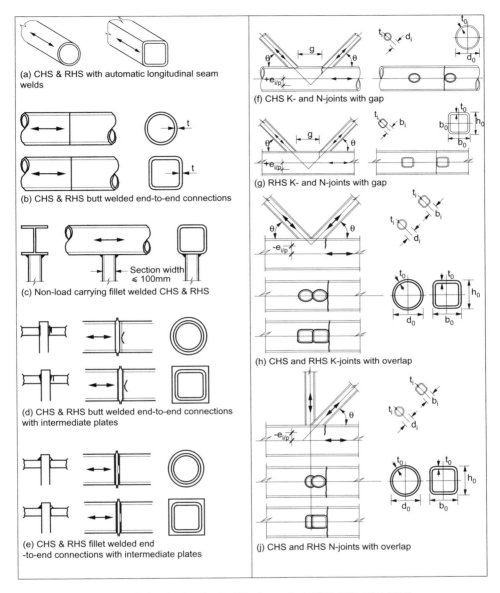

Figure 1. Types of Connections designed using the classification method (CEN 2003, SAA 1998).

to the weld bead (Nihei et al. 1997). The fatigue life of the joint in this method is related to the maximum (hot spot) geometrical stress occurring in the joint. This is the location where the cracks usually initiate leading to propagation and final fracture of the joint. The hot spot stress can be determined mathematically by multiplying a nominal stress by the relevant stress concentration factor (SCF). The stress concentration factor (SCF) is the ratio between the hot spot stress at the joint and the nominal stress in the member due to a basic member load which causes this hot spot stress (IIW 2000). Stress concentration factors are determined experimentally or numerically around the weld toes of a joint. A variation in stiffness around a joint causes differences in SCFs.

The hot spot stress method has been studied by the following researchers, among others, resulting in the development of parametric equations for stress

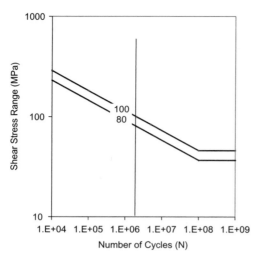

Figure 2. S-N curves for shear stress ranges (CEN 2003).

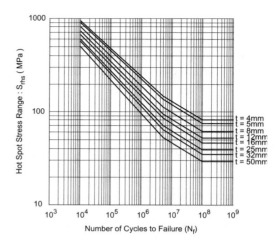

Figure 3. Fatigue strength curves for according to the hot spot stress method.

concentration based on non-dimensional parameters and the loading mode: (a) Efthymiou and Durkin (1985): SCFs for T- or Y-joints and gap and overlap K-joints made up of circular hollow sections, (b) van Wingerde (1992): SCFs for T- and Y- joints made up of square hollow sections, (c) Romeijn (1994): SCFs for multiplanar K-joints made up of circular hollow sections, and (d) Panjeh Shahi (1994): SCFs for multiplanar K-joints made up of square hollow sections.

3.1 Existing fatigue design guidelines

The hot spot stress method has been adopted by recent standards such as CIDECT Design Guide No. 8 (Zhao et al. 2000) and IIW Subcommision XV-E recommendations (IIW 2000). Other standards such as EC3 Part 1.9 also propose the hot spot stress method in plated details, for toes of butt welds, fillet welded attachments and fillet welds in cruciform joints (CEN 2003). Other older standards such the Department of Energy (1990) also proposed the use of the hot spot stress method using the T curve for fatigue design of tubular joints. The hot spot stress S-N curves that are currently used for the design of tubular nodal joints in CIDECT Design Guide No. 8 are shown in Figure 3. Figure 3 shows the dependence of fatigue life on tube wall thickness the hot spot stress is used for design. Figure 4 shows the types of tubular connections details that are currently covered in IIW (2000) and CIDECT Design Guide No. 8 (Zhao et al. 2000). These details include both uniplanar and multiplanar nodal joints.

3.2 Limitations/advantages of the method

There are limitations to the use of the hot spot stress method for design which is apparent in CIDECT Design Guide No. 8 (Zhao et al. 2000). At present this design guide, for example, does not have guidance for the fatigue design of overlapped CHS K-joints and overlapped KK-joints. CIDECT Design Guide No. 8 also does not cover out-of-plane bending in RHS T-/X-, K(gap)-, K(overlap)- and KK(gap)-joints among others. The use of the parametric equations for determining SCFs is also limited to a given validity range for each type of specimen.

3.3 Procedures in existing fatigue design guidelines

The following procedure is used for fatigue design of tubular joints for the hot spot stress method: (a) Determine the nominal stress range from analysis or from rainflow counting of stress-time histories for the appropriate basic loads (axial, in-plane bending, out-of-plane bending) for the brace and chord weld toes, (b) Determine the stress concentration factors for "hot spots" around the welded interface between the brace(s) and the chord(s) where cracking is likely to happen for the appropriate basic loads, (c) Determine the total hot spot stresses at "hot spot" locations around the welded interface, (d) Determine the maximum hot spot stress in the chord and brace at the different hot spots of the welded connection, (e) Determine the corresponding fatigue life in the chord and brace member using design curves for the appropriate tube wall thickness.

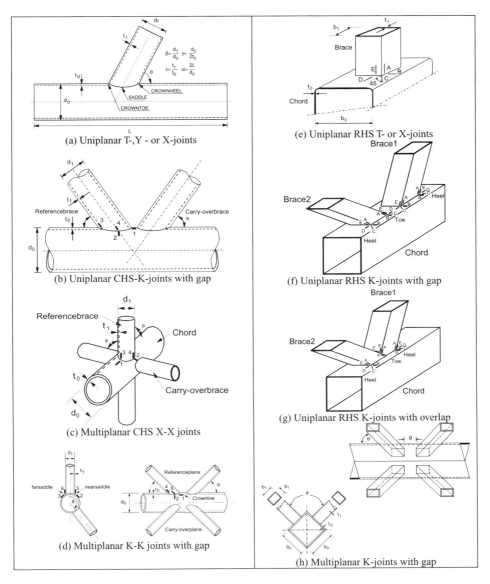

Figure 4. Types of joints designed using the hot spot stress method (IIW 2000, Zhao et al. 2000).

4 MESH-INSENSITIVE STRUCTURAL STRESS METHOD

The structural stress definition used by Dong (2001) was based on an equilibrium argument and implemented numerically by taking advantage of internal nodal force definitions in displacement-based finite element methods. Although the term "structural stress" has been used for many years, e.g., earlier work by Radaj (1990) and others within IIW community, their calculation methods have been based surface extrapolation procedures from a series of pre-determined locations. In the mesh-insensitive structural stress method, a stress state at a fatigue-prone location, e.g., at a weld toe (Fig. 5a), the normal structural stress is presented in the form of membrane and bending components that satisfy equilibrium conditions (Fig. 5b) along a hypothetical cut. Both the transverse shear (Fig. 5) and in-plane shear (perpendicular to the paper, not present in the 2D problem)

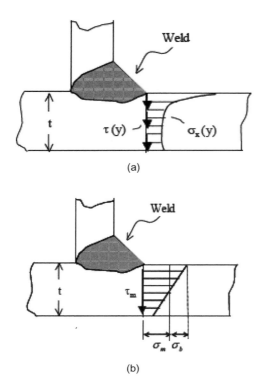

Figure 5. (a) Local through thickness stress, (b) Equilibrium-equivalent membrane and bending stresses (Dong 2001).

at the same cut can be represented in the same way (Dong, 2005 & 2008). The imposition of the equilibrium arguments as embodied by using nodal forces minimizes the mesh sensitivity in structural stress calculations. Healy (2004) showed that this method is reliable in its determination of stress concentration factors and its reliability is not affected by element type as well as element size for weld lines (along weld toe) that are continuous.

The structural stress, σ_s at a location of interest such as the weld toe (Dong et al. 2005):

$$\sigma_s = \sigma_m + \sigma_b \qquad (2)$$

where σ_m and σ_b is the membrane component and bending representation of actual local through-thickness stress distribution, shown in Figure 5(a).

The structural stress method shows that the structural stress SCF remains essentially the same for structural stress distributions in finite element models of relatively refined mesh say, $0.25t \times 0.25t$ to a very coarse mesh, say $1.0t \times 1.0t$.

Dong (2001) pointed out that nominal stresses and SCFs can be problematic to determine from FE models because of their dependency on mesh sensitivity. Dong (2001) stressed that dependency on mesh sensitivity means that there is a need to calibrate FE results with the experimental results to ensure that FE models capture what happens in the real tests. It can also be difficult to select an appropriate S-N curve for design because the determination of fatigue design curves is based on the testing a constructional detail with a specific geometry and loading mode. In the hot spot stress method, there are also issues that relate to the uncertainties in extrapolation especially in thin-walled joints (Dong 2001). Based on this background, Dong (2001) proposed that to improve fatigue design using S-N curves for both the classification and hot spot stress methods, the stress parameter must be mesh insensitive in finite element solutions and must have the ability to differentiate stress concentration effects in different joints.

4.1 *Existing fatigue design guidelines*

Based on the mesh-insensitive structural stress method, a master S-N curve has been produced based on the equivalent structural stress parameter, ΔS_s, which reflects the effects of the stress concentration ($\Delta\sigma_s$), the thickness (t) and loading mode (r) (Dong et al. 2005):

$$\Delta S_s = \frac{\Delta\sigma_s}{t^{*\frac{2-m}{2m}} \cdot I(r)^{\frac{1}{m}}} \qquad (3)$$

where t^* is a relative thickness with respect to a unit thickness, say 1 mm (i.e. $t^* = t/1$ mm), $I(r)$ is a dimensionless function of bending ratio, $r = \sigma_b/\sigma_s$. $I(r)$ is obtained by integrating the two-stage crack growth law expressed by the structural stress based stress intensity factor solutions. $I(r)$ also depends on whether crack aspect ratios in elliptical cracks approximate load-controlled or displacement controlled conditions as shown in Figure 6. The crack propagation exponent in the conventional Paris Law, m = 3.6. $\Delta\sigma_s$ is the structural stress range.

The master S-N curve shown in Figure 7 was first reported by Dong et al. (2005) and has been recently adopted by ASME (ASME, 2007) and API A579 (API 2007). The database supporting the master S-N curve contain over 1000 tests from plate joints, plate to tube joints, and tubular joints. The design master curve is a mean-minus-two standards deviation S-N curve (Dong et al. 2005) as follows:

$$\Delta S_s = C \times N^h \qquad (4)$$

where h is the slope of the S-N curve in the log-log scale, $h = -0.32$, C is a constant equal to 13875.8.

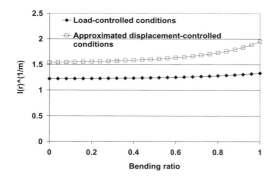

Figure 6. I(r) as a function of bending ratio for displacement and load controlled conditions (Dong et al. 2005).

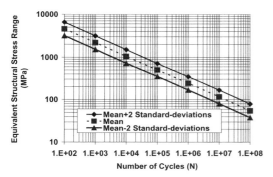

Figure 7. Master S-N curve (Dong et al. 2005).

5 DISCUSSION

5.1 Classification method

The S-N curves for the classification method are determined from the statistical analysis of tested specimens. According to EC3 Part 1.9 (CEN 2003) at least 10 specimens are required as a basis for determining a reliable S-N curve. This means that the development of S-N curves is a relatively expensive exercise. In the hot spot stress method, S-N curves for a variety of connections have been shown to depend on the wall thickness of the tube undergoing fatigue failure allowing different connections details to be designed using one S-N curve for a given thickness. In contrast, the classification method requires, the identification of an existing constructional detail in the standards that is comparable a detail of interest if its fatigue strength is to be estimated.

Design of tubular connections in the classification method, is therefore heavily depended on tested details and the validity ranges used in the tests.

At present the classification method in standards such AS4100 (SAA 1998) and EC3 Part 1.9 (CEN 2003) cover CHS and RHS K- and N-joints but there are no guidelines for dealing with T- and multiplanar joints. For multiplanar joints, this means that in the classification method, judgements have to be made at the discretion of the design engineer on how to subdivide a multiplanar joint into uniplanar components whose details are currently covered in the standards. AWS (1996) Annex L provides an ovalizing parameter which modifies the normal punching shear to account for these considerations in CHS.

5.2 Hot spot stress method

Although the use of the hot spot stress method also depends on the fatigue testing of connection details, it is the knowledge of stress concentration factors that is more important. The S-N curves that can be used for different connection types are the same but depended on the thickness of the tube in which fatigue failure occurs. This means that as long as the stress concentration factors can be determined, for example, through finite element modelling or on physical experimental models then fatigue life of the connection can be determined as long the tube thickness under consideration falls within the range of thicknesses that are currently covered by standards such as CIDECT Design Guide No. 8 (Zhao et al. 2000) and IIW (2000). At present hot spot stress fatigue design curves cover the following thicknesses, (a) for CHS joints, 4 mm $\leq t \leq$ 50 mm and (b) for RHS joints, 4 mm $\leq t \leq$ 16 mm, see Figure 3.

At present CHS K-joints with verlap, for example, are not covered by the standards that employ the hot spot stress method such as CIDECT Design Guide No. 8 (Zhao et al. 2000) and IIW (2000). AWS (2008) proposals would provide coverage where T × T or finer FEM is employed.

5.3 Mesh-Insensitive structural stress method

The mesh-insensitive structural stress method is a relatively new method that was proposed by Dong (2001).

As demonstrated by Dong et al. (2005), the accuracy of the master S-N curve which has been derived based on the mesh-insensitive structural stress method, depends on the amount of tests that are included in statistical determination of the design curve. Dong et al. (2005) showed that the lower bound S-N curve can shift depending on the amount of data included in the analysis. This is not unique to the master S-N curve but is an inherent characteristic of S-N curves determined from fatigue test data using statistical methods.

Doerk et al. (2003) pointed out that the use of structural stress is restricted to fatigue strength analysis of weld toes, where cracks start from the surface of the structure. The disadvantage of the structural stress method is that it is not suitable for cracks propagating from the weld roots. However, the above statement is not true. The same structural stress definition shown in Figure 5(b) can be directly applied to weld throat cracking as demonstrated by Dong (2001) and is being currently used in both the 2007 ASME and API A579 Codes (ASME 2007 and API 2007).

Analyses carried out by Doerk et al. (2003) concluded that for 2D problems, the structural stress that is evaluated at the weld toe remained the same regardless of the mesh refinement. Differences in structural stresses were however observed in 3D models. These differences were thought to be due to the fact that stresses at the transverse element sides were not taken into account in the equilibrium equations. It must be noted that the observations by Doerk et al. (2003) was based on their calculations using the integration of stresses, rather than using nodal forces which is the key essence of the mesh-insensitive structural stress method. The same example (a circular doubling plate) investigated by Doerk et al. (2003) was analysed by Dong et al. (2005), demonstrating a remarkable mesh-insensitivity for all mesh sizes investigated.

Marshall and Wardenier (2005) pointed out that despite some concerns about the Battelle Structural Stress method in its treatment of 3D corner singularities, it has the potential to handle both tubular and non-tubular applications. A large number of calculation examples for tubular joints including K joints with overlap and T-joints with internal stiffening have been investigated by using the mesh-insensitive structural stress method, as reported in Dong and Hong (2008). In addition to the mesh-insensitivity in the structural stress calculations, all these tubular joint tests are shown within plus/minus two standards derivations of the master S-N curve in Fig. 7.

Future research is needed to compare the pre-dicted fatigue life using the three approaches de-scribed in this paper.

REFERENCES

AISC 1993: Load and resistance factor design specification for structural steel buildings, American Institute of Steel Construction, Chicago, USA.

API 2005, "Recommended practice for planning, designing and constructing fixed offshore platforms", API Recommended Practice 2 A (RP 2 A), 21 st Edition, Washington, USA.

API 2007: API 579-1/ASME FFS-1, Fitness for Service Codes and Standards, August 2007, American Petroleum Institute, Houston, Texas, USA.

ASME 2007: Boiler and Pressure Vessel Code Sec. VIII Div 2 (2007), New York, USA.

AWS 1996, Structural Welding Code-Steel, ANSI/AWS D1.1–96, American Welding Society, Miami, USA.

AWS 2008: American Welding Society D1A Design Task Group proposed fatigue revisions to D1.1.

BSI 1993: *Code of Practice for Fatigue design and assessment of steel structures*, BS 7608: 1993, British Standards Institution, London, UK.

CSA 1989: Limit State Design of Steel Structures, CAN/CSA-S16.1-M89, Canadian Standards Association, Ontario.

CEN 2003, Eurocode 3, Design of Steel Structures-Part 1.9: Fatigue, PrEN 1993-1-9, European Committee for Standardisation.

Department of Energy, 1990, " Offshore Installations: Guidance on design, construction and certification", Fourth Edition, London, HMSO, UK.

Doerk O., Fricke W. and Weissborn C. 2003, "Comparison of different calculation methods for structural stresses at welded joints", Int. Journal of Fatigue, Vol. 25, pp. 359–369.

Dong P. 2001, "A structural stress definition and numerical implementation for fatigue analysis of welded joints", International Journal of Fatigue, Vol. 23, No. 10, pp. 865–876.

Dong P., Hong J.K., Osage D., and Prager M. 2002, "Master S-N curve approach for fatigue evaluation of welded components", WRC Bulletin No. 474, New York, USA.

Dong P. 2005, "A robust structural stress method for fatigue analysis of offshore/marine structures", J. of Offshore Mechanics and Arctic Eng. February 2005, Vol. 127, pp. 68–73.

Dong P. Hong J.K. and De Jesus A.M.P. 2005, "Analysis of recent fatigue data using the structural stress procedure in ASME Div. 2 Rewrite", Proc. of PVP2005, Paper No. PVP2005-71711, 17–21/07/2005, Denver, Colorado, USA.

Efthymiou M. and Durkin S. 1985, "Stress concentrations in T/Y and gap/overlap K-joints", *Proc. Int. Conf. on Behaviour of Offshore Struc.*, Elsevier, pp. 429–441.

Fricke W. 2003, "Fatigue analysis of welded joints: state of development", Marine Structures, Vol. 16, pp.185–200.

Gurney T.R. and Maddox S.J. 1973, "A re-analysis of fatigue data for welded joints in steel", WRC, 3(4), pp. 1–54.

Healy B. 2004, "A case study comparison of surfaceextrapolation and Battelle Structural Stress methodologies", Paper No. OMAE 2004-51228, Proc. of the 23rd OMAE International Conference, June 2004, Vancouver, Canada.

Hobbacher A. 1996, "Recommendations for fatigue strength of welded components", Cambridge: Abington Publishers.

IIW 2000: Fatigue Design Procedures for Welded Hollow Section Joints, IIW Doc. XV-1035-99, Eds. Zhao & Packer, Abington Publishing, Cambridge, UK.

Japanese Society of Steel Construction (JSSC) 1995: *Fatigue Design Recommendations for Steel Structures*. Japan.

Nihei K, Inamura F. and Koe S. 1997, "Study on Hot Spot Stress for Fatigue Strength Assessment of Fillet Welded Structure" Proc. 7th Int. Offshore and Polar Eng. Conf. Vol. IV, Honolulu, USA, May 25–30, 1997.

Marshall P.W. and Wardenier J. 2005, "Tubular vs Nontubular Hot Spot Stress Methods", Proc. 15th Int. Offshore and Polar Eng. Conf., Seoul, Korea, 19–24 June 2005, pp. 254–263.

Olivier R. and Ritter D. 1979, "Catalogue of S-N curves of welded joints in structural steel", Vol. 1–5, Report 56, Dusseldorf, DVS-Verlag.

Panjeh Shahi E. 1994, "Stress and strain concentration factors of welded multiplanar joints between square hollow sections", PhD Thesis, Delft University Press, Netherlands.

Romeijn A. 1994: Stress and strain concentration factors of welded multiplanar tubular joints, PhD Thesis, Delft University Press, Delft, The Netherlands.

Radaj D. 1990, "Design and analysis of fatigue resistant welded structures", Abington Publishers 1990.

SAA 1998: *Steel Structures, Australian Standard AS 4100-1998*, Standards Association of Australia, Sydney, Australia.

van Wingerde A.M. 1992: The fatigue behaviour of T- and X- joints made of SHS, *Heron*, Vol. 37, No. 2, pp. 1–180.

van Wingerde A.M., Packer J.A. and Wardenier J. 1997c, "IIW fatigue rules for tubular joints", IIW Int. Conf. on Performance of Dynamically Loaded Welded Structures, July 1997, San Francisco, USA, pp. 98–107.

Zhao X.L., Herion S., Packer J.A., Puthli R., Sedlacek G., Wardenier J., Weynand K., Wingerde A., and Yeomans N. 2000, "Design Guide for Circular and Rectangular Hollow Section Joints under Fatigue Loading", Verlag TUV Rheinland, Koln, Germany.

Tests of welded cross-beams under fatigue loading

F.R. Mashiri
School of Engineering, University of Tasmania, Australia

X.L. Zhao
Department of Civil Engineering, Monash University, Australia

ABSTRACT: This paper summarises the tests performed on welded cross beams joints made up of Rectangular Hollow Sections (RHS), channel and angle sections. Four different types of connections were tested: RHS-RHS, RHS-Angle, RHS-Channel and Channel-Channel connections. The specimens were subjected to constant stress amplitude cyclic load. The basic load that was applied to the specimens is cyclic bending load in the bottom member. A total of 35 specimens were tested among the RHS-RHS, RHS-Angle, RHS-Channel and Channel-Channel cross-beam connections. The paper summarises the failure modes, SCFs and resultant fatigue S-N data obtained. Existing fatigue design S-N curves are recommended based on the deterministic method for the classification method.

1 INTRODUCTION

Cold-formed high strength tubes and open sections have become readily available in steel markets around the world including in Australia (Hancock 1999). This has led to the study of cold-formed steel connections and members under static loading (Zhao 1993, Zhao & Hancock 1995). Some of this research has now been incorporated into Australian Standards.

To compliment previous research on static strength, research has been undertaken to understand the behaviour of cold-formed steel structural connections under fatigue loading (Mashiri 2001). This research concentrated on the fatigue behaviour of tubular nodal joints of circular hollow section and square hollow sections as well as tube-to-plate joints that are typical in trussed structures (Mashiri et al. 2002a, 2002b, 2004). This research was driven by the potential application of cold-formed tubular and tube-to-plate joints in the manufacture of structural systems subjected to cyclic loading in service such as communication towers, and tower cranes. There is also potential use in the road transport and agricultural industry in the manufacture of trailers, swing ploughs and haymakers.

For fatigue design of welded steel structures, there is also a lack of fatigue design rules in existing standards (SAA 1998, Zhao et al. 2000, IIW 2000) for structural details made up of cold-formed sections. This is because cold-formed sections are relatively thin-walled, with some having wall thicknesses less than 4 mm. Most of the early research on fatigue of welded connections focused on relatively thick-walled joints that were applicable in offshore structures (Marshall 1992).

In this paper, the so called 'cross-beam' connections have been investigated under fatigue loading. Cross-beam connections are made up through the welding of one member on top of another as shown in Figure 1. This connection detail is suitable for structural systems in undercarriages of motor vehicles and trailers.

Four different cross beam connection details were tested. The constructional details were made of RHS-RHS, RHS-Angle, RHS-Channel and Channel-Channel cross-beam connections. The welded connections were tested under constant stress amplitude cyclic loading to failure. A total of 35 specimens tested.

The failure modes are described. The cracking patterns observed in the different cross-beam connection details are supported by the stress concentration factors that were determined based on the measurement of strains at hot spot locations. Fatigue test data was also obtained based on the nominal applied stress range and the number of cycles to failure.

The fatigue test data is compared to the family of fatigue design curves in the existing standard for cold-formed steel structures (SAA 2005). Fatigue design of the cross beams is recommended based on the deterministic method and the classification method.

2 SPECIMENS

The specimens were manufactured from cold-famed high strength steel of grade C350LO and C450LO (SAA 1991). The hollow sections and open sections used were in-line galvanized DuraGal sections of Grade 350LO and C450LO. Grade C350LO steel has a specified minimum yield strength of 350 MPa and an ultimate tensile strength of 430 MPa. Grade C450LO on the other hand, has a specified minimum yield strength and ultimate tensile strength of 450 MPa and 500 MPa respectively (SAA 1991). The measured yield stress ($f_{y,m}$) and the measured ultimate tensile strength ($f_{u,m}$) of the steel sections used in this investigation are given in Table 1.

The cross-beams consist of a top member welded onto a bottom member as shown in Figure 1. The specimens that were manufactured and tested in this investigation are: (a) RHS-RHS cross beams, see Figure 2, (b) RHS-Angle cross-beams, see Figure 3, (c) RHS-Channel cross-beams, see Figure 4 and (d) Channel-Channel cross-beams, see Figure 5.

The gas metal arc welding method was used to join the bottom member of the cross beam to the top member. A single run fillet weld was used. Note that the resultant welds produced on these thin-walled joints are oversized. This is due to the fact that it is practically impossible to deposit weld metal in proportion to the weld thickness as the steel sections become thinner using manual metal arc welding methods. Similar observations were reported by Zhao and Hancock (1995) and Mashiri and Zhao (2004). The fact that oversized welds are produced in the welding of thin-walled joints means that the weld toes are located further away from the top and bottom member junction, resulting in relatively lower stresses at the weld toes (Mashiri et al. 2004).

Hardness tests were performed on the macros of the welded connections to check compliance with the Australian standards for welding AS1554-1 and its 1554.5 (SAA 2004a, 2004b). The hardness tests were reported by Mashiri and Zhao (2005). The hardness comparison test between the weld metal and the parent metal is designed to ensure that when failure occurs it is not concentrated in parent metal heat affected zone (HAZ) (WTIA 1998). The difference in hardness between the weld metal and parent metal is recommended to be less than 100HV10. Another hardness measurement that is carried out is that of the heat affected zone and is designed to ensure that there is less likelihood of heat affected cracking occurring. There is less likelihood of heat affected zone cracking when the value of the Vickers' hardness in the heat affected zone is less than 350HV10 (WTIA 1998).

The nature of the cross-beam connections is that it is unavoidable to locate start-stop positions at the corners of the connected sections since one section sits on top of another, see Figures 2 to 5.

For this investigation the validity range of the specimens that were tested are shown in Table 1.

Table 1. Validity range of specimens.

Connection	Top member $d_1 \times b_1 \times t_1$ (mm)	$f_{y,m}$ (MPa)	$f_{u,m}$ (MPa)	Bottom member $d_0 \times b_0 \times t_0$ (mm)	$f_{y,m}$ (MPa)	$f_{u,m}$ (MPa)	Validity range
RHS-RHS	75 × 50 × 3RHS	459	533	50 × 50 × 3SHS	477	531	$8.3 \leq b_0/t_0 \leq 21.9$
							$0.5 \leq t_0/t_1 \leq 1.0$
				50 × 50 × 1.6SHS	401	438	$1 \leq d_0/b_0 \leq 2$
RHS-Angle	100 × 50 × 4CC	453	554	35 × 35 × 3SHS	533	594	$8.3 \leq b_0/t_0 \leq 21.9$
							$0.4 \leq t_0/t_1 \leq 0.8$
				35 × 35 × 1.6SHS	492	545	$1 \leq d_0/b_0 \leq 2$
RHS-Channel	75 × 75 × 4CA	469	549	50 × 25 × 3SHS	519	576	$8.3 \leq b_0/t_0 \leq 21.9$
							$0.4 \leq t_0/t_1 \leq 0.8$
				50 × 25 × 3SHS	398	430	$1 \leq d_0/b_0 \leq 2$
Channel-Channel	100 × 50 × 4CC	453	554	100 × 50 × 4CC	453	554	$b_0/t_0 = 13.2$
							$t_0/t_1 = 1.0$
							$d_0/b_0 = 2$

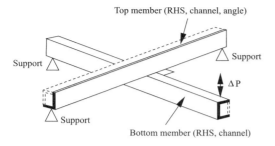

Figure 1. Schematic diagram of cross-beam connections.

Figure 2. RHS-RHS cross-beam.

Figure 3. RHS-Angle cross-beam.

Figure 4. RHS-Channel cross-beam.

Figure 5. Channel-Channel cross-beam.

The validity range is based on the dimensions of the bottom member with subscript 0 and those of the top member with subscript 1 as shown in Figure 1. In the cross beam specimens, b_0 and b_1 are the widths of the bottom and top members respectively, d_0 and d_1 are the depths of the bottom and top members respectively and t_0 and t_1 are the thicknesses of the bottom and top members respectively.

3 FATIGUE TESTS AND FAILURE MODES

Fatigue tests of the cross-beam specimens were carried out using a multiple fatigue testing rig (Mashiri 2001). The test rig has capabilities to apply constant

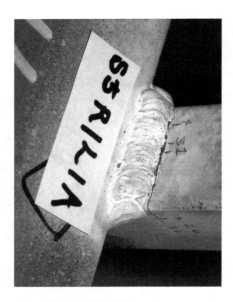

Figure 6. Failure in RHS-RHS cross-beam.

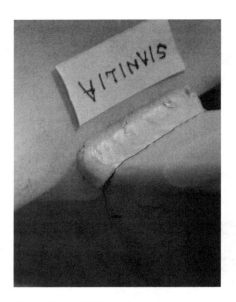

Figure 7. Failure in RHS-Angle cross-beam.

stress amplitude loading, to perform cycle counting of applied stress cycles and to detect failure through brake detectors.

The load that was applied to the specimens was bending moment through the bottom member as shown in Figure 1. The bending moment was applied as constant stress amplitude cyclic loading.

For the RHS-RHS, RHS-Angle, RHS-Channel cross-beams failure occurred through crack initiation and propagation at weld toes in the RHS bottom member. This failure is referred to as bottom member cracking (BMC) (Mashiri & Zhao 2005, 2006, 2007). Figure 6 shows failure in an RHS-RHS cross-beam specimen. Figure 7 shows failure in an RHS-Angle cross-beam. Figure 8 shows failure in an RHS-Channel cross-beam. Crack initiation in RHS-RHS, RHS-Angle and RHS-Channel cross-beams occurred at the weld toes around the corners of the bottom RHS member. This was followed by cracks propagating from the corners to the mid width location of the bottom RHS member resulting in a crack along the full width of the bottom member. A definition for end of test failure similar to that adopted by van Wingerde (1992) was adopted for this investigation. Failure was defined as the number of cycles corresponding to a crack length equal to the width of the bottom member plus twice the weld leg length in the bottom member. Crack initiation for the RHS-RHS, RHS-Angle, RHS-Channel started at the hot spot locations with the highest stress concentration factors. Details of the determination of experimental stress concentration factors are given in a later section.

Figure 8. Failure in RHS-Channel cross-beam.

For the Channel-Channel cross-beams, failure in the majority of specimens tested occurred through crack initiation and propagation at weld toes in the top member. This is refereed to as top member cracking (TMC) (Mashiri & Zhao 2008), see Figure 9.

Figure 9. Top member cracking (TMC) in channel-channel cross-beams.

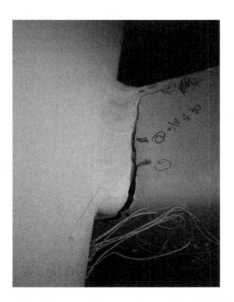

Figure 10. Bottom and top member cracking (BTMC) in channel-channel cross-beams.

Crack initiation in the 9 out of 10 Channel-Channel cross beams tested occurred at the weld toes around the corner of the Channel top member, as shown in Figure 9. However, in one of the Channel-Channel specimens tested, failure occurred through cracking at the bottom and top member weld toes. This can be referred to as bottom and top member cracking (BTMC) (Mashiri & Zhao 2008), see Figure 10. Subsequent propagation of the cracks away from the corners resulted in failure of the Channel-to-Channel cross-beams. Crack initiation occurred at the hot spot location with the highest stress concentration factors. Details of the measurement of strains and determination of stress concentration factors in Channel-Channel cross-beams are given later.

4 STRESS CONCENTRATIONS AND FATIGUE CRACKING

Under static loading conditions, strains were measured at hot spot locations at weld toes in the bottom and top members using strip strain gauges. The strip strain gauges were located at the weld toes and placed within the extrapolation region for hot spot stresses as recommended by CIDECT Design Guide No. 8 (Zhao et al. 2000) and IIW (2000). CIDECT Design Guide No. 8 (Zhao et al. 2000), recommends that the first point of extrapolation be located at a distance equal to $0.4t$ from the weld toe but not less than 4 mm. The second point of extrapolation is at a distance $1.0t$ further away from the first point of extrapolation for RHS connections, where t is the tube wall thickness. Similar distances for extrapolation were used at corners of open sections. Although open sections are not included in the recommendation for extrapolation distances in CIDECT Design Guide No. 8, the fact that stress concentrations are due to a similar change in direction of the section at the corners means that this should be a reasonable first approximation.

The strip strain gauges used had five strain sensitive strips located at 2 mm centres. Figure 11 shows the strip strain gauges installed on an RHS-Angle cross-beam. The strains were converted to stresses and used to determine the stress gradient in the determination of hot spot stresses through extrapolation to the weld toe. Quadratic extrapolation was used since the hot spot locations are at corners of members such as RHS, Angle and Channel. CIDECT Design Guide recommends quadratic extrapolation at hot spot locations in RHS connections and linear extrapolation of weld toes in CHS members (Zhao et al. 2000).

The specimens used in the determination of stress concentration factors were subjected to static loading at load levels within the elastic response range of the connection. Four or more different load levels were used to determine the strains and hence stress distribution at weld toes. The resultant stress distributions were used to determine the hot spot stresses at weld toe as shown in Figure 12. The load applied at the four different load levels was within the elastic response

Figure 11. Strain gauge location on cross-beam connections.

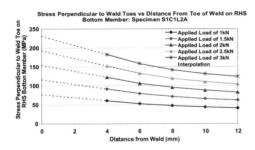

Figure 12. Typical stress distribution and resultant hot spot stresses at weld toe.

Table 2. Stress concentration factors.

Connection	Specimen	SCF_{bm}*	SCF_{tm}**
RHS-RHS	S1R1 L2 A	1.5	1.0
RHS-Angle	S2 AN1 L1 A	1.8	–
RHS-Channel	S1C1 L2 A	1.4	0.2
Channel-Channel	C1C1 L1B	1.3	1.6

* SCF_{bm} = stress concentration factor in bottom member.
** SCF_{tm} = stress concentration factor in top member.

limit of the connection resulting in the stress concentration factors of equal magnitude at the four different loads.

The stress concentration factors at the hot spot locations were defined as the ratio of the hot spot stress to nominal stress due to the basic load causing that hot spot stress (Zhao et al. 2000). For this investigation, the basic load is the "bending moment in the bottom member".

Stress concentration factors were determined at weld toes in both the bottom and vertical members. The resultant stress concentration factors for the bottom member (SCF_{bm}) and top member are (SCF_{tm}) are shown in Table 2.

Figures 6, 7 and 8 show the cracking that occurred in RHS-RHS, RHS-Angle and RHS-Channel cross beams. All the cracking in these connections occurred in the RHS bottom member in which the bending moment was applied. Table 2 shows a summary of the SCFs in both the bottom and top members. Table 2 shows that for the RHS-RHS and RHS-Channel cross beams, where SCFs were determined in both the top and bottom members, the maximum stress concentration occurred in the bottom member. This is consistent with the crack initiation that was observed in RHS-RHS and RHS-Channel cross-beams. Table 2 also shows that the SCFs in the top member of RHS-RHS and RHS-Channel cross-beams are significantly smaller than that in the bottom member, confirming the reason why no cracks were observed in the top member.

For the Channel-Channel cross beams, the maximum SCF was determined in the top member. Figure 9 shows that crack initiation and propagation occurred at the weld toes in the top member for the Channel-Channel cross-beams. This was the predominant mode of failure for Channel-Channel cross-beams. However, out of a total of 10 Channel-Channel specimens tested, one specimen C1C1 L1C failed through cracking at weld toes in both the bottom and top members (Mashiri & Zhao 2008) as shown in Figure 10. Table 2 shows that although the maximum stress concentration factor was found in the top member, the SCF in the bottom member is of a similar magnitude but slightly smaller. This means that after the top member has cracked there is a possibility of stress redistribution resulting in bottom member cracking. Alternatively, a comparatively shorter weld leg length in the bottom member could cause the notch stress at the weld toes in the bottom member to become larger than that in the top member, resulting in cracking occurring at the weld toes in the bottom member for Channel-Channel cross-beams.

The measured SCFs in this investigation can be used in future analysis of specimen fatigue data to produce fatigue design curves in the hot spot stress method. The magnitude of SCFs at weld toes in the top and bottom members were all less than 2.0. Fatigue design standards such as CIDECT Design Guide No. 8 (Zhao et al. 2000) and IIW (2000) recommend the use of a minimum SCF of 2.0. Therefore a stress concentration factor of 2.0 can be recommended for cross-beam connections under the load cyclic bending in the bottom member.

5 FATIGUE TEST DATA AND RECOMMENDATIONS FOR DESIGN

Fatigue tests were carried out under constant stress amplitude cyclic loading. For each test, nominal stress range ($S_{r\text{-nom}}$) and the number of cycles to failure (N) were recorded.

Ten (10) RHS-RHS, 7 RHS-Angle, 8 RHS-Channel and 10 Channel-Channel cross-beam connections were tested in this investigation. The fatigue test data for the different types of specimens is shown in Figure 13.

Eurocode 3 Part 1.9 (CEN 2003) recommends that at least 10 data points must be used in the determination of a design S-N curve.

It can be shown that RHS-RHS, RHS-Angle and RHS-Channel cross beam data fall within the same scatter band. Most of the RHS-RHS, RHS-Angle and RHS-Channel cross-beam fatigue data lie above the Class 55 curve except for one specimen.

For the Channel-Channel cross beams, all the fatigue test data expressed in terms of the classification method, lies above the Class 36, as shown in Figure 13.

The slope of the design S-N curves in Figure 13 are as follows: (a) 1 in 3 for fatigue life between 10^5 and 5×10^6 cycles (a region determined through constant stress amplitude loading) and (b) 1 in 5 for fatigue life between 5×10^6 cycles and 10^8 cycles (a region applicable to variable amplitude only).

5.1 Design for RHS-RHS, RHS-Angle and RHS-Channel cross beams

In deterministic analysis, a deterministic lower bound curve represents an eyeball fit through the lowest failure data (Wallin 1999). The advantage of this method is that when used in conjunction with the existing fatigue design guidelines, it allows for the use of existing fatigue design S-N curves. Using the deterministic method and the scatter of the S-N data for RHS-RHS, RHS-Angle and RHS-Channel cross beams in Figure 13 and the existing family of S-N curves in AS4600 (SAA 2005), the Class 55 S-N curve is recommended for the design of these cross-beam connection types for the classification method.

5.2 Design for channel-channel cross beams

Figure 13 also shows that based upon the deterministic method, the scatter of the S-N data for Channel-Channel cross beams and the existing family of S-N curves in AS4600 (SAA 2005), the class 36 curve is recommended for design of Channel-Channel cross beams for the classification method.

6 CONCLUSIONS

Four different types of cross-beams, RHS-RHS, RHS-Angle, RHS-Channel and Channel-Channel were tested under cyclic loading. The load applied to the specimens was cyclic bending on the bottom member. A total of thirty five (35) specimens were tested. The following conclusions can be reached based on the observations and results from this investigation.

Welding Method: Welding using the gas metal arc welding method showed that the hardness values for the heat affected zone were lower than the maximum value of 350HV10 recommended by the Australian Standards for prevention of heat affected zone cracking. Measurement of hardness values in the parent metal and weld metal showed that the differences in hardness between the parent metal and the weld metal was less than 100HV10 as recommended by Australian Standards for limiting the localisation of deformation in the parent metal.

Failure Modes: Failure in the RHS-RHS, RHS-Angle and RHS-Channel cross-beams occurred through fatigue crack initiation and propagation at weld toes in the RHS bottom member. Failure in the Channel-Channel cross-beams occurred predominantly through cracking at the weld toes in the top member. However, in one Channel-Channel specimen failure occurred through cracking at weld toes in both the bottom and top members. The cracking was in agreement with the measured maximum experimental stress concentration factors in the specimens.

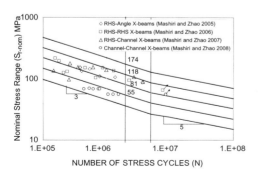

Figure 13. Test data and fatigue design curves.

Stress Concentration Factors (SCFs): The magnitude of SCFs at weld toes in the top and bottom members were all less than 2.0. Fatigue design standards such as CIDECT Design Guide No. 8 normally recommend the use of a minimum SCF of 2. Therefore a stress concentration factor of 2.0 can be recommended for cross-beam connections under the load, cyclic bending in the bottom member.

Fatigue Design: The Class 55 S-N curve is recommended for the fatigue design of RHS-RHS, RHS-Angle and RHS-Channel cross-beams. On the other hand, the Class 36 S-N curve is recommended for the fatigue design of Channel-Channel cross-beams.

ACKNOWLEDGEMENTS

The authors would like to thank staff in the Department of Civil Engineering laboratory at Monash University for their assistance in this project. This Project is sponsored by OneSteel Market Mills and a Monash University SMURF2 fund.

REFERENCES

CEN 2003: Eurocode 3, Design of Steel Structures-Part 1.9: Fatigue, PrEN 1993-1-9, European Committee for Standardisation.

Hancock G.J. 1999, "Recent research and design developments in cold-formed open section and tubular members", Advances in Steel Structures Vol. I, Proceedings of The Second International Conference on Advances in Steel Structures, ICASS'99, Editors Chan S.L. and Teng G.J., Elsevier Science Ltd, Hong Kong, 15–17 December 1999, pp. 25–37.

IIW 2000: Fatigue Design Procedures for Welded Hollow Section Joints, IIW Doc. XIII-1804-99, IIW Doc. XV-1035-99, Recommendations for IIW Subcommission XV-E, Edited by X.L. Zhao and J.A. Packer, Abington Publishing, Cambridge, UK.

Marshall P.W. 1992, "Design of Welded Tubular Connections, Basis and Use of AWS Code Provisions", Shell Oil Company, Houston, Texas, USA, Elsevier Science Publishers B.V. 1992.

Mashiri F.R. 2001, "Thin-Walled Tubular Connections under Fatigue loading", PhD Thesis, Monash University, Melbourne, Australia.

Mashiri F.R., Zhao X.L. and Grundy P. 2002a, "Fatigue Tests and Design of Thin Cold-Formed Square Hollow Section-to-Plate T-Connections under In-Plane Bending". *Journal of Structural Engineering, ASCE*, 128(1), January 2002: 22–31.

Mashiri F.R., Zhao X.L. and Grundy P. 2002b, "Fatigue Tests and Design of Welded T-Connections in Thin Cold-Formed Square Hollow Sections under In-Plane Bending". *Journal of Structural Engineering, ASCE*, 128(11), November 2002: 1413–1422.

Mashiri F.R., Zhao X.L. and Grundy P. 2004, "Stress Concentration Factors and Fatigue Behaviour of Welded Thin-Walled CHS-SHS T-Joints under In-Plane Bending", *Engineering Structures*, 26(13), Elsevier Science Ltd, 2004: 1861–1875.

Mashiri F.R. and Zhao X.L. 2005, "Fatigue Behaviour of Welded Thin-Walled RHS-to-Angle Cross-Beams under Bending", *Structural Engineering-Preserving and Building into the Future, Australian Structural Engineering Conference (ASEC2005)*, Editors: Stewart M.G. and Dockrill B, 11–14 September 2005, Newcastle, Australia, 10 pp. CDROM. ISBN: 1-877040-37-1.

Mashiri F.R. and Zhao X.L. 2006, "Welded Thin-Walled RHS-to-RHS Cross-Beams under Cyclic Bending", *Tubular Structures XI, Proceedings of The 11th International Symposium and IIW International Conference on Tubular Structures*, ISTS11 August 31 to September 2, 2006, Quebec City, Canada, Editors: Packer J.A. & Willibald S., pp. 97–104.

Mashiri F.R. and Zhao X.L. 2007, "Thin RHS-Channel Cross-Beams under Cyclic Bending", *Proceedings of The 5th International Conference on Advances in Steel Structures*, 5–7 December 2007, Singapore, Editors: Liew J.Y.R and Choo Y.S., pp. 991–996.

Mashiri F.R. and Zhao X.L. 2008, "Thin Channel-Channel Welded Cross-Beams under Cyclic Bending", *The 5th International Conference on Thin-Walled Structures*, 18–20 June 2008, Brisbane, Australia, Editor: Mahendran M. (*In Press*).

SAA 1991: Structural Steel Hollow Sections, Australian Standard AS1163-1991, Standards Association of Australia, Sydney, Australia.

SAA 1998, Steel Structures, Australian Standard AS 4100-1998, Standards Association of Australia, Sydney, Australia.

SAA 2004a: Structural Steel Welding, Part 1: Welding of Steel Structures, Australian Standard AS/NZS 1554. 1-2004, Standards Association of Australia, Sydney, Australia.

SAA 2004b: *Structural Steel Welding, Part 5: Welding of Steel Structures subject to High Levels of Fatigue Loading*, Australian Standard AS/NZS 1554. 5-2004, Standards Association of Australia, Sydney, Australia.

SAA 2005, Cold-Formed Steel Structures (Draft), Australian Standard AS 4600-2005, Standards Association of Austra-lia, Sydney, Australia.

van Wingerde A.M. 1992: The fatigue behaviour of T- and X-joints made of SHS, Heron, Vol. 37, No. 2, pp. 1–180.

Wallin K. 1999, "The probability of success using deterministic reliability", Fatigue Design and Reliability, Editors: G. Marquis and J. Solin, ESIS Publication 23, Elsevier Science Ltd., 1999.

WTIA 1998, "Commentary on the Structural Steel Welding Standard AS/NZS 1554", TN 11-98, WTIA Technical Note No. 11, Published jointly by the Welding Technology Institute of Australia and the Australian Institute of Steel Construction, NSW, Australia.

Zhao X.L. 1993, "The Behaviour of Cold-Formed RHS Beams Under Combined Actions", PhD Thesis, The University of Sydney, Sydney, Australia.

Zhao X.L and Hancock G.J. 1995: Butt Welds and Transverse Fillet Welds in Thin Cold-Formed RHS Members, *Journal of Structural Engineering, ASCE*, 121 (11): 1674–1682.

Zhao X.L., Herion S., Packer J.A., Puthli R., Sedlacek G., Wardenier J., Weynand K., Wingerde A., & Yeomans N. 2000, "Design Guide for Circular and Rectangular Hollow Section Joints under Fatigue Loading", Verlag TUV Rheinland, Koln, Germany.

*Parallel session 3: Static strength of joints
(review by G.J. van der Vegte & B. Young)*

Parametric finite element study of branch plate-to-circular hollow section X-connections

A.P. Voth & J.A. Packer
Department of Civil Engineering, University of Toronto, Canada

ABSTRACT: A parametric finite element study has been undertaken on plate-to-circular hollow section X-connections, with either longitudinal or transverse branch plates and with the branches loaded in either tension or compression. The finite element models employed were validated by comparison with the behaviour of seven plate-to-CHS connections previously tested. The 210 "numerical tests" generated illustrated that current limit states design recommendations for such connections are conservative, and especially so for tension-loaded branch plates. Distinct resistance equations are proposed for the latter, which will ultimately enable the capacity of through plate-to-CHS connections to be quantified.

1 INTRODUCTION

The limit states design resistance of plate-to-hollow structural section connections subjected to branch plate axial load is generally relatively low, due to an imposed deformation limit of 3% of the connecting face width for rectangular hollow section (RHS) connections or 3% of the diameter for circular hollow section (CHS) connections (Lu et al. 1994, Wardenier et al. 2008a). Though a deformation limit is practical and necessary, the potential strength of the hollow section member is being under-utilized, particularly with the increase in use of heavily-loaded plate-to-hollow structural section connections in tubular arch bridges and cable-stayed roofs. The need to develop a means to strengthen these connections becomes apparent. One method of strengthening these connections is to pass the branch plate through the chord member and weld the plate to both sides of the tube, producing a "through plate connection". Through plate-to-RHS connections have previously been studied (Kosteski & Packer 2003) and design recommendations have since been incorporated into CIDECT Design Guide No. 9 (Kurobane et al. 2004). However, design recommendations for through plate-to-CHS connections have been absent in published literature, indicating the need for further research.

An experimental and numerical investigation has been previously undertaken to investigate the behaviour of through plate-to-CHS connections by comparison with their branch plate counterparts (see Fig. 1), as well as to examine the current design methodology for T- and X-type branch plate-to-CHS connections.

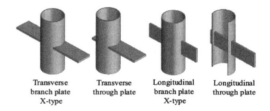

Figure 1. Plate-to-CHS connections.

The previous experimental program consisted of a total of 12 plate-to-CHS connections: six T-type connections tested under branch plate tensile load, four T-type connections tested under branch plate compressive load and two X-type connections tested under branch plate tensile load (Willibald et al. 2006, Voth & Packer 2007, 2008). Rigorous testing and measurement of both material and geometric properties was undertaken. Some of the important conclusions from the experimental work were (Voth & Packer 2007, 2008):

- T-type through plate-to-CHS connections have approximately 1.6 times the capacity of similar branch plate connections, at the 3% d_0 deformation limit.
- Current design guidelines for T- and X-type plate-to-CHS connections are conservative.
- T-type branch plate-to-CHS connections have significantly different behaviour in tension and compression; however, T-type through plate-to-CHS connections in tension and compression

behave in almost the same manner. In addition, the summation of the load-displacement characteristics of T-type branch plate-to-CHS connections in tension and compression produces similar behaviour to a T-type through plate-to-CHS connection.

2 FINITE ELEMENT MODELLING STUDY

2.1 General FE modelling

Seven finite element (FE) models of previously tested experimental specimens have been constructed and analyzed using the commercially available software package ANSYS 11.0 (Swanson Analysis Systems 2007). The models were constructed to replicate all geometric properties of the experimental test specimens, including experimental chord end conditions and weld fabrication details. Either eight-node solid brick elements (SOLID45) or 20-node solid brick elements (SOLID95) were used; each with three translational degrees of freedom per node and reduced integration with hourglass control. All FE models were analysed using non-linear time step analysis which incorporated non-linear material properties, large deformation allowance and a full Newton-Raphson frontal equation solver. At each time step an incremental displacement was applied to the nodes at the branch plate end to reproduce the displacement control loading used in the previous tests. Material properties, geometric and analysis considerations and model calibration are highlighted in the following sections.

2.2 Material properties

A multi-linear true stress (σ_T)—true strain (ε_T) curve was used to describe the behaviour of the CHS, branch plate and weld material properties. Material properties of both the CHS, ASTM A500 Grade C (ASTM 2003) and branch plate (CSA 2004), in the form of engineering stress-strain (σ-ε) relationships, were determined through a series of tensile coupon tests until the point of necking and for a single point at coupon rupture (see Table 1). Prior to necking the engineering stress-strain curve was converted to true stress-true strain by using (Boresi & Schmidt 2003):

$$\sigma_T = \sigma(1+\varepsilon) \qquad (1)$$

$$\varepsilon_T = \ln(1+\varepsilon) \qquad (2)$$

Post necking, these relationships are no longer valid as the stress distribution at the point of necking changes from a simple uniaxial case to a more complex triaxial case (Aronofsky 1951). A method developed

Table 1. Measured material properties.

	E (GPa)	f_y (MPa)	f_u (MPa)	ε_u (%)
CHS	211.5	389	527	30.0
Plate	210.5	326	505	37.7

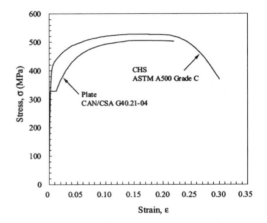

Figure 2. Engineering stress-strain properties for plate & CHS.

by Matic (1985) and modified by Martinez-Saucedo et al. (2006) to determine the post necking material behaviour is described and adopted herein.

To describe the post necking material behaviour, Matic (1985) outlines a procedure that generates a quadratic expression for the full true stress-true strain behaviour. The engineering stress-strain behaviour resulting from a FE analysis of a tensile coupon with Matic material properties is compared with the average material experimental engineering stress-strain behaviour and the coupon rupture value. Through an iterative process the Matic true stress-true strain curve that produces the best fit engineering stress-strain relationship is used as the FE material property. Martinez-Saucedo et al. (2006) suggest that the Matic curve be utilized only in the post necked region of the true stress-true strain curve. The material property curves are given in Figure 2. Weld material, for the purposes of this study, is given the same properties as the plate; both ignore the post necking response as they have been shown to remain in the pre-necked region.

2.3 Connection modelling

Branch plate connections were modelled using the exact geometric and material properties of each of seven experimental test specimens. Using connection symmetry, only one eighth or one quarter of X- and T-type connections respectively was modelled and symmetrical boundary conditions were applied. All models were developed with increased mesh density around areas associated with large deformation and ultimate fracture to better capture the connection behaviour. A gap of 1.5 mm was left between the CHS member and the branch or through plate to ensure force transfer through the weld only, in accord with fabrication and experimental observation (Fig. 3). To account for ultimate fracture, a maximum equivalent strain (ε_{ef}) was imposed to initiate the element "death feature" whereby an element stiffness and stress are reduced to near-zero allowing the element to deform freely.

To determine the best suited mesh layout, element type, restraint conditions and fracture strain, a model sensitivity study was preformed. From this study it was determined that a fine mesh layout with three elements through the CHS member thickness, constructed from element type SOLID45, produced the best agreement between experimental load-displacement response, spot strain values, the load at the 3% deformation limit ($N_{3\%}$) and overall behaviour. The CHS chord end condition for T-type connections was modelled in two ways: the first with no end plate and CHS chord end fully fixed, and the second with an end plate and fixed only at the inside surface of the bolt holes which more closely emulates experimental test geometry (Figs. 3–4). The chord end plates were modelled for the X-type connection, but no end restraint was applied to match the experimental test. The exclusion of the end plate for T-type connections resulted in stiffer overall connection behaviour that differed from experimental results, showing that the end plate should be modelled. As the experimental and FE models have relatively short CHS chord lengths (free chord lengths (L_0^*) between 420 mm and 500 mm for longitudinal and transverse connections respectively or approximately d_0 between end plate and branch plate), the chord end conditions significantly affect the connection behaviour and thus the type of end restraint provided becomes extremely important.

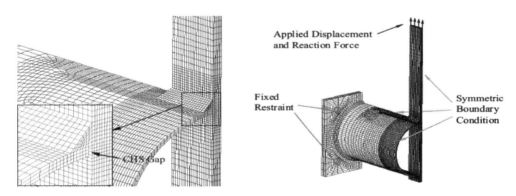

Figure 3. Mesh and restraint arrangements used in sensitivity study.

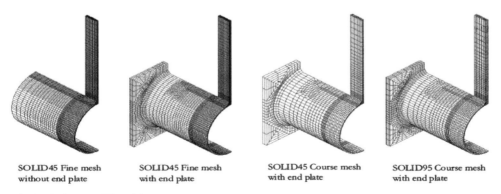

Figure 4. Typical quarter FE models.

The maximum equivalent strain value for all FE models was determined through empirical correlation between the experimental test results and the FE model results. An initial value of the maximum equivalent strain was chosen to be equal to that used for tensile coupon models ($\varepsilon_{ef} = 1.1$); however, for connection FE models, a high deformation was required to initiate fracture. All connections were therefore analysed with various ε_{ef} values and FE analysis and experimental fracture results were compared resulting in an average best fit value of $\varepsilon_{ef} = 0.20$. The lack of direct transferability between two models suggests that differences in local material boundary conditions, connection geometry, mesh arrangement and loading have a significant impact on the equivalent strain values of individual elements and thus connection failure using this method. Caution must therefore be observed if element boundary conditions or failure modes change when applying this method to further studies. This remark is further emphasized by the work of Martinez-Saucedo et al. (2006) where a similar method was used to determine ε_{ef} which produced a significantly different value of 0.60 for a slotted end plate connection.

2.4 Evaluation of FE models against experiments

In general, the FE connection models correlated well with the experimental tests with respect to overall load-displacement behaviour, 3% deformation limit load, spot displacements, fracture location, ultimate failure load and ultimate failure mode. Due to slightly unsymmetrical loading and plate initial out-of-straightness in experimental tests, the spot strain values for some FE models did not match those recorded in the experimental tests. As all other characteristics correlate well, the spot strain values were not used as a primary method of validation. A comparison of experimental and FE results is shown in Table 2 along with experimental specimen configuration shown in Table 3.

The load-displacement response of the FE models for T-type branch plate-to-CHS connections tested in tension correlates well with the experimental test results, as shown in Figure 5. Both the longitudinal (CB0EA) and transverse (CB90EA) FE models exhibited a high concentration of stress around the weld perimeter in the CHS. The location of the killed element in both FE models was very similar to the initial crack location in the experiments.

The through plate-to-CHS connection FE models tested in tension (CT0EA longitudinal; CT90EA transverse) correlate reasonably well with experimental results with respect to load-displacement behaviour (see Fig. 6). Though these FE models do not correspond as closely to the experimental results as their branch plate counterparts, they reproduce the overall trend and experimental observations such as localized deformation at the bottom plate connection, significant plastic deformation and CHS ovalization. The FE models further emulate the experimental results by the initial killed element being at the location of experimental initial crack.

Figure 7 shows the load-displacement behaviour of branch (CB0EB) and through (CT0EB) longitudinal

Table 2. Comparison of $N_{3\%}$ for experiments and FE models.

Specimen ID	Failure mode	$N_{3\%}$ (kN)	$N_{3\%FE}$ (kN)	$N_{3\%}/N_{3\%FE}$	N_u (kN)	N_{uFE} (kN)	N_u/N_{uFE}
CB0EA	PS	161	174	0.93	286	292	0.98
CB90EA	PS/TO	283	288	0.98	320	313	1.02
CT0EA	PS	259	237	1.09	406	396	1.03
CT90EA	PS/TO	447	409	1.09	459	460	1.00
CB0EB	PS	−84.9	−85.2	1.00	−258	−251	1.03
CT0EB*	PS	−273	−242	1.13	−387*	−413	0.94*
XB90EA	PS	124	131	0.95	226	216	1.05

PS = Punching shear failure and TO = Tear-out failure
*failure of overall experimental setup before fracture.

Table 3. Experimental specimen configuration.

Specimen ID	Plate orientation	Plate type	Plate load
CB0EA	Longitudinal	Branch	Tension
CB90EA	Transverse	Branch	Tension
CT0EA	Longitudinal	Through	Tension
CT90EA	Transverse	Through	Tension
CB0EB	Longitudinal	Branch	Compression
CT0EB	Transverse	Branch	Compression
XB90EA	Longitudinal	Branch	Tension

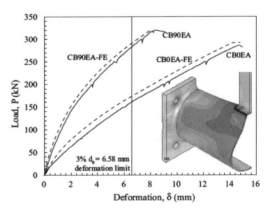

Figure 5. Load-displacement behaviour for T-type branch plate-to-CHS connections in tension: FE vs. Experimental.

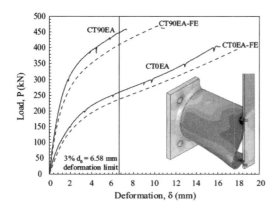

Figure 6. Load-displacement behaviour for T-type through plate-to-CHS connections in tension: FE vs. Experimental.

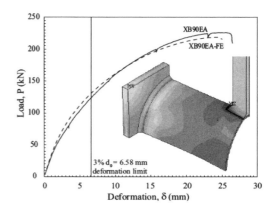

Figure 8. Load-displacement behaviour for X-type longitudinal plate-to-CHS connection in tension: FE vs. Experimental.

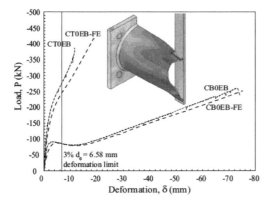

Figure 7. Load-displacement behaviour for T-type longitudinal plate-to-CHS connection in compression: FE vs. Experimental.

T-type connections tested in compression. Correlation between the FE and the experimental results for CB0EB is very good; the FE model predicts the load-displacement curve, overall deformed shape, stress concentrations and facture in the CHS. The through plate connection (CT0EB), while emulating the experimental load-displacement curve, does not match the experimental ultimate load or fracture due to overall test setup failure. The model, however, reflects the concentration of stress around the weld, the deformation and ovalization of the CHS and location of initial cracking.

Finally, for the X-type longitudinal branch plate-to-CHS connection tested in tension, FE and experimental behaviour again correlate well (Fig. 8).

3 FINITE ELEMENT PARAMETRIC STUDY OF X-TYPE BRANCH PLATE CONNECTIONS

For the first phase of a study into the behaviour of branch plate-to-CHS connections, a parametric analysis of transverse and longitudinal X-type connections, with branch members loaded in both tension and compression, was undertaken using the FE methodology established through the validation study. The study set out to examine the influence of the parameters in the current design equations: chord diameter-to-thickness ratio (2γ), nominal connection width ratio (β) and nominal branch member depth-to-chord diameter ratio (η). A total of 210 FE connections were modelled with a CHS chord diameter (d_0) of 219.1 mm, varying values of β from 0.2 to 1.0 or η from 0.2 to 4.0, and 2γ from 13.80 to 45.84, as outlined in Table 4.

A constant plate thickness (t_1) of 19.01 mm and nominal fillet weld size (w) of 13.44 mm were used in all FE models. The weld size was chosen to provide adequate resistance in shear based on the plate thickness, in accordance with CAN/CSA-S16-01 (CSA 20-01). When the affect of the weld size on connection behaviour is incorporated, both β and η are converted to effective values (β' and η'). The values of η' used in this study are 0.32, 0.53, 0.72, 0.92, 1.12, 1.62, 2.12, 2.62, 3.12 and 4.12. The values of β' used are 0.32, 0.51, 0.69, 0.87 and 1.00. It should be noted that the horizontal component of the weld size changes with the changing curvature of the CHS circumference, affecting the β' value, but with little affect on weld or connection capacity.

Table 4. Geometric parameters investigated.

t_0 (mm)	2γ	β or η					η				
		0.2	0.4	0.6	0.8	1.0	1.5	2.0	2.5	3.0	4.0
15.88	13.80	1	2	3	4	5	6	7	8	9	10
12.70	17.25	11	12	13	14	15	16	17	18	19	20
11.10	19.74	21	22	23	24	25	26	27	28	29	30
9.53	23.99	31	32	33	34	35	36	37	38	39	40
7.95	27.56	41	42	43	44	45	46	47	48	49	50
6.35	34.50	51	52	53	54	55	56	57	58	59	60
4.78	45.84	61	62	63	64	65	66	67	68	69	70

3.1 Connection modelling

The finite element analysis was carried out using the same general characteristics proven to produce effective results from the FE model validation study including material properties, mesh density, element type (SOLID45), incremental displacement control and analysis procedure. Though only one experimental test was directly validated for X-type branch plate connections, the general modelling procedure has been shown to be valid and is suited for application to other X-type connection geometries and loading conditions.

Both the experimental and FE validation models have relatively short chord lengths allowing the chord end boundary or restraint conditions to significantly influence the overall connection behaviour. Van der Vegte & Makino (2006, 2007) suggest that to reduce the influence of chord end boundary conditions a total chord length of at least $10d_0$, for both T-type and X-type CHS-to-CHS connections, is needed. It is also shown that, for X-type CHS-to-CHS connections, the use of free chord end conditions results in lower strength behaviour than if rigid end conditions are adopted. Based on these findings a free chord length $(L_0^*/2)$ of $5d_0$ with a free end condition was adopted herein, to ensure a chord length (L_0) of at least $10d_0$ and a conservative strength behaviour if some end effects remain. The configuration and geometric parameters for this parametric study are illustrated in Figure 9.

For the majority of connections, the branch plate was welded along the entire plate perimeter; however, special consideration was given to weld modelling for transverse connections with β values near 1.00 where a fillet weld over the ends of the plate is impossible. As typically fabricated, the FE models were constructed with a small portion of the plate clipped at the extremes and filled with a groove weld. Thus, the nominal and effective β values are the same for transverse connections with $\beta = 1.00$ as no additional thickness or perimeter is added. For a few connections with low β or η and 2γ values the branch plate thickness (t_1) was increased to allow connection behaviour to govern over branch plate yielding. In these cases a groove weld was utilized over the entire plate perimeter thus making the nominal and effective β or η values the same.

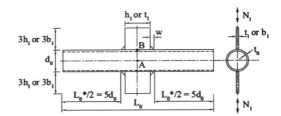

Figure 9. Geometric connection properties for FE models.

3.2 Results and observations

For each X-type connection under branch plate axial load, the load-displacement curve was determined with the displacement defined as the change in distance between point A in Figure 9 and a point at the crown of the CHS chord (point B in Figure 9). From these curves the connection ultimate load was determine by the minimum of: (i) $N_{3\%FE}$, (ii) N_{uFE} and (iii) branch plate yielding. $N_{3\%FE}$ governed over N_{uFE} in all cases. Figure 10 depicts the normalized ultimate load $(N_{3\%FE}/f_{y0}t_0^2)$ as a function of η' for all 2γ values and both tension and compression branch plate loading conditions, for longitudinal X-type connections. Figures 11 and 12 similarly depict the normalized ultimate load as a function of β' for all 2γ values and both tension and compression branch plate loading for transverse X-type connections. Figures 10, 11 and 12 have FE results compared with the current CIDECT connection design equations (Wardenier et al. 2008a), calculated using effective geometric properties to better evaluate the effectiveness of the design equations. Figure 13 compares the effective

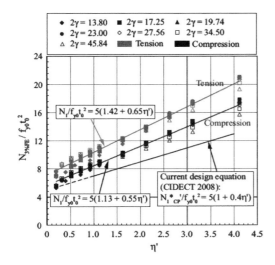

Figure 10. Parametric FE results for longitudinal X-type branch plate-to-CHS connections.

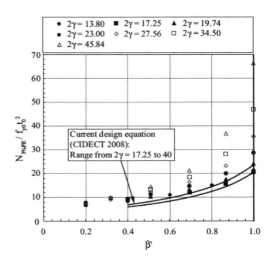

Figure 12. Parametric FE results for transverse X-type branch plate-to-CHS connections in tension.

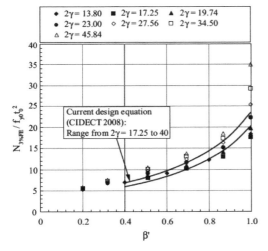

Figure 11. Parametric FE results for transverse X-type branch plate-to-CHS connections in compression.

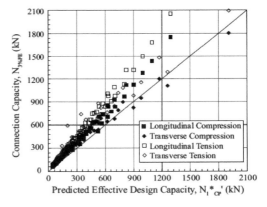

Figure 13. Comparison of parametric FE results to current design equations.

design capacity for chord plastification ($N_{1\,CP}^{*}{}'$) and the ultimate capacity from FE analysis ($N_{3\%\,FE}$).

Figure 10 shows that, like the experimental results for T-type branch plate-to-CHS, the connection capacity in tension is significantly higher than for connections tested in compression. The results for both tension and compression sets are grouped close together, regardless of 2γ value and form approximately linear trends. The current CIDECT design equation is plotted and extrapolated beyond the lower validation limit of $\eta = 1$ (dashed line). Compared to the current CIDECT design equation for X- and T-type longitudinal plate-to-CHS connections, which has been validated by Wardenier et al. (2008b) against existing limited experimental and numerical data, both compressive and tensile FE results have higher capacities. The CIDECT design equations, which are based on the compression test results only, in an aim to be conservative, eliminate a significant potential increase in design capacity for tension-loaded longitudinal plate-to-CHS connections. Provisional equations based on the FE data trends are shown in Figure 10, for comparison with the current CIDECT design equation.

Figure 11 reveals that the current CIDECT design equation (Wardenier et al. 2008a) for transverse plate-to-CHS X-type connections closely follows the FE results for connections loaded in compression, with the exception of connections with high β' values and high 2γ values. As the current design equation for X-type branch plate connections has a validity limit of $2\gamma = 40$, the thinnest CHS chord with a 2γ value of 45.8 is outside this limit. For transverse connections tested in tension (Fig. 12) the CIDECT design equation is very conservative, especially for connections with high 2γ and mid to high values of β', indicating that the behaviour of transverse plate-to-CHS connections is again significantly different in tension and compression. For the transverse connection tested in tension with $\beta = 1.0$ and $2\gamma = 45.84$ the FE ultimate capacity, governed by the 3% deformation limit, is almost three times that of the effective design capacity predicted by the current CIDECT design equation (although outside the range of 2γ validity).

All 210 FE X-type connection models, for which the chord ultimate deformation limit governed the ultimate limit state, are summarized in Figure 13. This plot clearly illustrates that current design guidelines do not reflect the ultimate capacity of X-type plate-to-CHS connections tested in tension.

4 CONCLUSIONS

Finite element models of laboratory experiments have been created, using commercially available finite element software and utilizing the non-linear geometric and material properties of the experimental test specimens. The FE model results, with regard to load-displacement curves, overall connection deformation, local stress concentrations, failure modes and initial fracture location, correlate well with the experimental connection results. The X-type validated model was then modified for use in a parametric study on X-type longitudinal and transverse plate-to-CHS connections, loaded in either branch plate tension or compression.

210 X-type connections were "tested numerically" with nominal β from 0.20 to 1.0, η values from 0.20 to 4.0 and 2γ values from 13.80 to 45.84. The results of the FE parametric analysis yielded the following insights:

- Longitudinal connections tested in tension have significantly higher ultimate capacity, based on a deformation limit of 3% d_0, over connections in compression.
  Longitudinal connections tested in both tension and compression follow the same trend as the current CIDECT design equation, but have higher capacities suggesting that the current design equation is conservative over the parameter range studied. Thus, the current CIDECT design equation could be extrapolated safely to a lower η limit of 0.2.

- Transverse plate-to-CHS connections tested in compression have similar FE ultimate capacity results to the current CIDECT design equation. However, there is a significant difference between the connection capacity results obtained by FE analysis and the current design recommendation for connections in tension. FE analysis showed connection capacities up to nearly times the CIDECT design capacity, for connections tested in tension with high β and 2γ values.

ACKNOWLEDGEMENTS

Financial support has been provided by CIDECT (Comité International pour le Développement et l'Étude de la Construction Tubulaire), the Steel Structures Education Foundation (SSEF, Canada), the Natural Sciences and Engineering Research Council of Canada (NSERC) and Deutsche Forschungsgemeinschaft (DFG, Germany). Generous donations of CHS, plate and fabrication of test specimens were made by Atlas Tube Inc., IPSCO Inc. and Walters Inc. (Ontario, Canada) respectively. The authors are grateful to Dr. Silke Willibald for assistance with the project experimental phase.

NOMENCLATURE

CHS = circular hollow section
RHS = rectangular hollow section
E = modulus of elasticity
K = rate of change of the tangent modulus
L_0 = total chord length
L_0^* = effective chord length (excludes plate and weld dimensions along chord length)
$N_{1\,CP}^{*}{}'$ = CIDECT design capacity using effective geometric parameters
$N_{3\%}$ = branch member load at a connection displacement of 3% d_0
$N_{3\%FE}$ = FE branch member load at a connection displacement of 3% d_0
N_u = fracture load
N_{uFE} = FE fracture load
b_0 = external width of RHS chord member
b_1, b_1' = nominal, effective external width of branch
d_0 = external diameter of CHS chord member
f_u = ultimate stress
f_y, f_{y0} = yield stress, yield stress of CHS
h_1, h_1' = nominal, effective depth of branch
t_0 = thickness of hollow section member
t_1 = thickness of branch
w = nominal weld size (leg length) = $1.41(t_1/2)$
β, β' = nominal, effective connection width ratio ($\beta = b_1/d_0, \beta' = b_1'/d_0$)

2γ = chord diameter-to-thickness ratio = d_0/t_0
$\varepsilon, \varepsilon_T$ = engineering, true strain
ε_{ef} = maximum equivalent strain
ε_u = fracture strain
η, η' = nominal, effective branch member depth-to-chord diameter ratio
$(\eta = h_1/d_0, \eta' = h_1'/d_0)$
σ, σ_T = engineering, true stress

REFERENCES

Aronofsky, J. 1951. Evaluation of stress distribution in the symmetrical neck of flat tensile bars. *Journal of Applied Mechanics* 3: 75–84.

ASTM 2003. *Standard specification for cold-formed welded and seamless carbon steel structural tubing in rounds and shapes, ASTM A500–03*. West Conshohocken, Pennsylvania, USA: ASTM International.

Boresi, A.P. & Schmidt, R.J. 2003. *Advanced mechanics of materials, 6th edition*. New Jersey, USA: John Wiley & Sons, Inc.

CSA 2001. *Limit states design of steel structures, CAN/CSA-S16-01*. Toronto: Canadian Standards Association.

CSA 2004. *General requirements for rolled or welded structural quality steel/structural quality steel, CAN/CSA-G40.20-04/G40.21-04*. Toronto: Canadian Standards Association.

Kosteski, N. & Packer, J.A. 2003. Longitudinal plate and through plate-to-hollow structural section welded connections. *J. Structural Engineering, ASCE* 129 (4): 478–486.

Kurobane, Y., Packer, J.A., Wardenier, J. & Yeomans, N. 2004. *Design guide for structural hollow section column connections*. Köln, Germany: Verlag TÜV Rheinland GmbH.

Lu, L.H., Winkel, G.D. de, Yu, Y. & Wardenier, J. 1994. Deformation limit for the ultimate strength of hollow section joints. In P. Grundy, A. Holgate & B. Wong (eds), *Proc. 6th intern. symp. on tubular structures*: 341–347. Rotterdam: Balkema.

Martinez-Saucedo, G., Packer, J.A. & Willibald, S. 2006. Parametric finite element study of slotted end connections to circular hollow sections. *Engineering Structures* 28: 1956–1971.

Matic, P. 1985. Numerically predicting ductile material behavior from tensile specimen response. *Theoretical and Applied Fracture Mechanics* 4: 13–28.

Swanson Analysis Systems 2007. ANSYS release 11.0, Houston, USA.

Vegte, G.J. van der & Makino, Y. 2006. Ultimate strength formulation for axially loaded CHS uniplanar T-joints. *Intern. J. Offshore and Polar Engineering* 16(4): 305–312. ISOPE.

Vegte, G.J. van der & Makino, Y. 2007. The effect of chord length and boundary conditions on the static strength of CHS T- and X-joints. In Y.S. Choo & J.Y.R. Liew (eds), *Proc. 5th intern. conf. on advances in steel structures*: 997–1002. Singapore: Research Publishing.

Voth, A.P. & Packer, J.A. 2007. *Branch plate connections to round hollow sections, CIDECT Report 5BS-5/07*. Toronto: University of Toronto.

Voth, A.P. & Packer, J.A. 2008. Experimental study on stiffening methods for plate-to-circular hollow section connections. *Proc. ECCS/AISC intern. workshop, connections in steel structures VI, Chicago, USA*.

Wardenier, J., Kurobane, Y., Packer, J.A., Vegte, G.J. van der & Zhao, X.-L. 2008a. *Design guide for circular hollow section (CHS) joints under predominantly static loading*. Köln, Germany: Verlag TÜV Rheinland GmbH.

Wardenier, J., Vegte, G.J. van der & Makino, Y. 2008b. Joints between plates or I sections and a circular hollow section chord. *Proc. 18th intern. offshore and polar engineering conf.* Vancouver, Canada.

Willibald, S., Packer, J.A., Voth, A.P. & Zhao, X. 2006. Through plate joints to elliptical and circular hollow sections. In J.A. Packer & S. Willibald (eds), *Proc. 11th intern. symp. on tubular structures*: 221–228. London, UK: Taylor & Francis Group.

Load-carrying capacity study on T- and K-shaped inner-stiffened CHS-RHS joints

G.F. Cao & N.L. Yao
Shanghai Institute of Architectural Design & Research Co., LTD., Shanghai, China

ABSTRACT: It is well known that stiffened joints have higher ultimate strength. Inner-stiffened joints were recently applied to some important projects in Shanghai. The main model of these joints in this paper is made up of CHS brace and RHS chord which comes from actual projects. The failure mode of unstiffened and inner-stiffened joints here is flexural failure of the chord face by controlling joint parameters. Through systematic Finite Element Analysis (FEA) on load-carrying capacity of simple T-shaped inner-stiffened joints under axial load, in-plane and out-of-plane bending in the brace, a set of formulae used to structural design were presented. Based on the study of T-joints, ultimate strength formulae of K-shaped inner-stiffened joints were obtained. At last the applications of inner-stiffened joints in two projects are introduced. Despite complexity of the inner-stiffened joints in actual projects, quantitative results were acquired in this paper contributed to the basic research of inner-stiffened joints.

1 INTRODUCTION

1.1 Background

Along with the development of large-span-weight spatial structures, stiffened joints are more and more applied to actual projects. The study on hollow structural sections is always including unstiffened joints and stiffened joints. The externally reinforced joints (Packer et al. 1997) are often applied, because of more difficult weld inside small size hollow sections than large ones. There is much study on ring-stiffened tubular joints in offshore structures (Myers et al. 2001, Rama Raju et al. 1995, Lee et al. 1999), and ring-stiffened joints were applied to some projects in China (Zhang et al. 2004).

Inner-stiffened joints were used to the megatrusses of Shanghai International Circuit and the Vierendeel trusses of Shanghai's Naihui Administration Center, successfully resolved the strength lack of unstiffened joints. In process of the project design, FEA and test study were performed aimed to particular joints. The study on representative inner-stiffened joints is necessary for future project need, including mechanical model, positioning of inner-stiffener and ultimate strength formulae. The main joint model in this paper is made up of CHS brace and RHS chord (large size welded box section). Based on systematic FEA, the load-carrying capacity formulae of T-shaped inner-stiffened joints under axial load, in-plane and out-of-plane bending in the brace were derived. The formulae of K-shaped ones under axial load were also obtained.

1.2 Failure modes

Failure modes of HSS joints have been summarized below by Chen et al. (2002).

- Flexural failure of the chord face
- Local buckling of the RHS chord walls
- Punching shear failure of the chord face
- weld failure between the brace and the chord

He defined local buckling of the compression brace near the chord face, tension failure and buckling of the compression brace as member failure.

The main failure mode of RHS joints is flexural failure of the chord face when the brace to chord width ratio (β) is less than 0.85 (Packer et al. 1997). In general, the diameter of circular brace is small in actual projects and the value of β does not exceed 0.85. So this failure mode was discussed mainly in this paper.

The early significant test research on T-shaped ring-stiffened joints was due to Sawada et al. (1979). These tests show that ring-stiffener can significantly enhance the strength of unstiffened joint, and both unstiffened and ring-stiffened joints attain the maximum load at approximately the same deformation levels, i.e. stiffener has no adverse effect on ductility. In the comparison tests (Cao et al. 2007) of unstiffened and inner-stiffened joints ($\beta = 0.6$), the same failure mode (flexural failure of the chord face) was found.

1.3 Deformation limit and ultimate strength

Compared with ultimate load, the concept of ultimate deformation limit was proposed by Yura et al. (1978), aimed to a joint which does not have a pronounced peak load. Lu et al. (1994) proposed the following conclusions:

- For a joint with an obvious peak load at a deformation around 3% b_o (chord width), the peak load or the load at 3% b_o deformation is considered to be the ultimate load.
- For a joint without a pronounced peak load, the ultimate deformation limit depends on the ratio of the load at 3% b_o to the load at 1% b_o. If the ratio is greater than 1.5, the deformation limit is 1% b_o, i.e. serviceability is in control, and the ultimate strength is taken as 1.5 times the load at 1% b_o. If the ratio is less than 1.5, the deformation limit is 3% b_o, i.e. strength is in control, and the ultimate strength is taken as the load at 3% b_o.
- A validity range of β and 2γ (chord width to thickness ratio) is given to determine whether the design is governed by serviceability or by strength.

The upper proposal was mainly based on tests of hot-rolled sections. The same conclusions were proposed by Zhao (2000) based on tests of cold-formed RHS sections. The chords of the joints in this paper are welded box sections. There is different stress distribution at the corner of the section, but the ultimate deformation limit is same based on flexural failure of the chord face. The concept of serviceability and strength limit was used to the unstiffened joints in this paper. Based on FEA, Von Mises stress distribution on the chord face of both unstiffened and inner-stiffened joints at 3% b_o deformation level is uniform. So the ultimate deformation limit standard of the unstiffened joints was applied to the inner-stiffened joints in this paper.

2 FEA OF T-SHAPED CHS-RHS JOINTS

2.1 FEA model

The FEA model of T-shaped joints in this paper is shown in Figure 1, which boundary conditions are simply supported at the ends of the chord, and loads are applied at the brace ends simulating the joint load conditions of axial load, in-plane and out-of-plane moment. The symmetrical boundary conditions were applied to the joint models under axial load and in-plane bending. In the finite element models, restraints and loading are applied to the center nodes

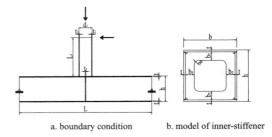

a. boundary condition   b. model of inner-stiffener

Figure 1. FEA model of T-shaped joints.

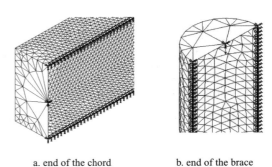

a. end of the chord   b. end of the brace

Figure 2. Details of finite element meshes of the end plates (under axial load and in-plane bending).

in the end plates. One end of the chord is pinned while the other pin-rolled (Fig. 2a). The chord length is set to five times the width or height of the chord and the brace length is set to three times the brace diameter to minimize end effects. The plates' thickness at the ends of the members is set to five times the wall thickness of every member to facilitate the application of boundary restraints and external loading. The model of inner-stiffeners is shown in Figure 1b. The corner radius (R) is 50 mm.

The choice of the size of the chord and brace depended on actual projects. The value of β is between 0.5 and 0.75 in this paper. The FEA showed that the thickness of the brace is irrelative to the ultimate strength, so the choice of the thickness of the brace only ensured that buckling of the brace do not appear before attaining maximum load. The FEA also showed that whether the chord is square or rectangular tubular has little effect on the ultimate strength, so square section was used in this paper. The ultimate strength mainly depend on β, t (thickness of the chord), b_r and t_r (width and thickness of the inner-stiffener).

2.2 Analysis program and elements

The FEA was carried out using the non-linear finite element program ANSYS (Ver 9.0). A four-node quadrilateral shell element (SHELL 143 from ANSYS library) with six degrees of freedom per node was used to model the joints. The deformation shapes are linear in both in-plane directions. For the out-of-plane motion, it uses a mixed interpolation of tensorial components. The triangular-shaped element was used to model the joints conveniently, which is formed by defining the same node number for nodes K and L. An elastic-perfectly plastic stress-strain curve with a yield stress of 345 N/mm^2, a Young's modulus of 206 kN/mm^2 and a Poisson's ratio of 0.3 were used. The material model is bilinear Kinematic hardening (rate-independent plasticity). The large deflection was taken into account.

2.3 Numerical study on T-joints under axial load

In this team of numerical study, the model parameters are shown in Table 1: the length of the chords is 3 m, the outside width and depth of the chords is 600 mm, the thickness of the chords is 20 or 25 mm. The number of joints including the unstiffened joints total 32.

The parameter δ is the ratio of ultimate strength of stiffened joints to that of unstiffened ones at 3% b_o, which shows increasing rate to the unstiffened joints with different stiffeners. The results are shown in Figures 3a, b. The parameter τ_r of the figures is the ratio of the thickness of the stiffener to that of the chords. From these figures and the results, the following conclusions can be drawn:

- The ultimate strength of the inner-stiffened T-joints is significantly higher than that of the unstiffened ones, and δ increases with τ_r almost linearly.
- δ decreases with larger β when same τ_r. When less β, the deformation of the upper face of the unstiffened joints is larger, so the stiffener can more significantly restrict the deformation and increase the strength.
- δ increases with γ when same τ_r. When larger γ, the deformation of the upper face of the unstiffened joints is larger, so the stiffener can more significantly restrict the deformation and increase the strength.
- On the early loading, the load-displacement curves are linearly. Along with plastic development, the slope of the curves become smaller and smaller.

Table 1. Geometrical parameters of T-joints.

Brace size mm	β	b_r mm	t_r mm	t mm
300 × 12 CHS	0.500	100	10	20
350 × 12 CHS	0.583		15	25
400 × 12 CHS	0.667		20	
450 × 12 CHS	0.750			

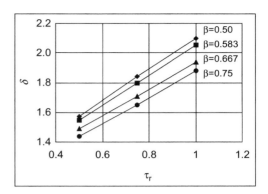

Figure 3a. Relationship between δ and τ_r ($2\gamma = 30$).

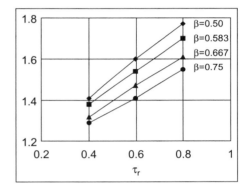

Figure 3b. Relationship between δ and τ_r ($2\gamma = 24$).

2.4 Numerical study on T-joints under out-of-plane bending

The faces of the chords bear tensile and compressive state under out-of-plane bending which have different deformation levels, so the whole joint models are used. For avoiding large rigid torsional displacement, the nodes of the ends of the chords were restricted (Fig. 4a). The numerical study showed that this method does not affect the relative deformation of crown and saddle points.

In this team of numerical study, the models are same as Table 1 except for the thickness of the braces (20 mm) to avoid the buckling of the braces. The results which have same characteristic as axial load, are shown in Figures 5a, b.

2.5 Numerical study on T-joints under in-plane bending

The inner-stiffener on the upper analysis is only one in middle of the chords. This stiffened mode has no enhance on ultimate strength under in-plane bending,

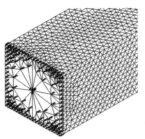

a. end of the chord   b. end of the brace

Figure 4. Details of finite element meshes of the end plates (under out-of-plane bending).

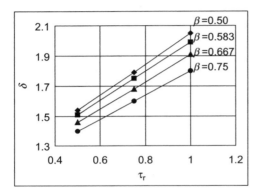

Figure 5a. Relationship between δ and τ_r ($2\gamma=30$).

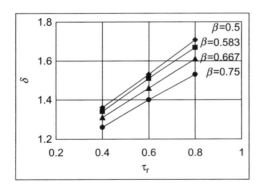

Figure 5b. Relationship between δ and τ_r ($2\gamma=24$).

which has been confirmed in many papers. So under in-plane bending two stiffeners are necessary, which lie on the crown points symmetrically (Fig. 6). In this team of numerical study, the models are same as Table 1 except for the thickness of the braces (20 mm) to avoid the buckling of the braces.

The results are shown in Figures 7a, b. It has the different trait: δ changes little with different β on the same other parameters, which is different from the results under axial load and out-of-plane bending.

2.6 Parametric study — material constitutive relationship

The representative material constitutive relationships are shown as follows: elastic-perfectly plastic material, bilinear hardening material which second tangent modulus is often 1% times to elastic modulus, et al. Based on the numerical study, the effect of the upper two material properties on the ultimate strength is drawn: difference of the ultimate strength of the unstiffened joints is less than 20%, that of the inner-stiffened joints is less than 30%. The choice of elastic-perfectly plastic material in this paper is appropriate and conservative.

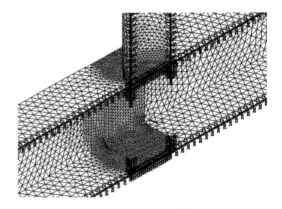

Figure 6. Details of finite element meshes of the two stiffeners (under in-plane bending).

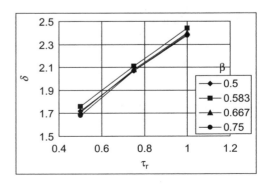

Figure 7a. Relationship between δ and τ_r ($2\gamma=30$).

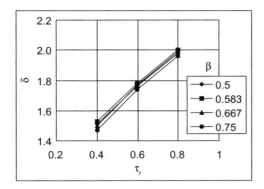

Figure 7b. Relationship between δ and τ_r ($2\gamma = 24$).

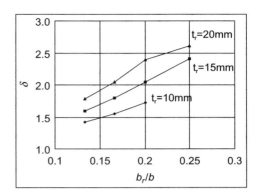

Figure 8a. Relationship between δ and b_r/b (under axial load).

Table 2. Geometrical parameters of T-joints.

Brace size mm	b_r mm	t_r mm	t mm
350 × 12 CHS	80	10	20
CHS	100	15	
CHS	120	20	

2.7 Parametric study —the size of stiffeners

The size of the stiffeners (b_r, t_r) is important factor to ultimate strength of the inner-stiffened joints. In this team of numerical study, the models are shown in Table 2. The results are shown in Figures 8a, b, c.

From the upper figures, δ increases with b_r and t_r. For the joints with only one middle stiffener under axial load, increasing with b_r and t_r, the maximum deformation position changes from saddle point to crown point. In Figure 8a, the below two curves are linearly. But increasing with b_r, the increasing ratio becomes small, and the upper curve becomes nonlinear. This shows that there is an appropriate width and thickness of inner-stiffeners. On the other hand, width-thickness ratio should less than 15 for avoiding the plate buckling.

3 ULTIMATE STRENGTH OF T-JOINTS

Sawada et al. (1979) proposed the strength formula of the ring-stiffened joints which might be considered as the sum of the strength of an unstiffened joint and the shearing strength of a stiffening ring. This method is called "cut and try". This method will be used in this paper, and the formulae can be shown as follows:

$$N_{T,s} = N_{T,u} + N_s \quad (1)$$

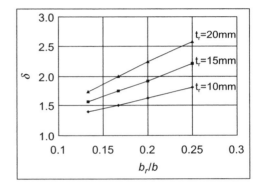

Figure 8b. Relationship between δ and b_r/b (under out-of-plane bending).

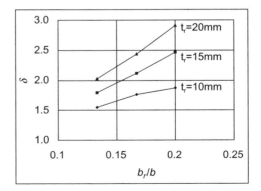

Figure 8c. Relationship between δ and b_r/b (under in-plane bending).

$$M_{TOB,s} = M_{TOB,u} + M_{s,OB} \quad (2)$$

$$M_{T,s} = M_{T,u} + M_s \quad (3)$$

Where $N_{T,u}$, $M_{TOB,u}$, $M_{T,u}$ are the ultimate strength of the unstiffened joints respectively under axial load, out-of-plane bending and in-plane bending, N_s, $M_{s,OB}$, M_s are the ultimate strength of the inner-stiffeners, $N_{T,s}$, $M_{TOB,s}$, $M_{T,s}$ are the total ultimate strength of the unstiffened joints and stiffeners.

3.1 Strength study on inner-stiffened joints under axial load

Based on numerical study, the following results were found:

- N_s increases with t_r linearly when b_r and β do not change.
- N_s increases with β nonlinearly when b_r and t_r do not change.
- The effect of t is very small, so can be ignored.

From the upper characteristic and mechanical traits of the inner-stiffeners, N_s may be assumed as:

$$N_s = f_y t_r b_r \beta^{\alpha} \cdot F_1(\eta_r) \quad (4)$$

Where $\eta_r = b_r/b$. Based on numerical results, the formula was obtained as:

$$N_s = f_y t_r b_r \beta^{0.5521} \cdot (6.0875\eta_r^2 + 2.3050\eta_r + 1.3815) \quad (5)$$

The upper formula has its appropriate parametric ranges. Beyond this range, the formula is not suitable. The parametric ranges are shown as follows:

- β is between 0.5 and 0.75.
- η_r is between 0.1 and 0.2.
- t_r is between 0.4 and 1.0.
- b_r/t_r must be less than 15, should be less than 10.

3.2 Strength study on unstiffened joints under out-of-plane bending

CIDECT (Packer et al. 1997) gave ultimate strength of RHS unstiffened T-joints, but that of CHS-RHS ones is not given. Mashiri et al. (2004) gave ultimate strength formula of CHS-RHS unstiffened T-joints under in-plane bending, which was obtained by the yield line model from the tests. In this paper, the yield line theory was used to derive ultimate strength formula of CHS-RHS unstiffened T-joints under out-of-plane bending. The failure mode is flexural failure of the chord face, i.e. β is less than 0.85. The yield line model was obtained from Von Mises stress distributing of FEA. When the deformation is up to 3% b_o, the boundary of plastic stress distributing is approximately a circular, i.e. $AB = CD$ (Fig. 9).

The energy due to the deformation of yield lines is equated to the energy due to the external moment, shown as:

$$M_{TOB,u} \cdot \varphi = m_1 \cdot \sum n_i \cdot l_i \cdot \phi_i \quad (6)$$

Where ϕ is the rotation of the brace under in-plane bending, l_i is the length of a yield line, n_i is the number of the same yield line, Φ_i is the rotation of the yield line, m_i is the plastic moment per unit length of the yield lines, which is given below:

$$m_1 = f_y t^2 / 4 \quad (7)$$

There are seven different yield line types in the model shown in Figure 9. After the parameters of every line are derived, the formula can be shown as:

$$M_{TOB,u} = f_y t^2 d_1 \cdot F(\theta, \beta) \quad (8)$$

Where $F(\theta,\beta)$ is a function of double parameters. For each value of β, a corresponding value of θ can minimize the function, so Equation 8 can be given by:

$$M_{TOB,u} = f_y t^2 d_1 \cdot \min[F(\theta,\beta)] \quad (9)$$

So a curve relating to β can be obtained, and the formula can be given by:

$$M_{TOB,u} = f_y t^2 d_1 \cdot (16.5751\beta^2 - 16.5356\beta + 8.0335) \quad (10)$$

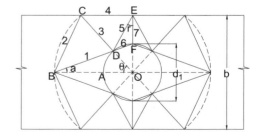

Figure 9. Yield line model under out-of-plane bending.

The study showed that there is a good agreement between the formula and the FEA results. Especially for small value of β, the difference is less than 6%. When β is equal to 0.75, the maximum difference is 12.25%. This shows that the proposed yield line model based on FEA is appropriate.

3.3 Strength study on inner-stiffened joints under out-of-plane bending

Based on numerical study, $M_{s,OB}$ increases with t_r and β linearly when b_r do not change. So $M_{s,OB}$ may be assumed as:

$$M_{s,OB} = f_y t_r b_r^2 \beta \cdot F_2(\eta_r) \qquad (11)$$

Using the same method, the equation can be given in Equation 12. The appropriate parametric ranges for Equation 5 are also suitable for the below formula.

$$M_{s,OB} = f_y t_r b_r^2 \beta \cdot (88.3740\eta_r^2 - 53.9426\eta_r \\ + 11.5234) \qquad (12)$$

3.4 Strength study on inner-stiffened joints under in-plane bending

Based on numerical study, M_s increases with t_r and β linearly when b_r do not change. So M_s may be assumed as:

$$M_s = f_y t_r b_r^2 \beta \cdot F_3(\eta_r) \qquad (13)$$

Using the same method, the equation can be given in Equation 14. The appropriate parametric ranges for Equation 5 are also suitable for the below formula.

$$M_s = f_y t_r b_r^2 \beta \cdot (6.98 + 20.8103\eta_r \\ - 133.5769\eta_r^2) \qquad (14)$$

3.5 Effective width

In the upper derived formulae, β is an important parameter. Mashiri et al. (2004), CAO et al. (1998), Kosteski et al. (2003) used the concept of "effective width", i.e. d_1 is replaced by $d_1' = d_1 + 2w$, b is replaced by $b' = b - t$, and β is replaced by $\beta' = d_1'/b'$, where w is the weld size of the brace. In fact, there are backings at the corner of the welded box section. The backing also affects the value of β', which can be verified by the tests (Cao et al. 2007).

4 STUDY ON K-SHAPED INNER-STIFFENED CHS-RHS JOINTS

The FEA model of K-shaped joints in this paper is shown in Figure 10a, which boundary conditions are simply supported at the ends of the chord. Axial loads are applied at the brace ends. One load is tension, and the other is compression. The model of inner-stiffeners is shown in Figure 10b. The corner radius (R) is 50 mm. The stiffeners lie against the middle of the braces. In this team of numerical study, the models are shown in Table 3: the length of the chords is 3 m, the outside width and depth of the chords is 600 mm, the thickness of the chords is 20 or 25 mm. The number of joints including the unstiffened joints total 42. The sections and angles between the braces and chords are same, and all of the joints are gapped K-joints.

From the FEA, the following conclusions can be drawn:

- The ultimate strength of the inner-stiffened K-joints is significantly higher than that of the unstiffened ones, and δ increases with τ_r almost linearly.
- δ decreases with larger β when same τ_r.
- δ increases with γ when same τ_r.
- On the early loading, the load-displacement curves are linearly. Along with plastic development, the slope of the curves become smaller and smaller.
- For different θ, $N_s \cdot \sin\theta$ almost changes.

Based on flexural failure of the chord face, the gapped inner-stiffened K-joints have higher ultimate strength than the unstiffened ones like T-joints. The action of the inner-stiffeners is same as T-joints, so the ultimate strength may be shown as:

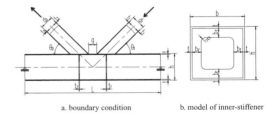

a. boundary condition   b. model of inner-stiffener

Figure 10. FEA model of K-shaped joints.

Table 3. Geometrical parameters of K-joints.

Brace size mm	β	b_r mm	t_r mm	t mm	θ_i
250 × 12 CHS	0.417	100	10	20	45°
300 × 12 CHS	0.500		15	25	50°
350 × 12 CHS	0.583		20		

$$N_{K,s} = N_{K,u} + N_s / \sin\theta \qquad (15)$$

Where $N_{K,s}$ is ultimate strength of the gapped inner-stiffened K-joints, $N_{K,u}$ is that of the unstiffened ones, and N_s is that of the stiffeners (Eq. 5). From the numerical analysis, the results are very close to Equation 15.

5 PROJECT APPLICATION OF INNER-STIFFENED JOINTS

5.1 *Inner-stiffened joints of the megatrusses of shanghai international circuit*

The large-span spindle-shaped truss system was applied to the restaurant and the press centre of Shanghai International Circuit (Figure 12). The span of trusses is 91.3 m, and the cantilever spans at each ends are 26.91 m and 17.41 m. The total length is 135.62 m, and the maximum width is 30.43 m, the maximum height of the trusses is 12.4 m. One side of megatrusses is supported on the two columns, and the other side is supported on the transfer truss which transfers the load to the below columns. The chords of megatrusses and transfer truss are welded box section. The braces of the megatrusses are CHS or welded RHS. The applied inner-stiffened joints are shown in Figures 11a, b.

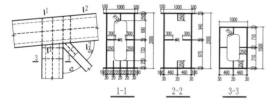

Figure 11a. Inner-stiffened joints (RHS chord and brace).

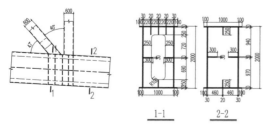

Figure 11b. Inner-stiffened joints (RHS chord and CHS brace).

Figure 12. Exterior view of the megatrusses.

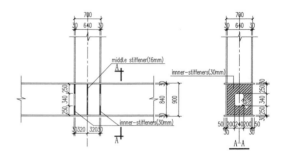

Figure 13. Moment-resisting inner-stiffened joints (RHS chord and brace).

5.2 *Inner-stiffened joints of the vierendeel trusses of shanghai's naihui administration center*

The Vierendeel truss system of Shanghai's Naihui Administration Center is composed of four trusses which spans are from 43.2 m to 81.6 m (Figure 14). The height of the trusses is 5.4 m. The supporting system of each side includes one concrete-filled CHS column and eleven concrete-filled RHS columns distributing in a triangle plan. The four trusses are linked rigidly by H-shaped beams. The section of the chords is 700 × 900 × 30 or 700 × 900 × 22 mm, and that of the braces is 700 × 700 × 30 or 700 × 700 × 22 mm. The applied inner-stiffened joints are shown in Figure 13.

6 CONCLUSIONS

FEA has been carried out in this paper. Through the analysis of loading-carrying capacity of simple T-shaped inner-stiffened joints under each kind of load condition, a set of formulae used to structural design were quantitatively summarized. Based on the

Figure 14. Photo of the Vierendeel trusses being bulit.

analysis of T-shaped joints, loading-carrying capacity formulae of K-shaped inner-stiffened joints were presented. At last application of inner-stiffened joints in two projects are introduced. The future work will be include:

- The formulae derived from numerical analysis will be verified and modified by corresponding tests.
- The ultimate strength was proposed. It is important to confirm the design strength in actual design, because the more plastic zones are dangerous.
- The rigidity of inner-stiffeners is important for moment-resisting joints. The effect of inner-stiffeners on rigidity will be studied.
- The modes of inner-stiffeners in actual projects are very complicated. It is necessary to study farther.

REFERENCES

Cao, G.F. & Yao, N.L. 2007. Load-carrying Capacity of Welded Square Inner-stiffened T-joints Under Axial Loads. *Journal of Tongji University* 35(10): 1333–1338.

Cao, J.J. & Packer, J.A. & Yang, G.J. 1998. Yield line analysis of RHS connections with axial loads. *Journal of Constructional Steel Research* 48: 1–25.

Chen, Y.Y. & Chen, Y.J. 2002. Present research of steel tubular joints. *Building structure* 32(7): 52–55.

Kosteski, N. & Packer, J.A. & Puthli, R.S. 2003. A finite element method based yield load determination procedure for hollow structural section connections. *Journal of Constructional Steel Research* 59: 453–471.

Lee, M.M.K. & Llewelyn-Parry, A. 1999. Strength of ring-stiffened tubular T-joints in offshore structuresóa numerical parametric study. *Journal of Constructional Steel Research* 51: 239–264.

Lu, L.H. & de Winkel, G.D. & Yu, Y. et al. 1994. Deformation limit for the ultimate strength of hollow section joints. In: P. Grundy & A. Holgate & M.B. Wong (eds), *Tubular structures VI*: 341–347. Rotterdam: Balkema.

Mashiri, F.R. & Zhao, X.L. 2004. Plastic mechanism analysis of welded thin-walled T-joints made up of circular braces and square chords under in-plane bending. *Thin-walled Structures* 42: 759–783.

Myers, P.T. & Brennan, F.P. & Dover, W.D. 2001. The effect of rack/rib plate on the stress concentration factors in jack-up chords. *Marine Structures* 14: 485–505.

Packer, J.A. & Henderson, J.E. & CAO, J.J. 1997. *Design manual of hollow structural section*. Beijing: Science Press.

Rama Raju, K. & Keshava Rao, M.N. 1995. Elastoplastic behaviour of unstiffened and stiffened steel tubular T-joints. *Computers & Structures* 55(5): 907–914.

Sawada, Y. & Idogaki, S. & Sketia, K. 1979. Static and fatigue tests on T-Joints stiffened by an internal ring. *Offshore Technology Conference*, paper OTC 3422. Houston: USA.

Yura, J.A. & Howell, L.E. & Frank, K.H. 1978. Ultimate load tests on tubular connections. *Civil Engineering Structural Research Laboratory, University of Texas*, Report No.78–1.

Zhang, F. & Chen, Y.J. & Chen, Y.Y. et al. 2004. Effects of ring-stiffeners on the behaviour steel tubular joints. *Spatial structures* 10(1): 51–56.

Zhao, X.L. 2000. Deformation limit and ultimate strength of welded T-joints in cold-formed RHS sections. *Journal of Constructional Steel Research* 53: 149–165.

Experimental study on overlapped CHS KK-joints with hidden seam unwelded

X.Z. Zhao & Y.Y. Chen
Tongji University, Shanghai, China

G.N. Wang
Shanghai Municipal Engineering Design Institute, China

ABSTRACT: Overlapped circular-hollow-section double K-joints, either overlapped in longitudinal or in transverse direction, are widely used in large span structures in China. In practice, the hidden toe of this type of joints is normally left unwelded due to the difficulty resulted from the fabrication sequence. It is thus necessary to investigate how the unwelded hidden seam may affect the behaviour of overlapped CHS KK-joints. This paper reports three static tests on multiplanar overlapped CHS KK-joints under axial loading. Among these 3 specimens, one KK-joint is designed to be overlapped in longitudinal direction, and the other two are overlapped in transverse direction, to study the effect of pattern of overlap on the joint behaviour. In order to investigate the effect of the hidden seam on the behaviour of the connection, one specimen with the hidden seam welded were also designed and tested for comparison purposes. Results of these tests show that the pattern of overlap leads to completely different failure mode of the joints; both the pattern of overlap and the welding situation of the hidden seam have significant effects on the joint ultimate capacity. Comparison of the joint capacities obtained from the tests with strength prediction from current design codes was presented.

1 INTRODUCTION

Multiplanar circular hollow section double K-joints, formed from member lying in two planes, are widely used in large span structures in China. This largely owes to the excellent properties of the CHS shape, the development of CHS end preparation machine and the geometrical stability of spatial truss with double K-joints. In the past decades, both experimental study and numerical simulation have been carried out on multiplanar gap KK-joint in order to investigate the behaviour of the joints and to establish formulations for static strength calculation. However, little substantial information on the static behaviour of multiplanar overlapped CHS KK-joints is available from the current international database of tubular connections, especially for the KK-joints considering the welding situation of hidden seam. For partially overlapped CHS double K-joints, it is a common practice to mount the CHS members on an assembly rig and tack welds them in fabrication workshops. Final welding is then carried out in a following separated operation. This sequence makes it impossible to weld the hidden toe of overlapped KK-joints. In tubular structures, this type of joints is widely used. It is thus necessary to investigate how the pattern of the overlap and how the unwelded hidden seam affects the behaviour of overlapped CHS KK-joints.

2 DESCRIPTION OF OVERLAPPED CHS KK-JOINT

Overlapped CHS KK-joints are one of the fundamental joint configurations used in spatial truss structures. The configuration and notation describing the geometry of overlapped KK-joints are shown in Figure 1 and Figure 2. The through brace is welded entirely to the chord, whereas the overlapping (lap) brace is welded to both the chord and the through brace. In an overlapped joint, all or part of the load is transferred directly between the two braces through their common weld, which may lead to a change in the chord wall thickness required.

Compared with gap KK-joints, more factors, including the pattern of overlap (the longitudinal overlap, i.e. lap-in-plane and gap-out-of-plane LIGO, or the transverse overlap, i.e. gap-in-plane and overlap-out-of-plane GILO) and the presence or absence of the hidden weld, need to be considered for overlapped KK-joints, in addition to the joint geometry parameters (the ratio of the mean diameter of the brace member

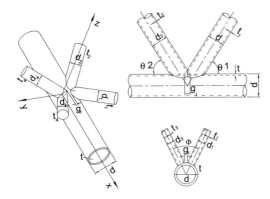

Figure 1. Configuration of overlapped CHS KK-joint.

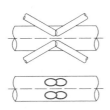

Figure 2a. Longitudinal overlap.

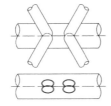

Figure 2b. Transverse overlap.

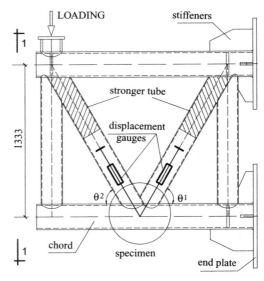

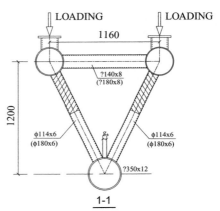

Figure 3. Overlapped CHS KK-joint specimen.

to that of the chord β, the ratio of the chord width to twice its wall thickness γ, and the ratio of the mean thickness of the brace member to that of the chord τ), the relative material strengths of the brace member and the chord, and the loading hierarchy reversal as Dexter (Dexter & Lee, 1999) listed.

3 TESTS OF OVERLAPPED KK-JOINTS

3.1 *Test specimens*

In total three specimens of multiplanar overlapped CHS KK-joints were tested under the static axial loading. Two of them are absence of the hidden seam; and for comparison purpose, the other one is presence of the hidden seam to study the effect of hidden seam on the behaviour of the connection.

A typical CHS overlapped KK-joint specimen is shown in Figure 3. To simulate the loading pattern that may occur within a KK-joint in the framework, the specimen was designed to be part of a cantilever truss. To ascertain the internal force of each brace members, the parts of the members far away the joint to be tested are manufactured by different tubes with higher steel strength and larger thickness to keep them elasticity in the whole loading process.

To minimize the effects of the secondary moments in the joints, the brace length is designed to be at least nine times of the brace diameter. End profiles of braces were prepared by automatic flame cutting. Weld details were in accordance with the Technical Specification for Welding of Steel Structures of Building JGJ81-2002 (China); but the throat thickness of the fillet weld was varied for different specimens. The details of all the specimens are given in Table 1. The material properties of the chord and braces, as well as the welding material, are listed in Table 2.

As shown in Table 1, specimen SJ1 is a multiplanar overlapped CHS KK-joint with the longitudinal overlap and the hidden seam unwelded; specimen SJ2 and SJ3 with the transverse overlap. Specimens

Table 1. Details of test specimens.

No.	d × t	$d_i × t_i$	θ_i	ϕ	β	γ	τ	g_l (mm)	g_t (mm)	Throat thickness	Condition of hidden seam	Pattern of overlap
SJ1	ϕ350 × 12	ϕ114 × 6	58°	52°	0.326	15	0.5	−41	41	12 mm	Unwelded	LIGO*
SJ2	ϕ350 × 12	ϕ180 × 6	50°	52°	0.514	15	0.5	59	−31	10 mm	Welded	GILO**
SJ3	ϕ350 × 12	ϕ114 × 6	50°	52°	0.514	15	0.5	59	−31	10 mm	Unwelded	GILO**

* LIGO = lap-in-plane and gap-out-of-plane, the longitudinal overlap.
** GILO = gap-in-plane and overlap-out-of-plane, the transverse overlap.

Table 2. Results of the tensile coupon tests.

Specimen nominal size	Yield strength (N/mm²)	Ultimate strength (N/mm²)	Elongation (%)	Grade of Steel
Chord (ϕ350 × 12)	324.5	487.4	16.7	Q235B
Brace (ϕ180 × 6)	382.6	544.9	12.2	Q235B
Brace (ϕ114 × 6)	344.6	511.9	20.5	Q235B
Welding material	450.4	583.4	23.1	N/A

Figure 4. Test setup.

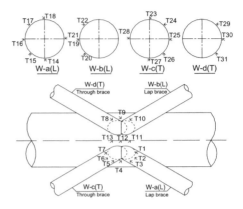

Figure 5a. Arrangement of rosette strain gauges of SJ1.

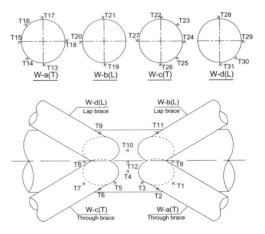

Figure 5b. Arrangement of rosette strain gauges of SJ2 & SJ3.

SJ2 and SJ3 have the same size of chord/braces, but the former has the hidden seam welded and the latter unwelded. The fillet weld in SJ1 had a relatively larger throat thickness.

3.2 Test setup

The specimens were tested using the long truss-beam system facility housed in the State Key Laboratory for Disaster Reduction in Civil Engineering (Tongji University, China), as shown in Figure 4. The overlapped CHS KK-joint specimen is designed to be part of a cantilever truss, which was fixed to the long truss beam through its end plates. The vertical loads were applied to the end of the cantilever truss by two 320 ton hydraulic jacks with a spreader beam, which is balanced by four columns anchored to the test floor.

3.3 Measurement program

In total 31 rosette strain gauges were applied at the intersections of chord and braces to trace the stress distribution, as shown in Figure 5. In order to obtain the actual loads applied in the chord and the braces, 15 strain gauges were attached on the chord/braces surfaces along the length of each member.

The deformations of the chord wall under axial brace loadings were measured using four wire displacement gauges (D1–D4) attached to bolts welded at the intersection of axes of the chord/braces and at the central line of each brace (as shown in Figure 3). Two displacement gauges were placed at the free end of the cantilever truss vertically and two gauges at the fixed end of the truss horizontally to obtain the actual vertical displacement under the hydraulic jacks. Another two displacement gauges were placed horizontally to obtain the out-of-plane deflections of the truss for safety reason.

4 TEST RESULTS OF KK-JOINTS

4.1 Brace load-displacement relationships

Figure 6 shows typical brace load-displacement plots. The load, N, is the internal force in the braces calculated from the measured strains in the braces; the displacement is the deformation of the chord wall under the respective through/lap brace loading in the direction of the original brace axis, calculated from the measured value of the wire displacement gauges deducting the elastic deformation of the braces. Tensile force in the brace is nominated to be positive and compressive force negative.

It can be seen that: (1) the two through (or lap) braces in SJ1 with LIGO have the same ultimate capacity, yet the two braces in SJ2 and SJ3 with GILO have different capacities; (2) both the ultimate capacity and the deformation capacity of SJ2 are larger than that of SJ3 with the hidden seam unwelded.

4.2 Strain distribution

Strain distribution near the welds between braces and chord are shown in Figure 7 in the form of the equivalent strain versus the rosette strain gauge number illustrated in Figure 5.

It can be seen that: (1) for SJ1, the intersection point (IP) of the through brace, lap brace and the chord in one plane, reaches the yield stress of the steel first, and the equivalent strain at the IP, obtained from the rosette stain stuck on the outer surface of the tube, remains to be the largest and the strain at the transverse gap to be the smallest; yet for SJ2 and SJ3, the equivalent strain at the

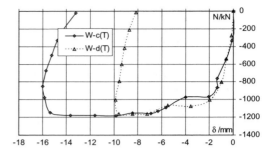

Figure 6a. Compressive brace load-deformation plot of SJ1.

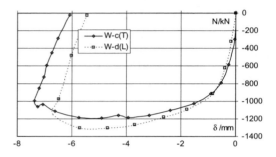

Figure 6b. Compressive brace load-deformation plot of SJ2.

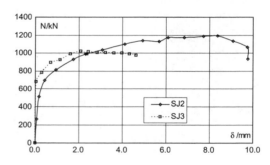

Figure 6c. Brace W-a(T) load-deformation plot of SJ2 & SJ3.

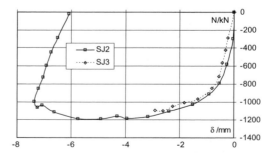

Figure 6d. Brace W-c(T) load-deformation plot of SJ2 & SJ3.

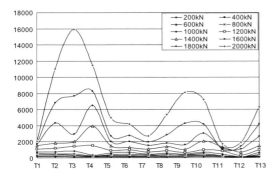

Figure 7a. Strain distribution of the chord of SJ1.

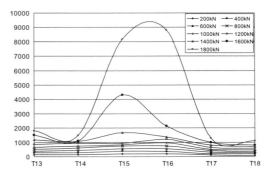

Figure 7c. Strain distribution of the brace W-a(T) of SJ2.

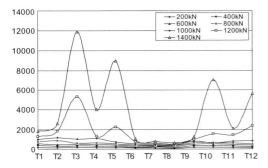

Figure 7b. Strain distribution of the chord of SJ3.

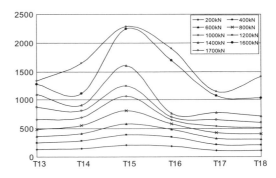

Figure 7d. Strain distribution of the brace W-a(T) of SJ3.

longitudinal gap in each plane yields first and remains largest during the whole loading process, as the uniplanar gap K-joint behaves. (2) for SJ1, the strain at the saddle spot is larger than that at the crown spot; but it is reverse for SJ2 and SJ3. (3) the equivalent strain on the chord/braces of SJ2 with hidden seam welded is smaller/larger than that of SJ3, which indicates the presence of the hidden weld significantly affects the joint behaviour.

4.3 Failure modes

For specimen SJ1, of which the overlap lies in one plane, local buckling of the compressive brace near the common weld and the necking of the tensile brace were observed (Figure 8a), then the cracks, starting from the IP occurred and propagated along the welds while the joint stiffness decreased, as shown in Figure 8b. For SJ2 and SJ3, the obvious plastic deformation of the chord at the longitudinal gap was first found (Figure 8c), then the crack at toe of tensile braces occurred and extended; the crack of SJ3 is severer than that of SJ2 with hidden seam welded (Figure 8d).

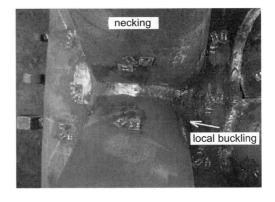

Figure 8a. Failure mode of SJ1 (local buckling).

4.4 Ultimate capacity of overlapped KK-joints

Three potential failure criteria were considered and checked for each joint to get the ultimate capacity: (1) the peak load on the brace load-displacement plot; (2) deformation reaches its limit, i.e. 3% of the chord

Figure 8b. Failure mode of SJ1 (crack propagating).

Figure 8c. Failure mode of SJ2 & SJ3 (chord plasticity).

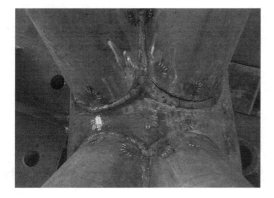

Figure 8d. Failure mode of SJ3 (crack extends).

Table 3. Measured and predicted specimen ultimate capacity.

No. of specimen	Ultimate capacity Test /kN	Ratio of prediction to test ultimate capacity			
		GB50017	EN1993	AWS	M.LEE
SJ1	916	0.78	0.82	0.66	0.75
SJ2	1197	0.73	0.73	0.84	0.88
SJ3	1023	0.84	0.83	0.96	1.01

diameter; (3) visually observable crack initiates. The ultimate capacity of joints is defined as the lowest load in all braces obtained from (1), (2) and (3). The ultimate capacity of each specimen, as well as the comparison with the formulae predictions from GB50017 (2003), EN1993 (1992), AWS (2004) and M.LEE (1996) are listed in Table 3. The measured dimensions and material properties were used for calculating the ultimate capacities by these code predictions with no safety factor considered. Note that (1) for specimen SJ1 with LIGO, the ultimate capacity of the joint increased due to the load transfer between the two braces through the common weld, resulting a larger measured joint capacity than code predictions; (2) the ultimate capacity of SJ2 is 17 percent larger than that of SJ3; (3) the measured joint capacity of SJ2 & SJ3 agrees well with the AWS predictions, due to AWS taking account of the parameter of transverse gap.

5 NUMERICAL STUDIES AND DISCUSSION

5.1 Finite element analysis

Elasto-plastic large displacement analysis was carried out on all specimens tested using the general purpose FE package ABAQUS to obtain the stress distribution and load-carrying capability of the joints. The contact situation of hidden seam, as well as the weld configuration and material property, were also taken into account.

Figure 9 illustrates the failure mode of the local buckling of compressive brace, the necking of the tensile brace and the chord plasticity obtained from FE analysis. Good agreement is achieved and thus the finite element model generated is validated.

5.2 LIGO KK-joint

As for the CHS KK-joint with the configuration of lap-in-plane and gap-out-of-plane, the failure mode observed in the test appeared quite similar to the CHS uniplanar overlapped K-joint (Zhao, 2006). Thus parameters, including brace-to-chord thickness ratio τ, transverse gap g_t and the presence of hidden welds, were selected and parametric studies were then performed on 24 models to investigate KK-joint behaviour using

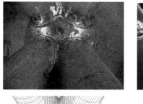
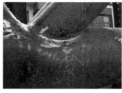
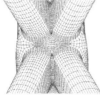
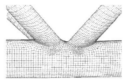

Figure 9. Failure modes by numerical simulation.

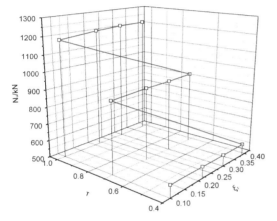

Figure 10a. Joint capacity vs. τ & ξ_t (hidden seam unwelded).

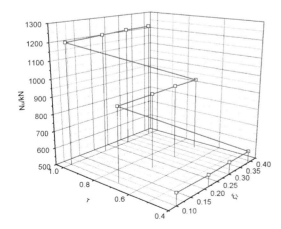

Figure 10b. Joint capacity vs. τ & ξ_t (hidden seam welded).

the above validated FE model. Figure 10 illustrates the results. It can be seen that the brace-to-chord thickness ratio has significant effect on the LIGO KK-joint ultimate capacity, but the transverse gap and the presence of hidden weld has little effect on it.

5.3 GILO KK-joint

As for the CHS KK-joint with the configuration of gap-in-plane and lap-out-of-plane, the failure mode observed in the test appeared similar to the CHS uniplanar gap K-joint, but the two compressive (or tensile) braces behaved differently due to the two braces contacting the chord asymmetrically, which thus causes the deformation and cracks near through brace are larger than that near lap brace, as shown in Figure 8. This also indicates that the presence of the hidden welds will significantly increase the ultimate capacity of this kind of joint and the hidden welds have to be made.

6 CONCLUSIONS

Experimental and numerical studies on multiplanar overlapped CHS KK-joint with various patterns of overlap and welding situation were carried out and presented in this paper. The LIGO CHS KK-joint performs as uniplanar overlapped K-joint and thus the absence of hidden weld has little effect on the joint ultimate capacity; the GILO KK-joint appears similar to the uniplanar gap K-joint and the presence of hidden weld will greatly affect the joint behaviour.

REFERENCES

AWS D1.1/D1.1 M: 2004. *Structural welding code—Steel (19th Edition)*. American Welding Society.

BS EN 1993-1-8: 2005. *Eurocode 3: Design of steel structures—Part1–8: Design of joints*. British Standards Institution.

Dexter, E. & Lee, M.M. 1999. Static strength of axially loaded tubular K-joints I: Behaviour. *Journal of structural engineering, ASCE*, 125(2): 194–201.

GB50017: 2003. *Code for design of steel structures*. National Standards of P.R.China.

Lee, M. & Wilmshurst, S.R. 1996. Parametric study of strength of tubular multiplanar KK-joints. *Journal of structural engineering, ASCE*, 122(8): 893–904.

Makino, Y. & Kurobane, Y. 1994. Tests on CHS KK-joints under anti-symmetrical loads. In R. Puthli & S. Herion (ed.), *Proceedings of the Sixth International Symposium on Tubular Structures, Australia, 14–16 December 1994*. Rotterdam: Balkema.

Wardenier, J. 2002. *Hollow sections in structural applications*. CIDECT.

Zhao, X.Z., Chen, Y.Y & Chen, Y. 2006. Experimental study on overlapped CHS K-joints with hidden seam unweldedd. In J.A. Packer & S. Willibald (ed.), *Proceedings of the eleventh International Symposium on Tubular Structures, Canada, August 2006*. Rotterdam: Taylor & Francis.

Behaviour of semi-rigid steelwork connections of I-section beams to tubular columns

A.H. Orton
Corus Tubes, UK

ABSTRACT: This paper reviews the use of semi-rigid site connections between I-section beams and tubular columns and also presents new analyses and test data on the behaviour of a particular type of semi-rigid connection, the double angle bolted connection. The connections reviewed are those made on site and can be characterised as plastic, semi-continuous and partial strength. They have important applications in practice because of their economy and because of the many cases where tubular columns are required to develop frame action, for example because cross-bracing has to be excluded.

Although much of this data already exists for open section columns, such as H-section columns, very little data exists on the behaviour of tubular columns. Similar types of semi-rigid connections, using H-section columns, have been used in traditional American and British practice to develop frame action against wind forces. In the present paper, the use of frames with tubular columns is presented not only as a means of resisting wind forces but also of absorbing seismic forces in zones of moderate activity.

1 INTRODUCTION

1.1 Design background

Current steel design practice favours the use of so-called 'simple' beam to column connections, in the vast majority of steel framed buildings that are erected worldwide. In this context, the use of the word 'connection' implies that this is a connection made on site externally, usually with bolts; a 'joint' is used as a generic term but also often refers to a joint made under controlled conditions, usually by means of welding. For the purposes of calculation, 'simple' connections are assumed to behave structurally like hinges. In a minority of cases, rigid joints are used. Semi-rigid joint design has a role in those cases, where frame action is required because there is no cross-bracing but where full rigidity is not required. However it has yet to establish itself as a valid general method in steel frame design but there are important reasons for thinking that it will do so in the future (Nethercot 2007).

Firstly, the cost of making semi-rigid connections is the same or very little more than that of the so-bcalled simple connections, because these 'simple' connections in fact require some rigidity in order to facilitate erection of the frame. Taking account of the rigidity of the connection, however, will produce savings both in height and weight of the beam elements, particularly where deflection is a critical factor. Secondly, many of the analytical difficulties in calculating the effects of the connection restraint, including the effect of this on column design, are being eliminated, largely by finite element analysis and methods based on the finite element analysis. Compared with rigid joint design, the use of semi-rigid joint can be justified by its greater reliability and ductility compared with the highly restrained and complex stress states associated with rigid welded joints.

1.2 New design requirements

Almost all steelwork joints in buildings in non-seismic areas are designed so as to resist the static and wind loads that might be applied to the structure in its normal configuration. In seismic areas, joints in new structures are designed to resist, to some specified damage level, the dynamic loads caused by the seismic event. However a new factor now presents itself: to an increasing degree, owners and the building authorities require that buildings, including those in non-seismic areas, be able in some degree to resist exceptional loadings such as those due to impacts, gas and other kinds of explosions, as well as those due to fire and seismic events, if relevant. In some countries these requirements have already been translated into legislation. This development has the potential to completely transform the existing practice of connection design because additional properties will be required of the connection particularly greater ductility. These new connection designs, if successfully achieved, will work as semi-rigid joints, and should obtain superior performance for very little additional cost. Generally, in the case of tubular columns, the site connections between the tube and the open-section beam can only be designed as pinned, because of a lack of data on semi-rigid connections; however some welded tubular joints do have known semi-rigid joint characteristics (Kurobane et al. 2004).

2 NEW CONNECTION DESIGNS FOR TUBE

2.1 Generic connection types

There are two principal methods of connecting steel I-section beams to steel hollow sections: either by use of a blind bolting system, in which the fixing bolt is secured against the wall of the hollow section on the inside by an expansion of the main body of the bolt (Figures 1 and 2) or by welding of an attachment to the hollow section column to which the beam may be fixed by normal structural bolts (Figure 3). It is the object of this paper to propose two general types of semi-rigid joint and to examine one of these types in more detail. The aim is that the site connection detail provide, economically, semi-rigid joint properties, including sufficient bending stiffness and ductility to meet the new requirements for better performance under exceptional loads.

2.2 Particularities of connections to tube

A special feature, and particularly difficulty, of connection design using tubes is the relative thinness of the tube wall, as compared, say, with that available when designing with open sections. The economy of designing with tubes will most often dictate that the tube has the smallest possible column wall thickness and this means that reinforcement is generally necessary for moment connections. Recent research, supported by Corus Tubes, on semi-rigid connections for tubular columns has been aimed at those connections which have the necessary stiffness and strength, as well as good ductility to meet demands for performance under extreme loading conditions. Work at present under review has centred on two particular connection types: the first, the Reverse Channel connection, arising out of work testing the robustness of different types of connection under fire conditions (Ding 2007); the second, the Double Angle connection, arising from work on partial strength connections for H-section columns in seismic areas (Elghazouli 2006).

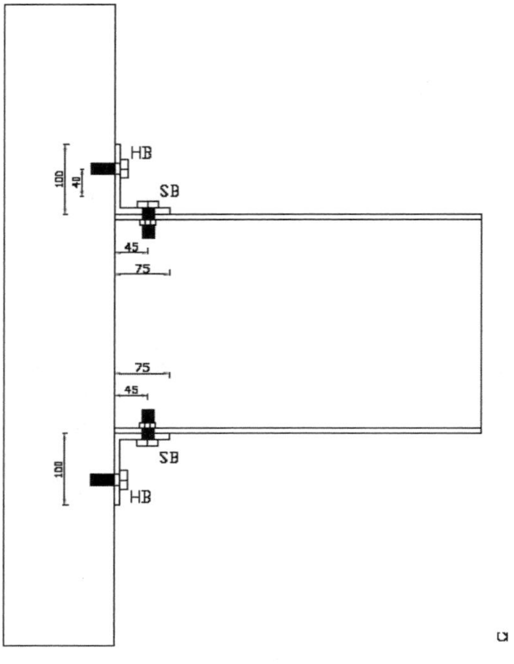

Figure 1. Double angle connection of I-beam to hollow section column with 100 mm vertical legs of angle attached by blind bolts to the column face.

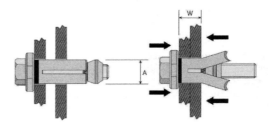

Figure 2. Action of blind bolts in Hollo-Bolt system.

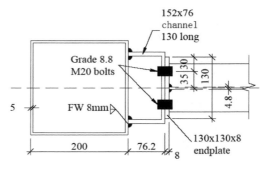

Figure 3. Plan view on reverse channel connection for I-beam to hollow section column.

3 DOUBLE ANGLE CONNECTION

3.1 Analysis of overall response of connection

A double angle connection consisting of top and seat angles attached to a tubular column, with a web plate as necessary, was selected for analysis as this detail had already been examined with open section columns and found to perform well under wind and moderate seismic forces (Figure 1). The shelf angles at the top and bottom of the beam are expected to work in double bending under wind or seismic loads, at the limit forming two plastic hinges in the angle. Previous work with open sections showed that this type of connection had good rotation capacity and could achieve a yield moment of about 40% of the moment capacity of the beam (Elghazouli 1998).

A numerical investigation at Imperial College London was undertaken on a T-frame consisting of a 2 metre long cantilever beam attached at mid-height to one side of a hollow section column with an overall height of 2 metres (Figure 4). The column is assumed fixed at its ends. This frame was used to examine the interaction between the top angle and the tube wall in contact with it (Ezequiel 2007). The angle chosen for the finite element analyses was a $100 \times 75 \times 8$ for both the top- and seat-angle; the beam was a $305 \times 102 \times 25$ I-beam and the column a $150 \times 150 \times 5$ square hollow section. The seat-angle was connected to the column face at points along the horizontal and half of its depth (shown at left in Figure 5), while the top-angle was assumed bolted to the column face by 2 bolts 61 mm apart (shown at right in Figure 5); both angles are also assumed bolted to the beam flanges. The connection between points on the angle and the corresponding points on the beam and column was achieved in the model by coupling of nodes, which allows for two or more nodes to be given the same degrees of freedom. Two analyses were run; a real case and, for comparison, an idealised case in which the top-angle was given a yield strength ten times that of the other elements so that it remained elastic at all times. Figure 6 gives the moment-rotation curves and it can be seen that the top angle elastic case gives higher moments, indicating the influence of the angle in the real case. The column wall also has a large influence, as shown by the plastic action that occurs in the top angle elastic case. Figure 7 shows the column face response for this same arrangement, the nearly identical yield points and similar plastic action for the two cases indicating the influence of the column wall thickness. Figures 8 and 9 treat the angle response and show the horizontal displacement of the top angle, defined as the difference between

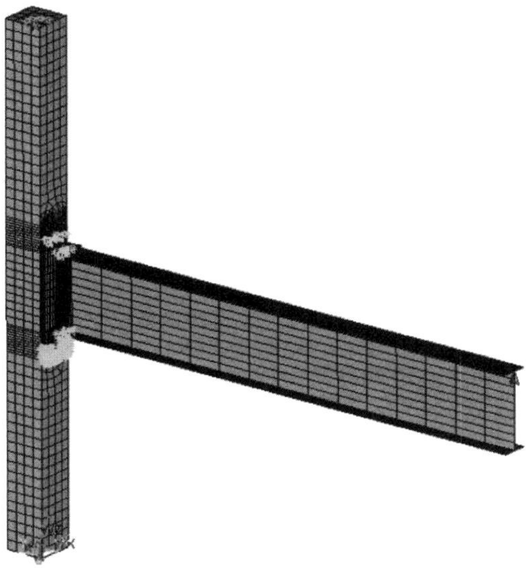

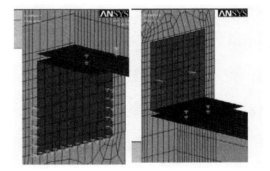

Figure 5. Detail of meshing at seat angle, left, and top angle, right, showing coupled points in finite element analysis.

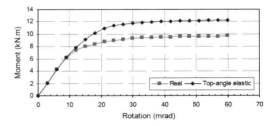

Figure 4. T-frame with 2 m long beam and 2 m high column showing elements and meshing.

Figure 6. Moment versus rotation curve of double angle connection for real and top angle elastic case.

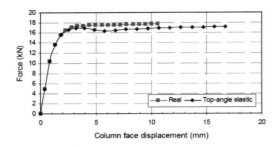

Figure 7. Force on one bolt versus displacement at coupled bolt positions of top angle for real and top angle elastic case.

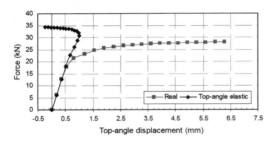

Figure 8. Force on bolts in top angle versus horizontal displacement of top angle for real and top angle elastic case.

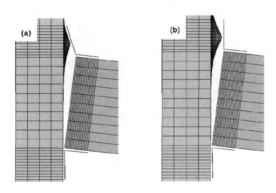

Figure 9. Displacement of top angle for (a) real case and (b) top angle elastic case.

the displacement of the heel of the angle and that of the column face at the bolt position, against the horizontal force measured at the coupled bolt nodes. This demonstrates that essentially column face is controlling the capacity and this may be compared with the test results given in Section 5.

3.2 Parametric studies

A series of parametric studies was also conducted using the same finite element model. The parameters investigated were column wall thickness, column width, column depth, beam depth and bolt spacing. As expected the most sensitive parameters were column thickness and width. The results for the study of column wall thickness are shown in Figure 10. A 150 × 150 square hollow section was analyzed with wall thicknesses varying from 5 mm to 12.5 mm and with the same bolt spacing of 61 mm. The three graphs showing overall rotation, column face and top angle displacement for the real case clearly show that there is a critical thickness, just above 6.3 mm, at which behaviour is essentially linear and at which the maximum moment is governed by the angle thickness.

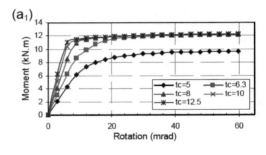

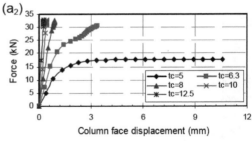

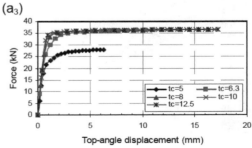

Figure 10. Effect of column wall thickness on behaviour of connection for real case.

Examination of the data for the same column but in which the column used was 150 deep, with a 5 mm column wall thickness, with the column width parameter varying from 150 mm to 250 mm showed an approximate decrease in moment capacity of 10% for every 50 mm increase in width. By contrast varying the column depth had very little effect on moment capacity and deformation, indicating that data with only different depth columns could be compared with very little error.

4 TESTS ON DOUBLE ANGLE CONNECTION

Following on from the analytical investigations, tests were undertaken on a double angle connection between a 305 × 102 × 25 I-beam and a 150 × 150 × 10 hollow section column using 100 × 75 × 8 angles (Elghazouli 2007). These are the same element sizes as those used in the finite element analysis, except that a 10 mm column wall thickness was selected for the test, as the parametric analysis had indicated that at this thickness the angle would control the maximum moment developed. The analysis (Section 3) had assumed that there would be no bolt pull out but this was not always achieved in the tests.

The experimental programme comprised seven tests on open beam to tubular column sub-assemblages. For this purpose a test rig was built, as illustrated in Figure 11. As shown, the hydraulic actuator attached to the rig is aligned with the end of the beam. On the other hand, the column is fixed at both ends by utilization of a supplementary rig. The hydraulic actuator, which has maximum stroke capacity of 250 mm, is used to apply a vertical displacement to the end of the beam. The vertical displacement and corresponding force are recorded by the load cell and displacement transducer incorporated in the actuator. Due to the nature of the of semi-rigid connections considered, it was anticipated that the beam end would observe significant vertical displacements and rotations and in order to accommodate those rotations at the loading point, it was decided to construct a loading mechanism that would allow the rotation of the beam to develop without any restraint provided by the actuator.

Of the total of seven tests, four were undertaken using M16 Hollobolts to fix the angles to the column with the longer 100 mm length of the angle fixed against the vertical face of the column (Figure 1). Loading was under deformation control, either monotonic or cyclic, using either 8.8. grade or 10.9 grade Hollobolts. The moment-rotation curve with 10.9 Hollobolts under monotonic load is shown in Figure 12. An ultimate moment of 36 kNm was

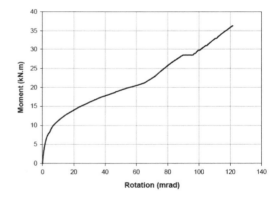

Figure 12. Moment versus rotation for double angle connection with 100 mm vertical leg.

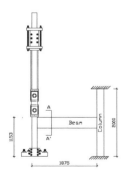 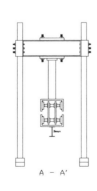

Figure 11. Elevation and cross-section through test rig with free end of cantilever on left hand side (obscured by frame).

Figure 13. Deflection of top angle showing formation of plastic hinges in the angle, at root radius and in vertical leg.

achieved which is 32% of the beam capacity; the test had to be stopped at this point because the actuator ran out of travel. The connection is classed as semi-rigid, according to the definitions of EC3, based on the initial stiffness. For capacity, the connection is classed as partial-strength. The non-linear response of the curve, after the elastic stage, is due to formation of two plastic hinges in the angle, one at its root radius and one at the centre line of the Hollo-Bolt (Figure 13). Most of the displacement at the heel of the top angle, relative to the column face, was due to bending of the vertical leg of the angle; however in other tests, with shorter vertical legs, there was significant axial displacement at the bolts caused by yielding and slippage of the Hollo-Bolts so that the top angle was not put into double bending. Because of these axial displacements, the test results could not be compared with the prior finite element analysis.

In another test with the same arrangement but this time under cyclic loading, applied according to

Figure 14. Concentration of plasticity leading to fracture in heel of angle at end of cyclic loading.

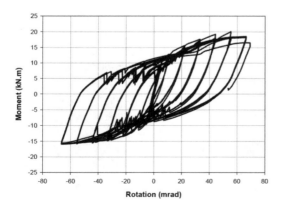

Figure 15. Plot of hysteretic curve for double angle connection with 100 mm vertical leg.

ECCS procedures (ECCS 1986), it was found that, at each deformation level, the applied load was in close agreement with that for the monotonic test and the inelastic response was similar, with very little axial deformation at the bolts, eventually leading to fracture at the root radius of the angle at the last load cycle (Figure 14) . The cyclic load gives rise to stable hysteretic loops but with some drops in load due to slippage in the bolts from sudden release of friction forces (Figure 15). Some pinching was observed. As expected there were extensive residual deflections, of the same general shape, in both top and seat angles resulting in separation of the beam from the column face.

5 TESTS ON ENDPLATE CONNECTIONS

Tests previous to this were undertaken on beams and columns of a larger size. In these tests, also supported by Corus Tubes, the test rig was in a cruciform arrangement (Figure 16). The beams were connected to the hollow section column by endplates.

Figure 17, from an unpublished report (SCI 2002), shows the result from one of the nine tests undertaken in this series to determine typical moment-rotation plots of semi-rigid connections of I beams to tubular columns. For this test a $305 \times 127 \times 48$ I-beam with a $150 \times 490 \times 15$ extended endplate, was connected to the narrow face of a $250 \times 150 \times 8$ RHS column by 8 M20 bolts threaded into sockets in the tube wall formed by the Flowdrill process. This process provides a thread in the wall of the tube so that a normal bolt may be used to attach the beam to the column. The moment-rotation curves for cycled loading is superimposed on those for which there was only one cycle of loading.

The test rig was loaded symmetrically so that the flexibility of the column web in shear is not taken into account. The measured initial test stiffness was only 11.8 KNm/mrad, less than that calculated by an elastic analysis assuming the column face to be fixed

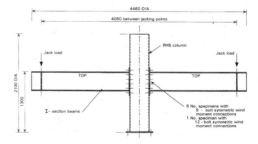

Figure 16. Test rig for beams with endplates.

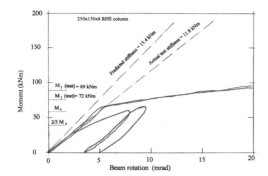

Figure 17. Moment-rotation curve for 250 × 150 column.

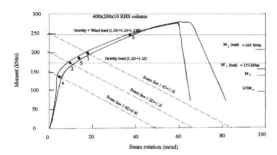

Figure 18. Moment-rotation curve, with beam lines drawn on, for 400 × 200 column.

along its vertical edges. Also shown are the moments M_1 and M_3 at which there is 1% and 3% deformation of the column face; Mc is a predicted nominal yield point, calculated as 83% of the ultimate moment based on an assumed yield pattern for the column face in tension. Figure 17 suggests that the initial stiffness is a reasonable estimate of the stiffness of the connection after shakedown.

Figure 18 shows similar results for a 406 × 178 UB 74 beam with a 200 × 595 × 15 long endplate attached to the narrow face of a 400 × 200 × 10 RHS column, this time taken to clear failure points. For a practical perspective, Beam Lines are drawn on for some typical gravity load combinations. These are based on loadings for a 4-bay long, 4-storey high office. The initial equilibrium point for the 1.2D + 1.2I (dead load + imposed load) is satisfactory (point 2) but with typical wind load moments added on (point 3), the factored moment easily exceeds that for the 3% deformation limit (M_3). Deformation at the serviceability state going from gravity load only (point 4) to full load with wind (point 5) was similarly unsatisfactory, that is greater than M_1.

These tests also indicated, as demonstrated by the finite element analysis, that the column depth only had a minimal influence on the connection performance.

6. SUMMARY

6.1 Design specification

The aim of the current research programme, supported by Corus Tubes, is to examine suitable economic types of semi-rigid connections between I-beams and tubular columns that give satisfactory performance across the complete range of the loading spectrum, covering both the serviceability states and the ultimate limit states, including those due to extreme loading. The connection requires suitable strength, stiffness and ductility for the applied load. In most economic arrangements, however, optimizing for ductility also reduces stiffness at the serviceability state and this requires some compromise in the arrangement.

The two key limiting factors on the use of tube for semi-rigid connections are: the thin walls, that are necessary for economic design, and the related issue of fixing to the tube walls, either by blind bolts or by welding of attachments to the tube for use with normal bolts. Only connection systems using blind bolting systems are considered here.

6.2 Tube wall thickness

The results presented indicate that in many applications, it will be satisfactory to use unreinforced tube walls but that in the majority of applications it will prove more economic to strengthen the walls at the connection rather than make a general increase in tube wall thickness. Work is at present taking place to devise suitable strengthening and other alternative details.

6.3 Tube wall fixings

In both the blind bolt fixing systems discussed in this paper, the axial stiffness and strength of the blind bolts were critical influences in the overall behaviour of the connection in most cases. In the endplate tests, using the Flowdrill system, structural failure in 7 of the 9 cases tested was due to sudden pull-out of 2 or 4 bolts in the upper bolt group due to stripping of threads (Figure 16). In the cantilever tests using the Hollo-Bolt blind bolting system, the pull-out force and the axial deformation of the bolts were critical in determining the deflection and ultimate moment achieved in some of the tests.

In the case of the double angle connection system, excessive axial deformation of the bolts will prevent development of the double bending in the angle that is required, so that high axial stiffness, comparable to that of a normal bolted system, is essential where significant bending moments need to be developed. However both blind bolting systems, used as recommended, are suitable for applications in which there are only limited tensile forces present. Note that the large axial deflections of the blind bolts that were found in some cases, while reducing bending stiffness, can also provide ductility, if this is required for extreme loading conditions and assuming that sufficient axial force is developed. Work is at present continuing on further improving the properties of the Hollo-bolt system, both in terms of its initial stiffness and its ultimate strength (Tizani 2003), and on general improvements in the axial stiffness of blind bolt systems by filling the hollow section column with concrete.

ACKNOWLEDGEMENTS

The free use of research work undertaken by Dr Walid Tizani of Nottingham University and by Dr Ahmed Elghazouli of Imperial College London in the preparation of this paper is gratefully acknowledged.

REFERENCES

Ding, J. (2007) Behaviour of restrained concrete filled tubular (CFT) columns and their joints in fire, PhD thesis, University of Manchester.

ECCS (1986) Recommended testing procedure for assessing the behaviour of structural steel elements under cyclic loads, Brussels.

Elghazouli, A.Y. (1998) Seismic design of steel frames with partial strength connections, *Proceedings of Sixth SECED Conference on seismic design practice into the next century, Oxford, UK*. Rotterdam: Balkema.

Elghazouli, A.Y. (2007) *Tests on top and seat angle open beam to tubular column connections* (unpublished report) Imperial College London.

Ezequiel, D. (2007) Rotational behaviour of semi-rigid top and seat angle open beam to tubular column connections, M. Sc. thesis (unpublished report) Imperial College London.

Kurobane, Y., Packer, J.A., Wardenier, J., & Yeomans, N.F. (2004). *Design guide for structural hollow section column connections*. Köln, Germany: CIDECT/Verlag TUV, Rheinland, Design Guide 9, pp. 64–72.

Nethercot, D. A. (2007) Semi-rigid and partial strength joint action: Then, now, when ? *Proceedings of 3rd International Conference on Steel and Composite Structures, Manchester, UK: Taylor and Francis* pp. 3–9.

SCI (2002). *Report on Wind-Moment Method for Frames with RHS Columns* (unpublished). Document RT934.

Tizani, W. & Ridley-Ellis, D.J. (2003) The performance of a new blind bolt for moment resisting connections. In: Jaurietta, M.A., Alonso, A., Chica, J.A., eds. *Tubular Structures X: Proceedings of the tenth symposium on tubular structures, Madrid, Spain*. Rotterdam: A.A. Balkema, pp. 395–400.

Shear capacity of hollow flange channel beams in simple connections

Y. Cao & T. Wilkinson
The University of Sydney, Sydney, NSW, Australia

ABSTRACT: A new hollow flange channel section has recently been manufactured in Australia. This paper summarises the results of a series of tests on simple shear connections of these beams to a column via the webside plate connection. A new failure mode was observed, relating to the opposing shear flows in the two hollow flange portions. A simple design model incorporating a plastic mechanism in the web was developed. Reliability analyses were performed and indicated that the design model had an acceptable degree of reliability.

1 INTRODUCTION

Smorgon Steel Tube Mills (SSTM) launched its new product—the LiteSteel Beam (LSB) in 2005. The LSB is a hollow flange channel section manufactured from a single steel strip. The base steel strip, (nominal yield stress of 380 MPa), is cold formed and welded using the unique inline Dual Electrical Resistance Welding process to create the desired shape. The flanges of the LSB, due to extensive cold-forming, have an increased nominal yield stress of 450 MPa, whilst the web remains at 380 MPa. Figure 1 shows the shape and designation of the range of LSBs.

The hollow cells of the flanges give the LSB a higher torsional rigidity compared to traditional hot-rolled I-sections of similar size. The web is generally more slender than that of a comparable I-section. Being a partially open and partially closed cold-formed section, plus the different yield stress between the web and flanges, the characteristics of LSB differ from those covered in the current design standards. It is thus necessary to study the behaviour of the LSB to verify whether the current design rules are suitable for LSB and, in cases where they are not, develop appropriate design rules. This paper investigates the behaviour of the LSB beam connected to rigid supports by single row bolted web side plate (WSP) connections. Figure 2 shows a LSB beam connected by a WSP connection.

Traditionally, WSP connections have been considered as simple (or shear) connections which transfer only shear force without producing moment at the connection (Hogan & Thomas (1994)). In the case of beam to column WSP connections, it is thought that such a connection, when the web plate is placed close to the top flange of the supported beam, provides full restraint to the beam, i.e. lateral deflection and twist restraint against flexural torsional buckling (Trahair et al. (1993)).

The current LSB Connection Design Manual (CDM) (SSLST (2005b)), which provides a range of WSP connection configurations involving different sizes and number of bolts and different web side plates, is largely based on the Hogan & Thomas (1994) design model which is based on research mainly on open sections (typically I-sections). The Hogan & Thomas (1994) design model involves calculating the strength of the elements of the connection and using the smallest as the design capacity of the connection. For the purposes of this paper, the relevant elements are the design capacities of the weld in shear, the bolt group, the web side plate in shear and bending, and the supported member in shear.

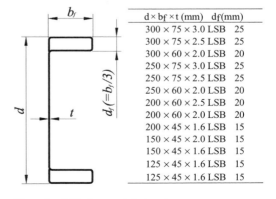

Figure 1. LSB shape and designation.

$d \times b_f \times t$ (mm)	d_f (mm)
300 × 75 × 3.0 LSB	25
300 × 75 × 2.5 LSB	25
300 × 60 × 2.0 LSB	20
250 × 75 × 3.0 LSB	25
250 × 75 × 2.5 LSB	25
250 × 60 × 2.0 LSB	20
200 × 60 × 2.5 LSB	20
200 × 60 × 2.0 LSB	20
200 × 45 × 1.6 LSB	15
150 × 45 × 2.0 LSB	15
150 × 45 × 1.6 LSB	15
125 × 45 × 1.6 LSB	15
125 × 45 × 1.6 LSB	15

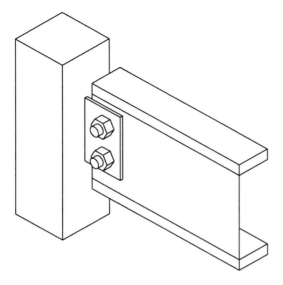

Figure 2. WSP connection of LSB beam.

SSTM also provides design manuals for residential construction (residential) (SSLST (2005c)) and industrial & commercial floors (industrial) (SSLST (2005d)) using LSB. These manuals specify WSP connections slightly different to that of the CDM and from each other. In addition, for connections between primary and secondary beams, the CDM offers an extended WSP connection to avoid coping of the secondary beams. A comparison of the various configurations, typically for a 200 LSB using two M16 bolts except where specified, is illustrated in Figure 3.

The design of LSB beams were investigated in the Queensland University of Technology (Mahaarachchi & Mahendran (2006)) and the results of the research have since been implemented in the current Australian standard for the design of cold-formed steel sections AS/NZS 4600:2005 (SA/NZS (2005)). Accordingly, the current design model for the design section and member capacities of LSB beams can be found in Clauses 3.3.2.2 and 3.3.3.3(b) of the standard respectively.

This paper investigates the behaviour of all the relevant configurations of WSP connections by experimental tests. By comparing the test results, the factors affecting the capacity of the connections and the effect of the connection on the supported beam are considered.

2 TEST PROGRAM

Tests were conducted on beams at pre-selected lengths. A range of lengths were chosen to model the

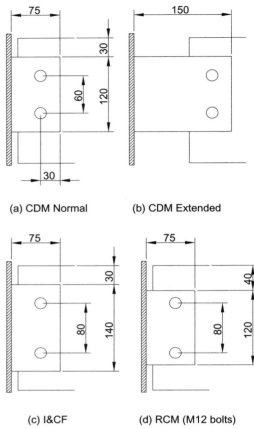

Figure 3. LSB WSP configurations.

combination of the end rotation, connection shear and beam lengths that would be seen in practice. Due to the mono-symmetric shape of the LSB section, back-to-back tests were performed for simplicity of loading and data verification.

Specially designed testing rig as shown in Figure 4 was used for the tests. Detailed information of the test program can be found in Cao & Wilkinson (2007).

3 TOP FLANGE SHEAR BUCKLING MODE

In all the tests conducted in this thesis, a tendency of the LSB flanges to displace laterally was observed (Figure 5). At the connection, the top flange of the LSB tended to displace laterally towards the heel side of the flange while the bottom flange would displace towards the opposite side (the toe side). This occurred regardless of whether the web plate was placed outside of the LSB or inside of the LSB.

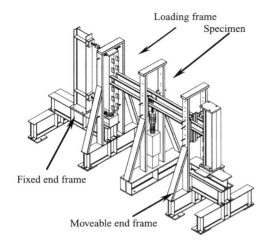

Figure 4. Testing rig.

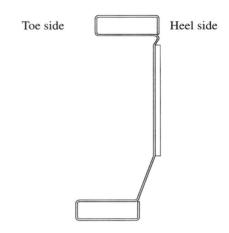

Figure 5. LSB flange lateral displacement.

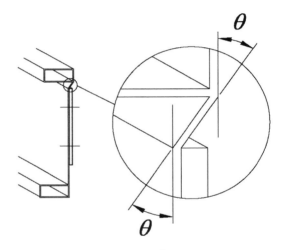

Figure 6. Top flange shear buckling.

Where extra lateral supports were provided to the LSB top flanges at the connections, the LSB top flanges were effectively prevented from lateral displacement at the connections. The bottom flanges lateral displacement were nevertheless observed in these tests and the degree of the displacement seemed to vary and was relative to factors such as the shear force at the connection, the configuration of the connection and the type of lateral support provided at the connection. Typically the increase of the shear force at the connection and/or the distance from the bottom flange to the bottom bolt of the connection would cause larger displacement of the LSB bottom flange. In relation to the different lateral supports at the connections, it was observed that the continuous flooring support seemed to result in a smaller bottom flange lateral displacement than the tests in which RHS sections were placed between the top flanges.

In tests 9, 16, 17, 18, 21, 22 and 26, no lateral support was provided at the connections and the specimens failed by the mode shown in Figure 6.

In the failure mode plastic yield lines formed in the LSB web at the position of the top edge of the web plate and the junction of the web and the flange. Consequently the top flange of the LSB displaced both laterally and downwards to eventually rest on the top edge of the web plate (Figure 7) (in test 22, flooring was fixed to the top flanges of the LSB beam specimens by screws however screw pull out failure occurred thus in effect the specimen failed as if no lateral support was provided).

The nature of the deformation observed as shown in Figure 5 was consistent with those seen in lateral distortional buckling (LDB) and therefore this failure mode was investigated first. There are, however, two factors that seem to suggest that the deformation is unlikely to be a result of LDB.

Firstly, the test results indicated that the bending moment in the connection was small. In all tests the calculated position of the point of contra-flexure (POC), at which point the bending moment was zero, were reasonably close to the centre of the bolt line (the detailed discussion on this is provided in the next section of this chapter). Thus it was evident that this failure mode could only be relevant to perhaps an extension of the LSB beam LDB at the loading point (where the bending moment was the maximum).

Secondly, although there were bending moment in the beam segment, a comparison of the test results against the AS/NZS 4600 LDB curve (Figure 8)

Figure 7. Top flange shear buckling.

Figure 9. Shear flow in LSB.

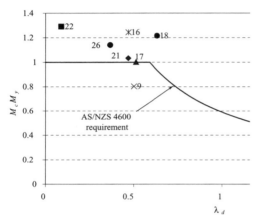

M_c = moment capacity for LDB; M_c = yield moment;
$\lambda_d = \sqrt{(\text{yield moment/elastic LDB moment})}$

Figure 8. Test results comparing to LDB.

showed that most beams in the tests (all but Test 18) were in the range where the bending moment capacity was not affected by LDB. This indicated that it was unlikely that LDB was the primary cause in this failure.

A different failure mode, resulting from the horizontal shear stresses in the hollow flanges, was therefore considered. This was being termed as "Top Flange Shear Buckling" (TFSB).

Due to the shear force present in the LSB, the shear flow in the section, as a result of an analysis by a software developed in The University of Sydney—ThinWall, is shown in Figure 9.

To simplify the calculation of the TFSB effect, the following assumptions are made:

1. All the corners of the LSB are 90 degrees.
2. The slight difference of the shear flow between the top and bottom horizontal segments of the top hollow flange of the LSB is ignored.
3. The effect of the shear flow in the vertical segments of the top hollow flange is ignored.

As shown in Figure 10(c), suppose that a vertical downward shear force V is applied to the LSB, the maximum shear flow in each of the hollow flanges would approximately be $V/2d_1$ where d_1 is the depth of the web. Thus the shear force in each of the horizontal segments of the top hollow flange (each with a triangular shear flow distribution) can be expressed as the area of the shear flow triangle. As a result the total shear force in the top hollow flange V_f because of the shear force V can be calculated as:

$$V_f = 2 \times \left(\frac{1}{2} \times \frac{V}{2d_1} \times b_f \right) = \frac{Vb_f}{2d_1} \qquad (1)$$

where: b_f = the width of the LSB flange.

If yield lines with a length of l_1 are formed in the LSB web at the top of the web plate and at the root of the top hollow flange as shown in Figure 10(d). And the portion of the LSB web between the top of the web plate and the top hollow flange of the LSB (denoted g_1 in Figure 10(a)) rotates an angle θ, from energy conservation:

$$V_f \times g_1 \times \theta = 2 \times M_p \times \theta \times l_1 \qquad (2)$$

where: M_p = the plastic moment of the web segment and may be calculated by:

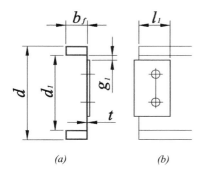

Table 1. Experimental values for γ.

Test	g_1 Nominal (mm)	g_1 actual (mm)	γ
9	15	12	0.80
16	10	6	0.60
17	10	7	0.70
18	15	10	0.67
21	20	16	0.80
22	15	13	0.87
26	10	6	0.60
		Mean	0.72
		Std. Dev.	0.11

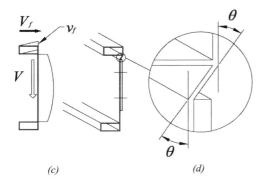

Figure 10. Shear lag effect calculation.

$$M_p = \frac{t^2}{4} f_{yweb} \quad (3)$$

where: t and f_{yweb} = the thickness and yield stress of the LSB web respectively. Therefore the shear force V required to cause the TFSB failure can be calculated as:

$$V = (t^2 f_{yweb}) \frac{l_1}{g_1} \frac{d_1}{b_f} \quad (4)$$

In the application of the equation (4) to the actual tests, it was noticed that, while some dimensions were easy to determine, the dimensions l_1 and g_1 might be difficult to define or determine. Firstly l_1 might be defined as the length of the web supported by the web plate (Figure 10(b)), however it was observed in the tests that the yield line would usually extend beyond that length as shown in Figure 7. Secondly g_1 would typically be smaller than the nominal value because of the factors such as in reality the bolt holes in both the web plate and the LSB web were 2 mm larger in diameter than the bolts, position of the bolt holes might be slightly inaccurate, etc. It was thus thought necessary to take into account of such uncertainty by adding factors α and γ for l_1 and g_1 respectively in equation (4). Consequently the equation became:

$$V = (t^2 f_{yweb}) \frac{\alpha l_1}{\gamma g_1} \frac{d_1}{b_f} \quad (5)$$

To determine the factors α and γ, experimental data was used. Firstly the actual values of g_1 were measured in tests 9, 16, 17, 18, 21, 22 and 26, where shear lag failure mode eventuated. The experimental value for γ were then determined as in Table 1 and taken as the mean ratio of the actual g_1 and the nominal g_1.

The experimental value of α was then determined as the mean ratio of the maximum shear reached in tests V_{test} and the predicted shear load V_{pred} calculated by using equation (5) with $\alpha = 1$ and $\gamma = 0.83$, as listed in Table 2.

An initial value of 1.4 and 0.8 for α and γ respectively were selected as a result. To finalise the values of α and γ, a reliability analysis, based on the First Order Second Moment (FOSM) method was carried out using different combinations of α and γ.

The FOSM method, details of which will not be discussed here, primarily involves the calculation of a safety index β. In this thesis the computation of such index was based on the mean ratio of the ultimate test loads to the predicted failure loads using different combinations of α and γ, factors determined pursuant to AS/NZS 4600 (2005) and AS/NZS 4600 Commentary (1998) (including a safety factor ϕ of 0.8) and those provided by Smorgon Steel Tube Mills from their long term tests. The calculated safety index β was then plotted against a term $D_n / (D_n + L_n)$, which considered the effects of different load combinations ranging from "dead load only" to "live load only".

Figure 11 and Table 3 show the safety index calculated.

The AS/NZS 4600 Commentary (SA/SNZ (1996b)) specifies, in Section C1.6.2.1, a target value of the safety index of 2.5 and 3.5 for the design of members and connections respectively. For the design of the

Table 2. Experimental values for α.

Test	V_{test} (kN)	V_{pred} (kN)		α
9	28.6	24.7		1.16
16	46.6	38.6		1.20
17	38.4	38.1		1.01
18	30.8	24.1		1.28
21	42	19.0		2.22
22	33.3	24.3		1.37
26	56.8	38.3		1.48
			Mean	1.39
			Std. Dev.	0.37

Table 3. Safety index.

Factor	$\alpha = 1.3$	$\alpha = 1.4$	$\alpha = 1.3$	$\alpha = 1.3$	$\alpha = 1.2$
Combination	$\gamma = 0.8$	$\gamma = 0.8$	$\gamma = 0.7$	$\gamma = 0.9$	$\gamma = 0.8$
β minimum	2.29	2.05	1.86	2.67	2.55

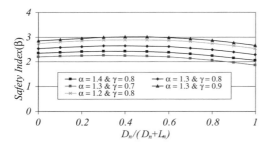

Figure 11. Safety index.

(a) TFSB

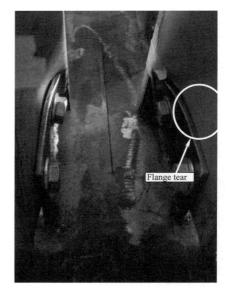

(b) Flange Tear

Figure 12. Flange tear after TFSB.

TFSB, it is recommended that a target value of 2.5 should be used. There are two main reasons for the recommendation. Firstly the failure mode is a failure of the supported member (LSB beam) and is caused by the shear stresses within the member. Secondly, the failure mode does not appear to be a sudden failure that causes the immediate collapse of the structure. After the occurrence of TFSB, typically at a later stage the top flange of the LSB section tears away from the LSB web from the flange-web junction as shown in Figure 12.

Ultimately local buckling occurs in the LSB beam resulting in its collapse. Test results showed that in the period between TFSB and the final collapse of the beam, substantially more lateral displacement of the LSB top flange could be sustained without significant loss in capacity. For example Figure 13 plots the shear—top flange lateral displacement curve of Test 21. It can be seen that after TFSB (point A), substantially more lateral displacement occurred in the top flange (from 3.75 mm to 20.7 mm) with approximately 29% loss in capacity (from 42 kN to 30 kN). As a result, a target safety index of 2.5 is deemed appropriate for TFSB design.

It can be seen from Table 4–7 that there are three combinations produced results which satisfy that

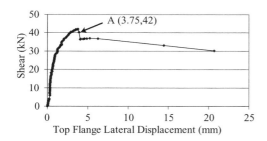

Figure 13. Shear—Top flange lateral displacement curve in test 21.

Table 4. Comparison of the predicted results using design model.

Test	LSB Size (mm)	Web Side Plate	V_{pred} (kN)	V_{test} (kN)	V_{test}/V_{pred}
9	200 × 45 × 1.6	Extended	26.7	28.6	1.07
16	200 × 60 × 2.0	Normal	41.6	46.4	1.11
17	200 × 45 × 1.6	Extended	41.1	38.4	0.93
18	200 × 60 × 2.0	Normal	26.1	30.8	1.18
21	200 × 45 × 1.6	Residential	20.5	42.0	2.05
22	200 × 60 × 2.0	Normal	26.3	33.3	1.27
26	200 × 60 × 2.0	Extended	41.4	56.8	1.37
				Mean	1.28
				Std. Dev.	0.36

requirement, namely, "$\alpha = 1.2$ & $\gamma = 0.9$", "$\alpha = 1.0$ & $\gamma = 0.8$" and "$\alpha = 1.1$ & $\gamma = 0.9$". The "$\alpha = 1.0$ & $\gamma = 0.8$" combination is thought more favourable because firstly it produces a reasonable result of minimum β, and secondly it simplifies the equation by having one of the factors being 1.0. As a result, the recommended equation for the consideration of the shear lag effect becomes:

$$V = (t^2 f_{yweb}) \frac{l_1}{0.8 g_1} \frac{d_1}{b_f} \quad (6)$$

This equation may be recommended for the design of LSB beam with WSP connections, where there is no lateral support to the LSB beam at the connection, for the checking of the possibility of the TFSB failure mode. A comparison of the predicted capacities using the design model (V_{pred}) and the ultimate load reached in the tests that failed by TFSB mode (V_{test}) is given in Table 4.

4 PREVENTION OF TOP FLANGE SHEAR BUCKLING

In addition to the web plate configurations specified by the SSTM manuals, trying to improve the current web plates, 'tall' web plates were used in some tests. Such plates were the same with the CDM normal web plates except the upper edge of the plates were extended so that it was nearly flush with the top flange of the LSB beams (Figure 14).

It was found that the use of the tall web side plate effectively prevented the lateral displacement of the LSB top flange. This is demonstrated in the comparison of the top flange lateral displacement in Test 15 (tall WSP), Test 18 (normal WSP) and Test 8 (normal WSP with lateral support) in Figure 15. The figure shows that in Test 15, the use of the tall web side plate significantly reduced the top flange lateral displacement and prevented the TFSB failure, which occurred

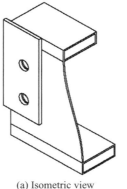

(a) Isometric view

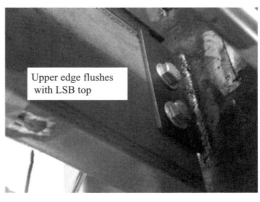

(b) Tall WSP

Figure 14. Tall web side plate.

in Test 18. As a result the member moment capacity was critical in Test 15. The top flange lateral displacement and failure load achieved (32.7 kN) in Test 15 was similar to that in Test 8 (32.5 kN), where a short

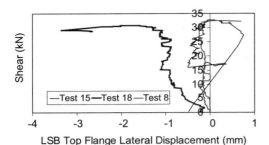

Figure 15. Top flange lateral displacement comparison.

SHS section was placed between the specimens to provide lateral support at the connection.

Hence it can be concluded that the tall web side plate could eliminate the TFSB failure mode and offered similar lateral support to placing SHS sections between the specimens. Due to its obvious practical advantages comparing to placing extra lateral support between the LSB beams (for example, in floor bearers), the tall web side plate is a favourable improvement to the current web side plate used in the LSB WSP connections.

5 CONCLUSION

A total of 30 tests were performed to investigate the behaviour of the new cold-formed structural steel section LiteSteel Beam with web side plate connections.

A new failure mode was observed where lateral support was not provided to the LSB beams at the connections. The failure mode was termed as the top flange shear buckling and was characterised by the lateral displacement of the LSB flanges at the connection. Design model was developed for this failure mode and the reliability analysis indicated that the new design model was reliable for the design of LSB top flange shear buckle.

The use of tall web side plate where the top edge of the web side plate was extended to nearly flush with the top of the LSB beam significantly reduced the top flange lateral displacement and prevented the TFSB mode. Due to its obvious practical advantages comparing to placing extra lateral support between the LSB beams, the tall web side plate is a favourable improvement to the current web side plate used in the LSB WSP connections.

REFERENCES

Cao Y & Wilkinson T, 2007, "Shear Connection Tests for Hollow Flange Channels", *Proceedings*, 6th International Conference, Steel & Aluminium Structures, ICSAS 07, (R.G. Beale, edition), Oxford Brookes University: 799–807.

Hogan TJ & Thomas IR, 1994, *Design of Structural Connections*, 4th edition, Australian Institute of Steel Construction, Sydney.

Mahaarachchi D & Mahendran M, (2006), "Lateral Distortional Buckling Behaviour of a New Cold-formed Hollow Flange Channel Section", *Proceedings*, eighteenth International Specialty Conference on Cold-Formed Steel Structures, Orlando, 2006: 47–175.

SA/SNZ (1998), *Supplement 1 to AS/NZS 4600:1996 Cold-formed steel structures—commentary*, Standards Australia, Sydney, Australia and Standards New Zealand, Wellington, New Zealand.

SA/SNZ (2005), *Australian/New Zealand Standard AS/NZS 4600 Cold-formed steel structures*, Standards Australia, Sydney, Australia and Standards New Zealand, Wellington, New Zealand.

SLST (2005c), *Residential Construction Manual for LiteSteel Beams*, Smorgon Steel LiteSteel Technologies, Queensland, Australia.

SSLST (2005a), *Design Capacity Tables for LiteSteel Beam*, Smorgon Steel LiteSteel Technologies, Queensland, Australia.

SSLST (2005b), *Connection Design Manual for LiteSteel Beams*, Smorgon Steel LiteSteel Technologies, Queensland, Australia.

SSLST (2005d), *Industrial and Commercial Flooring Design Manual for LiteSteel Beams*, Smorgon Steel LiteSteel Technologies, Queensland, Australia.

Trahair NS, Hogan TJ & Syam AA, (1993), "Design of Unbraced Beams", *Journal of the Australian Institute of Steel Construction*, AISC, Vol 27, No 1, 1993: 2–26.

Parallel session 4: Seismic

Comparative study of stainless steel and carbon steel tubular members subjected to cyclic loading

K.H. Nip, L. Gardner & A.Y. Elghazouli
Department of Civil and Environmental Engineering, Imperial College London, UK

ABSTRACT: The material properties of stainless steel differ from those of carbon steel due to the influence of the alloying elements. Of particular significance to the performance of structural members under high-amplitude cyclic loading are the ratio of ultimate-to-yield strength and ductility at fracture of the material, both of which are greater for stainless steel than carbon steel. In this paper, results of cyclic axial and bending material tests on the two materials are presented. A method based on microplastic strain energy is adopted to correlate data from these two test configurations. Despite significantly different behaviour in monotonic tensile coupon tests, the two materials were found to perform similarly in terms of fatigue life and energy dissipation under high strain amplitude cyclic loading. In order to study the influence of material properties on the response of structural members, numerical models were developed to simulate tubular braces under cyclic axial loading. A damage model, in conjunction with output from the finite element analysis of cyclic axial loading tests, was used to predict the fatigue life of the braces. It was found that stainless steel members were more resistant to local buckling and exhibited a larger number of cycles to failure than their carbon steel counterparts. This is believed to be due to a higher post-yield stiffness.

1 INTRODUCTION

The behaviour of a structural member is controlled by its geometry and by the characteristics of the constituent material. When a member is subjected to large cyclic displacements, the ratio of ultimate-to-yield strength and fatigue life of the material are of particular importance. This paper focuses on comparing the fatigue lives of stainless steel and carbon steel by means of cyclic material testing and assessing the influence of cyclic material properties on the behaviour of structural members.

A number of previous studies have included extensive testing on both stainless steel and carbon steel to assess fatigue performance. Testing can be broadly divided into two categories: strain-controlled low cycle fatigue testing and stress-controlled high cycle fatigue testing. Low cycle fatigue testing, which reveals the relationship between strain amplitudes and fatigue lives, is more relevant to the material behaviour of members subjected to large deformation. This relationship can be expressed by the Coffin-Manson rule as:

$$\frac{\Delta \varepsilon}{2} = \frac{\Delta \varepsilon_e}{2} + \frac{\Delta \varepsilon_p}{2} = \frac{\sigma'_f}{E}(2N_f)^b + \varepsilon'_f(2N_f)^c \quad (1)$$

where $\Delta \varepsilon_e/2$ = elastic strain amplitude; $\Delta \varepsilon_p/2$ = plastic strain amplitude; $2N_f$ = number of reversals to failure; σ'_f = fatigue strength coefficient; ε'_f = fatigue ductility coefficient; b = fatigue strength exponent; c = fatigue ductility exponent; and E = Young's modulus.

The parameters in Equation 1 can be derived from the results of fully reversed strain-controlled axial tests. Bhanu Sankara Rao et al. (1993), Bergengren et al. (1995), Wong et al. (2001) and Ye et al. (2006) have determined these parameters for grades 1.4301, 1.4310, 1.4432 austenitic stainless steel and some other duplex stainless steel grades. The strain amplitudes applied to these specimens ranged from ±0.25% to ±2.0%. Similarly, various carbon steel grades were tested by Roessle & Fatemi (2000) and Meggiolaro & Castro (2004) at strain amplitudes of ±0.15% to ±1.5%.

Existing strain-life relationships reported in the literature were derived from tests at cyclic strain amplitudes of less than ±2.0%. However, strain amplitudes encountered by structural members under extreme loading, such as earthquakes, may significantly exceed this value. A strain-based approach to predict low cycle fatigue life using these data must therefore rely on extrapolation of the strain-life relationships to high strain amplitudes. The validity of using data at low amplitudes for determining high amplitude fatigue lives is questionable. Therefore, cyclic material tests at high strain amplitudes were carried out to obtain the material parameters in the low cycle-high strain domain.

2 MATERIAL TESTS

2.1 Materials

Material properties may be altered during the production process of tubular members. In order to reflect the actual material properties of complete members, test coupons were cut directly from the centrelines of the flat faces opposite to the weld of cold-formed and hot-rolled hollow sections. The tests covered 4 cold-formed 1.4301 stainless steel (SS) sections, 5 cold-formed S235 and 5 hot-rolled S355 carbon steel (CS) sections (Table 1). Material properties of a typical coupon from each material are shown in Table 2. For the cold-formed carbon steel and stainless steel, the yield strength was defined as the 0.2% proof strength.

Figure 1 shows the dimensions of the cyclic material test coupons, including the central necked region to constrain the location of fracture. Drilled and reamed holes 20 mm from each end were provided to accommodate the pins in the jaws of the testing machine.

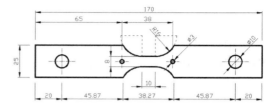

Figure 1. Dimensions of a test coupon (in mm).

Figure 2. Set-up of axial material test.

2.2 Axial material tests

Axial material tests were performed to investigate the relationship between strain amplitudes and the corresponding fatigue life, which is required for determining the material parameters in the Coffin-Manson

Table 1. Sections from which test coupons were obtained.

Cold-formed SS	Cold-formed CS	Hot-rolled CS
50 × 50 × 3	40 × 40 × 3	40 × 40 × 3
60 × 60 × 3	40 × 40 × 4	40 × 40 × 4
100 × 100 × 3	60 × 60 × 3	60 × 60 × 3
60 × 40 × 3	100 × 100 × 4	100 × 100 × 4
	60 × 40 × 4	60 × 40 × 4

Table 2. Material properties under monotonic tensile load.

Material	Yield strength N/mm²	Ultimate strength N/mm²	Elongation at fracture %
Cold-formed SS	419	725	66
Cold-formed CS	482	500	29
Hot-rolled CS	488	570	33

rule. Coupons were subjected to fully-reversed constant strain amplitude cycles in the set-up shown in Figure 2. Test coupons were gripped by friction as well as positively held by pins at each end and close to the necked region. The jaws provided restraint to the coupons to prevent buckling and guides prevented any relative movement between the top and bottom jaws. Since only the middle 3 mm of the coupon was exposed, deformation was concentrated in that area, which was covered by strain gauges.

Tests were carried out at ±1%, ±3%, ±5% and ±7% strain amplitudes at a frequency of 0.025 Hz. Figure 3 shows the measurement of strain during the first few cycles of a typical test at ±5% strain amplitude. Measurement of load during the test is shown in Figure 4. Cyclic hardening during the first 5 cycles was observed and fracture was defined as a 10% drop in load.

Preliminary results of the axial material tests are presented in Table 3 and the parameters in the Coffin-Manson rule were determined as $\sigma'_f = 1120$, $\varepsilon'_f = 0.2893$, $b = -0.0424$ and $c = -0.6922$ for stainless steel. Those for cold-formed carbon steel were $\sigma'_f = 445$, $\varepsilon'_f = 0.2079$, $b = -0.1104$ and $c = -0.4882$. Figure 5 indicates that data from the current study follow the trend extrapolated from existing data.

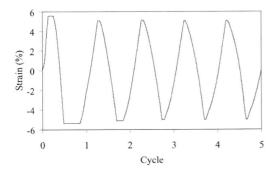

Figure 3. Measurement of strain during the first few cycles of a typical test at ±5% amplitude.

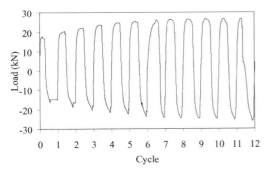

Figure 4. Measurement of load during a typical test at ±5% amplitude.

Table 3. Results of axial material tests.

Section*	Strain amplitude (%)	Fatigue life	Energy to fracture (Nmm/mm³)
50 × 50 × 3a (S)	1	336	4098
50 × 50 × 3b (S)	3	20	1749
50 × 50 × 3c (S)	5	37	4864
50 × 50 × 3d (S)	7	3	410
100 × 100 × 3a (S)	1	267	1923
100 × 100 × 3b (S)	3	45	3913
100 × 100 × 3c (S)	5	8	1121
60 × 40 × 3a (S)	3	28	1936
60 × 40 × 3b (S)	5	13	2236
60 × 40 × 3c (S)	5	5	708
40 × 40 × 3a (C)	1	522	7816
40 × 40 × 3b (C)	3	17	691
40 × 40 × 3c (C)	5	9	933
40 × 40 × 3d (C)	7	4	564
60 × 40 × 4a (C)	5	21	2100

*(S) = cold-formed stainless steel; (C) = cold-formed carbon steel.

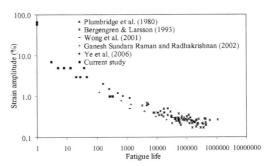

Figure 5. Data from strain-controlled cyclic axial material tests on stainless steel.

Table 4. Results of bending material tests.

Section*	Strain amplitude (%)	Fatigue life	Energy to fracture (Nmm/mm³)
60 × 60 × 3 (S)	4.9	33	1789
100 × 100 × 3b (S)	10.0	13	1284
100 × 100 × 3a (S)	10.9	12	1361
60 × 60 × 3 (C)	6.3	53	2395
60 × 60 × 3 (H)	7.2	45	2482
100 × 100 × 4a (H)	15.7	8	1372
100 × 100 × 4b (H)	16.0	9	1616

*(S) = cold-formed stainless steel; (C) = cold-formed carbon steel; (H) = hot-rolled carbon steel.

2.3 *Bending material tests*

Bending material tests are an alternative means of determining the material parameters for the Coffin-Manson rule. Since the coupons used in this study were thin plates, testing them by bending about their minor axes avoided the problem of buckling. Due to the non-uniform distribution of strain through the thickness, the strain amplitudes and fatigue lives obtained from the tests cannot be used directly in the Coffin-Manson rule. A method to correlate the results from axial and bending tests will be introduced in Section 4.

A four-point bending configuration was adopted to produce constant bending moment at the necked area of the test coupon (Figs. 6–7). Each end support consisted of a block sliding freely on rollers to minimise any axial force in the coupon. Strain gauges were attached to both the upper and lower faces of the coupons. Tests were carried out at various strain amplitudes at a frequency of 0.025 Hz. Fracture was defined as a 10% drop of the load.

Preliminary results of the bending material tests are presented in Table 4 and the strain-life relationships of the three materials are shown in Figure 8.

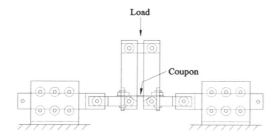

Figure 6. Set-up of bending test.

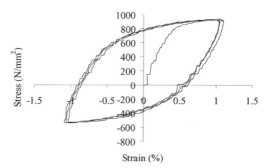

Figure 9. Stress-strain hysteresis loops of 50 × 50 × 3a (S) in the first 3 cycles.

Figure 7. Set-up of bending test at maximum displacement.

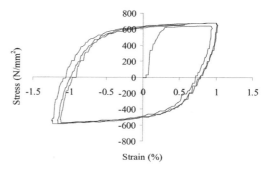

Figure 10. Stress-strain hysteresis loops of 40 × 40 × 3a (C) in the first 3 cycles.

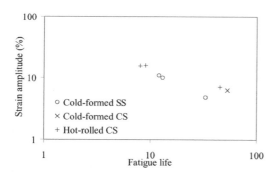

Figure 8. Strain-life relationships of the three materials in cyclic bending material tests.

Stainless steel does not show the superior ductility that it exhibits in monotonic tensile tests.

3 FATIGUE LIFE PREDICTION USING AN ENERGY CRITERION

Prediction of fatigue fracture by energy criteria has been studied by Morrow (1965) and Lefebvre & Ellyin (1984). Both inelastic strain energy per cycle and total inelastic strain energy to fracture have been used as fatigue fracture criteria. Figures 9 and 10 show the stress-strain hysteresis loops of a stainless steel and carbon steel specimen under cyclic axial loading. The area of under each hysteresis loop is the inelastic strain energy dissipated by the material in a loading cycle.

Morrow (1965) proposed the following expression for predicting fatigue life from energy dissipated per cycle:

$$(2N_f)^{b+c} = \frac{\Delta W}{4\sigma_f' \varepsilon_f' \left(\dfrac{c-b}{c+b}\right)} \quad (2)$$

where ΔW = energy dissipated per cycle and other terms are defined as the same as in Equation 1.

Since the energy per cycle is nearly constant throughout fatigue tests (Morrow & Tuler, 1964), the total inelastic strain energy to fracture may be approximated as:

$$W_f = \Delta W N_f \quad (3)$$

By substituting ΔW and rearranging terms, Morrow (1965) obtained the total inelastic strain energy to fracture in the form of:

$$2N_f = \left(\frac{2W_f}{W_f'}\right)^{1/(1+b+c)} \quad (4)$$

where $W_f' = 4\sigma_f' \varepsilon_f' \left(\frac{c-b}{c+b}\right)$.

It should be noted that Equation 4 implies that the total energy to fracture increases as the fatigue life increases.

4 COMPARISONS BETWEEN AXIAL AND BENDING MATERIAL TESTS

4.1 Correlation between axial and bending material tests

It has been observed that specimens from the same material tested under different configurations yield different fatigue lives. Axial and bending tests differ from each other in the variation of strain through the thickness. Since the underlying fatigue process at a localised area is the same regardless of the distribution of strain through the thickness, there have been attempts to correlate the fatigue lives obtained from these two testing configurations (Esin, 1980). Previous studies (Feltner & Morrow, 1961, Esin, 1968) indicated that fatigue was an energy conversion process and microplastic strain energy was introduced as an index of fatigue damage. This criterion is valid for any loading configuration and therefore was used to correlate axial and bending fatigue tests.

In order to predict fatigue fracture, the plastic hysteresis energy for a cycle of reversed stress amplitude is firstly obtained as:

$$\Delta W = 2\int_0^{\Delta \varepsilon_p} \sigma \, d\varepsilon_p \quad (5)$$

Assuming

$$\sigma = K\varepsilon^n \quad (6)$$

Equation 5 becomes

$$\Delta W = 2\int_0^{\Delta \varepsilon_p} K\varepsilon^n \, d\varepsilon_p \quad (7)$$

Every material has a nominal stress-strain response. This mechanical property holds true for even very small elements. These elements, which conform to the classical mathematical assumptions and repeat the nominal properties of the material, are called macro-elements. Every macro-element is made up of micro-elements. They are infinitesimal elements having a range of values of material constants and stress-strain response. The material properties of a macro-element represent a statistical average of the contributions of micro-elements. Therefore, the microplastic strain hysteresis energy dissipated by a macro-element in a cycle is given by:

$$u_i = 2\sum_{(m)} P_k \int_0^{\Delta \varepsilon_k} K_k \varepsilon^n \, d\varepsilon \quad (8)$$

where the subscript k = randomness of the stresses, strains and the material constants of the micro-elements; $\Delta \varepsilon_k$ = plastic strain amplitude; P = probabilistic number of micro-elements of equal strain amplitude and; K and n are strength coefficient and strain hardening exponents respectively.

The total plastic hysteresis energy dissipated by a unit volume of metal in a cycle can be expressed as the sum of the contributions of all of the macro-elements which contain plastic micro-elements:

$$W = \sum_{(m)} T_i u_i \quad (9)$$

where T_i is the number of macro-elements at the same strain level.

According to the microplastic strain energy criterion of fatigue, failure of a metal occurs when the plastic hysteresis energy accumulates to the value of the true fracture energy of the metal. This can be expressed as:

$$N = \frac{UT_t}{W} \quad (10)$$

where N = the number of cycles to failure; U = true fracture energy determined as the area under the true stress-true strain curve and; T_t = total number of macro-elements of the section with unit depth.

When a strain gradient exists in a cross-section, the total plastic hysteresis energy dissipated by the specimen is not equal to that of an axially-loaded one at the same surface strain amplitude. An equivalent strain amplitude is defined so that it would then be possible to correlate the strain-life curves with one another:

$$\varepsilon_{eq} = \frac{\sum \varepsilon_i A_i}{\sum A_i} \quad (11)$$

where A_i is area of macro-elements with equal strain amplitude.

When the concept of equivalent strain is employed, the plastic strain energy, i.e. u_i in Equation 8, dissipated by the contributing elements becomes equal. Equation 6 can be rewritten as:

$$W = u_i \sum T_i \quad (12)$$

Substituting Equation 12 into Equation 10, we obtain

$$N = \left(\frac{U}{u_i}\right)\left(\frac{T_t}{\sum T_i}\right) \quad (13)$$

The last term of the equation represents the ratio of the total area of the section to the area of the elements contributing to the fatigue process.

4.2 Comparisons of fatigue life between axial and bending material tests

Due to the strain gradient through the thickness of the specimen in bending tests, the energy dissipation is lower than that by a specimen with the same amplitude of surface strain in an axial test. Consider a rectangular cross-section (Fig. 11), where the macro-elements subjected to fatigue are indicated by the shaded area. Assuming a linear strain distribution, the equivalent strain is given by:

$$\varepsilon_{eq} = \frac{\int_{h-x}^{h} \varepsilon_s \frac{y}{h} w dy}{wx} = \varepsilon_s \left(1 - \frac{x}{2h}\right) \quad (14)$$

Assuming all material in a cross-section is subjected to fatigue (i.e. $x = h$), Equation 14 can be simplified to:

$$\varepsilon_{eq} = 0.5\varepsilon_s \quad (15)$$

Figure 12 compares the fatigue lives of stainless steel specimens from axial and bending tests. The dotted line is the axial fatigue lives predicted from bending test data by using the method outlined in Section 4.1.

4.3 Comparisons of energy dissipation between axial and bending material tests

The energy dissipation by a coupon was calculated from the area under the stress-strain hysteresis loops for axial tests and moment-curvature hysteresis loops for bending tests. Figures 13 and 14 show that the energy dissipation of the specimens follows the same trend regardless of the loading configuration. As it has been observed in other experimental programmes (Morrow, 1965, Lefebvre & Ellyin, 1984), the total energy dissipation to fracture increases as the fatigue life increases, in both axial and bending material tests. On the other hand, the energy dissipation per cycle decreases as the fatigue lives increases due to the smaller strain amplitude applied to the specimen in higher cycle fatigue tests.

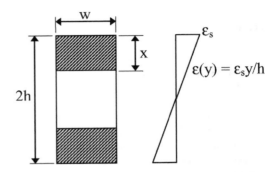

Figure 11. Distribution of strain through a rectangular cross-section subjected to bending.

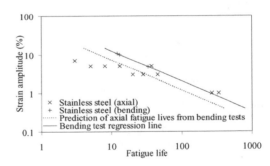

Figure 12. Prediction of fatigue lives under axial loading from bending test results.

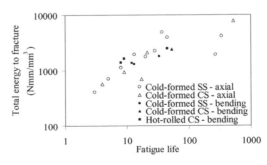

Figure 13. Relationship between total energy to fracture and fatigue life in axial and bending material tests.

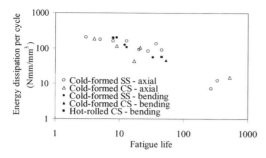

Figure 14. Relationship between energy dissipation per cycle and fatigue life in axial and bending material tests.

In Figures 12–13, the three materials follow a similar trend in terms of energy dissipation. This is in contrast with the significant differences observed in the area under the monotonic stress-strain curves of tensile coupon tests of the three materials. More low cycle fatigue tests are needed to confirm the trend and identify the strain amplitude at which the convergence of behaviour occurs.

5 INFLUENCE OF CYCLIC MATERIAL PROPERTIES ON BEHAVIOUR OF BRACING MEMBERS

5.1 Numerical study of bracing members under cyclic loading

Concentrically braced frames are commonly employed as lateral load resisting systems. Under significant seismic loading conditions, their response largely depends on the behaviour of the diagonal braces, which represent the key energy dissipating zones. Numerical study was performed to assess the influence of material properties on the behaviour of tubular steel braces under cyclic loading.

5.2 Finite element model

The general purpose finite element (FE) package ABAQUS (2007) was used to develop models to replicate cyclic loading tests on SHS and RHS braces. The models were subjected to cyclic axial displacements with large plastic deformation within non-linear analyses. The numerical results were initially validated against tests, after which, a series of parametric studies were performed (Nip et al. 2007).

Cyclic axial loading tests on 1800 mm long members with four different cross-sections (40 × 40 × 1.5, 40 × 40 × 2.5, 40 × 40 × 3.0, and 40 × 40 × 4.0) were simulated. Material properties of stainless steel and carbon steel were used in the models.

5.3 Damage prediction

The bracing members examined in the current study were subjected to large cyclic displacements, resulting in high localized plastic deformation which would, in reality, lead to low cycle fatigue failure. However, since fracture was not explicitly modeled within the numerical models, an indirect approach to assess the damage and predict the occurrence of fracture was needed. A strain-based approach was deemed most suitable for this purpose.

The adopted procedure for predicting damage first requires the monitoring of strains during each cycle of loading. A fatigue life, adjusted for the effects of multi-axial loading, is then obtained through a Coffin-Manson relationship (Coffin, 1954, Manson, 1960). Finally, the damage caused by each cycle is accumulated by means of Miner's rule and fracture is predicted from this damage value. The results of the material tests in this study were used to derive the parameters in the Coffin-Manson relationship.

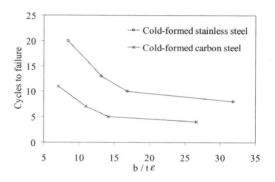

Figure 15. Local slenderness against numbers of cycles to failure for stainless steel and carbon steel braces.

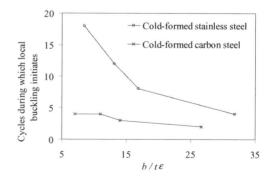

Figure 16. Local slenderness against numbers of cycles when local buckling initiated for stainless steel and carbon steel braces.

5.4 Comparative study

The results of the numerical study are presented in Figure 15. Models of both materials show a decrease in cycles to failure as local slenderness increases. The local slenderness is defined as $b/t\varepsilon$ where b = the flat width of the wider face of the section; t = the thickness of the section; and $\varepsilon = \sqrt{235/f_y}$.

In this study, the stainless steel members sustained approximately twice the number of cycles in comparison to carbon steel members of the same local slenderness. Since stainless steel did not exhibit superior fracture life over carbon steel in the cyclic material tests, the difference in member behaviour may be due to other factors, such as the initiation of local buckling. Figure 16 shows that stainless steel members buckle locally at larger applied displacements than carbon steel members. Since local buckling induces high localised strain and fracture follows soon after, stainless steel members have a higher number of cycles to failure than carbon steel members. The higher resistance to local buckling of stainless steel members may be due to more substantial strain hardening of the material and maintenance of higher stiffness after yielding, as shown in Figures 9 and 10.

6 CONCLUSIONS

Cyclic axial and bending material tests were carried out on cold-formed stainless steel, cold-formed carbon steel and hot-rolled carbon steel specimens. No significant difference was found in terms of fatigue life and energy dissipation among these materials. The correlation between axial and bending tests was established through an energy criterion. In order to investigate the influence of material properties on the behaviour of structural members, tubular braces under cyclic loading were modelled numerically. A damage model, in conjunction with output from the finite element analysis of cyclic axial loading tests, was used to predict the fatigue life of the members.

It was shown that stainless steel members were more resistant to local buckling and had a larger number of cycles to failure than their carbon steel counterparts. However, better performance at member level may not be attributed to the higher ductility of stainless steel demonstrated in monotonic tensile tests. Instead, it may be due to delayed local buckling of stainless steel members resulting in lower strain amplitudes being experienced and a higher number of cycles to failure. Factors influencing resistance to local buckling, including ability to maintain stiffness after yielding and degree of strain hardening, may have contributed to the superior performance of stainless steel braces over carbon steel braces under cyclic loading.

REFERENCES

ABAQUS 2007. *Analysis User's Manual I-V. Version 6.7.* USA: ABAQUS, Inc., Dassault Systèmes.

Bergengren, Y. & Larsson, M. 1993. *Fatigue properties of austenitic and duplex stainless sheet steels in air at room temperature.* Stockholm, Sweden: Swedish Institute for metals Research.

Bergengren, Y., Larsson, M. & Melander, A. 1995. Fatigue properties of stainless sheet steels in air at room temperature. *Materials Science and Technology* 11: 1275–1280.

Bhanu Sankara Rao, K., Valsan, M., Sandhya, R., Mannan, S.L. & Rodriguez, P. 1993. An Assessment of Cold Work Effects on Strain-Controlled Low-Cycle Fatigue Behavior of Type 304 Stainless Steel. *Metallurgical Transactions* 24 A: 913–924.

Coffin, Jr. L.F. A study of the effects of cyclic thermal stresses on a ductile metal. 1954. *Transactions of ASME* 76: 931–50.

Esin, A. The microplastic strain energy criterion applied to fatigue. 1968. *Journal of Basic Engineering, ASME* 90(1): 28–36.

Esin, A. A Method for Correlating Different Types of Fatigue Curve. 1980. *International Journal of Fatigue* 2: 153–158.

Feltner, C.E. & Morrow, J.D. 1961. Microplastic strain hysteresis energy as a criterion for fatigue fracture. *Journal of Basic Engineering, ASME* 88(1): 15–22.

Lefebvre, D. & Ellyin, F. 1984. Cyclic response and inelastic strain energy in low cycle fatigue. *International Journal of Fatigue* 6(1): 9–15.

Manson S.S. 1960. Thermal stress in design. Part 19, cyclic life of ductile materials. *Machine design* 139–144.

Meggiolaro, M.A. & Castro, J.T.P. 2004. Statistical evaluation of strain-life fatigue crack initiation predictions. *International Journal of Fatigue* 26: 463–476.

Morrow, J.D. 1965. Cyclic plastic strain energy and fatigue of metals. *Internal Friction, Damping and Cyclic Plasticity* ASTM STP 378: 48–84.

Morrow, J.D. & Tuler, F.R. 1964. Low Cycle fatigue evaluation of Inconel 713C and Waspaloy. *AWS-ASME Metals Engineering Conference, Detroit, Michigan, May 1964.*

Nip, K.H., Gardner, L. & Elghazouli, A.Y. 2007. Numerical modelling of tubular steel braces under cyclic axial loading. *Proceedings of the Third International Conference on Steel and Composite Structures, Manchester, U.K., July, 2007.*

Plumbridge, W.J., Dalski, M.E. & Castle, P.J. 1980. High strain fatigue of a type 316 stainless steel. *Fatigue of Engineering Materials and Structures* 3: 177–188.

Roessle, M.L. & Fatemi, A. 2000. Strain-controlled fatigue properties of steels and some simple approximations. *International Journal of Fatigue* 22: 495–511.

Wong, Y.K., Hu, X.Z. & Norton, M.P. 2001. Low and high cycle fatigue interaction in 316 L stainless steel. *Journal of Testing and Evaluation* 29: 138–145.

Ye, D., Matsuoka, S., Nagashima, N. & Suzuki, N. 2006. The low-cycle fatigue, deformation and final fracture behaviour of an austenitic stainless steel. *Materials Science and Engineering* A 415: 104–117.

Seismic response of circular hollow section braces with slotted end connections

G. Martinez-Saucedo & R. Tremblay
École Polytechnique, Montréal, Canada

J.A. Packer
University of Toronto, Toronto, Canada

ABSTRACT: During severe earthquakes, brace members supply lateral resistance to structural frames while also dissipating energy through cyclic inelastic deformations. However, this performance may be negated by a sudden fracture at an end connection, or at the brace mid-length. As a result, the use of a structural section that shows excellent behaviour under such loading is invaluable and Circular Hollow Sections (CHS) have shown promising characteristics. However, the lack of a simple end connection detail has limited their use to date. A novel connection detail that is capable of preventing premature brittle failure at the brace end, and thus turning all attention to the brace mid-length, is presented here. The capacity of this detail has been previously demonstrated by means of two large-scale specimen tests under pseudo-dynamic loading. A Finite Element (FE) analysis of these connections has been undertaken and is presented herein. Results from the numerical analysis compared well with the experimental tests. A failure criterion is also included as a means of successfully predicting the location and time of crack initiation, validating the use of the FE models for a subsequent parametric analysis.

1 INTRODUCTION

Hollow structural sections have been studied extensively for use as braces subject to cyclic loading since the late 1970s. As a result, the parameters defining the behaviour of hollow section braces are well known nowadays. In most studies, the brace cross-section was fully connected at the supports as a means of preventing connection failure. Although this support is frequently used in laboratories, simplified connection details, reducing the time and materials used during fabrication, are preferred for conventional design conditions. Slotted end connections exemplify the simplest method of connecting CHS braces. The gusset plate or the CHS may be slotted (resulting in several connection details) which has allowed their extensive use in structures under static loading. Despite these advantages, strict requirements in current North American design provisions (AISC 2005a, AISC 2005b and CSA 2001) restrict their use for seismic applications. These provisions demand that the connection must withstand the maximum tension load when the brace gross cross-sectional area (A_g) yields (i.e. $R_y A_g F_y$, where R_y represents the ratio between the expected and the specified minimum yield stress, F_y). Since the CHS cross-section is only partially attached at the connection, the load transfer creates an uneven stress distribution that reaches its maximum value at the beginning of the weld (see Fig. 1).

Due to shear lag, the tube may crack at that location leading to fracture of the connection. To consider this phenomenon, the tube net area (A_n) is reduced by an efficiency factor (U). Hence, to avoid the connection fracture, AISC (2005b) requires that:

$$R_y A_g F_y \leq U A_n R_t F_u \quad (1)$$

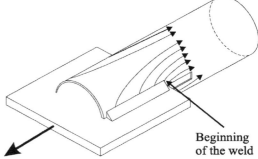

Figure 1. Shear lag in slotted CHS connections.

where all resistance factors (or partial safety factors) are set equal to unity and Rt represents the ratio between the expected and the specified minimum ultimate tensile strength (F_u).

Eq. 1 may be satisfied when A_n exceeds A_g. This alternative has been investigated by Yang & Mahin (2005) and Haddad & Tremblay (2006) adding cover plates to the hollow section in the connection region. Nevertheless, this solution may become impractical for CHS. Another option requires that the presence of shear lag be minimized and the ratio $UA_n R_t F_u / R_y A_g F_y$ exceeds unity, which may require a rigorous control of the material properties. Martinez-Saucedo et al. (2006) have suggested that full connection efficiency ($U = 1.0$) may be achieved under static loading when the connection length (L_w) exceeds the distance between the welds (w, measured around the tube perimeter). Moreover, slotted gusset plate connections produced necking at the tube mid-length when a ratio of $L_w/w > 1.0$ was used. Hence, a connection merging the advantages of slotted gusset plate and slotted CHS connections, herein called a "Modified-Hidden-Gap" (MHG) connection, was regarded as the optimal solution. Figure 2 shows the fabrication drawing for this connection type. Both the gusset plate and the CHS are slotted, but during the assembly the slotted plate is pushed into the CHS slot leaving a small (hidden) gap. Two gap sizes have been considered experimentally: 6 and 30 mm for the details MHG-1 and MHG-2 respectively. Once together, the pieces are attached only by longitudinal welds.

This paper focuses on the FE analysis of two CHS braces, exhibiting MHG connections at either end, tested under pseudo-dynamic cyclic axial loading.

2 FINITE ELEMENT MODELLING

2.1 Material properties

The data used to develop the FE and material models are reported in Martinez-Saucedo & Packer (2007) and Martinez-Saucedo et al. (2008). A non-linear material, multi-linear true stress-true strain (Tσ-Tε) curve was employed to reproduce the gusset plate, CHS and weld material properties. Weld material was represented by the CHS material properties. The Tσ-Tε curves were generated based on the engineering (σ-ε) relations (from coupon tests under monotonic loading) and a method suggested by Martinez-Saucedo et al. (2006) to deal with coupons having a rectangular cross-section. Whereas isotropic hardening was assumed for monotonic analysis, kinematic hardening has been assumed for cyclic analysis. A preliminary FE analysis of the braces, using curve material properties from monotonic tests, resulted in loads exceeding those from the laboratory. The reason for this is that the hardening function in a cyclic hardening model is different to the monotonic case, due to the effects of consolidation induced by the cyclic loading (Lemaitre & Chaboche 1994). In the absence of coupon test results under low cycle fatigue cyclic loading, to capture the hardening function, the static material curves were scaled down to represent the anticipated cyclic hardening response. It was found that scaling the static material curves to 94% of their original value (see Fig. 3) was sufficient to allow the FE models to closely predict the response measured for the test specimens.

2.2 Modelling and analysis considerations

ANSYS (Swanson 2007) FEA software was used for the analyses. The fillet welds connecting the tube and plates were fully modelled. The gusset plate, tube and weld materials were modelled with a reduced

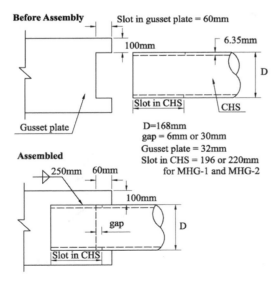

Figure 2. Connections tested experimentally.

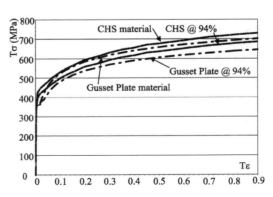

Figure 3. Material Tσ-Tε curves.

integration eight-noded brick element (SOLID45) with hourglass control. This element has large deflection, large strain capabilities and each node has three translational degrees of freedom. An elastic buckling analysis was undertaken to determine the critical buckling mode of the braces; this was then used to modify the original geometry and include an initial geometrical imperfection. The use of Symmetry Boundary Conditions (SBC) along the longitudinal axis allowed the modelling of half of the test specimens (see Fig. 4). In addition, strain readings were acquired through the FE analysis at the connection region. The ends of the gusset plates were fixed. While displacements were applied at the top, the reaction forces were calculated at the bottom of the brace.

A refined meshing was used in areas expected to potentially develop a fracture; i.e. the tube mid-length and near the beginning of the welds. Based on a sensitivity analysis, it was decided to use three and six elements through the tube and gusset plate thickness respectively. To emulate the displacement-control loading used during the tests, a non-linear time step analysis was performed by applying incremental displacements. A frontal equation solver and full Newton Raphson method were used. Geometrical nonlinearities were considered by allowing large deformations as well as shape changes.

2.3 Material fracture

Material fracture was emulated by the activation of the element "death feature" that was triggered by comparing the accumulated equivalent plastic strain (EPEQ) in the element to the critical failure strain (ε_f). Material was assumed to fracture and the element was "killed" when the ratio EPEQ/ε_f exceeded unity. The critical failure strain, which is a function of the triaxiality, is computed as suggested by Hancock and Mackenzie (1976):

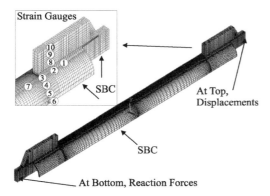

Figure 4. Typical FE model.

$$\varepsilon_f = \alpha \exp\left(-1.5\frac{\sigma_m}{\sigma_e}\right) \quad (2)$$

where α = material constant, σ_m = mean stress and σ_e = effective stress. Kanvinde & Deierlein (2006) have suggested a method to calculate α which is based on the FE models used to calibrate the Tσ-Tε curves. Since the FE model closely matches the load-displacement response of the test coupons, it may be assumed that the FE model emulates the stress distribution in the coupon test. Hence, at the point of failure in the test coupon (defined by a load and displacement), ε_f, σ_m and σ_e are computed at the centre of the FE cross-section, at the same load and displacement. Hence, rearranging Eq. 2 one can obtain α by:

$$\alpha = \frac{\varepsilon_f}{\exp\left(-1.5\frac{\sigma_m}{\sigma_e}\right)} \quad (3)$$

2.4 Evaluation of FE models against experiments

2.4.1 Connection detail MHG-1

The loading protocol recommended by ATC-24 (1992) was used during the laboratory test and the FE analysis of this connection detail. During the initial cycles, the FE model closely followed the response of the test specimen. At a displacement of −2.0 δ_y (where δ_y corresponds to the specimen yield deformation), the FE model attained overall buckling as seen during the test (see Fig. 5). A displacement of −3.0 δ_y was associated with the start of an inelastic local buckle in the tube. At −4.0 δ_y, the tube showed an ovalization of its cross-section with the presence of a local buckle at the tube mid-length, and this was accurately captured by the FE Model (see Fig. 5, insert). At −5.0 δ_y, the ovalization and the local buckling increased. It was close to the end of the second excursion (at ±5.0 δ_y) that material cracking started in corners of the buckle. As a result, the transition from −5.0 δ_y to 6.0 δ_y (which was not attained) was marked by the gradual propagation of the crack and failure of the brace. The FE model was able to capture this behaviour as a gradual material cracking occurred at the tube mid-length during this transition. An insert in Figure 5 shows the FE model at the end of the analysis, where the killed elements have been removed to clearly illustrate the crack in the tube. Even though the material fracture model was implemented throughout the entire FE model, fracture only occurred in elements at the tube mid-length, as seen in the test.

The behaviour of the FE model at the connection was also verified by a comparison of Strain Gauge (SG) readings from the test and the FE analysis.

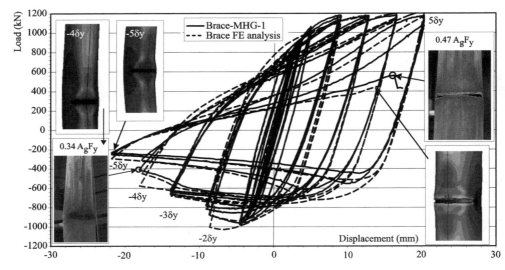

Figure 5. Load-displacement response of CHS brace with MHG-1 (test specimen and FE model).

It was found that FE strain readings along the connection (i.e. SG-1, SG-2, SG-3 and SG-8 in Fig. 4) exhibited differences for the initial cycles (see Fig. 6). This has been associated with geometrical irregularities in the weld that were related to the fabrication process and not included in the FE model. These may have disrupted the connection strain distribution during the load transfer, which could not be re-produced by the FE model. Nevertheless, this anomaly decreased as the strain increased. Moreover, this effect vanished as the SGs were located further from the connection.

Figure 7 shows the reading from SG-5 (as no data is available for SG-6 due to a failure of this SG during the test). Here the FE model shows a better response, especially for the initial cycles. Moreover, the influence of the weld region was also decreased at SG-7 (see Fig. 8), located 50 mm in front of the weld.

In the gusset plate, strains readings at the gap region (i.e. SG-8, SG-9 and SG10) remained within the elastic range throughout the test and the FE analysis. During the connection design stage, it was decided to minimize the deformation here as the bowing of the wings of the gusset plate can increase the demand in the connection region, which may lead to weld fracture. Figure 9 illustrates the strain readings at SG-9. Despite the strains being very low, the test and FE model exhibited a similar trend.

2.4.2 Connection detail MHG-2

A modified loading protocol was applied to this brace as a means of testing the connection while the demand was minimized at the tube mid-length. Fig. 10 shows

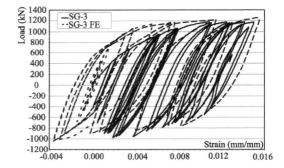

Figure 6. Load-strain readings SG-3 (MHG-1).

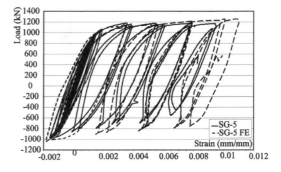

Figure 7. Load-strain readings SG-5 (MHG-1).

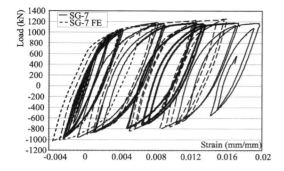

Figure 8. Load-strain readings SG-7 (MHG-1).

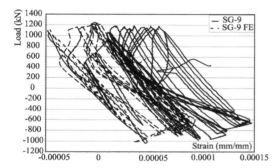

Figure 9. Load-strain readings SG-9 (MHG-1).

the load-displacement response of the test specimen and the FE model. After two initial cycles with a deformation of δ_y, tension was applied until a displacement of 12.0 δ_y was attained. Then, the specimen was reloaded until it reached a displacement of $-3.0\,\delta_y$ from the zero load point. This displacement in compression was sufficient to reach the buckling load in the brace while it avoided damaging the tube at its mid-length. This protocol (i.e. $+1.0\,\delta_y$ and $-3.0\,\delta_y$) continued until $+16.0\,\delta_y$ was reached. At this stage, it was decided to increase the displacements to $+18.0\,\delta_y$, $+20.0\,\delta_y$ and $+22.0\,\delta_y$ (with $-7.0\,\delta_y$, $-8.0\,\delta_y$ and $-8.0\,\delta_y$ respectively). During these last cycles, the load drop due to overall buckling in the brace became more evident. The FE model was able to emulate this behaviour as well as the generalized yielding along the tube beyond the connection region (see Fig. 10, insert). At this point, it was decided to continue with a tensile displacement to failure. Close to the end, a neck developed at the tube mid-length; it was here where the accumulation of plastic strain continued until fracture. The material fracture ability was implemented throughout the entire FE model. However, fracture only occurred in elements located in the necking region at the tube mid-length as observed during the test.

As seen in test MHG-1, the strain readings at the beginning of the weld (see Fig. 11) also exhibited disruptions during the initial cycles. However, subsequent FE strain readings exceeded the values achieved during the test. The difference between these responses has been related to the loading protocol used herein.

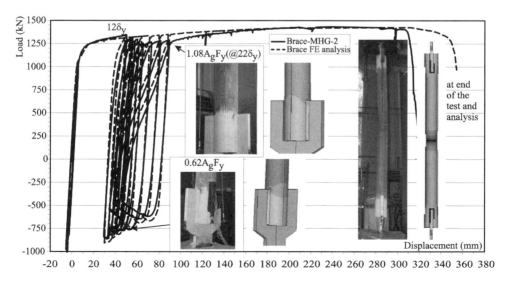

Figure 10. Load-Displacement response of CHS brace with MHG-2 (test specimen and FE model).

231

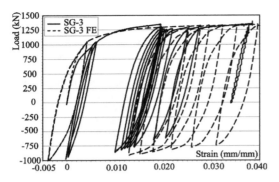

Figure 11. Load-strain readings SG-3 (MHG-2).

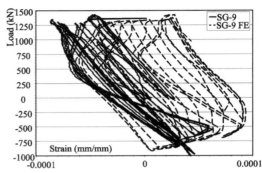

Figure 14. Load-strain readings SG-9 (MHG-2).

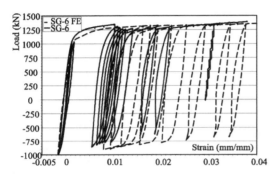

Figure 12. Load-strain readings SG-6 (MHG-2).

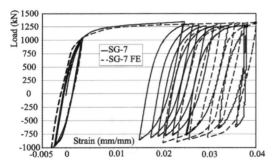

Figure 13. Load-strain readings SG-7 (MHG-2).

Whereas steel naturally adapts to changes in the loading, i.e. modifying its hardening model and hardening rules as the load speed and history changes, the FE model used herein lacked this capability. After two initial cycles at 1.0 δ_y, the attainment of +12.0 δ_y under a monotonic load would require switching the FE hardening model from cyclic to monotonic, allowing the attainment of a higher load and smaller strains. From this point, a switch to cyclic modelling would allow a better representation of the gradual strength deterioration during the cyclic load. Finally, from +22.0 δ_y, a last model change would permit it to exhibit the gradual load increase and the load decrease due to tube necking. Although this may seem a natural course to follow, the simpler approach used herein was deemed acceptable as it could provide adequate results without increasing the FE analysis time.

As seen previously, the strain disruption at the weld region also decreased in SGs away from the weld. The FE strain readings in SG-6 (at 90° away from the weld, see Fig. 12) and SG-7 (see Fig. 13) exhibited a better response. Although the FE strain readings do not match the test results for the later cycles, for the reasons previously discussed, these clearly follow a similar trend.

Despite the fact that this connection MHG-2 had a larger hidden gap (30 mm) and the brace sustained a larger demand, strains in the gusset plate remained within the elastic range, decreasing the bowing effect of the plate in the connection. Moreover, the FE model emulated this behaviour as FE readings in SG-9 exhibited a similar trend (despite being very small) to the test specimen (see Fig. 14).

3 CONCLUSIONS

The FE models presented have shown their capacity to emulate the overall brace response as well as the strain distribution at the connection region. Moreover, the suggested fracture model allowed the FE models to replicate the time and the location of the fracture observed during the tests. This validates their use for further studies where geometrical dimensions, as well as loading protocols, will be modified.

As seen during the tests, the FE models confirmed the ability of the so-called "Modified Hidden Gap" (MHG) connections to prevent brittle fracture in the connection region by effectively decreasing the influence of shear lag there. Because of this, and the high stiffness associated with this connection detail, most of the inelastic deformation was located away from the connection at the CHS brace mid-length. Thus, the MHG detail should enable the use of unreinforced CHS brace connections in seismic regions.

It was shown that the use of material $T\sigma$-$T\varepsilon$ curves from monotonic coupon tests can provide fairly good results for FE analysis under cyclic inelastic loading when they can be properly adjusted to fit available specimen test data. However, accurate blind numerical predictions would require the use of $T\sigma$-$T\varepsilon$ curves obtained from cyclic coupon tests.

ACKNOWLEDGEMENTS

Financial support for this project has been provided by CIDECT (Comité International pour le Développement et l'Etude de la Construction Tubulaire) and NSERC (the Natural Science and Engineering Research Council of Canada). IPSCO Inc. and Atlas Tube Inc. generously donated steel material. For fabrication services, the authors gratefully acknowledge Walters Inc. (Ontario, Canada).

NOMENCLATURE

A_g = gross cross-sectional area of CHS
A_n = net cross-sectional area of CHS
CHS = Circular Hollow Section
D = outside diameter of CHS
EPEQ = accumulated equivalent plastic strain
F_u = ultimate tensile strength
F_y = yield tensile stress
L_w = connection weld length
MHG = Modified Hidden Gap
R_t = ratio of expected tensile strength to specified minimum tensile strength
R_y = ratio of expected yield stress to specified minimum yield stress
SBC = Symmetry Boundary Condition
SG = Strain Gauge
U = Efficiency factor
w = distance between welds (around tube)
α = material constant
δ_y = specimen yield deformation
ε_f = critical fracture strain
σ-ε = engineering stress–strain curve
σ_e = effective stress
σ_m = mean stress
$T\sigma$-$T\varepsilon$ = uniaxial true stress–true strain curve

REFERENCES

AISC. 2005a. *Specification for structural steel buildings, ANSI/AISC 360-05.* Chicago: American Institute of Steel Construction.

AISC. 2005b. *Seismic provisions for structural steel buildings, ANSI/AISC 341-05.* Chicago: American Institute of Steel Construction.

ATC. 1992. *Guidelines for cyclic seismic testing of components of steel structures, ATC-24.* Redwood: Applied Technology Council.

CSA. 2001. *Limit states design of steel structures, CAN/CSA-S16-01.* Toronto: Canadian Standards Association.

Haddad, M. & Tremblay, R. 2006. Influence of connection design on the inelastic seismic response of HSS steel bracing members. In J.A. Packer & S. Willibald (eds), *Proc. 11th international symposium and IIW international conference on tubular structures*: 639–649. London, UK: Taylor & Francis Group.

Hancock, J.W. & Mackenzie, A.C. 1976. On the mechanisms of ductile failure in high-strength steels subjected to multi-axial stress-states. *Journal of the Mechanics and Physics of Solids (24)*: 147–169.

Kanvinde, A.M. & Deierlein, G.G. 2006. Void growth model and stress modified critical strain model to predict ductile fracture in structural steels. *Journal of Structural Engineering (12)*: 1907–1918.

Lemaitre, J. & Chaboche, J.L. 1994. *Mechanics of Solid Materials.* New York: Cambridge University Press.

Martinez-Saucedo, G., Packer, J.A. & Willibald, S. 2006. Parametric finite element study of slotted end connections to circular hollow sections. *Engineering Structures (28)*: 1956–1971.

Martinez-Saucedo, G. & Packer, J.A. 2007. Slotted tube end connections: seismic loading performance. *Proc. 8th pacific structural steel conference*: Vol. 2: 91–96. New Zealand: Heavy Engineering Research Association.

Martinez-Saucedo, G., Packer, J.A. & Christopoulos C. 2008. Gusset plate connections to circular hollow section braces under inelastic cyclic loading. *Journal of Structural Engineering*: in press.

Swanson Analysis Systems 2007. ANSYS release 11.0. Houston, USA.

Yang, F. & Mahin, S. 2005. Limiting net section fracture in slotted tube braces. *Steel Tips Technical Information and Product Service—Structural Steel Educational Council*: 1–34.

Seismic performance of encased CFT column base connections

W. Wang, Y.Y. Chen & Y. Wang
Tongji University, Shanghai, China

Y.J. Xu & X.D. Lv
China Northwest Building Design Research Institute, Xi'an, China

ABSTRACT: Concrete-filled tubular (CFT) column has been quite often used in the high-rise buildings, because of its high strength and good performance against seismic loading. In a seismic design of CFT structures, column bases form highly important connections which must ensure smooth transferring of internal forces from column to foundation. So far, encased column base connection is becoming popular in the CFT buildings of China. The rigidity of this type of column base has been widely studied under static loads, but few investigations have been reported for the hysteretic behavior of the connections subjected to cyclic loading. This paper examines the cyclic performance of two encased concrete-filled tubular (CFT) column to base connections. Among the two, one is an encased CFT base plate connection with traditional detail, the other is a new type of encased CFT column base proposed by the authors which is strengthened by central reinforcing bars going through base plate into foundation concrete. Each specimen was subjected to increasing horizontal drift amplitude under a predetermined condition of axial compression. The experimental results clearly show that (1) stable restoring force characteristics are obtained for these two column bases, (2) use of central reinforcing bars can yield larger deformation capacity and higher strength than that of traditional encased CFT column bases, and (3) encased CFT column bases of these two types may be expected to have excellent properties for progressive collapse prevention since they all keep a stable residual load-bearing capacity though drastic encased concrete crush and reinforcing bar rupture occur. Finally, an analytical model for base strength estimation is proposed for design references.

1 INTRODUCTION

Concrete-filled tubular (CFT) members possess high compressive strength and significant ductility and thus have been found to be effective structural forms for earthquake-resistant purposes. The excellent seismic performance of such structural system can be guaranteed only when adequate connection performance is achieved. Current studies on the behavior of CFT designs under earthquake excitation are focused primarily on the response of CFT members under axial load, bending or a combination (Shanmugam 2001, Hsu 2003). Information on the CFT column to beam or base design details and the effects of the connection details on the seismic performance is still limited (Hajjar 2002, Nishiyama 2004, Kawaguchi 2006) and therefore should be further investigated.

To fully develop understanding of seismic behavior of CFT connections, a number of experimental programs were funded through national grants to provide technical support for the seismic improvement of CFT constructions in China. These test programs included the assessment of the seismic response of beam-to-column connections and column base connections. In particular, the authors of the present work investigated the inelastic seismic response of column base connections. There are mainly three types of column base used in the practical engineering: exposed type connection, encased type connection, and embedded type connection (Akiyama 1985). Among them, encased base plate connection is becoming more and more popular in the high-rise buildings of China. In doing so, two encased column base connections, with different connection details, were tested at Tongji University. The results of the cyclic tests are analyzed and discussed in the following sections.

2 EXPERIMENTAL PROGRAM

2.1 *Specimens*

The experiments were carried out on two types of encased CFT column bases: ZJ-1 and ZJ-2 (see Figure 1). Among them, the former is an encased

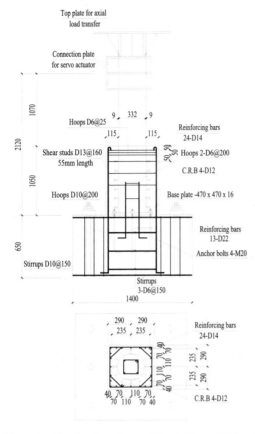

Figure 1. Details of specimens: ZJ-1 (without CRB) and ZJ-2 (with CRB).

CFT base plate connection with traditional detail. Reinforcement in the encased concrete is designed so that the force transferred from the column will be resisted by the RC foundation under the base plate. Shear studs are welded on the steel tube to intensify composite action between the CFT column and the encased concrete. In each cross section twelve shear studs are arranged around the diameter at equal arc length. The encased length is three times the diameter of the steel tube. The latter is an alternative or innovative type joint in which the column is strengthened by central reinforcing bars (CRB) going through base plate into foundation concrete. A typical configuration of specimens is shown in Figure 1. The principal properties of specimens are listed in Table 1. Table 2 summarizes the measured material properties of different components of the specimens.

2.2 Test setup and loading

Figure 2 shows the general arrangement for the connection tests. The specimens were tested under combined constant axial and cyclic lateral loads. Two sets of load apparatus were used to generate the required loads. The axial load N was applied to the head of the specimen which was mounted onto the strong floor with six high strength rods, by the hydraulic jack attached to the reaction frame. The lateral load P was applied to the top of the cantilever column at the level about 2120 mm above from the surface of the foundation, by the servo-controlled hydraulic actuator through a series of prescribed cyclic displacements.

The magnitude of the constant axial load for two specimens was set to 600 kN, i.e. 20% of the CFT

Table 1. Specimens and test parameters.

Specimen	CFT column (mm)	Encased CFT column (mm)	RB	CRB	Anchor bolt	N/N_y
ZJ-1	Φ350 × 9	580 × 580	24-D14	NA	4-M20	0.2
ZJ-2	Φ350 × 9	580 × 580	24-D14	4-D12	4-M20	0.2

Table 2. Material properties of the specimens.

Specimen	Component	f_y (MPa)	f_u (MPa)	ζ (%)	f_y/f_u (MPa)	f_{cu} (MPa)
ZJ-1	Reinforcing barD14	402	543	–	0.74	–
	Steel tube	382	561	29.9	0.68	–
	Anchor bolt	235	317	–	0.74	–
	Encased concrete	–	–	–	–	31
ZJ-2	Reinforcing bar D14	446	605	–	0.74	–
	Reinforcing bar D12	364	546	–	0.67	–
	Steel tube	345	478	32.1	0.72	–
	Anchor bolt	242	321	–	0.72	–
	Encased concrete	–	–	–	–	35

member's compressive strength (N_y), which could be calculated as follows:

$$N_y = A_t f_{yt} + A_c f_{cu} \qquad (1)$$

in which A_t and A_c are the cross-sectional areas of the steel tube and the in-filled concrete, and f_{yt} and f_{cu} are the yield strength of the tube and the compressive strength of the concrete, respectively.

2.3 Instrumentation

Figure 3 provides a detailed view of the arrangement of the displacement transducers and strain gauges for the specimens.

Figure 2. Test setup.

The strain gauges were positioned on the main reinforcing bars, central reinforcing bars as well as steel tubes at different levels of the column height. Displacement transducers were located at the loading point and the top of the foundation. They were mainly used to measure the specimen responses.

3 OBSERVATIONS

Figure 4 shows the failure patterns for two specimens. It can be observed from the figure that whether for ZJ-1 or for ZJ-2, they all exhibited cracking and peeling in the encased concrete of the column. Reinforcing bars buckled and concrete crushed under the compression caused by the combination of bending moment and axial load. Note that several tension bars of ZJ-2 ruptured due to the extremely large lateral deformation in the end.

4 TEST RESULTS AND DISCUSSIONS

4.1 Hysteretic properties

The lateral force-drift hysteretic curves obtained for two tests are presented in Figure 5. The horizontal load P has been plotted against the horizontal relative displacement δ_0, which is defined as the displacement detected at the loading point minus the displacement detected at the top of the foundation beam.

From the results for two specimens, peak force is obviously shown, accompanied by deterioration in capacity and stiffness at the next several cycles close

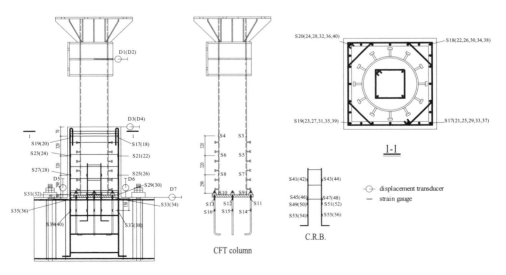

Figure 3. Arrangement of displacement transducers and strain gauges.

(a) ZJ-1

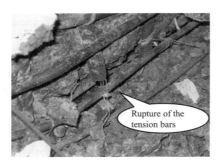

(b) ZJ-2

Figure 4. Typical failure modes of specimens.

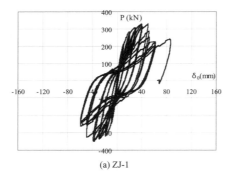
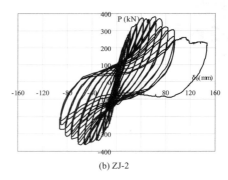

(a) ZJ-1   (b) ZJ-2

Figure 5. Lateral force-drift curves for two types of column base connections ($N = 600$ kN).

to the end. The hysteretic loops were observed to pinch as the displacement amplitude was gradually magnified. This phenomenon revealed that the column base connections experienced an increase in deformation without a significant increase in load, thus resulting in a loss in stiffness of the connection. From the comparison between the hysteretic curves of two specimens, it is founded that the ultimate strength of ZJ-2 is a little higher than that of ZJ-1. This is caused by the strengthening scheme adopting central reinforcing bars in the former specimen. Considering the difference in material strengths, the degree of connection strength enhancing is very limited. It is easily understood if the smaller distance relative to the section width between tensile and compressive bars is taken into account.

The rigidities in the early stage are almost same for two specimens. In addition, two specimens all keep a stable residual load-bearing capacity though drastic crushing of encased concrete and rupture of reinforcing bars occur. It may be expected that they have excellent properties for progressive collapse prevention.

4.2 Ductility of column base connections

To effectively assess the seismic performance of encased CFT column base connections, the ductility must first be adequately identified. The ductility ratio is defined as $\mu = \delta_u/\delta_y$, where δ_u is assumed to be the displacement corresponding to the post-peak load equal to 85% maximum load, and δ_y is the displacement when the reinforcing bar in the encased concrete is observed to yield for the first time.

The comparison of the skeleton curves for two specimens is shown in the Figure 6. Ductility ratio values of the specimens are calculated and listed in Table 3, where μ^+ and μ^- are the ductility ratio in the positive half-cycle and in the negative half-cycle, respectively.

It can be observed from Figure 6 and Table 3 that significant improvement in ductility was achieved when the strengthening scheme was adopted with central reinforcing bars going through base plate.

4.3 Load transferring mechanism

For encased CFT segments of the base connections, the member strength can be evaluated by summing the component strength of the CFT and encased reinforced concrete. Accordingly, the moment distribution between these two components is a major concern in the effective CFT base connection design. Figure 7 shows the comparison of the moments resisted by the CFT and the Encased RC along the column height in the elastic range schematically. The load induced on the CFT was calculated using the strain measurements installed on the CFT surface.

It can be found from the figure that the moments born by CFT reach the maximum at the boundary between the encased and the unencased column portion. In a small range below the boundary level, the moments born by CFT decrease dramatically and the moments born by encased RC increase rapidly. The latter is significantly larger than the former when it comes to the end of the column height. This observation can be explained by composite action due to load transferring from CFT to encased RC by shear studs welded onto the steel tube.

Figure 8 shows strain distribution of reinforcing bars of specimen ZJ-2 in a certain occasion of the elastic range, which was measured with the strain gauges shown in Figure 3. Vertical axis H denotes the column height away from the top of foundation beam. The value of H is positive for above the foundation beam and negative for below the top surface of foundation beam. Whether for main reinforcing bars or for central reinforcing bars, the longitudinal strains become larger with the decrease in the distance away from distance close to the base plate. It is inferred that the central reinforcing bars partly participate in moment resisting and load transferring.

4.4 Ultimate strength evaluation

Figure 9 shows a model for the analysis of the ultimate moment strength M_u of the column base subjected to the compressive load N simultaneously. It is assumed

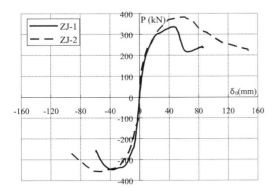

Figure 6. Skeleton curves for two specimens.

Table 3. Ductility ratio of column base connections.

Specimen	δ_y^+ (mm)	δ_y^- (mm)	δ_u^+ (mm)	δ_u^- (mm)	μ^+	μ^-
ZJ-1	10	10	55	55	5.5	5.5
ZJ-2	-10	-10	-75	-78.5	7.5	7.85

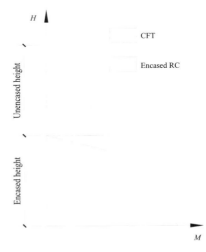

Figure 7. Moment distribution between CFT and Encased RC.

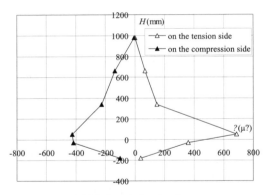

(a) Main reinforcing bar

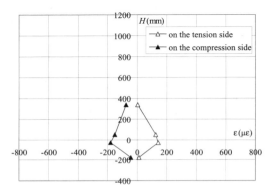

(b) Central reinforcing bar

Figure 8. Strain distribution of reinforcing bars along the column height in specimen ZJ-2.

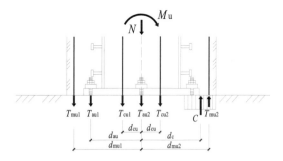

Figure 9. Analytical model for ultimate moment.

that tensile forces T_{mu1}, T_{au}, and T_{cu}, are generated in the outmost main reinforcing bars, anchor bolts and central reinforcing bars, respectively, which are all in the yielding or ultimate stage. A rectangular stress block with the stress equal to the cube strength f_{cu} forms in the foundation concrete, and the anchor bolt located in the bending compression side does not contribute to the ultimate strength. It is also assumed that the concrete stress distributes uniformly in the area with x distance from the column surface, and thus the resultant compression force acts at the center of this area. These assumptions were based on the observations in the test. The following equilibrium conditions of axial force and bending moments is obtained:

$$N + T_{mu1} + \sum T_{mui} + T_{au1} + T_{au2} + T_{cu1} + T_{cu2} = C + T_{mu2} \quad (2)$$

$$M_u = T_{mu1} \cdot d_{mu1} + T_{mu2} \cdot d_{mu2} + \sum T_{mui} \cdot d_{mui} + T_{au1} \cdot d_{au} + C \cdot d_c \quad (3)$$

Where d_c is the distance between the center and the acting point of C, and T_{mui} is the tension force in the inner main reinforcing bar. At the ultimate stage, the outmost main reinforcing bars and anchor bolts fractured, but the strains in the other main reinforcing bars and central reinforcing bars exceeded the yield value a little. Therefore, it is assumed for the calculation of the lower limit of M_u that the value of each component force is given as follows:

$$T_{mu1} = f_u \cdot A_s, \quad T_{mui} = f_y \cdot A_s, \quad T_{au1} = f_{um} \cdot A_{sm},$$
$$T_{au2} = f_{ym} \cdot A_{sm}, \quad T_{cu1} = f_{yx} \cdot A_{sx}, \quad T_{cu2} = f_{yx} \cdot A_{sx} \quad (4)$$

where f_u and f_{um} are ultimate stresses of the main reinforcing bars and anchor bolts, respectively f_y, f_{ym}, f_{yx} are yield stresses of the main reinforcing bars, the anchor bolts and the central reinforcing bars. A_s, A_{sm} and A_{sx} are cross-sectional areas of main reinforcing bars within tensile region, anchor bolts and central reinforcing bars, respectively.

Eliminating C from Equation 2 and Equation 3 in view of Equation 4 leads to the ultimate moment M_u.

The ultimate moment strength M_u was calculated for each specimen and compared with the experimental value M_{ue} in Table 4. The accuracy of M_u was good with the maximum error 6% for two specimens.

Table 4. Comparison of experimental and analytical strength.

Specimen	M_{ue} (kN-m)	M_u (kN-m)	M_{ue}/M_u
ZJ-1	699.6	744	0.94
ZJ-2	819.2	854	0.96

5 CONCLUSIONS

The aim of this paper was to provide information regarding the cyclic performance of encased CFT column base connections. Based upon the analysis on the test results of two specimens with different details, the following conclusions can be drawn:

1. Restoring force characteristics were obtained for these two column bases. It can be expected that they have excellent properties for progressive collapse prevention since they all keep a stable residual load-bearing capacity though drastic encased concrete crush and reinforcing bar rupture occur.
2. Significant improvement in ductility was achieved when the strengthening scheme was adopted with central reinforcing bars going through base plate.
3. It was also found that the base connections possess higher strength when central reinforcing bars were added to the encased segment, but the increase in strength is limited.
4. An analytical model for base strength estimation is proposed for design references. The accuracy of this method to evaluate the ultimate bending strength of the specimens was good.

ACKNOWLEDGEMENTS

The reported work is supported by National Research and Development Foundation of New Technology in Construction (CECEC-2005-Z). The experimental investigation was conducted in the Structural Engineering Laboratory, Department of Building Engineering, Tongji University.

REFERENCES

Akiyama, H. 1985. *Seismic design of steel column base for architecture*, in Japanese. Tokyo: Gihodo Shuppan.

Hajjar, J.F. 2002. Composite steel and concrete structural systems for seismic engineering. *Journal of Constructional Steel Research* 58(5–8): 702–723.

Hsu, H.L. & Yu, J.L. 2003. Seismic performance of concrete-filled tubes with restrained plastic hinge zones. *Journal of Constructional Steel Research* 59(5): 587–608.

Kawaguchi, J. & Morino, S. 2006. Experimental study on strength and stiffness of bear type CFT column base with central reinforcing bars. *Composite Construction in Steel and Concrete* V(N).

Nishiyama, I. & Morino, S. 2004. US–Japan cooperative earthquake research program on CFT structures: achievements on the Japanese side. *Prog. Struct. Engng Mater.* 6: 39–55.

Shanmugam, N.E. & Lakshmi, B. 2001. State of art report on steel-concrete composite columns. *Journal of Constructional Steel Research* 57(10): 1041–1080.

Tubular steel water tank tower dynamic analysis

J. Benčat, M. Cibulka & M. Hrvol
University of Žilina, Faculty of Civil Engineering, Department of Structural Mechanics

ABSTRACT: The subject of the paper is the dynamic behaviour and expected wind response of the steel water tank tower. Full scale dynamic measurements were realised on the lattice tower structure with empty and full water tank. The used sources of vibration were ambient ones and portable mechanical exciter of vibration. The analytical studies included comparison of different calculation models as to their sensitivity to natural modes and frequencies. The calculation of forced harmonic vibration proved acceptable agreement with measured values.

1 INTRODUCTION

The water tanks belong to structures that should be verified with regard to their wind response and resistance. The progress in development of new National and International Standards activates verifications of analytical models, design and execution of new and existing structures. The Eurocode 1 covers problems of wind loading in Part 1–4 (prEN 1991-1-4 Actions on Structures Part 1–4 Wind Actions) and those devoted to water tanks in Part 4 (prEN 1991–4 Actions on Structures Part 4 Silos and Tanks). The rules for the design of steel structure and its details are included in Eurocode 3. The design procedure includes also the dynamic analysis of the respective structure. In order to create the appropriate amount of data and limit statements necessary for simplified or more sophisticated analytical models, any knowledge of actual existing structure dynamic behaviour is welcome.

2 STRUCTURE DESCRIPTION

The analysed steel water tank was designed and constructed at Drahňov in Eastern Slovakia and now serves for field watering. The tower is 97 m high constructed like lattice tower with three corner tubes that create the basic carrying system. Two of tubes have diameter 1.34 m and the third one has 2.54 m. The thickness of tube wall is variable from 20 mm at the foundation and it decreases in top direction through 16, 14, 12 up to 10 mm. The corner tubes are anchored into R/C foundation slab. The cylindrical water tank is situated on the top of the tower and has diameter 15.5 m, height 6.8 m and its content can reach 500 m³. The brace is from welded I profiles 2 × 300/12 + 322/8. The bottom of the tank is the orthotropic plate with primary and secondary welded I-beams. The sheet of plate is 10 mm thick, the wall of tank has thickness 6 mm. The service content of the tank is limited to 470 m³. In view of expected stresses the steel S 355 for the lower part of the tower and S 235 for the upper part were used. The tower steel structure is welded, the bracing is connected to corner tubes by high strength bolts (Fig. 2). The view of R/C foundation slab is in Figure 1, general soil profile indicates clays, sands, sandy gravel and rather high underground water table. The added artificial gravel layer below the foundation is 0.5 m thick.

3 EXPERIMENTAL OBSERVATIONS

In the case of field measurements it is advantageous and reasonable to use all available sources of excitation, either random sources or artificial ones. Small vibration exciters have the advantage of being portable and easily operated. The applied mechanical exciter is similar to that described by Juhásová (1991). Firm bracing between the load carrying elements of the structure provides efficient transfer of the exciting force to the structure.

The relatively small total mass of the exciter and possibility of simple installation into the tested structure enable measurements of different systems of buildings, towers and bridges. Beside the measurements of ambient vibrations the mechanical exciter was used like the basic source of excitation of investigated water tank tower. The position of exciter during the tests was inside of the largest corner tube at the mid-height of the first brace. The set of accelerometers was used for measurement of global and

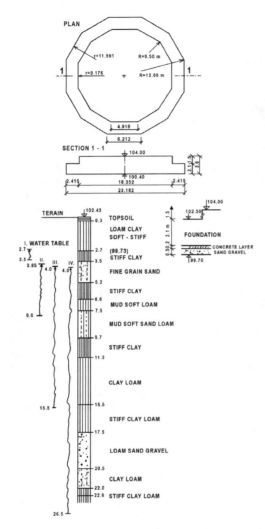

Figure 1. R/C foundation slab below the water tank tower and soil profile beneath.

Figure 2. General view of steel water tank tower Drahňov.

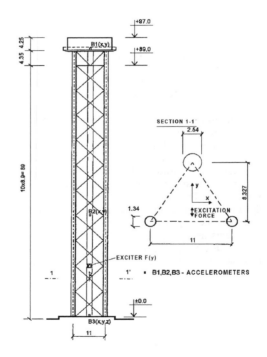

Figure 3. Position of accelerometers in the tower (B1x,y, B2x,y, B3x,y,z).

local vibrations in the most representing sections and points (Figure 3). For the recording both analogue and digital methods were used including A/D converters, filters, on/off line evaluation and reasonable data acquisition (Beneat et al. 2000).

The damping of tower determined from free tail vibration and that from steady harmonic vibration varied through damping ratio $\zeta = 0.024$ for empty tank and $\zeta = 0.021$ for the full tank. Local vibration damping ratio of the largest tube was determined from impact tests and reached the value up to 0.05.

Represented harmonic response of the tower to mechanical exciter is in Figure 4 for empty tank and in Figure 5 for full tank, Juhásová et al. (2002). The main experimental natural frequencies via spectral analysis (Figs. 6–7) of the tested water tank tower are in Table 1.

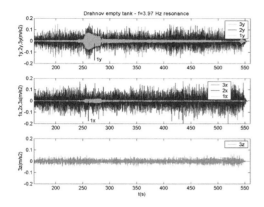

Figure 4. Harmonic response of empty water tank tower to mechanical exciter.

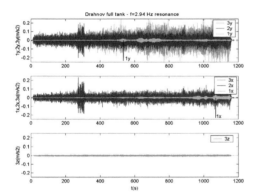

Figure 5. Harmonic response of full water tank tower to mechanical exciter.

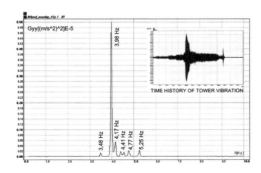

Figure 6. PSD at point B1(y)—Empty Tank.

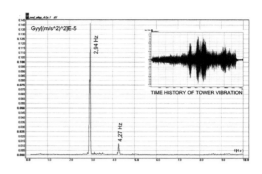

Figure 7. PSD at point B1(y)—Full Tank.

4 NATURAL FREQUENCIES AND MODES

The original calculations were performed before any test started. After the results of tests were known, the sensitivity study included few analytical model modifications (Benčat et al. 2000, Juhásová et al. 2002). They are as follows:

a. FEM model that consists of set of beam elements and stiff connection to the base;
b. FEM model that considered boundary springs and beam elements;
c. FEM model with liquid in the tank considered like stiff body, beam elements and stiff connection to the base
d. FEM model with liquid in the tank considered like stiff body, beam elements and boundary springs;
e. FEM model with liquid modelled according Equations 1–5 (Juhásová et al. 2002), beam elements and boundary springs.

Calculated modes and frequencies of empty tank when using model b) are in Figure 8, review of calculated and measured frequencies is in Table 1.

5 FORCED HARMONIC RESPONSE

Dynamic response to mechanical exciter creates harmonic response that could be transferred from measured acceleration to respective deflections. The comparison with calculated response is influenced by differences in calculated and measured modes and frequencies and by the contribution of spurious motions to the measured response. Illustrative results for empty water tank are in Table 2. Calculated response corresponds to analytical model b) at calculated resonant frequency 3.72 Hz and damping ratio $\zeta = 0.024$. Measured response corresponds to steady vibration at measured resonant frequency 3.97 Hz for direction y.

6 COMMENTS TO EXPECTED SEISMIC RESPONSE

Considering usual standard approach the seismic response can be calculated through application of seismic response spectra. The practice in earthquake engineering and dynamics of structures in Slovakia is interconnected with the demands of valid *National Standards STN 73 0036:1997 Seismic Actions on Structures and STN 73 2044:1983 Dynamic Tests of Structures*. New demands are those of STN P ENV 1998. The execution of dynamic tests on existing structures drawn down necessary computations, dynamic analysis and assessment of expected seismic response (Krištofovič 2001). The obtained data and subsequent analyses give more precise basis for accomplishment of appropriate seismic resistance of tank structure. The full-scale measurement of structure vibrations is important both for empty and full tank. The received results should be compared with data from previous static and dynamic computations and assumptions used in the design.

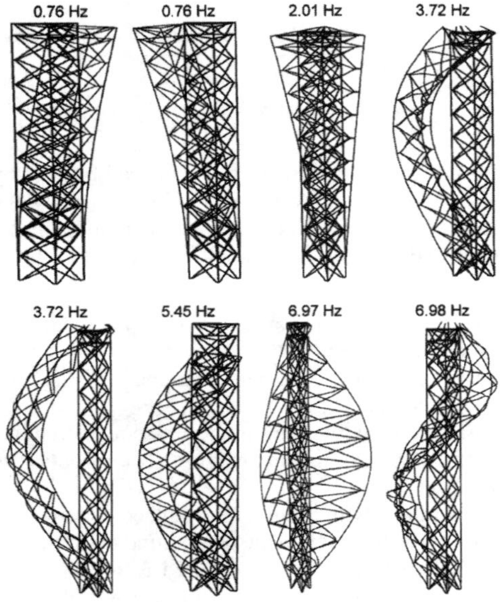

7 CONCLUSION

The structure of the water tower tank was subjected to the verification of its dynamic properties and dynamic resistance capacity before its full service operation. The original calculation results were compared to those obtained from dynamic calculations with and without consideration of water sloshing

Figure 8. Calculated natural modes of empty water tank tower-model b).

Table 2. Dynamic response to mechanical exciter—empty water tank.

Point and direction	Measured deflection at 3.97 Hz mm	Calculated deflection at 3.72 Hz for model b) mm
B1-y	−0.075	−0.096
B2-y	+0.157	+0.137
B3-y	−0.032	−0.007

Table 1. Experimental and calculated natural frequencies of water tank tower.

Mode No.	Exper im. empty Hz	Calculated-empty a. Hz	b. Hz	Exper Im. full Hz	Calculated-full tank c. Hz	d. Hz	e. Hz
water				0.23			0.18
water				0.26			0.20
1	0.74	0.85	0.76	0.61	0.34	0.31	0.56
2	0.87	0.89	0.76	0.67	0.36	0.31	0.61
3	2.03	2.07	2.01	2.94y	0.86	0.84	1.51
4	3.97y	4.09	3.72	3.50	3.42	3.11y	3.23y
5	4.27	4.14y	3.72y	4.27	3.48y	3.12	3.25
6	6.45	5.64	5.45	5.01	4.34	4.14	4.47
7	8.67	8.67	6.97	6.30	4.74	4.58	5.47
8	8.91	8.78	6.98	7.30	7.17	6.63	6.50
9	9.22	9.62	7.12	7.50	7.22	6.70	6.68

and test results. The detailing and resistance capacity of the most sensitive connections were checked and proved that the structure complies with new standards demands and conditions. Experiences obtained from this case study give also the indications of existing uncertainties and assessment variance in the design period comparing to actual dynamic properties and the expected structure performance in the case of "design seismic effects". It was confirmed, the experimental verification of dynamic properties and expected response to any accidental dynamic action contributes to the appropriate determination of the reliability and safety of civil engineering structures including towers and water tanks.

REFERENCES

Benčat, J., Juhásová, E., Podhorský, P. & Půbal, J. 2000. The results of dynamic test of water tank Drahňov. *FCE University of Žilina, Report*. No.-76-SvF-02.

Juhásová, E. Benčat, J. Krištofovič, V. & Kolcún, Š. 2002. Expected seismic response of steel water tank. *12th European Conference on Earthquake Engineering*. No. 592. London: UK Paper Ref.

Juhásová, E.1991. *Seismic effects on structures*. Amsterdam—Oxford-NewYork-Tokyo:Elsevier.

Krištofovič, V. 2001. Dynamic interaction of steel towers and chimneys with subsoil under earthquake excitation. *Proc. Seminar on New trends in Steel Construction, FCE TU Košice, 2001, Jahodná, Slovakia: pp 163–168.*

Elasto-plastic behavior of circular column-to-H-shaped-beam connections employing 1000 N/mm²-class super-high-strength steel

N. Tanaka & T. Kaneko
Kajima Technical Research Institute, Kajima Corporation, Japan

H. Takenaka
Technical Research Institute, Toda Corporation, Japan

S. Sasaki
Technology Development Division, Fujita Corporation, Japan

ABSTRACT: Super-high-strength steels with yield points of 800–1000 N/mm² are planned to be used in the near future for building frames. Thus, a design method needs to be established that takes into account the poor welding performance as well as the larger elasticity of these materials. To this end, cruciform circular-column-to-beam connection specimens were tested and analyzed. Frame connections can be made by partial butt welding or super-high-strength bolts using an ordinal outer or ring diaphragm and a horizontal beam haunch. Tests and analytical results confirmed that the elastic region extends further than that of an ordinary frame made of mild steel by about 1% of story deflection angle. Initial stiffness, yield strength and maximum strength due to local beam buckling can be assessed by ordinary calculation methods. However, the slip and fracture strength of bolted joints using super-high-strength steel and bolts needs further investigation.

1 INTRODUCTION

Severe damage that occurred to buildings during the 1995 Hanshin-awaji earthquake in Japan had a great impact on us. Many countermeasures, such as horizontal beam-haunches (Tanaka & Sawamoto 2003) and new structural steels, SN400 and SN490, have since been employed in construction. A guideline to prevent brittle fracture in beam-to-column welded connections has also been recommended (BCJ 2003). However, these methods can only prevent collapse of buildings to save lives in a great earthquake. Buildings that narrowly escaped collapse would stand like Stone henge but be of no practical use. A performance-based seismic design method has been proposed in the USA (SEAOC 2000). According to this method, if a building is designed for a higher performance level, it could survive a severe earthquake without damage such as residual deflection, allowing rapid return to service after an earthquake. To this end, a national project was started to develop a structure with the required high performance after an earthquake up to an intensity of 7 on the JMA seismic scale. Because of its high elasticity, super high-strength-steel (SHSS), 800–1000 N/mm² strength class used as a construction material is expected to show no residual deflection. However, it has poor site-welding performance and is vulnerable to brittle fracture.

To seek a feasible design and to verify the higher elasticity of frames made from SHSS, five cruciform H-beam-to-circular-column connection specimens were tested and then analyzed by an FEM analysis.

2 EXPERIMENTS

2.1 Specimens

The specimens are listed in Table 1 and described in Figure 1. Table 2 shows the mechanical properties of the material used in the specimens. The parameters are for the column filled, or not, with high-strength concrete (nominal compression strength is 100 MPa), column and beam steel strength (YS880 and YS650), and shape of a diaphragm (outer or ring).

Specimens BC1 to BC3 had concrete-filled columns. BC1 had outer diaphragms frictionally jointed to the beam by super-high-tension bolts (SHTBs, F14T; nominal tensile strength larger than 1400 MPa), while BC2 and BC3 had ring diaphragms partially butt welded to the beam flange. The steel strength of the

Table 1. List of specimens.

Specimen	Column Section	Column Strength class	Column Filled concrete	Diaphragm Type	Diaphragm Thickness (mm)	Diaphragm Height (mm)	Diaphragm Strength class	Beam Section	Beam-Haunch Strength class	Joint bolt Width (mm)	Joint bolt Length (mm)	Joint bolt Flange	Joint bolt Web
BC1				Outer	12	63	YS650		YS650			8-F14T, M16	4-F14T, M16
BC2			Fc100					H-280×90×6×9		140	103.4		
BC3	O-300×9	YS880		Ring	22	40	YS880		YS880			null	8-F14T, M16
BC4			null	Outer	12	60	YS650	H-280×80×6×9	YS650	130	165	6-F14T, M16	5-F14T, M16
BC5		YS650											

Table 2. Mechanical material properties.

Strength Class	Thickness (mm)	Yield strength (MPa)	Tensile strength (MPa)	Yield strength ration (%)	Elongation (%)	Compression strength (MPa)	Young's modulus (GPa)	Remarks
YS880	22	1019	1052	97	15.2		–	Ring for BC3
	9	944	1038	91	15.1		176	Column for BC1~BC4
	9	933	1000	93	16.9		196	Beam flange for BC3
	–6	933	993	94	15.9		187	Beam web for BC3
YS650	22	807	851	95	15.5		–	Ring for BC2
	12	773	832	93	19.1		193	Outer diaphragm
	9	760	835	91	19.2		175	Column for BC5
	9	801	836	96	20.1		195	Beam flange for BC1,BC2, BC5 and Splice Plate for BC4, BC5
	6	786	829	95	19.2		203	Beam Web for BC1, BC2, BC4, BC5 and splice plate for BC4, BC5
SHTB	16	≧1235	1437~1460	–	≧14		–	Super high tension bolt
Fc100	–	–	4.8	–	–	99	38.2	Obtained before test
			4.2			105	41.7	Obtained after test

diaphragm and beam of BC3 was higher than that of BC2. BC4 and BC5 had no concrete in the column, but had attached outer diaphragms jointed to the beam by SHTBs. The steel strength of BC5's column was the least of all. All the diaphragms were also welded to the column by partial butt welding, which was expected to decrease thermal input and deterioration.

For the design, the specimens satisfied two criteria of the AIJ recommendations (AIJ 1990, AIJ 2003, AIJ 2006). First was beam yielding before beam joint slip. The second was maximum joint strength exceeding full plastic strength of the beam. To meet the criteria using high yield ratio of the SHSS (91–97%, see Table 2), the beam section was smaller than that of a conventional frame made of mild steel. Horizontal haunches were also used to increase the joint strength.

2.2 *Loading procedure*

All specimens were loaded cyclically up to fracture. The load was applied at the tip of the upper column for BC1 to BC3 and at both beam tips for BC4 and BC5, by a hydraulic jack with a controlled lateral deflection

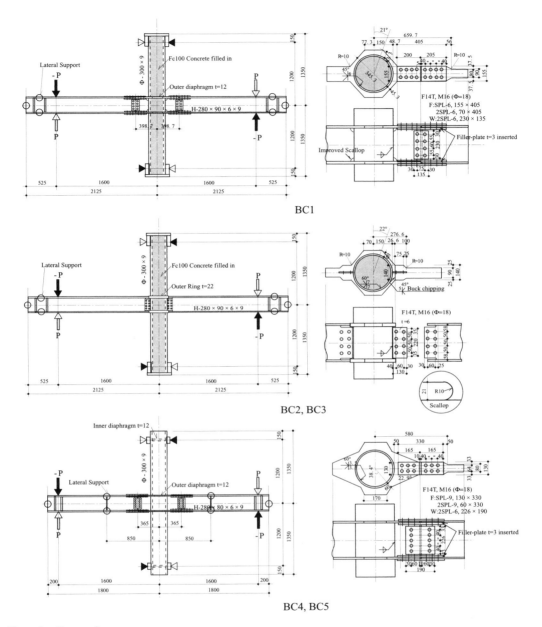

Figure 1. Test specimens.

angle (R). The difference of the loading was due to the apparatus used in the two laboratories. The jack load was converted to lateral shear force (Q_c). To prevent out-of-plane beam deflection, constraining rigs were set up on both sides.

The load, displacements, and strains of the specimens were automatically measured during the test.

2.3 *Test results and discussion*

2.3.1 *Load-deflection relationships*
Lateral shear force (Q_c) and lateral deflection angle (R) relationships are shown in Figure 2. The Q_c-R relations up to main slip of the bolted joints are also aded for BC4 and BC5.

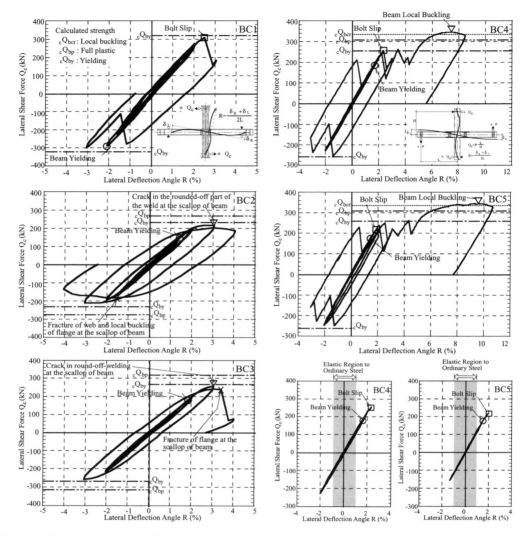

Figure 2. Load-deflection curves obtained by test.

For BC1, under negative loading, a strain gauge attached to the beam flange yielded (strain gauge yielding: SGY) near the start point of the horizontal haunch at $R = 2.08\%$. Under positive loading, bolted joint slip (BJS) occurred at about 2.5% of R, accompanied by a rapid drop of Q_c. The same drop of strength due to BJS was observed under negative loading, so the test was stopped. The SGY of BC2 occurred at 1.42% followed by reduction of stiffness, which made the Q_c-R relations spindle-shaped. At 3.03%, cracking occurred in the rounded-off part of the weld of the web at the scallop. Finally, fracture of the web root and buckling of the flange were cogenerated near the scallop, as shown in Figure 3, at about 1.5% in the following negative loading. For BC3, almost the same pass as BC2 was observed up to 2.83%. However, at 3.3% the beam flange fractured in the scallop, thus ending the test. For BC2 and BC3, no BJS occurred owing to the ring diaphragm welded to the beam flange. For BC4, SGY occurred at 1.62% and BJS at 2.3%, causing a rapid drop of Q_c, which was also observed in the following negative loading. Further loading caused a small drop of Q_c, so that the bolted joint mechanism changed from friction to bearing with increase in Q_c. Finally, local beam flange buckling occurred at 7.31% at the start point of the

Table 3. Test results of the initial stiffness, yielding and slipping load and the corresponding calculations.

Specimen	Initial stiffness Test (kN/%)	Initial stiffness Cal.	Test/Cal.	Yielding Strength Test[1] (SGY) (kN)	Yielding Strength Test[2] (GYP)	Yielding Strength Cal.[3]	Test/Cal.	Story deflection angle[4] (%)	Slip Strength Test[5] (kN)	Slip Strength Cal.[6]	Test/Cal.	Story strength angle (%)	Full plastic strength Cal.(FPS) (kN)	Maximum Strength Test (kN)	Maximum Strength Cal.[8] (LBS)	Test/Cal.	Story deflection angle (%)	Ductility factor[9] (DF) (%)
BC1	137	102	1.24	290	–	320	–	2.26	310	413	0.75	2.5	373	310[7]	411	0.75	3.07	–
BC2	101	102	0.99	144	204	232	0.88	2.02	–	300	–	–	270	215	298	0.72	3.03	0.51
BC3	96.5	102	0.95	179	243	265	0.92	2.52	–	338	–	–	309	257	331	0.78	2.83	0.12
BC4	118	91.9	1.28	187	–	260	–	2.18	257	300	0.86	2.37	306	346	339	1.02	7.31	2.25
BC5	122	91.9	1.33	178	–	260	–	1.81	221	300	0.74	1.97	306	344	339	1.02	9.31	2.09

*1: Value is determined by strain gauge, *2: Value is determined by G. Y. P. method, *3: Value is calculated at the start point of the beam-haunch, *4: Value is obtained from test[2] or test[5] divided by initial stiffness, *5: Value is obtained at bolt first slipping, *6: Slip factor is 0.45, *7: Slip strength, *8: Value is when beam locally buckles, *9: Value is calculated from story deflection angle of maximum divided by that of yielding.

Figure 3. Fracture and local buckling (BC2).

haunch. The progress of BC5 was almost the same as that of BC4, except that the 9.31% of R at maximum strength was greater than that of BC4.

2.3.2 *Initial stiffness*

Test and calculation results of initial stiffness and yield strength are compared in Table 3.

The initial stiffness is Q_c divided by R when Q_c is 100 kN in the test. BC1's initial stiffness was the largest due to the concrete in the column, while those of BC2 and BC3 were the smallest. This was because the relatively long scallops of these two specimens, shown in Figure 1, deteriorated the unity of bending stiffness of the flange and web in the beam. This lack of unity was confirmed by the strain distribution of the beam web shown in the Figure 4, where the strain near the beam-to-column connection (section A) was small and its distributions were nonlinear when R was 1% or 2%. Thus, an improved scallop or no scallop is recommended in the welding of the beam-to-column connections. Except for BC2 and BC3, the test results were underestimated by the calculations, which were obtained by considering bending and shear deflection in the beam and column as well as shear deflection in the connection panel with the beam haunch and the diaphragm neglected. However, neglecting the beam haunch and diaphragm could lead to discrepancies between the test and calculated values.

2.3.3 *Yield strength*

Because a strain gauge indicated local yielding of the beam flange near the haunch start point, the lateral shear force obtained by the General Yield Point (GYP) method shown in Figure 5 is selected as the yield strength (GYP strength). However, GYP was not

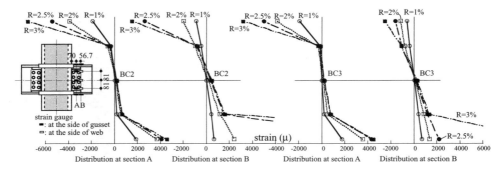

Figure 4. Strain distributions in web at near beam-to-column connections for BC2 and BC3.

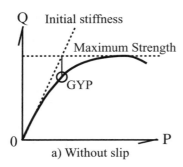

Figure 5. GYP method to obtain yield strength.

applied to BC1, BC4 and BC5 for their bolted joint slip (BJS) precedence. The calculation result was Q_c when the beam flange yielded at the start point of the haunch. The test results were smaller than the calculation results for BC2 and BC3. This discrepancy was due to their larger scallops, as discussed previously. Dividing the GYP strength or BJS strength mentioned later by the initial stiffness obtained by the test gave the story yield deflection angle. These values varied from 1.81% to 2.52% and were about 1% greater than those in an ordinary frame using mild steel, that is, almost double. This exemplifies the superiority of SHSS as a material with larger elasticity leading to less residual deflection.

2.3.4 Slip strength

The slip strength (BJS strength) is defined as the strength when the first and major slip occurs in the beam joints. All test results were smaller than the calculation results, which were obtained with a slip factor of 0.45 on the friction surface. One reason for this is considered to be insufficient slip surface due to filler plates for ensuring the slip factor of 0.45, as shown in Figure 1. The filler plate was employed to adjust the difference between the thicknesses of the diaphragm and the beam flange to be jointed together. However, further investigation is needed to develop a brand new friction joint employing SHSS and SHTB.

2.3.5 Maximum strength

The full plastic strength (FPS) is Q_c when the beam reaches calculated full plastic bending moment at the start point of the haunch. Local buckling strength (LBS), referring to Kato B. 2000, is the calculated maximum strength. For BC2 and BC3, the test results were smaller than those for both FPS and LBS, owing to the co-occurrences of fracture and local buckling in the larger scallop region shown in Figure 3. However, the calculation results showed good accordance with the test results for BC4 and BC5. Although the calculation for LBS was derived from many experimental studies exclusively based on mild steel, this method could be applied to higher yield strength steel like SHSS.

2.3.6 Plastic deformation capacity

Ductility factor (DF) is defined as plastic deformation capacity with the story deflection angle at maximum strength divided by that at the yield strength. For BC2 and BC3, it is considered to have hardly any plastic deformation capacity (PDC) because their DFs were 0.51 and 0.12, respectively. However, some PDC is expected for BC4 and BC5 with DF being a little more than 2.

2.3.7 Comparison between test parameters

2.3.7.1 Column concrete

Although their beam sections differ, comparison of BC1 with BC4 shows an increase in initial stiffness and a decrease in strain in the connection panel due to the concrete infill, as shown in Figure 6.

2.3.7.2 Type of diaphragm

Major slip and rapid drop of the beam joints is observed depending on the joining method, i.e., bolted or welded, rather than difference of diaphragm type.

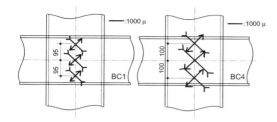

Figure 6. Strain distributions in connection panel at R = 2%.

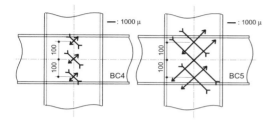

Figure 7. Strain distributions in connection panel at beam yielding.

2.3.7.3 *Difference of beam steel strength in BC2 and BC3*

Increase in strength and story deflection angle at yielding is recognized because of the high strength of the steel for BC3 compared with that for BC2. However, rupture in the welded part of BC3 suggested the vulnerability of SHSS to welding and stress concentration.

2.3.7.4 *Difference of column steel strength in BC4 and BC5*

Strain distributions are shown in Figure 7 for BC4 and BC5 at beam yielding. It is clear that the strain of BC4 is smaller than that of BC5 because of its higher steel strength, although there was no apparent difference in the Q_c-R relation.

3 NUMERICAL SIMULATION

3.1 *Analytical model*

Targets for analysis were BC2 to BC5.

Figure 8 shows one of the analytical models representing half of the specimen cut along the plane of geometrical symmetry. To simplify the model and to emphasize the effect of steel strength, the bolted joints were assumed to be a unified body. The steel part was modeled as a plane shell member consisting of three or four nodes, while the concrete part was modeled as a solid member with six or eight nodes. BC2 and BC3 comprised 3,968 members and 3,183 nodes, while BC4 and BC5 comprised 1,624 members and 1,683 nodes.

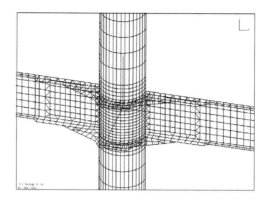

Figure 8. Analytical model.

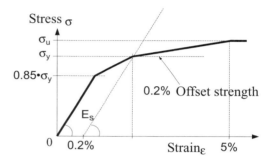

Figure 9. Stress and strain relationship for steel.

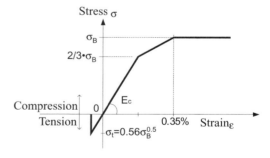

Figure 10. Stress and strain relationship for concrete.

3.2 *Analytical method*

Monotonic elasto-plastic analytical loading was carried out using the finite element method (FEM) provided by Marc Corporation (Marc 2005).

Stress (σ) and strain (ε) relationships are shown in Figure 9 for steel and Figure 10 for concrete. A quadric-linear σ-ε relation was used in both tension and compression for steel, where σ_y is yield strength of 0.2% offset and σ_u is tensile strength. The

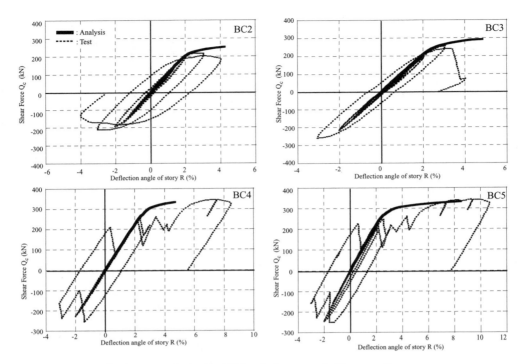

Figure 11. Load-deflection curves obtained by FEM analysis compared with test results.

reason why $0.85\sigma_y$ is given as a braking point is that a yield stress of 0.2% offset has nonlinearity from about $0.8\sigma_y$ to $1.0\sigma_y$. Strain of 5% at σ_u is based on the uniform elongation obtained from coupon tests. Mises's yield criterion is also applied to determine yielding of steel whose Young's modulus is 205 kN/mm² and Poisson's ratio is 0.3.

For the concrete, a tri-linear σ-ε relation was used in compression assuming that the concrete was elastic up to 2/3 of axial compression strength, σ_B. The strain of 0.35% at σ_B is based on the coupon tests. Equation 1 was used for the breaking point in tension (AIJ 1999). Buyukozturk's destruction criterion given by Equation 2 was used (Marc 2005). The Young's modulus given by Equation 3 and a Poisson's ratio of 1/6 were used. Adhesion between steel and concrete was neglected.

$$\sigma_t = 0.56\sqrt{\sigma_B} \quad (1)$$

$$\left(J_2 + \frac{\beta \cdot \sigma_B \cdot J_1}{\sqrt{3}} + \frac{\gamma}{3}J_1^2\right)^{\frac{1}{2}} - \frac{\sigma_B}{\sqrt{3}} \quad (2)$$

where J_1, J_2 are invariant of stress and $\beta = \sqrt{3}$, $\gamma = 0.2$

$$E_c = 33500\kappa_1 \cdot \kappa_2 \left(\frac{\gamma_c}{24}\right)^2 \cdot \left(\frac{\sigma_B}{60}\right)^{\frac{1}{3}} \quad (3)$$

Where κ_1, κ_2 are coefficients for coarse aggregate (0.95 was used for both) and γ_c is weight per unit volume (24 was used).

3.3 Analytical results and discussion

The Q_c-R relationships obtained from the analysis are shown along with the test results in Figure 11. All specimens showed elastic behavior near 0.2% of R in the analysis, which is almost the same as that in the test results. However, the calculations after 0.2% were greater than those of the tests for BC2 and BC3. The reason is that early co-occurrence of rupture of welded parts and local buckling near the scallop prevented the beam from showing bending capacity. However, for BC4 and BC5, it can be seen that the analyses closely modeled the test results, although it could not model the jagged parts due to bolt-slipping.

4 CONCLUDING REMARKS

Loading tests and numerical simulation by FEM were carried out for beam-to-column sub-assemblages made from super-high-strength steel (SHSS) to seek the feasible design and to verify its large elasticity. The conclusions are as follows.

a. Ordinary techniques can be used to connect beams and columns of SHSS in the following conditions even if site-welding can not be used.
   1. An outer or ring diaphragm should be used for shop welding of beam-to-column connections with partial-butt grooves, which enable decreased heat input during welding.
   2. Super-high-tension bolts (SHTB) should be used for beam-to-beam and diaphragm-to-beam joints.
   3. A horizontal beam haunch should be used to decrease stress in the welded parts and to enable joining by SHTB compact.
b. Tests and analytical results verified that the elastic region is about double that of an ordinary frame made of mild steel.
c. Among the parameters, the concrete filling and higher steel strength of the column works well, showing decreased stress in the connection panel. However, there is no clear difference with beam steel strength or diaphragm type, although the joining method of the latter to the beam had a great impact on the Q_c-R relations depending on bolting or welding.
d. Initial stiffness, yield strength and maximum strength due to the local buckling of the beam can be assessed by ordinary calculation methods. However, the slip and fracture strength of the bolted joints comprising SHSS and SHSB need further investigation.
e. The overall Q_c-R relationship can be roughly assessed by the FEM analysis.

ACKNOWLEDGEMENTS

This research was carried out under the Collaborative Project by Cabinet and Ministries: Development of New Building Systems Using Innovative Structural Materials, whose promoters are the Association of New Urban Housing Technology (ANUHT), the Japan Iron and Steel Federation (JISF), and Japanese Society of Steel Construction (JSSC). The authors would like to thank the members of the New Building Systems Research Committee supervised by ANUHT. We owe special thanks to Dr. S. Sakamoto of Shimizu Corporation for his valuable suggestions for the design of specimens.

REFERENCES

AIJ 1990. *Recommendations for the design and fabrication of tubular structures in steel (in Japanese)*, Architectural Institute of Japan.
AIJ 1999. *Standard for structural calculation of reinforced concrete structures-based on allowable stress concept- (in Japanese)*, Architectural Institute of Japan.
AIJ 2003. *Guidebook on design and fabrication of high strength bolted connections (in Japanese)*, Architectural Institute of Japan.
AIJ 2006. *Recommendation for design of connection in steel structures (in Japanese)*, Architectural Institute of Japan.
BCJ 2003. *Guideline for prevention of brittle fracture of welding parts in steel beam-to-column connections (in Japanese)*, Building Center of Japan.
Kato B. 2000. Prediction of the vulnerability of premature rupture of beam-to-column connection (in Japanese). *Journal of structural and construction engineering*. No. 527: 155–160, No. 529: 175–178.
MARC 2005. *Background information Volume F Part II*. Marc analysis research corporation.
SEAOC 2000. *Vision 2000-A framework for performance based seismic engineering buildings*, SEAOC Vision 2000 Committee.
Tanaka & Sawamoto. 2003. Behavior of steel H-shaped-beam-to-hollow-column connection with horizontal haunch and its optimum length, *Tubular Structure X*: 385–393.

Plenary session B: Specification & code development

New IIW (2008) static design recommendations for hollow section joints

X.L. Zhao
Department of Civil Engineering, Monash University, Australia

J. Wardenier
Delft University of Technology, The Netherlands
National University of Singapore, Singapore

J.A. Packer
University of Toronto, Canada

G.J. van der Vegte
Delft University of Technology, The Netherlands

ABSTRACT: The first edition of the IIW sub-commission XV-E design recommendations on static strength of tubular joints was published in 1981. The second edition was published in 1989, which was the basis for the CIDECT Design Guides on static strength of tubular joints in the early 90s. Extensive research on tubular joints has been carried out all over the world in the last 18 years. There is a need to update the current static design rules. This paper summarizes the 3rd edition of the IIW Static Design Procedure for Welded Hollow Section Joints, recently completed by the IIW sub-commission XV-E. The paper will also briefly describe the relevant background information.

1 INTRODUCTION

The first edition of IIW sub-commission XV-E design recommendation on static strength of tubular joints (IIW 1981) was published in 1981. The second edition (IIW 1989) was published in 1989, which was the basis for the CIDECT Design Guides 1 and 3 on the static strength of tubular joints (Wardenier et al. 1991, Packer et al. 1992). Extensive research on tubular joints has been carried out all over the world in the last 18 years. There is a need to update the current static design rules. This paper summarizes the 3rd edition of the IIW Static Design Procedure for Welded Hollow Section Joints, recently completed by the IIW sub-commission XV-E (IIW 2008). The paper also briefly describes the major changes made to the previous edition and the relevant research background in terms of yield stress, moment resistance, chord stress function, overlap joints, multiplanar joints, plate-to-tube joints and formulae harmonization.

2 OVERVIEW OF THE NEW RECOMMENDATIONS

2.1 Layout

The new recommendations contain 18 sections and three annexes. The first 6 sections cover the scope, normative references, terms and definitions, symbols, limit state design and partial safety factors. Section 7 outlines the design procedures followed by two sections on design member forces and design criteria. The design resistance formulae are presented in the remaining nine sections. The annexes give information related to quality requirements for hollow sections, weld details and partial safety factors.

2.2 Scope

The recommendations deal with the static resistances of welded uniplanar and multiplanar joints in lattice structures of hollow sections. The recommendations are valid for both hot-finished and cold-formed steel hollow sections. The manufactured hollow sections should comply with the applicable national manufacturing specification for structural hollow sections. The nominal yield strength of hot finished hollow sections and the nominal yield strength of the cold-formed hollow section finished product should not exceed 460 MPa (N/mm^2). The nominal wall thickness of hollow sections is limited to a minimum of 2.5 mm. The nominal wall thickness of a steel hollow section chord should not be greater than 25 mm unless special measures have been taken to ensure that the through thickness properties of the material will be adequate.

Table 1. Validity ranges for the new IIW recommendations.

(a) uniplanar CHS to CHS joints (T, Y, X and K-gap).

General		$0.2 \leq d_i/d_0 \leq 1.0$	
		$g \geq t_1 + t_2$ $e/d_0 \leq 0.25$	
Chord	tension	$d_0/t_0 \leq 50$	For X joints:
	compression	class 2, but also $d_0/t_0 \leq 50$	$d_0/t_0 \leq 40$
Braces	tension	$d_i/t_i \leq 50$	
	compression	class 2 but also $d_i/t_i \leq 50$	

(b) gusset plate, I or RHS sections to CHS chord joints.

Same as those given in Table 1(a) with the following additional limits:
Transverse plate: $\beta \geq 0.4$
Longitudinal plate: $1 \leq \eta \leq 4$

(c) uniplanar RHS or CHS to RHS chord joints (T, Y, X and K-gap).

Type of joint		T, Y or X	K gap	Circular brace member
Brace-to-chord ratio		$b_i/b_0 \geq 0.1 + 0.01\, b_0/t_0$ but ≥ 0.25	as left, but also $d_i/b_0 \leq 0.80$	
Chord	tension	b_0/t_0 and $h_0/t_0 \leq 40$		
	compression	class 2, but b_0/t_0 and $h_0/t_0 \leq 40$		
Braces	tension	b_i/t_i and $h_i/t_i \leq 40$	$d_i/t_i \leq 50$	
	compression	class 2, but b_i/t_i and $h_i/t_i \leq 40$	class 2, but $d_i/t_i \leq 50$	
Aspect ratio		$0.5 \leq h_i/b_i \leq 2.0$	N/A	
Gap		N/A	$0.5(1-\beta) \leq g/b_0 \leq 1.5(1-\beta)$ [1)] but $g \geq t_1 + t_2$	

Note (1):
For $g/b_0 > 1.5(1-\beta)$, check the joint also as two separate T or Y joints.
For multiplanar joints $\phi = 90°$.

(d) additional conditions (to Table 1 (c)) for square RHS or CHS to square RHS chord joints in order to use simple rules.

Type of brace	Type of joint	Joint parameters	
Square hollow section	T, Y and X	$b_i/b_0 \leq 0.85$	
	K gap or N gap	$0.6 \leq (b_1 + b_2)/(2b_i) \leq 1.3$	$b_0/t_0 \geq 15$
Circular hollow section	K gap or N gap	$0.6 \leq (d_1 + d_2)/(2d_i) \leq 1.3$	$b_0/t_0 \geq 15$

(e) gusset plate to RHS chord joints.

chord class 2 and $b_0/t_0 \leq 35$
$\beta \geq 0.4$ (transverse plate)
and $1 \leq \eta \leq 4$ (longitudinal plate)

(f) uniplanar CHS or RHS to I or H section chord joints.

Type of joint	X	T or Y	K gap	Circular brace
h_i/b_i	$0.5 \leq h_i/b_i \leq 2.0$		1.0	N/A
d_w/t_w	class 1 and $d_w \leq 400$ mm	class 2 and $d_w \leq 400$ mm	as left	
b_0/t_0	class 2			
Brace tension	$h_i/t_i \leq 40$, $b_i/t_i \leq 40$			$d_i/t_i \leq 50$
Brace compression	class 1			class 1

(g) uniplanar overlap joints with a CHS, RHS or I-section chord.

General	f_{yi} and f_{yj} $\leq f_{y0}$	t_i and $t_j \leq t_0$ $t_j \leq t_i$	Ov $\geq 25\%$	
	$d_i/d_0 \geq 0.2$	$b_i/b_0 \geq 0.25$ $d_i/b_0 \geq 0.25$	$d_i/d_j \geq 0.75$ $b_i/b_j \geq 0.75$	
Chord	tension	$d_0/t_0 \leq 50$; b_0/t_0 and $h_0/t_0 \leq 40$		
	compression	class 2 but $d_0/t_0 \leq 50$; b_0/t_0 and $h_0/t_0 \leq 40$		
Braces	tension	$d_i/t_i \leq 50$; b_i/t_i and $h_i/t_i \leq 40$		
	compression	class 2 but $d_i/t_i \leq 50$; b_i/t_i and $h_i/t_i \leq 40$		

The joints covered in the recommendations consist of circular or rectangular hollow sections as used in uniplanar trusses or girders, such as T-, Y-, X- and K-joints and their multiplanar equivalents, gusset plates to CHS or RHS joints, open sections to CHS and hollow sections to open sections. Some examples are shown in Figure 1 where geometric dimensions are defined. The ranges of validity are summarized in

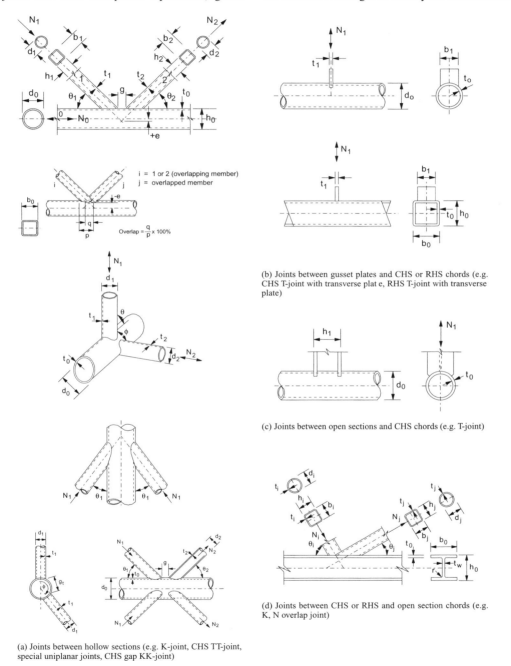

(a) Joints between hollow sections (e.g. K-joint, CHS TT-joint, special uniplanar joints, CHS gap KK-joint)

(b) Joints between gusset plates and CHS or RHS chords (e.g. CHS T-joint with transverse plate, RHS T-joint with transverse plate)

(c) Joints between open sections and CHS chords (e.g. T-joint)

(d) Joints between CHS or RHS and open section chords (e.g. K, N overlap joint)

Figure 1. Types of joints covered in the new IIW recommendations (some examples).

Tables 1 (a) to (g) for the types of joints covered in the recommendations, where the section class definition can be found in Eurocode 3 (CEN 2005).

2.3 Design procedures

The static design procedures can be summarized in the following three steps.

2.3.1 Step 1: determine the design member forces in brace and chord

For welded hollow section structures, design member forces must be obtained by analysis of the complete structure, in which nodal eccentricity of the member centrelines at the joint is taken into account.

For lattice girders or triangulated trusses with joints within the range of validity given, simplified analysis methods can be used: (a) Pin-jointed analysis, where moments due to eccentricity need to be taken into account for chord design, (b) Analysis with continuous chords and pin-ended braces. This produces axial forces in the braces, and both axial forces and bending moments in the chord. This modelling assumption is particularly appropriate for loads on the chord members which are away from the node points or panel points. For two- or three-dimensional Vierendeel girders rigid frame analysis should be used.

2.3.2 Step 2: determine the design resistance of the joint

Design resistance formulae are presented in Sections 10 to 18 of the new recommendations for various joint types. Partial safety factors (γ_M) or resistance factors (ϕ) for hollow section joints have already been incorporated into the design resistance formulae. For information purposes, the partial safety factors used in the various joint resistance formulae are given in Annex C of the new recommendations. The partial safety factors for applied loading, for the ultimate (γ_F) limit state, if applicable, should be taken from the relevant building code or specification being used.

2.3.3 Step 3: apply design criteria to assess if the joint is sufficient

Design criteria are given in Section 9 of the new recommendations to assess if the joint is sufficient by comparing design action and design resistance. Six types of failure modes are considered. They are:

a. chord face failure or chord plastification (plastic failure of the chord face or plastic failure of the chord cross-section),
b. chord side wall failure (or chord web failure) by yielding, crushing or instability (crippling or buckling of the chord side wall or chord web) under the relevant brace member,
c. chord shear failure,
d. chord punching shear failure of a hollow section chord wall (crack initiation leading to rupture of a brace member from the chord member),
e. brace failure with reduced effective width (cracking in the weld or in a brace member, or local buckling of a brace member),
f. local chord member yield failure (local buckling of the chord connecting face in an overlapped joint), and
g. brace shear failure in an overlap joint.

For uniplanar joints within the given range of validity, only failure modes listed in resistance tables need be considered. The design resistance of a joint should be taken as the minimum value for these criteria. For joints outside the given range of validity, all the seven failure modes mentioned above should be considered. In joints with the brace member(s) subject only to axial forces, the design axial force N_i should not exceed the design axial resistance of the welded joint N_i^*, expressed as an axial force in the brace member. For multiplanar joints, in each relevant plane the design criteria given for uniplanar joints should be satisfied but with correction factors as specified for multiplanar behaviour. Criteria are also given for combined bending and axial force cases, and special uniplanar joints.

3 MAJOR CHANGES MADE TO THE PREVIOUS EDITION

The previous edition (IIW 1989) only covers uniplanar CHS to CHS chord joints, RHS to RHS chord joints and CHS or RHS to I- or H-section chord joints. The major changes are summarized in Table 2 with some comments.

4 RESEARCH BACKGROUND REGARDING THE MAJOR CHANGES

4.1 Yield stress

The yield stress of steel tubes was limited to 355 MPa in the previous edition (IIW 1989). The new recommendations allow the design with steel tubes with a yield stress up to 460 MPa provided that a reduction factor is applied to the design resistance. For material with a nominal yield strength (referred to the finished tube product) exceeding 355 MPa, the joint resistances should be multiplied by 0.9. In addition, if the nominal yield strength (f_y) exceeds 0.8 of the nominal ultimate strength (f_u) then the design yield strength should be taken as $0.8f_u$. The main reasons are:

Table 2. Major changes made to the previous edition.

Design aspects	Exist in the IIW (1989)?	Comments
Yield stress	Yes	Limit increased from 355 MPa to 460 MPa
Uniplanar CHS to CHS chord joints	Yes	Axial resistance formulae modified, moment resistance added, harmonization in format done
Chord stress function for CHS joints	Yes	Completely revised
Uniplanar RHS to RHS chord joints	Yes	Axial resistance formula for K joints modified, moment resistance added, harmonization in format done
Chord stress function for RHS joints	Yes	Completely revised
Uniplanar CHS or RHS to I- or H-section chord joints.	Yes	Moment resistance added
Overlap joints	Yes	Modified to include local chord member yield and brace shear resistance, 100% overlap added. Uniform approach established for both CHS and RHS joints
Multiplanar joints	No	New clauses added for CHS or RHS TT, XX and KK joints
Plate-tube joints	No	New clauses added for Plate-CHS and Plate-RHS joints
Thick walled sections	No	Validated

a. a 3% deformation limit (Lu et al. 1994), as well as the maximum load, is used to define the ultimate capacity in calibrating the resistance formulae, because relatively larger deformation is observed in joints with nominal yield strength around 450 to 460 MPa when plastification of connecting chord face occurs,
b. the deformation capacity may be lower with yield strengths exceeding 355 MPa,
c. sufficient connection ductility is required in cases where punching shear failure or "brace or plate effective width" failure govern, since formulae for these failure mode checks are based on the yield stress. The maximum reduction in joint resistance is about 15% for 460 MPa steel tubes due to the reduction factor of 0.9 and the limitation on f_y to $0.8 f_u$. More details are given in Liu & Wardenier (2004).

4.2 Moment resistance

Moment design resistance was not covered in the previous edition (IIW 1989). Reanalyses were conducted by van der Vegte et al. (2008) and for thick-walled joints by Qian et al. (2008), to calibrate CHS-to-CHS joints for chord plastification and chord punching shear failure. For CHS joints loaded by out-of-plane bending moments, similar equations are used as in the CIDECT Design Guide No. 1 (Wardenier et al. 1991) but with reduced constants. The CIDECT (1991) equations are based on a relation with the capacity of axially loaded X joints and for in-plane bending on a modified Gibstein formula (Wardenier 1982).

For uniplanar RHS to RHS joints, the equations are similar to those included in the CIDECT Design Guide No. 3 (Packer et al. 1992).

4.3 Chord stress function for joints between hollow sections and for joints between open sections to CHS chords

The effect of chord stress in tubular joints is taken into account using the so-called chord stress function (Q_f). In the previous edition (IIW 1989), the chord stress function for CHS joints and that for RHS joints were not consistent. The chord stress function was based on the chord pre-stress for CHS joints whereas the maximum chord stress was used for RHS joints.

Research was conducted (van der Vegte et al. 2002, 2003, van der Vegte & Makino 2006) to reanalyze the effect of chord stress on the axial strength of CHS T, X, and K gap joints using the concept of maximum stress. A basic chord stress function expression was proposed by van der Vegte et al. (2007) as:

$$Q_f = (1-|n|)^{(B+C\beta+D\gamma)} \qquad (1)$$

where n is the non-dimensional chord stress ratio, and B, C and D are constants.

Further work was conducted by Wardenier et al. (2008c) on CHS-to-CHS joints, gusset-plate, I-, H- or RHS-to-CHS joints.

Research was also carried out by Liu et al. (2004), Liu & Wardenier (2006) and Wardenier et al. (2007a, b) to reanalyze the chord stress function for RHS joints aiming to come up with an expression consistent with the new function for CHS joints. In the new IIW recommendations all chord stress functions are expressed in the format of Equation 1, using the concept of maximum stress.

The recent work by Choo & Qian (2005) and Qian et al. (2008) has validated the validity of the equations for very low d_0/t_0 values for CHS joints.

4.4 Overlap joints

In the previous edition (IIW 1989) for RHS joints only an effective width criterion was considered whereas for CHS joints only chord plastification failure was considered. Chord stress functions were not consistent for RHS and CHS joints.

Extensive numerical simulations were conducted by Liu et al. (2005) on 50% RHS overlap joints and by Chen et al. (2005) on 100% RHS overlap joints. The ultimate strength was defined as the lesser of the ultimate peak load and the load at a deformation limit of 3%b_0. Generalized formulae were proposed by Wardenier & Choo (2005) for RHS overlap joints with Ov between 25% and 100%. Three failure criteria were considered in the analysis, namely:

a. brace effective width criterion for the overlapping brace,
b. ultimate shear capacity of the connection between the brace(s) and the chord, and
c. local chord member yield capacity at the joint location based on a linear interaction between axial load and bending moment.

Qian et al. (2007) and Wardenier (2007) reanalysed CHS overlap joints using the same principles as for RHS overlap joints. Experimental and FE databases used in the calibration were Makino et al. (1996), Dexter & Lee (1998) and Qian (2005). For the brace effective width and shear capacity criteria, the so-called conversion method (Wardenier 1977, Packer et al. 2007) was used to convert the formulae for RHS joints to those for CHS joints, by multiplying by a factor of $\pi/4$ and replacing b and h by d. For the chord member local yield criterion, the same interaction formula (see Eq. 2) was adopted except that an exponent c of 1.7 was used for CHS joints whereas c = 1.0 was used for RHS joints.

$$(N_0/N_{p1,0}) + (M_0/M_{p1,0})^c \leq 1.0 \quad (2)$$

For circular and square brace members, there is no need to check the shear failure if the overlap ratio is less than a certain limit Ov_{limit}. The limit is taken as 60% if the hidden seam of the overlapped brace is not welded or 80% if the hidden seam of the overlapped brace is welded. In the case of rectangular braces with h_i/b_i or $h_j/b_j < 1.0$ brace shear can be governing at lower overlaps.

4.5 Multiplanar joints

The design of multiplanar joints was not covered in the previous edition (IIW 1989). Extensive research work has been carried out on CHS TT and XX joints (Mitri et al. 1987, Paul et al. 1989, Paul 1992, van der Vegte 1995), CHS KK joints (Makino et al. 1984, Paul 1992), RHS TT and XX joints (Davies & Morita 1991, Liu et al. 1993, Davies & Crockett 1996, Yu 1997) and RHS KK joints (Bauer & Redwood 1988, Liu & Wardenier 2003). The design resistance for each relevant plane of a multiplanar joint is determined by applying an appropriate multiplanar factor (μ) to the strength of the corresponding uniplanar joint. The same expression for μ has been adopted in the new IIW recommendation (IIW 2008) for both CHS and RHS joints, as shown in Equation 3, except that μ in Equation 3b should not be greater than 1.0 for RHS XX joints.

$$\mu = 1.0 \quad \text{for TT joints} \quad (3a)$$

$$\mu = 1.0 + 0.35\, N_2/N_1 \quad \text{for XX joints} \quad (3b)$$

$$\mu = 1.0 \quad \text{for KK joints} \quad (3c)$$

where $|N_1| \geq |N_2|$. N_1 and N_2 are axial forces in the braces with tension taken as positive and compression as negative.

4.6 Plate-to-tube joints

The design of plate-to-tube joints was not covered in the previous edition (IIW 1989). For plate-to-CHS joints, the design formulae in the new recommendations (IIW 2008) are very similar to those for CHS-to-CHS joints. The calibration was mainly based on the test database of Kurobane (1981), Wardenier (1982) and Makino et al. (1991). Modifications are made based on the recent study by Wardenier et al. (2008c). For example, the constant in Q_u is reduced from 2.6 to 2.2, the term $(1 + 0.25\eta)$ is changed to $(1 + 0.40\eta)$, and a reduction factor of 0.8 was applied to in-plane bending of the joints with a longitudinal plate. The chord stress function Q_f was based on the work of de Winkel (1998) for chord compression loading while the function is the same as for CHS-to-CHS joints for chord tension loading.

Longitudinal plate-to-RHS joints tend to have excessive distortion or plastification of the RHS connecting face. Cao et al. (1998) developed design formulae for such connections using a plastic mechanism analysis. The chord stress function was modified by Wardenier et al. (2008c), to the same format as that shown in Equation 1. Transverse plate-to-RHS joints are less flexible than those with longitudinal plates because of a higher β value, which leads to different failure modes (Wardenier et al. 1981, Davies & Packer 1982). There are four basic failure modes for such joints:

a. chord face plastification,
b. punching shear,
c. chord side wall failure, and
d. branch yielding.

The chord stress function, Q_f, for chord face failure and chord side wall failure is based on the work of Lu (1997) and is converted to the same format as in Equation 1.

4.7 Formulae harmonization

Great efforts have been made by the IIW XV-E subcommission to produce a user-friendly design recommendation. The criteria used to adopt design approaches are:

a. easy to use,
b. reasonable accuracy,
c. consistency among different types of joints,
d. consistent between RHS and CHS joints, and
e. simplified format.

For example, a general expression is established (see Eq. 4) to describe the chord capacity, expressed as an axial resistance in the brace:

$$N_1^* = (f_{y0} t_0^2 / \sin\theta_1) Q_u Q_f \quad (4)$$

where the first term is related to the plastic moment capacity of a steel plate, $(1/\sin\theta_1)$ is used to consider the influence of the brace angle, the term Q_u is a function of non-dimensional geometries of the joints (e.g. β, 2γ), and the term Q_f is the chord stress function discussed in Section 4.3 of this paper.

The chord capacity is also expressed in a similar manner (see Eq. 5), as a moment resistance in the brace:

CHS chord in-plane and out-of-plane moment resistance:

$$M_1^* = (f_{y0} t_0^2 / \sin\theta_1) d_1 Q_u Q_f \quad (5a)$$

RHS chord in-plane moment resistance:

$$M_{ip,1}^* = (f_{y0} t_0^2 / \sin\theta_1) h_1 Q_u Q_f \quad (5b)$$

RHS chord out-of-plane moment resistance:

$$M_{op,1}^* = (f_{y0} t_0^2 / \sin\theta_1) b_1 Q_u Q_f \quad (5c)$$

In regression analyses, both the mean and COV (coefficient of variation) values have been considered. Comparison of the new recommendations (IIW 2008) and the previous edition (IIW 1989) and the new API (2007) are given in Wardenier et al. (2008a). Design examples based on the new IIW (2008) are given in the new edition of CIDECT Design Guide No. 1 (Wardenier et al. 2008b).

5 CONCLUSIONS

This paper has summarized the 3rd edition of the IIW Static Design Recommendations for Hollow Section Joints. The major changes made to the previous edition, as well as the relevant research background and references, have been given to assist the designers.

It should be pointed out that information is available for other types of welded joints that are not covered in the recommendations, for example gusset plate to slotted hollow sections (Packer 2006, Martinez-Saucedo & Packer 2006, Ling et al. 2007a, b) and reinforced joints (Lee & Llewelyn-Parry 1998, Willibald 2001, Choo et al. 2004, 2005).

ACKNOWLEDGEMENTS

The authors wish to thank Professors Y. Makino, Y. Kurobane and Y.S. Choo, Dr. X.D. Qian and the late Dr. D.K. Liu for their contributions. The financial support provided by CIDECT on many of the research projects is very much appreciated.

REFERENCES

American Petroleum Institute 2007. Recommended practice for planning designing and constructing fixed offshore platforms—Working stress design. API RP-2 A, 21st Edition, Supp. 3, American Petroleum Institute, USA.

Bauer, D. & Redwood, R.G. 1988. Triangular truss joints using rectangular tubes. *Journal of Structural Engineering*, ASCE, Vol. 114, No. 2, pp. 408–424.

Cao, J.J., Packer, J.A. & Kosteski, N. 1998. Design guide for longitudinal plate to HSS connections. *Journal of Structural Engineering*, ASCE, Vol. 124, No. 7, pp. 784–791.

CEN. 2005. Eurocode 3: Design of steel structures—general rules—Part 1–8: design of joints, EN1993-1-8:2005(E). Brussels: European Committee for Standardization.

Chen, Y., Liu, D.K. & Wardenier, J. 2005. Design recommendations for RHS K-Joints with 100% overlap. *Proc. 15th Intern. Offshore and Polar Engineering Conf.*, Seoul, Korea, Vol. IV, pp. 300–307.

Choo, Y.S., Liang, J.X. & Vegte, G.J. van der 2004. An effective external reinforcement scheme for circular hollow section joints. *Proc. ECCS-AISC Workshop 'Connections in Steel Structures V'*, Bouwen met Staal, Zoetermeer, The Netherlands, pp. 423–432.

Choo, Y.S., Vegte, G.J. van der, Zettlemoyer, N., Li, B.H. & Liew, J.Y.R. 2005. Static strength of T joints reinforced with doubler or collar plates—I: Experimental investigations. *Journal of Structural Engineering*, ASCE, Vol. 131, No. 1, pp. 119–128.

Choo, Y.S. & Qian, X.D. 2005. Recent research on tubular joints with very thick-walled chords. *Proc. 15th Intern. Offshore and Polar Engineering Conf.*, Seoul, Korea, Vol. IV, pp. 272–278.

Davies, G. & Packer, J.A. 1982. Predicting the strength of branch plate-RHS connections for punching shear. *Canadian Journal of Civil Engineering*, Vol. 9, No. 3, pp. 458–467.

Davies, G. & Morita, K. 1991. Three dimensional cross joints under combined axial branch loading. *Proc. 4th Intern. Symp. on Tubular Structures*, Delft, The Netherlands, pp. 324–333.

Davies, G. & Crockett, P. 1996. The strength of welded T-DT joints in rectangular and circular hollow sections under variable axial loads. *Journal of Constructional Steel Research*, Vol. 37, No. 1, pp. 1–31.

Dexter, E.M. & Lee, M.M.K. 1998. Effect of overlap on the behaviour of axially loaded CHS K-joints. *Proc. 8th Intern. Symp. on Tubular Structures*, Singapore, pp. 249–258.

IIW 1981. Design recommendations for hollow section joints-Predominantly statically loaded. 1st Edition, IIW Doc. XV-491–81, IIW Annual Assembly, Lisbon, Portugal.

IIW 1989. Design recommendations for hollow section joints-Predominantly statically loaded. 2nd Edition, IIW Doc. XV-701–89, IIW Annual Assembly, Helsinki, Finland.

IIW 2008. IIW static design procedure for welded hollow section joints-Recommendations. IIW Doc. XV-1281-08, IIW Annual Assembly, Graz, Austria.

Kurobane, Y. 1981. New developments and practices in tubular joint design. IIW Doc. XV-488–81, IIW Annual Assembly, Oporto, Portugal.

Lee, M.M.K. & Llewelyn-Parry, A. 1998. Ultimate strength of ring stiffened T-joints-A theoretical model. *Proc. 8th Intern. Symp. on Tubular Structures*, Singapore, pp. 147–156.

Ling, T.W., Zhao, X.-L., Al-Mahaidi, R. & Packer, J.A. 2007a. Investigation of shear lag failure in gusset-plate welded structural steel hollow section connections. *Journal of Constructional Steel Research*, Vol. 63, No. 3, pp. 293–304.

Ling, T.W., Zhao, X.-L., Al-Mahaidi, R. & Packer, J.A. 2007b. Investigation of block shear tear-out failure in gusset-plate welded connections in structural steel hollow sections and very high strength tubes. *Engineering Structures*, Vol. 29, No. 4, pp. 469–482.

Liu, D.K., Wardenier, J., Koning, C.H.M. de & Puthli, R.S. 1993. Static strength of multiplanar DX-joints in rectangular hollow section. *Proc. 5th Intern. Symp. on Tubular Structures*, Nottingham, UK, pp. 419–428.

Liu, D.K. & Wardenier, J. 2003. The strength of multiplanar KK-joints of square hollow sections. *Proc. 10th Intern. Symp. on Tubular Structures,* Madrid, Spain, pp. 197–206.

Liu, D.K. & Wardenier, J. 2004. Effect of the yield strength on the static strength of uniplanar K-joints in RHS. IIW Doc. XV-E-04-293.

Liu, D.K., Wardenier, J. & Vegte, G.J. van der 2004. New chord stress functions for rectangular hollow section joints. *Proc. 14th Intern. Offshore and Polar Engineering Conf.*, Toulon, France, Vol. IV, pp. 178–185.

Liu, D.K., Chen, Y. & Wardenier, J. 2005. Design recommendations for RHS K-Joints with 50% overlap. *Proc. 15th Intern. Offshore and Polar Engineering Conf.*, Seoul, Korea, Vol. IV, pp. 308–315.

Liu, D.K. & Wardenier, J. 2006. Effect of chord loads on the strength of RHS uniplanar gap K-joints. *Proc. 11th Intern. Symp. on Tubular Structures*, Quebec, Canada, pp. 539–544.

Lu, L.H., Winkel, G.D. de, Yu, Y. & Wardenier, J. 1994. Deformation limit for the ultimate strength of hollow section joints. *Proc. 6th Intern. Symp. on Tubular Structures*, Melbourne, Australia, pp. 341–348.

Lu, L.H. 1997. The static strength of I-beam to rectangular hollow section column connections. *PhD Thesis*, Delft University of Technology, Delft, The Netherlands.

Makino, Y., Kurobane, Y. & Ochi, K. 1984. Ultimate capacity of tubular double K-joints. *Proc. IIW Conf. on Welding of Tubular Structures*, Boston, USA, pp. 451–458.

Makino, Y., Kurobane, Y., Paul, J.C., Orita, Y. & Hiraishi, K. 1991. Ultimate capacity of gusset plate-to-tube joints under axial and in plane bending loads. *Proc. 4th Intern. Symp. on Tubular Structures*, Delft, The Netherlands, pp. 424–434.

Makino, Y., Kurobane, Y., Ochi, K., Vegte, G.J. van der & Wilmshurst, S.R. 1996. Database of test and numerical analysis results for unstiffened tubular joints. IIW Doc. XV-E-96-220.

Martinez-Saucedo, G. & Packer, J.A. 2006. Slotted end connections to hollow sections. CIDECT Final Report 8G-10/4, University of Toronto, Toronto, Canada.

Mitri, H.S., Scola, S. & Redwood, R.G. 1987. Experimental investigation into the behaviour of axially loaded tubular V-joints. *Proc. CSCE Centennial Conf.*, Montreal, Canada, pp. 397–410.

Packer, J.A., Wardenier, J., Kurobane, Y., Dutta, D. & Yeomans, N. 1992. *Design guide for rectangular hollow section (RHS) joints under predominantly static loading.* CIDECT Design Guide No. 3, 1st Edition, TÜV-Verlag, Köln, Germany.

Packer, J.A. 2006. Tubular brace member connections in braced steel frames. Houdremont Lecture, *Proc. 11th Intern. Symp. on Tubular Structures*, Quebec, Canada, pp. 3–14.

Packer, J.A., Mashiri, F.R., Zhao, X.L. & Willibald, S. 2007. Static and fatigue design of CHS-to-RHS welded connections using a branch conversion method. *Journal of Constructional Steel Research*, Vol. 63, No. 1, pp. 82–95.

Paul, J.C., Valk, C.A.C. van der & Wardenier, J. 1989. The static strength of circular multi-planar X joints. *Proc. 3rd Intern. Symp. on Tubular Structures*, Lappeenranta, Finland, pp. 73–80.

Paul, J.C. 1992. The ultimate behaviour of multiplanar TT and KK joints made of circular hollow sections. *PhD Thesis*, Kumamoto University, Japan.

Qian, X.D. 2005. Static strength of thick-walled CHS joints and global frame behaviour. *PhD Thesis*, National University of Singapore, Singapore.

Qian, X.D., Wardenier, J. & Choo, Y.S. 2007. A uniform approach for the design of 100% CHS overlap joints. *Proc. 5th Intern. Conf. on Advances in Steel Structures*, Singapore, Vol. II, pp. 172–182.

Qian, X.D., Choo, Y.S., Vegte, G.J. van der & Wardenier, J. 2008. Evaluation of the new IIW CHS strength formulae for thick-walled joints. *Proc. 12th Intern. Symp. on Tubular Structures*, Shanghai, China.

Vegte, G.J. van der 1995. The static strength of uniplanar and multiplanar tubular T- and X-joints. *PhD Thesis*, Delft University of Technology, Delft, The Netherlands.

Vegte, G.J. van der, Makino, Y. & Wardenier, J. 2002. The effect of chord pre-load on the static strength of uniplanar tubular K-joints. *Proc. 12th Intern. Offshore and Polar Engineering Conf.*, Kitakyushu, Japan, Vol. IV, pp. 1–10.

Vegte, G.J. van der, Liu, D.K., Makino, Y. & Wardenier, J. 2003. New chord load functions for circular hollow section joints. *CIDECT Report* 5BK-04/03, Delft University of Technology, Delft, The Netherlands.

Vegte, G.J. van der & Makino, Y. 2006. Ultimate strength formulation for axially loaded CHS uniplanar T-joints. *Intern. Journal Offshore and Polar Engineering*, ISOPE, Vol. 16, No. 4, pp. 305–312.

Vegte, G.J. van der, Makino, Y. & Wardenier, J. 2007. Effect of chord load on ultimate strength of CHS X-joints. *Intern. Journal Offshore and Polar Engineering*, ISOPE, Vol. 17, No. 4, pp. 301–308.

Vegte, G.J. van der, Wardenier, J., Qian, X.D. & Choo, Y.S. 2008. Re-analysis of the moment capacity of CHS joints. *Proc. 12th Intern. Symp. on Tubular Structures*, Shanghai, China.

Wardenier, J. 1977. Design of hollow section joints (in Dutch). *Proc. Symp. on Tubular Structures*. Delft, Staalbouwkundig Genootschap, Rotterdam, The Netherlands.

Wardenier, J., Davies, G. & Stolle, P. 1981. The effective width of branch plate to RHS chord connections in cross joints. *Stevin Report* No. 6-81-6, Delft University of Technology, Delft, The Netherlands.

Wardenier, J. 1982. *Hollow section joints*. Delft University Press, Delft, The Netherlands.

Wardenier, J., Kurobane, Y., Packer, J.A., Dutta, D. & Yeomans, N. 1991. *Design guide for circular hollow section (CHS) joints under predominantly static loading*, CIDECT Design Guide No. 1, 1st Edition, TÜV-Verlag, Köln, Germany.

Wardenier, J. & Choo, Y.S. 2005. Some developments in tubular joints research. *Proc. 4th Intern. Conf. on Advances in Steel Structures*, Shanghai, China, pp. 31–40.

Wardenier, J. 2007. A uniform effective width approach for the design of CHS overlap joints. *Proc. 5th Intern. Conf. on Advances in Steel Structures*, Singapore, Vol. II, pp. 155–165.

Wardenier, J., Vegte, G.J. van der & Liu, D.K. 2007a. Chord stress function for rectangular hollow section X and T joints. *Proc. 17th Intern. Offshore and Polar Engineering Conf.*, Lisbon, Portugal, Vol. IV, pp. 3363–3370.

Wardenier, J., Vegte, G.J. van der & Liu, D.K. 2007b. Chord stress functions for K gap joints of rectangular hollow sections. *Intern. Journal Offshore and Polar Engineering*, ISOPE, Vol. 17, No. 3, pp. 225–232.

Wardenier, J., Vegte, G.J. van der, Makino, Y. & Marshall, P.W. 2008a. Comparison of the new IIW (2008) CHS joint strength formulae with those of the previous IIW (1989) and the new API (2007). *Proc. 12th Intern. Symp. on Tubular Structures*, Shanghai, China.

Wardenier, J., Kurobane, Y., Packer, J.A., Vegte, G.J. van der & Zhao, X.L. 2008b. *Design guide for circular hollow section (CHS) joints under predominantly static loading*. CIDECT Design Guide No. 1, 2nd Edition, TÜV-Verlag, Köln, Germany.

Wardenier, J., Vegte, G.J. van der & Makino, Y. 2008c. Joints between plates or I sections and a circular hollow section chord. *Proc. 18th Intern. Offshore and Polar Engineering Conf.*, Vancouver, Canada.

Willibald, S. 2001. The static strength of ring-stiffened tubular T- and Y-joints. CIDECT Student Prize Paper, *Proc. 9th Intern. Symp. on Tubular Structures*, Düsseldorf, Germany, pp. 581–588.

Winkel, G.D. de 1998. The static strength of I-beam to circular hollow section column connections. *PhD Thesis*, Delft University of Technology, Delft, The Netherlands.

Yu, Y. 1997. The static strength of uniplanar and multiplanar tubular connections in rectangular hollow sections. *PhD Thesis*, Delft University of Technology, Delft, The Netherlands.

SYMBOLS

f_{yi} = yield stress of the brace member i (i = 1 or 2)
f_{y0} = yield stress of the chord member
M_0 = design value of the moment in the chord member
$M_{pl,0}$ = plastic moment capacity of the chord
N_0 = design value of the axial force in the chord member
$N_{pl,0}$ = design axial resistance of a chord member
β = ratio of the mean diameter or width of the brace members, to that of the chord, e.g. for T, Y and X joints, $\beta = d_1/d_0$ or d_1/b_0 or b_1/b_0 for K and N joints, $\beta = (d_1 + d_2)/(2d_0)$ or $(b_1 + b_2)/(2b_0)$ or $(b_1 + b_2 + h_1 + h_2)/(4b_0)$ for plate to CHS or RHS, $\beta = b_1/d_0$ or b_1/b_0
γ = ratio of the chord width or diameter to twice the thickness
η = ratio of the brace member depth h_i to the chord diameter d_0 or width b_0

Evaluation of the new IIW CHS strength formulae for thick-walled joints

X. Qian & Y.S. Choo
Center for Offshore Research and Engineering, National University of Singapore, Singapore

G.J. van der Vegte
Faculty of Civil Engineering and Geosciences, Delft University of Technology, The Netherlands

J. Wardenier
Faculty of Civil Engineering and Geosciences, Delft University of Technology, The Netherlands
Department of Civil Engineering, National University of Singapore, Singapore

ABSTRACT: This paper presents the comparison between the new IIW strength equations (proposed by the Sub-commission XV-E) for circular hollow section X-, T- and K-joints and the finite element (FE) data for joints with very thick chord members. The chord diameter-to-wall thickness ratio 2γ in this investigation ranges from 7 to 40, which extends beyond the validity limit of most design equations. This study examines the applicability of the proposed IIW equations for thick-walled CHS joints under brace axial loads, under brace in-plane bending (IPB) and under brace out-of-plane bending (OPB). The comparison procedure filters out joints with member (brace or chord) failure and punching shear failure. The numerical data obtained from the extensive, calibrated nonlinear FE study demonstrate the applicability of the new IIW equations to thick-walled joints.

1 INTRODUCTION

Extensive numerical analyses and re-assessment of existing experimental results (van der Vegte et al., 2007, 2008a, 2008b) lead to more detailed understanding of the static behaviour of circular hollow section (CHS) joints under monotonic brace axial or moment loads. This resulted in the development of new, updated IIW design recommendations (IIW, 2008), which aim to cover a wide range of geometries for CHS joints with improved accuracy. Sub-commission IIW-XV-E has proposed a new set of strength equations for CHS joints under brace axial loads, in-plane bending (IPB) and out-of-plane bending (OPB), including the interaction with chord stresses. Wardenier et al. (2008) present a detailed comparison of the new IIW (2008) design equations with the previous IIW (1989) equations and the revised API (2007) equations.

Previous research efforts on the static strength of CHS connections focus primarily on joint configurations with a chord diameter-to-thickness ratio 2γ larger than 20. Recent design and construction of both onshore bridges (Schumacher et al., 2001), mining equipments (Zhao et al., 2007) and offshore jack-up platforms adopt circular hollow sections with 2γ ratios much less than 20, e.g., as low as 7.0. The industrial application of the thick-wall connections requires a validated formulation for these CHS joints under different brace loading conditions. To foster a better understanding of the static response of CHS joints under overloading conditions, the offshore research group from the National University of Singapore has performed an extensive numerical study on the static strength of thick-walled joints through large-deformation, elastic-plastic analyses.

This paper compares the new strength equations proposed by Sub-commission IIW-XV-E with the finite element (FE) data for thick-walled CHS X-, T- and K-joints (generally with $7 \leq 2\gamma \leq 20$) under brace axial/moment loads and subjected to chord axial stresses.

This paper starts with a brief description of the new IIW ultimate strength formulation. The next section describes the numerical research on thick-walled joints, followed by a detailed comparison between the new IIW strength formulation and the thick-walled joint data.

2 NEW IIW FORMULATION

2.1 *Brace axial loads*

Table 1 lists the non-dimensional ultimate strength parameter Q_u, for X-, T- and K-joints under brace axial loads, adopted in the new IIW (2008) equations.

Table 1. IIW (2008) ultimate strength Q_u functions for axially loaded CHS joints.

Joint	Q_u	mean	CoV	No. of data
X	$3.16\left(\dfrac{1+\beta}{1-0.7\beta}\right)\gamma^{0.15}$	1.01	5.8%	23
T	$(3.1+21\beta^2)\gamma^{0.2}$	1.01	4.5%	33
K	$(2+16\beta^{1.6})\gamma^{0.3}Q_g^*$	1.00	4.2%	122

$$^*Q_g = \left[1 + \dfrac{1}{1.2 + \left(\dfrac{g}{t_0}\right)^{0.8}}\right].$$

Table 2. IIW (2008) Q_f functions for axially loaded CHS joints.

Joint	Q_f (compression) *	mean	CoV	No. of data
X	$(1-|n|)^{(0.25-0.25\beta+0.01\gamma)}$ $(1-|n|)^{(0.45-0.25\beta)}$ (lower bound for $2\gamma=40$)	1.03	6.8%	156
T	$(1-|n|)^{(0.20-0.25\beta+0.01\gamma)}$ $(1-|n|)^{(0.45-0.25\beta)}$ (lower bound for $2\gamma=50$)	1.00	6.0%	141
K	$(1-|n|)^{0.25}$	1.00	4.2%	122

* For chord tension $Q_f = (1-n)^{0.2}$.

The mean and CoV values refer to the comparison of the equations against the original data used in the development of the formulation reported by van der Vegte et al. (2007, 2008a, 2008b).

The new IIW equations include a simple γ dependence in the Q_u factor. The generally used theoretical model (e.g., the simplified ring model proposed by Togo, 1967) assumes a high diameter-to-thickness ratio of the chord and therefore does not include the γ dependence in the model. In contrast, a more refined theoretical model (e.g., the exact ring model proposed by van der Vegte, 1995) includes the effect of transverse shear in the plastic moment capacity, but evolves an expression which is too complex for design purposes.

The equation for gapped K-joints in Table 1 replaces the complex expression for the gap ratio in the current IIW (1989) equation, with a simple Q_g factor.

For overlapped K-joints, previous studies (Wardenier, 2007; Qian et al., 2007) summarize the new IIW approach based on the effective width concept originally developed for RHS K-joints.

2.2 *Moment loaded joints*

For moment loaded joints, the comparison is made with the IIW (2008) design strengths, because for both X- and T-joints, the same equations are used, however both have different means and CoV's. Therefore, the FE data are first compared with the design equations, followed by an evaluation of the results with regard to the mean values obtained.

For X- and T-joints under brace in-plane bending, the IIW (2008) design strength M^*_{ipb} is given by:

$$M^*_{ipb} = 4.3\beta\,\gamma^{0.5}d_1\dfrac{f_{y0}\,t_0^2}{\sin\theta}Q_f \qquad (1)$$

where Q_f defines the chord stress function (see Table 2).

For joints subjected to brace out-of-plane bending, the IIW assumes that the resistance of the joint can be derived from a couple force acting in parallel to the brace axis. The moment arm between the couple equals approximately to the brace outer diameter, while the magnitude of the couple force depends on the axial capacity of the joint, which, for both X- and T-joints, is approximated by 50% of the axial capacity of the X-joint. The *design strength* M^*_{opb} for X- and T-joints under brace out-of-plane bending thus follows:

$$M^*_{opb} = 1.3\left(\dfrac{1+\beta}{1-0.7\beta}\right)\gamma^{0.15}d_1\dfrac{f_{y0}\,t_0^2}{\sin\theta}Q_f \qquad (2)$$

where Q_f defines the effect of chord stresses, as illustrated in the following section.

2.3 *Chord stress function*

Extensive numerical evidence in the last decade demonstrates significant dependence of the ultimate strength on the presence of tensile chord stresses, which has been neglected in most of the previous design equations.

Recent research efforts further prove that the chord stress effect predicted by the chord stress functions in the current design codes underestimate the reduction in the joint capacity at high compressive chord stresses. The new IIW (2008) chord stress function Q_f for compressive chord stress, shown in Table 2, incorporates these effects through a more rational maximum chord stress ratio n instead of the conventionally used chord pre-stress ratio n'. The mean and CoV values in Table 2 indicate the comparison of the IIW equations with the original data used in the development of the new formulation (van der Vegte et al., 2008a).

The chord stress ratio n denotes the maximum chord stress ratio in the chord, including the contribution from the brace induced chord stresses.

For T- and K-joints, the chord stress ratio includes both the axial chord stress and the in-plane bending stresses caused by the end reactions of the chord member or the eccentricity between the two braces. The chord stress ratio n follows,

$$n = \left|\frac{N_0}{N_p}\right| + \left|\frac{M_0}{M_c}\right|^a \quad (3)$$

where N_0 and M_0 define the axial and moment loads in the chord member. N_p refers to the axial capacity of the chord member, and M_c denotes the plastic or elastic moment capacity depending on the section classification. The exponent a in Equation 3 equals to 1 for compressive chord load and takes a value of 2 for tensile chord loads. As compared to the chord axial tension, Equation 3 includes a higher contribution from the chord bending moment when the chord member experiences axial compression, due to the potential local instability in the chord wall.

3 FE ANALYSIS ON THICK-WALLED JOINTS

3.1 Numerical procedure

The FE model for the thick-walled joints adopts 20-node brick elements with a reduced integration scheme (Qian, 2005). The FE model utilizes four layers of elements across the wall thickness of the chord member.

The nonlinear material properties follow the Von Mises constitutive relationship with J_2 flow theory. The uni-axial true stress-log strain relationship utilizes the stress-strain curve reported by van der Vegte (1995). The element response adopts the finite-strain formulation.

The loading conditions follow a displacement-controlled brace load, and a load-controlled chord load for cases with axial chord stresses. The chord stresses are applied through a sequential loading procedure with the chord loads introduced first and maintained subsequently during loading of the braces.

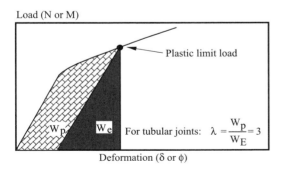

Figure 1. Definition of joint strength following the plastic limit load.

3.2 Scope of analysis

The numerical analysis covers a wide range of geometric parameters, with β varying from 0.4 to 1.0 and γ varying from 3.5 to 20 for T- and K-joints. For X-joints, the γ ratio extends from 3.5 to 25.4. The additionally applied chord stresses for X-, T- and K-joints include 8 levels of magnitude: $-0.9f_{y0}$, $-0.7f_{y0}$, $-0.5f_{y0}$, $-0.3f_{y0}$, $0.3f_{y0}$, $0.5f_{y0}$, $0.7f_{y0}$ and $0.9f_{y0}$.

3.3 Strength definition

Thick-walled CHS joints exhibit frequently ductile responses without determinative peaks in the load-displacement (or moment-rotation) curves. The strength definition for the thick-walled joints used here follows the plastic limit load approach, originally proposed by Gerdeen (1980). This approach defines the joint strength as the ratio of the plastic work over the elastic work (both done by the applied load) when a pre-defined, semi-empirical limit is reached, as shown in Figure 1.

The semi-empirical limit λ takes a value of 3, which provides consistent estimations with Lu's deformation limit of 3% of the chord diameter (Lu et al., 1994) and peak load values observed in thin-walled joints (Choo et al., 2003). For moment loaded joints, Lu imposes an additional limit corresponding to a maximum brace rotation of 0.1 radian.

4 COMPARISION WITH THE NEW IIW (2008) EQUATIONS

4.1 Axially loaded joints

Table 3 compares the non-dimensional ultimate joint strength factor Q_u with the FE data for thick-walled X- and T-joints under brace axial loads, without the presence of chord stresses. For T-joints, compensating moments balance the chord bending moments at the brace-to-chord intersection induced by the chord end reactions.

The FE data for thick-walled X-joints indicate close agreement with the new IIW equation in Table 1. At ultimate load levels, the thin-walled joints experience significant plastic deformations in the chord wall near the brace-to-chord intersection. However, the very high rigidity of a thick-walled section limits the

Table 3. Comparison of the thick-walled FE data with the new IIW Q_u functions for axially loaded CHS joints.

Joint	FE/IIW	No. of data	mean	CoV
X	All data	54	0.98	6.3%
T	All data	15	0.93	20%
T	Excluding chord yield	9	1.05	6.8%

development of the same degree of plastic deformations observed in the thin-walled joints. In addition, the plastic limit load yields 2–3% lower estimations of the joint strength as compared to Lu's deformation limit, which is used in the development of the IIW formulation in Table 1.

Figure 2 compares the ratio of the numerical ultimate joint strength and the new IIW ultimate strength equation for thick-walled X-joints with respect to the parameters β and γ. The joints with a large β ratio show relatively lower joint capacities compared to the IIW (2008) recommendations.

For T-joints under brace axial loads, the compensating moments in six very thick-walled joints cause yielding in the chord section away from the brace-to-chord intersection. Figure 3 compares the ultimate joint strength obtained from the FE data of thick-walled T-joints with the IIW ultimate strength equation as a function of β. Two joints for β = 0.7 with γ = 7 and 9 show slightly lower joint strength than the IIW equation. The other joints have capacities exceeding the predictions by the Q_u factor in the new IIW formulation. The ultimate strength data for X- and T-joints, excluding chord member failures, demonstrate a good agreement with the new IIW ultimate strength equation.

Table 4 compares the chord stress effect obtained from the FE analyses and the chord stress function Q_f proposed by IIW (2008). For X-joints, the comparison procedure simply calculates the ratio of the joint strength under the influence of chord axial stresses and that without the influence of chord axial stresses. For T- and K-joints, the comparison procedure includes the effect of the chord stresses into the estimated ultimate joint strength, since the maximum chord stress ratio in T- and K-joints depends on the magnitude of the brace loads.

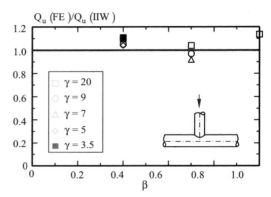

Figure 3. Comparison of the numerical T-joint strength data (excluding chord yield failure) with the IIW (2008) ultimate strength equation.

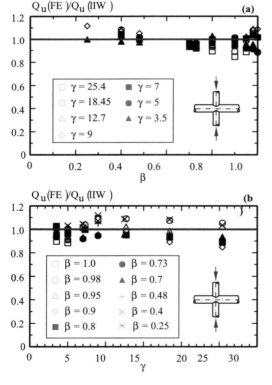

Figure 2. Comparison of thick-walled X-joint data obtained from the FE study with the new IIW (2008) ultimate strength equation: (a) as a function of β, and (b) as a function of γ.

Table 4. Comparison of the thick-walled FE data (plastic limit load and Lu's deformation limit) with the new IIW Q_f functions for axially loaded CHS joints.

Joint	FE/IIW	No. of data	mean	CoV
X	All data	108	0.99	12.5%
X	All data (lower bound)*	108	1.05	15.5%
X	$|n| \leq 0.85$ (lower bound)*	84	1.06	8.5%
X	*$|n| \leq 0.85$ (lower bound)***	*84*	*1.06*	*7.2%*
T	All data	41	1.08	14.6%
T	All data (lower bound)*	41	1.15	10.4%
T	$|n| \leq 0.85$ (lower bound)*	25	1.16	8.4%
T	*$|n| \leq 0.85$ (lower bound)***	*25*	*1.16*	*8.3%*
K	All data	159	0.91	30.3%
K	Excl. member failures	78	1.07	14.4%
K	$|n| \leq 0.85$ Excl. member failures	57	1.06	11.3%
K	*$|n| \leq 0.85$** Excl. member failures*	*57*	*1.04*	*14.4%*

* Lower-bound estimations using $(1-|n|)^{(0.45-0.25\beta)}$.
** based on Lu's deformation limit.

Table 4 shows that the lower-bound chord stress function for X- and T-joints demonstrates a conservative estimation of the chord stress effect for the thick-walled joints, with more conservatism for the T-joint. The K-joint strength equation indicates a conservative estimate for most joints excluding brace and chord yielding.

Figure 4 illustrates the chord stress function for X-joints under brace axial loads. All FE data for the thick-walled joints with chord compression loads show a less significant chord stress effect, compared to the prediction by the Q_f expression adopted by IIW. The three continuous curves in Figure 4 represent the lower bound Q_f function in IIW for the three β ratios considered (β = 0.4, 0.7 and 1.0). The relatively thin-walled FE data are in good agreement with the IIW curves. For chords loaded under very high axial tension, some thick-walled FE data show a more significant chord stress effect, but in general the agreement is good.

Figure 4 further indicates an increasing scatter for large chord stress ratios (as shaded in Figure 4 for $|n| > 0.85$). As $|n|$ approaches to unity, the IIW Q_f function has a very steep slope, causing significant sensitivity of the data. A small deviation from the Q_f function may introduce a large difference measured in percentage, and hence a pronounced scatter in the comparison.

The comparison with the new IIW Q_f function for FE data with $|n| < 0.85$ shows significantly reduced CoV values, as indicated in Table 4. The plastic limit load approach shows very similar comparisons with the IIW equation for X-, T- and K-joints.

Figure 5 compares the lower bound IIW chord stress function Q_f with the FE data for thick-walled T-joints. Based on the plastic limit load criterion, the thick-walled joints show a significantly smaller chord load effect than estimated by the new IIW equations, which is caused by the smaller deformations for thick-walled joints. The relatively thin-walled FE data are in good agreement with the IIW curves.

Figure 6 compares the IIW (2008) equation with the thick-walled K-joint data, excluding joints with chord/brace failure. Most of the thick-walled joints experience member failures, i.e. yielding of the brace under axial loads and yielding of the chord under combined chord axial load and eccentric moment. For very thick-walled joints ($\gamma < 7$) with lower β ratios ($\beta \leq 0.7$), yielding of the brace member occurs prior to extensive plastic deformation near the brace-to-chord intersection.

4.2 CHS joints under brace IPB

Table 5 compares the FE *ultimate strength* data and the IIW (2008) *design* equation for thick-walled X- and T-joints under brace IPB, using three different strength definitions. The plastic limit load approach corresponds to the smallest deformation level (at $\lambda = 3$) among the three strength definitions. Lu's deformation limit defines a maximum brace rotation of 0.1 radian and Yura's deformation criterion (Yura

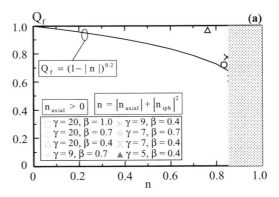

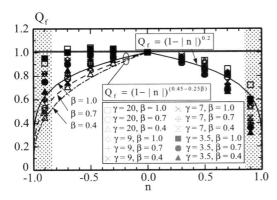

Figure 4. Comparison of the lower bound chord stress function Q_f for X-joints subjected to brace axial loads.

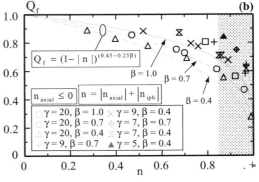

Figure 5. Comparison of the lower bound IIW equation Q_f with the FE data for thick-walled T-joints: (a) chord axial tension ($n_{axial} > 0$); and (b) chord axial compression ($n_{axial} < 0$).

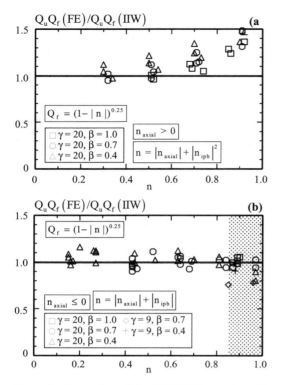

Figure 6. Comparison of the IIW (2008) equation with thick-walled K-joint data: (a) chord under axial tension; and (b) chord under axial compression.

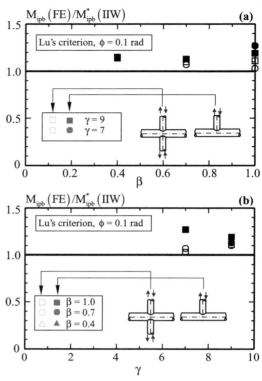

Figure 7. Comparison of the thick-walled FE data, X(2) and T(2) in Table 5, with the design strength for X- and T-joints under brace IPB (Equation 1): (a) as a function of β; and (b) as a function of γ.

et al., 1980) equals to $80 f_{y0}/E$, or 0.139 radian for f_{y0} = 355 MPa and E = 205 GPa.

For thick-walled X- and T- joints, excluding data with punching shear failure and brace yielding based on $f_{yi} \leq f_{y0}$, the plastic limit load approach yields a mean value of 1.01 to 1.07. Yura's deformation limit, corresponding to a large deformation level, indicates the largest mean value of 1.15 to 1.26 among the three strength definitions. Lu's deformation limit shows a mean value in-between the plastic limit load approach and Yura's deformation limit for both X- and T-joints, as shown in Table 5. The comparison of the three different deformation definitions indicate that, using Lu's deformation limit as reference, Yura's deformation limit yields 5–7% higher and the plastic limit load approach 7–12% lower estimations of the joint capacity for X- and T-joints under brace IPB.

Figure 7 compares, based on Lu's deformation limit, the thick-walled joint FE data with the *design strength* for X- and T-joints under brace IPB (Equation 1) vs. β and γ ratios. The FE data exclude joints governed by punching shear failure and brace yielding.

Table 5. Comparison of the thick-walled FE data with the new IIW design equations for CHS joints under brace IPB.

Joint	No. of data	mean			CoV (%)		
		$\lambda = 3$	Lu	Yura	$\lambda = 3$	Lu	Yura
X (1)	12	0.96	1.03	1.08	9.2	9.7	10.0
X (2)	5	1.01	1.09	1.15	4.7	3.6	4.9
T (1)	12	1.05	1.12	1.19	8.5	7.8	8.0
T (2)	4	1.07	1.19	1.26	2.5	5.4	5.3
X(2) + T(2)	9	1.04	1.13	1.20	4.7	6.2	6.8

* All data.
** Data excluding punching shear failure and brace yielding.

The new IIW recommendations (2008) adopt the same strength equation (Equation 1) for CHS X- and T-joints under brace IPB. The comparison of the combined X- and T-joint FE data (defined by Lu's deformation limit) with Equation 1 shows a mean value of 1.13 with a CoV equal to 6.2%.

Considering the CoV and the evaluation procedure described by van der Vegte et al. (2008a), the mean value for the thick-walled FE results for X- and T-joints loaded by brace in-plane bending and based on Lu's rotation limit of 0.1 radian, should have been 1.17 to be in line with the ultimate strength equation on which the IIW (2008) design strength formula is based.

4.3 CHS joints under brace OPB

For joints under brace OPB, Equation 2 adopts the brace outer diameter d_1, to estimate the yield line location in the chord. For joints with very thick-walled braces, this gives an overestimation of the capacity. On the other hand, for the joints with very thick-walled chords the braces will generally have a smaller thickness.

Table 6 shows the comparison between the FE data, based on different deformation criteria, and Equation 2 for thick-walled X- and T-joints under brace OPB.

Similar to X- and T-joints under brace IPB, Lu's deformation limit corresponds to a deformation level less than the Yura's deformation criterion and larger than the deformation level defined by the plastic limit load approach.

Figure 8 compares the FE data (excluding joints with punching shear failure and brace yielding based on $f_{yi} \leq f_{y0}$) and Equation 2 for X- and T-joints subjected to brace OPB. The joint strength obtained from the FE data follows Lu's deformation limit.

The comparison between Equation 2 and the combined X- and T-joint data (following Lu's deformation limit) shows a mean value of 1.15 with a CoV equal to 8.0%, more or less in line with the values obtained for X- and T-joints under brace IPB.

Considering the CoV and the evaluation procedure described by van der Vegte et al. (2008a), the mean value for the thick-walled FE results for X- and T- joints loaded by a brace out-of-plane bending moment and based on Lu's rotation limit of 0.1 radian, should have been 1.22 to be in line with the ultimate strength equation which served as basis for the IIW

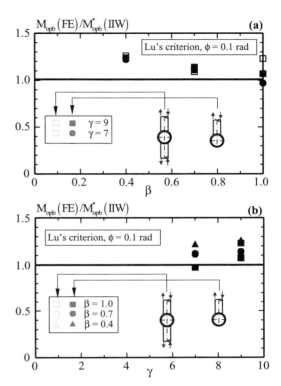

Figure 8. Comparison of the FE data, X(2) and T(2) in Table 6, for thick-walled X- and T-joints subjected to brace OPB: (a) as a function of β; and (b) as a function of γ.

(2008) design strength formula. Thus, also here the FE results based on the plastic limit load approach and Lu's criterion are slightly too low, whereas those based on Yura's rotation criterion are completely in line with the new equations adopted by IIW.

4.4 Evaluation

The FE data for the axially loaded thick-walled X- and T-joints give, compared to the ultimate strength equations on which the IIW design strength equations are based, Q_u mean values between 0.98 and 1.05 with CoV values between 6.3 and 6.8%. Thus, they show an excellent agreement. For the Q_f chord stress effect, the IIW lower bound solution provides a conservative estimate for the thick-walled joints under chord axial compression.

For K-joints, the $Q_u Q_f$ FE data for thick-walled joints agree for small chord loads with the IIW equation, and are for larger chord loads generally higher than the IIW predictions.

Table 6. Comparison of the thick-walled FE data with the new IIW (2008) equations for CHS joints under brace OPB.

Joint	No. of data	mean $\lambda=3$	Lu	Yura	CoV (%) $\lambda=3$	Lu	Yura
X (1)	12	1.06	1.14	1.21	14.1	12.9	11.9
X (2)	4	1.09	1.17	1.23	7.8	7.2	6.6
T (1)	12	1.02	1.03	1.09	17.2	15.5	16.0
T (2)	6	1.09	1.13	1.18	11.6	8.8	11.0
X(2) + T(2)	10	1.09	1.15	1.19	9.8	8.0	9.2

(1) All data.
(2) Data excluding punching shear, brace yielding.

The FE results for thick-walled X- and T- joints loaded by a brace in-plane bending moment and based on Lu's rotation limit of 0.1 radian, are about 4% too low to be in line with the ultimate strength formula on which the IIW (2008) design strength equation is based.

For X- and T- joints loaded by brace out-of-plane bending moment and again based on Lu's rotation limit of 0.1 radian, the FE results give a mean value which is 7% too low to be in line with the ultimate strength formula which served as basis for the IIW (2008) design strength equation.

Based on Yura's rotation criterion, for in-plane bending and out-of-plane bending, the FE results would be in line with the new IIW equations. It should further be mentioned that the design strength for joints loaded by brace bending moments is limited by the brace plastic bending moment capacity with $f_{yi} \leq f_{y0}$.

5 CONCLUSIONS

This study examines the application of the new IIW (2008) strength equations for thick-walled CHS X-, T- and K-joints. The thick-walled joint data have been derived from an extensive FE investigation at the National University of Singapore, using large deformation, elastic-plastic analyses.

Based on the plastic limit load approach, the FE results for thick-walled X-, T- and K-joints under brace axial loads support the following conclusions:

– The Q_u factors in the ultimate strength equations, which are the basis for the IIW design strength equations, provide a close prediction of the FE results.
– The Q_f factors generally provide a conservative estimation for the joints subjected to brace axial loads and chord stresses.

Based on Lu's maximum rotation limit of 0.1 radian, which is larger than that according to the plastic limit load approach but smaller than Yura's rotation limit, the FE results for thick-walled X- and T- joints under brace bending moment support the following conclusion:

– The FE results for thick-walled joints, loaded by brace in-plane bending or brace out-of-plane bending (and excluding data governed by punching shear and brace yielding) are 4–7% lower than the predictions by the equations which served as basis for the IIW design strength formulae. FE results based on Yura's rotation criterion are more in line with the IIW recommendations.

The design strength for joints loaded by brace bending moments is further limited by the brace plastic bending moment capacity with $f_{yi} \leq f_{y0}$.

REFERENCES

American Petroleum Institute 2007. Recommended practice for planning, designing and constructing fixed offshore platforms - Working stress design. API RP 2A, 21st Edition, Revision Supp. 3, American Petroleum Institute, USA.

Choo, Y.S., Qian, X.D., Liew, J.Y.R. & Wardenier, J. 2003. Static strength of thick-walled CHS X-joints—Part I: New approach in strength definition. *J. Constr. Steel. Res.*, 59, 1201–1228.

Gerdeen, J.C. 1980. A critical evaluation of plastic behavior data and a united definition of plastic loads for pressure components. *Welding Res. Council Bulletin.*

IIW 1989. Design recommendations for hollow section joints – Predominantly statically loaded, 2nd Edition. International Institute of Welding, Sub-commission XV-E, Annual Assembly, Helsinki, Finland, IIW Doc. XV-701-89.

IIW 2008. Static design procedure for welded hollow section joints - Recommendations. 3rd Edition, International Institute of Welding, Subcommission XV-E, IIW Doc. XV-1281-08.

Lu, L.H., Winkel, G.D. de, Yu, Y. & Wardenier, J. 1994. Deformation limit for the ultimate strength of hollow section joints. *Proc. 6th Int. Symp. Tubular Struct.*, Tubular Structures VI, Melbourne, Australia, 341–347.

Qian, X. 2005. Static strength of thick-walled CHS joints and global frame behavior. PhD thesis. National University of Singapore, Singapore.

Qian, X., Wardenier, J. & Choo, Y.S. 2007. A uniform approach for the design of 100% CHS overlap joints. *Proc. 5th Int. Conf. Adv. Steel Struct.*, Singapore.

Schumacher, A., Nussbaumer, A. & Hirt, M. 2001. Fatigue behavior of welded circular hollow section (CHS) joints in bridges. *Proc. 9th Int. Symp. Tubular Struct.* Tubular Structures IX, 291–297.

Togo, T. 1967. Experimental study on mechanical behavior of tubular joints. PhD thesis, Osaka University, Osaka, Japan.

Vegte, G.J. van der 1995. The static strength of uniplanar and multiplanar tubular T and X joints. PhD thesis, Delft University Press, Delft, The Netherlands.

Vegte, G.J. van der, Makino, Y. & Wardenier, J. 2007. New ultimate strength formulation for axially loaded CHS K-joints. *Proc. 5th Int. Conf. Adv. Steel Struct.*, Singapore.

Vegte, G.J. van der, Wardenier, J., Zhao, X.-L. & Packer, J.A. 2008a. Evaluation of the new CHS strength formulae to design strengths. *Proc. 12th Int. Symp. Tubular Struct.*, Shanghai, China, Tubular Structures XII, Taylor & Francis Group, London, UK.

Vegte, G.J. van der, Wardenier, J., Qian, X.D. & Choo, Y.S. 2008b. Re-analysis of the moment capacity of CHS joints. *Proc. 12th Int. Symp. Tubular Struct.*, Shanghai, China, Tubular Structures XII, Taylor and Francis Group, London, UK.

Wardenier, J. 2007. A uniform effective width approach for the design of CHS overlap joints. *Proc. 5th Int. Conf. Adv. Steel Struct.*, Singapore.

Wardenier, J., Vegte, G.J. van der, Makino, Y. & Marshall, P.W. 2008. Comparison of the new IIW (2008) CHS joint strength formulae with those of the previous IIW (1989) and the new API (2007). *Proc. 12th Symp. Tubular Struct.*, Shanghai, China, Tubular Structures XII, Taylor and Francis Group, London, UK.

Yura, J.A., Zettlemoyer, N. & Edwards, I.E. 1980. Ultimate capacity equations for tubular joints. OTC 3690, USA.

Zhao, X.L., Dayawansa, P. & Price, J. (2007). Dragline structures subjected to fatigue loading. *Proc. 5th Int. Conf. Adv. Steel Struct.*, Singapore.

NOMENCLATURE

CHS	circular hollow section
CoV	coefficient of variation
FE	finite element
IPB	in-plane bending
OPB	out-of-plane bending
RHS	rectangular hollow section
C_{ipb}	coefficient for joint capacity under brace IPB
C_{opb}	coefficient for joint capacity under brace OPB
M	bending moment
M_c	moment capacity of the member depending on the section classification
M^*_{ipb}	joint in-plane design bending moment capacity
M^*_{opb}	joint out-of-plane design bending moment capacity
$M_{ipb,u}$	joint in-plane ultimate bending moment capacity
$M_{opb,u}$	joint out-of-plane ultimate bending moment capacity
N	axial load
N_0	axial chord load
N_p	axial chord capacity
Q_f	chord stress function
Q_g	gap function for gapped K-joints
Q_u	geometric function for ultimate joint strength
W_E	elastic work by external load
W_P	plastic work by external load
d_0	external chord diameter
d_1	external brace diameter
f_{y0}	yield stress of the chord
g	gap size
n	maximum chord stress ratio
n_{axial}	maximum chord axial stress ratio
n_{ipb}	maximum chord in-plane bending stress ratio
t_1	wall thickness of the brace
t_0	wall thickness of the chord
β	diameter ratio between braces and chord
δ	brace displacement
φ	brace rotation
γ	half diameter to thickness ratio of the chord, $\gamma = d_0/2t_0$
λ	semi-empirical plastic limit load factor
θ	angle between the brace member and the chord
τ	brace to chord wall thickness ratio

Comparison of the new IIW (2008) CHS joint strength formulae with those of the previous IIW (1989) and the new API (2007)

J. Wardenier
Delft University of Technology, Delft, The Netherlands National University of Singapore, Singapore

G.J. van der Vegte
Delft University of Technology, Delft, The Netherlands

Y. Makino
Kumamoto University, Kumamoto, Japan

P.W. Marshall
National University of Singapore, Singapore
MHP Systems Engineering, Houston, USA

ABSTRACT: Recently Sub-commission IIW-XV-E has drafted new design recommendations for hollow section joints. The joint strength equations for Circular Hollow Section (CHS) joints in these new recommendations are the result of extensive reanalyses discussed in companion papers. This paper deals with uniplanar axially loaded T, X, K-gap and K-overlap joints made of CHS, and CHS joints loaded by in-plane or out-of-plane bending moments. The capacities of the new IIW (2008) equations and those of the previous IIW (1989) recommendations and the API (2007) are compared with each other. The background and differences are discussed and explained. In general, the capacities of the new IIW (2008) design equations for CHS joints are between the predictions of the previous IIW (1989) or CIDECT (1991) recommendations and those of the API (2007).

1 INTRODUCTION

Recently Sub-commission IIW-XV-E has drafted new design recommendations for hollow section joints, designated as IIW (2008). The joint capacity equations for CHS joints in these new recommendations are the result of extensive reanalyses discussed in companion papers (van der Vegte et al. 2008a, b, Qian et al. 2008, Zhao et al. 2008).

One of the main differences with the previous IIW (1989) design recommendations is that in the new IIW (2008) recommendations the influence function for the chord stress effect on the joint capacity is based on the maximum chord stress instead of a so-called prestress, i.e. based on the chord load excluding the brace load components (van der Vegte et al. 2007). Further, for tension loaded chords, a chord stress reduction is included whereas this was not the case in the previous IIW (1989) recommendations. Also, the equations are presented in a format similar to that in the API, i.e. for axially loaded joints:

$$N_i^* = Q_u Q_f \frac{f_{y0} t_0^2}{\sin\theta_i} \quad (1)$$

where Q_u is a function of the diameter ratio β, the chord diameter to thickness ratio 2γ and for gap joints, the relative gap g'. The function Q_f is the chord stress function based on the maximum chord stress; in the previous IIW (1989) recommendations called $f(n')$ but based on the chord prestress.

For CHS overlap joints, a new approach is followed, i.e. similar to that for overlap joints of rectangular hollow sections (RHS) or joints with an I section chord.

For CHS joints loaded by in-plane or out-of-plane bending, the same philosophy is used as for the CIDECT recommendations (Wardenier et al. 1991).

2 BACKGROUND

2.1 New IIW (2008) recommendations

In the framework of the re-evaluation of the design recommendations of IIW Sub-commission XV-E, many finite element (FE) analyses have been carried out on CHS X, T and K-gap joints (van der Vegte et al. 2008b), using the FE package Abaqus/Standard. For the analyses of T- and X joints, 8 noded, quadratic shell elements were employed. For gap K joints, which require a more accurate representation of the gap area and weld geometry, 20 noded, quadratic solid elements with two layers of elements modelled through the wall thickness, were used.

For failure, either the peak in the load-deformation diagram is used or Lu's 3% d_0 deformation criterion (Lu et al. 1994). Further, to avoid cracking, the strain in an element is limited to 20%. However, this may not capture ductile fracture in mode II (punching shear) or mode III (Mohr 2003) at the toe of the weld.

The FE analyses for the gap K joints with $d_0 = 406$ mm were calibrated against the experiments carried out by de Koning & Wardenier (1981).

Since the welds are modelled according to AWS with the minimum requirements for butt welds, the results of the FE calculations are in general, a lower bound for the strength of the experiments, especially for those with fillet welds.

The basic proposed joint strength functions are developed using the parameters of the analytical ring model as basis. Multi-regression analyses of the FE results are conducted to derive the coefficients of the strength formulations. These equations are further simplified, resulting in equations with a CoV and mean values (m) as given in Table 1. For X and T joints, as further simplification, a lower bound function is adopted for Q_f.

The characteristic strength $N_{u,k}$ is determined taking account of the tolerances ($s_{t_0} = 0.05$, $s_g = 0.05$ and $s_{f_{yo}} = 0.0075$) and scatter in numerical data, followed by a correction $1/0.85$ being $f_{y0,mean}/f_{y0,k}$.

The design capacities N^* in Table 2 are determined after dividing $N_{u,k}$ by a partial safety factor $\gamma_M = 1.1$. The resulting equations are further checked with the available experimental data. After careful screening of published data, a good agreement is found with the proposed equations (van der Vegte et al. 2008b).

For CHS overlap joints, the criteria are completely new and fully consistent with those for an RHS or I section chord (Wardenier 2007, Qian et al. 2007). In case of joints with overlaps Ov ≤ 60%, the joint

Table 2. New IIW (2008) design capacity equations.

Joint	Q_u	Q_f (comp.) *
X	$2.6 \left(\dfrac{1+\beta}{1-0.7\beta} \right) \gamma^{0.15}$	$(1-\|n\|)^{C_1}$ with $C_1 = 0.45 - 0.25\beta$
T	$2.6(1+6.8\beta^2) = \gamma^{0.2}$	
K-gap	$1.65(1+8\beta^{1.6})\gamma^{0.3}\left[1+\dfrac{1}{1.2+\left(\dfrac{g}{t_0}\right)^{0.8}}\right]$	$(1-\|n\|)^{0.25}$

* For chord tension $Q_f = (1-n)^{0.2}$

capacity is governed by the brace effective width criterion or by chord member failure due to axial load and bending. For larger overlaps, brace shear between braces and chord may be governing.

For CHS joints loaded by in-plane bending, it is found that the equation in the CIDECT Design Guide No. 1 (Wardenier et al. 1991) fits the data well, provided that the constant is reduced by 13%. This reduction is a result of newly derived data with larger dimensions and relatively smaller welds.

For CHS joints loaded by out-of-plane bending, as in CIDECT Design Guide No. 1, the capacity is related to the capacity of axially loaded X joints using a factor $c.d_1$.

2.2 *Previous IIW (1989) recommendations*

The joint strength equations for CHS X, T and K joints in the previous IIW recommendations (IIW 1989), shown in Table 3, are mainly based on work of Kurobane (1981). The governing parameters are partly based on the ring or punching shear model and the statistical analysis of experimental results.

For CHS joints loaded by out-of-plane bending moments, the CIDECT (1991) equations are based on a relation with the capacity of axially loaded X joints and for in-plane bending, on a modified Gibstein formula, see Wardenier (1982).

The design equations for all CHS joints are determined taking account of the scatter in test results, the tolerances/variations in dimensions (t_0, g) and mechanical properties (f_y, f_y/f_u), see Wardenier (1982).

The equations of the previous IIW (1989) recommendations are among others included in the current Eurocode 3 (CEN 2005), CIDECT Design Guide No. 1 (Wardenier et al. 1991) and many other international codes.

Table 1. CoV and mean values for the basic strength equations.

Joint	Q_u m and CoV	data	Q_f (compression) m and CoV	data
X	m = 1.01 CoV = 5.8%	23	m = 1.03 CoV = 6.8%	156
T	m = 1.01 CoV = 4.5%	33	m = 1.00 CoV = 6.0%	141
K-gap	m = 1.00 CoV = 4.2%	122	m = 1.00 CoV = 4.2%	122

Table 3. Previous IIW (1989) design capacity equations.

Joint	Previous IIW (1989) design strength functions	
	Q_u	$f(n')$
X	$\dfrac{5.2}{1 - 0.81\beta}$	
T	$(2.8 + 14.2\,\beta^2)\,\gamma^{0.2}$	
K-gap	$(1.8 + 10.2\beta)\,\gamma^{0.2} \left[1 + \dfrac{0.024\gamma^{1.2}}{\exp\left(0.5\dfrac{g}{t_0} - 1.33\right) + 1} \right]$	See note below

* $f(n') = 1 + 0.3n' - 0.3n'^2$ for chord compression pre-stress; no reduction for chord tension stress.

Table 4. API (2007) characteristic capacity equations.

Joint	New API (2007) characteristic strength functions	
	Q_u*	Q_f
X	$[2.8 + (12 + 0.1\gamma)\,\beta]\,Q_\beta$	See Figs. 1 to 3
T	$2.8 + (20 + 0.8\gamma)\,\beta^{1.6}$	
K-gap	$(16 + 1.2\gamma)\,\beta^{1.2}\left[1 + 0.2\left(1 - 2.8\dfrac{g}{d_0}\right)^3\right]$**	

* for T and K-gap joints limit $\gamma \leq 20$ value;

for $\beta \geq 0.6$: $Q_\beta = \dfrac{0.3}{\beta(1 - 0.833\beta)}$ and for $\beta < 0.6$: $Q_\beta = 1.0$

** Q_g for $\dfrac{g}{d_0} \geq 0.05$ and $Q_g \geq 1.0$.

2.3 API (2007) recommendations

Recently the equations for the capacity of CHS joints in the API (2007) have been revised (Marshall 2004). These new API equations (Table 4) are mainly based on numerical work carried out by Pecknold et al. (2007). In the numerical models, no welds are included, which for small β ratios and small gaps, gives lower capacities. In general, this approach not only results in ultimate capacities lower than the values found in experiments, but also lower than those derived for the IIW (2008) recommendations (van der Vegte et al. 2007, 2008b). Compared to the previous edition of the API, substantially lower capacities are given for joints loaded by in-plane or out-of-plane bending.

The API (2007) equations represent characteristic strength functions for which, for working load design, a safety factor of 1.6 has to be applied. This 1.6 is about equivalent to the load factor multiplied by the partial safety factor $\gamma_M = 1.1$, for example used in the Eurocode and IIW recommendations. The API characteristic strength equations were selected to provide a sharp lower bound to the data, based on a visual fit. For individual type combinations of joint/loading/database, the characteristic strength ranged from 1.3 to 3.9 standard deviations below the mean, influenced mainly by the amount of scatter on the safe side. The composite for all data was 2.3 standard deviations. For offshore jacket members with 100% Gulf of Mexico wave loading, the lifetime safety index is 2.7, exceeding previous RP2A targets, and representing considerable improvement over older API joint design equations.

3 AXIALLY LOADED X, T AND K-GAP JOINTS

For a realistic comparison, the characteristic joint capacity functions of the IIW ($1.1 \times N^*$) are compared with the characteristic functions of the API (2007). (Note: values $2\gamma < 20$ are outside the validity range of the API (2007)). Both functions Q_f and Q_u are first separately considered and then combined.

3.1 Chord stress function Q_f

Figures 1a and 1b show for X and T joints and Figures 3a and 3b for K-gap joints, a comparison between the chord axial stress functions in the new IIW (2008), the previous IIW (1989) and the API (2007) recommendations. Figures 2a and 2b show the comparisons for the in-plane bending chord stress functions for T joints.

3.1.1 New IIW (2008) chord stress functions vs. those in the previous IIW (1989) recommendations

3.1.1.1 General
For tension loaded chords, the function in the new IIW (2008) recommendations gives a reduction in capacity whereas the previous IIW (1989) function does not give a reduction.

3.1.1.2 X and T joints
The previous IIW (1989) chord stress function (which was based on K joint test results) is a lower bound for the new IIW (2008) chord stress functions for T and X joints. Hence, the new IIW (2008) function gives a somewhat smaller reduction in capacity.

3.1.1.4 K-gap joints

The new IIW (2008) chord stress function for K-gap joints (based on the maximum chord stress ratio n) gives, for $n = n'$, a some what smaller reduction in capacity than that in the previous IIW (1989) recommendations (based on the chord compression prestress n'. However, n is always larger than n'.

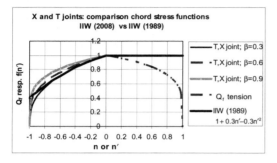

Figure 1a. X joint: new IIW (2008) chord axial stress functions vs previous IIW (1989).

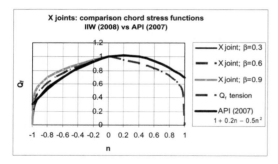

Figure 1b. X joint: new IIW (2008) chord axial stress functions vs API (2007).

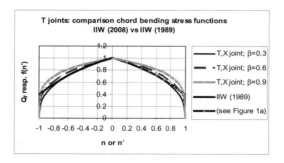

Figure 2a. T joint: new IIW (2008) chord bending stress functions vs previous IIW (1989).

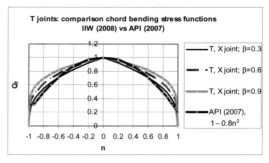

Figure 2b. T joint: new IIW (2008) chord bending stress functions vs API (2007)—for axially loaded brace.

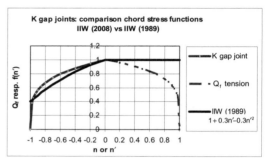

Figure 3a. K-gap joint: new IIW (2008) chord axial stress function vs previous IIW (1989).

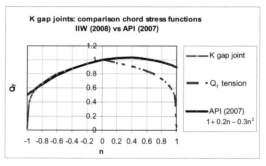

Figure 3b. K-gap joint: new IIW (2008) chord axial stress function vs API (2007).

3.1.2 *New IIW (2008) chord stress functions vs. those in the API (2007) recommendations*

3.1.2.1 X joints

The API (2007) chord axial compression stress function for X joints is a lower bound for the new IIW (2008) chord stress functions for T and X joints.

For tension loaded chords, the new IIW (2008) function gives a larger reduction in capacity than that in the API (2007).

3.1.2.2 T joints

The API (2007) chord bending stress function for T joints gives about the same reduction due to chord bending than the new IIW (2008) chord compression stress function for T and X joints. However, for chord axial compression loading, the new IIW (2008) chord stress function is less severe than that according to the API (2007).

3.1.2.3 K-gap joints

The new IIW (2008) function for chord compression stress gives up to a compression chord stress of 90% of the yield load, about the same reduction in capacity as that of the API (2007). However, the two Q_u functions on which Q_f is normalized are substantially different, and the chord stress to be included is not exactly the same; API (2007) is based on a model where the brace load components are divided over both sides of the chord at the joint.

The new IIW (2008) function for chord tensile stress gives a considerably larger reduction in capacity than that included in the API (2007). In offshore jacket design, where $2\gamma > 20$, tensile chord load reduces the tripping tendency of chord shell in the gap region. The new IIW (2008) function is a simplified mean function, conservative for data with a high γ ratio and slightly unconservative for data with low γ ratio (van der Vegte et al. 2007).

3.2 Geometric function Q_u

Figures 4a and 4b show for X joints and Figures 5a and 5b for T joints, a comparison between the Q_u functions in the new IIW (2008), the previous IIW (1989) and the API (2007) recommendations. Figures 6a to 7b show these comparisons for K-gap joints with $g' = 2$ and 20.

3.2.1 X joints

The Q_u function in the previous IIW (1989) recommendations gives, for small and large β ratios, a larger capacity and for medium β ratios with medium to high γ ratios, an equal or lower capacity than that in

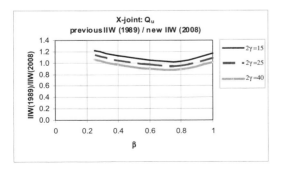

Figure 4a. X joint: new IIW (2008) chord Q_u function vs previous IIW (1989).

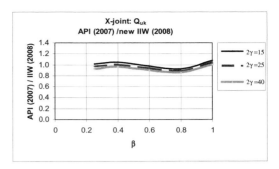

Figure 4b. X joint: new IIW (2008) Q_{uk} function vs API (2007).

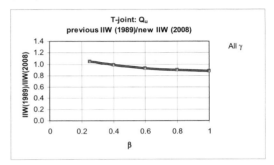

Figure 5a. T joint: new IIW (2008) Q_u function vs previous IIW (1989).

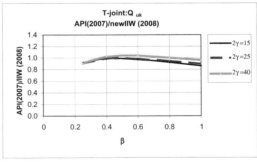

Figure 5b. T joint: new IIW (2008) Q_{uk} function vs API (2007).

the new IIW (2008) recommendations. It was known that the previous IIW (1989) function gave too high values for low β ratios.

The Q_{uk} function in the API (2007) gives, with exception for β about 0.8, about the same capacity as that in the new IIW (2008) function.

3.2.2 T joints

The Q_u function in the previous IIW (1989) recommendations gives, compared to the new IIW (2008) recommendations, for large β ratios a lower capacity and for small β ratios an equal capacity. This is caused by the chord bending stress effect included in the previous IIW (1989) Q_u function, which is larger for high β ratios and very small for low β ratios.

The Q_u function in the API (2007) gives for low and high β ratios, a slightly lower capacity than the IIW (2008) function. For medium β ratios, the capacity is the same.

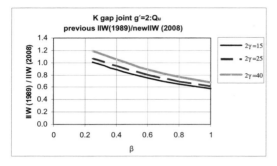

Figure 6a. K-gap joint with small gaps g' = 2: new IIW (2008) Q_u function vs previous IIW (1989).

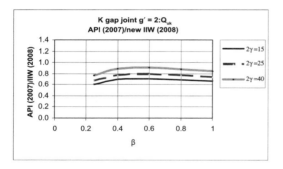

Figure 6b. K-gap joint with small gaps g' = 2: new IIW (2008) Q_{uk} function vs API (2007).

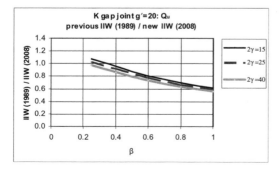

Figure 7a. K-gap joint with large gaps g' = 20: new IIW (2008) Q_{uk} function vs previous IIW (1989).

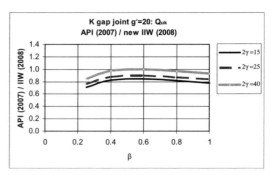

Figure 7b. K-gap joint with large gaps g' = 20: new IIW (2008) Q_{uk} function vs API (2007).

3.2.3 *K-gap joints*

For small and larger gaps, the Q_u function in the previous IIW (1989) recommendations gives for low β ratios, a larger or equal capacity and for medium to high β ratios, a much lower capacity. This is caused by the chord stress effect due to the brace load components included in the previous IIW (1989) recommendations which is larger for medium to high β ratios.

The Q_{uk} function in the API (2007) gives, especially for β < 0.4 ratios, considerable lower capacities than the new IIW (2008) function. For larger β ratios, the capacity according to API (2007) is, especially for the smaller gaps, lower than that according to the new IIW (2008) function. This is partly offset by the chord stress effect of half of the brace load components included in the Q_{uk} function of the API.

3.3 *Comparisons of gross capacity*

3.3.1 *X joints*

Figure 8 shows for $2\gamma = 25$ and 40 in relation to the diameter ratio β, a comparison between the new IIW (2008), the previous IIW (1989) and the API (2007) $Q_{uk} \cdot Q_f$ functions for $Q_f = 1$, which is the case for no chord preload. Thus, in principle the comparison is similar as that in Figures 4a and 4b.

There is a good agreement between the gross capacity of the new IIW (2008) formula for X joints with compression loaded chords and that of the previous IIW (1989) and the API (2007) recommendations. However, for X joints with tension loaded chords, the new IIW (2008) equation gives a lower gross capacity than that of the previous IIW (1989) and that of the API (2007), which is mainly driven by the chord tension stress effect for X joints with thick walled chords.

3.3.2 *T joints*

Figure 9 shows for $2\gamma = 25$ and 40 in relation to the diameter ratio β, a comparison between the new IIW

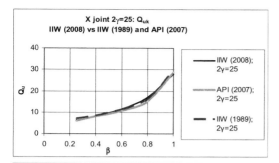

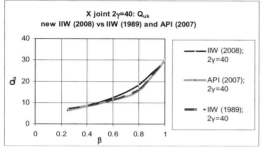

Figure 8. Comparison Q_{uk} for the new IIW (2008), previous IIW (1989) and API (2007) characteristic strength for axially loaded X joints.

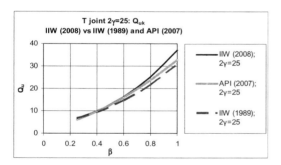

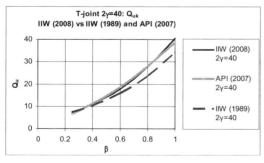

Figure 9. Comparison Q_{uk} for the new IIW (2008), the previous IIW (1989) and API (2007) characteristic strength for axially loaded T joints (excluding Q_f effect, although chord bending effect included in IIW (1989).

(2008), the previous IIW (1989) and the API (2007) Q_{uk} functions for axially loaded T joints. Also here, the comparison is similar to that in Figures 5a and 5b and the same conclusions apply.

Considering the Q_{uk} with the Q_f functions together (see Figs. 2 and 5 or 9), the capacity of the new IIW (2008) equation for T joints is somewhat lower than that according to the previous IIW (1989) (bending effect included) and equal or larger than that according to API (2007) (Q_f in the API is a lower bound for the Q_f in the new IIW (2008)).

3.3.3 K-gap joints

Figures 10 and 11 show a comparison between the new IIW (2008), the previous IIW (1989) and API (2007), for K-gap joints with g' = 2 and 20 in relation to the diameter ratio β, based on the combined $Q_{uk} \cdot Q_f$ functions for the case of no chord prestress.

In general, the capacity according to the new IIW (2008) equations for K-gap joints with small gaps and small β ratios will be smaller or equal and for large gaps with medium or high γ ratios, considerably larger than that of the previous IIW (1989) recommendations. This is mainly caused by the gap function in the IIW (1989), which gives for large g' with high γ ratios, a considerable lower capacity.

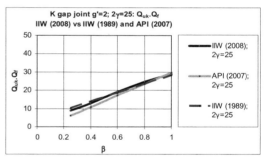

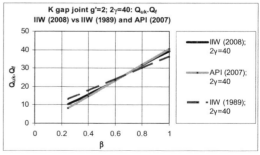

Figure 10. Comparison of the characteristic strength ($Q_{uk} \cdot Q_f$) of the new IIW (2008), the previous IIW (1989) and API (2007) for K-gap joints with g' = 2 (small gaps)

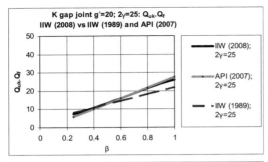

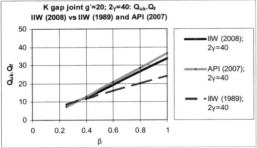

Figure 11. Comparison of the characteristic strength ($Q_{uk} \cdot Q_f$) of the new IIW (2008), the previous IIW (1989) and API (2007) for K-gap joints with g' = 20 (large gaps).

Compared to API (2007), the capacity of the new IIW (2008) formula for K-gap joints gives for small β ratios, larger capacities than that according to API (2007); for larger β ratios the capacity is about the same as that of API (2007). This effect may be caused by the FE model used for API (2007) where no welds are included. For very small β ratios and small gaps, this results in lower capacities.

In general, the formulae in the new IIW (2008) recommendations give results between those of the previous IIW (1989) and the API (2007).

4 K-OVERLAP JOINTS

As mentioned in Section 2, the criteria for CHS overlap joints are completely new and fully consistent with those for joints with an RHS or I section chord (Wardenier 2007, Qian et al. 2007). If the joint capacity is not governed by chord member failure due to axial load and bending or brace shear, for overlaps Ov ≤ 60% the joint capacity is governed by the brace effective width criterion.

For indication only, Table 5 shows a comparison between the joint capacities in terms of efficiency (joint capacity/brace yield load) for a 50% overlap joint, calculated according to the three recommendations. The differences can be considerable.

Table 5. Joint efficiencies for a 50% overlap joint ($\beta = 0.6$; $2\gamma = 25$; $\tau = 1$ or 0.5; $\theta_i = 45°$ and equal brace dimensions).

Joint efficiency	IIW (2008)	IIW (1989)	API (2007)/1.1
$\beta = 0.6$; $2\gamma = 25$; $\tau = 1$	0.81	0.63	$0.63 Q_f$
$\beta = 0.6$; $2\gamma = 25$; $\tau = 0.5$	0.84	>1.0	$1.32 Q_f \leq 1.0$

5 JOINTS LOADED BY IN-PLANE AND OUT-OF-PLANE BENDING MOMENTS

5.1 Chord stress functions Q_f

Table 6 shows the chord stress functions for joints with the braces loaded by in-plane or out-of-plane bending moments.

For simplicity, the new IIW (2008) recommendations adopt the same Q_f function as for axially loaded X and T joints. In API (2007), the Q_f function was derived from the database and Q_{uk}, but forced to be the same for all bending cases, giving a milder reduction than that for axially loaded joints. Depending on the β ratio and the chord stress ratio n, the value of the bending Q_f in the API (2007) for n > 0.8 is about 1.05 to 1.5 times the blanket Q_f in IIW (2008). However, n > 0.8 rarely occurs in offshore jacket design with thick joint cans.

5.2 Geometric functions Q_u (or Q_{uk}) for in-plane bending

As shown in Figure 12, the Q_u function in the CIDECT (1991) recommendations gives for all β and γ ratios, a 13% larger capacity than that in the new IIW (2008) recommendations.

As shown in Figure 13, the Q_{uk} function in the API (2007) gives a considerably lower capacity than the new IIW (2008), especially for moderate to low β. The API strength function for in-plane bending was forced to be the same for all joint types, and was driven down by finite element results for symmetrical K joint bending, a common case for offshore jacket braces loaded by wave action. The resulting conservatism for other joint types is partially offset by a lesser penalty in Q_f.

5.3 Geometric functions Q_u (or Q_{uk}) for out-of-plane bending

As shown in Figure 14, the Q_u function in the CIDECT (1991) recommendations gives, compared to that in the IIW (2008) recommendations for small and large β ratios especially in combination with lower γ ratios, a considerable larger capacity and for medium β ratios

Table 6. Chord stress functions (for in-plane bending) for joints loaded by brace bending moments.

	Chord stress function
IIW (2008)	$Q_f = (1 - |n|)^{(0.45 - 0.25\beta)}$
CIDECT (1991)	$f(n') = 1 + 0.3n' - 0.3n'^2$
API (2007)	$Q_f = (1 - 0.4n^2)$

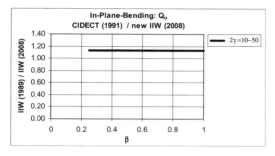

Figure 12. Comparison of the in-plane bending moment capacities according to the new IIW (2008) and CIDECT (1991).

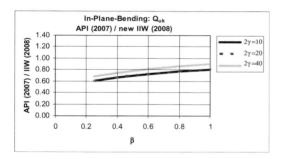

Figure 13. Comparison of the in-plane (characteristic) bending moment capacities according to the new IIW (2008) and API (2007). (Note: $2\gamma = 10$ is outside the validity range of the API).

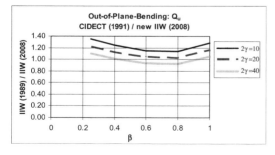

Figure 14. Comparison of the out of plane bending moment capacities according to the new IIW (2008) and CIDECT (1991).

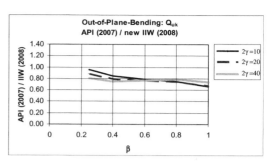

Figure 15. Comparison of out-of-plane (characteristic) bending moment capacities according to the new IIW (2008) and API (2007). (Note: $2\gamma = 10$ is outside the validity range of the API).

with medium γ ratios, an equal capacity. For medium β ratios with high γ ratios, the CIDECT (1991) recommendations give a somewhat lower capacity.

The Q_{uk} function in API (2007) (see Fig. 15) gives, with exception for low β values, about 20% lower capacities than the new IIW (2008) function. Again, this is influenced by the symmetrical K joint loading case in offshore jackets.

5.4 Combined functions $Q_u \cdot Q_f$ (or $Q_{uk} \cdot Q_f$)

Considering the combined $Q_f \cdot Q_u$ functions for in-plane and out-of-plane bending, it can be concluded that the differences between the new IIW (2008) and API (2007) become smaller, although depending on the parameters β and n, still considerable differences may occur.

Generally, the IIW (2008) recommendations give capacities between that of CIDECT (1991) and that of API (2007).

6 CONCLUSIONS

From the comparisons between the CHS joint capacity equations in the new IIW (2008) recommendations, those in the IIW (1989) and the API (2007), the following conclusions can be drawn:

6.1 X joints

A good agreement exists between the capacity of the new IIW (2008) equation for X joints with compression loaded chords and that of the previous IIW (1989) and the API (2007) recommendations.

However, for X joints with tension loaded chords, the new IIW (2008) equation gives a lower capacity than that of the previous IIW (1989) but also compared to that of API (2007), which is mainly caused by tension chord stress effect for the X joints with thick walled chords.

6.2 T joints

The capacity of the new IIW (2008) equation for T joints gives slightly lower capacities than those according to the previous IIW (1989) (bending effect included) and larger capacities than those according to API (2007) (Q_f of API is a lower bound for the new Q_f of IIW (2008)).

6.3 K-gap joints

In general, the capacity predictions of the new IIW (2008) formula for K-gap joints are somewhat smaller or equal to and only in some cases somewhat larger than those of the previous IIW (1989) recommendations.

Compared to API (2007), the new IIW (2008) equation for K-gap joints gives, for small β ratios, considerably larger capacities than those according to API (2007); for larger β ratios, the capacity is about the same as that of API (2007).

The equation in the new IIW (2008) recommendations gives results between those of the previous IIW (1989) and API (2007).

6.4 K-overlap joints

The approaches for overlap joints are totally different in the three considered recommendations. IIW (1989) uses a continuous function with gap joints, API (2007) takes account of the yield-thickness ratio between chord and brace, whereas IIW (2008) takes not only account of the yield-thickness ratios between chord and braces but also between the braces itself. This may result in considerable differences, depending on the yield-thickness ratios.

6.5 Joints loaded by in-plane bending moments

The new IIW (2008) recommendations give for all β and γ ratios, a lower capacity than that according to the CIDECT (1991) recommendations, but generally larger capacity than that in API (2007).

In general, it can be concluded that the new IIW (2008) gives capacities between those of the CIDECT (1991) recommendations and API (2007).

6.6 Joints loaded by out-of-plane bending moments

The IIW (2008) recommendations give, compared to that of the CIDECT (1991) recommendations, for small and large β ratios especially in combination

Table 7. Range of validity for the various recommendations.

IIW (2008)	IIW (1989)	API (2007)
$0.2 \leq d_1/d_0 \leq 1.0$	$0.2 \leq d_1/d_0 \leq 1.0$	$2 \leq d_1/d_0 \leq 1.0$
$2\gamma \leq 50$ *	$2\gamma \leq 50$ *	$20 \leq 2\gamma \leq 100$
	$30° \leq \theta \leq 90°$	
$g \geq t_1 + t_2$	$g \geq t_1 + t_2$	$g > 50$ mm
$Ov \leq 100\%$	$Ov \leq 100\%$	$g/d_0 > -0.6$
$f_y \leq 460$ MPa	$f_y \leq 355$ MPa	$f_y \leq 500$ MPa

* for X joints $2\gamma \leq 40$

with lower γ ratios, a considerable lower capacity and for medium β ratios with medium γ ratios, an equal capacity. For medium β ratios with high γ ratios, it gives a somewhat larger capacity.

The new IIW (2008) recommendations give a considerable larger capacity than that according to API (2007). Wave loading on offshore jacket bracing presents a bending pattern not often found onshore.

In general, it can be concluded that the new IIW (2008) gives capacities between those of the previous IIW (1989) or CIDECT (1991) recommendations and those of API (2007). Thus the two newer standards agree on the direction of change.

6.7 Range of validity

The ranges of validity are not discussed in the previous sections, but are as given in Table 7.

6.8 General conclusion

In most cases, the CHS joint capacities according to the new IIW (2008) design recommendations for hollow section joints are between those according to the previous IIW (1989) recommendations and those according to API (2007).

ACKNOWLEDGEMENT

The continuous discussions with our colleagues Prof. J.A. Packer from University of Toronto, Canada and Prof. X.-L. Zhao from Monash University, Australia are gratefully acknowledged.

REFERENCES

American Petroleum Institute 2007. Recommended practice for planning, designing and constructing fixed offshore platforms—Working stress design. API RP 2 A, 21st Ed, Supp. 3, American Petroleum Institute, USA.

CEN 2005. Eurocode 3: Design of steel structures—Part 1.8: Design of joints. EN 1993-1-8:2005, Comité Européen de Normalisation, Brussels, Belgium.

IIW 1989. Design recommendations for hollow section joints—Predominantly statically loaded. 2nd Edition, International Institute of Welding, Subcommission XV-E, Annual Assembly, Helsinki, Finland, IIW Doc. XV-701–89.

IIW 2008. Static design procedure for welded hollow section joints—Recommendations. 3rd Edition, International Institute of Welding, Subcommission XV-E, IIW Doc. XV-1281–08.

Koning, C.H.M. de & Wardenier, J. 1981. The static strength of welded joints between structural hollow sections or between structural hollow sections and H sections, Part 1: Joints between circular hollow sections. Stevin Report 6-84-18. Delft University of Technology, The Netherlands.

Kurobane, Y. 1981. New developments and practices in tubular joint design (÷ Addendum). International Institute of Welding, Annual Assembly, Oporto, Portugal, IIW Doc. XV-488–81.

Lu, L.H., Winkel, G.D. de, Yu, Y. & Wardenier, J. 1994. Deformation limit for the ultimate strength of hollow section joints. *Proc. 6th Intern. Symp. on Tubular Structures*, Melbourne, Australia, Tubular Structures VI, Balkema, Rotterdam, The Netherlands.

Marshall, P.W. 2004. Review of tubular joint criteria. *Proc. ECCS-AISC Workshop 'Connections in Steel Structures V'*, Bouwen met Staal, Zoetermeer, The Netherlands.

Mohr, W.C. 2003. Strain based design of pipelines. Minutes of JIP meeting on March 6. Edison Welding Institute.

Pecknold, D.A., Marshall, P.W. & Bucknell, J. 2007. New API RP2A tubular joint strength design provisions. *Journal of Energy Resources Technology*, American Society of Mechanical Engineers, USA, Vol. 129, No. 3, pp. 177–189.

Qian, X.D., Wardenier, J. & Choo, Y.S. 2007. A uniform approach for the design of 100% CHS overlap joints. *Proc. 5th Intern. Conf. on Advances in Steel Structures*, Singapore.

Qian, X.D., Choo, Y.S., Vegte, G.J. van der & Wardenier, J. 2008. Evaluation of the new IIW CHS strength formulae for thick-walled joints. *Proc. 12th Intern. Symp. on Tubular Structures*, Shanghai, China. Tubular Structures XII, Taylor & Francis Group, London, UK.

Vegte, G.J. van der, Wardenier, J. & Makino, Y. 2007. New ultimate strength formulation for axially loaded CHS K-joints. *Proc. 5th Intern. Conf. on Advances in Steel Structures*, Singapore.

Vegte, G.J. van der, Wardenier, J., Qian, X.D. & Choo, Y.S. 2008a. Reanalysis of the moment capacity of CHS joints. *Proc. 12th Intern. Symp. on Tubular Structures*, Shanghai, China, Tubular Structures XII, Taylor & Francis Group, London, UK.

Vegte, G.J. van der, Wardenier, J., Zhao, X.-L. & Packer, J.A. 2008b. Evaluation of the new CHS strength formulae to design strengths. *Proc. 12th Intern. Symp. on Tubular Structures*, Shanghai, China, Tubular Structures XII, Taylor & Francis Group, London, UK.

Wardenier, J. 1982. *Hollow section joints*. Delft University Press, The Netherlands, ISBN 90-6275-084-2.

Wardenier, J., Kurobane, Y., Packer, J.A., Dutta, D. & Yeomans, N. 1991. *Design guide for circular hollow section (CHS) joints under predominantly static loading*. Verlag TUV Rheinland GmbH, Köln, Germany, ISBN 3-88585-975-0.

Wardenier, J. 2007. A uniform effective width approach for the design of CHS overlap joints. *Proc. 5th Intern. Conf. on Advances in Steel Structures*, Singapore.

Zhao, X.-L., Wardenier, J., Packer, J.A. & Vegte, G.J. van der 2008. New IIW (2008) static design recommendations for hollow section joints. *Proc. 12th Intern. Symp. on Tubular Structures*, Shanghai, China, Tubular Structures XII, Taylor & Francis Group, London, UK.

SYMBOLS

CHS	circular hollow section
FE	finite element
RHS	rectangular hollow section
A_0	cross sectional area of the chord
M_1^*	bending joint moment capacity
$M_{u,k}$	characteristic joint moment capacity
N_1^*	axial load capacity of the joint based on load in the compression brace
$N_{u,k}$	characteristic joint capacity
N_0	axial chord load
N_{0p}	external axial chord load
Ov	overlap
Q_f	chord stress function
Q_u	geometric function for the limit design joint strength
$Q_{u,k}$	geometric function for the characteristic joint strength
$W_{pl,0}$	plastic section modulus
c	constant
d_0	external chord diameter
d_1	external brace diameter
e	eccentricity
f_{y0}	yield stress of the chord
$f_{y0,k}$	characteristic value for the yield stress of the chord
$f_{y0,mean}$	mean value for the yield stress of the chord
$f(n')$	chord (pre)stress function in the IIW (1989) equations
g	gap size
g'	relative gap g/t_0
n	non-dimensional chord stress ratio based on maximum chord load
n'	non-dimensional chord pre-stress ratio
s	standard deviation
t_i	wall thickness of the brace i (i = 1 or 2)
t_0	wall thickness of the chord
β	diameter ratio between braces and chord
γ	half diameter to thickness ratio of the chord, $γ = d_0/2t_0$
$γ_M$	partial safety factor for the joint capacity
$θ_i$	angle between brace member i and the chord

Extending existing design rules in EN1993-1-8 (2005) for gapped RHS K-joints for maximum chord slenderness (b_0/t_0) of 35 to 50 and gap size g to as low as $4t_0$

O. Fleischer & R. Puthli
Research Centre for Steel, Timber and Masonry, University of Karlsruhe, Germany

ABSTRACT: Many cold formed hollow sections with a width to thickness ratio $2\gamma > 35$ can be found in the product standard EN10219-2 (2006). However, due to the design restrictions in EN1993-1-8 (2005), they have normally to be excluded from use. For joints with these slender sections, experimental investigations on K-joints with gap have therefore been carried out at Karlsruhe University to extend the limiting range. The results of these investigations are statistically evaluated according to EN1990 (2002) with reference to the ob-served failure modes. In this paper, the existing design models are suitably adapted for the slender sections not covered by existing rules. The influence of reducing the gap size limitation on slender chords is also investigated.

1 INTRODUCTION

Experimental investigations on axially loaded K-joints have been carried out at Karlsruhe University to extend the limitations in EN1993-1-8 (2005) of the chord width to thickness ratio from $2\gamma = b_0/t_0 \leq 35$ to 50. The lower bound of the permitted gap size in EN1993-1-8 (2005) is either $g_{min} \geq 0.5 \cdot b_0 \cdot (1 - b_i/b_0)$ or $g_{w,min} \geq t_1 + t_2$, whichever is larger. For the investigated range of the chord slenderness, the first criterion almost always governs. However, for the present work, the minimum gap size $g_{w,min}$ is allowed to go down up to $g_{w,min} \geq 4 \cdot t_0$, which is the minimum gap size for practical purposes of welding these thin walled sections. This proposed gap, below that permitted above, not only affects the yield mechanism for which the governing gap size is given in existing rules, but also the deformation capacity.

Because the design rules in EN1993-1-8 (2005) are based on maximum loads $N_{i,max}$ (overall ultimate loads), these maximum loads achieved are used for the first statistical evaluations for the failure modes observed in the tests (Fleischer & Puthli 2006).

Since chord face failure cannot be observed visually, the deformation criterion from previous work (Lu et al. 1993), which limits the indentation of the chord flange to 3% of the chord width b_0, has been used for further evaluation (of chord face failure only) on the assumption that the chord face failure mode would best represent the behaviour at lower deformations. With the experimentally determined loads $N_{i,u}$, based on the deformation criterion, a statistical evaluation for chord face failure based on the corresponding design model used in EN1993-1-8 (2005) has been carried out.

Details of the test specimens, the measured dimensions and material properties as well as the results of the tests and the statistical evaluations are reported in Fleischer & Puthli (2006).

2 ADAPTION OF THE EXISTING DESIGN MODELS IN EN1993-1-8 (2005)

The design models of failure modes visually observed in the tests and determined by using a de-formation criterion (Lu et al., 1993) have been statistically evaluated according to specifications in EN1990 (2002). The results of these evaluations show that the existing design rules in EN1993-1-8 (2005) for gapped K-joints using rectangular hollow sections with chord slenderness $2\gamma \leq 35$ and gap size $g > g_{min}$ cannot be extended for the parameter range considered in the present work without modifications (Fleischer & Puthli, 2006).

For punching shear and brace failure modes, which could clearly be observed in the experiments, the effective lengths are to be reduced to $l_{e.p.red}$ (punching shear) and $l_{e.red}$ (effective width) to overcome the deviations between the statistically determined design resistances and the design models of EN1993-1-8 (2005) that are otherwise observed.

Chord web failure was also observed in the tests, which is not considered in EN1993-1-8 (2005) for

K-joints, but only for X- and Y-joints. However, the design equation for X- and Y-joints was not found to be suitable for use with K-joints, particularly as it did not correlate with the experimental results for specimens that failed by chord web failure.

For chord face failure, which was determined on the basis of a maximum indentation of 3% of the chord width b_0, basic yield line models are found to be necessary, in lieu of the semi-analytical simplified equation in EN1993-1-8 (2005), so that the parameter range could be extended to include chord slenderness $35 < 2\gamma \leq 50$ and gap sizes $g < g_{min}$. This function will be further calibrated in subsequent work by using results of numerical investigations that are already carried out but not reported here.

2.1 Chord face failure

For the statistical evaluation of chord face failure in the present work "ultimate" loads $N_{i,u}$, which are obtained by using a deformation criterion (Lu et al., 1993) are used.

2.1.1 Design resistance of EN1993-1-8 (2005)

If the design resistance in EN1993-1-8 (2005) given by Eq. 1 below is used, a reduction factor of $\xi_d = \gamma 0.71$ has to be applied for connections with chord slenderness $35 < 2\gamma \leq 50$ and small gap size ($4t_0 \leq g < g_{min}$) (Fleischer & Puthli, 2006).

$$r_t = \frac{8.9 f_{y0} b_0^{0.5} t_0^{1.5}}{2^{0.5} \sin\Theta_i} \left(\frac{b_i + h_i}{2 b_0} \right) \cdot f(n') \quad (1)$$

where f_{y0}, b_0, t_0 = yield strength, width and wall thickness of the chord; Θ_i = brace angle; b_i, h_i = width and height of the braces, respectively; $f(n')$ = chord stress function.

The semi-empirical design formula for chord face failure (Eq. 1) is partly based on the yield line theory, however, based on the maximum loads recorded in the experimental databases. The validity of Eq. 1 is restricted to chord slendernesses $2\gamma \leq 35$ and gap sizes $g \geq g_{min}$. Therefore, as expected, the reduction factor of $\xi_d = 0.71$ is smaller than 1.0, particularly since the limiting range is extended beyond that specified for Eq. 1.

2.1.2 Yield line model

To obtain a better agreement of resistances r_t calculated by an analytical model to the experimental results $N_{i,u}$, the basic yield line model for K-joints with gap (Figure 1) simplified as a push-pull joint is used for determining the joints resistances r_t in Eq. 2 (Wardenier, 1982).

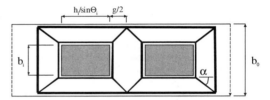

$$r_t = \frac{f_{y0} t_0^2}{(1-\beta)\sin\Theta_i} \left(\frac{2\frac{h_i}{b_0}}{\sin\Theta_i} + \frac{g}{b_0} + \frac{b_0}{2g}(1-\beta) \right) \quad (2)$$

where $\beta = b_i/b_0$; g = gap size.

The statistical evaluation of the joint resistances r_t (Eq. 2), calculated with measured dimensions and material properties, is given in Figure 2.

Since this analytical model does not account for stress biaxiality and shear stresses, which are more dominant for small gap sizes, and membrane and strain hardening effects are also ignored, the lower bound design resistance r_t from yield line theory marginally underestimates the ultimate loads $N_{i,u}$

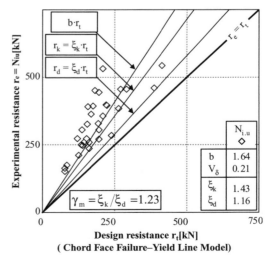

Figure 1. Yield lines for K-joint with gap.

Figure 2. Statistical evaluation of chord face failure acc. to the yield line model.

based on the deformation criterion. For the extended limiting range of the chord slenderness $35 < 2\gamma \leq 50$ and the gap size $4t_0 \leq g < g_{min}$, Eq. 2 requires a factor of $\xi_d = 1.16$ for design purposes on the basis of the statistical analysis of the experimental data.

2.1.3 Modified yield line model

If the dominant influence of shear stress and biaxial stress for small gaps is to be included, a modified yield line theory presented by Partanen (1991) and further refined by Partanen & Björk (1993) is found to be suitable.

With modifications to the classic yield line theory stress biaxiality in the plastic hinges and shear stress correction in the biaxial stress state can be considered (Figure 3, Mode A) (Partanen 1991).

The more refined approach by Partanen & Björg (1993) allows the possibility of a plastic hinge in the brace wall of a fillet welded joint giving the lowest shear capacity (Figure 3, Mode B). The resulting failure is called effective gap reduction failure.

The axial load capacity r_t of the connection is obtained by adding half of the Knife Edge Capacity $\frac{1}{2} \cdot N_{ke}$ (capacity of the heel sides) and the capacity of the transverse and the longitudinal gap side (Eq. 3) (Partanen & Björk, 1993). Full details of the derivation of the Knife Edge Capacity and the transverse and longitudinal shear capacities are given by Partanen (1991) and Partanen & Björk (1993). The formulae and derivation are too lengthy and complex to be presented here, so that only the basic approach is presented below.

$$r_t = \frac{1}{\sin\Theta_i}\left(\frac{1}{2}N_{Ke} + q_t b_0 + q_l l_i\right) \quad (3)$$

where (see Figure 4): $\frac{1}{2} \square N_{Ke}$ = Knife Edge Capacity of one heel side, q_t = capacity of the transverse gap side per unit length; q_l = capacity of the longitudinal sides per unit length; $l_i = 2h_i/\sin\Theta_i + g_t$ (for g_t see Figure 3) = length of the shear yield lines in the longitudinal direction; q_t and q_l are taken as a minimum value determined from three criteria (Partanen & Björk 1993).

The results of the statistical evaluation for the design resistance r_t from the approach by Partanen & Björk (1993) are presented in Figure 5. Measured dimensions and material properties are used in the calculations and compared against the ultimate joint resistance $r_e = N_{i,u}$ from the experimental work. Although the mean variation is b = 1.47, the factor to be applied to Eq. 3 is 0.99 from the statistical evaluation, so that Eq. 3 could effectively be used without any correction factors for design.

Although 31 tests are evaluated for the modified yield line model of Partanen et al. (1993), a large scatter is obtained (Figure 5). Therefore this analytical model needs further investigations.

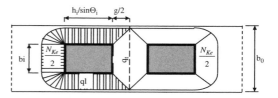

Figure 4. Modified yield line theory acc. to Partanen & Björk (1993).

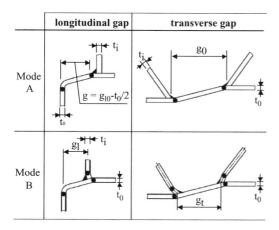

Figure 3. Yield mechanism (Partanen & Björk, 1993).

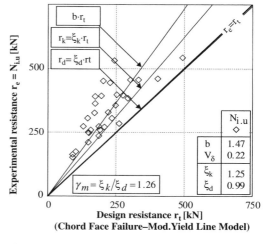

Figure 5. Statistical evaluation of chord face failure acc. to the modified yield line model of Partanen & Björk (1993).

2.2 Punching shear failure

The effective length for punching shear $l_{e.p}$ in EN1993-1-8 (2005) is as given in Eq. 4.

$$l_{e.p} = \frac{2h_i}{\sin\Theta_i} + b_i + b_{e.p} \quad (4)$$

where

$$b_{e.p} = \frac{10}{b_0/t_0} \cdot b_i \leq b_i \quad (5)$$

$l_{e.p}$, $b_{e.p}$ = effective length and effective width for punching shear failure.

For small gap sizes, when $g < g_{min}$, the disproportionately higher stiffness in the gap region requires a reduced effective length $l_{e.p.red}$ for punching shear failure. If the design formula for punching shear failure in EN1993-1-8 (2005) is used without modification to include all cases where $g < g_{min}$, the design resistance has to be reduced by using a factor of $\xi_d = 0.47$ on the basis of statistical analysis according to EN 1990 (2002).

By taking into account that the brace width b_i near the gap is effective for all gap sizes $g < g_{min}$ and the brace walls $h_{e.p.red}$ are effective partially depending upon the b_i/b_0-ratio. Figure 6a and Figure 6b show the reduced length for punching shear failure $l_{e.p.red}$.

If the exact geometry including the corner radii of the braces $r_{o,i}$ were to be considered, a complex formula would be the result. Therefore, a simplified formula is provided for the reduced effective length $l_{e.p.red}$ for punching shear failure, as given by Eq. 6.

$$l_{e.p.red} = b_i + h_{e.p.red} \quad (6)$$

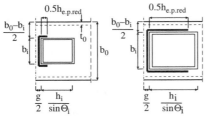

a) effective length $l_{e.p.red}$ for a small b_i/b_0-ratio

b) effective length $l_{e.p.red}$ for a high b_i/b_0-ratio

Figure 6. Reduced effective length $l_{e.p.red}$ for punching shear failure for small gap sizes in relation to the b_i/b_0-ratio.

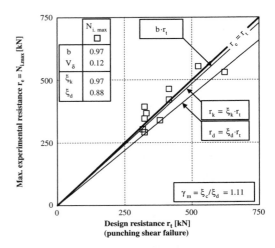

Figure 7. Statistical evaluation of punching shear failure with reduced effective length $l_{e.p.red}$

where

$$h_{e.p.red} = 2\beta \cdot \frac{h_i}{\sin\Theta_i} \quad (7)$$

$l_{e.p.red}$ = reduced effective length for small gap sizes;
$h_{e.p.red}$ = effective part of the brace wall for punching shear failure; β = brace to chord width ratio.

In Figure 7 a comparison of the maximum experimental joint resistance $r_e = N_{i,max}$ from tests that failed by punching shear is made with the design resistance r_t in Eq. 8 determined by using measured dimensions, material properties and the reduced effective length $l_{e.p.red}$ (Eq. 6).

$$r_t = \frac{f_{y0} \cdot t_0}{\sqrt{3}\sin\Theta_i} \cdot l_{e.p.red} \quad (8)$$

The statistical evaluation of tests which failed by punching shear on the basis of using the modified effective length $l_{e.p.red}$ in the design equation for punching shear in EN1993-1-8 (2005) shows that the modified equation gives a mean variation of $b = 0.97$ and a coefficient of variation of $V_\delta = 12\%$. For the determination of design resistance a reduction factor of $\xi_d = 0.88$ has to be applied to Eq. 7.

2.3 Effective width failure

As in punching shear failure, the effective length for brace failure modes l_e (Eq. 9) of EN1993-1-8 (2005) has also to be reduced because of the disproportionately higher stiffness in the gap region for small gaps. If the design formula for effective width failure in EN1993-1-8 (2005) is used without modification to include all cases where $g < g_{min}$, the design resistance has to be reduced by using a factor of $\xi_d = 0.79$ on the basis of statistical analysis according to EN 1990 (2002) (Fleischer & Puthli, 2006).

$$l_e = 2 \cdot h_i - 4 \cdot t_i + b_i + b_e \tag{9}$$

where

$$b_e = \frac{10}{\frac{b_0}{t_0}} \cdot \frac{f_{y0} \cdot t_0}{f_{yi} \cdot t_i} \cdot b_i \leq b_i \tag{10}$$

l_e, b_e = effective length and effective width for brace failure modes; f_{yi}, t_i = yield strength and wall thickness of the braces.

The brace width b_i near the gap is effective for all gap sizes $g < g_{min}$. The effective part of the brace side walls $h_{e.red}$ (Eq. 11) is only partially effective depending on the thickness ratio $\tau = t_i/t_0$, the yield strength ratio f_{y0}/f_{yi} and the b_i/b_0-ratio (Figure 8a and Figure 8b), see Wardenier (2002).

According to the simplifications already used to determine the effective punching shear height $h_{e.p.red}$, the effective parts of the brace walls $h_{e.red}$ are obtained (Eq. 11).

$$l_{e.p.red} = b_i + h_{e.red} \tag{11}$$

where

$$h_{e.red} = 2\beta \cdot \frac{f_{y0} \cdot t_0}{f_{yi} \cdot t_i} \cdot h_i \tag{12}$$

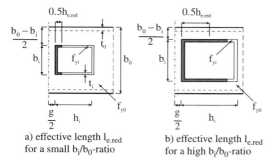

a) effective length $l_{e.red}$ for a small b_i/b_0-ratio
b) effective length $l_{e.red}$ for a high b_i/b_0-ratio

Figure 8. Reduced effective length $l_{e.red}$ for brace failure for small gap sizes in relation to the b_i/b_0-ratio.

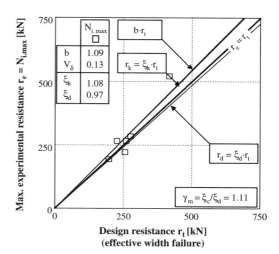

Figure 9. Statistical evaluation of brace failure modes (effective width failure) with reduced effective length $l_{e.red}$

$l_{e.red}$ = reduced effective length of brace failure modes for small gap sizes; $h_{e.red}$ = effective side wall height of the braces; β = brace to chord width ratio.

The comparison of the design resistances r_t (Eq. 13), calculated by using measured dimensions, material properties and the reduced effective length $l_{e.red}$ in

$$r_t = f_{yi} \cdot l_{e.red} \cdot t_i \tag{13}$$

with the maximum experimental joint resistances $r_e = N_{i.max}$ of the tests that failed by effective width failure is given in Figure 9.

On the basis of the 6 experimental results, the statistical evaluation of the modified model for brace failure give a mean variation of b = 1.09 and a coefficient of variation of V_δ = 13%. Design resistance for these 6 cases can be determined by using a reduction factor of $\xi_d = 0.97$ in Eq. 13.

2.4 *Chord web failure*

Since chord web failure is not considered for K-joints in EN1993-1-8 (2005), the design resistance r_t (Eq. 14) is obtained by using the design model for Y-joints of EN1993-1-8 (2005) given in Eq. 14.

$$r_t = \kappa \cdot \frac{t_0 \cdot f_{y,0}}{\sin \Theta_i} \cdot \left(\frac{2 \cdot h_i}{\sin \Theta_i} + 10 \cdot t_0 \right) \tag{14}$$

where κ = reduction factor for buckling obtained from European buckling curve "a".

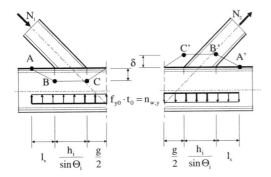

Figure 10. Modified "4-hinge" yield line and web buckling mechanism for K-joints.

The statistical evaluation according to EN 1990 (2002) for this failure mode gives an increase of the design resistance of $\xi_d = 1.13$. Since this evaluation gives a large mean variation of $b = 1.73$ and a high coefficient of variation $V_\delta = 0.23$, it was concluded that the design model of EN1993-1-8 (2005) for Y-joints cannot be used without basic modifications (Fleischer & Puthli, 2006).

The 4-hinge yield line and web buckling model for axially X-joints (Yu, 1997), which gives a more accurate solution for chord web failure of X-joints with $\beta = 1.0$ (Wardenier, 2002) is therefore taken as a basis and modified for the use of gapped K-joints (Figure 10).

The model considers the plastic hinges in the chord flange and the yielding of the chord webs determined by an elastically supported beam consisting of the chord web for a full width joint $b_i/b_0 = 1.0$.

By equating the external work $E_a = N_1 \cdot \sin\Theta_1$ to the internal energy dissipation (Eq. 15), the length l_x (Eq. 16) is obtained for the minimum failure load $(dN_1/dl_x = 0)$.

$$N_1 = \frac{2}{\sin\Theta_1}\left[b_0 m_p\left(\frac{1}{l_x} + \frac{1}{g}\right) + \left(\frac{h_1}{\sin\Theta_1} + \frac{l_x}{2} + \frac{g}{4}\right)n_{w.y}\right] \quad (15)$$

where m_p = plastic bending moment of the chord flange $m_p = f_{y0} t_0^2/4$; $n_{w.y}$ = reaction force in the chord web at yield $n_{w.y} = f_{y0} t_0^2$.

$$l_x = t_0 \cdot \sqrt{\gamma} \quad (16)$$

where: γ = chord slenderness $\gamma = b_0/(2t_0)$.

By substituting l_x and rearranging Eq. 15, the chord web bearing capacity is given in Eq. 17 as:

$$N_{1.u} = \frac{t_0^2 f_{y0}}{\sin\Theta_1}\left(2\sqrt{\gamma} + \frac{b_0}{2g} + \frac{2h_1}{t_0 \sin\Theta_1} + \frac{g}{2t_0}\right) \quad (17)$$

Since stability of the chord webs may become critical for an increasing chord web slenderness λ, the yield strength f_{y0} has to be reduced by the buckling reduction factor κ, which is based on the European buckling curve a to the critical buckling stress $n_{w.b} = \kappa \cdot f_{y0}$. Therefore, the reaction force $n_{w.y}$ is replaced by $n_{w.b} = \kappa \cdot n_{w.y}$ in Eq. 15.

The chord web bucking resistance r_t is therefore given by:

$$r_t = \frac{t_0^2 f_{y0}}{\sin\Theta_1}\left[\sqrt{\gamma}(1+\kappa) + \frac{b_0}{2g} + \kappa\left(\frac{2h_1}{t_0 \sin\Theta_1} + \frac{g}{2t_0}\right)\right] \quad (18)$$

The slenderness λ of the chord web and therefore the critical buckling stress $n_{w.b}$ is mainly influenced by the assumed supports of chord webs, which are idealized as struts. For X joints with $\beta = 1.0$, where the rotation of the chord web edges is restrained by the chord corners and the walls of the braces, Yu (1997) proposed to use half of the strut length $(L/l) = 0.5$ as buckling length. For K-joints with $\beta < 1.0$, the support behaviour is influenced by the width ratio β of the connection. For $\beta = 0$ (an impracticable but theoretical width ratio) the upper chord web support almost behaves as unrestrained and it is proposed to use $(L/l) = 1.5$. For $\beta = 1.0$ the assumption of Yu for a fixed ended strut $(L/l) = 0.5$ is used. For interim width ratios $0.35 \le \beta \le 1.0$, the factor (L/l) is obtained by linear interpolation (Eq. 19):

$$\left(\frac{L}{l}\right) = 1.5 - \beta \quad (19)$$

where L = System length of the strut; (L/l) = Factor with reference to the statical system of the strut, giving the buckling length l.

Using this proposal to determine the buckling length l, the slenderness λ is obtained by (Eq. 20):

$$\lambda = \frac{l \cdot \left(\frac{L}{l}\right)}{i} = 3.46 \frac{(h_0 - 2r_{o,0})(1.5 - \beta)}{\sqrt{}} \quad (20)$$

where i_0 = radius of inertia of the chord section; $r_{o.0}$ = outer corner radius of the chord section.

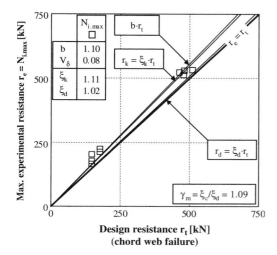

Figure 11. Statistical evaluation of the chord web failure mode using the modified 4-hinge model and the β-dependent buckling length.

In Figure 11 a comparison of the maximum experimental joint resistance $r_e = N_{i,max}$ from tests that failed by chord web buckling is made with the design resistance r_t in Eq. 18, determined by using measured dimensions, material properties and the reduction factor for buckling κ obtained from the European buckling curve "a" and a buckling length according to Eq. 20.

The statistical evaluation of the modified "4-hinge model" for chord web failure on basis of the maximum loads $N_{i,max}$ give a mean variation of b = 1.11 and a coefficient of variation of $V_δ$ = 8%. Since the obtained reduction factor is almost equal to $ξ_d$ = 1.02, Eq. 20 can directly be used for design.

3 DEFORMATION CAPACITY

The non-uniform stiffness distribution around the brace perimeters result in secondary bending moments M_{II}, which cause additional stresses in the brace and chord sections. In order to ignore these effects for the determination of joint resistance, it has to be ensured that the connection offers sufficient deformation capacity, so that the stress redistribution allows the secondary bending moments to disappear with increasing loads.

3.1 Determination of the deformation capacity

The strains in the braces have not been measured during the experimental investigations, so that the

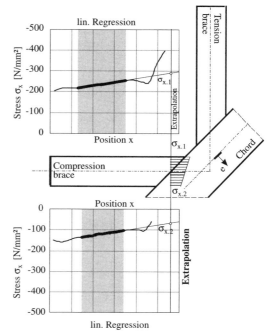

Figure 12. Determination of bending moment (typically on the compression brace for one load increment).

verification of adequate deformation capacity can only made through numerical investigations for cases where cracking can be excluded. Details of the numerical model can be found in Fleischer & Puthli (2006).

The moment loading in the braces and chord is obtained from the differences of the axial stresses $σ_{x,i}$ on the outer surfaces of the brace and chord sections. Since the geometrical stresses near the gap (welds) cannot be determined due to stress concentration, only stresses in unaffected areas such as shown shaded in Figure 12 are taken into account and extrapolated to the positions shown in Figure 12.

Assuming a linear moment distribution in the sections and therefore linear stress distributions in the braces and in the chord, the moments in the braces M_1 (compression brace), M_2 (tension brace) and the chord M_0 can be determined from the differences of the axial stresses at the outer surfaces of the sections. The stresses obtained by extrapolation from the unaffected areas (shown shaded in Figure 12) to the meeting point between brace system line and chord face is used to determine the moments.

$$M = \frac{σ_{x,1} - σ_{x,2}}{2} = \frac{Δσ_x}{2} \cdot W_{el} \qquad (21)$$

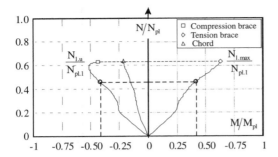

Figure 13. Typical moment rotation capacity ($\beta = 0.50$, $2\gamma = 50$, $g = 4.t_0$, $e = -0.06.h_0$

where $\sigma_{x.1}$, $\sigma_{x.2}$ = stresses obtained by linear extrapolation; W_{el} = elastic section modulus according to the formulae in EN10210-2 (2006) or EN10219-2 (2006).

For all load increments N_i in the numerical investigation, the moment M is calculated. By plotting the moment M (normalized by the plastic moment capacity M_{pl}) against the axial load N_i (normalized by the plastic load capacity N_{pl}), it is determined whether the deformation capacity is sufficient and if the secondary bending moments could be redistributed (Figure 13).

If the joint offers sufficient deformation capacity, the ratio M_i/M_{pl} will decrease with increasing loads N_i and will finally almost disappear.

3.2 Results

For the typical K-joint for which this procedure is presented, a width ratio of $\beta = (b_1 + h_1)/(2b_0) = 0.50$, a chord slenderness of $2\gamma = 50$ and a gap size of $g = 4.t_0$ is taken. It can be observed that the secondary moments cannot redistribute as desired. For the compression brace, redistribution only partially occurs at and near the maximum load N_{max}, but no redistribution occurs for the tension brace and the chord member.

The evaluation from ongoing numerical investigations (Fleischer, 2008) will investigate in detail, if secondary bending moments can be neglected or if insufficient deformation capacity occurs for certain parameter ranges.

4 CONCLUSIONS

The existing design rules given in EN1993-1-8 (2005) for RHS gapped K-joints are modified to include slender chord members with b/t up to 50 and connections with gap sizes down to $g = 4t_0$, which is smaller than the permitted value of $g_{min} = 0.5.b_0.(1 - b_1/b_0)$. Except the chord face failure model, the modifications are found to give good agreement with experimental tests having the corresponding failure modes.

For punching shear and brace failure modes the use of reduced effective lengths $l_{e.p.red}$ (punching shear) and $l_{e.red}$ (effective width) required for the gap sizes below g_{min} in the existing rules give good agreement with the corresponding experiment results.

For chord face failure, the ongoing numerical work (Fleischer, 2008) will show if the design model given here and based on yield lines (Wardenier, 1982) gives adequate overall accuracy, or whether the more accurate and complicated analytical design model by Partanen & Björk (1993), which allows for shear stress and stress biaxiality, has to be used.

For the determination of the design resistance for chord web failure of RHS K-joints with gap, the 4-hinge yield line and web buckling model of Yu (1997) for full width $b_1/b_0 = 1.0$ X- and Y-joints is revised accordingly. If the buckling length of the chord webs are determined as a function of the width ratio, the agreement to the experimental results is good.

5 FUTURE WORK

The results presented in this paper will be used in conjunction with the FE parameter studies that have already been performed to re-evaluate the data for chord face failure.

On the basis of the re-evaluation, modifications to the design rules given in EN1993-1-8 (2005) will be presented, which will cover an extended range for the chord slenderness b_0/t_0 and gap (function $g_0 = g/t_0$). This will include modifications to existing formulae for chord face yield, punching shear, chord web failure and effective width failure. These modifications, supported by the presented statistical evaluations, are intended to extend the existing rules.

Finally, the secondary bending moments may have to be considered for certain parameter ranges. This may result in complex formulae, which are more appropriate for computational programming.

ACKNOWLEDGEMENTS

This paper is dedicated to the memory of Mr. Reijo Ilvonen, who sadly died unexpectedly, shortly after the last ISTS Symposium in Quebec City, Canada, in 2006. He was a member of the ECSC research project (Salmi et al., 2003), which forms the basis of the work presented here.

The authors would also like to thank the ECSC (European Community for Steel & Coal) for financial support and Rautaruukki Metform, Finland and Voest Alpine, Austria for the supply of material.

REFERENCES

EN1990. 2002. Basis of structural design. Brussels: European Committee for Standardisation.

EN1993-1-8. 2005. Design of steel structures – Design of joints. Brussels: European Committee for Standardisation

EN 10210-1. 2006. Hot finished structural hollow sections of non-alloy and fine grain structural steels – Part 1: Technical delivery requirements. Brussels: European Committee for Standardisation.

EN 10210-2. 2006. Hot finished structural hollow sections of non-alloy and fine grain structural steels – Part 2: Tolerances, dimensions and sectional properties. Brussels: European Committee for Standardisation.

EN 10219-1. 2006. Cold formed welded structural hollow sections of non-alloy and fine grain steels – Part 1: Technical delivery requirements. Brussels: European Committee for Standardisation.

EN 10219-2. 2006. Cold formed welded structural hollow sections of non-alloy and fine grain steels – Part 2: Tolerances, dimensions and sectional properties. Brussels: European Committee for Standardisation.

Fleischer, O. & Puthli, R. 2003. RHS K-joints with b/t ratios and gaps not coverd by Eurocode 3. In M.A. Jaurrieta & A. Alonso & J. Chica J.A. (eds), *Proc. intern. Symp on Tubular structures., Madrid, 18–20 September 2003.* Rotterdam: Balkema.

Fleischer, O. & Puthli, R. 2006. Evaluation of experimental results on slender RHS K-gap joints. In J.A. Packer & S. Willibald (eds), *Proc. intern. Symp. and IIW intern. Conference on Tubular Structures, Québec City, 31. August to 2. September 2006.* London: Taylor & Francis Group.

Fleischer, O. 2008. Symmetrische K-Knoten aus RHS mit Spaltweiten und Gurtschlankheiten außerhalb der aktuellen Normung unter statischer Beanspruchung. Dissertation (under preparation). Karlsruhe.

Lu, L.H., de Winkel, G.D., Yu, Y. & Wardenier, J. 1993. Deformation limit for the ultimate strength of hollow section joints. In P. Grundy & A. Holgate & B. Wong (eds), *Proc. intern. Symp. on Tubular Structures, Melbourne, 14–16 December 1994.* Rotterdam: Balkema.

Partanen, T. 1991. On Convergence of Yield Line Theory and Nonlinear FEM Results in Plate Structures. In J. Wardenier & E. Panjeh Shahi (eds), *Proc. intern. Symp. on Tubular Structures, Delft 26 to 28 June 1991.* Delft: Delft University Press.

Partanen, T. & Björk, T. 1993. On Convergence of Yield Line Theory and Experimental Test Capacity of RHS K- and T-joints. In M.G. Coutie & G. Davies (eds), *Proc. intern. Symp. on Tubular Structures, Nottingham 25–27 August 1993.* London: E & FN Spon.

Salmi, P., Kouhi, J., Puthli, R., Herion, S., Fleischer, O., Espiga, F., Croce, P., Bayo, E., Goñi, R., Björk, T., Ilvonen, R. & Suppan, W. 2003. Design rules for cold formed structural hollow sections. *Final Report 7210-PR/253.* Research programme of the Research Fund for Coal and Steel, Brussels: European Commission.

Wardenier, J. 1982. Hollow Section Joints. *Ph.D. Thesis.* Delft: University Press.

Wardenier, J. 2002. Hollow Sections in Structural Applications. Bouwen met Staal, Zoetermeer, The Netherlands.

Yu, Y. 1997. The Static Strength of Uniplanar and Multiplanar Connections in Rectangular Hollow Sections. *Ph.D. Thesis.* Delft: Delft University Press.

Plenary session C: Design rules & evaluation

Recent development and applications of tubular structures in China

Z.Y. Shen, W. Wang & Y.Y. Chen
Tongji University, Shanghai, China

ABSTRACT: To meet the requirement of economic booming, a large number of mega-structures for commercial, industrial and infrastructure uses have been built in China in recent years. Tubular structures have thus been developed rapidly and used widely in such mega-structures as large span spatial structures (such as stadiums, airport terminals, station, theatre, and exhibition centre), high-rise buildings, bridges, and offshore platforms. Reliable design of these tubular structures requires a sound understanding of their behaviors under various working conditions. Considerable research has been carried out on tubular structures especially on tubular joints in China, and the achievements are valuable and encouraging. In this paper, tubular structures are classified into three large groups: truss-type structures, frame-type structures and lattice shell structures. A number of typical applications of tubular structures in China are described firstly. The latest developments on behavior study of tubular joints in China are reviewed, together with background of practical applications. Emphases are given to CHS joints, which are the most commonly used types. Research on static, seismic, and fatigue behavior of tubular joints has been addressed, as well as the performance with different details. Based on results obtained from both research and practical applications, Chinese technical specification for design of tubular structures have been built and introduced.

1 INTRODUCTION

To meet the requirement of economic booming, a large number of buildings for commercial, industrial and infrastructure uses have been constructed in China in recent years. Steel tubular structures are well recognized for their pleasing appearance, light weight, easy fabrication and rapid erection. They have thus been developed rapidly and used widely in such mega-structures as large span spatial structures (such as stadiums, airport terminals, station, theatre, and exhibition centre), high-rise buildings, and offshore platforms. According to official statistics, the consumption of steel tubes for structural use in building construction of China is increasing at a rate of 20% per year, covering a total weight of one million tons. By the end of 2007, steel tubes account for 30% of the total consumption of steel used in the newly built stadiums and gymnasiums matching with Beijing Olympic Games. Currently there are more than 120 fabricators of tubular structures in China, representing a new manufacturing industry.

In this paper a number of typical engineering applications of tubular structures in China are described, the latest developments on behavior study of tubular joints in China are reviewed, together with background of practical applications, and the technical specification for design of tubular structures recently completed by the China Association for Engineering Construction Standardization are overview-ed in the last section.

2 TYPES AND ENGINEERING APPLICATIONS OF TUBULAR STRUCTURES IN CHINA

In this paper, tubular structures are classified into three large groups: truss-type structures, frame-type structures and lattice shell structures.

2.1 *Truss-type tubular structures*

Greater load-bearing capacity and the increase in span can be achieved with the use of planar or spatial truss-type structures. For planar tubular trusses, a Warren type truss with K-joints, a Pratt type truss with N-joints and a Vierendeel type truss with T-joints are three main types of trusses which are mostly used in practice.

The large-span shuttle-shaped trussed system applied to the restaurant and the press centre of Shanghai F1 International Circuit is an example of the

Pratt type tubular trusses (Fig. 1). The chord members adopted welded box sections and the web members adopted circular hollow sections (CHS) or welded rectangular hollow sections (RHS). One side of this megatruss is supported on two columns, and the other side is supported on the transfer truss which transfers the load to the bottom columns. The span between the supports is 91.3 m, and the cantilever spans from each support are 26.91 m and 17.41 m, respectively. The maximum height of the truss is 12.4 m.

A Vierendeel truss is a type of truss without diagonals, in which shear forces are resisted by the vertical web members and chords, acting as an moment-resisting frame. It may have diagonals in some bays in some designs, but may also be designed to rely totally on the verticals. The CHS roof truss of Chongqing Jiangbei International Airport Terminal (Fig. 2) is an example of this type of tubular structures.

The same construction can be done with spatial system.

No. 3 Terminal of Beijing Capital International Airport (Fig. 3) is a project matching with the Beijing 2008 Olympic Games. The roof structure of the project belongs to a slightly bending hyperboloidal vacuum triangle pyramid type space truss. The length from south to north is 958 m, the width from east to west is 775 m, and the elevation of the highest top reaches 42 m. The space truss owns over 12,600 tubular joints and over 50,000 components with different sizes, including about 4,300 bolted spherical joints and over 8,300 welded hollow spherical joints. Solid steel pin nodes were adopted in the supports.

CHS space trusses are commonly used as the main structural system in super high steel TV towers in China, such as the 336 m high Heilongjiang Dragon Tower (Fig. 4a). It is the highest steel tower in Asia and the 2nd highest in the world. Besides, in the recent years, space tubular trusses are being widely used in high electricity transmission towers in China. By the end of 2007, approximately one hundred high electricity transmission towers have been built in China, 50 percent of which are space tubular trusses. Fig. 4b gives a typical example of high electricity transmission tower with 215 m high. The maximum tube diameter at the tower foot reaches 1,580 mm.

The main reasons of adopting CHS members in high steel towers are: (1) wind load is the dominated load for high towers and the wind pressure coefficient for CHS is lower than any other types of cross

Figure 3. No.3B Terminal of Beijing Capital International Airport.

Figure 1. Shanghai F1 International Circuit.

Figure 2. Chongqing Jiangbei International Airport Terminal.

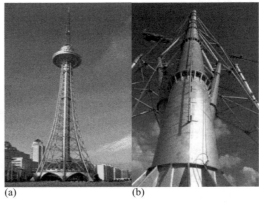

Figure 4. Space tubular truss towers, (a) Heilongjiang Dragon Tower and (b) Yamen electricity transmission tower.

sections; and (2) axial compression force is the dominated force in the main structural members. CHS is central symmetric and has the same radius of gyration about any axis of the cross section.

2.2 Frame-type tubular structures

Frame-type tubular structures can be divided into two subclasses, i.e. high-rise moment resisting tubular frames and space frame tubular structures.

High-rise moment resisting tubular frames are defined as rigid frames adopting continuous tubular columns or concrete filled tubular (CFT) columns as main bearing elements. Guangzhou New TV Tower is a high-rise tubular frames under construction in China (Fig. 5). It consists of a main tower of 454 m height and an antenna mast of 156 m, totaling 610 m in height. After completion of the project, it will be the highest tower around the world. The tower adopts 24 oblique straight cone-shaped steel pipe columns filled with concrete, 46 ring tubular beams and 24 oblique tubular supporting. The sections of cone-shaped steel pipe are $\Phi 2,000 \times 50 \sim \Phi 1,200 \times 30$, the steel material includes Q345GJC and Q460GJC. The consumption of the steel for the project totals about 50,000 tons.

West Tower of Guangzhou Twin Towers (Fig. 6) is also a representative moment resisting tubular frame. The project covers an area of 449,000 m². There are 103 floors on the ground for the main tower with 432 m height. The tube-in-tube structural system was adopted. The outer tube is composed of oblique grid CHS columns and the inner tube is a concrete core tube. The plane size is a centrosymmetric equilateral triangle with round-corners. CHS tube ranges from $\Phi 1,800 \times 55 \sim \Phi 700 \times 20$.

Compared with high-rise moment-resisting frames, space frames are another class of frame-type tubular structures. They usually have irregular configuration and demonstrate a frame behavior with significant spatial load transferring mechanism. National Stadium is the main stadium of Beijing 2008 Olympic Games (Fig. 7), adopting frame-type structure formed by 48 huge tubular trusses. The stadium has a plane projection size of 323.3 m × 296.4 m, maximum elevation of 68.5 m and minimum elevation of 42.8 m. The structure is composed of bending-torsion members and straight members with welded box section. The maximum size of the box sections of frames is 1,200 mm, and the maximum thickness is 110 mm. Most materials are Q345D and Q345GJD-Z25, and the material with highest strength is Q460E-Z35. The column base uses GS 20Mn5V casting steel joints.

Polyhedron space frame structure with EFTE cladding is adopted in the National Swimming Center

Figure 6. West Tower of Guangzhou Twin Towers.

Figure 5. Guangzhou New TV Tower.

Figure 7. National Stadium (nicknamed Bird's Nest).

Figure 8. National Swimming Centre (nicknamed Water Cube).

Figure 9. National Grand Theater (nicknamed Big Egg).

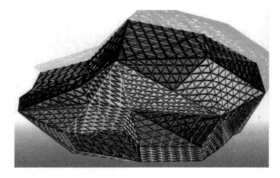

Figure 10. Guangzhou Opera.

"Water Cube" (Fig. 8). In which, rectangular hollow sections are used as chords, circular hollow sections are used as web members and welded hollow balls are used as joints. Although the form of the new structure is like space truss, the member behavior is entirely different from the truss members which mainly bearing axial force. The joints of the polyhedron space frame are rigidly connected and the members are space frame elements. Element forces contain bending moment, axial force, shear force and torsional moment. The moment stress is larger than the axial stress.

2.3 Lattice shell tubular structures

Like space trusses, lattice shell structures are three-dimensional skeletal structures. They are now the most widely used tubular structures for long-span buildings in China.

The main building of the National Grand Theater consists of an exterior shell, within which are a 2,416-seat opera house, a 2,017-seat concert hall, a 1,040-seat theater, a lobby and other supporting facilities (Fig. 9). The exterior is a shell in the shape of a half ellipsoid with a long east-west axis of 212.20 meters and a short south-north axis of 143.64 meters. It is 46.285 meters tall, and it reaches to a maximum underground depth of 32.50 meters. Most of the shell is covered by titanium panels, and it is broken in the middle by a curtain of glass that opens gradually from top to bottom.

Guangzhou Opera will be a landmark architecture in Guangzhou City of China (Fig. 10). It adopts "Double-Gravel"-based irregular geometric configuration, covering an area of 42,000 m^2. The periphery of the building adopts three-dimensional grids folded plate and single layer reticulated shell structures. All members of this structure are thin-walled box girder. Rigid joints made of cast steel are used where planes intersect. The highest weight of the single joint is 37 tons and the longest one is 12 m. The consumption of steel totals 10,000 tons, including 1,200 tons of steel casting.

3 RECENT ADVANCES IN THE RESEARCH OF TUBULAR JOINTS IN CHINA

One important difference of tubular structures from other steel structures lies in the design of tube-to-tube connections. Connections between hollow section members are usually configured by welding one member directly to the surface of the other, where possible, without any exposed stiffening or reinforcing element. Conventional design procedures applicable to welded plate-to-plate connections can not be applied to the design of these connections. With the vigorous application of tubular structures in practical engineering in China, fruitful research of both theoretical and practical significance has been carried out on tubular joints. Some of these are described below.

3.1 Non-rigid behavior of tubular joints and interaction with global performance of tubular structures

In engineering practice, tubular joints are usually assumed pinned or rigid. Recent research showed that tubular joints may exhibit non-rigid behavior under axial or bending loads (Fig. 11). It may have significant influence on the behavior of Vierendeel truss and single-layer lattice shell structures.

Recent investigations on non-rigid behavior of tubular joints include experimental investigations on CHS X- and KK-joints under in-plane bending or out-of-plane bending loads on the brace (Chen & Wang 2003, Wang et al. 2007). Test results showed that the joints primarily failed by excessive deformation and punching shear cracking at the weld toe. It is found that the stiffness ratio of the connection to the brace member is a critical factor to classify this joint to be rigid or semi-rigid. Parametric formulae for predicting tubular joint rigidities are proposed, which are based on numerical investigations through systematic variation of the main geometric parameters (Wang & Chen 2005). By considering the deformation patterns of respective parts of Vierendeel lattice girders, the boundary between rigid and semirigid tubular connection is built in terms of joint bending rigidity. In order to include characteristics of joint rigidity in the global structural analysis, a type of semirigid element which can effectively reflect the interaction of two braces in K joints is introduced and validated. The numerical example of a Warren lattice girder with different joint models shows the great effect of tubular joint rigidities on the internal forces, deformation and secondary stresses (Wang & Chen 2005). A further study is in process on a new classification system of tubular joints based on the global performance of CHS lattice shell structures.

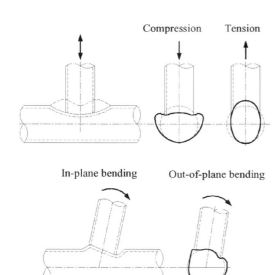

Figure 11. Non-rigid deformation under axial or bending loads.

3.2 Seismic behavior of tubular joints

Many large-span tubular truss applications now exist in regions of high seismic risk in China. Unlike common beam-column connections in moment resisting frames, unstiffened tubular joints sometimes have lower capacity than adjacent tubular members. In order that the inelastic seismic performance of truss structures can be evaluated properly, it is necessary to study the hysteretic behavior of tubular joints.

A series of cyclic loading tests of CHS T-, K-, and KK-joints were performed to understand the seismic properties of these joints (Wang & Chen 2007, Chen & Zhao 2007) (Fig. 12). Some of the new findings were as follows: (1) Monotonic load-deformation curves can cover the skeleton of the hysteretic curves for the joints having the same geometry, but the ductility will deteriorate under cyclic loading. (2) Both member failure mechanism in the form of brace yielding and joint failure mechanism in the form of chord wall plasticity demonstrate good energy dissipation capacity, if weld cracking can be prevented or delayed. For proper design of weld, stress distribution around intersection welds should be further studied, and seismic design criteria of tubular joints should be established pertain to structural deformability.

3.3 Behavior of inner-stiffened tubular joints

Along with the development of large-span spatial structures, inner-stiffened tubular joints are more and more applied to engineering projects of China. It helps to improve both strength and rigidity of the joint, as well as to keep its original appearance.

A number of experimental and analytical investigations into inner-stiffened tubular joints are found in recent years, including CHS joints with multiple braces (Tong et al. 2007), RHS T-joints (Yan et al. 2007), and CHS brace to RHS chord joints (Cao & Yao 2008).

Figure 12. Cyclic loading test of a multiplanar CHS KK-joint.

The present study proposed some basic design guidelines for the ring stiffeners in the complicated CHS joints with multiple brace members. A set of formulae used to predicting ultimate strength of inner-stiffened CHS-RHS T- and K-joints were also recommended.

3.4 Behavior of tubular joints with curved chords

In tubular structures of China, welded trusses with curved circular hollow sections have been increasingly used in order to meet the needs of more attractive architectural appearance. It confronts structural engineers with the question that how to check the strength of tubular joints with curved chord members.

Nine static tests on CHS joints with curved and straight chord members were carried out (Tong et al. 2006) (Fig. 13). Behavior in failure modes, load-deformation curves and stress distributions is compared between joints with curved and straight chord members. Test results show that the curved circular chord members do not exert more significant influence upon the behavior of joints. Both the experimental and numerical results indicated that the strength of the joints with curved chords at a wide range of curvature radiuses used in practical engineering can be checked like the joints with straight chords based on the current design specifications.

3.5 Behavior of overlapped tubular K-joints

For partially overlapped CHS K-joints (Fig. 14), it is a common practice to mount the CHS members on an assembly rig and tack welds them in fabrication workshops. Final welding is then carried out in a following separated operation. This sequence makes it impossible to weld the hidden toe of overlapped K-joints. In tubular structures, this type of joints is widely used. It is thus necessary to investigate what parameters and how they affect the behavior of overlapped CHS K-joints with unwelded hidden seam.

In total twelve specimens of uniplanar overlapped CHS K-joints were tested under axial loading (Zhao et al. 2006). The study concentrated on the effect of varying the joint geometry, the loading hierarchy as well as the presence or absence of the hidden weld on the behavior of the joints. Results show that the welding situation of the hidden seam has some effect on the stress distribution and failure mechanism, but the static ultimate capacity of the overlapped CHS K-joints is not affected significantly given that the through brace is under compression. When the brace-to-chord thickness ratio is smaller, the local failure is the main failure mechanism observed.

3.6 Fatigue behavior of tubular joints

Welded trusses composed of the brace members with CHS and the chord members with concrete filled CHS (CFCHS) have come into increasing use in large span highway arch bridges in China. It is worth while to pay attention to the fatigue problem of this new kind of welded joints under repeated loads.

Stress concentration factors (SCF) of CHS-to-CFCHS T- and K-joints subjected to axial loading and in-plane bending were experimentally studied by Tong et al. (2007, 2008) (Fig. 15). Fatigue assessment of the welded CFCHS joints can be performed based on one of three approaches, namely classification method, hot spot stress method and fracture mechanics method. The last one is the advanced approach in which the parameter of dominating fatigue crack propagation, stress intensity factor (SIF), needs to be determined first. Gu et al. (2008) proposed a simplified fracture mechanics model, and successfully predicted stress intensity factors (SIF) of this joint with semi-elliptical

Figure 13. A CHS joint with curved chord under loading.

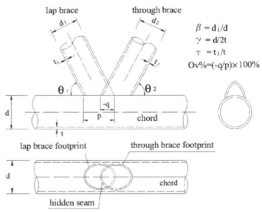

Figure 14. Configuration of overlapped CHS K-joint.

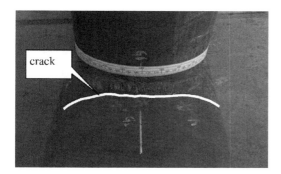

Figure 15. Fatigue crack propagation of CFCHS T-joints.

Figure 16. Test set-up of a tubular flange joint.

surface cracks under axial brace loading through the FE methodology.

3.7 Behavior of tubular flange joints

Steel poles are more and more used in transmission line systems especially in urban areas of China in recent years. Bolted flange joints subjected to axial forces and bending moments are one most common type of connection of steel poles.

Experimental study were performed on 3 flange joints with stiffened ribs of steel transmission poles for Baihuadong transmission lines being constructed in Guangdong province in China (Deng, Huang & Jin 2008) (Fig. 16). The tubes are made of Q420B steel. The biggest diameter of the tube is 2,100 mm. It is found that the flange plates have enough strength and prying force does not exist in the flange plates connected to the tensile zone of the tube. A new rotational axis for calculating bolt force is recommended for large-scale flange joints, and maximal bolt force will be increased by 17.3% compared with the conventional calculating method. This can be used as a reliable reference in engineering practice.

4 CODE DEVELOPMENT OF TUBULAR STRUCTURES IN CHINA

The aim of all realistic research projects should be to provide new or improved design criteria. Indeed, many researchers consider the acceptance of their proposals into the codes to be the highest form of peer recognition and achievement.

The first edition of design provisions for steel tubular structures in China was published as a section of Code for Design of Steel Structures (MCC 1988) in 1988. The second edition (MCC 2003) was revised and published in 2003. Extensive research on steel tubular structures has been carried out all over the world in the last 5 years. There is a need to build a special design specification for tubular structures in China. This section summarizes the Chinese Technical Specification for Structures with Steel Hollow Sections (CECS 2008), recently completed by the China Association for Engineering Construction Standardization (CECS).

The new Chinese specification contains 9 sections and one annex. The first two sections cover the scope, symbols, terms and definitions. Section 3 gives the requirements for steel tube materials, connection materials, and infilled concrete materials. Section 4 prescribes the basic design principle, material strength value, service limit state design indices and some detail requirements for steel tubular structures. The design procedures and the resistance formulae of tubular members and joints are outlined in the following Section 5, 6 and 7. Section 8 gives the information about the calculating method of predicting the fatigue strength of tubular joints. In the last section 9, some detailed recommendations of fabrication, assembly and erection aspects of tubular structures are presented.

5 CONCLUSIONS

The aim of this paper was to provide information regarding the state of the art of research, design and applications of tubular structures in China. There are some issues that are awaiting intensive future study, including a new classification system of tubular joints based on the global structural performance, seismic design criteria of tubular joints and welds, and resistance formulae of overlapped tubular joints.

It should be pointed out that because many intensive studies have been conducted, studies dealing with beam-to-column connections in concrete filled tubular frame are excluded from this review for the sake of brevity.

ACKNOWLEDGEMENTS

The reported work is supported by Natural Science Foundation of China (50578117).

REFERENCES

Cao, G.F. & Yao, N.L. 2008. Load-Carrying Capacity Study on T- and K-Shaped Inner-Stiffened CHS-RHS Joints. *Proc. 12th Intern. Symp. on Tubular Structures*, Shanghai, China.

China Association for Engineering Construction Standardization 2008. *Technical Specification for Structures with Steel Hollow Sections*. CECSXX: 2008, to be published, China Association for Engineering Construction Standardization, Beijing, China.

Chen, Y.Y. & Wang, W. 2003. Flexural behavior and resistance of uni-planar KK and X tubular joints. *Steel & Composite Structures*, Vol.3, No.2: 123–140.

Chen, Y.Y. & Zhao, X.Z. 2007. Experimental study on hysteretic behavior of CHS overlap K-joints and Gap KK-joints. *Proceedings of the 5th International Conference on Advances in Steel Structures*, Singapore, Vol.2: 207–217.

Deng, H.Z., Huang, Y. & Jin, X.H. 2008. Experimental Research on large-scale flange joints of steel transmission pole. *Proc. 12th Intern. Symp. on Tubular Structures*, Shanghai, China.

Fan, Z., Fan, X.W., Peng, Y., Wang, Z. & Sun, H.L. 2008. Design and research of welded thin-wall box components for the national stadium. *Proc. 12th Intern. Symp. on Tubular Structures*, Shanghai, China.

Gu, M., Tong, L.W., Zhao, X.L. & Lin, X.G. 2008. Stress intensity factors of surface cracks in welded T-joints between CHS brace and concrete-filled CHS chord. *Proc. 12th Intern. Symp. on Tubular Structures*, Shanghai, China.

Ministry of Construction of China. 1988. Code for design of steel structures. GB17-88, Ministry of Construction of China, Beijing, China.

Ministry of Construction of China 2003. *Code for design of steel structures*. GB50017-2003, Ministry of Construction of China, Beijing, China.

Tong, L.W., Wang, B., Chen, Y.Y. & Chen, Z. 2006. Study on effect of curved chords on behavior of welded circular hollow section joints. *Proc. 11th Intern. Symp. on Tubular Structures*, Quebec, Canada: 153–160.

Tong, L.W., Sun, C.Q., Chen, Y.Y., Zhao, X.L., Shen, B. & Liu, C.B. 2008. Experimental comparison in hot spot stress between CFCHS and CHS K-joints with gap. *Proc. 12th Intern. Symp. on Tubular Structures*, Shanghai, China.

Wang, W. & Chen, Y.Y. 2007. Hysteretic behavior of tubular joints under cyclic loading. *Journal of Constructional Steel Research*, Vol.63, No.10: 1384–1395.

Tong, L.W., Sun, J.D., Zhou, L.Y., Chen Y.Y. & Gu M. 2007. Behavior of circular hollow section joints with multiple brace members and internal stiffeners. *Proceedings of the 5th International Conference on Advances in Steel Structures*, Singapore, Vol.3: 1009–1015.

Tong, L.W., Wang, K., Shi W.Z. & Chen Y.Y. 2007. Experimental investigation on stress concentration factors of CHS-to-CFCHS T-joints subjected to in-plane bending. *Proceedings of the 5th International Conference on Advances in Steel Structures*, Singapore, Vol.3: 1003–1007.

Tong, L.W., Wang, K., Shi W.Z. & Chen Y.Y. 2007. Experimental study on stress concentration factors of concrete—filled Circular hollow Section T-joints under axial loading. *Proceedings of Eighth Pacific Structural Steel Conference*, Wairakei, New Zealand: 153–158.

Wang, W. & Chen, Y.Y. 2005. Modeling & classification of tubular joint rigidity and its effect on the global response of CHS lattice girders. *Structural Engineering and Mechanics*, Vol.21, No.6: 677–698.

Wang, W., Chen, Y.Y. & Du, C.L. 2007. Study on the static behavior of CHS X-joints under out-of-plane bending. *Proceedings of Eighth Pacific Structural Steel Conference*, Wairakei, New Zealand, Vol.2: 53–58.

Yan, S., Wang, W. & Chen, Y.Y. 2007. Study on the static behavior of vierendeel truss joints reinforced with internal stiffeners under in-plane bending. *Proceedings of the Second International Conference on Advances in Experimental Structural Engineering*, Shanghai, China, Vol.1: 117–122.

Zhao, X.Z., Chen, Y.Y., Chen, Y., Wang, G.N. & Xu, L.X. 2006. Experimental study on overlapped CHS K-joints with hidden seam unwelded. *Proc. 11th Intern. Symp. on Tubular Structures*, Quebec, Canada: 125–132.

Evaluation of new CHS strength formulae to design strengths

G.J. van der Vegte
Delft University of Technology, Delft, The Netherlands

J. Wardenier
Delft University of Technology, Delft, The Netherlands
National University of Singapore, Singapore

X.-L. Zhao
Monash University, Melbourne, Australia

J.A. Packer
University of Toronto, Toronto, Canada

ABSTRACT: In recent years, extensive numerical analyses were conducted on axially loaded uniplanar K-gap, T- and X-joints made of Circular Hollow Sections (CHS), considering a wide range of geometric parameters and different types of chord load. Based on the large amount of numerical data, new joint capacity and chord stress equations were derived for CHS joints. This article presents the statistical evaluation procedure to convert the mean strength equations into design equations. Further, as examples, detailed comparisons are shown between the newly proposed (mean) strength equations and available experimental data for T- and X-joints. Similar comparisons are made for K-gap joints but these will be published in the near future. The new design equations are included in the 3rd edition of the IIW Static Design Procedure for Welded Hollow Section Joints (IIW 2008).

1 INTRODUCTION

In recent years, extensive finite element (FE) analyses were conducted on axially loaded uniplanar K-, T- and X-joints between CHS for a wide range of geometric parameters (van der Vegte et al. 2006, 2007a, b). In addition, chord load was varied between axial load, in-plane bending moments and combinations of axial load and in-plane bending moments. Although the initial purpose was to evaluate the influence of chord load on the joint strength for K-joints (as part of CIDECT programme 5BK), it was decided, after having generated a numerical database of this size, to check whether a new set of ultimate strength equations in combination with new chord stress functions would be desirable, not only for K-joints but also for T- and X-joints. Emphasis was put on the following criteria: the new strength equations should (i) be accurate, (ii) show relationships between the various joint types, (iii) be easy to understand for designers and (iv), where possible, be based on an analytical model.

This article first presents the mean strength equations derived for axially loaded T-, X- and K-gap joints, followed by the statistical evaluation procedure to convert these equations into design equations. Further, detailed comparisons between the strength equations and available experimental data are shown for T- and X-joints; for K-gap joints these will be published in another article.

As mentioned in a companion paper describing the reanalysis of the capacity equations for joints under (brace) moment loading (van der Vegte et al. 2008), experimental data inevitably include a certain amount of scatter, even if duplicate tests are carried out under seemingly identical conditions. Because FE results are supposed to be free of such scatter, numerical data provide a reliable tool for identifying outliers in experimental programmes. The current study not only identifies and highlights trends within certain series of experiments but also makes comparisons between series of experiments obtained from different researchers.

The new design equations are included in the 3rd edition of the IIW Static Design Procedure for Welded Hollow Section Joints, recently completed by IIW sub-commission XV-E (IIW 2008). A companion paper by Zhao et al. (2008) summarizes the revised edition of the IIW recommendations and briefly describes the major changes made to the previous edition (IIW 1989).

2 NEW BASIC JOINT STRENGTH FUNCTIONS

As mentioned, the new joint capacity equations derived for axially loaded T-, X- and K-gap joints are primarily based on numerical data. For the numerical analyses, the FE package Abaqus/Standard was used. The numerical model generated for T- and X-joints included 8 noded, quadratic shell elements. For gap K-joints, which require a more accurate representation of the gap area and weld geometry, 20 noded, quadratic solid elements with two layers of elements modelled through the wall thickness, were employed. Welds were modelled according to AWS with the minimum requirements for butt welds. Both models (i.e. with shell or solid elements) were found to give accurate predictions of the load-deformation behaviour observed in experimental programmes.

The proposed mean strength functions are presented in a format similar to that in the API (2007), i.e. for axially loaded joints:

$$\frac{N_1 \sin\theta_1}{f_{y0} t_0^2} = Q_u Q_f \qquad (1)$$

where Q_u represents the basic joint strength or reference strength function, while Q_f accounts for the influence of the chord stress. Table 1 gives the expressions for Q_u, while Table 2 shows the equations for Q_f. Tables 1–2 further list the coefficient of variation (CoV) and the number of data used for the derivation of the expressions.

3 EVALUATION TO DESIGN STRENGTHS

Various approaches exist to derive design equations based on experimental and numerical data. Two approaches are briefly described here and illustrated for uniplanar X-joints, while a third procedure is only mentioned.

3.1 Evaluation based on scatter of the data

A simple approach, used in some codes, solely uses the scatter of the data V_δ without taking account of tolerances and other variables:

$$N_{u,k} = N_{u,m}(1 - 1.64 V_\delta) \qquad (2)$$

where $N_{u,k}$ = characteristic strength; and $N_{u,m}$ = mean strength.

$$N_1^* = \frac{N_{u,k}}{\gamma_M} = \frac{N_{u,k}}{1.1} = 0.81 N_{u,m} \qquad (3)$$

where N_1^* = design strength; and γ_M = partial safety factor (γ_M = 1.1 for CHS joints with sufficient rotation/deformation capacity).

In this analysis, the highest value for CoV and the lowest value for m (i.e. the numbers given in Tables 1–2) are used. For example, for X joints, the lowest value for m = 1.01 and the highest value for CoV = 6.8%. Hence:

$$N_{u,k} = N_{u,m}(1 - 1.64 \times 0.068) = 0.89 N_{u,m} \qquad (4)$$

Table 2. Chord stress functions Q_f (for chord compression) with Mean (m) and CoV derived from the FE database.

Joint	Q_f	Mean	CoV	No. of data
X	$(1-|n|)^{(0.25-0.25\beta+0.01\gamma)}$	1.03	6.8%	156
T	$(1-|n|)^{(0.20-0.25\beta+0.01\gamma)}$	1.00	6.0%	141
K-gap	$(1-|n|)^{0.25}$	1.00	4.2%	122

* For chord tension: $Q_f = (1-n)^{0.2}$

Table 1. Reference strength functions Q_u of the proposed mean strength equations with Mean (m) and CoV derived from the FE database.

Joint	Q_u	Mean	CoV	No. of data
X	$3.16\left(\dfrac{1+\beta}{1-0.7\beta}\right)\gamma^{0.15}$	1.01	5.8%	23
T	$(3.1+21\beta^2)\gamma^{0.2}$	1.01	4.5%	33
K-gap	$(2+16\beta^{1.6})\gamma^{0.3}\left[1+\dfrac{1}{1.2+(g/t_0)^{0.8}}\right]$	1.00	4.2%	122

$$N_1^* = \frac{N_{u,k}}{\gamma_M} = \frac{N_{u,k}}{1.1} = 0.81 N_{u,m} \quad (5)$$

with m = 1.01, this results in:

$$N_1^* = 2.6 \left(\frac{1+\beta}{1-0.7\beta}\right) \gamma^{0.15} \frac{f_{y0} t_0^2}{\sin\theta_1} Q_f \quad (6)$$

3.2 Evaluation based on the same procedure as used for current IIW design strength functions (1989)

The procedure adopted for the current IIW design strength functions (1989) is described by Wardenier (1982). The characteristic strength $N_{u,k}$ of the tests is determined by taking account of the tolerances, mean values and scatter in data, followed by a correction of $f_{y,mean}/f_{y,k}$ assumed to be 1/0.85.

Although for the evaluation a log-normal distribution is preferred, for low values of the CoV as here, the difference with a normal distribution is negligible. The procedure is then as follows:

If the variables are not correlated, the variance VAR(N_u) is given by:

$$VAR(N_u) = \left(\frac{\partial f}{\partial x_1} s_{x_1}\right)^2 + \cdots + \left(\frac{\partial f}{\partial x_n} s_{x_n}\right)^2 \quad (7)$$

$$V_{N_u} = \frac{[VAR(N_u)]^{0.5}}{E(N_u)} \quad (8)$$

For the standard deviations, the values of the variables marked by "important" in Table 3 are used, which are the same as for the current recommendations. It should be noted that for the mean value $t_0/t_{0,nominal} = 1.0$ is assumed, which may depend on the delivery standard.

Similar to the evaluation in the previous section, the highest value for CoV and the lowest value for m, obtained from the analyses for Q_u and Q_f, are used. For X joints, the procedure is as follows:

Table 3. Standard deviations used.

Variable	Standard deviation	Relevance
V(f_{y0})	0.075	important
V(t_0)	0.05	important
V(d_0)	0.005	negligible
V(d_1)	0.005	negligible
V(θ_1)	1°	negligible
V(δ) data	See Tables 1–2	important

$$N_1^* = 3.16 \left(\frac{1+\beta}{1-0.7\beta}\right) \gamma^{0.15} \frac{f_{y0} t_0^2}{\sin\theta_1} Q_f \quad (9)$$

$$VAR(N_u) = \left(\frac{\partial N_u}{\partial f_{y,0}} s_{f_{y,0}}\right)^2 + \left(\frac{\partial N_u}{\partial t_0} s_{t_0}\right)^2 + \left(\frac{\partial N_u}{\partial \delta} s_\delta\right)^2 \quad (10)$$

$$VAR(N_u) = N_u^2 \left[\left(\frac{s_{f_{y,0}}}{f_{y0}}\right)^2 + \left(1.85\frac{s_{t_0}}{t_0}\right)^2 + \left(\frac{s_\delta}{\delta}\right)^2\right] \quad (11)$$

$$V_{N_u} = \frac{[VAR(N_u)]^{0.5}}{E(N_u)} \quad (12)$$

$$(V_{N_u})^2 = (0.075)^2 + 3.42 \times (0.05)^2 + (0.068)^2 = 0.0188 \quad (13)$$

$$V_{Nu} = 0.14$$

The characteristic strength for more than 20 test data (factor 1.68 instead of 1.64) with 5% probability of lower strengths, is:

$$N_{u,k} = N_{u,m}(1 - 1.68 \times 0.14) = 0.76 N_{u,m} \quad (15)$$

$$N_{u,k} = \frac{0.76}{0.85} N_{u,m} = 0.90 N_{u,m} \quad (16)$$

where f_{y0} represents the design value instead of the mean value.

$$N_1^* = \frac{N_{u,k}}{\gamma_m} = \frac{N_{u,k}}{1.1} = 0.82 N_{u,m} \quad (17)$$

with m = 1.01, this results in:

$$N_1^* = 2.6 \left(\frac{1+\beta}{1-0.7\beta}\right) \gamma^{0.15} \frac{f_{y0} t_0^2}{\sin\theta_1} Q_f \quad (18)$$

For T and K-gap joints, the procedure is similar. However, for these joints it is assumed that $Q_f = 1.0$. Inclusion of Q_f makes this method very complicated, since the chord stress parameter n is correlated to t_0 and f_{y0}.

3.3 Evaluation based on Eurocode 1—Basis of structural design

The current IIW (1989) strength functions were previously analysed according to the procedure in Eurocode 1 (Sedlacek et al. 1991). Since the proposed equations generally give either equal or lower strengths than those incorporated in Eurocode 3 (which are the same as in the current IIW and CIDECT recommendations), no further evaluations are carried out here.

3.4 Recommended design strength formulae

Based on the evaluations in the previous section, the design strength functions for Q_u and Q_f in Table 4 are recommended. For simplicity, a single, lower bound function is adopted for the Q_f function for T and X joints.

Comparisons between the proposed equations and the strength predictions of the current IIW formulae (1989) and API recommendations are described by Wardenier et al. (2008a).

4 COMPARISON WITH EXPERIMENTAL DATABASE

4.1 General

A major point of discussion in the past was the database to be used. For example, the database developed by Kumamoto University (Kurobane 1981) differs from the API database with respect to the dimensions of joints included. Small-scale specimens tend to have relatively larger welds, which may increase the capacity of the joints. Hence, in the API database, all experiments with chord diameters smaller than 100 mm are excluded from the database. Yura et al. (1980) were the first researchers to remove small-scale specimens, even with chord diameters smaller than 150 mm. On the other hand, Kurobane's database does not impose limitations on the dimensions. In this study, the database compiled by Makino et al. (1996) is adopted. Similar to Kurobane's database (1981), no restrictions are imposed with respect to the joint sizes.

The following sections present, for X- and T-joints only, comparisons between the experimental database and the mean strength predictions of the new formulae summarized in Tables 1–2. The number of observations described here is limited. Attention is primarily focussed on the results with either high or low test/prediction ratios. Similar comparisons between the database and the mean strength formula are also made for K-joints; however, these will be published in the near future.

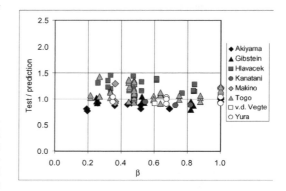

Figure 1. Comparison between experiments and mean strength equation for uniplanar X-joints.

Table 4. Recommended design strength formulae (compression loaded chords).

	New IIW design strength formulae	$N_1^* = \dfrac{f_{y0} t_0^2}{\sin \theta_1} Q_u Q_f$
	Q_u	Q_f
X	$2.6 \left(\dfrac{1+\beta}{1-0.7\beta} \right) \gamma^{0.15}$	$(1-|n|)^{(0.45-0.25\beta)}$
T	$2.6 \left(1+6.8\beta^2 \right) \gamma^{0.2}$	
K gap	$1.65 \left(1+8\beta^{1.6}\right) \gamma^{0.3} \left[1 + \dfrac{1}{1.2 + (g/t_0)^{0.8}} \right]$	$(1-|n|)^{0.25}$

* For chord tension: $Q_f = (1-n)^{0.2}$.

4.2 X-joints

Figure 1 illustrates the ratios between the strength values obtained in the experiments and the mean strength predictions for uniplanar X-joints. Table 5 summarizes the statistical data of the comparison between the various series of experiments and the proposed mean strength equation. The row indicated by "all tests" presents the statistical data of all experiments.

Sixty of the 118 available experiments on uniplanar X-joints were carried out by Togo (1967). In general, Togo (1967) conducted various series of experiments, thereby varying only one parameter within each series. It is of interest to note that some series match well with the proposed strength formula, while others give strength values considerably higher than the predicted values. For example, the largest test/prediction ratio for Togo is 1.43. Examination of this data point (XC-24) makes clear that this test is part of a series of three almost identical tests ($\alpha = 15.7$ and $\beta = 0.27$) with slightly different values for the f_{y0}/f_{u0} ratio (0.87 for XC-22 and XC-23 vs. 0.81 for XC-24), as shown in Table 6. However, for unknown reasons, the non-dimensional strength of test XC-24 is significantly higher than that of XC-22 and XC-23, causing the large test/prediction ratio for this data point.

Further evaluation of Togo's tests reveals more peculiarities. For example, tests XC-7 and XC-17 are comparable with respect to the chord length parameter α, and the chord material properties are almost identical for both joints. The differences between the two tests are (i) the β and 2γ values, which are slightly lower for XC-7, and (ii) the chord diameter d_0 ($d_0 = 76.3$ mm for XC-7 and $d_0 = 101.6$ mm for XC-17). Hence, it would be expected that the non-dimensional strength of XC-17 is higher than that of XC-7. As presented in Table 7, the experiments do not show this. On the contrary, the non-dimensional strength of XC-17 is considerably lower than that of XC-7, causing the different values for the test/prediction ratios of both tests. Both XC-7 and XC-17 are part of a series of experiments with the chord length α as variable.

Figure 2 plots the results, both Togo's data and the predicted strength values as a function of α. Because the proposed strength equation does not account for the effect of the chord length, the predicted strength is equal for the joints within each series. The results of the six tests for XC-17 (with $2\gamma = 34.2$) are in line with each other and with the predictions of the

Table 5. Test/prediction ratios for uniplanar X-joints.

	No. of data	Mean	CoV	Min.	Max.
Akiyama	14	0.889	0.097	0.76	1.03
Gibstein	9	0.935	0.083	0.79	1.05
Hlavacek	16	1.269	0.087	1.12	1.46
Kanatani	5	0.945	0.071	0.88	1.03
Makino	5	1.121	0.156	0.92	1.30
Togo	60	1.113	0.104	0.91	1.43
Yura	7	0.985	0.062	0.90	1.04
All tests	118	1.079	0.145	0.76	1.46

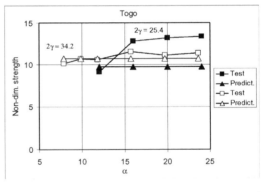

Figure 2. Comparison between selected experiments of Togo (1967) and the mean strength equation for uniplanar X-joints.

Table 6. Details of three uniplanar X-joints tested by Togo (1967).

	2γ	f_{y0} N/mm²	f_{u0} N/mm²	Non-dim. strength test	Non-dim. strength Predict.	Test /predict. ratio
XC-22	32.2	400	462	7.90	7.48	1.06
XC-23	32.2	400	462	7.85	7.48	1.05
XC-24	33.4	365	448	10.76	7.52	1.43

Table 7. Details of two uniplanar X-joints tested by Togo (1967).

	α	β	2γ	Non-dim. test	Non-dim. predict.	Test / predict. ratio
XC-7	20.0	0.446	25.4	13.18	9.72	1.355
XC-17	19.7	0.478	34.2	11.13	10.75	1.035

proposed formulae. However, for the series of XC-7 (with $2\gamma = 25.4$), only the result with $\alpha = 12$ matches well with the new strength equation, while the other three results seem odd.

From Figure 1, it is observed that Akiyama et al. (1974) gives test/prediction ratios close to 1.0 for $\beta = 1.0$. For all other β ratios, the test/prediction ratios are lower than 1.0. It should be pointed out that in general, the test data of Akiyama et al. (1974) are lower than the results found by other researchers, not only for CHS X-joints but also for CHS T-joints as described in the section hereafter. Wardenier et al. (2008b) came to a similar conclusion for CHS TP-joints under axial load. The lowest test-to-prediction ratios are found for the joints with $2\gamma = 95$, although this value is beyond the range of validity. For uniplanar X-joints, Akiyama's low ultimate strength values may be due to the relatively short chords used in the tests ($\alpha = 10$) in combination with the thin cap plates. This would explain why the test results and predicted strength values match well for $\beta = 1.0$. For this β value, the effect of the chord length and cap plates is small for X-joints.

In general, the data of Kanatani (1965), Yura (Boone et al. 1982, Weinstein & Yura 1986) and Gibstein (1973) match well with the predicted strength values. The lowest test/prediction ratio for Gibstein (1973) is noticed for the joint with $\beta = 0.82$ and $2\gamma = 29.8$. Figure 3 illustrates Gibstein's nine test data as a function of β. It is observed that the particular test is not in line (i.e. too low) with the data for $2\gamma = 20.6$ and $2\gamma = 40.5$ and $\beta = 0.82$.

From Figure 1 and Table 5, it is found that Hlavacek's tests for X-joints (1970) are relatively high for all β values considered which might be due to the effect of the chord length. No further information is available with regard to the chord length (and chord end conditions) applied in Hlavacek's tests.

4.3 T-joints

Figure 4 illustrates the test/prediction ratios for uniplanar T-joints. Figure 4a presents the results for joints with a chord diameter $d_0 < 150$ mm, while Figure 4b shows the data with $d_0 > 150$ mm. Table 8 summarizes the statistical data for the various series of experiments, whereby only series with three or more experiments are included. The row indicated by "all tests" includes the statistical data of all experiments.

From Table 8 and Figure 4, it is noticed that the lowest test/prediction ratio (0.62) is found for one of Akiyama's tests. Akiyama (1974) considers 14 tests of which at first sight, three are similar (TC-10, TC-15 and TC-51). Table 9 lists details of these three tests. Two tests consider identically sized specimens (TC10 and TC15). However, for unknown reasons, the non-dimensional failure load of both tests (7.29 for TC-15 vs. 11.92 for TC-10) varies significantly.

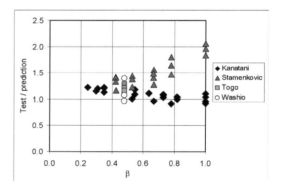

Figure 4a. Comparison between experiments (with $d_0 < 150$ mm) and mean strength equation for uniplanar T-joints.

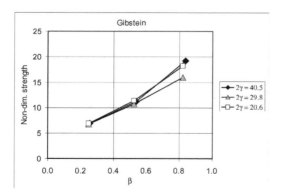

Figure 3. Selected Gibstein's tests (1973) on X-joints as a function of β.

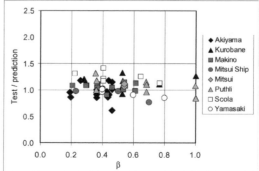

Figure 4b. Comparison between experiments (with $d_0 > 150$ mm) and mean strength equation for uniplanar T-joints.

Table 8. Test/prediction ratios for uniplanar T-joints.

	No. of data	Mean	CoV	Min.	Max.
		Specimens with $d_0 < 150$ mm			
Kanatani	21	1.077	0.090	0.91	1.22
Stamenkovic	17	1.523	0.170	1.16	2.06
Togo	9	1.223	0.034	1.16	1.28
Washio	3	1.145	0.193	0.97	1.39
		Specimens with $d_0 > 150$ mm			
Akiyama	14	0.960	0.155	0.62	1.18
Kurobane	10	1.122	0.132	0.86	1.33
Makino	20	1.065	0.055	0.91	1.15
Mitsui Ship Corp.	5	0.945	0.105	0.77	1.03
Puthli	8	1.081	0.134	0.85	1.32
Scola	6	1.251	0.084	1.13	1.42
Yamasaki	5	0.952	0.076	0.85	1.02
All tests	122	1.138	0.194	0.62	2.06

Table 9. Details of selected tests on uniplanar T-joints by Akiyama et al. (1974).

	α	β	2γ	θ	Non-dim. Strength test	Test / predict. ratio
TC-10	10	0.462	35.2	90°	11.92	1.01
TC-15	10	0.462	35.4	90°	7.29	0.62
TC-12	10	0.439	70.8	90°	12.08	0.92
TC-14	10	0.361	93.3	90°	10.54	0.92
TC-51	10	0.462	35.2	45°	13.51	1.15
TC-52	10	0.439	72.4	45°	15.56	1.18
TC-53	10	0.361	93.3	45°	11.56	1.01

Further, Akiyama's test TC-51 with identical parameters and dimensions as specimen TC-10 but with a brace angle $\theta = 45°$, even gives a higher non-dimensional failure load (13.51) than specimen TC-10. As shown in Table 9, the same observation can be made for specimens TC-12 and TC-14 with 90° vs. comparable specimens TC-52 and TC-53 with 45°. A possible explanation could be the chord end restraint being located at the tensile side of the chord, causing the ultimate strength to increase. Unfortunately, Akiyama's report does not provide additional details with regard to the chord end restraints employed.

As a result of the large scatter in experimental results, the test/prediction ratios for Akiyama's tests also show large variations.

Another data point which gives a relatively low test/prediction ratio (0.77) is a test conducted by Mitsui Shipping Corporation (1978). Figure 5 shows that three of the four data match the predicted strength values very well. However, the result found in test TC-63 seems suspicious, see also Table 10. In the range TC-61, TC-62 and TC-63, β increases from 0.4 to 0.5 to 0.7. For TC-63, the non-dimensional

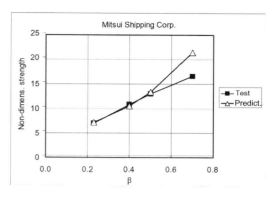

Figure 5. Selected tests by Mitsui Shipping Corporation (1978) on T-joints.

Table 10. Details of selected tests on uniplanar T-joints by Mitsui Shipping Corporation (1978).

	α	β	2γ	Non-dim. Strength test	Test / predict. ratio
TC-60	7.5	0.23	39.4	−6.99	0.98
TC-61	9.5	0.40	39.4	−10.81	1.03
TC-62	9.5	0.50	39.4	−13.01	0.98
TC-63	7.5	0.70	39.4	−16.56	0.77

strength does not follow the strength increments seen from TC-61 to TC-62. Instead the curve presented in Figure 5 flattens. Because the α value of TC-63 ($\alpha = 7.5$) is lower than that of TC-61 and TC-62 ($\alpha = 9.5$), an even larger strength increase would have been expected.

Table 8 shows that the largest test-to-prediction ratio for the experiments of Kurobane et al. (1991) is 1.33. A closer examination reveals that this data point (TC-96) is part of a series of three experiments with approximately the same dimensions but with different chord material properties (TC-96, TC-97 and TC-98—see Table 11). The ultimate strength of the three joints seems to have a very strong dependency on the strength ratio f_{y0}/f_{u0}, as shown in Figure 6 where the test/prediction ratios are plotted as a function of f_{y0}/f_{u0}. Only the joints with the higher yield strengths (TC-97 and TC-98) match well with the proposed strength formula. However, Kurobane's results (1991) regarding the yield strength ratio f_{u0}/f_{y0} are not confirmed by tests TC-68 and TC-69 conducted by Yamasaki et al. (1979), as shown in Figure 6.

It is further noticed that the four tests of Kurobane et al. (1991) with $2\gamma = 48$ give higher ultimate strengths than the values predicted by the proposed equations, especially for $\beta = 1.0$ (see results for TC-92 in Table 11). The strength of thin-walled joints in combination

Table 11. Details of tests on uniplanar T-joints by Kurobane et al. (1991) and Yamasaki et al. (1979).

	α	β	2γ	f_{y0} N/mm²	f_{u0} N/mm²	Non-dim. strength test.	Test/predict. ratio
				Kurobane			
TC-96	6.4	0.527	34.5	298	519	−19.24	1.33
TC-97	6.4	0.525	32.3	580	650	−14.96	1.05
TC-98	6.4	0.526	32.6	832	875	−13.32	0.93
TC-92	6.4	0.999	48	377	522	−51.14	1.27
				Yamasaki			
TC-68	9.3	0.398	40.1	366	524	−10.17	0.96
TC-69	9.3	0.398	40.2	728	827	−10.71	1.02

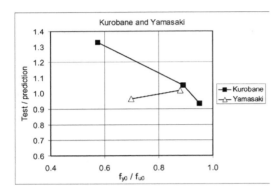

Figure 6. Effect of f_{y0}/f_{u0} according to tests of Kurobane et al. (1991) and Yamasaki et al. (1979).

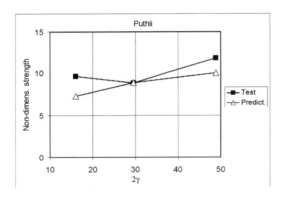

Figure 7. Selected tests by Puthli (Stol et al. 1984) on T-joints with β = 0.36.

with α = 6.4 is possibly affected by short chord effects. Makino's unpublished results (1976–1979) and Kurobane's tests (1991) for 2γ = 27 are much better in line with the proposed strength equations, even though the α values of Makino's tests are as small as in Kurobane's tests. However, Makino (1976–1979) did not consider β = 1.0 and only tested specimens with chord thickness ratios 22.5 ≤ 2γ ≤ 37.0.

Another series of experiments for which a rather high test/prediction ratio (of 1.32) is found, are the tests conducted by Puthli (Stol et al. 1984). The particular data point (TC-108 with 2γ = 16) can be compared with two other tests (TC-110 and TC-113) with similar α and β values. Figure 7 shows the experimental and predicted non-dimensional strengths of these three joints as a function of the chord thickness ratio 2γ. The experiments (in particular the data with 2γ = 16) do not confirm the anticipated 2γ dependency i.e. increasing non-dimensional strength values as a result of increasing 2γ values. On the other hand, it is recognized that the chord stress function presented in Table 2 is conservative for low 2γ values. This might explain why for a similar series of experiments with β = 0.68, joint TC-109 with 2γ = 16 also gives the largest test/prediction ratio (1.15). Unfortunately, the three joints with β = 0.68 do not have the same α value, making a direct comparison as shown in Figure 7 impossible.

From Figure 4 and Table 8, it is observed that Scola's tests (1989) are relatively high for all β values considered, which is caused by the short chords used in the tests. A chord length parameter of α = 4.8 is too short to exclude the effect of the chord end plates on the strength of T-joints.

The largest test/prediction ratios are observed for the tests of Stamenkovic & Sparrow (1983). In these tests, 12.5 mm thick cap plates on the chord ends were bolted to a closed steel frame and hence fixed against rotations. These boundary conditions are different from the restraints commonly used (i.e. pin-supported chord ends), causing the significant strength enhancement observed.

5 SUMMARY AND CONCLUSIONS

This study first presents the mean strength equations derived for axially loaded T-, X- and K-gap joints, followed by the statistical evaluation procedure to convert these equations into design equations.

Further, detailed comparisons between the strength equations and available experimental data are shown for T- and X-joints. Similar comparisons are made for K-gap joints but because of paper length limitations, these will be published in the near future.

An important conclusion is that the available experimental data exhibit a large scatter due to a variety of reasons, e.g. effect of (i) chord length, (ii) chord end restraints, (iii) chord material properties (i.e. f_{y0}/f_{u0} ratio), (iv) size of specimens (i.e. small scale specimens versus commonly sized specimens, and (v) definition of the end of a test, etc. Even within series of experiments conducted by the same researcher, considerable but inexplicable differences are found. All these effects contribute to a significant amount of scatter in the experimental data and hence, increase the CoV of the test/prediction ratios.

As far as FE analyses are concerned, attention should be given to the modelling of the welds, especially for joints with small gaps between the in-plane or out-of-plane braces. Excluding the weld geometry in such cases may grossly underpredict the ultimate strength. Another point of attention is the definition of ultimate load if fracture is expected to be the failure mode; different values of strain levels will result in different ultimate load values.

Notwithstanding the large scatter obtained at first sight, it can be concluded that after careful screening, examination of, and comparisons between the available experimental data, in general, a good agreement is observed between the experimental results and the predictions of the new strength equations.

REFERENCES

Akiyama, N., Yajima, M., Akiyama, H. & Ohtake, A. 1974. Experimental study on strength of joints in steel tubular structures. *Journal of Society of Steel Construction*, Vol. 10, No. 102, pp. 37–68, (in Japanese).

American Petroleum Institute 2007. Recommended practice for planning, designing and constructing fixed offshore platforms—Working stress design. API RP 2 A, 21st Ed, Supp. 3, American Petroleum Institute, USA.

Boone, T.J., Yura, J.A. & Hoadley, P.W. 1982. Chord stress effects on the ultimate strength of tubular joints—Phase 1. PMFSEL Report No. 82–1, Phil M. Ferguson Structural Engineering Laboratory, University of Texas, Austin, USA.

Gibstein, M.B. 1973. Static strength of tubular joints. Det Norske Veritas Report No. 73–86-C, Oslo, Norway.

Hlavacek, V. 1970. Strength of welded tubular joints in lattice girders. *Construzioni Metalliche*, No. 6.

IIW 1989. Design recommendations for hollow section joints—Predominantly statically loaded. 2nd Edition, Intern. Institute of Welding, Subcommission XV-E, Annual Assembly, Helsinki, Finland, IIW Doc. XV-701–89.

IIW 2008. Static design procedure for welded hollow section joints—Recommendations. 3rd Edition, Intern. Institute of Welding, Subcommission XV-E, IIW Doc. XV-1281–08.

Kanatani, H. 1965. Experimental study of welded tubular connection—Part 1, 2 and 3. *Transactions Architectural Institute of Japan*, No. 108–110, (in Japanese).

Kurobane, Y. 1981. New developments and practices in tubular joint design (+ Addendum). International Institute of Welding, Annual Assembly, Oporto, Portugal, IIW Doc. XV-488–81.

Kurobane, Y., Makino, Y., Ogawa, K. & Maruyama, Y. 1991. Capacity of CHS T-joints under combined OPB and axial loads and its interactions with frame behavior. *Proc. 4th Intern. Symp. on Tubular Structures*, Delft, The Netherlands, pp. 412–423.

Makino, Y. 1976–1979. Unpublished test results.

Makino, Y., Kurobane, Y., Mitsui, Y. & Yasunaga, M. 1978. Experimental study on ultimate strength of tubular joints with high strength steel and heavy walled chord. Research Report, Chugoku-Kyushu Branch of Architectural Institute of Japan, No. 4, pp. 255–258, (in Japanese).

Makino, Y. 1982. Unpublished test results.

Makino, Y., Kurobane, Y., Ochi, K., Vegte, G.J. van der & Wilmshurst, S.R. 1996. Database of test and numerical analysis results for unstiffened tubular joints. IIW Doc. XV-E-96–220.

Mitsui, Y. 1973. Experimental study on local stress and strength of tubular joints in steel. *PhD Thesis*, Osaka University, Osaka, Japan, (in Japanese).

Mitsui Engineering and Shipbuilding Co. Ltd. 1978. Development of computer program for fatigue design of tubular joints for floating offshore structures. Unpublished report, (in Japanese).

Scola, S. 1989. Behavior of axially loaded tubular V-joints. *MSc Thesis*, McGill University, Montreal, Canada.

Sedlacek, G., Wardenier, J., Dutta, D. & Grotmann, D. 1991. Evaluation of test results on hollow section lattice girder connections. Background Report to Eurocode 3 'Common unified rules for steel structures', Document 5.07, Eurocode 3 Editorial Group.

Stamenkovic, A. & Sparrow, K.D. 1983. Load interaction in T-joints of steel circular hollow sections. *Journal of Structural Engineering*, ASCE, Vol. 109, No. 9, pp. 2192–2204.

Stol, H.G.A., Puthli, R.S. & Bijlaard, F.S.K. 1984. Static strength of welded tubular T-joints under combined loading. TNO-IBBC Report B-84–561/63.6.0829, The Netherlands.

Togo, T. 1967. Experimental study on mechanical behavior of tubular joints. *PhD Thesis*, Osaka University, Osaka, Japan, (in Japanese).

Vegte, G.J. van der & Makino, Y. 2006. Ultimate strength formulation for axially loaded CHS uniplanar T-joints. *Intern. Journal Offshore and Polar Engineering*, ISOPE, Vol. 16, No. 4, pp. 305–312.

Vegte, G.J. van der, Wardenier, J. & Makino, Y. 2007a. New ultimate strength formulation for axially loaded CHS K-joints. *Proc. 5th Intern. Conf. on Advances in Steel Structures*, Singapore, Vol. II, pp. 218–227.

Vegte, G.J. van der, Makino, Y. & Wardenier, J. 2007b. Effect of chord load on ultimate strength of CHS X-joints.

Intern. Journal Offshore and Polar Engineering, ISOPE, Vol. 17, No. 4, pp. 301–308.

Vegte, G.J. van der, Wardenier, J., Qian, X.D. & Choo, Y.S. 2008. Reanalysis of the moment capacity of CHS joints. *Proc. 12th Intern. Symp. on Tubular Structures*, Shanghai, China.

Wardenier, J. 1982. *Hollow section joints*. Delft University Press, The Netherlands, ISBN 90-6275-084-2.

Wardenier, J., Vegte, G.J. van der, Makino, Y. & Marshall, P.W. 2008a. Comparison of the new IIW (2008) CHS joint strength formulae with those of the previous IIW (1989) and the new API (2007). *Proc. 12th Intern. Symp. on Tubular Structures,* Shanghai, China.

Wardenier, J., Vegte, G.J. van der & Makino, Y. 2008b. Joints between plates or I sections and a circular hollow section chord. *Proc. 18th Intern. Offshore and Polar Engineering Conf.*, Vancouver, Canada.

Washio, K. & Mitsui, Y. 1969. High stress fatigue tests of tubular T-joints. Summary Papers, *Annual Conf. of Architectural Institute of Japan*, pp. 1155–1156, (in Japanese).

Weinstein, R.M. & Yura, J.A. 1986. The effect of chord stresses on the static strength of DT tubular connections. *Proc. Offshore Technology Conf.*, OTC 5135, Houston, USA.

Yamasaki, T., Takizawa, S. & Komatsu, M. 1979 Static and fatigue tests on large-size tubular T-joints. *Proc. Offshore Technology Conf.*, OTC 3424, pp. 583–591.

Yura, J.A., Zettlemoyer, N. & Edwards, I.E. 1980. Ultimate capacity equations for tubular joints. *Proc. Offshore Technology Conf.*, OTC 3690, USA.

Zhao, X.-L., Wardenier, J., Packer, J.A. & Vegte, G.J. van der 2008. New IIW (2008) static design recommendations for hollow section joints. *Proc. 12th Intern. Symp. on Tubular Structures,* Shanghai, China.

Structural design rules for elliptical hollow sections

L. Gardner
Imperial College London, UK

T.M. Chan
University of Warwick, UK

ABSTRACT: Elliptical hollow sections have been recently added to the familiar family of hot-finished tubular sections, featuring square, rectangular and circular hollow sections. Their usage has attracted considerable interest from industry, and significant recent research towards development of comprehensive structural design rules has been carried out. Investigations have been performed at both cross-section and member levels and in a range of structural configurations. On the basis of full-scale tests, non-linear numerical simulations and experience gained from studies on other tubular sections, design rules for elliptical hollow sections have been developed. This paper describes the background to and validation of these rules, which include, at cross-section level, a set of slenderness parameters and classification limits for compression and bending, shear resistance formulations and an interaction curve for combined shear and bending. At member level, column buckling curves have been proposed in harmony with those for structural steel circular hollow sections.

1 INTRODUCTION

Recent investigations into the structural use of elliptical hollow sections have become increasingly wide-ranging. Concerns with connection methods have been investigated by Bortolotti et al. (2003), Choo et al. (2003), Pietrapertosa & Jaspart (2003) and Willibald et al. (2006). The composite behaviour of filled elliptical hollow sections has been studied by Zhao et al. (2007) and Roufegarinejad and Bradford (2007). With regards to cross-section and member resistance, recent studies have been conducted by Gardner & Chan (2007, in press) and Chan & Gardner (2008a, 2008b, submitted). Numerical studies were reported by Zhu & Wilkinson (2006). This paper aims to summarise the proposed design rules which have been developed on the basis of full-scale testing, numerical simulations and comparisons with other tubular sections.

2 AXIAL COMPRESSION

Axial compression represents a fundamental loading arrangement for structural members. For cross-section classification under pure compression, of primary concern is the occurrence of local buckling in the elastic material range. Cross-sections that reach the yield load are considered Class 1–3 (fully effective), whilst those where local buckling of the slender constituent elements prevents attainment of the yield load are Class 4 (slender). For uniform compression, the cross-section slenderness parameter has been determined by consideration of the elastic critical buckling stress.

For an elliptical section, the maximum radius of curvature lies at the ends of the cross-section major (y-y) axis, and may be shown to be equal to a^2/b (Fig. 1). Under pure compression, this is the point at which local buckling initiates. To account for the continuously varying curvature of an EHS, the concept of an equivalent diameter is introduced and the cross-section slenderness is defined as:

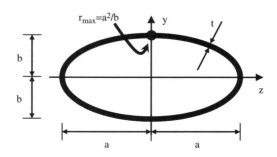

Figure 1. Location of maximum radius of curvature in an elliptical hollow section.

$$\frac{D_e}{t\varepsilon^2} = 2\frac{\left(\frac{a^2}{b}\right)}{t\varepsilon^2} \quad (1)$$

where D_e is an equivalent diameter $(2a^2/b)$, t is the section thickness and $\varepsilon^2 = 235/f_y$ to allow for the influence of varying yield strengths.

Alongside the theoretical derivations, a series of precise full-scale laboratory tests on EHS (grade S355), manufactured by Corus Tubes (2006), was performed at Imperial College London. The first series of tests comprised a total of 25 material tensile coupon tests and 25 cross-section capacity stub column tests. The primary objective of the tensile coupon tests was to determine the basic engineering stress-strain behaviour of the material for each of the tested section sizes. Results were used to facilitate the numerical study and the development of design rules. Stub column tests were conducted to develop a relationship between cross-section slenderness, deformation capacity and load-carrying capacity for elliptical hollow sections under uniform axial compression. A typical test arrangement is shown in Figure 2 and the typical load-end shortening curves for two stub columns (EHS 150 × 75 × 8-SC1 and SC2) are depicted in Figure 3.

A numerical modelling study, using the finite element (FE) package ABAQUS (2006), was carried out in parallel with the experimental programme.

The primary aims of the study were to replicate the experimental tests and, having validated the models, to perform parametric studies. Replication of test results was found to be satisfactory (Chan & Gardner 2008a) with the numerical models able to successfully capture the observed stiffness, ultimate load, general load-end shortening response and failure patterns. Having verified the general ability of the FE models to replicate test behaviour for EHS with an aspect ratio of 2, a series of parametric studies, aiming to investigate the influence of cross-section slenderness and aspect ratio (from 1 to 3) on the ultimate load carrying capacity, were conducted.

The relationship between test N_u/N_y, where N_u is the maximum test compressive load and N_y is the yield load, and the corresponding cross-section slenderness parameter $2(a^2/b)/t\varepsilon^2$ is plotted in Figure 4. The results of the FE parametric study on elliptical hollow sections with aspect ratios, a/b of 1, 2 and 3 are included in the figure. A value of N_u/N_y greater than unity represents meeting of the Class 1–3 requirements, whilst a value less than unity indicates a Class 4 section where local buckling prevents the yield load from being reached. Figure 4 demonstrates the anticipated trend of reducing values of N_u/N_y with increasing slenderness. For comparison, existing compressive test data from circular hollow sections (CHS) have also been added. Overall, the FE results mirror the experimental findings and suggest

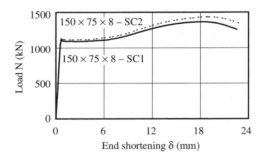

Figure 3. Stub columns load-end shortening curves.

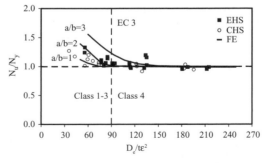

Figure 4. Normalised compressive resistance versus cross-section slenderness.

Figure 2. Stub column test arrangement.

that adopting the Class 3 CHS slenderness limit of 90 from *Eurocode 3* (EN 1993-1-1 2005) for EHS, based on the proposed slenderness parameter, is suitable. Further details presented in this section have been reported by Gardner & Chan (2007) and Chan & Gardner (2008a).

Further research on the elastic buckling of elliptical tubes has been conducted by Ruiz-Teran & Gardner (in press), where a modified definition of equivalent diameter (and hence slenderness parameter) was proposed. A simple analytical model was developed to (1) describe the boundary behavior (circular shells and flat plates) and intermediate behaviour of elliptical sections and (2) examine the influence of the aspect ratio and relative thickness of the section on the structural behaviour. Numerical simulations using the finite element package ABAQUS (2006) were also carried out. Based on the results from the analytical and numerical studies, for the current practical range of elliptical hollow sections, a simple expression for the equivalent diameter was deduced and is given by Equation 2.

$$D_e = 2a\left[1 + f\left(\frac{a}{b} - 1\right)\right] \quad (2)$$

where f is obtained from Equation 3,

$$f = 1 - 2.3\left(\frac{t}{2a}\right)^{0.6} \quad (3)$$

With reference to Equations 2–3, the corresponding cross-section slenderness of an elliptical hollow section is therefore defined as:

$$\frac{D_e}{t\varepsilon^2} = 2a\frac{\left[1 + f\left(\frac{a}{b} - 1\right)\right]}{t\varepsilon^2} \quad (4)$$

where D_e is the equivalent diameter and $\varepsilon^2 = 235/f_y$ to allow for a range of yield strengths.

A comparison of CHS and EHS test data in compression is shown in Figure 5. For EHS, the results are plotted on the basis of the equivalent diameters proposed by Gardner & Chan (2007, Equation 1) and by Ruiz-Teran & Gardner (in press, Equation 4). Regression curves have also been added for the three data sets in the figure. The results demonstrate that both slenderness parameters for EHS are conservative in comparison to CHS, but the proposal by Ruiz-Teran & Gardner (in press) yields closer agreement between the two section types, and increases the number of sections from the current range of EHS

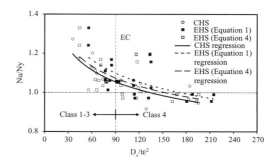

Figure 5. Comparison of different equivalent diameters employed in EHS slenderness parameters.

being fully effective, and is thus more accurate and appropriate for design.

On this basis, it is recommended that in order to achieve structural efficiency, EHS may be classified in compression using current CHS slenderness limits and the equivalent diameter and slenderness parameter defined by Equations 2–4.

3 BENDING

The second phase of this research focused on the in-plane bending response of EHS. For minor (z-z) axis bending, similar to axial compression, the maximum compressive stress coincides with the point of maximum radius of curvature. It is therefore proposed that the same cross-section slenderness parameter given by Equation 1 can be adopted. This proposal is supported by the test data. For bending about the major (y-y) axis, buckling would initiate in general neither at the point of maximum radius of curvature (located at the neutral axis of the cross-section with negligible bending stress) nor at the extreme compressive fibre, since this is where the section is of greatest stiffness (i.e. minimum radius of curvature). A critical radius of curvature r_{cr} was therefore sought to locate the point of initiation of local buckling. This was achieved by optimizing (i.e. finding the maximum value of) the function composed of the varying curvature expression and an elastic bending stress distribution. The result was modified to provide better prediction of observed physical behaviour, to yield r_{cr} equal to $0.4a^2/b$ (Fig. 6).

For an aspect ratio, a/b of less than 1.155, where the section is approaching circular, elastic buckling was found to occur at the extreme of the major axis and r_{cr} would therefore be equal to b^2/a. Thus, the proposed slenderness parameters for major axis bending are given by Equations 5–6. Note that the transition between Equations 5–6 is at $a/b = 1.357$ for the modified case, as compared to 1.155 in the elastic case.

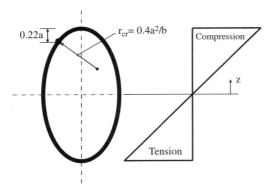

Figure 6. Location of critical radius of curvature in major axis bending.

$$\frac{D_e}{t\varepsilon^2} = 0.8\frac{\left(\dfrac{a^2}{b}\right)}{t\varepsilon^2} \quad \text{for } a/b > 1.357 \qquad (5)$$

$$\frac{D_e}{t\varepsilon^2} = 2\frac{\left(\dfrac{b^2}{a}\right)}{t\varepsilon^2} \quad \text{for } a/b \leq 1.357 \qquad (6)$$

Note that for the special case of an EHS with an aspect ratio of unity, the cross-section slenderness defined by Equation 6 reverts to that for CHS in *Eurocode 3*.

A total of 18 in-plane bending tests were performed in three-point and four-point bending configurations in order to study the structural behaviour of EHS under a moment gradient and constant moment scenarios respectively (Figs. 7–8).

Results from in-plane bending tests formed the basis of the experimental performance database for the development of the cross-section classification system. In parallel with the experimental programme, a numerical modelling study was carried out. The scope of this study was to extend the structural performance database numerically by exploring cases not examined experimentally. The numerical models were validated against the test results by using measured geometric and material properties. Local imperfections were included in the models by means of the lowest local eigenmode, with a range of imperfection amplitudes being considered to assess sensitivity. Comparisons between test and FE results were made to ensure that the FE models were capable of mimicking not only the peak response, but also the general load-displacement histories and failure patterns from the tests. The validated models were utilized for parametric studies to

Figure 7. Four-point bending test arrangement.

Figure 8. Three-point bending test arrangement.

examine the influence of aspect ratio and cross-section slendernesses. The results of the numerical study were used to verify the applicability of using the proposed cross-section slenderness parameters for EHS in conjunction with the current slenderness limits for CHS.

Distinction between Class 3 and 4 cross-sections in bending is made by assessing to their ability to reach the elastic moment resistance $M_{el,Rd}$. The relationship between test $M_u/M_{el,Rd}$, where M_u is the maximum moment reached in the test, and the proposed cross-section slenderness parameters given by Equations 1, 5–6 is plotted in Figures 9–10 for bending about the minor axis and major axis respectively. A value of $M_u/M_{el,Rd}$ greater than unity represents meeting of the Class 3 requirement, whilst a value less than unity indicates a Class 4 cross-section where local buckling prevents the yield moment being reached. A lower bound to the experimental results in both figures suggests that the *Eurocode 3* slenderness limit of 90, representing the boundary between Class 3 and 4 cross-sections may be safely adopted.

For comparison, existing CHS bending test data have been plotted. In addition to the experimental results, results from the described parametric studies on elliptical hollow sections with aspect ratios, a/b of 1, 2 and 3 have also been plotted complementing the experimental results in illustrating the appropriateness of adopting the Eurocode limit. Further analysis of the results indicates that the Class 3 slenderness limit for both CHS and EHS (bending about either axis)

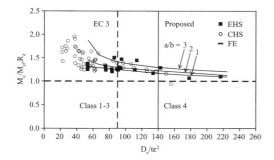

Figure 9. Normalized minor-axis bending resistance versus cross-section slenderness.

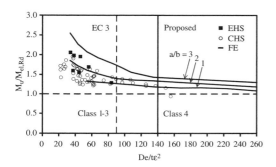

Figure 10. Normalized major-axis bending resistance versus cross-section slenderness.

Figure 11. Shear test arrangement (three-point bending configuration).

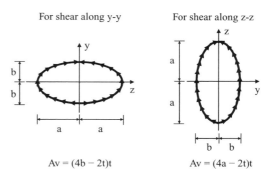

Figure 12. Derivation of plastic shear area.

may be relaxed to $140\varepsilon^2$. This limit, together with the approach of using separate Class 3 (yield) slenderness limits for CHS in compression and bending is broadly in line with the provisions of the Australian and North American Standards. Comparison of test results with the plastic moment and consideration of rotation capacity has demonstrated that, using the proposed measures of slenderness, the Class 1 and 2 limits for CHS (50 and 70, respectively) may also be adopted for EHS (Gardner & Chan 2007, Chan & Gardner 2008b).

4 SHEAR

Having established a cross-section classification system for axial compression and bending, the next investigation focussed on the influence of shear. A total of 24 shear tests in a three-point bending configuration were performed (Fig. 11). A range of span lengths was chosen in order to study the degradation effect on bending resistances with increasing shear. Numerical simulations covering a wider range of cross-section slendernesses and aspect ratios were also conducted to support and complement the experimental results.

With regards to the plastic shear resistance, for an EHS of uniform thickness subject to transverse loading along the y-y direction (Fig. 12), assuming a uniform shear stress distribution around the cross-section (acting circumferentially), the applied shear force is balanced by the summation of the vertical components of the shear stresses × dA. The resulting shear force may be found to be equal to twice the product of the vertical projection of the elliptical section (measured to the centreline of the thickness) and the thickness given by $(2b - t)t$ multiplied by the shear stress τ. Likewise, when the transverse load is applied in the z-z direction (Fig. 12), the corresponding projected area is equal to $2(2a - t)t$. Therefore, for an elliptical hollow section of constant thickness, the shear area A_v may be defined by Equations 7–8.

$$Av = (4b - 2t)t \quad \text{for load in the y-y direction} \quad (7)$$

$$Av = (4a - 2t)t \quad \text{for load in the z-z direction} \quad (8)$$

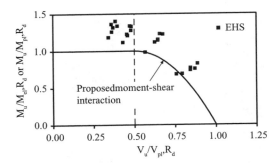

Figure 13. Moment-shear interaction.

Based upon the structural performance data, the normalised test results are plotted in Figure 13. The results demonstrate that where the shear force is less than half the plastic shear resistance ($V_{pl,Rd}$), the effect of shear on the bending moment resistance is negligible. Conversely, for high shear force (greater than 50% of $V_{pl,Rd}$), there is a degradation of the bending moment resistance. The proposed moment-shear interaction derived from *Eurocode 3* has also been plotted in Figure 13 and shows good agreement with the test results. Further details on shear buckling are reported by Gardner & Chan (in press).

5 COLUMN BUCKLING

Complementary to the structural performance study at the cross-section level, research into member column buckling behavior was also carried out. This behavioral study again employed the key methodologies demonstrated in the cross-section investigations. Carefully conducted experiments were performed on a total of 24 columns, buckling about the major and minor axes (Fig. 14).

A range of column lengths was chosen to cover a spectrum of member slendernesses. Comparison between the experimental results and elastic and plastic column buckling theories revealed satisfactory agreement, demonstrating that the theoretical models were capable of capturing the observed load-deformation behavior (Fig. 15).

Numerical models were also developed alongside the experimental programme and validated against the test results. Local and global imperfections were included in the numerical models by means of superimposed eigenmodes. Following satisfactory replication of the tests, parametric studies which aimed to assess the influence of different aspect ratios, thicknesses and lengths on the column buckling resistances were carried out. The resulting structural performance data demonstrated a similar trend to their circular counterparts (Fig. 16). The current design column curve for hot-finished

Figure 14. Column test arrangement.

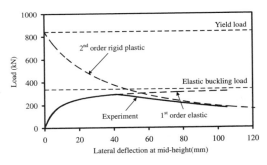

Figure 15. Comparison between experimental results and theoretical models.

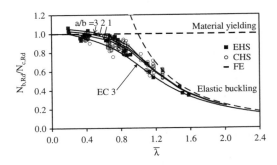

Figure 16. Normalized column test results versus member slenderness.

hollow sections from *Eurocode 3* (EN 1993-1-1 2005) provides, in general, a lower bound to the structural performance data and it is therefore recommended that this column curve can also be adopted for elliptical hollow sections. Further details have been reported by Chan & Gardner (submitted).

6 CONCLUSIONS

This paper summarises proposed design rules for elliptical hollow sections at both cross-section and member levels on the basis of theoretical derivations, experimental investigations and numerical simulations. At cross-section level, a total of 92 experimental tests have been conducted, a cross-section classification system has been developed with the associated slenderness parameters derived on the basis of an equivalent diameter emerging from theoretical and experimental observations. Shear area formulations have also been proposed on the basis of a uniform shear stress distribution acting tangentially around the section to mimic the shear area formulations for circular hollow sections. At member level, 24 column tests have been performed, and the column buckling curve from *Eurocode 3* for hot-finished hollow sections has been validated against experimental and numerical results and thus proposed to be applicable to hot-finished EHS columns.

ACKNOWLEDGEMENTS

The authors are grateful to the Dorothy Hodgkin Postgraduate Award Scheme for the project funding, and would like to thank Corus Tubes for the supply of test specimens and for funding contributions, An- drew Orton (Corus Tubes) for his technical input and Ron Millward, Alan Roberts and Trevor Stickland for their assistance in the laboratory works.

REFERENCES

Hibbitt, Karlsson & Sorensen, Inc. 2006. *ABAQUS, Version 6.6*. USA: Pawtucket.

Bortolotti, E., Jaspart, J.P., Pietrapertosa, C., Nicaud, G., Petitjean, P.D. & Grimault, J.P. 2003. Testing and modeling of welded joints between elliptical hollow sections. *Proceedings of the 10th International Symposium on Tubular Structures*: 259–266. Madrid, Spain.

Chan, T.M. & Gardner, L. 2008a. Compressive resistance of hot-rolled elliptical hollow sections. *Engineering Structures* 30(2): 522–532.

Chan, T.M. & Gardner, L. 2008b. Bending strength of hot-rolled elliptical hollow sections. *Journal of Constructional Steel Research* 64(9): 971–986.

Chan, T.M. & Gardner, L. Submitted. Flexural buckling of elliptical hollow section columns. *Journal of Structural Engineering*. ASCE.

Choo, Y.S., Liang, J.X. & Lim, L.V. 2003. Static strength of elliptical hollow section X-joint under brace compression. *Proceedings of the 10th International Symposium on Tubular Structures*: 253–258. Madrid, Spain.

Corus. 2006. Celsius® 355 Ovals—Sizes and resistances Eurocode version. Corus Tubes—Structural & Conveyance Business.

Gardner, L. & Chan, T.M. 2007. Cross-section classification of elliptical hollow sections. *Steel and Composite Structures* 7(3): 185–200.

Gardner, L. & Chan, T.M. In press. Shear resistance of elliptical hollow sections. *Structures and Buildings*. ICE.

EN 1993-1-1 2005. Eurocode 3: Design of steel structures— Part 1–1: General rules and rules for buildings. CEN.

Pietrapertosa, C. & Jaspart, J.P. 2003. Study of the behaviour of welded joints composed of elliptical hollow sections. *Proceedings of the 10th International Symposium on Tubular Structures*: 601–608. Madrid, Spain.

Roufegarinejad, A. & Bradford, M.A. 2007. Local buckling of thin-walled elliptical tubes containing an elastic infill. *Proceedings of the 3rd International Conference on Steel and Composite Structures*: 943–948. Manchester, United Kingdom.

Ruiz-Teran, A.M. & Gardner, L. In press. Elastic buckling of elliptical tubes. *Thin-Walled Structures*.

Willibald, S., Packer, J.A. & Martinez-Saucedo, G. 2006. Behaviour of gusset plate connections to ends of round and elliptical hollow structural section members. *Canadian Journal of Civil Engineering* 33(4): 373–383.

Zhao, X.L., Lu, H., & Galteri, S. 2007. Tests of elliptical hollow sections filled with SCC (self-compacting concrete). *Proceedings of the 5th International Conference on Advances in Steel Structures*, Singapore.

Zhu, Y. & Wilkinson, T. 2006. Finite element analysis of structural steel elliptical hollow sections in pure compression. *Proceedings of the 11th International Symposium on Tubular Structures—ISTS11*: 179–186. Quebec City, Canada.

Application of new *Eurocode 3* formulae for beam-columns to Class 3 hollow section members

N. Boissonnade & K. Weynand
Feldmann +Weynand GmbH, Aachen, Germany

J.P. Jaspart
ArGEnCo Department, University of Liège, Liège, Belgium

ABSTRACT: This paper presents a new set of design formulae for Class 3 hollow section beam-column members. The specificity of the proposed rules consists in offering a fully continuous model along the Class 3 range, filling the existing gap between plastic and elastic resistance in *Eurocode 3*. Through comparisons with both experimental tests and extensive FEM numerical results, the proposal is found accurate and safe. Moreover, a second parametric study dedicated to the determination of the level of safety led to a fully satisfactory safety factor γ_M.

1 INTRODUCTION

This paper reports on investigations led towards the application of the new *Eurocode 3* formulae for hollow section beam-columns. More precisely, it (i) focuses on the behavior of semi-compact "Class 3" cross-section members, and (ii) presents the essentials of a new design model for the resistance of Class 3 cross-sections.

The new set of design rules for checking the stability of beam-column members proposed in the latest version of *Eurocode 3* (EN 1993-1-1 2005) differ significantly from the previous ENV version set (ENV 1993-1-1 1992). Two different sets of formulae are now available, according to either Method 1 or Method 2. Based on second-order principles, both methods provide a high level of accuracy and emphasize a clear physical background, which is of prime importance for a safe day-by-day application.

Besides this, recent research efforts on the behavior and design of Class 3 cross-sections have been undertaken, within RFCS research project "Semi-Comp". The outcome of the project mainly consisted in an improved analytical formulation for cross-section resistance that enables a fully continuous transition between the plastic bending resistance $M_{pl,Rd}$ at the Class 2–3 border and the elastic bending resistance $M_{el,Rd}$ at the Class 3–4 border: indeed, as Figure 1 shows, such a gap still exists in *Eurocode 3*, associated with the principles of cross-section classification. Present paper shows how this new concept for semi-compact cross-sections can be successfully incorporated in the more general context of beam-column members with Class 3 cross-sections.

It should be noted that only three classes are defined in other standards such as the *Australian Standard AS4100* (Standards Australia 1998) and *AISC LRFD Specifications* (AISC 2000), namely compact, non-compact and slender sections.

In a first step, the paper proposes a brief review of the new beam-column formulae with a particular emphasis on their main features; then, the basic principles of the proposed Class 3 cross-section resistance model are detailed in Section 2. Section 3 focuses on the validation of a purposely-derived FEM tool towards experimental tests; this numerical tool is extensively used in the parametric studies described in Section 4, where comparisons with the results provided by the design formulae are presented.

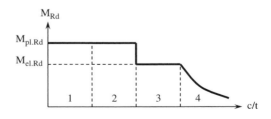

Figure 1. Distribution of bending resistance along cross-section class ranges according to Eurocode 3.

2 NEW DESIGN FORMULAE FOR CLASS 3 HOLLOW SECTION MEMBERS

2.1 New beam-column formulae in Eurocode 3

As already mentioned, the new beam-column rules in *Eurocode 3* consist of two different sets, Method 1 and 2, given separately in Annexes A and B, respecttively (each country has the possibility to recommend one of the formulae set, by means of the National Annexes).

Method 1 has been basically built on the principles of generality, consistency and continuity with the other formulae of the code; it provides a high level of accuracy, and is applicable to any type of cross-section. Reference is especially made here to Method 1 set of design rules that have been derived in order to fulfill the following requirements: accuracy, generality, consistency, continuity, transparency and mechanical basis (Boissonnade et al. 2006, CIDECT 2007).

Method 2 is dedicated to a more simple and practical use, through global factors that cover several physical effects; specific formulae for hollow cross-sections are also proposed.

The basic format of both methods relies on elastic second-order considerations. Spatial behavior and elastic-plastic behavior have been progressively incorporated in the elastic in-plane format (Boissonnade et al. 2006), and the general formulae for Method 1, rearranged in a suitable format, are as follows (for Class 1 and 2 cross-sections):

$$\frac{N_{Ed}}{\chi_y N_{pl,Rd}} + \mu_y \left[\frac{C_{my} M_{z,Ed}}{\left(1 - \frac{N_{Ed}}{N_{cr,y}}\right) C_{yy} M_{pl,y,Rd}} \right.$$

$$\left. + \gamma^* \mu_y \frac{C_{mz} M_{z,Ed}}{\left(1 - \frac{N_{Ed}}{N_{cr,z}}\right) C_{yz} M_{pl,z,Rd}} \right] \leq 1 \quad (1)$$

$$\frac{N_{Ed}}{\chi_z N_{pl,Rd}} + \delta^* \mu_z \left[\frac{C_{my} M_{z,Ed}}{\left(1 - \frac{N_{Ed}}{N_{cr,y}}\right) C_{zy} M_{pl,y,Rd}} \right.$$

$$\left. + \mu_z \frac{C_{mz} M_{z,Ed}}{\left(1 - \frac{N_{Ed}}{N_{cr,z}}\right) C_{zz} M_{pl,z,Rd}} \right] \leq 1 \quad (2)$$

As can be seen, the formulae presented here make use of the well-known amplification factor concept. Also, the concept of equivalent moment factor C_m is employed; it can be shown that the definitions of both of them are intrinsically dependent on each other (Boissonnade et al. 2006).

The use of an equivalent moment factor is a first important feature of the proposed formulae. It brings significant simplification since it allows substituting the (general) actual bending moment distribution on the member for an equivalent one that leads to the same maximum second-order bending moment on the member M^{max}, see Figure 2.

In the particular case of a linear bending moment, which is of prime importance in practice, Method 1 proposes a new expression for C_m:

$$C_m = 0.79 + 0.21\psi + 0.36(\psi - 0.33)\frac{N_{Ed}}{N_{cr}} \quad (3)$$

where ψ is the ratio between the applied end moments ($-1 \leq \psi \leq 1$). This last definition, when compared to well-known definitions such as the ones of Austin (Austin 1981) or Massonnet (Massonnet & Campus 1955), significantly improves the accuracy of the beam-column formulae, where the determination of C_m is decisive.

Besides this, Method 1 proposes to take due account of elastic-plastic effects by means of (i) C_{ii} and C_{ij} factors for $M - N$ interactions (*i* and *j* indexes are relative to either y – y or z – z axes, see Equation 4) and (ii) γ^* and δ^* factors for $M_{y,Ed}$–$M_{z,Ed}$ biaxial bending interaction.

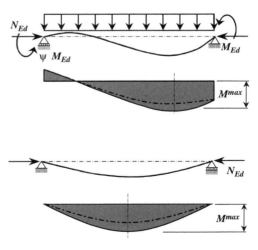

Figure 2. Illustration of equivalent moment factor C_m concept.

Since instability effects influence the extent of yielding in the member, the plastic cross-section resistance $M_{pl,Rd}$ of a Class 1 or 2 section might not be reached, but an elastic-plastic $C\,M_{pl,Rd}$ one, where C is so that $C\,M_{pl,Rd}$ is comprised between $M_{pl,Rd}$ and $M_{el,Rd}$. When multi-axial behavior and loading must be contemplated, a distinction has to be made between what concerns strong axis and weak axis, therefore the need to introduce distinctive C_{ii} and C_{ij} coefficients (Boissonnade et al. 2006). The general writing of the C_{ii} factor is as follows:

$$C_{ii} = 1 + (w_i - 1)\left[2 - \frac{1.6}{w_i} C_{mi}^2 (\bar{\lambda}_i + \bar{\lambda}_i^2)\right] \frac{N_{Ed}}{N_{pl,Rd}} \geq \frac{W_{el,i}}{W_{pl,i}} \quad (4)$$

with $w_i = W_{pl,i}/W_{el,i} \leq 1.5$. W_{pl} and W_{el} designate respectively the plastic and elastic modulus of the member cross-section in bending; $\bar{\lambda}_i$ represents the relative column slenderness for flexural instability.

The definition of C_{ii} must obviously depend on $\bar{\lambda}_i$, so that the behavior of the beam is plastic for small slenderness, and becomes elastic as the member slenderness and the axial compression increase. In Equation 4, the factor $(w - 1)$ represents the maximum available bending potential between pure elasticity and pure plasticity, and must be multiplied not only by a function of $\bar{\lambda}_i$, as explained before, but also by a function of C_m, because the member cannot develop the same elastic-plastic effects whatever the transverse loading is. This calibrated coefficient then clearly permits a smooth physical transition between plasticity and elasticity; the bending moment of the column cross-sections being always greater than the elastic moment resistance, the C_{ii} coefficient must be bounded by W_{el}/W_{pl} for Class 1 and 2 cross-sections.

As explained before, γ and δ are factors to account for biaxial bending elastic-plastic interaction; they are respectively chosen as $0.6\sqrt{w_z/w_y}$ and $0.6\sqrt{w_y/w_z}$.

As one of the most outstanding features of the Method 1 set of design formulae, continuity aspects have been receiving specific attention. It can be observed that the different formulae provide continuity:

– from elastic to plastic behavior;
– from cross-section to member resistance when the member slenderness increases;
– with the other formulae of the code;
– between simple loading situations;
– from spatial to in-plane behavior.

Finally, it has to be mentioned that the proposed formulae have been validated for Class 1 and 2 cross-section members (both open and hollow section shapes) on about 15 000 non-linear GMNIA FEM simulations results as well as on 370 test results (Boissonnade et al. 2006, CIDECT 2007).

2.2 *Proposal for a design model for Class 3 hollow section members*

As Equations 1–2 show, the new beam-column formulae have been mainly built for plastic Class 1 and 2 cross-sections. However, Class 3 and 4 cross-sections situations are obviously covered, since the basis of the formulae is elastic behavior (Boissonnade et al. 2006).

As already mentioned, the formulae have been derived so that to offer a maximum level of continuity; nevertheless, a lack of continuity in bending resistance can be observed at the Class 2 to 3 border, as Figure 1 shows. Indeed, in accordance with *Eurocode 3* assumptions, the cross-section resistance for pure bending drops from $M_{pl,Rd}$ to $M_{el,Rd}$ when the cross-section passes from Class 2 to Class 3. More than 50% of resistance may be lost due to the presence of this gap, which is physically not acceptable; the loss of resistance may rise up to 65% when biaxial bending is of concern.

In order to propose a solution to this problem, a RFCS research project named "Semi-Comp" has been initiated, where a mechanical model for a continuous cross-section and member buckling resistance along the Class 3 field has been developed. The basic idea on which the model relies on consists in offering an intermediate "elastic-plastic" resistance of the semi-compact cross-section in bending, between pure plastic and pure elastic behavior (see Figure 3). In this way, the distribution of stresses of Figure 4 is adopted. The total bending resistance then consists of two distinctive contributions (Equation 5):

– a plastic contribution $W_{3,pl} f_y$ that corresponds to the yielded fibers of the section;
– an elastic contribution $W_{3,el} f_y$ arising from the other fibers of the section that have not reached the yield stress (fibers within h_{ep}, see Figure 4).

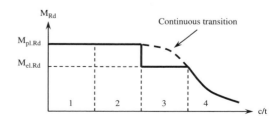

Figure 3. Proposed continuous distribution of bending resistance along cross-section class ranges.

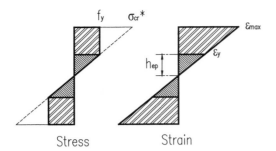

Figure 4. Elastic-plastic distribution of stresses.

$$M_{3,Rd} = (W_{3,pl} + W_{3,el})f_y \quad (5)$$

Obviously, the key aspect here lies in the correct determination of h_{ep}, that must be so that the cross-section reaches a full plastic resistance at the Class 2–3 border (i.e. $h_{ep} \to 0$), and that the cross-section exhibits its sole elastic resistance at the Class 3–4 border (i.e. $h_{ep} = h$).

This is achieved in assuming that whenever the stress distribution were still linear beyond f_y, the maximum stress reached would be equal to σ_{cr}, at a strain level $\varepsilon = \varepsilon_{max} > \varepsilon_y$, ε_y being the maximum elastic strain. The critical plate stress σ_{cr} appears as a convenient stress measure here, since the local (plate) instability effects play the key role here.

$$\sigma_{cr} = \frac{k_\sigma \pi^2 E}{12(1-\nu^2)}\left(\frac{t}{b}\right)^2 \quad (6)$$

However, in order to fulfill the continuity aspects at the ends of the Class 3 field, it is necessary to bring modifications to the original definition of σ_{cr}. Indeed in accordance with Figure 4, σ_{cr} must be so that:

– $\sigma_{cr} = \infty$ at the limit between Class 2 and Class 3;
– $\sigma_{cr} = f_y$ at the limit between Class 3 and Class 4.

In the particular case of a simply supported plate in compression, such a modified σ_{cr}^* writes:

$$\sigma_{cr}^* = 1.616\, E\left(\frac{1}{9.5c/t - 323\varepsilon}\right)^2 \quad (7)$$

It is to be noted that this definition stands for a Class 3 field defined for so-called c/t ratios comprised between 34ε and 38ε, unlike the 38ε to 42ε one prescribed by Eurocode 3. Place is missing here to provide full justifications for this fact (cf. Semi-Comp 2007).

Finally, the design model resorts to the plastic cross-section design checks of *Eurocode 3* for combined loading, adequately modified to allow for continuous transitions along the Class 3 range:

$$M_{N,3,y,Rd} = M_{3,y,Rd}\frac{1-n}{1-0.5a\left(1-\dfrac{f_y}{\sigma_{cr}^*}\right)} \leq M_{3,y,Rd} \quad (8)$$

$$M_{N,3,z,Rd} = M_{3,z,Rd}\frac{1-n}{1-0.5a\left(1-\dfrac{f_y}{\sigma_{cr}^*}\right)} \leq M_{3,z,Rd} \quad (9)$$

$$\left(\frac{M_{y,Ed}}{M_{N,3,y,Rd}}\right)^{\alpha^*} + \left(\frac{M_{z,Ed}}{M_{N,3,z,Rd}}\right)^{\beta^*} \leq 1 \quad (10)$$

where

$$\alpha^* = \beta^* = \frac{1.66}{1-1.13n^2}\left(1-\frac{f_y}{\sigma_{cr}^*}\right) + \frac{f_y}{\sigma_{cr}^*} \leq 6 \quad (11)$$

These equations ensure full continuity with the simple loading cases and within the Class 3 range.

The implementation of this cross-section model is straightforward in the Method 1 set of beam-column design formulae. Obviously, one must replace $M_{pl,y,Rd}$ and $M_{pl,z,Rd}$ by $M_{3,y,Rd}$ and $M_{3,z,Rd}$ in the different factors, such as for w_y and w_z in the C_{ii} factors for example. In addition, the γ^* and δ^* factors need to be adjusted to:

$$\gamma^* = 0.6\sqrt{\frac{w_z}{w_y}}\left(1 + \frac{1-0.6\sqrt{\dfrac{w_z}{w_y}}}{0.6\sqrt{\dfrac{w_z}{w_y}}}\frac{f_y}{\sigma_{cr}^*}\right) \quad (12)$$

$$\delta^* = 0.6\sqrt{\frac{w_y}{w_z}}\left(1 + \frac{1-0.6\sqrt{\dfrac{w_y}{w_z}}}{0.6\sqrt{\dfrac{w_y}{w_z}}}\frac{f_y}{\sigma_{cr}^*}\right) \quad (13)$$

3 ASSESSMENT OF FEM NUMERICAL TOOLS

3.1 Tests on Class 3 hollow section members

Within the frame of "Semi-Comp" project, a series of 24 tests on beam-column members has been undertaken.

The main goal of the tests was to help validating a purposely-derived FEM model that, if found satisfactory, would be used extensively in parametric studies (see section 4).

Amongst the 24 member buckling tests, 6 were performed on RHS $200 \times 120 \times 4$ (S275) members, and 6 on SHS 180×5 (S355) members; all profiles were about 4000 mm long.

In order to fully characterize the different specimens, in view of further comparison with the FEM calculations, several additional measurements and tests have been conducted, mainly regarding dimensions, material characteristics and initial imperfections.

Accordingly, several tensile coupon tests have been completed, including in the corner regions, to provide the necessary information on material behavior (see Table 1).

The coupon tests first showed that the σ–ε curves were not exhibiting a typical "yield plateau", as is usual for constructional steel; consequently, a specific implementation has been made in the FEM models, and the determination of the design yield stress has been made possible through the "$f_{y\,0.2\%}$ procedure".

Secondly, these tests revealed, as expected, a much higher level of yield stress in the corner zones, due to strain hardening (resp. 548 N/mm² for RHS cross-sections and 520 N/mm² for SHS sections).

Besides this, the so-called "cutting strip technique" was used for the determination of initial residual stresses. This method consists in measuring the length of several strips all over the cross-section (see Figure 5) before and after cutting of the strips. The differences in length are directly linked to membrane stresses, while additional measurements of the variation of curvature provide further information on the flexural initial residual stresses.

Figure 5 presents some of the obtained results, where a rather low level of initial membrane residual stresses is observed, in comparison with the level of initial flexural residual stresses, as was expected.

Information on the initial geometrical imperfections is also decisive for the correct interpretation of member buckling tests: it is indeed known to have a non negligible influence on the carrying capacity of real members. In the present project, it was necessary to determine both the "global" longitudinal imperfections and the "local" (plate) defaults, since the occurrence of local buckling is a key feature in the behavior of Class 3 cross-section members.

Accordingly, imperfections on several faces of the tested members have been measured all along the specimens' length, at three different points in a cross-section (see Figure 6).

Three measurement devices have then been fixed on a small trolley that displaces all along the specimen's length, recording the longitudinal defaults along three different lines. Through appropriate geometrical treatment, it becomes possible to separate the member global imperfection from the local imperfection. Indeed, a measurement close to the corner zone of a tubular profile is assumed to be free from local imperfections, i.e. representative of the global imperfection, while a measurement in the middle of a face contains both local and global defaults. Figure 7 shows an example of result where global and local imperfections have been separated.

It was finally found that the relative level of both global and local initial imperfections was small, though highly variable from one specimen to another. These data (shape and amplitude) have been further implemented in the FEM models for validation purposes.

Table 1. Results for tensile coupon tests.

Section	E GPa	f_y N/mm²	f_u N/mm²
RHS $200 \times 120 \times 4$ (S275)	177	378	486
SHS 180×5 (S355)	178	413	538

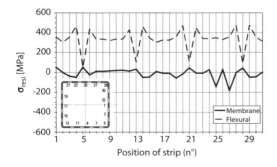

Figure 5. Measured initial residual stresses on SHS180×5 section (cold formed).

Figure 6. Measurement of local and global geometrical imperfections on a face.

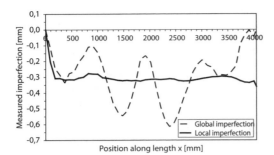

Figure 7. Geometrical global and local imperfections (RHS 180 × 5 profile, strong axis).

Figure 8. General view of test setup—Support conditions and instrumentation in the mid-height cross-section and at supports.

The member buckling main tests consisted in mono-axial or biaxial bending with axial compression tests. For convenient purposes, primary bending was applied through eccentrically applied thrust. As a consequence, the primary bending moment diagrams applied on the members were linear (with $\psi = 1$ or $\psi = 0$ values).

The support conditions of the members during the tests may be assumed to be "pinned conditions" (Figure 8). The end-sections bending rotations can be assumed to be free, since the specimen was fixed at its bottom and top to half-spheres lying on a film of oil ensuring a low level of friction; the centre of these half-spheres were exactly located at the member's end sections. In addition, torsional rotation of the member ends was prevented. So-called end plates were also welded to the specimens end sections, so that to fix them in the test rig. Axial compression was applied through controlling the pressure of oil in the system.

For each test, the following data have been recorded: applied load, axial shortening, and at top and bottom of the specimens, rotations in both principal planes. For the mid-span cross-section, major and minor axis displacements were measured. This information has been further used in comparisons with the FEM results.

3.2 Comparison with FEM models and design proposal

Adequate FEM numerical models (see Figure 9) have been developed, that resorts to shell elements: indeed, the key aspect here consists in characterizing the early or late occurrence of local buckling on the behavior of the semi-compact cross-section member, thus the need for shell elements.

Specific attention has been paid to the meshing of the corner regions, and mesh-density tests have been performed for each possible type of analysis.

Additional fictitious nodes have been defined at the centroids of the end cross-sections for the definition of the support conditions.

Realistic boundary conditions (i.e. with welded plates at the ends of the members) as well as measured characteristics (material law, dimensions, and geometrical imperfections) have been introduced in the numerical models, in order to perform a comparison with the experimental tests as accurate as possible.

It is to be noted that no initial residual stresses have been introduced in the numerical models, since they mainly consist in flexural residual stresses (see section 3.1) that are known to have a lesser influence than the membrane residual stresses which are negligible here.

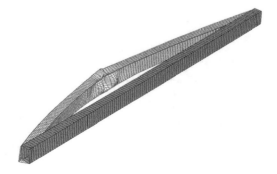

Figure 9. Shell FEM model (amplified deformation at failure and initial shape).

Table 2. Results for hollow section member buckling tests.

Specimen	e_y mm	e_z mm	P_{test} kN	P_{FEM} kN	P_{model} kN	$\%_{FEM}$	$\%_{model}$
R275_BU_1 ($\psi=1$)	55	0	404	378	294	7.0	37.2
R275_BU_2 ($\psi=0$)	55	0	451	453	331	−0.4	36.3
R275_BU_3 ($\psi=1$)	0	45	261	239	211	9.2	23.9
R275_BU_4 ($\psi=0$)	0	45	331	296	240	12.0	38.1
R275_BU_5 ($\psi=1$)	55	45	268	225	183	19.3	46.7
R275_BU_6 ($\psi=0$)	55	45	307	282	217	8.8	41.5
S355_BU_1 ($\psi=1$)	55	0	563	546	449	3.1	25.4
S355_BU_2 ($\psi=0$)	55	0	656	660	509	−0.6	29.0
S355_BU_3 ($\psi=-0.45$)	55	0	708	700	532	1.2	33.1
S355_BU_4 ($\psi=1$)	55	55	460	453	328	1.6	40.3
S355_BU_5 ($\psi=0$)	55	55	600	568	382	5.5	57.1
S355_BU_6 ($\psi=-0.45$)	55	55	629	608	419	3.4	50.0

Table 2 presents the results obtained for both RHS and SHS test series.

In addition to the experimental maximum axial loads, Table 2 reports the results of the FEM calculations and the analytical results yielded by the application of the model described in section 2.2. It also provides information on the applied loading: monoaxial or biaxial bending, and constant ($\psi = 1$) or triangular ($\psi = 0$) first order bending moment.

First, a very good agreement between experimental and numerical results can be observed: the difference is usually less than 5% (except for two results where the experimental load exceeds the FEM load up to about 20%), and is always on the safe side.

Considering the uncertainties inherent to the experimental setup and material behavior, the results presented herein are assumed to be satisfactory enough to validate the ability of the FEM models to simulate the behavior of such Class 3 cross-section members up to collapse. Accordingly, the developed FEM models have been intensively used in parametric studies, as detailed in the next section.

Besides this, Table 2 also shows rather conservative results obtained through the application of the proposed model. This can first be explained by the difference between the assumed levels of imperfections: indeed, the design model implicitly accounts for a rather high level of imperfections, while the measured one was relatively small, which leads to over-conservative design results.

Secondly, due to the delivery of test specimens exhibiting a higher yield stress than expected (see Table 1), all cross-sections are classified as Class 4 sections. As a consequence, a reduced (effective) analytical resistance can only be reached, while the members exhibited a certain amount of elastic-plastic effects, therefore the high level of safety of the design formulae.

4 PARAMETRIC STUDIES

The numerical models being shown to be accurate and appropriate, extensive parametric studies have been performed. The following parameters have been taken into account in a first parametric study:

- 2 different cross-section shapes: RHS 250 × 150 × 6 and SHS 180 × 5;
- 2 different steel grades: $f_y = 235$ N/mm² and $f_y = 355$ N/mm²;
- 2 primary linear bending moments distributions: $\psi_y = \psi_z = 1$ and $\psi_y = \psi_z = 0$ ($M_{y,Ed}$ and $M_{z,Ed}$ end-moments applied on the same side);
- 2 relative slenderness $\bar{\lambda}_z = 0.5$ and $\bar{\lambda}_z = 1.0$;
- 4 different values of relative axial compression $n = N_{Ed}/N_{b,Rd}$: $n = 0$, 0.30, 0.50 and 0.70, where $N_{b,Rd}$ represents the flexural buckling load under pure compression;
- For each fixed values of the previous parameters, 9 combinations of $M_{y,Ed}$ and $M_{z,Ed}$ values have been investigated, so that to allow for the determination of the biaxial bending interaction.

Steel grade S355 have been studied as it has now become the "standard" one, and S235 was also kept in the calculations in order to exhibit a more plastic behavior. Indeed, the influence of instability phenomena (i.e. local buckling) is more pronounced for high strength steel, and as a consequence, the expected elastic-plastic behavior of Class 3 member may be more pronounced for S235 than for S355 steel. In the same way, the relative low values of $\bar{\lambda}_z$ have been chosen so as the buckling behavior of members to be influenced by elastic-plastic effects, since the influence of member instabilities (i.e. flexural buckling or lateral torsional buckling) makes the behavior of members nearly elastic for high relative member slenderness ($\bar{\lambda} > 1.0$).

For each plate of the different cross-sections, a local geometrical imperfection has been accounted for, along the clear width of the considered plate. It consists in a doubly sinusoidal shape (i.e. in both longitudinal and transversal directions) with maximum amplitude equal to $b/200$, b being the clear width.

In addition, a global imperfection has been introduced, that consists in a combination of sinusoidal out-of-straightness for both strong and weak axis (maximum amplitude $L/1000$ at mid-span). Considering such a level of initial geometrical imperfection in both principal axes together with the adopted local one appears to be rather severe, and should therefore lead to safe results.

The values of axial buckling loads $N_{b,Rd}$ have been obtained through separate calculations. The level of relative axial compression n has been applied trough a first load sequence where N_{Ed} applies until the expected level; a second load sequence followed where N_{Ed} is fixed and both $M_{y,Ed}$ and $M_{z,Ed}$ were rising simultaneously. Doing so affects the expansion of second-order effects, since the coupling between lateral displacements caused by primary bending and the axial force may not be the same as in a situation where all the internal (1st order) forces rise proportionally.

In total, 576 non-linear FEM-shell calculations have been performed.

Besides these FEM simulations, their equivalent analytical counterparts have been calculated, in order to investigate the accuracy of the proposed model for Class 3 members. It is to be noted that the results presented herein do not allow for a rigorous determination of the benefits brought by either the semi-compact cross-section model or the new beam-column formulae, since the global response of the member is studied here. However, within the frame of the Semi-Comp project, the cross-section model has been shown accurate and safe through other specific parametric studies (Semi-Comp 2007), while the new beam-column formulae have been deeply investigated and validated, cf. section 2.1 (Boissonnade et al. 2006).

Table 3 gives a summary of results on the comparison between numerical and analytical results. The values reported refer to so-called R_{simul} values, which represent the ratio of the $M_{y,Ed}$–$M_{z,Ed}$ loading leading to failure according to the FEM result to the proportional $M_{y,Ed}$–$M_{z,Ed}$ loading leading to failure according to the proposed formulae. Consequently, a value higher than unity means safety, while a value lower than one indicates an unsafe result. The R_{simul} values also form a good indicator of the level of accuracy of the proposal: a value of 1.10 indicates a 10% resistance reserve.

As can be seen in Table 3, the mean m and standard deviation s values of the R_{simul} ratios are quite satisfactory, highlighting the good accuracy of the proposal. Table 3 indicates in addition minimum and maximum values of R_{simul}, together with an $R_{simul} < 0.97$ criterion that intends at collecting the "real" unsafe situations,

i.e. that cannot be charged to the interaction formulae (it is known that *Eurocode 3* recommendations for flexural buckling can be up to 3% on the unsafe side compared to FEM results, thus the 0.97 value).

Figure 10 also further illustrates the ability of the proposed model to lead to intermediate results between full plastic behavior and pure elastic behavior. It also clearly shows that in this particular case, it would appear unnecessarily conservative to restrict the resistance of such members to their sole *elastic* carrying capacity. The tendency of exhibiting a certain level of plastic behavior has been generally observed on the cases studied, even in situations where the cross-section is classified as Class 4, as Figure 11 shows. Such results are especially responsible for the high R_{simul} ratios reported in Table 3.

In addition to this "accuracy" parametric study, a second "safety" parametric study has been undertaken, aiming at the determination of so-called γ_{M1} safety factors. In this respect, some 576 additional cases have been studied, where all relevant geometrical and material data have been generated by random (using the Monte-Carlo simulation principles) and introduced in both the FEM models and the analytical calculations. Table 4 gives further information on the statistical data used to generate the different parameters needed. A rigorous and strict application of *EN 1990 Annex D* (EN 1990 Annex D 2003) led to a $\gamma_{M1} = 1.03$ value,

Table 3. Results for accuracy parametric study.

	RHS 200 × 120 × 4		SHS 180 × 5	
	$\psi = 1$	$\psi = 0$	$\psi = 1$	$\psi = 0$
m	1.263	1.316	1.275	1.340
s	0.141	0.183	0.229	0.307
min	0.916	0.907	0.903	0.917
max	1.552	1.756	1.820	2.092
<0.97	4	7	7	15
tests	144	144	144	144
Class 4	22	22	16	16

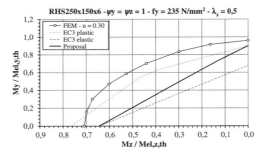

Figure 10. Illustration of an intermediate behavior.

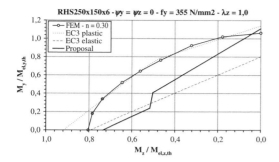

Figure 11. Situation where plastic reserve is observed while cross-section is Class 4 or nearly (first 3 points only are Class 4).

Table 4. Statistical data for hollow specimens.

Parameter	Distribution	σ	V
h	Normal	1	0.005
b	Normal	1	0.01
t	Normal	1	0.05
f_y	Log-normal	1	0.06
$e_{loc.\,web}$	Normal	1	0.324
$e_{loc.\,flange}$	Normal	1	0.324
$e_{glob.\,y}$	Uniform	1	0.1
$e_{glob.\,z}$	Uniform	1	0.1

which complies with the recommended value of 1.10 in *Eurocode 3*.

5 CONCLUSION

This paper has presented the essentials of a new design proposal for semi-compact hollow section beam-columns, that provides a fully continuous transition between Class 2 sections (i.e. exhibiting a plastic behavior) and Class 4 cross-sections (i.e. having an elastic resistance limited by effective properties).

Through extensive comparisons with both experimental tests and adequately-performed FEM simulation results, the new proposal was found accurate and safe. In addition, a safety factor $\gamma_{M1} = 1.03$ has been determined following the specifications of *EN 1990*, on the basis of FEM results obtained from another specific parametric study where all data have been generated by random; this γ_{M1} value complies with *Eurocode 3* recommendations.

Whereas the new beam-column formulae have been incorporated in the latest version of *Eurocode 3*, the design proposal for Class 3 cross-sections results from recently achieved research projects, and is therefore not included in the Eurocode rules. Since the level of reliability of the model is found to be fully satisfactory, the model may be further proposed for implementation in the next draft of *Eurocode 3*.

ACKNOWLEDGEMENTS

The Research Fund for Coal and Steel (RFCS), by means of research project "Semi-Comp" n°RFS-CR-04044, as well as CIDECT, by means of research project 2X, are gratefully acknowledged.

REFERENCES

AISC. 2000. *Load and Resistance Factor Design Specifications for Steel Hollow Structural Sections*. Chicago, Illinois, USA: American Institute of Steel Construction.

Austin, W.J. 1981. Strength and design of steel beam-columns. *Proceedings of ASCE, Journal of the Structural Division* (87).

Boissonnade, N. Greiner, R. Jaspart, J.P. & Lindner, J. 2006. *Rules for Member Stability in EN 1993-1-1-Background Documentation and Guidelines*. ECCS publication N 119.

Campus, F. & Massonnet, C. 1955. Recherches sur le flambement de colonnes en acier A37, à profil en double té, sollicitées obliquement. *Bulletin du CERES*. Liège: Tome VII.

Jaspart, J.P. & Weynand, K. 2007. Resistance and stability of structural members with hollow sections under combined compression and bending. *CIDECT Research Project 2X*.

1992. *Eurocode 3: Design of Steel Structures, Part 1-1: General Rules and Rules for Buildings (ENV 1993 1-1)*. Brussels: CEN (Comité Européen de Normalisation).

2003. *EN 1990 Eurocode–Annex D. Basis of Design*. Brussels: CEN (Comité Européen de Normalisation).

2005. *Eurocode 3: Design of Steel Structures, Part 1-1: General Rules and Rules for Buildings (EN 1993 1-1)*. Brussels: CEN (Comité Européen de Normalisation).

2007. Plastic member capacity of semi-compact steel sections–a more economic design ("Semi-Comp"). *Final report (01/01/05–30/06/07)–RFCS–Steel RTD (Contract RFS-CR-04044)*.

1998. *Standards Australia*. Australian Standard AS4100, Steel Structures. Sydney, Australia: Standards Australia.

Re-evaluation of fatigue curves for flush ground girth welds

Y.H. Zhang & S.J. Maddox
TWI, Cambridge, United Kingdom

A. Stacey
HSE, London, United Kingdom

ABSTRACT: The fatigue performance of girth welded steel pipes with the weld cap and root flush-ground is designated as Class C in the BS 7608 fatigue design rules. However, this is not based directly on experimental data. Since fatigue test results for flush-ground girth welds in steel pipes have now become available, a comprehensive examination of published work was carried out to re-evaluate this and other design $S-N$ curves for such welds on the basis of the relevant experimental data. The data considered included some from tests on full-scale welded pipes and other from strip specimens cut from girth-welded pipes. In addition, the opportunity was taken to consider the larger database available from butt-welded plate specimens for comparison.

An important condition on the use of relatively high design curves for flush-ground butt welds is that the weld should be proved free from significant welding defects. Concern about the detectability of small defects that could result in fatigue performance below such design curves has limited their acceptance. In the case of volumetric flaws, relevant fatigue data obtained from girth welds containing reportable embedded flaws are available and they were used, together with data obtained from butt-welded plate specimens, to address such concerns.

1 INTRODUCTION

Flush-grinding of a butt weld is an established method for improving its fatigue performance. This eliminates the stress concentration created by the weld profile and removes the inherent weld toe flaws at which fatigue cracks typically initiate. Consequently a fatigue performance far superior to that typically assigned to conventional structural welds in the as-welded condition is expected. However, the current fatigue design rules in various standards and codes for such welds are not based on fatigue test data from flush-ground girth welds but on data for joints between flat plates, much of which were obtained many years ago. Furthermore, the welding procedures, types and consumables might not be representative of those currently used in offshore structures. However, since data from flush-ground girth welds in steel pipes have now become available, there is the opportunity to re-evaluate the design $S-N$ curve(s) for such welds on the basis of relevant experimental data.

To qualify for BS 7608 Class C (1993), flush-ground welds must be free from defects that could reduce that potentially high fatigue strength. However, in the great majority of situations it is found that a fatigue strength higher than Class D cannot be justified. This is attributed mainly, from evidence from plate specimen tests, to the possible presence of flaws which are too small for reliable detection using current non-destructive testing (NDT) methods but which could be of sufficient size to reduce the fatigue strength of the joint. Therefore, it is necessary to examine the defect acceptance criteria for flush-ground welds using data from girth welds, particularly for embedded defects since they become the weak link in the fatigue performance and the minimum defect sizes need to be assessed against the detection capability of current NDT methods.

2 FLUSH-GROUND WELDS WITH NO REPORTABLE FLAWS

2.1 *Full scale pipe specimen*

Published data from fatigue tests on full-scale pipes with flush-ground girth welds are very scarce and only four relevant test series were found (Wirsching et al.

1995, Salama 1999, Maddox et al. 2002, Wastberg & Salama 2007). Fatigue test results for seven API-5L X60 UOE steel pipe specimens with outer diameter (OD) 610 mm and wall thickness (WT) 20 mm were reported by Wirsching et al. (1995). Details of the test results, NDT examination records and failure locations were not reported and only the statistical analysis of fatigue endurances was published.

The results reported by Wastberg and Salama (2007) were also from UOE X60 pipe specimens with an OD of 610 mm and WT of 20.6 mm. They included the data previously published by Salama (1999). All ground welds were inspected by two independent organizations and were found to be either free from any defects or to contain only small defects of acceptable sizes according to API 1104 (2005). Unfortunately, the NDT results were not reported. A total of eight pipe specimens were tested, each containing three girth welds. All fatigue tests were carried out under axial loading at a stress ratio $R = 0.1$. It was found through post-test examinations that fatigue crack had occurred at lack of fusion defects, located close to the inside or outside surface. The defect sizes, which caused significant reduction in fatigue lives, were around 2×15 mm and were considered to be at the limits of reliable detection by NDT.

Maddox et al. (2002) reported results obtained from flush ground girth welds in 609 mm OD and 21.4 mm WT API 5L X60 steel pipe for tendons and 273 mm OD and 12.6 mm WT API 5L X80 pipe for risers. Two tendon specimens containing three welds each were fabricated in the 1G position, using single sided submerged arc welding (SAW) throughout. The weld root was made onto backing tape. Three riser specimens containing two welds each were fabricated in the 1G position, using a single sided GTAW root pass followed by SAW fill and cap. All weld root beads and caps were ground flush with the pipe surfaces after welding. Comprehensive NDT including radiography testing (RT), ultrasonic testing (UT) and magnetic particle inspection (MPI) did not reveal any defects. Six full-scale girth welds in the two tendon specimens and six full-scale girth welds in the three riser specimens were fatigue tested under tension-tension axial loading at a mean stress of ~175 MPa and 125 MPa, respectively. All the tests gave run-outs with no evidence of fatigue cracking in any of the welds.

These results, or just the reported S-N curves in the case of Wirsching et al, are plotted in Figure 1 together with the BS 7608 Class C design curve. As will be seen, the fatigue test results for full-scale specimens do not agree well. Although the mean curve from Wirsching et al. (1995) was reported to be slightly above the BS 7608 C design curve, the design curve derived is below it. This was due to the large standard deviation in the original data and the small

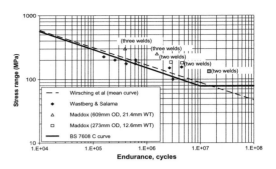

Figure 1. Comparison of fatigue endurance of full-scale specimens with the BS 7608 C curve. All specimens tested by Maddox et al. (2002) were run-outs.

number of results. For the test results reported by Wastberg and Salama (2007), the endurances of four of the eight tests were below the BS 7608 design C curve. Failure of these girth welds started from planar defects of about 2×15 mm. They were missed during NDT by two independent organisations. For the four pipe specimens which achieved the Class C curve, failure started from defects of $\leq 1 \times 8$ mm, which is considered to be too small to be reliably detected. On the other hand, all the results from Maddox et al were run-outs to endurances that exceeded the Class C curve significantly. This highlights the importance of careful NDT and stringent requirement on the elimination of defects to qualify for the Class C curve. For this reason, the test results from Wirsching et al should be taken with caution. The work reported by Wastberg and Salama raised a question about whether or not a lack of fusion defect, which could significantly reduce the fatigue lives of ground welds, can be reliably detected by relevant NDT.

2.2 Strip specimen

In the early 1990's, a large testing programme (Razmjoo et al. 1998) was carried out in three laboratories to investigate the fatigue performance of strip specimens extracted from flush-ground girth welds in pipes for applications in the Heidrun TLP. The tendons were made from 12 m lengths of 1118 mm OD by 38 mm WT pipes in steel equivalent to API 5L Grade X70 specification. They were fabricated using a number of welding procedures.

Waisted fatigue test specimens with a minimum width of around 100 mm were extracted from the pipes. All welds were flush-ground both inside and outside before fatigue testing. The fatigue tests were conducted under constant amplitude axial loading at $R = 0.1$ in most cases. The tests were run until failure occurred or to target lives based on the BS 7608 Class C curve.

From the total of 68 specimens, 18 specimens were tested at 6 °C in seawater with cathodic protection (−1050 mV). Others were tested at room temperature in air.

In general, establishing the fatigue strengths of the flush-ground butt welds was a challenge. In the absence of weld toes and any significant embedded flaws, fatigue cracking could initiate at various locations in the specimens other than the weld. In the event, a test was terminated for one of four reasons:

- Failure in the weld (11 specimens)
- Failure in the weld but from the edge of the specimen (11 specimens)
- Failure from the machine grips or in parent plate (22 specimens)
- Run-out (i.e. specimen did not fail) (24 specimens)

The test results are compared with the Class C curve in Figure 2. It can be seen that all the test data are above the C curve, including those from tests carried out in seawater cathodic protection (CP) (Razmjoo et al. 1998). Furthermore, these data from strip specimens exhibit a much shallower slope in the long life regime (>10^6 cycles) when compared with the Class C curve, and a higher fatigue limit than that of the full-scale pipes shown in Figure 1.

2.3 Plate specimen

As the S-N curve classification of welded plates is the same as that for girth welded pipes, the review of the fatigue performance of flush-ground butt welded steel plate specimens conducted by Maddox (1997) is useful, see Figure 3. Data in the low-cycle fatigue regime were expressed in terms of the pseudo-elastic stress range, as suggested in PD 5500 (2000). As will be seen in Figure 3, the plate specimen data are more widely scattered than the girth weld strip specimen data. The majority of the results qualified for the C curve with only a few exceptions where the results were only slightly below the C curve. The larger scatter is to be expected since the database is much larger than that for strip specimens. However, it might also be associated with inconsistent welding quality in the plate specimens obtained from many different sources, in contrast with the comparatively good welding control and stringent NDT acceptance criteria in girth welding. Consequently, some of the plate specimens might have contained defects larger than the acceptable limit for girth-welded pipes. Furthermore, the tests on these plate specimens were carried out at different stress ratios, some even with $R < 0$. Overall, the results from plate specimens suggest that the C curve is applicable to flush-ground butt welds.

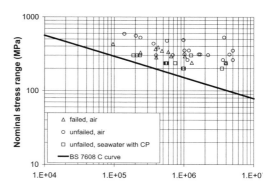

Figure 2. Comparison of the fatigue performance of flush-ground strip specimens cut from girth welds (Razmjoo et al. 1998) with the BS 7608 C curve.

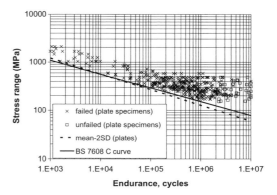

Figure 3. Fatigue endurance data for flush-ground butt-welded plate specimens (Maddox 1997).

2.4 Comparison of fatigue design rules for flush ground welds

Generally, flush-ground butt welds are recognised as having enhanced fatigue performance compared with as-welded joints. This is reflected in the different design standards. A graphic presentation of the corresponding design S-N curves is shown in Figure 4, where it can be seen that the Class C curve in BS 7608 has a slope of 3.5, whereas 3.0 is adopted by DNV (2005) and IIW (Hobbacher 1996). In comparison, Class C is more conservative for endurances below 2×10^5 cycles but less conservative above this endurance. The AWS designation for flush-ground butt welds is Class B which has a shallow S-N curve ($m = 4.3$). As noted earlier, a shallow slope with $m > 3$ would be expected to be appropriate for flush-ground

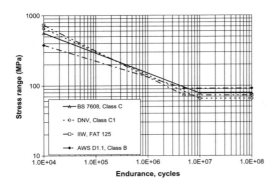

Figure 4. Comparison of fatigue design curves for flush ground butt welds from different standards.

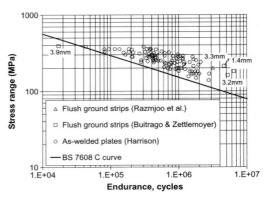

Figure 5. Fatigue performance of strip and plate specimens containing porosity less than 4%.

butt welds as a result of crack initiation occupying a significant proportion of the fatigue life. Compared to other standards, the AWS B curve is conservative at fatigue endurances below one million cycles, especially in the regime of high stress ranges. It should be noted that the AWS C1 curve has been reported to be more appropriate as the design curve for flush-ground girth welds. (Buitrago & Zettlemoyer 1999).

It must be emphasized that the above comparison is based on the S-N curves at the reference thickness for each code. Since some codes apply a thickness correction for flush ground welds for thickness above the reference value and recommend some degree of penalty, care must be taken when such a comparison is made at a specimen thickness beyond the reference thickness for a certain code.

3 FLUSH GROUND GIRTH WELDS WITH REPORTABLE FLAWS

3.1 Surface breaking and internal planar defects

There do not appear to have been any investigation specifically directed at studying the effect of surface breaking or internal planar defects, revealed during NDT, on fatigue performance of flush ground welds. The lack of fusion defects reported by Wastberg and Salama (2007) were only revealed after the post test examinations of the fracture surfaces of the failed welds. In fact, to qualify for the fatigue design curves for flush-ground girth welds, the criteria of defect acceptance are very strict in relevant standards (BS 7910 2005, ASME 2007). Surface breaking and internal planar defects, such as lack of fusion or lack of penetration, are particularly important since they are much more deleterious to the fatigue performance of the weld than embedded volumetric defects such as pores or slag inclusions. In recognition of this, it has been proposed that they cannot be accepted in flush-ground welds in TLP tendons designed to AWS C1 (Buitrago & Zettlemoyer 1999).

3.2 Effect of porosity

Recently the effect on fatigue performance of porosity and slag inclusions in flush-ground girth welds has been investigated (Razmjoo et al. 1998, Buitrago & Zettlemoyer 1999). Figure 5 shows the fatigue test results from the strip specimens containing porosity. The maximum pore sizes have been characterised by RT and UT in these specimens. As it is likely that fatigue crack initiation may occur from internal flaws, the fatigue performance of the plate specimens (12 mm in thickness) containing porosity density <4% (Harrison 1972a) are also included for comparison. It will be seen that when porosity density was less than 4% and the maximum reported individual flaw size was up to 4.8 mm, the fatigue performance could still be qualified as Class C, except in the high stress range/low endurance (<10^5 cycles) regime. The results suggest that, in the medium and long fatigue life regimes, a pore size of around 4 mm can be tolerated, a size which can be detected reliably by RT or UT.

However, when porosity density was increased to 8% the test data for many plate specimens fell below the C curve, Figure 6 (Harrison 1972a). This suggests that when pore density is low (below 4%), the individual pore size plays an important role. When the pore density is high, however, the fatigue performance is significantly reduced regardless of the individual pore size.

3.3 Effect of slag inclusions

Fatigue test results from flush-ground girth welds in strip specimens with slag inclusions (Razmjoo et al.

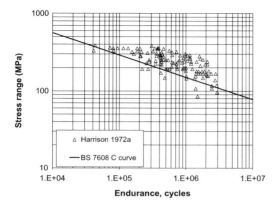

Figure 6. Fatigue performance of plate specimens containing porosity less than 8% (Harrison 1972a).

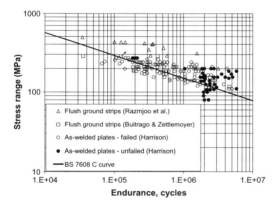

Figure 7. Fatigue performance of strip and plate specimens containing slag inclusions.

1998, Buitrago & Zettlemoyer 1999) are compared in Figure 7. To illustrate the effect of inclusions, the fatigue data from as-welded plates containing inclusions less than 10 mm long (Harrison 1972b) were also included for comparison. It can be seen that, although the data from Razmjoo et al. (1998), which had defect lengths between those reported in two other references (Buitrago & Zettlemoyer 1999, Harrison 1972b), qualify for Class C, many of the test results from Buitrago & Zettlemoyer and Harrison fall below the C curve even from the plate specimens. This suggests that the fatigue life depends not only on the inclusion length, as specified in BS 7910 (2005), but also on other factors such as the ligament and the inclusion height.

4 DISCUSSION

When considering the fatigue design of girth weld in pipes or tubes, results obtained from full-scale specimens are highly recommended in order to exclude uncertainties associated with other types of specimen. However, the requirements for the special testing facility, time and cost make this difficult to achieve. In practice, small-scale specimens have often been used to simulate the behaviour of large-scale components. When the test results from these specimens are used for design purposes, however, care must be taken to ensure they represent the fatigue performance of the large-scale components.

Available fatigue data from flush-ground welds in strip specimens cut from girth welded pipes confirm the qualification of the BS 7608 C curve. Even those specimens that were originally rejected due to the presence of unacceptable defects (Razmjoo et al. 1998) gave lives consistent with Class C. This suggests that the fatigue performance of strip specimens free from defects can be comfortably classified as Class C. However, some of the results of the flush-ground girth welded pipes were lower than the corresponding strip specimen results and could not be qualified for the C curve (Wirsching et al. 1995, Wastberg & Salama 2007). Although the data from the above full-scale specimens should be viewed with caution, as described before, the difference in fatigue performance between the strip and the full-scale specimens was not unexpected as the same has been observed in tests on as-welded girth welds (Razmjoo et al 1998). In this case, the difference was attributed to the possible effects of size and residual stress.

With regard to size, as a strip specimen contains only a small proportion of the length of a girth weld, it is unlikely to contain the most severe defect present or to introduce the same level of stress concentration as in some locations in the pipe. With respect to residual stress effects, it is speculated that the welding-induced residual tensile stress could be partly or even fully released during extraction of a strip specimen. Even if this were not the case, residual stress measurements in girth welds have shown that they can be widely scattered (Maddox et al. 2006), even for a single weld. Thus, a strip specimen may coincide with a region of low residual stress in the original girth weld. A consequence of having lower residual stresses in the strip specimens is that the effective mean stress, resulting from the superposition of the applied and residual stress, would have been lower than in the full-scale specimens.

Another issue, which could affect the direct comparison of the fatigue results between strip and full-scale specimens, is the quality of flush-grinding at weld roots. For single sided welds in pipes, grinding the weld roots is not as easy as for strip specimens. Consequently cracking might preferentially initiate at

Table 1. Defect acceptance criteria for quality category Q1 (BS 7910 2005) and flush-ground girth welds with thickness limited to 50 mm (ASME 2007).

Standard	Porosity, % area on radiograph	Flaw acceptance limit			
		Individual pore size, mm	Slag inclusion length, mm	Undercut, (depth/ wall thickness)	Planar defects
BS 7910 -Q1	3.0	thickness/4 or 6.0	2.5	0.025	not applicable*
ASME VIII Div.1	not defined	6.3	6.0	no undercut	not allowed

* can be calculated by fracture mechanics but considered to be undetectable by current NDT methods.

the weld root, resulting in a lower fatigue endurance. Thus, to make the fatigue results of strip specimens representative of large-scale specimens, the same quality of grinding both inside and outside must be assured.

To qualify for the Class C design curve, detailed NDT of welds must be conducted to ensure that they are free from significant defects. Thus, any test data for which the NDT results are lacking, e.g. the data reported by Wirsching et al. from full-scale fatigue testing, should be taken with caution.

The work carried out by Maddox et al (2002) provided full details of the NDT examination and testing conditions. All twelve welds qualified with respect to the C curve without failure. The results therefore strongly support the BS 7608 C curve. Furthermore, all the results from the strip specimens also support the adoption of the C curve for flush-ground girth welds, even for those tests undertaken in seawater with cathodic protection (Razmjoo et al. 1998).

The defect acceptance criteria in two codes, which might be used for flush-ground girth welds, are compared in Table 1. One is based on fitness-for-service (BS 7910 2005) and the other on fabrication limits (ASME 2007). It should be noted that BS 7910 only provides defect limits up to quality category Q1 (equivalent to design Class D) on the basis that beyond this limit NDT cannot be relied upon to detect critical defects. The defect acceptance criteria in ASME, Section VIII, Division 1 are those currently used for some tendons (Buitrago & Zettlemoyer 1999).

The results obtained from strip specimens cut from girth welded tendons reported by Razmjoo et al (1998) provided a direct comparison between specimens containing reportable defects, some of which were measured, and others with no reportable defects under the same production procedures. A pore size up to 8 mm and slag inclusions up to 18 mm long were reported. The comparable fatigue endurances of these specimens containing reportable defects with those containing no reportable defects suggest that these defects can be tolerated, providing confidence in detecting the limiting defect sizes using current NDT methods. However, this finding was based on a comparatively small sample and more such tests are required to determine the critical defect sizes.

5 CONCLUSIONS

- There is uncertainty in the fatigue performance of flush ground girth welds in full-scale pipe specimens. Although most of the results reviewed strongly support the BS 7608 C curve, some data, not reported in full (lacking details of NDT records, failure locations and individual test results) suggest that Class C can be unsafe. They highlight the importance of careful NDT inspection and stringent requirement on the elimination of defects to qualify for the Class C curve. It is apparent that there is a need for further effort to establish the missing details and for more full-scale fatigue testing of flush-ground girth welds with full details of the welds.
- BS 7608 Class C is comfortably qualified on the basis of fatigue data obtained from strip specimens cut from girth welded joints.
- The optimistic fatigue performance of small specimens can be attributed partly to their reduced residual stresses compared to large–scale pipe specimens. Fatigue testing of such specimens at high stress ratios is required in order to predict the behaviour of actual girth welds conservatively.
- Data from flush-ground plate specimens also support Class C classification. However, because of deficiencies, for example, NDT records, weld quality and misalignment, their fatigue behaviour may not be truly representative of that of girth welds.
- The limited database on flush-ground girth welds containing reportable embedded flaws suggests that the flaw sizes present in welds which achieved Class C can be detected using current NDT methods.

REFERENCES

API 1104. Welding of pipelines and related facilities. *American Petroleum Institute*, November 2005.

ASME Section VIII, Division 1. 2007. Boiler and Pressure Vessels and Code. *The American Society of Mechanical Engineers*, New York.

AWS D1.1/D1.1M:2002: Structural Welding Code—Steel, *American Welding Institute*, August 31 2001.

BS 7608:1993. Fatigue design and assessment of steel structures, BSI, London, UK.

BS 7910:2005. Guide on methods for assessing the acceptability of flaws in metallic structures. BSI, London, UK.

Buitrago, J., Weir, M.S. & Kan, W.C. Fatigue design and performance verification of deepwater risers. In Proc. of OMAE2003, Paper No. OMAE 2003-37492, Cancum, Mexico, June 2003.

Buitrago, J. & Zettlemoyer, N. 1999. Fatigue of tendon welds with internal defects. In Proc. of OMAE99, Paper No.: Mat-2001, St. Johns, Newfoundland, Canada.

DNV RP C203:2005: Fatigue strength analysis, *Det Norske Veritas*, May 2005.

Harrison, J.D. 1972a. The Basis for An Acceptance Standard for Weld Defects, Part 1: Porosity. *Metal Construction* 4: 99.

Harrison, J.D. 1972b. The basis for an acceptance standard for weld defects, Part 2: Slag inclusions. *Metal Construction* 4: 262.

Hobbacher, A.: Fatigue design of welded joints and components, Recommendations of IIW Joint Working Group XIII–XV, Abington Publishing, 1996.

Maddox, S.J. 1997. Developments in fatigue design codes and fitness-for-service assessment methods. *In Performance of Dynamically Loaded Welded Structures, IIW 50th Annual Assembly Conference*, San Francisco, Ed: S J Maddox, M Prager. Publ: New York, NY 10017, USA; *Welding Research Council, Inc.*: 22–42.

Maddox, S.J., Speck, J.B., Lockyer, S.A. & Razmjoo, G.R. 2002. Fatigue performance of girth welds made from one side. *TWI Report No. 5680/26/02*, confidential to membership companies.

Maddox, S.J., Razmjoo, G.R. & Speck, J.B. 2006. An investigation of the fatigue performance of riser girth welds. *In Proceedings of Conf. Offshore Mechanics and Arctic Engineering, ASME, Paper No. OMAE2006-92315*.

PD 5500:2000: 'Unfired fusion welded pressure vessels', BSI, January 2000.

Razmjoo, G.R., Maddox, S.J. & Hayes, B. 1998. Fatigue performance of flush-ground TLP tendon girth welds. *TWI Report No. 5680/13/98*, confidential to membership companies.

Salama, M.M. 1999. Fatigue design of girth welded pipes and the validity of using strips. *In Proc. of OMAE99, Paper No.: Mat-2003*, St. Johns, Newfoundland, Canada.

Wirsching, P., Karsan, D.I. & Hanna, S.Y. 1995. Fatigue/fracture reliability and maintainability analysis of the Heidrum TLP tether system. *In Proc. of OMAE95*, pp.187, Copenhagen, Denmark.

Wastberg, S. & Salama, M.M. 2007. Fatigue testing and analysis of full scale girth welded tubulars. *In Proc. of OMAE07, Paper No.: OMAE2007-29399*, June 10–15 2007, USA.

*Parallel session 5: Fatigue & fracture
(reviewed by P. Marshall & F. Mashiri)*

Experimental determination of stress intensity factors on large-scale tubular trusses

A. Nussbaumer & L. Borges
Swiss Federal Institute of Technology Lausanne, ICOM—Steel Structures Laboratory, Lausanne, Switzerland

ABSTRACT: The fatigue design of tubular space trusses for bridge applications requires additional knowledge with respect to their fatigue resistance. In order to be able to model thoroughfully the fatigue crack initiation and propagation phases, it is important to have at disposal as much as possible information on crack development. The paper presents the results of crack depth measurements using an Alternate Current Potential Drop (ACPD) system. The originality is that the system was installed on the nodes of two truss specimens, 9m by 2m, with K-joints made out of CHS. This system has, to our knowledge, never been applied to large-scale specimens such as those. The multiple electrical current passes possibilities made it difficult to predict if the system would actually measure correctly. The paper describes the tests carried out and the equipment installed. A total of eight probes where installed at the most likely crack locations. The results obtained from the tests are given in terms of S-N data, of crack depth versus number of cycles, of crack propagation rate and of stress intensity factor range, obtained using the Paris law with constants calibrated on previous tests on similar elements. Comparisons with opened crack surface after testing show good agreement with crack depth deduced from ACPD measurements. The paper also contains plots with both crack propagation from numerical calculations and deduced from measurements, which show similar features. Finally, for the test conditions used in this study and from the ACPD measurements, it can be concluded that crack propagation starts early and thus that the crack initiation period is small. This is an important result for proper modelling of the fatigue behavior in this type of tubular joints

1 INTRODUCTION

In welded tubular truss bridges, rectangular or circular hollow sections (RHS, CHS) with low values of the ratio diameter to thickness are usually preferred. This results from the will to best satisfy both the constraints of high forces, fatigue loadings and truss transparency for keeping good aesthetics. Since fatigue is a dominant design criteria for welded joints in CHS, significant work has gone into the development of design methodologies and guidelines for the fatigue behaviour of welded tubular joints (CIDECT 2000, IIW 2000).

The fatigue strength of a tubular structure depends both on the absolute as well as on the relative size of its members. Currently, this fatigue size effect is dealt with in analogy to plates in tension with welded details, that is using a size correction formula in which only the main plate thickness appears. The use of this formula for tubes in various structural applications has shown the limitations of this approach, in particular due to differences in the loading cases and can be very penalizing for the thick-walled tubular joints used in tubular truss bridges.

In order to improve the understanding and predictions of the fatigue cracking processes in tubular joints, researches have been carried out (Schumacher 2003, Schumacher & Nussbaumer 2006) and currently continue (Borges & Nussbaumer 2008). In order to be able to model thoroughfully the fatigue crack initiation and propagation phases, it is important to have at disposal as much as possible information on crack development. This paper focusses on the experimental determination of the crack depth in large-scale tubular truss specimens and illustrate how this information helps in the understanding of the behavior and validation of numerical models.

Several methods for get the crack depth and crack shape at different numbers of cycles during fatigue testing exist for many years and are described in literature (Marsh 1991):

- Heat-tinting of crack surface (but heat may change loading behavior during subsequent fatigue crack growth);
- Crack marking with overload (but again this induce crack depth and shape modifications during subsequent crack growth);

- Ink marking (may induce corrosion at crack tip and thus modification in subsequent crack growth rate);
- Beach-marking by applying a different stress amplitude for a limited number of cycles (if not made properly, may induce modification in subsequent crack growth rate);
- Various direct AC or DC potential difference methods (not as sensitive as the next methods);
- Alternating current potential drop (ACPD) method, indirect method providing continuous information at discrete points along surface crack (but crack may start at probe wires);
- Ultrasonic testing (using unidirectional or phased-array systems);
- Acoustic emission systems (direct measure of crack growth rate, not depth);

In our case, it was chosen to use the ACPD method because the location of the crack initiation was well defined, the method has proven accurate, and we believed the system could be applied on our large specimens. The large-scale tubular truss specimens tested in fatigue in this project (Borges et al. 2006) were designed to complete the database comprising about 180 test results worldwide (van Wingerde et al. 1997).

2 DESCRIPTION OF EXPERIMENTS

2.1 Test specimens

The test specimens are two uni-planar tubular CHS welded truss beams that were tested under fatigue loading. These beams constitute series S5 (S5-1 and S5-2) and follow four series using similar specimens previously tested by Schumacher (Schumacher 2003, Schumacher & Nussbaumer 2006).

Truss members are made of steel S355 J2 H conforming to EN10210-1:2006 and EN10210-2:2006. This refers to a weldable steel with a minimum tensile yield stress, f_y, of 355 N/mm² (for nominal thicknesses up to 16 mm) or 345 N/mm² (for 16 mm up to 40 mm), and a minimum ultimate tensile stress, f_u, between 490 and 630 N/mm² at 22% elongation. The minimum toughness of the steel is defined as 27 J at −20°C. The nominal dimensions of the specimens are given in Figure 1. Brace members were cut to fit the contour of the chord using computer control technology. Bevels were prepared at angles ranging from 30° to 45°. Backing-rings were used to facilitate the welder's task and make sure complete penetration of the weld is achieved. A MAG tubular cored metal arc welding with active gas shield (process 136) was used. Non-destructive testing controls of the welds were done by the fabricator following the Swiss code SIA 263/1 (SIA263 2003) rules (requirement weld quality B: 100\% VT + 50\% UT).

The test set-up static system corresponded to a simple beam with a concentrated load at mid-span as shown in Figure 2. Three steel blocks were machined to fit the top chord circular shape with the support blocks and to allow for proper introduction of the load at mid-span. For the sake of safety, and to prevent possible instability, blocks with a Teflon layer were put in place in each side of the beam to provide lateral support.

Weld size has an important effect in the stress concentration at the weld toes of tubular joints. In order to measure the weld size of joints, a mould impression of the welds was done using Rhodorsil RTV 3535 (Rodia Chimie 2002). This modelling material presents very low shrinkage, fast hardening and easiness for application.

Real weld dimensions were then compared to the American Welding Society (AWS) recommendations (see Fig. 3). AWS recommendations are globally fulfilled for both joint 1 and joint 2 and actual weld size is found to be normally much higher then AWS minimum recommendation.

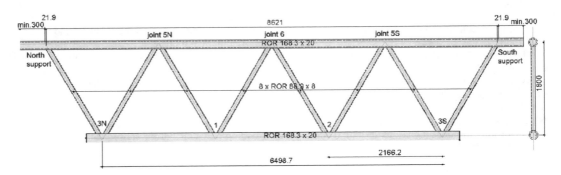

Figure 1. Elevation of a truss beam with dimensions.

Figure 2. Tubular truss fatigue test set-up.

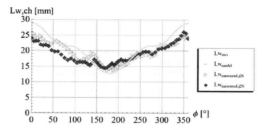

Figure 3. Comparison of weld leg lengths on chord.

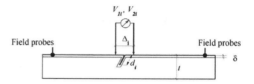

Figure 4. Alternate current potential drop theory and notation (adapted from Chiew (2004)).

2.2 Instrumentation

Each one of the two trusses has seven welded K-joints and two Y-joints, and there are 6 interesting K-joints simultaneously tested. Out of these six, four of them, namely j1, j2, j5N and j5S (see Fig. 1) were selected and instrumented because of the respective loading conditions. In particular, j1 and j2 were equipped with an ACPD system.

Other instrumentation comprised LVDT for displacement measurements of the truss, uni-axial strain gages to measure the nominal strains in the members and gages strips to evaluate the hot-spot strains and stresses (for beam S5-2).

2.3 Testing

Static tests were carried out in order to: verify the linearity of the loading/unloading response, check symmetry, verify that the out-of-plane bending remains negligible, determine the nominal stress in main joints and evaluate the hot-spot stresses (for specimen S5-2). Nominal stresses and hot-spot strains and stresses were then used to validate the numerical model.

The fatigue test then started, with periodic stop for NDE and static tests. A sinusoidal fatigue load was applied at a frequency of about 0.7 Hertz, with a load ratio $R = Q_{min}/Q_{max}$ of $R = 0.1$. A single Hydrel™ 1200 kN (static capacity) actuator under load control was used. The prescribed force range $\Delta Q = 549$ kN, was chosen to obtain the same chord nominal stress range as in previously tested series S3, namely $\Delta\sigma_{nom,ch} = 35$ N/mm².

3 ACPD SYSTEM

3.1 Principle

Alternating Potential Drop (ACPD) techniques have been widely used for crack depth monitoring in laboratory experiments for many years. The ACPD system (Fig. 4) was conceived for the detection of cracks by measuring the evolution of the electrical impedance of the welds where a stress concentration occurs.

The basic principle is to introduce an alternating current (AC) to flow between two electrodes (field probes) and measure the local voltage drop over the area adjacent to the weld and over the crack by means of voltage probes (Marsh 1991) (see Fig. 4).

For an infinitely long crack of constant depth, ACPD estimation of the crack depth is given by:

$$d_i = \frac{\Delta_i}{2}\left(\frac{V_{2i}}{V_{1i}} - 1\right) \quad (1)$$

where d_i is the crack depth at probe location i, V_{2i} is the potential difference measured with the probe i, V_{1i} is a reference potential difference, Δ_i is the spacing between probe contacts. The reference measure, V_{1i} was taken as the average of the first 30 measures on the uncracked joint at location i. Accurate determination of the probe contacts distance is specially hard because of the irregular geometry across the weld toe. This difficulty can be overcome using ACPD systems with independent reference measures for each probe site immediately before the weld toe, however the complexity of the tubular joint in the weld toe, can lead to other inaccuracies.

This one-dimensional interpretation of the ACPD measurements to give the crack depth is accurate for low crack aspect ratios ($a/2c < 0.1$) and is always used as a first estimation of the crack depth. As the geometry of the specimen and the weld become more

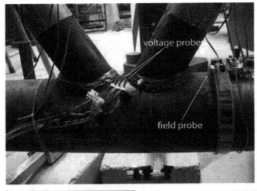

Figure 5. Alternating current potential drop system—joint with probes (top)—Acquisition box (bottom, left) and AC 120W generator (bottom, right).

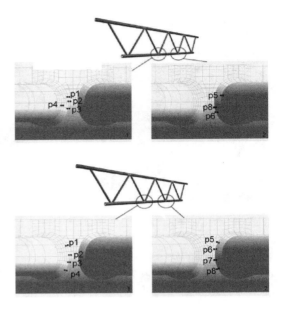

Figure 6. Alternating current potential drop probe locations for S5-1 (top) and S5-2 (bottom).

complex or the crack aspect ratio higher ($a/2c > 0.1$) this estimation is likely to be less accurate, underestimating the crack size. In this case it is recommended to use a multiplier to calibrate the final result. For the case of tubular joints, the aspect ratio remains low, and it was demonstrated by Michael et al (1982) that correction factor would be less than 1.05. Therefore Equation 1 can provide a satisfactory estimation for tubular joints.

3.2 Application to our specimens

The ACPD equipment used was developed at ICOM in collaboration with the manufacturer (Lavanchy 2007). Two different acquisition boxes were used, with slightly different characteristics.

Compared to other ACPD equipments described in literature, the AC frequency (5 to 6 Hz) is about 10000 lower, but compensated by the introduction of higher intensity of 100 to 150A. The ACPD acquisition box used on specimen S5-2 showed improved signal/noise ratio and thus more precise measurement results. In total, 8 (7 for S5-1) probes were disposed in potential crack sites corresponding to the crack toe, near imperfections found after visual inspection (Fig. 6).

A problem with ACPD estimations in our case is the impossibility to determine through thickness crack penetration. This happens because the depth estimation represents the length along the crack faces and the crack is not vertical but is inclined; and the angle of inclination remains unknown until the joint is opened

up. The reported fatigue tests presented two main challenges regarding the use of the ACPD technique: the size and complex geometry of the tubular truss.

In order to keep the monitored areas as close to the field probes (primary current introduction) as possible to reduce resistance, only joints j1 and j2 were equipped with voltage probes. The actuator and the supports were electrically insulated. Figure 5 shows one of the joints, the other being the same but with the field probe on the left side of the joint. Before trying, we did not know if the system would actually work since the current has multiple paths, actually two, to go from one field probe to the other. It can go through the joints and the bottom chord (that is what we want), but it can also go up one brace and down the next one.

For beam S5-2, to obtain even more information on crack shape and propagation, ink was sprayed into detected cracks. Also, Beach-marks by difference stress amplitude marking was made following the procedure suggested by Husset et al. (1985): Load amplitude is decreased by half, maximum load is kept at the same level, and these loading conditions are applied during 1000 cycles.

3.3 Measurements analysis

ACPD results were analysed following the flowchart in Figure 7. Raw potential results were filtered using Mathematica (Wolfram, 1988) moving average algorithm with a sample of 300 points (corresponding to a

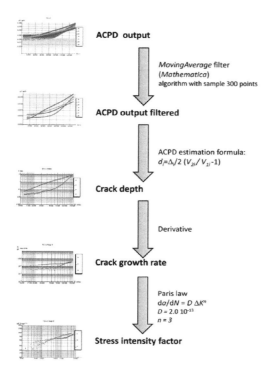

Figure 7. ACPD measurements analysis flowchart.

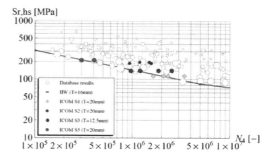

Figure 8. S-N test results.

lapse of 6000 cycles) in order to eliminate the noise. This was done at cost of losing information at the beginning and end of the fatigue test.

Afterwards, using Equation 1, the estimated crack depths, d_i, at probe sites were computed. Then, an estimation of the crack growth rate can be obtained every 50 cycles (Equation 2), considering a 750 cycles interval ($2\Delta N$).

$$\frac{dd}{dN} = \frac{d(N + \Delta N) - d(N - \Delta N)}{2\Delta N} \quad (2)$$

The experimental values of the stress intensity factor range were then obtained using the Paris law (Equation 3) with $C = 2.0 \times 10^{-13}$ (mm/cycle) (N/mm$^{-3/2}$)^m and m = 3 as often found as reference values in literature for Ferrite/Pearlite steels (Gurney 1979, Barsom & Rolfe 1999) and also found by calibration with previous fatigue tests on similar CHS truss beams (Walbridge 2005).

$$\frac{da}{dN} = C \cdot \Delta K^m \quad (3)$$

4 RESULTS

4.1 S-N results

The results are expressed in the form of hot-spot stress range versus number of cycles. In order to compare the results, one has to define a coherent definition of joint failure. Here, all results are expressed as the life corresponding to complete loss of strength of the joint and identified as N_4. Thus, the test results corresponding to through-thickness cracking were multiplied by a mean factor equal to 1.49 found in previous studies (van Wingerde et al. 1997, Schumacher et al. 2006). Figure 8 shows the results from these tests, together with the ones from previous similar tests and those from the existing IIW database (van Wingerde et al. 1997). The size of the symbols is proportional to the chord thickness.

Our new data shows on the graph as only point, but in fact represents 10 fatigue cracks. On both test beams, fatigue cracks were obtained in joints j1, j2, j5N, j5S, and j6. All the cracks occurred at hot-spots 1 or 1c (compression side in the case of elements in compression). It was observed that crack growth, at least up to half the chord thickness, occurred at about the same rate both in tension or compression joints, with crack shapes (a/c) being also similar.

No crack growth was detected at hot-spot 11 or from the weld root. Size effects were observed when considering all 5 series carried out at ICOM, due to proportional or non-proportional sizing effects between the different series (Borges et al. 2006). More tests needed to confirm the non-proportional size effects and propose improved design rules (compared to Gurney's or CIDECT) (Gurney 1979, IIW 2000).

4.2 Crack depth deduced from ACPD

The measurements analysis is carried out for each location according to the method described in section 3.3. The results can be plotted in terms of crack depth at each probe location, d, in function of the number

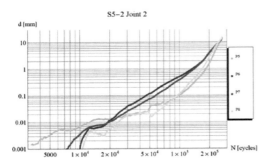

Figure 9. Number of cycles vs. crack depth for series S5-2 joints (filtered results).

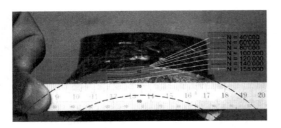

Figure 10. Comparison between ACPD predicted depths and real crack for S5-2, joint j2.

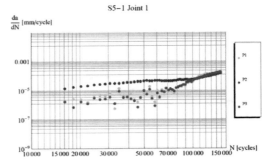

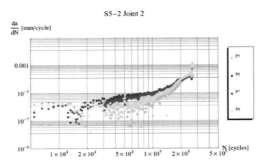

Figure 11. Crack growth rate vs. number of cycles deduced from ACPD measurement for S5-1, joint j1 and S5-2, joint j2.

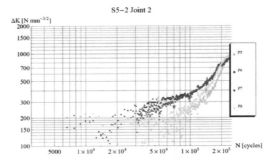

Figure 12. Stress intensity factor range vs. number of cycles deduced from ACPD measurement for S5-2, joint j2.

of cycles. A typical example of the plot is given in Figure 9 for joint 2 of beam S5-2.

One can observe that the first signs of cracking were detected by probe P5, which is on the side of hot spot 1. However, such small crack sizes can be within the tolerances of the instrumentation. After 20000 cycles, the probes the closest to, besides hot spot 1, namely P6 and P7, show larger crack depths. But Figure 10, which shows the comparison between ACPD predicted depths and real crack, confirm the fact that the cracking started from the left side (the side of probe P5).

Indeed, the other measurements also showed that the first detectable crack could be on the side of hot spot 1 (not belonging to the vertical symmetry plane). Also, the first detectable crack happened after less than 10% of the test fatigue life. For comparison, in field inspection by MPI, smallest detectable crack depth is 0.5 mm, that is between 25% and of the total fatigue life (Miki et al. 1989). All measurements made were stable and with the good results and comparisons obtained, we can say that the size of the specimen and the multiple current paths did not disturb the functioning of the ACPD system.

4.3 Crack propagation rate

Using the information on the crack depth, again using the method described in section 3.3, one can find the crack propagation rate and plot the results as in Figure 11. For joint S5-1, j1, the results from probe P4 are not shown as no propagation was measured (probe at hot spot 11, see Fig. 6). Crack propagation rates from 10^{-7} to 10^{-3} mm/cycle could be measured. For C-Mn

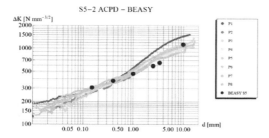

Figure 13. Comparison between "measured" and computed stress intensity factor range vs. crack depth for S5-2.

steels, the value 10^{-7} mm/cycle corresponds to stress intensity factor ranges near the threshold value.

4.4 SIF ranges

Experimentally evaluated SIF ranges can be found once a crack propagation law, see section 3.3, has been assumed. A plot of the result is given in Figure 12.

One should mention that the evaluated SIF ranges do not correspond with the deepest point along the crack front but depends on the location of the probe. Very low ΔK values, between 100 and 200 Nmm$^{3/2}$ could be evaluated from measurements after only 5000 fatigue cycles, which is remarkable. For comparison, threshold SIF range values, ΔK_{th}, for C-Mn steels, can for example approximately be expressed as follows (Zheng 1987):

$$\Delta K_{th} = 320(1-R) \quad for \ R \leq 0.4 \quad (4)$$

This means that we are able to measure threshold values. It further could be said that due to high tensile residual stresses from welding, it is logical that SIF ranges values below the ΔK_{th} for $R = 0.1$, as predicted by Formula 4, could be measured.

4.5 Comparison between experimental and numerical SIF ranges

The plot of the comparison between experimentally evaluated SIF ranges and computed ones for both joints of beam S5-2 is given in Figure 13. The computed values come from a numerical boundary element model, made with the software BEASY (BEASY 2002). There, among other assumptions, a constant crack shape with a/c = 0.2 was assumed in the computations. It was chosen as it matched with the average crack shape seen during the tests.

Further information on the numerical model can be found in Borges & Nussbaumer (2008). Since the SIF range values plotted here were computed for the deepest crack point only, one can say that a good agreement exist between the computed and measured values.

5 CONCLUSIONS

From the instrumented fatigue tests on tubular truss beams, the following conclusions can be drawn:

– All the fatigue cracks occurred at hot-spots 1 for joints on the tension chord and at 1c (between compression chord and compression brace) for the joints on the compression chord.
– With regard to the influence of welding residual stresses, fatigue cracks were observed both in joints with the chord in tension (joints 1 and 2) as well as in joints with the chord in compression (joints 5N, 5S and 6).
– No crack growth was detected at hot-spot 11 (weld toe on the side of the braces) or at weld root.
– During the two tests carried out, we were able to successfully apply an ACPD (Alternative Current Potential Drop) measuring system on a large-scale tubular truss beam with multiple current paths. Previously, ACPD systems had only been used on isolated joints or small specimens.
– The ACPD system was found very sensitive and gave very precise measurements of the crack depth. Also, from the measurements, crack propagation rates in the range from 10^{-7} to 10^{-3} mm/cycle could be deduced as well as low ΔK values, between 100 and 200 Nmm$^{3/2}$, which is close to threshold values.
– Tests confirm that welded joints in fatigue, also tubular welded joints, show mainly crack propagation because of initial welding imperfections.

ACKNOWLEDGMENTS

The research carried out is part of projects from FOSTA and the Swiss National Science Foundation (SNF). The FOSTA project P591 is supervised by the Versuchsanstalt für Stahl, Holz und Steine at the Technische Universität Karlsruhe, which is supported financially and with academic advice by the Forschungsvereinigung Stahlanwendung e. V. (FOSTA), Düsseldorf, within the scope of the Stiftung Stahlanwendungsforschung, Essen. The SNF support was given for the thesis work of L. Borges under grant no 200021-112014. Other support was given by Vallourec & Mannesmann Tubes, Germany, who provided the tubes for the laboratory tests and Zwahlen & Mayr SA, Switzerland, who fabricated the test specimens.

REFERENCES

Barsom, J.M. & Rolfe, S.T. 1999. *Fracture and fatigue control in structures: application of fracture mechanics, Third edition*. Prentice-Hall, New Jersey, ISBN 0-8031-2082-6.

Borges, L., Schumacher, A., & Nussbaumer, A. 2006. Size effects on the fatigue behavior of welded CHS bridge joints. *International Conference on Fatigue and Fracture in the Infrastructure*, Philadelphia, USA, August 6 to 9, 2006, CD-Rom.

BEASY (2003). *BEASY User Guide*. Computational mechanics BEASY Ltd, Ashurst, Southhampton, UK.

Borges, L., & Nussbaumer, A. 2008. Advanced numerical modelling of fatigue size effects in welded CHS K-joints, *12th International Symposium on Tubular Structures (ISTS 12)*, Shanghai.

Chiew, S.P., Lie, S.T., Lee, C.K. & Huang, Z.W. 2004. Fatigue performance of cracked tubular T joints under combined loads. I: Experimental. *Journal of Structural Engineering*, April: 562–571.

CIDECT no 8 (2000), Zhao, X. L., Herion, S., Packer, J. A. et al. *Design guide for circular and rectangular hollow section joints under fatigue loading*. CIDECT, Comité International pour le développement et l'étude de la construction tubulaire 8, TüV-Verlag Rheinland, Köln.

EN10210-1 (2006). *Hot finished structural hollow sections of non-alloy and fine grain steels*. Technical delivery requirements, CEN, Brussels.

EN10210-2 (2006). *Hot finished structural hollow sections of non-alloy and fine grain steels*. Tolerances, dimensions and sectional properties, CEN, Brussels.

Gurney, T. R. 1979. *Fatigue of Welded Structures*. Cambridge University Press, Cambridge, UK.

Husset, J., Lieurade, H., Maltrud, F. & Truchon, M. 1985. Fatigue crack growth monitoring using a crack front marking technique. *Welding in the World*: 276–282.

IIW (2000), Zhao, X.L., Packer, J.A., *Fatigue design procedure for welded hollow section joints*, Doc. XIII-1804-99, XV-1035-99, IIW, Cambridge, Abington.

Lavanchy Electronique (2007). *Mesureur de fissure-7422*, Av. de la Rochelle 12, CH-1008 Prilly, Switzerland.

Michael, D.H., Waechter, R.T. & Collins, R. 1982. The measurement of surface cracks in metals by using a.c. electric fields. *Proceedings of the Royal Society of London. Series A, Mathematical and Physical Sciences* (1934–1990), 381(1780), May: 139–157.

Marsh, K.J., Smith, R.A. & Ritchie, R.O. 1991. *Fatigue Crack Measurement: Techniques and applications*. EMAS Engineering Materials Advisory Service LTD, ISBN 0947817468.

Miki, C., Fukazawa, M., Katoh, M, & Ohune, H. 1989. Feasibility study on non-destructive methods for fatigue crack detection in steel bridge members, Doc. IIS/IIW-990-88, *Welding in the World* 27(9/10): 248-266.

Rhodia Chimie. 2002. *fiche technique SIL 01 021 1*, Silicones Europe, 55 Av. des frères Perret, BP 60, 69192, St-Fons-Cedex, France.

Schumacher, A. 2003. *Fatigue behaviour of welded circular hollow section joints in bridges*. PhD thesis EPFL n°2727, Swiss Federal Institute of Technology (EPFL), Lausanne.

Schumacher, A. & Nussbaumer, A. 2006. Experimental study on the fatigue behavior of welded tubular k-joints for bridges. *Engineering Structures* 28:745–755.

SIA263 (2003). Steel structures, Schweizerischer Ingenieur- und Architektenverein (SIA), Zürich, Switzerland.

Van Wingerde, A.M., van Delft, D.R.V., Wardenier, J. & Packer, J.A. 1997. *Scale Effects on the Fatigue Behaviour of Tubular Structures*. WRC Proceedings.

Walbridge, S. 2005. *A Probabilistic Fatigue Analysis of Post-Weld Treated Tubular Bridge Structures*. PhD thesis EPFL n°3330, Ecole polytechnique fédérale de Lausanne (EPFL).

Wolfram, S. 1998. Mathematica, A System for Doing Mathematics by Computer, U.S.A.

Zheng Xiulin. 1987. A simple formula for fatigue crack propagation and a new method for the determination of ΔK_{th}, *Engng Fracte Mech* 27(4): 465–475.

Stress intensity factors of surface cracks in welded T-joints between CHS brace and concrete-filled CHS chord

M. Gu & L.W. Tong
School of Civil Engineering, Tongji University, Shanghai, China

X.L. Zhao
School of Civil Engineering, Tongji University, Shanghai, China
Department of Civil Engineering, Monash University, Clayton, Australia

X.G. Lin
School of Urban Construction, Zhejian Shuren University, Hangzhou, China

ABSTRACT: With the increasing use of arch bridges made of welded T-joints between CHS brace and concrete-filled CHS chord (CFCHS) in China, fatigue of the welded CFCHS joints under vehicle loading is becoming an issue to be addressed. Like welded CHS joints, fatigue assessment of the welded CFCHS joints can be performed based on one of three approaches, namely classification method, hot spot stress and fracture mechanics. The last one is the advanced approach in which the parameter of dominating fatigue crack propagation, Stress Intensity Factor (SIF), needs to be determined first. This paper presents the simplified fracture mechanics model, the FE methodology predicting stress intensity factors of the welded CFCHS joint with semi-elliptical surface cracks under axial brace loading. In this method the software ANSYS was used to produce the finite element model of the T-joint. The special APDL programs were developed to generate the surface semi-elliptical weld toe cracks at the crown of the joint and to create wedge-shaped singularity elements around the crack tip since they are not available in ANSYS element library. In order to get reliable and convergent result, the nodes on the interior surface of the steel tube and the nodes on the external surface of the concrete must be in the same location and they shouldn't be merged together. Contact elements were introduced between them. The stress intensity factors were calculated respectively by the way of displacement extrapolation. The calculated results indicate that the finite element modeling is successful and the numerical simulation is an effective way to obtain the stress intensity factors of the complicated CFCHS joints.

1 INTRODUCTION

In recent eighteen years, concrete-filled circular hollow sections (CFCHS) have been increasingly used in Chinese bridge engineering, especially in large span arch bridges, such as the Wuxia Bridge in Chongqing as shown in Figure 1 which is 460 meters long. In these bridges, circular hollow section chord members and the brace members were welded together and then concrete is filled into the chords. The brace members are unfilled. For this type of structure, welded CFCHS joints are one of the fundamental components since the concrete can improve the stability of the chords near the intersection between the chords and the braces.

It's well known that cyclic loads result in fatigue failure at the weld toe of the CHS joints generally in the form of cracks due to the high stress concentration and welding defects. Consequently, much of effort has been directed towards the estimation of stress concentration factors and stress intensity factors of cracked tubular joints.

Figure 1. Wuxia Bridge in China.

The welded CFCHS joints of the arch bridges are under daily vehicle loads. Fatigue may be an important issue especially due to manufacturing defects such as welding slag, undercut and so on. However, the research in this field is quite limited. Previous research by Mashiri et al (2004) on T-joints with concrete-filled square hollow section chord revealed some advantages of using void-filled chord in terms of stress concentration factor and fatigue life.

A research project on the fatigue behavior of the welded CFCHS T-joints was conducted in Tongji University, which deals with experiments and numerical analysis of hot spot stresses, fatigue strength and crack propagation life prediction. In the part of crack propagation life prediction, initial defect assumptions, reliable stress intensity factors in conjunction with the appropriate crack growth law are necessary. The prediction of the stress intensity factors at the weld toe cracks of the CFCHS joint is not easy. This paper presents the simplified fracture mechanics model, the FE methodology predicting stress intensity factors of the welded CFCHS joint with semi-elliptical surface cracks under axial brace loading.

Figure 3. Crack propagation along weld toe of chord.

2 DESCRIPTIONS OF EXPERIMENTS

2.1 Characters of hot spot stresses

Ten static experiments on the welded CFCHS T-joints have been done in Tongji University (Tong & Wang 2007). Each specimen joint's chord was filled with concrete and its brace was kept empty. The test specimens of the CFCHS T-joints are shown in Figure 2.

In these specimens, different geometers and concrete grades were chosen to study their effect on joints behavior respectively. Both the brace and the chord were made of steel Q345. The brace was connected to the chord by manual arc welding with full penetration weld according to AWS 2000. These experiment results show that the concrete increases greatly the stiffness of the whole joint, restrains ovalisation of the chord and makes the stiffness distribution around the intersection more uniform so that the hot spot stresses in the joint become much smaller and the position of the maximum SCF changes from the chord saddle to chord crown after concrete infill.

2.2 Characters of fatigue behavior

Following these static experiments, fatigue strength experiments on the welded CFCHS T-joints under cyclic axial loading have been carried out in Tongji University with the same geometries and concrete grades of the static experiments (Zhu 2007). One of the experimental crack propagation pictures is shown in Figure 3. These experiment results indicated that the initial crack almost occurs at the chord crown. After propagating around the weld toe to some extent, the crack propagates along the chord circumference.

3 SIMPLIFIED MODEL OF FRACTURE MECHANICS

In order to numerically model the life of the welded CFCHS T-joint with defect, it's necessary to do some assumptions based on the experiment results of hot spot stresses, fatigue strength and some data of other researchers.

3.1 Location of initial cracks

Comparison has been done between the locations of the maximum SCF and the initial crack occurrence of the CFCHS T-joints under cyclic axial brace loading. It shows that the two locations are almost in accordance with each other and a large amount of parametric FE analyses present that most of the maximum SCFs occur at the crown of the CFCHS T-joints' chord

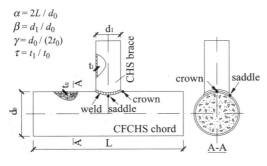

$\alpha = 2L/d_0$
$\beta = d_1/d_0$
$\gamma = d_0/(2t_0)$
$\tau = t_1/t_0$

Figure 2. Test specimen of CFCHS T-joint.

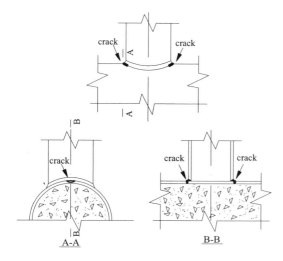

Figure 4. Assumption of crack initiation location.

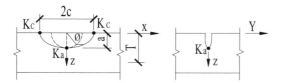

Figure 5. Model for semi-elliptical surface crack.

under axial brace loading. As shown in Figure 4, it's assumed that the deepest point of the crack locates at the crown of the chord in the fracture mechanics models and the number of the cracks is two.

3.2 Crack shape assumption

Based on the traditional fracture mechanics and crack propagation picture in Figure 3, the surface crack is simplified to a semi-elliptical crack with the depth a and the length 2c which surrounds the weld toe. When it is unwrapped in a plane, its equation is $x^2/c^2 + z^2/a^2 = 1$ as shown in Figure 5. The crack is perpendicular to the surface of the chord in the deep direction which needs the least amount of energy when it propagates. In this study the stress intensity factors of the deepest point K_a and the crack ends K_c are considered.

4 FINITE ELEMENT ANALYSIS

4.1 Model sequence

It's difficult to model the welded CFCHS T-joint with surface crack at the crown of the chord due to the crack, the weld and the concrete. Furthermore, there are contact nonlinear finite elements between the chord member which is metal and the concrete which is non-metal. It's very important that the nodes on the interior surface of the steel tube and the nodes on the external surface of the concrete must be in the same location and they shouldn't be merged together. Otherwise, it may lead to non-convergence or error results in software of ANSYS due to the initial gap or penetration between the two contact materials. In this paper, based on the mesh of the T-butt with surface crack at the weld toe, a good mesh was generated by the special APDL programs for the aim of fine mesh near the crack tip, good transition and match of the mesh and less time consumption. The model sequence is shown in Figure 6. The steps are shown below:

1. The model starts with a T-butt with a semi-elliptical crack with two non-dimensional parameters a/t_0, a/c.
2. Wrap the T-butt and generate a quarter of the brace due to the symmetry of the joint and loading about two planes.
3. Based on step 2, generate the unwrapped chord and merge the nodes on the board and the brace.
4. Wrap the board and generate the chord. Now, the parameters α, β, γ, τ, t_0 of the joint are defined.
5. Insert the concrete into the chord, make the mesh on the two contact areas matched and add contact elements between the two materials. The contact elements are shown in Figure 7.

In order to capture the bending stresses and stress concentrations around the intersection, more than four layers of finite elements are used over the chord wall thickness and three layers over the brace. The crack tip is surrounded by four rings of elements and each ring includes eight elements. In this method the mesh can be modified easily and be the basis of the investigation on the crack growth behavior.

4.2 Material property, element type and weld

In the analysis, most of the steel tube is simulated by SOLID45 excluding the crack tip which will be discussed later. The elasticity modulus of the steel is 206,500 MPa and the Poisson's Ratio is taken as 0.3.

The fatigue properties of the concrete in the chord member were directed much effort in the first stage of study. After the fatigue strength experiment, partial steel near the intersection of the brace and the chord was cut and removed to study the failure mode of the concrete. Almost no concrete was crushed. One representative specimen is shown in Figure 8. Associating with the following FE analysis, the influence of the concrete in the chord member is on

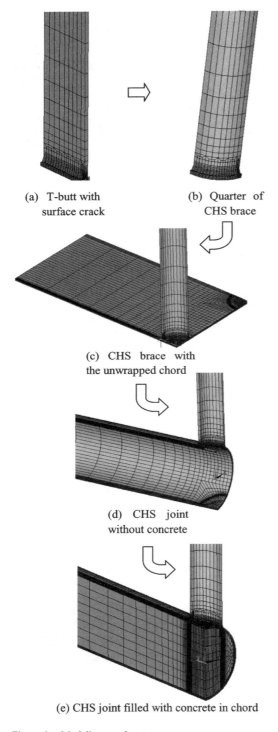

(a) T-butt with surface crack

(b) Quarter of CHS brace

(c) CHS brace with the unwrapped chord

(d) CHS joint without concrete

(e) CHS joint filled with concrete in chord

Figure 6. Modeling step by step.

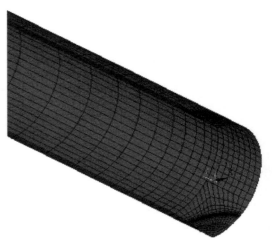

Figure 7. Contact elements used in FE analysis.

(a) Outer fibre of concrete under tension

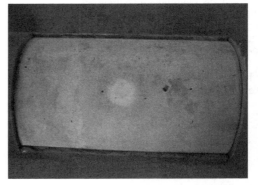

(b) Outer fibre of concrete under pressure

Figure 8. Concrete failure mode under pressure.

362

Table 1. Concrete fatigue deformation modulus (E^f_c).

Grade	C20	C25	C30	C35	C40
E^f_c (× 10⁴ MPa)	1.1	1.2	1.3	1.4	1.5
Grade	C45	C50	C55	C60	C65
E^f_c (× 10⁴ MPa)	1.55	1.6	1.65	1.7	1.75
Grade	C70	C75	C80	–	–
E^f_c (× 10⁴ MPa)	1.8	1.85	1.9	–	–

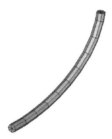

Figure 9. Singularity elements close to crack tip.

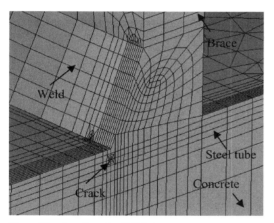

Figure 10. Local mesh of T-joint.

the hot spot stresses and then on the fatigue strength indirectly. When the grade of concrete changes, the hot spot stresses change too, but just a little. Although the grade of the concrete's contribution is insignificant, the fatigue property of the concrete can't be ignored. According to the *code for design of concrete structures (GB 50010-2002)*, the fatigue deformation modulus is listed in Table 1. The Poisson's Ratio is equal to 0.2. The concrete is simulated by SOLID65.

The weld is simulated according to the AWS 2000 recommendations and its properties were assumed to be the same as the parent metal.

The surface-to-surface contact elements (TARGE 170 and CONTA 174) were employed to simulate the flexible-flexible contact between the steel and concrete surfaces. The friction coefficient value was 0.4.

Due to the lack of them in ANSYS element library put forward by Barsoum (Barsoum, 1977) that mid-side nodes are moved to the 1/4 points to model the $1/\sqrt{r}$ singularity, the wedge-shaped singularity elements near the crack tip were created based on the SOLID95 in the APDL program (see Figure 9).

Up to now, good meshes are generated as shown in Figure 10.

5 EVALUATION OF STRESS INTENSITY FACTORS

5.1 Method of calculation

According to the elastic fracture mechanics (Zhu, 2000), the displacement near the crack tip and the stress intensity factors have the following relationships.

$$u = \frac{K_I}{4G}\sqrt{\frac{r}{2\pi}}\left[(2\kappa-1)\cos\frac{\theta}{2} - \cos\frac{3\theta}{2}\right]$$
$$- \frac{K_{II}}{4G}\sqrt{\frac{r}{2\pi}}\left[(2\kappa+3)\sin\frac{\theta}{2} + \sin\frac{3\theta}{2}\right] + O(r) \quad (1)$$

$$v = \frac{K_I}{4G}\sqrt{\frac{r}{2\pi}}\left[(2k+1)\sin\frac{\theta}{2} - \sin\frac{3\theta}{2}\right]$$
$$- \frac{K_{II}}{4G}\sqrt{\frac{r}{2\pi}}\left[(2k+3)\cos\frac{\theta}{2} + \cos\frac{3\theta}{2}\right] + O(r) \quad (2)$$

$$w = \frac{2K_{III}}{G}\sqrt{\frac{r}{2\pi}}\sin\frac{\theta}{2} + O(r) \quad (3)$$

and

$$k = \begin{cases} 3-4\gamma & \text{plane strain} \\ \dfrac{3\gamma}{1+\gamma} & \text{plane stress} \end{cases}$$

In these equations, u, v, w are the local radial, normal and tangential displacement (see Figure 11). K_I, K_{II}, K_{III} are the mode I, II, III stress intensity factors respectively. G is the shear modulus and γ the Poisson's ratio.

Based on these equations, displacement extrapolation was used to calculate the stress intensity factors because of the disappearance of the J-integral independence in the region where the crack meets the weld and plain strain assumption is used everywhere in the CFCHS T-joints analysis.

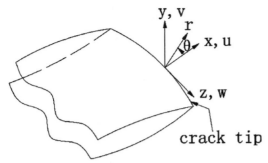

Figure 11. Crack tip coordinate system.

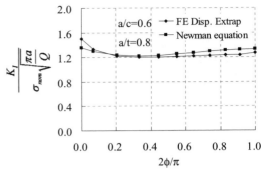

(a) under tension

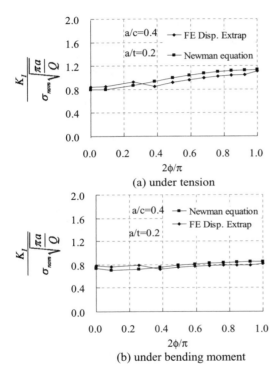

(a) under tension

(b) under bending moment

Figure 12. K_I distribution of a plate with shallow crack.

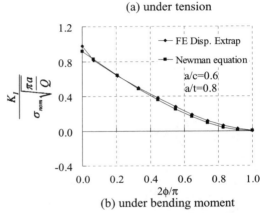

(b) under bending moment

Figure 13. K_I distribution of a plate with deep crack.

5.2 Validation of plane plate model

To validate the APDL programs and mesh, solutions for a plate with different crack sizes which mesh is compatible with the CFCHS T-joint near the crack tip were compared with Newman and Raju's (1981) which is well-known. The plate has a length, width and thickness of 200, 500 and 8 mm respectively. Two load cases are chosen. One is tensile load resulting in nominal membrane stress of 100 MPa, the other is bending moment resulting in nominal stress of 100 MPa in the outer fiber. The crack dimension is a/c = 0.4, 0.6, a/t = 0.2, 0.4, 0.6, 0.8. In these results, K_{II}, K_{III} are much less than one percentage of K_I. So they had been ignored. Comparison between the FE solution of eleven different points' $K_I/(\sigma_{nom}\sqrt{\pi a/Q})$ value on the crack tip and the Newman-Raju's solution shows that the trend agrees well with each other, where Q is the function of a/c in Newman-Raju' solution. The maximum deviation of the results for the deepest points is −9% and for the crack ends 12%. In these cracks, a shallow crack's K_I distributions are shown in Figure. 12 and a deep crack's K_I in Figure.13. It's clear that the FE model in case of a flat plate with surface crack can generate reliable SIF. Although in the absence of benchmark solutions, more confidence in the FE model and solution of the weld CFCHS T-joint which mesh is compatible to the plate near the crack tip is given.

5.3 Stress intensity factors of CFCHS T-joint

Based on the method and the APDL programs, the stress intensity factors of some CFCHS T-joints with surface cracks were calculated. One specimen in the hot spot stresses experiment with different

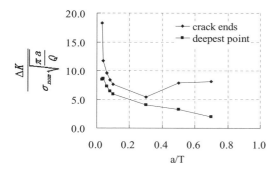

Figure 14. Calculated stress intensity factors at the deepest point and ends of the crack.

crack sizes is chosen as the example (see Figure 2). The dimensions are: $\alpha = 12.0$, $\beta = 0.54$, $\gamma = 15.31$, $\tau = 1.00$, $t_o = 8$ mm, concrete grade C70 and steel Q345. Tensile axial loading is applied on the brace and ΔF is equal to 80 KN. The results show that K_I, K_{II}, K_{III} are in the same magnitude and none of them can be ignored which is different from the plates. So the equivalent stress intensity factor K_{eff} is used. The results of the joint with different crack sizes are shown in Figure 14.

6 SUMMARY

The summaries are as follows :

1. Based on the experiment of hot spot stresses and fatigue strength, simplified fracture mechanics model of the weld CFCHS T-joint under axial brace loading is given which is different from the CHS T-joint due to the difference of the crack location, i.e. chord crown for CFCHS whereas the chord saddle for CHS T-joint.
2. The FE method predicting stress intensity factors of the welded CFCHS joint with semi-elliptical surface cracks at the crown under axial brace loading is presented. In the method, the fatigue parameter of concrete was discussed. Special APDL programs were developed to create the singularity elements around the crack tip and other model. In order to introduce contact elements between the concrete and the steel and to get reliable and convergent results, it was noted that the nodes on the interior surface of the steel tube and the nodes on the external surface of the concrete must be in the same location and they shouldn't be merged together. Although in the absence of benchmark solutions, much confidence was gained through the validation of a plate with different sizes of surface crack. In the calculation of stress intensity factors, displacement extrapolation was used and the calculated results indicate that the finite element modeling is successful.

ACKNOWLEDGEMENTS

The authors wish to thank the Natural Science Foundation of China for financially supporting the research in the paper through the grant NO. 50478108.

REFERENCES

A.W.S.: Structural Welding Code- Steel. ANSI/AWS D1.1-2000, 17th Edition, American Welding Society, Miami, Florida, USA, 2000.

Barsoum R.S. 1977. Triangular quarter-point elements as elastic & perfectly-plactis crack tip elements. Int. J. Num. Meth. Eng.: 85–98.

Bowness D. and Lee M.M.K. 1996. Stress intensity factor solutions for semi-elliptical weld-toe cracks in T-butt geometries. Fatigue Fract. Engng Mater. Struct, Vol.19: 787–797.

Cao J.J. et al. 1998. Crack modeling in FE analysis of circular tubular joints. Engineering Fracture Mechanics 61: 537–553.

Lee M.M.K. 1999. Strength, stress and fracture analyses of offshore tubular joints using finite elements. Journal of Constructional Steel Research, 51: 265–286.

Mashiri, F.R and Zhao, X.L. (2004), Fatigue Behaviour of Welded Composite Tubular T-Joints under In-Plane Bending, 4th International Conference on Thin-Walled Structures, UK, June.

Newman J.C. & Raju I.S. 1981. An Empirical Stress Intensity Factor Equation for the Surface Crack. Engineering Fracture Mechanics, 15: 185–192.

Oomens M. 2006. Numerical fatigue crack growth analysis of thick-walled CHS T-joints. Tubular structures XI, Taylor & Francis Group, London.

The Ministry of Construction of the People's Republic of China. GB 50010-2002. 2002. Code for design of concrete structures. Beijing: China Architecture & Building Press.

Tong L.W. & Wang K. et al. 2007. Experimental study on stress concentration factors of concrete-filled circular hollow section T-joints under axial loading. Proceedings of eighth pacific structural steel conference wairakei, New Zealand.

Zhu J. 2007. Fatigue behaviour of welded T-joints of concrete filled circular hollow sections. Master dissertation. Tongji University.

Failure assessment of cracked Circular Hollow Section (CHS) welded joints using BS7910: 2005

S.T. Lie
*Structures & Mechanics Division, School of Civil & Environmental Engineering,
Nanyang Technological University, Singapore*

B.F. Zhang
*Maritime Research Centre, School of Civil & Environmental Engineering,
Nanyang Technological University, Singapore*

ABSTRACT: The prediction of the residual strength of offshore tubular joints is based on the design equations of uncracked joints. However, many offshore structures installed in the open seas are found to contain cracks caused by severe cyclic loads. These fatigue cracks propagate and increase in size resulting in the reduction of the joint capacity. Therefore, there is a need to develop a systematic procedure for assessing the fracture strength of such cracked tubular joints in practice. The most widely used and accepted approach is the failure assessment diagram (FAD) method. Based on this approach, BS7910 (2005) gives guidance for assessing the acceptability of defects in welded structures. In this codes of practice, a specific assessment procedure for tubular joints in offshore structures has been proposed recently. In accordance with this procedure, a full-scale fatigue cracked tubular K-joint specimen, subjected under axial and in-plane loads, is analyzed and assessed illustrating the usage of this approach.

1 INTRODUCTION

In practice, the safety of any welded structure depends primarily on the usage of non-destructive inspection to detect crack before it develops to a critical size, and hence to permit component repair or replacement before catastrophic failure occurs. To determine the critical crack size, the structure should be assessed according to the knowledge of the service stresses and the knowledge of the fracture properties of the material.

Fracture mechanics is an indispensable tool for performing a critical assessment of a defect discovered in any steel structure. The main objective of this assessment is to establish the maximum tolerable defect size which would not compromise the service requirements. At the design stage, the ability to evaluate the tolerance of a structure to possible defects may be used to optimize the design with respect to properties of the material, geometric shape, and ease of inspection. During operation an assessment may be used to reassess a structure that has been found to contain defects, thus helping in making rational repair or no-repair decisions and improving the inspection strategy.

Recently, American Petroleum Institute (API) RP579 (2000), British Standard BS7910 (2005) and Central Electricity Generating Board (CEGB) R6 procedure (2001) give guidance for assessing the acceptability of defects in welded structures based on the failure assessment diagram (FAD) method. The FAD method was originally derived from the original two-criterion approach reported by Dowling and Townley (1975). This approach states that a structure can fail by either of two mechanisms, brittle fracture or plastic collapse, and that these two mechanisms are connected by an interpolation curve based on the strip yield model. If the service (assessment) point falls inside the assessment curve, the structure is considered safe, otherwise, the structure is deemed unsafe. This method enables the analyst to go directly from linear elastic fracture mechanics (LEFM) calculations to plastic instability calculations (Wiesner et al. 2000).

The assessment curves specified in the BS7910 (2005) are different for different materials and geometries. However, the lower bound curves are always used to assess all types of structures including the cracked circular hollow section (CHS) T, Y, K and KT-joints. In this paper, the standard Level 2A FAD curve is used to assess a typical cracked tubular CHS K-joint specimen. The plastic collapse correction factor F_{AR}, as recommended in Annex B of BS7910

(2005), is used to calculate the plastic collapse load P_c. Knowing the elastic stress intensity factor K_I and the fracture toughness K_{IC}, the corresponding values of K_r and L_r for different crack sizes are plotted in the FAD curve accordingly. The loading paths are then used to study the fracture assessment sensitivity analysis as the flaw (crack) increases with time.

2 STANDARD FAILURE ASSESSMENT DIAGRAM (FAD)

According to BS7910 (2005), any uni-planar cracked tubular CHS T, Y, K and KT-joints can be assessed using the normal assessment route. The standard FAD curve has two curves, namely Level 2A and 2B as shown in Figure 1 respectively.

The Level 2A and 2B curves can be described respectively by the following equations:

$$K_r = (1 - 0.14 L_r^2)[0.3 + 0.7 \exp(-0.65 L_r^6)] \quad (1a)$$

and

$$K_r = \left(\frac{E \varepsilon_{\text{ref}}}{L_r \sigma_Y} + \frac{L_r^3 \sigma_Y}{2 E \varepsilon_{\text{ref}}} \right)^{-\frac{1}{2}} \quad (1b)$$

where E is Young's modulus, ε_{ref} is reference strain and σ_Y is yield stress of the material, and

$$K_r = \frac{K_I}{K_{IC}} \quad (2)$$

$$L_r = \frac{\text{total applied load giving to } \sigma^P \text{ stresses } (P)}{\text{plastic collapse load of flawed structure } (P_c)} \quad (3)$$

ε_{ref} is the true strain obtained from the uniaxial tensile stress-strain curve at a true stress, $K_r \sigma_Y$. The application of Level 2B requires the knowledge of a complete stress-strain curve; in particular the region around the yield point has to be available in a detailed manner. However, there are many cases where this information is not available to the users. Therefore, Equation 1b is applied for a number of materials to generate a material independent lower bound curve of Level 2A as shown in Figure 1, which is the more conservative curve.

As it can be seen, this method adopts the assessment curve which uses the ratio of the stress intesity factor K_I to the fracture toughness K_{IC}, defined as K_r as the vertical (fracture) axis, and the ratio of the applied load P to the plastic collapse load P_c, defined as L_r as

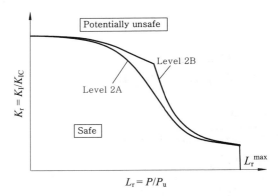

Figure 1. BS7910 (2005) Levels 2A and 2B FAD curves.

the horizontal (plasticity) axis. If the service (assessment) point falls inside the assessment curve, the structure is considered safe, otherwise, the structure is deemed unsafe.

It is also important to recognize that the K_r parameter of the assessment diagram uses the linear elastic stress intensity factor with no allowance for the effect of plasticity on the crack tip driving force. As L_r value increases, plasticity also increases the effective crack tip driving force. If it is considered that fracture actually occurs when the total effective crack tip driving force, the elastic plastic value of J_{ep} reaches a critical value equivalent to the fracture toughness, then this will occur at $\sqrt{EJ_{ep}} = K_{IC}$. Since the applied linear elastic stress intensity factor is equivalent to $\sqrt{EJ_e}$ where J_e is the linear elastic J-integral, then

$$K_r = \frac{K_I}{K_{IC}} = \sqrt{\frac{J_e}{J_{ep}}} \quad (4)$$

As plastic increases so the ratio $\sqrt{J_e / J_{ep}}$ reduces, and this defines the shape of the assessment curve with the increasing of L_r. The standard assessment curve is originally derived for the case of a large plate under tension loading with a central crack (Dawes & Denys 1984). They have been shown to represent a lower bound curves for other common simple geometries.

2.1 Fracture axis K_r

For assessing the safety and integrity of an existing cracked structure, the fracture parameter K_r given by Equation 2 is usually used in practice. The elastic stress intensity factors K_I along the 3D crack front of the K-joint specimen can be obtained from a finite element analysis (Lie et al. 2005), and the material fracture toughness K_{IC} can be determined from the standard CTOD or J values tests (BS7448-1, 1991). When the joint is subjected under a mixed mode condition, the effective stress intensity factor K_{eff} should be used to replace the K_I as

$$K_{eff} = \sqrt{K_I^2 + K_{II}^2 + K_{III}^2/(1-\nu)} \qquad (5)$$

where K_I, K_{II}, K_{III} are the Mode-I, II and III stress intensity factors respectively, and ν is the Poisson's ratio. However, it was shown that the Mode-I stress intensity factors are the dominant ones, and they are almost equal to K_{eff} for the K-joint subject under axial (AX) and in-plane bending (IPB) loads shown in Figure 2 (Lie et al. 2003).

2.2 Plasticity axis L_r

As it is very difficult to obtain the plastic collapse load of any offshore cracked tubular welded joint, BS7910 (2005) has recommended that the plastic collapse load of the cracked geometry P_c is determined by reducing the plastic collapse load for the corresponding uncracked geometry $P_{uncracked}$ using the correction factor F_{AR}. The plastic collapse load of an uncracked K-joint can be obtained from the Health and Safety Executive (HSE, 1995). The correction factor F_{AR} for axial load is given by

$$F_{AR} = \left[1 - \frac{\text{cracked area}}{(\text{intersection length} \times T)}\right]\left(\frac{1}{Q_\beta}\right)^{m_q} \qquad (6)$$

where T is chord thickness, Q_β is 1 for $\beta \leq 1.0$ and m_q is 1 for the K-joints.

For in-plane bending moment, the correction factor F_{AR} is given by

$$F_{AR} = \cos\left(\frac{\varphi}{2}\right) \cdot \left(1 - \sin\left(\frac{\varphi}{2}\right)\right) \qquad (7)$$

where the angle φ is defined in Figure 2.

In Annex B of BS7910 (2005), the L_r parameter for any tubular joint subjected under combined loads is given by the following equation:

$$L_r = \left(\frac{\sigma_F}{\sigma_Y}\right)\left\{\left|\frac{P_a}{P_c}\right| + \left(\frac{M_{ai}}{M_{ci}}\right)^2 + \left|\frac{M_{ao}}{M_{co}}\right|\right\} \qquad (8)$$

where σ_F and σ_Y are the flow and yield stresses; P_a, M_{ai} and M_{ao} are the applied axial load, in-plane bending and out-of-plane bending; and P_c, M_{ci} and M_{co} are the plastic collapse load in the cracked

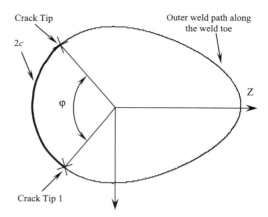

Figure 2. Definition of the angle φ in F_{AR}.

condition for axial load, in-plane bending and out-of-plane bending respectively. The plastic collapse load is obtained by reducing the plastic collapse load of the corresponding uncracked geometry on the basis of the net load-bearing area for axial load and the effect of the flaw area on the plastic collapse modulus for bending moments.

3 FRACTURE ASSESSMENT OF A CRACKED K-JOINT SPECIMEN

3.1 Specimen dimensions

A full-scale cracked tubular K-joint shown in Figure 3 containing fatigue cracks will be assessed in this paper. In the fatigue test carried out earlier (Lie et al. 2005), an axial (AX) and an in-plane bending (IPB) were applied at the brace end. As the hot spot stress was located at the crown of the chord, and the applied loads were symmetrical, the crack was found to initiate and propagate symmetrically from this position.

The notations used to describe the joint parameters, namely $\alpha = 2L/D$, $\beta = d/D$, $\gamma = D/2T$, $\tau = t/T$ and $2\zeta = g/D$ are given in Figure 4, and the overall dimensions are tabulated in Table 1.

To capture the crack details during the earlier test (Lie et al. 2005), an alternating current potential drop (ACPD) technique shown in Figure 5 was used in the fatigue test (Dover et al. 1995). The results showed that the captured crack profile by the ACPD technique agreed quite well with the actual crack shape. It is especially so at the deepest points where the ACPD measurements are capable of providing accurate and useful information such as the stress intensity factors.

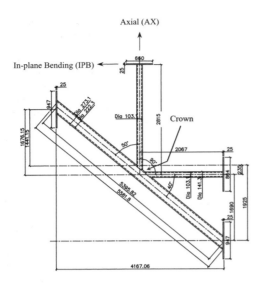

Figure 3. The full-scale K-joint specimen.

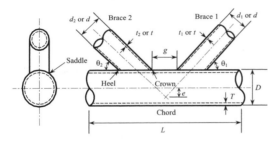

Figure 4. Parameters of the K-joint specimen.

Table 1. Overall dimensions of the K-joint specimen.

	Unit mm & dimensionless
Chord diameter (D)	273.1
Brace diameter (d)	141.3
Chord thickness (T)	25.4
Brace thickness (t)	19.1
Gap (g)	102
Brace to chord angle (θ_1, θ_2)	45°
Chord length (L)	5395.82
Ratio of $2L$ to D (α)	39.52
Ratio of d to D (β)	0.52
Ratio of D to $2T$ (γ)	5.38
Ratio of g to D (ζ)	0.37

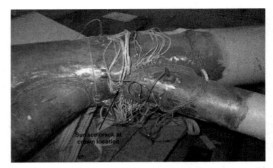

Figure 5. ACPD test set-up and location of the crack.

The measured crack depth a and the corresponding crack length 2c are tabulated in Tables 2 and 3 respectively.

The crack shapes at different propagation stages were plotted from the ACPD readings, and they were compared with the actual crack shape measured by manual method. The two types of crack shapes were then compared with a semi-elliptical shape having the same depth and length (Fig. 6). From the earlier fatigue test (Lie et al. 2005), it can be seen that it is reasonable to assume a semi-elliptical crack shape in the numerical model.

3.2 Finite element analysis

According to the joint dimensions and the 3D crack size, a finite element model shown in Figure 7 is generated automatically (Lie et al. 2005) to calculate the stress intensity factors at the critical locations, namely, at the deepest point and at the crack tips. The K-joint specimen is subjected with an axial load (AX) of 150 kN and an in-plane bending (IPB) of 38 kNm (13.5 kN × 2.815 m) respectively.

J-integral method is then used to obtain the stress intensity factors along the crack front and at the two crack tips. Although this method can not be used directly for the mixed mode problems, it can still be able to produce the Mode-I, II & III stress intensity factors through an indirect way by *introducing* an interaction integral method as proposed by Shih and Asaro (1998). The relationship between the J-integral and the SIFs can be written as

$$J = \frac{1}{8\pi} K^T \cdot B \cdot K \quad (9)$$

where $K = [K_I, K_{II}, K_{III}]^T$ and B is called the pre-logarithmic energy factor matrix.

Table 2. Calculated values of K_r and L_r at the crack depth.

a mm	c mm	$2c$ mm	K_I MPa·m$^{1/2}$	L_r P/P_c	K_r K_I/K_{IC}
3.30	35.20	70.40	25.12	0.210	0.171
5.33	44.82	89.64	25.51	0.390	0.173
8.13	58.98	117.96	26.39	1.335	0.180
10.41	66.12	132.24	28.21	2.988	0.192
13.46	75.33	150.66	30.10	10.947	0.205
15.75	82.67	165.34	30.94	41.518	0.210
18.03	88.19	176.38	32.51	148.174	0.221
20.83	95.89	191.78	35.42	1712.024	0.241

Table 3. Calculated values of K_r and L_r at the crack tips.

a mm	c mm	$2c$ mm	K_I MPa·m$^{1/2}$	L_r P/P_c	K_r K_I/K_{IC}
3.30	35.20	70.40	14.34	0.210	0.098
5.33	44.82	89.64	21.27	0.390	0.145
8.13	58.98	117.96	25.73	1.353	0.175
10.41	66.12	132.24	29.66	2.988	0.202
13.46	75.33	150.66	36.50	10.947	0.248
15.75	82.67	165.34	37.30	41.518	0.254
18.03	88.19	176.38	35.80	148.174	0.244
20.83	95.89	191.78	34.50	1712.024	0.235

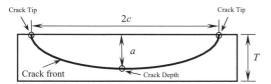

Figure 6. Semi-elliptical 3D surface crack.

Shih and Asaro (1998) gave the relationship between the SIF and the interaction J-integral, J_{int}, as follow:

$$K = 4\pi B \cdot J_{int} \qquad (10)$$

where $J_{int} = [J_{int}^I, J_{int}^{II}, J_{int}^{III}]^T$.

Therefore, once J_{int} is obtained, K can be easily calculated from Equation 10. The detailed calculations of J_{int} can be found in the paper published by Shih and Asaro (1998), and this method has been implemented in the ABAQUS (2006) general finite element software. The values of K_I corresponding to

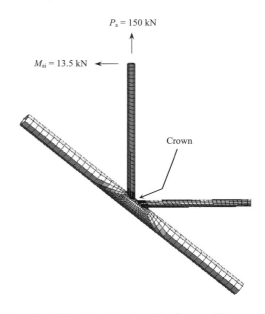

Figure 7. Finite element mesh and loading conditions.

the crack depth a and crack length $2c$ at the deepest point and the crack tips are tabulated in Tables 2 and 3 respectively.

4 ASSESSMENT POINTS K_r AND L_r

In order to assess the safety and integrity of this cracked tubular K-joint subjected under the combined loads, BS7910 (2005) Level 2A failure assessment diagram (FAD) method is employed in this study. FADs are used to consider failure by linear elastic fractures as one limiting criteria and failure by plastic collapse as the second criteria. When performing a structural integrity assessment of a flaw in a stressed structure, an assessment point is derived from two different calculations and plotted on the diagram. The structure is deemed unsafe if the point calculated lies on the curve or falls outside it, and it is safe if the point is within the curve.

For offshore tubular joints, the plastic collapse loads for the cracked geometry P_c are determined by multiplying the plastic collapse loads for the corresponding uncracked geometry $P_{uncracked}$ with the reduction factor F_{AR}. The plastic collapse loads of uncracked K-joint $P_{uncracked}$ can be obtained from the Health and Safety Executive (HSE, 1995), and the correction factors F_{AR} for axial load and in-plane bending are given by Equations 6–7 respectively.

The brace and chord members were fabricated from standard API 5L Grade B specifications pipes, and the fracture toughness measured at room temperature is approximately 147 MPa·m$^{1/2}$ (Somerday 2007). Because the material exhibits significant strain-hardening, the flow stress σ_F will be used and it is given by

$$\sigma_F = \frac{\sigma_Y + \sigma_u}{2} \quad (11)$$

to obtain the L_r parameter. From the standard coupon tests, the measured yield stress $\sigma_Y = 352$ MPa and $\sigma_u = 493$ MPa respectively.

Then, the corresponding values of K_r and L_r for different crack sizes can be computed and they are tabulated in Table 2 at the deepest points and Table 3 at the crack tips subsequently. The assessment points of this cracked K-joint are plotted in the assessment curve shown in Figures 8–9 respectively. For the surface cracks in the K-joint, the maximum crack driving force (SIFs or CTOD) seems to be at the crack tips. Therefore, the SIFs or CTOD at this point should be used to calculate the K_r values.

In accordance with the FADs shown in Figures 8–9, it can be seen that only three assessment points fall inside the standard Level 2A curve, i.e. for a/T is less

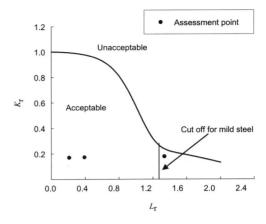

Figure 8. Plot of assessment points at the crack depth.

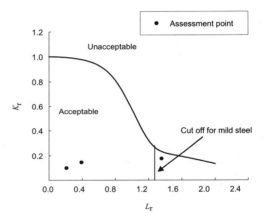

Figure 9. Plot of assessment points at the crack tips.

than 0.4. Therefore, the critical load value is less than the crack initiation load value and the crack K-joint is still safe. When a/T goes beyond 0.4, it is noted that the L_r values are substantially larger, and therefore the cracked K-joint is deemed to fail beyond this crack size.

The increase of K_r is more gradual compared to the increase of L_r as the stress intensity factors, both at the deepest points and crack tips, do not increase drastically as the crack size increases. An increased load or larger crack size will move the assessment point along the loading path toward the failure line as shown in Figures 8–9. The reliability of the method depends on how accurately the intersection zone is described by the failure curve which depends on the structural geometry, type of loading and crack size.

5 CONCLUSIONS

This paper demonstrates the usage of BS7910 (2005) to assess the safety and integrity of a typical cracked tubular CHS K-joint where failure is characterized by two criteria, namely, crack tip failure where failure occurs when the applied load equals the LEFM failure load, i.e.

$$K_r = \frac{K_I}{K_{IC}} = \frac{\text{load}}{\text{LEFM failure load}} = 1 \quad (12)$$

and failure by plastic collapse of the ligament given by

$$L_r = \frac{P}{P_c} = \frac{\text{load}}{\text{ligament area} \times \sigma_Y} = 1 \quad (13)$$

where σ_Y is the material's yield stress.

It enables the integrity of cracked CHS tubular joints to be assessed through two separate calculations based on the two extremes of fracture behaviour, linear elastic and fully plastic. A design curve is used to interpolate between the two failure criteria. The relative position of the assessment point on the diagram, derived from the two separate calculations, determines the integrity of the structure. If the assessment point falls inside the failure curve, the structure is deemed safe; if the assessment point is on or outside the curve, then failure is predicted to occur.

For the cracked K-joint considered, it is found that only three assessment points fall inside the standard Level 2A curve, and hence this damaged joint is still safe if a/T is less than 0.4 operating under the specified conditions.

REFERENCES

ABAQUS 2006. User's Manual. Hibbit, Karlsson & Sorensen Inc., Providence, USA.

American Petroleum Institute (API) RP579 2000. *Fitness-for-Service*. Washington, USA.

British Standard (BS) BS7448-1 1991. *Fracture Mechanics Toughness Tests—Part 1: Method for Determination of KIC, Critical CTOD and Critical J Values of Metallic Structures*. British Standards Institution, London, UK.

British Standard (BS) BS7910 2005. *Guide on Methods for Assessing the Acceptability of Flaws in Fusion Welded Structures*. British Standards Institution, London, UK.

British Electricity Generating Board (CEGB) R6 2001. *Assessment of the Integrity of Structures containing Defects. Revision 4*, British Energy, Gloucester, UK.

Dawes, M.G. & Denys, R. 1984. BS 5500 Appendix D: An Assessment Based on Wide Plate Brittle Fracture Test Data. *International Journal of Pressure Vessels & Pipings* 15: 161–192.

Dover, W.D., Dharmavasan, S., Brennan, F.P. & Marsh, K.J. 1995. *Fatigue Crack Growth in Offshore Structures. Engineering Materials Advisory Services (EMAS) Ltd.*. London, UK: Chameleon Press.

Dowling, A.B. & Townley, C.H.A. 1975. The Effect of Defect on Structural Failure: A Two-Criteria Approach. *International Journal of Pressure Vessels and Pipin* 3(2): 77–107.

Health and Safety Executive (HSE) 1995. *Offshore Installation: Guidance on Design, Construction and Certification*. 4th Edition, Third Amendment, Her Majesty's Stationary Office, London, UK.

Lie, S.T., Chiew, S.P., Lee, C.K. & Shao, Y.B. 2005. Validation of a Surface Crack Stress Intensity Factors of a Tubular K-joint. *International Journal of Pressure Vessels and Piping* 82(8): 610–617.

Lie, S.T., Lee, C.K., Chiew, S.P. & Shao, Y.B. 2005. Mesh Modelling and Analysis of Cracked Uni-planar Tubular K-joints. *Journal of Constructional Steel Research* 61(2): 235–264.

Lie, S.T., Lee, C.K., Chiew, S.P. & Shao, Y.B. 2003. Estimation of Stress Intensity Factors of Weld Toe Surface Cracks in Tubular K-joints. *Proceedings of the 10th International Symposium on Tubular Structures*, Madrid, Spain, 347–355.

Shih, C.F. & Asaro, R.J. 1998. Elastic–Plastic Analysis of Cracks on Bimaterial Interface: Part I—Small Scale Yielding. *Journal of Applied Mechanics* 51: 299–316.

Somerday, B.P. 2007. Technical Reference on Hydrogen Compatibility of Materials—Carbon Steel: C-Mn Alloys (Code 1100). Sandia National Laboratory, Livermore, California, USA.

Wiesner, C.S., Maddox, S.J., Xu, W., Webster, G.A., Burdekin, F.M., Andrews, R.M. & Harrison, J.D. 2000. Engineering Critical Analyses to BS 7910—the UK guide on methods for assessing the acceptability of flaws in metallic structures. *International Journal of Pressure Vessels and Piping* 77(14–15): 883–893.

Effect of chord length ratio of tubular joints on stress concentration at welded region

Y.B. Shao
Yantai University, Yantai, China

S.T. Lie & S.P. Chiew
Nanyang Technological University, Singapore

ABSTRACT: The stress distribution around the weld toe is critical in assessing the fatigue life of a tubular joint. Although there are many reported parametric equations in the literature for estimating the Stress Concentration Factor (SCF) of a tubular joint under basic load, the effect of the chord length ratio on the SCF values is generally neglected. This paper presents the Finite Element (FE) analysis of the effect of the chord length ratio on the SCF values of tubular T-joints under axial load. From the FE investigation, it is found that the chord length ratio has a remarkable influence on the SCF values at the weld toe when the geometry of a tubular T-joint is slightly outside the validity range specified in CIDECT. Therefore, the chord length ratio should be considered in calculating the hot spot stress and its distribution in practical design stage.

1 INTRODUCTION

In the assessment of fatigue life of tubular joints in offshore engineering, the hot spot stress range (HSS) is used to estimate the approximate loading cycles in S-N curves (Zhao et al. 2001). For a tubular joint subjected to certain basic load, i.e., axial load (AX), in-plane bending load (IPB) or out-of-plane bending load (OPB), the hot spot stress can be calculated from a parameter, namely stress concentration factor (SCF), which can be obtained from the parametric equations presented in many reported references or design guide. In these parametric equations, the SCF value of a tubular joint under one certain basic load is determined mainly by the joint geometry. Four commonly used geometrical parameters for tubular joints are: chord thickness ratio γ (chord radius/thickness), chord length ratio α (chord length/radius), chord/brace diameter ratio β (brace diameter/chord diameter) and chord/brace thickness ratio τ (brace thickness/chord thickness). In CIDECT (Zhao et al. 2001) and other references (Romeijn et al. 1994), it is believed that the SCF value of a tubular joint is influenced significantly by γ, β and τ, and the effect of the chord length ratio α on the SCF value is thought to be minor.

Finite element analysis on some tubular T-joints models subjected to axial load has been carried out in this study to investigate the influence of the chord length ratio α on the SCF values. In the FE analysis, solid elements are used to model the entire tubular structures and the proposed FE model is also benchmarked by experimental measurements. For brevity, only tubular T-joints under axial load are analyzed in this study. The conclusions obtained from the above study can be also used for other types of joints.

2 FINITE ELEMENT MODELLING

2.1 Finite element modeling of a tubular joint

In previous FE analysis of a tubular joint, shell elements are generally selected for modeling the entire structure (Hellier et al. 1990). There are some disadvantages for using shell elements to simulate a tubular joint. The most critical ones are: (1) shell elements can only model the mid-planes of the tubes, and the stress variation in the tube thickness direction can not be reflected accurately, especially when the tube thickness is very big; (2) shell elements are difficult in modeling the welding profile. To overcome the above problems, solid elements are an alternative element type. However, it is expensive in computer time and difficult to generate the mesh when solid elements are used to model the joint structure. To obtain more accurately FE results, solid elements are used in the FE analysis in this study.

To produce the FE mesh of a tubular T-joint more efficient and effective, sub-zone mesh generation method is generally used. In this method, the entire tubular structure is divided into several regions according to the requirements of both the accuracy of the FE results and the saving of the computer time. In the regions with high stress gradient, the mesh is refined to ensure the convergence of the FE results. However, relatively coarse mesh can be produced in the regions far away from the high stress gradient regions. The regions with refined mesh and those with coarse mesh are connected by the transition regions. This sub-zone mesh generation method is widely used in FE analysis, and thus it is also adopted in this study. The process of the sub-zone mesh generation method is shown in Figure 1.

To examine the accuracy of the FE results, it is necessary to carry out the convergence test. An automatic refining mesh generation for tubular T-joints is thus proposed in this study. In the mesh refining process, each element in the original mesh is divided into three equal segments in every length direction, and thus an element in the original mesh can be divided into 8 sub-elements (doubling process) or 27 sub-elements (tripling process). The original and the refined meshes of the region around the weld are shown in Figures 2a–2c.

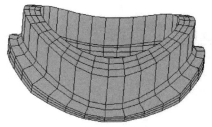

a. original mesh

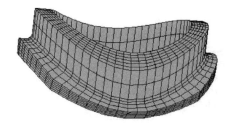

b. doubled mesh

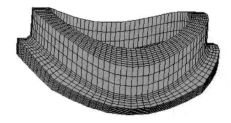

c. tripled mesh

Figure 2. FE mesh of the welded region.

2.2 *FE analysis of SCF*

The SCF of a tubular joint under axial load is obtained simply by dividing the hot spot stress at the weld toe with the nominal stress. The nominal stress is calculated by dividing the area of the cross section of the brace with the applied axial load. To find the peak stress, it is usually to calculate the stress distribution around the weld toe. The stresses at the weld toe, which is generally named as the hot spot stress, are calculated from linear extrapolation method. The hot spot stress is specified as the stress perpendicular to the weld toe in IIW (Zhao & Packer 2000) and CIDECT (Zhao et al. 2001). Such definition of the hot spot stress is more reasonable in assessing the fatigue life of a tubular joint using S-N curve method because the surface crack caused by cyclic

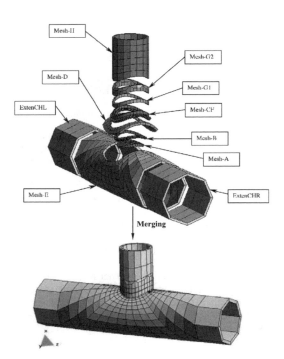

Figure 1. Sub-zone mesh generation method.

Table 1. Geometry of the T-joint specimen.

L (mm)	l (mm)	D (mm)	d (mm)	T (mm)	t (mm)
4130	2160	355.6	273	25.4	25.4

Table 2. Normalized parameters of the specimen.

α	β	γ	τ
23.23	0.77	6.4	1.0

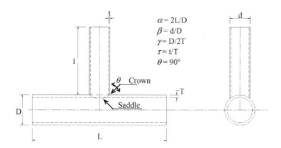

Figure 3. Definition of geometrical parameters.

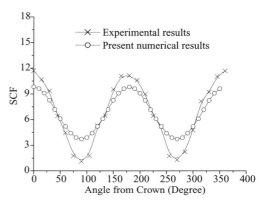

Figure 4. SCF distributions of the specimens.

load is initiating and propagating along the weld toe, and opening crack, which is perpendicular to the weld toe, is the principal crack mode in the fatigue failure mode. Therefore, the above definition of the hot spot stress is used in this study to calculate the SCF value of a tubular joint.

The reliability and accuracy of the presented FE model are critical to be evaluated from experimental measurements (Chiew et al. 2004). A full-scale tubular T-joint subjected to axial load is tested as benchmark to assess the FE model. For the T-joint specimen, axial load is applied at the end of the brace. Both ends of the chord are fixed for the T-joint specimen. The geometry of the T-joint specimen is tabulated in Table 1, and the definitions of the variables in Table 1 are shown in Figure 3.

For brevity, it is more convenient to describe the stress distribution along the weld toe with SCF values. The SCFs of a tubular joint under axial load are determined by the joint geometry, and some normalized geometrical parameters are used to define a particular joint. Figure 3 shows the definition of these parameters of a tubular T-joint.

The normalized geometrical parameters of the T-joint specimen are tabulated in Table 2. The value of the chord thickness parameter, γ, is slightly outside the specified values in some design guides, such as CIDECT (Zhao et al. 2001).

The SCF distributions along the weld toe for the T-joint specimen are obtained from FE analysis and experimental measurements. Such results on the chord surface are plotted together in Figure 4.

It can be found from Figure 4 that the proposed FE model in this study can produce reasonably accurate and reliable estimation of the stress distribution along the weld toe for the T-joint specimen. The FE model, therefore, is used to carry out parametric study to investigate the effect of the geometrical parameters on the SCF values for tubular T-joints.

3 EFFECT OF CHORD LENGTH RATIO ON SCF VALUES OF TUBULAR JOINTS

3.1 *Effect of α on SCFs*

In CIDECT (Zhao et al. 2001), the SCF of a tubular T-joint under axial load is thought to be determined mainly by parameters β, γ and τ. The chord length ratio parameter, α, is thought to have minor influence on such SCF value. To evaluate this conclusion, three T-joint models under axial load have been analyzed. The three T-joint models have different values of α, i.e. 10, 15 and 20 respectively. For the three T-joint models, the values of parameters β, γ and τ are the same, and their values are 0.7, 8 and 0.4 respectively. The stress distributions of the three T-joint models calculated using FE analysis are plotted in Figure 5. ϕ is the polar angle around the brace/chord intersection measured from the crown position in Figure 5.

It is clear that the stress distributions along the weld toe for the three tubular T-joints under axial load are influenced remarkably by the values of the chord length ratio α. When the value of α is smaller, the maximum stress is located at the saddle, and the posi-

377

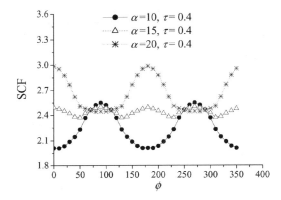

Figure 5. Stress distributions of three T-joint models.

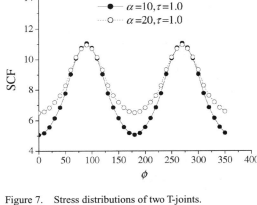

Figure 7. Stress distributions of two T-joints.

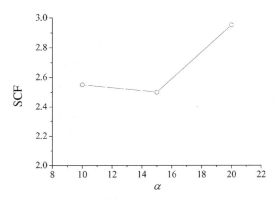

Figure 6. Effect of α on SCF values.

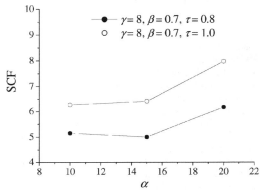

Figure 8. Effect of α on SCFs.

tion of the peak stress is then shifting toward to the crown when the value of α is increasing gradually. As the stress distribution around the weld toe is not consistent, the effect of parameter α on the SCFs is not monotonous, and this can be seen from Figure 6.

However, the effect of parameter α on the SCF values is much different from the FE results of two additional T-joint models. The values of parameter α of these two T-joint models are 10 and 20 respectively. The two models have the same values of $\beta(0.4)$, $\gamma(12)$ and $\tau(1.0)$. The stress distributions obtained from FE analysis are plotted in Figure 7. It is found that the peak SCF values are almost same even the values of α are much different. It seems that the effect of α on the SCF values of tubular T-joints under axial load is also influenced by other geometrical parameters. To make a clear understanding of such effect, further research work on parametric study must be carried out.

3.2 Parametric study

The effect of geometrical parameters on the SCFs of tubular T-joints under axial load has been reported by some researchers (Shao 2007). In previous study, three geometrical parameters, i.e., β, γ and τ are considered. However, the effect of the chord length α on the SCFs is not studied in details. The main purpose in this parametric study is to investigate the effect of α on the SCF values. In parametric study, the validity range of the four parameters are listed as follow

$10 \leq \alpha \leq 20$
$8 \leq \gamma \leq 27$
$0.4 \leq \beta \leq 0.9$
$0.2 \leq \tau \leq 1.0$

There are altogether 240 T-joints models have been analyzed in the parametric study. In the FE analysis, both chord ends of these 240 T-joints models are fixed, and a tensile load is applied at the end of the brace.

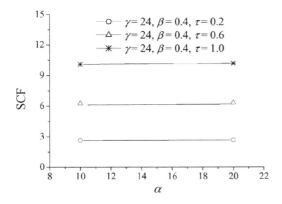

Figure 9. Effect of α on SCFs.

From the FE results, the general conclusion of the effect of parameter α on the SCFs of tubular T-joints under axial load can be drawn. Generally, such effect is remarkably influenced by the chord thickness ratio γ. When the value of parameter α is less than 12, α has great effect on the SCF values. This can be seen in Figure 8. Generally, the location of the peak stress is not fixed when the values of α become different. Such position variation of the peak stress influences the SCF values. It can be thus concluded that the location of the peak stress shifts between the saddle and the crown when the values of α changes.

However, the effect of parameter α on the SCFs is very small when the value of γ is greater than 12, and such effect can be neglected in this case. This conclusion can be illustrated in Figure 9. As can be seen, the SCF values almost keep constant when the values of α are different.

As the effect of α on the SCF values of tubular T-joints under axial load is influenced by parameter γ, such effect must be considered in design stage of practical engineering when chord thickness is very big. In this case, cautions must be taken when the parametric equations in CIDECT (Zhao et al. 2001) are used to calculate the SCF values.

4 CONCLUSIONS

The effect of the chord length ratio on the stress concentration at the weld toe for tubular T-joints under axial load is investigated from FE analysis. It is found from the investigation that the chord length ratio has very remarkable influence on the stress concentration when the chord thickness ratio is less than 12. Therefore, this effect has to be taken into account in design. Further work is needed to improve SCF equations in CIDECT guide.

REFERENCES

Chiew, S.P., Lie, S.T., Lee, C.K. & Huang, Z.W. 2004. Fatigue performance of cracked tubular T-joints under combined loads, I: experimental. *Journal of Structural Engineering, ASCE* 130(4): 562–571.

Hellier, A.K., Connolly, M.P. & Dover, W.D. 1990. Stress concentration factors for tubular Y- and T-joints. *International Journal of Fatigue* 12(1): 13–23.

Romeijn, A., Karamanos, S.A. & Wardenier, J. 1997. Effects of joint flexibility on the fatigue design of welded tubular lattice structures. *The 7th International Offshore and Polar Engineering Conference*, Honolulu, USA, IV: 90–97.

Shao, Y.B. 2007. Geometrical effect on the stress distribution along weld toe for tubular T- and K-joints under axial loading. *Journal of Constructional Steel Research* 63(9): 1351–1360.

Zhao, X.L., Herion, S., Packer, J.A., Puthli, R., Sedlacek, G., Wardenier, J., Weynand, K., van Wingerde, A. & Yeoman, N. 2001. *Design guide for circular and rectangular hollow section joints under fatigue loading*. Koln: TUV-Verlag.

Zhao, X.L. & Packer, J.A. 2000. *IIW recommended fatigue design procedure for welded hollow section joints*. Cambridge: Woodhead Publishing.

Welded KK-joints of circular hollow sections in highway bridges

U. Kuhlmann
Institute of Structural Design, University of Stuttgart, Germany

M. Euler
Institute of Structural Design, University of Stuttgart, Germany

ABSTRACT: Over the last 15 years steel-concrete composite highway bridges comprising spatial trusses made of Circular Hollow Sections (CHS) have become an innovative and aesthetic alternative to the conventional all-steel bridges in Europe. The so-called KK-joint is a major detail of the above mentioned trusses that normally consist of two planes of rising and falling (not crossing) braces and a continuous bottom chord forming together a KK at the intersection. From the viewpoint of economics and ease of construction the braces are preferred being welded directly on the bottom chord. To verify a welded spatial tubular truss against fatigue failure the well-known S-N method is not sufficient anymore due to the complexity of the welded joint geometry. The more sophisticated hot-spot stress approach has to be applied. In general, stress concentration factors are used in order to compute the hot-spot stresses at the truss joints. Unfortunately, the available SCF-values are not applicable to highway bridge geometries. In particular, the low γ-values (chord slenderness) being typical of highway bridges make an application of the existing SCF-values impossible. This paper aims to show the first results of a still on-going research project for developing recommendations concerning the fatigue design and construction of welded KK-joints in highway bridges.

1 INTRODUCTION

1.1 Objectives

In the last two decades an innovative type of highway bridge has been developed (Dauner 1998), (Schlaich 1999), (Seifried et al. 1999), (Bernhardt et al. 2003), (Denzer et al. 2006) etc. Architecturally appealing spatial trusses made of circular hollow sections have become more and more popular in steel-concrete composite highway bridge, see Figure 2. For this particular kind of composite bridges a spatial tubular truss is arranged under the reinforced concrete slab serving as bridge deck.

Preferably, the tubular trusses consist of two planes of rising and falling (not crossing) braces and a continuous bottom chord. Subsequently, the bottom chord forms in longitudinal direction together with two adjacent braces of each plane a lying K. Due to the spatial arrangement of the braces each truss connection is referred to as multi-planar K-joint or KK-joint, see Figure 1.

Up to now tubular truss joints in highway bridges have been carried out mainly as cast steel joints in order to avoid the otherwise complex geometry of the welded brace to chord intersections. With the introduction of the CNC controlled tube cut-off technology it is now possible to carry out the joint preparation for these complex intersections and to connect the braces directly to the chord tube by welding with a high amount of accuracy.

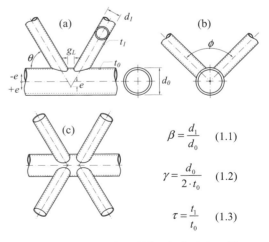

Figure 1. Welded KK-joints: (a) longitudinal view, (b) cross section, (c) top view.

$$\beta = \frac{d_1}{d_0} \quad (1.1)$$

$$\gamma = \frac{d_0}{2 \cdot t_0} \quad (1.2)$$

$$\tau = \frac{t_1}{t_0} \quad (1.3)$$

Figure 2. Existing bridges: (a) Lully, (b) Daettwil, (c) Nesenbachtal, (d) St. Kilian.

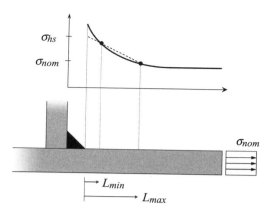

Figure 3. Concept of hot-spot stresses—stress extrapolation.

The welded KK-joint has got some essential advantages over the cast steel joint. Firstly, the direct connection of the braces and the chord saves the cast steel elements and, consequently, reduces the construction costs. Secondly, the fatigue behaviour of the welded joint has to be considered quite positive since the fatigue cracks are initiated on the outside of the truss members, especially the chord crown toe. Therefore, the cracks are well detectable through permanent checks. Furthermore, the fatigue resistance of the potential crack locations can be improved by post-weld treatment. Lastly, the design and the fabrication process are accelerated since interruptions caused by erroneous cast steel elements are eliminated. For these reasons a research project has been set up (still on-going) that focuses on welded KK-joints in highway bridges as an alternative to cast-steel joints.

1.2 The state of the art

Compared with the cast steel joints the fatigue verification of a welded KK-joint requires additional considerations. The stress distribution of the chord and the braces indicate an essential stress concentration (hot-spot stress) within the intersection. This stress-rising effect is caused by the sudden change of geometry (geometric notch). Additionally, the weld diminishes the fatigue resistance (metallurgic notch) at this region. In a preliminary study (Kuhlmann et al. 2002), (Stuba 2001) it has been shown that due to the aforementioned reasons for highway bridges the design of tubular trusses is governed by the fatigue verification of the KK-joints.

In order to verify a tubular truss against fatigue failure the S-N method (classification method) cannot be performed anymore because of the complex geometry of the welded connections.

The more sophisticated hot-spot stress method has to be applied. The hot-spot stress is determined by an extrapolation of the stresses outside the weld-influenced region toward the weld toe, see Figure 3. L_{max} and L_{min} are the limits of the extrapolation region. Unless a time-consuming Finite Element Analysis is performed the computation of the hot-spot stresses requires so-called SCF (stress concentration factors). If SCF values available the hot-spot stress is computed as:

$$\sigma_{hs} = \text{SCF} \cdot \sigma_{nom} \qquad (1.4)$$

where σ_{hs} is the hot-spot stress, also known as structural stress, σ_{nom} is nominal stress.

For KK-joints in offshore constructions, cranes etc SCF values are given in (Zhao et al. 2001), (Karamanos et al. 1997), (Romeijn 1994). However, the range of geometrical parameters of highway bridges comprises significantly lower γ-values ($\gamma < 12$) compared with the offshore and the building sector. Due to the strong dependency of the SCF values on the γ-values the available SCF values cannot be transferred to the scope of highway bridges.

2 INVESTIGATION

2.1 General

Within the research project the welded KK-joint in highway bridge design is considered in an integral view. The developed recommendations are derived both in an empirical and an analytical way. Firstly, the recommendations for the design have been summarized through an analysis of pilot bridge projects throughout Europe. Secondly, an intensive discussion with experts and construction companies has led to recommendations comprising important fabrication parameters. Lastly, based on the aforementioned identified parameter ranges an extensive parametric study has been carried out in order to obtain appropriate SCF values for highway bridges.

2.2 Range of geometrical design parameters

Generally, the design of tubular trusses requires an integral view. The classical approach to dimension a structure is performed step by step (Puthli 1998). Normally, it starts with the verification of the single members followed by the joint design. Since the member's geometry has a great influence on the geometry of the joint of a tubular truss and vice versa the design phase of tubular trusses has to account for this dependency. As an orientation and without exclusion of any other geometry the following range of geometrical parameters seems to be appropriate and therefore preferable for highway bridge design. The recommended range was identified through an analysis of the as yet been realized bridge projects, see Table 1:

$$0.50 \leq \beta \leq 0.6 \quad (2.1)$$

$$4 \leq \gamma \leq 1 \quad (2.2)$$

$$0.25 \leq \tau \leq 0.7 \quad (2.3)$$

$$\phi = 90° \quad \text{and} \quad 45° \leq \theta \leq 60°$$

See Figure 1 for the definition of the parameters.

One outcome of the research project is the recommendation to prefer KK-joints with gap rather than joints with overlapping braces for highway bridges. For the joints with gap it is known from the dynamic testing done by (Schumacher 2003) on uni-planar K-joints with highway bridge specific geometry that the fatigue failure is usually initiated at the weld toe of the chord in the gap region. Thus, a K-joint and KK-joint with gap is advantageous because of the possible visual inspection of the weld toes in the critical zones. Additionally, such a design offers the opportunity of a post-weld treatment of the welds to extend the service life.

A main issue of the design of KK-joints with gap is the minimum gap size g_L, see Figure 1. From the viewpoint of static strength the gap should be as small as possible. Nevertheless, the gap has to meet the requirements of the joint's weldability. On the other hand, the gap size is supposed to be sufficiently wide to ensure the determination of the hot-spot stress. In Figure 4 (a) the gap size is obviously too small to determine exactly the hot-spot stress because of the mutual interaction of the welds' structural stress fields.

To summarize, following minimum gap size is recommended:

$$g_{L,min} \geq 2 \cdot L_{max} + 2 \cdot w_{chord} \quad (2.4)$$

with

$$L_{max} = 0.40 \cdot \sqrt[4]{\frac{d_0 \cdot t_0 \cdot d_1 \cdot t_1}{4}} = 0.20 \cdot d_0 \cdot \sqrt[4]{\frac{\beta \cdot \tau}{\gamma^2}}$$

and w_{chord} as weld leg length on the chord. For design purposes two different gap sizes ($g_L = g_{L,min}$ and $g_L = 2 \cdot g_{L,min}$) have been chosen within the numerical investigation in which the realized gap size has to fall.

Table 1. Realized highway bridge projects with tubular trusses in Europe (selection).

Location	KK-joint	Member	Non-dimensional parameters				
			β	γ	τ	θ	ϕ
Lully, Switzerland, 1997 *)	welded	brace: ⌀ 267/11–25 chord: ⌀ 508/25–50	0.53	5.08–10.16	0.44–0.50	60°	69°
Daettwil, Switzerland, 2001 *)	welded	brace: ⌀ 267/11–25 chord: ⌀ 508/50	0.53	5.08	0.22–0.50	60°	69°
Nesenbach, Germany, 1999 **)	cast	brace: ⌀ 194/10–60 chord: ⌀ 324/16–80	0.60	2.03–10.13	0.63–0.75	46°	102°
Korntal-Muenchingen, Germany, 2002 **)	cast	brace: ⌀ 267/28–45 chord: ⌀ 457/45–65	0.58	3.51–5.07	0.62–0.69	60°	90°
St. Kilian, Germany, 2006 ***)	cast	brace: ⌀ 298.5 chord: ⌀ 610	≈0.50			≈52°	≈70°

*) Schumacher 2003, **) Kuhlmann et al. 2002, ***) Denzer et al. 2006.

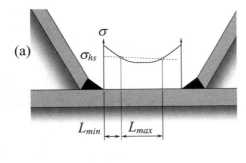

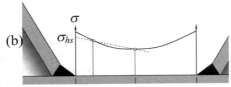

Figure 4. Gap size (a) too small, (b) appropriate.

2.3 Range of fabrication parameters

Secondly, the recommendations for the fabrication have been developed summarizing a couple of quality management linked rules that meet the European highway bridge design requirements for cyclically loaded welded structures. Particular qualification tests are suggested besides the basically required welding procedure specifications. For example, it is recommended to perform the suggested qualification tests on full-scale specimens in order to ensure a high quality of the welds. Furthermore, it is strongly recommended to reduce the geometric notch of the weld toe at the critical crown position. This can be achieved by post-weld treatment or additional weld passes. In brief, the aim of all quality management recommendations is to ensure that the requirements of the fatigue verification are realized in the real structure.

Among others the weld shape at the crown heel location has been a major part of discussion and has to be considered somewhat complicated from the viewpoint of fabrication, Figure 5. Undoubtedly, a full penetrated weld is to be preferred if possible to be realized. Up to now exclusively joints with full penetrated welds at the crown heel have been investigated for the scope of highway bridges (Schumacher 2003). Nevertheless, tests on KK-joints with higher γ-values ($\gamma > 12$) and fillet welds have revealed no significant deterioration of the fatigue resistance (Romeijn 1994).

That means that the location of fatigue failure does not shift from the crown toe to the crown heel. As a consequence, two different weld profiles have been considered: (i) full penetrated weld at all crown and saddle positions, (ii) full penetrated weld at crown toe and saddle with a smooth transition into a fillet weld

at crown heel. The second type appears reasonable especially for greater brace angles, see Figure 6.

2.4 Highway bridge specific SCF values

The fatigue verification of tubular truss connections on the level of hot-spot stresses is covered by various codes such as (Eurocode 3 2005)

$$\sigma_{hs} = \text{SCF} \cdot \sigma_{nom} \cdot \gamma_{Ff} \leq \frac{\Delta \sigma_c}{\gamma_{Mf}} \quad (2.5)$$

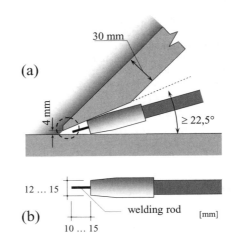

Figure 5. Welding detail: (a) joint preparation for full penetrated weld at crown heel according to (AWS 2004), (b) welding torch.

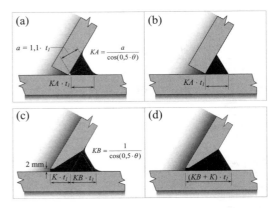

Figure 6. Weld shape at crown heel: (a) proposed fillet weld shape, (b) modeling, (c) proposed full penetration weld shape, (d) modeling.

where

σ_{hs} ... hot-spot stress
σ_{nom} ... nominal stress
$\Delta\sigma_c$ ... fatigue resistance on the level of hot-spot stresses
γ_{Ff}, γ_{Mf} ... partial safety factors

Up to now the hot-spot stresses for highway bridges have had to be obtained through a laborious Finite Element Analysis (FEA) because of the lack of appropriate SCF values. Therefore, one major issue of the still on-going research project has been to determine SCF values for KK-joints in order to simplify the computations of hot-spot stresses. The project aims at SCF values that serve as a powerful and fast tool to get reliable hot-spot stresses without a great loss of accuracy.

The SCF values are computed by means of a parameterized three-dimensional Finite Element model of a KK-joint using the software ANSYS 10.0. 36 combinations of β, γ and τ for two different gap sizes and types of weld profiles have been investigated within an extensive parametric study. The modeled KK-joint has been generated by 20-node solid elements (SOLID 95) with reduced integration scheme. The chord is restrained at one end, Figure 7. All other members are free and can be loaded separately depending on the considered load case.

All in all, eighteen load cases have been considered that comprise axial loading, in-plane bending (ipb) and out-of-plane bending (opb) of all truss members. It has been differentiated between balanced and unbalanced conditions of the brace loading in order to identify so-called carry-over effects. According to (Romeijn 1994) a compensation (N_{eq}, $M_{eq,ipb}$, $M_{eq,opb}$) for chord stresses has been used in order to get SCF values independent of the modeled boundary conditions, compare Table 2(e). The computation is entirely based on primary rather than principal stresses.

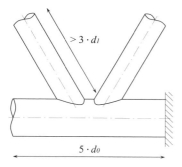

Figure 7. Boundary conditions of FE model.

The generated FE model has been verified by existing test series on uni-planar K-joints and comparative analysis. The verification is mainly focused on the test data of (Schumacher 2003) who did cyclic testing on uni-planar highway bridge like K-joints. Furthermore, the FEA results are compared with other available SCF series for low γ-values ($\gamma = 12$) (Romeijn 1994). Results of cyclic testing on KK-joints with highway bridge type geometries are not available.

In contrast to the uni-planar K-joint where the chord crown toe could be identified as the major hot spot, the most stressed location of the KK-joint is normally situated between the crown and the saddle depending on the particular load case. For this reason, the SCF values have been computed along the complete intersection for each considered load case. For illustration, a table of the background documentation (Kuhlmann, Euler 2008b) is printed, see Table 2. For design purposes the SCF values for eight particular points (see Figure 8) along the brace to chord intersection are tabled and will be summarized in the research report published by *Bundesanstalt fuer Strassenwesen (BASt)* (Kuhlmann, Euler 2008a).

The first results of the still on-going parametric study can be summarized as followed:

i. For chord loading the SCF values at chord crown heel are about 10% higher in case of a fillet weld at the crown heel than for the case of a full penetration weld.
ii. For axial brace loading the SCF values of the crown heel are increased by about 15% for the chord location and nearly unchanged for the brace location in case of a fillet weld at the crown heel.
iii. The influence of the gap size is far more pronounced. Under axial brace loading the SCF values for the chord location (especially crown toe) show an increase of up to 40%.

3 CONCLUSIONS

The design of highway bridges including tubular trusses is generally fatigue-controlled. The brace to chord intersections of the trusses result in high stress concentrations that have to be taken into account. Within a still on-going research project SCF values for KK-joints are computed that can be used to determine these stress concentrations on the level of hot-spot stresses. The computed SCF values are a part of a bundle of recommendations. Especially the integral view of the developed recommendations allows the efficient realization of aesthetic composite tubular truss bridges with welded KK-joints that meet the high requirements of the European highway bridge design.

Table 2. SCF KK-joint (chord location) under axial brace loading.

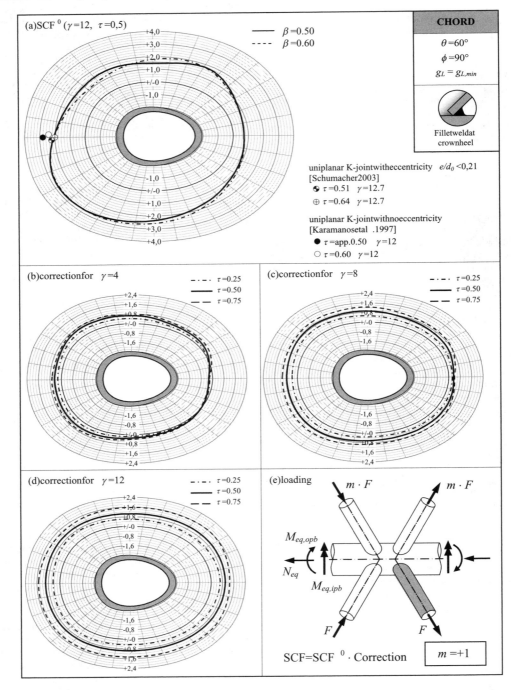

Example: $\beta = 0.60$, $\gamma = 8$, $\tau = 0.75$
maximum $SCF_0 = 2.8$ Corr = 1.20 SCF = $2.8 \times 1.20 = 3.36$

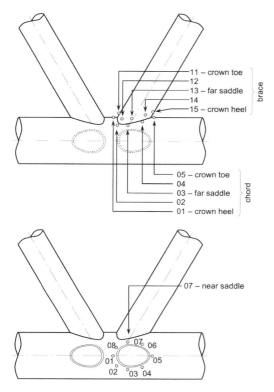

Figure 8. Full penetrated weld at heel according to (AWS 2004).

ACKNOWLEDGEMENT

The research project has been financed by the German Federal Highway Research Institute *Bundesanstalt fuer Strassenwesen* (*BASt*). We gratefully acknowledge the received support.

REFERENCES

AWS 2004. American Welding Society AWS D1.1/D1.1M Structural Welding Code—Steel. Miami, USA:2006.

Bernhardt, K., Mohr, B., Seifried, G., Angelmaier, V. 2003. Talbrücke Korntal-Münchingen innovativer Brückenentwurf als Rohrfachwerk-Verbundbrücke / Teil 1—Entwurf. Stahlbau 72 (2003), No. 2, p. 61–70.

Dauner, H.-G. 1998. Der Viadukt von Lully. Eine Neuheit im Verbundbrückenbau. Stahlbau 67 (1998), No. 1, pp. 1–14.

Denzer, G., Weyer, U., Dieckmann, C. 2006. Die Talbrücke St. Kilian—Entwurf und Ausführung. In: Stahlbau 75 (2006), No. 2, p. 105–116.

Eurocode 3 2005: Design of steel structures, Part 1–9: Fatigue, Release: May 2005.

Herion, S., Puthli, R. 2002. Brücken aus Stahlhohlprofilen—aktuelle Entwicklungen. In: Tagungsband 744 zum 2. Stahl-Symposium Werkstoffe, Anwendung, Forschung, 19. März 2002, Aachen, Düsseldorf: Verlags-und Vertriebsgesell-schaft mbH, 2002.

Karamanos, S. A., Romeijn, A., Wardenier, J. 1997. Stress concentrations and joint flexibility effects in multi-planar welded tubular connections for fatigue design. Research Report Delft 1997.

Kuhlmann, U., Euler, M. 2008a. Entwicklung von SCF-Werten für KK-Knoten im Straßenbrückenbau. Final report of BASt-Research project FE 15.0413/2005/CRB

Kuhlmann, U., Euler, M. 2008b. Entwicklung von SCF-Werten für KK-Knoten im Straßenbrückenbau. Background documentation of BASt-Research project FE 15.0413/2005/CRB

Kuhlmann, U., Günther, H.-P., Saul, R., Häderle, M.-U., Stuba, G. 2002. Zur Anwendung geschweißter Hohlprofilkonstruktionen im Brückenbau. Stahlbau 71 (2002), No. 7, pp. 507–515.

Puthli, R. 1998. Hohlprofilkonstruktionen aus Stahl nach DIN V ENV 1993 (EC 3) und DIN 18800 (11.90). Werner Verlag 1. Auflage 1998.

Romeijn, A. 1994. Stress and strain concentration factors of welded multiplanar tubular joints, PhD Thesis, Delft, The Netherlands, 1994.

Schlaich, J., Schober, H. 1999b. Bahnbrücke am Lehrter Bahnhof in Berlin—Die Humboldthafenbrücke. In: Stahlbau 68 (1999), No. 6, pp. 448–456

Schumacher, A. 2003. Fatigue Behaviour of Welded Circular Hollow Section Joints in Bridges. Dissertation, Thèse 2727, EPFL, Lausanne, 2003.

Seifried, G., Angelmaier, V., Wilhelm, G., Beschorner, K. 1999. Eisenbahnbrücke über den Humboldthafen in Berlin. In: Stahlbau 68 (1999), No. 7, pp. 511–519.

Stuba, G. 2001. Zur Anwendung geschweißter Hohlprofilkno-ten im Brückenbau. Universität Stuttgart, Institut für Kon-struktion und Entwurf, diploma thesis, September 2001, (No. 2001-34x).

Zhao, X.-L., Herion, S., Packer, J.A., Puthli, R.S., Sedlacek, G., Wardenier, J., Weynand, K., van Wingerde, A.M., Yeomans, N.F. 2001. Design guide for circular and rectangular hollow section welded joints under fatigue loading. CIDECT No. 8, TÜV-Verlag GmbH, Köln, 2001.

Experimental comparison in hot spot stress between CFCHS and CHS K-joints with gap

L.W. Tong, C.Q. Sun & Y.Y. Chen
School of Civil Engineering, Tongji University, Shanghai, China

X.L. Zhao
School of Civil Engineering, Tongji University, Shanghai, China
Department of Civil Engineering, Monash University, Clayton, Australia

B. Shen & C.B. Liu
Shanghai Baoye Construction Co. Ltd, Shanghai, China

ABSTRACT: For the purpose of fatigue assessment, the hot spot stresses of welded gap K-joints made of concrete-filled circular hollow sections (CFCHS) subjected to axial loads were investigated by means of the comparison with the same type joints made of circular hollow sections (CHS). The experiments on seven welded CHS K-joints with different non-dimensional joint geometric parameters were carried out firstly, and then these joints were filled with concrete in the chord so that they became CFCHS K-joints and were tested again under the same loading condition as the CHS joints. Hot spot stresses along the intersections between chords and braces for both kinds of joints were measured by means of both linear and quadratic extrapolation methods. The effects of non-dimensional geometric parameters on stress concentration factor (SCF) were presented. It was found that both the joints had similar varying configuration in hot spot stress and similar effects of geometric parameters. In general, the maximum hot spot stress was located at the crown toe, whereas the minimum was located at the crown heel. The hot spot stress at compression area was higher than that at tension area. However, both the joints had quite different magnitude of hot spot stress. The hot spot stress at the CFCHS joint significantly lower than that at the CHS joint. It is a effective method to decrease stress concentration of tubular joints and then improve fatigue behavior by filling a chord with concrete.

1 INTRODUCTION

Welded trusses composed of the brace members with circular hollow sections (CHS) and the chord members with concrete filled circular hollow sections (CFCHS) have come into increasing use in large span highway arch bridges in China (Zhou & Chen 2003). It is worth while to pay attention to the fatigue problem of the new kind of welded joints under repeated loads like the welded joints made of both brace and chord members with circular hollow sections (CHS). For the conveniences of discussion in the paper, the steel-concrete composite joints are called CFCHS joints, whereas the pure steel joints are called CHS joints (Fig. 1).

The fatigue behaviour of welded CHS joints such as T, Y, X and K types has been investigated well in the past twenty years and the resultant fatigue design guidelines have been proposed, in which hot spot stress method is used for their fatigue assessment (Zhao et al. 2000, Zhao et al. 2001). It is considered by the authors of the paper that the hot spot stress method is also suitable for the fatigue assessment of CFCHS joints. The study on behaviour of hot spot stress is first step for having an understanding of fatigue strength of CFCHS joints. The behaviour of hot spot stress at CFCHS T-joints subjected to both axial and bending loads on the brace was investigated recently by the authors (Tong et al. 2007a, b). Experimental comparison in hot spot stress between CHS joints and CFCHS joints of K-type with gap was performed in the paper in order to know their difference.

2 TEST SPECIMEN AND SETUP

Seven specimens of welded K gap joint with both braces at 45° with a chord were designed as shown in Figure.1. They had different non-dimensional geometric parameters β, γ and τ listed in Table 1, but they could be divided into three groups, namely K1, K5, K6 for the group one, K2, K3, K5 for the group two and K4, K6, K7 for the group three to reflect independently the

Table 1. Size of specimens.

No.	$d_0 \times t_0$ (mm)	$d_1 \times t_1$ (mm)	β*	γ*	τ*	g (mm)
K1	245 × 10	121 × 10	0.49	12.25	1.00	74
K2	273 × 8	121 × 8	0.44	17.06	1.00	102
K3	273 × 8	95 × 8	0.35	17.06	1.00	139
K4	273 × 7	133 × 5	0.49	19.50	0.71	85
K5	273 × 8	140 × 8	0.51	17.06	1.00	75
K6	273 × 7	133 × 7	0.49	19.50	1.00	85
K7	273 × 7	133 × 4.5	0.49	19.50	0.64	85

*Notes: $\beta = d_1/d_0$, $\gamma = d_0/(2t_0)$, $\tau = t_1/t_0$.

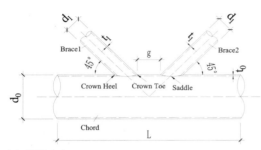

(a) CHS joint

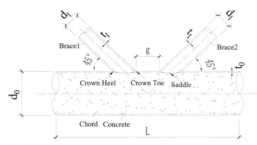

(b) CFCHS joint

Figure 1. Specimens of K-joint with gap.

effects of γ, β and τ on the hot spot stress respectively. All joints were made of hot-rolled seamless tubes of steel Q345 with yielding strength of 345 MPa.

Tests of seven CHS joints on the hot spot stress distribution along the connection between a brace and a chord were carried out at range of elasticity firstly. Afterwards, the chords of all CHS joints were filled with concrete C50 (cube strength of 50 MPa), so they became CFCHS joints, and then the seven CFCHS joints tested in the same loading condition as the CHS joints. Different strength grades of concrete were not considered, seeing that the effect of concrete strength on hot spot stress was investigated in CFCHS T-joints and no significant effect was found.

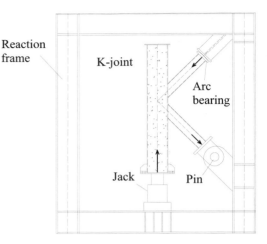

Figure 2. Test setup.

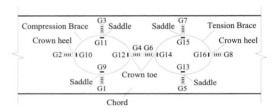

Figure 3. Hot spot location measured for K-joint with gap.

Both the CHS and CFCHS joints were experimented in the same way as shown in Figure 2. Each specimen was vertically installed inside a self-balanced frame. A jack was placed underneath the lower end of a chord, whereas its upper end was kept free. The end of a lower brace was connected to the reaction frame by a pin. The end of an upper brace was made to be arc bearing which was kept in touch with a bracket on the frame. When the jack exerted a compressive load to the chord, the lower and upper braces were subjected to tension and compression respectively, but both forces were equal numerically.

A strain set composed of four tiny strain gauges were fabricated to measure the hot spot strain at the joint as shown in Figure 3. Four hot spot positions such as crown toe, crown heel and two saddles were considered for each member. The strain sets G1 ~ G8 were placed in the chord, and G9 ~ G12 and G13 ~ G16 in the compression brace and the tension brace respectively. For the conveniences of discussion in the paper, the area including G1 ~ G4 and G9 ~ G12 is called a compression area, whereas the area including G5 ~ G8 and G13 ~ G16 is called a tension area.

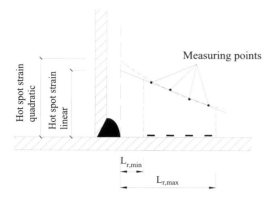

Figure 4. Extrapolation methods of hot spot stress.

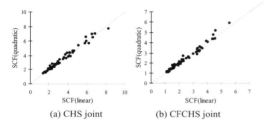

(a) CHS joint   (b) CFCHS joint

Figure 5. Comparison between linear and quadratic extrapolation.

As shown in Figure 4, both the linear and quadratic extrapolation methods recommend for CHS joints in CIDECT handbook were used for the determination of the hot spot strain to see which one was more suitable (Zhao X.L. et al. 2001). The hot spot strains measured were multiplied by a factor of 1.2 and then converted to the hot spot stresses. Moreover, some strain gauges were placed in the lower chord and in both the braces to measure their axial forces and make the chord loaded axially by means of position adjustment of the jack.

3 EXPERIMENTAL RESULTS AND DISCUSSION

The hot spot stress concentration factors measured using both the linear and quadratic extrapolation methods are presented in Figure 5. Among all measuring points totally, there were respectively 89% points in the CHS joints and 91% points in the CFCHS joints that showed less than 10% difference in SCF between the two methods used. This means that both kinds of joints had similar feature, not showing strong nonlinear stress gradient. In CIDECT handbook, the linear extrapolation method is recommended for CHS joints. So the method is considered to be suitable for CFCHS K-joints with gap.

The comparison in hot spot stress distribution between CHS and CFCHS joints subjected to 120 kN axial force on the chord is given in Figure 6 in which the abscissa is about the number of measuring points at the joints shown in Figure 3. Important information on behavior of hot spot stress can be observed:

As for similar feature, on the one hand, both kinds of joints had similar varying configuration in hot spot stress from the measuring points G1 to G16. Firstly, the maximum hot spot stress in the chord was located generally at the crown toe belonging to the gap area, namely at the point G4 for the chord in compression area and at the point G6 for the chord in tension area, but the maximum hot spot stress in the braces was located at the crown toe or at the saddle. Secondly, the hot spot stress at the crown heel was nearly always the lowest regardless of the braces or the chord. Finally, the maximum hot spot stress at the whole joint was located at the crown toe G4 of the chord within the compression area.

As for different feature, on the other hand, firstly, it is clear that the CFCHS joints had lower hot spot stress distribution than the CHS joints. Table 2 presents the ratio of the maximum hot spot stress at the CFCHS joint to that at the CHS joint for the chord and brace at both the compression and tension areas. The degree of decrease in hot spot stress was different for the joint with the different geometry. Their averages ranged from 0.64 to 0.84 for the chord and from 0.73 to 0.77 for the brace. Secondly, the hot spot stress in the chord was much higher than that in the braces for the CHS joints, but this effect became less for the CFCHS joints.

It was observed that the hot spot stress in the compression area was generally higher than that in the tension area. The situation was quite clear for the chord of the CHS joint. It can be explained by means of finite element analysis. Figure 7 illustrates the deformation of K5 joint. The largest bending deformation of the chord occurred at the compression area of the crown toe for the CHS joint and caused high hot spot stress thereby (Fig. 7a). After it was filled with concrete and became a CFCHS joint (Fig. 7b),

Table 2. Ratio of $S_{hs,CFCHS}$ to $S_{hs,CHS}$.

Specimen		K1	K2	K3	K4	K5	K6	K7	Avg.
Chord	Compression	0.73	0.46	0.50	0.70	0.87	0.48	0.72	0.64
	Tension	0.86	0.90	0.68	0.77	1.06	0.91	0.73	0.84
Brace	Compression	0.92	0.60	0.74	0.65	0.83	0.67	0.99	0.77
	Tension	0.73	0.53	0.81	0.74	0.76	0.87	0.69	0.73

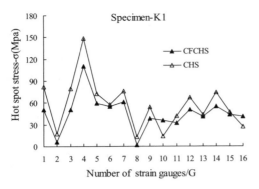
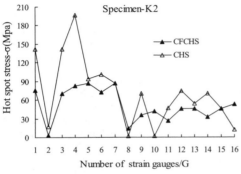
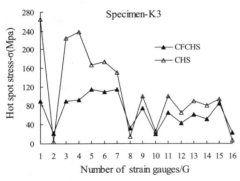
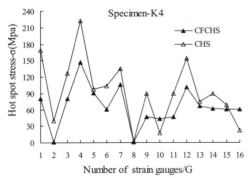
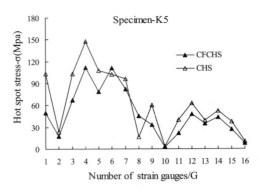
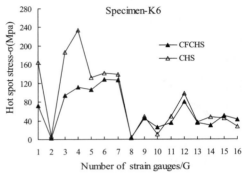
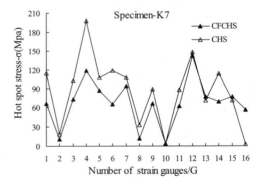

Figure 6. Hot spot stress distribution at CHS and CFCHS joints.

the bending deformation decreased significantly, but was still higher there in comparison with other position.

The stress concentration factor (SCF) can be calculated, which is the ratio of the hot spot stress at the joint to the nominal stress in the member due to a basic member load that causes this hot spot stress. Figures 8–10 presents the effects of non-dimensional geometric parameters β, γ and τ on the SCFs at crown toes of the chord and the brace under both the compression and tension areas. In each

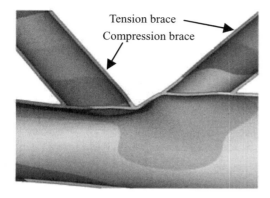

(a) CHS joint

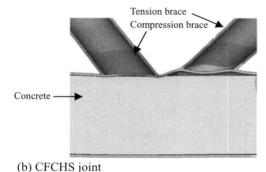

(b) CFCHS joint

Figure 7. Deformation of K5 joint from FE analysis.

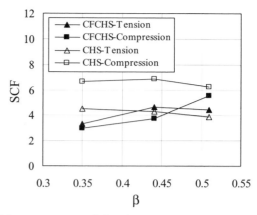

(a) at crown toe of chord

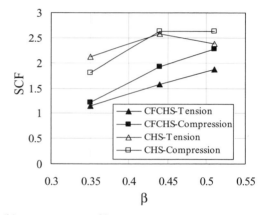

(b) at crown toe of brace

Figure 8. SCF vs. β.

figure, on one geometric parameter was variable, whereas the others were kept constant. It can be seen from these figures that both the CHS and CFCHS joints had similar effects of β, γ and τ on their SCFs. As for the range of geometric parameters tested in the paper, the effects of β, γ and τ on the SCF of the CFCHS joint still existed, and they were significant sometimes and slight sometimes. The further parameter study on the effects of β, γ and τ at a wide range will be needed by means of finite element analysis in order to get more information.

4 CONCLUSIONS

The hot spot stress distributions at both the CHS and CFCHS K-joints with gap were investigated experimentally and compared between them. The following conclusions can be drawn base on the geometric parameters of the joints tested:

1. The CFCHS joint did not show strong nonlinear stress gradient. The linear extrapolation method commonly used in CHS joints was also suitable for the CFCHS joints.
2. Both the joints had the similar varying configuration in hot spot stress and the similar effects of non-dimensional geometric parameters β, γ and τ on the stress concentration factor. However, the magnitude of hot spot stress at the CFCHS joint was lower or much lower than that at the CHS joint, which was attributed to the stiffness of CFCHS joint improved by concrete.
3. Both the joints showed that for a chord the maximum hot spot stress occurred at its crown toe, but for a brace the maximum hot spot stress occurred at its crown toe or saddle sometimes. For a member, regardless of chord or brace, the

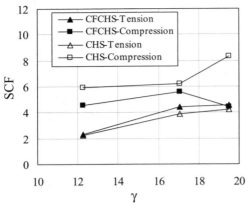

(a) at crown toe of chord

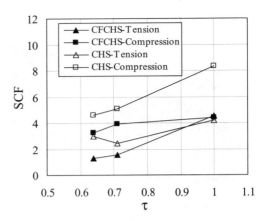

(a) at crown toe of chord

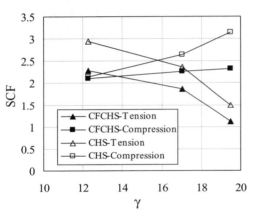

(b) at crown toe of brace

Figure 9. SCF vs. γ.

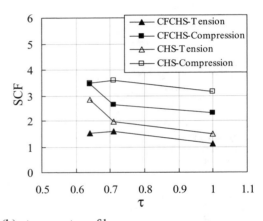

(b) at crown toe of brace

Figure 10. SCF vs. τ.

minimum hot spot stress occurred at its crown heel. As for the whole joints, the maximum hot spot stress was located at the crown toe of the chord in compression area.
4. Both the joints had such feature that the hot spot stress in compression area was generally higher than the hot spot stress in tension area. This was attributed to a large bending deformation taking place at the crown toe of the chord in compression area.
5. It is a good way to decrease stress concentration of tubular joints by filling a chord with concrete, which is expected to increase fatigue strength.

ACKNOWLEDGMENTS

The authors wish to thank the Natural Science Foundation of China for financially supporting the research in the paper through the grant No. 50478108.

REFERENCES

Tong, L.W. et al. 2007a. Experimental study on stress concentration factors of concrete-filled circular hollow section T-joints under axial loading. *Proceedings of 8th Pacific Structural Steel Conference*, Wairakei, New Zealand, 13–16 March 2007: 153–158.

Tong, L.W. et al. 2007b. Experimental investigation on stress concentration factors of CHS-to-CFCHS T-joints subjected to In-plane Bending. *Proceedings of 5th International Conference on Advances in Steel Structures*, Singapore, 5–7 December, 2007: 1003–1007.

Zhao, X.L. et al. 2000. Fatigue design procedures for welded hollow section joints. *IIW*, Cambridge, UK: Abington Publishing.

Zhao, X.L. et al. 2001. *Design guide for circular and rectangular hollow section welded joints under fatigue loading.* German: CIDECT and TUV-Verlag,

Zhou, S. & Chen, S. 2003. Rapid development of CFST arch bridges in China. Advances in Structures, *Proceedings of the International Conference on Advances in Structures: Steel, Concrete, Composite and Aluminium*, Sydney, Australia, Vol. 2. 23–25 June 2003: 915–920.

Parallel session 6: Composite construction
(reviewed by L.H. Han & M. Lefranc)

Behaviour and design of hollow and concrete filled stainless steel square sections subjected to combined actions

B. Uy & Z. Tao
University of Western Sydney, Australia

ABSTRACT: Modern structural engineering which considers whole of life costing of a project has illustrated that stainless steel can be utilised for more than just architectural applications. Major engineering structures such as the Hearst Tower in New York and the Stonecutters Bridge in Hong Kong are using stainless steel as one of the major structural elements for structural and durability purposes. This paper will consider the behaviour and design of concrete filled steel columns utilising stainless steel sections. The paper will involve both an experimental and theoretical component and will result in the development of design guidelines for practising structural engineers. The major innovation in this paper is the conduct of experimental and theoretical research and the development of design guidelines for the composite behaviour of stainless steel sections filled with concrete. Stainless steel columns with and without concrete were tested under the combined actions of axial force and bending moment. The results of these tests were compared with existing design procedures and recommendations have been made for utilising this material in an efficient design manner. It is believed that there will be significant applications in buildings, bridges, offshore and specialty structures utilising stainless steel in this form in the future.

1 INTRODUCTION

There has been a renaissance in the use of composite columns in structures throughout the developed world. This has in many ways been heavily influenced by the development of high-strength concrete, which enables these columns to be considerably economised. Further economies can also be achieved by the use of thin-walled fabricated steel columns (Bridge & Webb 1992).

Stainless steel is considered to be useful in seismic zones, because of its high energy absorption capability. Furthermore, its corrosion resistance is far superior to carbon steel and thus is very suitable to exterior environments. This is a major reason why the use of stainless steel has increased of late. Other advantages of stainless steel in comparison with carbon steel are its mechanical properties, weldability, chemical-physical compatibility with other materials, and life cycle cost (Di Sarno et al. 2003).

From a financial perspective, stainless steel attracts lower maintenance costs. This is because stainless steel does not require protective coating against corrosion. Also, structures, using stainless steel in their structural elements have considerable weight savings.

In addition, stainless steel has improved fire resistance in comparison with carbon steel. Consequently, stainless steel structural members are less likely to require protective fire coating to be applied for structural members (Gardner & Nethercot 2004). Furthermore, since the cost of stainless steel is approximately three times greater than carbon steel, efficient design seems to be essential and this justifies the use of a more complex design process (Gardner 2002).

Stainless steels are metal alloys which contain a high percentage of chromium (Cr). Normal grades are obtained by adding at least 12% by weight to low-alloy carbon steel (Standards Australia 2001a). The presence of Cr allows the formation of a protective oxide on the material surface (stainless metal). There are different types of stainless steel alloys available. Stainless steel alloys are classified as Austenitic (ASS), Ferritic (FSS), Austenitic-Ferritic (AFSS), Duplex, Martensitic (MSS), and Precipitation Hardening (PHSS). The most common grades for structural and architectural applications are the austenitic and duplex grades. Duplex stainless steel offers higher strength and wearing resistance than austenitic, but at a greater expense. Moreover, austenitic grades demonstrate the greatest non-linearity and strain hardening amongst other grades (Di Sarno et al. 2003).

The purpose of the study conducted herein is to provide an improved understanding of the behaviour of stainless steel tubes filled with concrete. This paper focuses on experiments for the short column behaviour of hollow and concrete filled stainless steel sections filled with concrete and subjected to the combined actions of axial compression and bending moment.

2 PREVIOUS APPLICATIONS

Previous applications for the use of stainless steel show mainly a preference for specialty structures, however with increased knowledge and research, this could lead to increased usage of this material for purely structural purposes.

2.1 *Gateway arch, St Louis, USA*

The gateway arch constructed in St Louis, Missouri was completed in 1966 and this was constructed of stainless steel. This included a 300 metre tall arch and is composed of a triangular annular cross-section which invoked a double skin construction technique. The inner and outer stainless steel skins were filled with reinforced concrete, thus forming a composite steel-concrete composite structure which would provide benefits for durability, as well as stiffness, strength and stability.

2.2 *Parliament house, Canberra, Australia*

The new federal parliament house building in Canberra, Australia was completed in time for the bicentennial celebrations in 1988. The 81 meters tall flag mast of the parliament house, shown in Figure 1 was constructed using stainless steel closed sections.

2.3 *Hearst tower, New York, USA*

Hearst Tower at 959 Eight Avenue, New York City, USA is a 46 storey building completed in 2006. The major lateral load resisting system for this building included concrete filled steel columns in mega columns of the diagrid exoskeleton (Fortner 2006). These columns incorporate a stainless steel skin as illustrated in Figure 2.

2.4 *Stonecutters bridge, Hong Kong, China SAR*

The Stonecutters bridge in Hong Kong will be the longest cable stayed bridge in the world upon its completion (1018 metres). The bridge consists of two 290 metre tall masts with their upper third (approximately 100 metres) comprised of a stainless steel section that will be filled with concrete. The major reason for the concrete infill is to ensure that minimal maintenance has to be conducted on this section as it would be difficult to access during the design life. The bridge was due for completion in 2007.

Figure 1. Parliament house, Canberra, Australia.

Figure 2. Hearst tower, New York City, USA.

3 PREVIOUS RESEARCH OF STAINLESS STEEL SECTIONS

Rasmussen (2000) considered recent research carried out on stainless steel tubular members and connections. Research was focused on square, rectangular and circular tube members.

Kouhi et al. (2000) summarised the significant research being conducted in Finland on stainless steel in construction and suggested that its use would be increased as life cycle costing was considered in building projects. Burgan et al. (2000) summarised UK findings for circular hollow sections made of stainless steel and suggested that the diameter to thickness slenderness limits used for normal carbon steel were far too conservative for stainless steel sections.

Nethercot & Gardner (2002) have looked at methods for exploiting the special features of stainless steel in structural design. The premise of their approach is to employ an ultimate stress for design based on the slenderness of the plate elements in tubular sections. For example in sections where very small plate slenderness is present, stresses much larger than the 0.2% proof stress may be employed in the design of these members.

Di Sarno et al. (2003) and Di Sarno & Elnashai (2003) have also recently shown that the use of stainless steel can provide up to 3 times the ultimate strain of carbon steels and they can exhibit improved post-local buckling giving them excellent application for seismic regions both for new structures and as braces for rehabilitating structures.

It is believed that there will be significant applications in buildings, bridges, and specialty structures utilising stainless steel as evidenced by the applications already outlined in this paper. The major benefit of the use of concrete infill is that due to restrained local buckling a concept developed by Uy & Bradford (1996), the thickness of the steel sections can be reduced considerably. Since stainless steel is approximately five times the cost of mild structural steel, this benefit is quite important in ensuring that wider application of the material is made possible. Initial development of a numerical model has been carried out by Roufegarinejad, Uy & Bradford (2004) to ascertain the behaviour of short concrete filled steel columns under combined axial force and bending moment. This paper will involve an experimental study to further investigate the issue.

4 BEHAVIOUR OF STAINLESS STEEL CONCRETE FILLED SECTIONS

The use of stainless steel with concrete also provides a very good combination of materials. When the stainless steel component is restrained from local buckling it will be able to achieve stresses which are significantly greater than the 0.2% proof stress which is generally used in design codes. This has the potential for the steel to significantly exceed the design stress and the steel will also subsequently confine the concrete. To date there has been limited research to consider these effects.

4.1 Axial strength

The axial strength of composite columns utilising stainless steel can thus be represented as:

$$N_u = N_{uc} + N_{uss} = f_c A_c + f_{ss} A_s \qquad (1)$$

Now the maximum concrete compressive strength f_c depends on the concrete strength chosen and the steel plate slenderness. For steel sections with compact slenderness the concrete is generally assumed to be adequately confined and thus the concrete strength is assumed to be higher than the mean cylinder strength. The steel compressive yield stress f_{ss} is also dependent on the concrete strength and steel plate slenderness. Generally if the concrete strength is low, then it is expected that the proof stress will be exceeded. However, if the concrete strength is high and the steel section is non-compact then the maximum stress may be lower than the 0.2% proof stress.

4.2 Combined strength

Initial numerical studies by Roufegarinejad, Uy & Bradford (2004) showed that when considering the

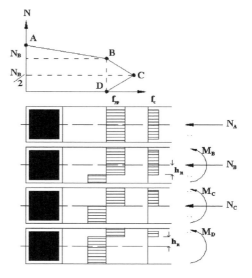

Figure 3. Eurocode 4 model—rigid plastic approach.

interaction between axial force and bending moment, the strength of a stainless steel concrete filled column will surpass the rigid plastic strength based on the 0.2% proof stress. This is due to the fact that stainless steel continues to increase in stress after the 0.2% proof stress. This is a strength issue which may prove useful in harnessing in future studies. As the strain increases, the stainless steel tube experiences significant strain hardening before the concrete crushes and this explains why the model gives a larger envelope to the Eurocode 4 method as illustrated in Figure 3 (British Standards Institution 2004).

5 EXPERIMENTAL PROGRAM

In order to investigate the strength of short composite stainless steel box columns under uni-axial loading combined with bending, an extensive experimental program has been carried out. Tests were undertaken on both hollow and composite short columns and hollow and composite beams as well. In the case of the composite columns, experiments were carried out on columns with a nominal concrete strength of 32 MPa.

5.1 Specimen preparation

Cold-formed hollow stainless steel sections were cut according to the requirements of the project. A total of sixteen specimens, divided into three series, were cut into their required dimensions. From that batch twelve were used for short box column tests and the other four were used for the beam tests.

5.2 Concrete pouring

A timber frame was constructed to restrain the specimens together and to stop any movement during casting. Care was taken during the casting of the concrete. The hollow sections were covered with small hardboards to ensure no concrete was poured inside the sections. The restraints in the timber framework ensured that there was no significant lateral deformation for the specimens during casting.

5.3 Testing of short hollow stainless steel box sections

Series 1 consisted of six hollow stainless steel box columns. The columns and their geometric properties are described in Table 1. The plate slenderness (b/t) of the specimens varied between 20 and 33, which indicates that all plates were considered to be compact or stocky. All columns are thus designed to avoid local buckling. The hollow columns were identified by the following code, H-t-e, where H signifies a hollow section, t is the nominal thickness of the section in mm and e is the eccentricity of the load from the plastic centroid.

End-plates with a restraining mechanism were designed to ensure that failure would occur at the centre of the columns and not at the end. It is important to prevent failure from occurring at the column ends, since this would affect the test results and the ability to monitor deformations in the failure zone. Sharp knife edges were designed and attached to the end plates, with grooves machined in the plates, to apply the bending moments to the specimens. Two grooves of 20 and 40 mm from the centre were machined into the plates to apply moments with different eccentricities. Experiments were undertaken using a 5000 kN capacity Denison Material testing system as shown in Figure 3. All columns were designed to fail at loads below the capacity of the machine. A total of eight strain gauges were attached to two sides of each column as shown in Figure 4. Four horizontal and four vertical strain gauges were used to measure the strains in the steel. In addition to the strain gauges, one Laser Linear Varying Displacement Transducers (LVDT) and one Midory LVDT were used to measure the shortening of specimens and the lateral deflection.

Table 1. Geometric properties for hollow columns.

Specimen	Nominal dimensions (mm)				
	b	d	L	t	b/t
H-3-0	100	100	300	3	33
H-3-20	100	100	300	3	33
H-3-40	100	100	300	3	33
H-5-0	100	100	300	5	20
H-5-20	100	100	300	5	20
H-5-40	100	100	300	5	20
H-3-0	100	100	300	3	33

Figure 4. Experimental set up for hollow sections.

They were positioned on the testing system as shown in Figure 4. The testing machine was operated under 'load control' until the load was close to the peak value. There was a need to use 'load control' until close to peak to be able to manually record the readings from the strain gauges.

5.4 Testing of short composite stainless steel box sections

Series 2 consisted of six composite stainless steel box columns filled with 33 MPa concrete and these are summarized in Table 2. The plate slenderness (b/t) of the specimens varies between 20 and 33, which indicates that all plates are compact or stocky. All columns are thus designed to avoid local buckling. The composite columns were identified by the following code, C-t-e, where C signifies a composite section, t is the nominal thickness of the section in mm and e is the eccentricity of the load from the plastic centroid. Six columns were cast with concrete with a compressive strength of 33 MPa and similarly four beams were cast with 33 MPa concrete. The columns were left to cure at room temperatures in the laboratory for a minimum of 28 days before the testing proceeded.

The same test method was used to test the composite columns as was described for the hollow stainless steel sections and this is illustrated in Figure 5.

5.5 Testing of hollow and composite stainless steel beams

Series 3 consisted of two hollow stainless steel beams and two composite stainless steel beams. The beams and their geometric properties are described in Table 3. A four point loading system was designed for the beam tests. Experiments were undertaken using a 5000 kN capacity Denison Material testing system as shown in Figure 6. All beams were designed to fail at loads below the capacity of the machine. A total of seven strain gauges were attached to four sides of each beam.

In addition to the strain gauges, one Laser Linear Varying Displacement Transducers (LVDT) was used to measure the deflection of specimens. It was placed on the testing system as shown in Figure 5. The testing machine followed 'load control' until the load was close to the peak value. There was a need to use 'load control' until close to the peak to be able to manually record the readings from the strain gauges.

Table 2. Geometric properties for composite columns.

Specimen	Nominal dimensions (mm)				
	b	d	L	t	b/t
C-3-0	100	100	300	3	33
C-3-20	100	100	300	3	33
C-3-40	100	100	300	3	33
C-5-0	100	100	300	5	20
C-5-20	100	100	300	5	20
C-5-40	100	100	300	5	20
C-3-0	100	100	300	3	33

Table 3. Geometric properties for beams.

Specimen	Nominal dimensions (mm)			
	b	d	L	t
H-3-Beam	100	100	800	3
H-5-Beam	100	100	800	5
C-3-Beam	100	100	800	3
C-5-Beam	100	100	800	5

Figure 5. Experimental set up for composite sections.

Figure 6. Experimental set up for beam tests.

6 RESULTS AND COMPARISONS

6.1 Column specimens

The ultimate loads of the hollow and composite columns are summarised in Tables 4 and 5. The hollow columns reached failure at a lower load than the composite columns. This indicates that the effect of the concrete core enhances the overall strength and stability of the columns, especially by denying the steel from buckling inwards and thus avoiding local buckling from occurring before the peak load is reached (Figures 7–9).

Table 4. 0.2% proof load, ultimate load and maximum load of hollow members in combined axial loading and bending.

Specimen	0.2% proof Load (kN)	Predicted ultimate load axial loading (kN)	Failure load for axial load combined with bending (kN)
H-3-20	410	453	340
H-3-40	410	453	247
H-3-0	410	453	447
H-5-20	1030	1213	833
H-5-40	1030	1213	625
H-5-0	1030	1213	1185

Table 5. 0.2% proof load, ultimate load and maximum load of composite members in combined axial loading and bending.

Specimen	0.2% proof Load (kN)	Predicted ultimate load axial loading (kN)	Failure load for axial load combined with bending (kN)
C-3-20	615	711	512
C-3-40	615	711	752
C-3-0	615	711	721
C-5-20	1200	1443	1043
C-5-40	1200	1443	389
C-5-0	1200	1443	1447

6.2 Beam specimens

The ultimate loads of the hollow and composite beams are summarised in Tables 6 and 7. The hollow beams reached failure at a lower load than the composite columns. This indicates that the effect of the concrete core enhances the overall strength and stability of the beams just as it was for columns (Figure 10).

Table 6. Ultimate load for beams.

Specimen	Nominal dimensions (mm)				Failure load (kN)
	b	d	L	t	
H-3-Beam	100	100	800	3	104
H-5-Beam	100	100	800	5	298
C-3-Beam	100	100	800	3	221
C-5-Beam	100	100	800	5	416

Table 7. Comparison of predicted and experimental values.

Specimen	e (mm)	P_{theory} (kN)	P_{expt} (kN)	(P_{expt}/P_{theory})
H-3-20	20	332	339.9	1.024
H-3-40	40	243	246.9	1.016
H-3-0	0	453	446.9	0.987
H-5-20	20	845	832.5	0.985
H-5-40	40	619	624.6	1.009
H-5-0	0	1213	1185	0.977
C-3-20	20	533	511.7	0.960
C-3-40	40	340	751.5	2.210
C-3-0	0	711	720.7	1.014
C-5-20	20	1120	1043	0.931
C-5-40	40	401	389	0.970
C-5-0	0	1443	1447	1.003
H-3-Beam	–	110	103.6	0.942
H-5-Beam	–	305	298.3	0.978
C-3-Beam	–	245	221.2	0.903
C-5-Beam	–	439	416.1	0.948
			Mean =	1.054
			Standard deviation =	0.310

6.3 Comparisons with design codes

In this section the prediction of the ultimate load has been compared with the experimental results. The Australian Standard AS3600 and AS4100 (Standards Australia 1998, 2001b) have been used to predict the ultimate load of both the hollow and the composite sections. The 0.2% proof stress and the ultimate load capacity obtained from the tensile coupon tests have been used in all predictions. Table 7 illustrates the difference between the experimental results and the model predictions for the hollow and composite specimens.

Almost all predictions from the design methods were conservative compared with the test results and they can thus be assumed to be safe for use in the design of composite stainless steel box sections. The main discrepancy in these results is the use of the 0.2% proof stress determined from the tensile

Figure 7. All the hollow columns after testing.

Figure 8. Composite sections (3 mm thickness) after testing.

Figure 9. Composite sections (5 mm thickness) after testing.

Figure 10. All beams after testing.

coupon tests, because of the manner in which the tests were undertaken. Also C-3-40 section showed a great deal more load capacity than expected. This should be further investigated.

7 CONCLUSIONS AND FURTHER RESEARCH

Utilizing the full strength of stainless steel sections has been previously alluded to by Di Sarno et al. (2003) and Gardner & Nethercot (2004). The attempt in this paper is to use stainless steel in composite sections rather than hollow sections. The paper highlights that stainless steel in this form can provide increased strength particularly when confinement is large. It is desirable to use the full benefits of stainless steel in design. This paper has presented the results of an extensive experimental program that considers the combined actions of axial force and bending moment. Some very preliminary theoretical comparisons have been made and this research will form part of an extensive research program at the University of Western Sydney.

ACKNOWLEDGEMENTS

The authors would like to acknowledge the assistance of Mr Tanvir Asgar and all the technical staff at the University of Wollongong where these tests were initially conducted. This program of research was sponsored by a University Research Council Small Grant at the University of Wollongong and further funding has been received through the Research Grant Scheme at the University of Western Sydney to continue this research. This support is gratefully acknowledged.

REFERENCES

Bridge, R.Q. & Webb, J. 1992. Thin-walled circular concrete filled steel tubular columns. *Proceedings, Engineering Foundation Conference, Composite Construction 11*, Potosi, USA.

British Standards Institution. 2004. Eurocode 4, EN 1994-1-1. *Design of composite steel and concrete structures, Part 1.1, General Rules and Rules for Buildings*.

Burgan, B.A., Baddoo, N.R. & Gilsenan, K.A. 2000. Structural design of stainless steel members—comparison between Eurocode 3, Part 1.4 and test results. *Journal of Constructional Steel Research, An International Journal* 54 (1): 51–73.

Di Sarno, L. & Elnashai, A.S. 2003. Special metals for seismic retrofitting of steel buildings, *Progress in Structural Engineering and Materials* 5 (2): 60–76.

Di Sarno, L., Elnashai, A.S. & Nethercot, D.A. 2003. Seismic performance assessment of stainless steel frames, *Journal of Constructional Steel Research* (59): 1289–1315.

Fortner, B. 2006. Landmark reinvented, *Civil Engineering, Magazine of the American Society of Civil Engineers*, 76 (4): 42–47.

Gardner, L. 2002. *A New Approach to Structural Stainless Steel Design*, PhD thesis, Imperial College of Science, Technology and Medicine, London, UK.

Gardner, L. & Nethercot, D. A. 2004. Experiments on stainless steel hollow sections—Part 1: material and cross-sectional behaviour, *Journal of Constructional Steel Research.*, 60 (9): 1291–1318.

Kouhi, J., Talja, A, Salmi, P. & Ala-Outinen, T. 2000. Current R&D work on the use of stainless steel in construction in Finland, *Journal of Constructional Steel Research, An International Journal* 54 (1): 31–50.

Nethercot, D.A. & Gardner, L. 2002. Exploiting the special features of stainless steel in structural design, *ICASS '02, 3rd International Conference on Advances in Steel Structures*, Hong Kong: 43–56.

Rasmussen, K.J.R. 2000. Recent research on stainless steel tubular structures, *Journal of Constructional Steel Research, An International Journal* 54: 75–88.

Roufegarinejad, A., Uy, B. & Bradford, M.A. 2004. Behaviour and design of concrete filled steel columns utilising stainless steel cross-sections under combined actions, *18th Australasian Conference on Mechanics of Structures and Materials*, Perth, Australia, December: 159–165.

Standards Australia. 1998. *Australian Standard, Steel Structures, AS4100–1998*, Sydney, Australia.

Standards Australia. 2001a. *Australian/New Zealand Standard. Cold-formed stainless steel structures, AS/NZS 4673:2001*, Sydney, Australia.

Standards Australia. 2001b. *Australian Standard, AS 3600–2001 Concrete Structures*, Sydney, Australia.

Uy, B. & Bradford, M.A. 1996. Elastic local buckling of steel plates in composite steel-concrete members. *Engineering Structures, An International Journal*, 18 (3): 193–200.

Test and analysis on double-skin concrete filled tubular columns

J.S. Fan, M.N. Baig & J.G. Nie
Department of Civil Engineering, Tsinghua University, Beijing, China

ABSTRACT: Concrete filled tubular columns are widely used in China for the attractions of higher strength and ductility. Under some conditions in bridges and high rise buildings, tubular columns with double skin have additional advantages of time saving, lower self-weight and arrangement of vertical equipment pipes. In this paper, a total of 32 Double Skinned Concrete Filled Tubular (DSCFT) columns are tested with different geometrical arrangements. The varied parameters of the specimens are diameter to thickness ratios, diameter of inner tubes and cross-section geometric shapes (square or circular). The test results on stress/strain of steel and compressive strength of the columns are described and discussed. The equation on axial loading capacity is suggested and compared with some design codes. This equation is proved to be workable for calculation of axial strength of all kinds of columns tested in this study.

1 INTRODUCTION

Concrete-filled steel tubular (CFT) columns have been used for earthquake-resistant structures, bridge piers subject to impact from traffic, columns to support storage tanks, columns in high-rise buildings, etc. According to past study on concentric compression behavior of CFT columns, the ultimate axial strength of CFT columns is considerably affected by the thickness of the steel tube, as well as by the shape of its cross section. Although a confining effect could be expected in circular CFT columns, square columns show only a small increase in axial strength even for those with large wall thicknesses. On the other hand, the axial load-deformation behavior of columns is remarkably affected by the cross-section shape, diameter/width-to-thickness ratio of the steel tube and strength of the filled concrete.

Double-skinned concrete filled steel tubular (DSCFT) columns consisting of two concentric thin steel tubes with concrete between them are studied in this paper under axial loads. DSCFTs have been used for over a decade for compression members in offshore construction. DSCFTs were first reported in the late 1980s (Shakir-Khalil & Illouli 1987). Since then, some research has been conducted. Compared to CFT columns, the DSCFT can reduce its own weight while have a high flexural stiffness. Due to these benefits, some researchers have investigated the DSCFT for high rise building and bridge applications. Different geometric arrangement have been studied by Wei et al. (1995), Lin et al, Tao et al. (2004). Most of these columns were composed of inner and outer tubes with circular shapes. Outer circular and inner square columns are studied by Elchalakani et al. (2002). Both outer and inner as square is studied by Zhao et al. (2002).

In this paper, a series of double skinned concrete filled tubular columns are tested. In addition, the test results are compared with some design code, and an equation for calculating the ultimate load of DSCFT is proposed.

2 TEST PROGRAM

A total of 32 specimens are tested. Figure 1 shows the cross sections of all series tested in this paper. The main parameters varied are thickness and geometry of inner and outer tubes. Length and outer diameter/side length are kept constant. The outer diameter of circular tube or side length of square tubes is kept as 240 mm for all specimens. The diameter or side length of inner tube is 120 mm or 80 mm respectively. The outer and inner tube thicknesses are 3 and 4 mm in two series. The height of all specimens is 720 mm. For each type of geometry, two specimens are fabricated and tested. All the test results listed in this paper are the average value of the two specimens. The details of the specimens are shown in Table 1, where t is the thickness of the outer and inner tubes, D_o/B_o is the diameter/edge length of the outer tube, D_i/B_i is the diameter/edge length of the inner tube, L is the height of columns.

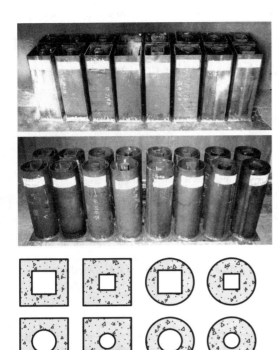

Figure 1. Pictorial views and cross sections of specimens.

Table 1. Details of specimens.

No.	Specimen	t mm	D_o or B_o	D_o/t or B_o/t	L/D_o or L/B_o	D_i or B_i	D_i/t or B_i/t	L/D_i or L/B_i
1	CC3 240-120	3	240	80	3	120	40	6
2	CC3 240-80	3	240	80	3	80	26.7	9
3	CS3 240-120	3	240	80	3	120	40	6
4	CS3 240-80	3	240	80	3	80	26.7	9
5	SS3 240-120	3	240	80	3	120	40	6
6	SS3 240-80	3	240	80	3	80	26.7	9
7	SC3 240-120	3	240	80	3	120	40	6
8	SC3 240-80	3	240	80	3	80	26.7	9
9	CC4 240-120	4	240	60	3	120	30	6
10	CC4 240-80	4	240	60	3	80	20	9
11	CS4 240-120	4	240	60	3	120	30	6
12	CS4 240-80	4	240	60	3	80	20	9
13	SS4 240-120	4	240	60	3	120	30	6
14	SS4 240-80	4	240	60	3	80	20	9
15	SC4 240-120	4	240	60	3	120	30	6
16	SC4 240-80	4	240	60	3	80	20	9

*SC3 240–120 means outer tube is Square "S", inner tube is Circular "C", both have 3 mm thickness, outer tube has side length of 240 mm and inner tube has diameter of 120 mm.

To avoid local buckling of steel members in composite columns, some design codes have been consulted for the minimum thickness of steel pipes to be incorporated in test specimens. These codes only give the thickness requirements for outer tubes of CFTs.

Table 2 shows the limiting and actual values of the specimens. Although some of the specimens do have values exceeding the limits for outer tubes, but it would be proved in succeeding paragraphs that this would not make any difference on the performance of the columns.

Testing is done in the structural engineering laboratory of Tsinghua University with a 500 tons capacity testing machine. The specimens are centered in the testing machine in order to avoid eccentricity effects. The vertical shortening is measured by four displacement transducers, two placed on sides for total shortening and two placed in middle one third of columns. Four strain gauges are also installed,

Table 2. Limiting values of D/t and B/t.

Geometry type	LRFD	EC4	ACI	Chinese	Specimen Outer	Specimen Inner
Circular	–	75.5	75.6	18.32~78	60~80	20~40
Square	37.8	47.63	46.3	–	60~80	20~40

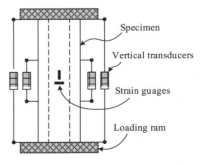

Figure 2. Test setup.

two for longitudinal and two for transverse strain measurements. The top and bottom surfaces of the specimens are made smooth and leveled to remove imperfections and to maintain uniformity of loading on the surface. The elastic modulus of steel E_s is 200,000 MPa, and the modulars of concrete E_c is 29,000 MPa. The average yield stress of steel is 280 MPa and the average cylinder strength of concrete is 29 MPa. The test setup is shown in Figure 2.

3 TEST RESULTS

The Load-axial shortening displacement curves are shown in Figure 3 and Figure 4. The curves of circular specimens show that the failure occurred at large shortening i.e. between 5 to 20 mm, whereas for square ones it is between 3 to 5 mm. The curves also show that the ductile behavior of circular columns and their load absorbing capacity are definite edges over the square ones. The square columns having inner tube as circular one also behave the same way as of square inner ones, leading to the conclusion that the behavior of DSCFTs is overall depicted by outer tube geometry irrespective of shapes of inner tube. Although the load-strain behavior is not drastically different from the CFTs and from previously studied DSCFTs, but it did vary for different series. Circular series show a ductile behavior whereas the square ones have a kind of brittle or slightly ductile behaviors.

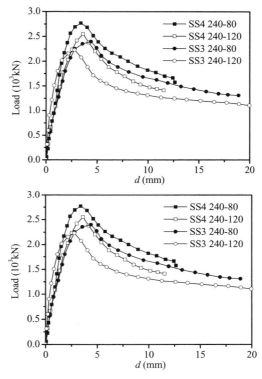

Figure 4. Load—vertical shortening displacement curves for specimens with square outer tubes.

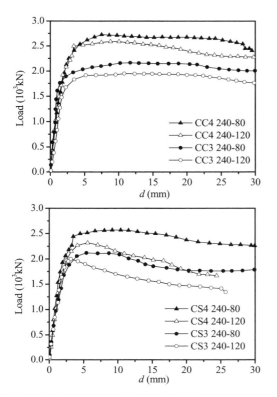

Figure 3. Load—vertical shortening displacement curves for specimens with circular outer tubes.

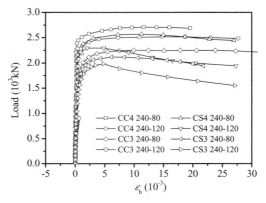

Figure 5. Load-hoop strain curves for specimens with circular outer tubes.

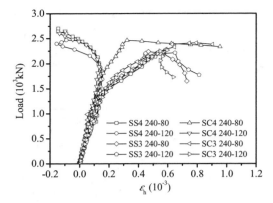

Figure 6. Load-hoop strain curves for specimens with square outer tubes.

Figure 5 and Figure 6 show the hoop strain behaviors of all specimens. It can be observed that the confinement effect for columns with square outer tubes is much less than that of circular ones.

4 ANALYSIS OF THE TEST RESULTS

The ACI and Australian Standards (AS) use the same formula for calculating the squash load of CFT. Neither code takes into consideration the concrete confinement. The limiting thickness of steel tube to prevent local buckling is almost the same for both codes. The squash load for square, rectangular and circular columns is determined by:

$$N_u = 0.85 A_c f'_c + A_s f_s \tag{1}$$

A modification for ACI and AS equation was proposed by Giakoumelis & Lam (2004) for CFTs. A coefficient is proposed for the ACI/AS equation to take into account the effect of concrete confinement on the axial load capacity of concrete filled circular steel tube. A revised equation was proposed as follows:

$$N_u = 1.3 A_c f'_c + A_s f_s \tag{2}$$

Thickness to diameter ratio of outer tube has a pronounced effect on the confinement effects of columns. It can be observed that with the increase in confinement factor there is a pronounced increase in compressive strength of circular columns but a moderate increase in square columns. Generally different levels of concrete confinement have been observed for concrete-filled tubes by a number of researchers. Load enhancement due to concrete confinement in circular hollow sections (CHS) is found to be larger than that in square hollow sections (SHS). The ultimate strength of CHS confined concrete is found to be several times of unconfined concrete strength. The confinement of concrete is considered for CHS by using equation (1995):

$$f_{cc} = f_c \left[1 + \eta_1 \frac{t_o \sigma_{yf0}}{D_o f_c} \right] \tag{3}$$

where η_1 is the confinement factor, and is taken as 4.9 for concrete-filled CHS stub columns. σ_{yf0} is the yield stress of the flat surfaces in the outer tube.

This equation is modified for concrete-filled double skin SHS stub columns by adopting η_1 as 1.0 and replacing D_o with B_o (2002), i.e.

$$f_{cc} = f_c \left[1 + \frac{t_o \sigma_{yf0}}{B_o f_c} \right] \tag{4}$$

It can be observed from here that D_o/t ratio of outer tube is the controlling factor for circular tubes. If all other values are kept constant on the right side of equation except the thickness of tube, it can be found that more the thickness of outer tube, higher will be the confined compressive strength of column.

The value of η_1 have been calculated for different geometries and also checked and verified for circular pipes having inner tube as circular too.

for circular columns $\quad \eta_1 = \left[\frac{f_{cc}}{f'_c} - 1 \right] \left(\frac{D_o}{t_o} \right) \left(\frac{f'_c}{f_y} \right) \tag{5}$

for square columns $\quad \eta_1 = \left[\frac{f_{cc}}{f'_c} - 1 \right] \left(\frac{B_o}{t_o} \right) \left(\frac{f'_c}{f_y} \right) \tag{6}$

The values of η_1 as 4.90 for CFTs may hold good, but for DSCFTs the same is not valid. Values for square columns of DSCFTs are lower than 5, where it is replaced by 1 by Zhao et al. (2002).

If the value of η_1 is kept as 2 for DSCFT square columns and 5 for circular ones, the ultimate axial force of all DSCFT columns can be calculated as:

$$N_u = 0.85 A_s f_y + A_c f'_c \left[1 + \eta_1 \frac{t f_y}{D_o f'_c} \right] \tag{7}$$

Table 3. Values given by different design codes and test results.

No.	Specimen	ACI kN	G. G. kN	Chinese code kN	Eq(7) kN	Test results kN
1	CC3 240-120	1652	2030	1875	1973	1990
2	CC3 240-80	1732	2185	2038	2156	2207
3	CS3 240-120	1685	2026	1832	1906	1997
4	CS3 240-80	1633	2070	1955	2139	2102
5	SS3 240-120	2123	2605	2399	2155	2153
6	SS3 240-80	2053	2630	2491	2318	2278
7	SC3 240-120	2087	2605	2428	2194	2211
8	SC3 240-80	2176	2769	2592	2323	2403
9	CC4 240-120	1887	2299	2161	2246	2517
10	CC4 240-80	2072	2566	2399	2606	2735
11	CS4 240-120	2016	2386	2157	2322	2314
12	CS4 240-80	1996	2472	2334	2460	2572
13	SS4 240-120	2526	3049	2819	2613	2505
14	SS4 240-80	2497	3127	2964	2676	2687
15	SC4 240-120	2380	2944	2791	2512	2656
16	SC4 240-80	2596	3243	3044	2794	2553

This equation is for circular columns and is valid for square columns by replacing t/D_o with t/B_o. Strength reduction factor of 0.85 for steel has been included due to effects of local buckling of both inner and outer tubes.

Values calculated by equation 7 and ACI method are listed in Table 3. It is observed that the capacities given by ACI code are too conservative whereas those calculated using equation 7 are more realistic, especially for circular columns. The values given by equation 7 are close to the actual ones. This is because of the inclusion of confinement effect. Eurocode and LRFD values give an error due to hollowness effects.

5 CONCLUSIONS

1. Cross sectional geometries significantly influence the axial behavior of columns. Circular columns behave in a ductile way as compared to square ones. The in filled concrete can delay the occurrence of local buckling of the steel tube, and steel tube improves the confinement of concrete. Thereby, these increase overall strength of columns.
2. No fracture is observed in any of the inner or outer pipes even at very large deformations, which indicates that no bursting effect was induced on pipes due to internal pressure of concrete.
3. The equations given in some design codes give conservative values and hence need to be revised for DSCFTs both for seismic and normal designs.

ACKNOWLEDGEMENT

The authors would like to acknowledge the financial support of the Chinese National Science Foundation, Grant No. 50438020.

REFERENCES

ACI Committee 318, Building code requirements for structural concrete (ACI 318-95). Detroit, American Concrete Institute, 1995.

CECS 28:90. Specification for design and construction of concrete-filled steel tubular structures. China Planning Press, 1992. (in Chinese).

Eurocode 4. Design of composite steel and concrete structures. Part 1: General rules and rules for buildings. 1994.

G. Giakoumelis, D. Lam. Axial capacity of circular concrete-filled tube columns. *Journal of Constructional Steel Research, 2004*, 60(7): 1049–1068.

H. Shakir-Khalil, S. Illouli. Composite columns of concentric steel tubes. *Proceedings of conference on the design and construction of non-conventional structures. London, 1987*, 1: 73–82.

LRFD. Load and resistance factor design: manual of steel construction. 3rd ed. American Institute of Steel Construction; 2001.

Min-Lang Lin, Keh-Chyuan Tsai. Mechanical behavior of double-skinned composite steel tubular columns. National Center for Research on Earthquake Engineering, Taipei, Taiwan. (http://www.ncree.gov.tw/ncree-jrc/CD/Poster/Min-LangLin.pdf).

Mohamed Elchalakani, Xiao-Ling Zhao, Raphael Grzebieta. Tests on concrete filled double-skin (CHS outer and SHS inner) composite short columns under axial compression". *Thin-Walled Structures, 2002*, 40(5): 415–441.

R. Bergmann, C. Matsui, C. Meinsma, D. Dutta. Design guide for concrete filled hollow section columns under static and seismic loading. *Serial No.5, Verlag TUV Rheinland,* Cologne, Germany, 1995.

S. Wei, S. T. Mau, C. Vipulanandan, S. K. Mantrala. Performance of new sandwich tube under axial loading. *Journal of Structural Engineering, 1995*, 121(12): 1806–1821.

Xiao-Ling Zhao, Byoungkee Han, R. H. Grzebieta. Plastic mechanism analysis of concrete-filled double-skin (SHS inner and SHS outer) stub Columns. *Thin-Walled Structures, 2002*, 40(10): 815–833.

Zhong Tao, Lin-Hai Han, Xiao-Ling Zhao. Behaviour of concrete-filled double skin (CHS inner and CHS outer) steel tubular stub columns and beam-columns. *Journal of Constructional Steel Research, 2004*, 60(8): 1129–1158.

Flexural limit load capacity test and analysis for steel and concrete composite beams with tubular up-flanges

C.S. Wang, X.L. Zhai, L. Duan & B.R. Li
Key Laboratory for Bridge and Tunnel Engineering of Shaanxi, Chang'an University, Xi'an, China

ABSTRACT: Steel and concrete composite girders with concrete-filled tubular up-flanges (SCCGCFTF) have been proposed in this paper. In order to study the bending capacity of SCCGCFTF, the static test under concentrated loads in vertical direction has been performed. Based on the test results, the bending capacity and mechanical characters of the specimen were obtained and the flexural failure mechanisms of SCCGCFTF were also summarized. Then the finite element software ANSYS was used to establish the nonlinear static analysis model of SCCGCFTF. By comparison of results from numerical analyze and test, it was showed that the analysis, considering material nonlinearities, was in great agreement with the test results. On the basis of analysis above, some requirements concerning the design of initial geometry sizes were recommended.

1 INTRODUCTION

Steel and concrete composite girders were used as bending components traditionally, which were realized by composition of steel girder and concrete through shear connections. The composite girder is characterized with advantages of lower architectural height, stronger capacity and rigidity (Nie & Yu 1999). The composition of steel tube and concrete was realized by fulfilling the steel tube with concrete (Han 2004). In this paper, the advised new type of steel and concrete composite girders with concrete filled tubular up-flanges (SCCGCFTF) can be formed when the upper flange of conventional I girder is substituted by steel tube, which is filled with self-compacting high performance concrete. Compared with composite girder with flat flange, SCCGCFTF have significant advantages, such as superior strength, rigidity, and stability under the same amount of steel. What's more, the height of web can be reduced when steel tube is used as the upper flange, therefore avoiding problems relating to larger slenderness limit of web (Sause et al. 2001). Compared with conventional I steel girders, SCCGCFTF have stronger limit load capacity and lower dead load. As a result, they are also superior in saving material and reducing height of girders, thus can be applied in city bridges where the architectural clearance is restricted strictly. Besides, this SCCGCFTF have better ductility, larger safety stockage, and superior seismic resistance, thus enlarging their application to regions with strict seismic resistance. So, SCCGCFTF can be better applied in the future because advantages of materials can be more effective utilized.

Currently, the research concerning SCCGCFTF is still insufficient. Smith (2001) in Lehigh University studied initially the behavior of the new composite girders. He analyzed the affection to the girder's stability caused by the distribution of diagrams and stiffers, and also conducted parametric analyze. Kim (2005) in Lehigh University conducted numerical analyze and experiment about the partial connection between up-flange and concrete deck. Yet, there is no relevant reports concerning concrete filled tubular flange girder in domestic region in China. In short, it is meaningful to study the bending capacity and conduct experiment of SCCGCFTF, because it is necessary to have an in-depth recognition of their behaviors, thereby serving as a preliminary study for their further application in bridge engineering.

2 EXPERIMENT OF BENDING CAPACITY

2.1 Design of test specimen

The specimen was designed after a seriers of numerical analyse considering equiments used in experiment, the objectives of test and other various factors. This new type of specimen is composed of concrete filled tube up-flange, stiffer, web and bottom flange.

The cross-section is illustrated in Figure 1. The specimen is 0.5 m in height and 4.3 m in length. The outer diameter of steel tube is 219 mm and the tube is 8 mm in thickness. The thickness of web is 6 mm. The bottom flange is 150 mm in width and 14 mm in thickness. In order to prevent local instablity of web, the stiffer, which is 12 mm in thickness, should be installed at the location of supporting and loading. The specimen is shown in Figure 2.

The specimen was fabricted by Q235. The straight welding steel tube served as the upper flange. The self-compacting high performance concrete filled in steel tube was made by 525 ordinary Portland cement, lime rock macadam and sand with 2.6 in modulus of fineness. The proportion of cement: flyash: sand: macadam: water was 360: 90: 700: 1050: 175. The parameter of water-reducing agent was 1.35%. The slump of self-compacting concrete was 250 mm. The degree of sprawl was 620 mm and the average speed was 52.5 mm per second when flowing through L shaped instrument. The normal cube, used for measuring the strength of concrete, shared the same maintenance condition with the specimen. The elastic modulus has been tested using prism coupon with dimension of 300 mm in height and 150 mm in length and width. The test result showed that the compressive strength of concrete cube after 28 days was 48.6 MPa and the elastic modulus was 3.47×10^4 MPa averagely.

2.2 Loading equipment and distribution of apparatus

The experiment was conducted in Laboratory for Bridge and Tunnel Engineering in Chang'an University. The 4-point loading method was used during the experiment. The jacking apparatus with oil press of 1500 kN was adopted and the TDS-602 static strain indicator was used to collect strain and displacement data in entire test.

In order to measure the strain distribution and certificate plane cross-section assumption more accurately, strain gauges were allocated to middle, quarter and loading sections, and strain rosettes were used at the web in the vicinity of supporting. At the same time, eight pairs of strain gauges were arranged around steel tube in the circumferential direction averagely in order to test circumferential strain and determine whether crippling caused by local buckling would occur. In the axis of concrete filled tube up-flange, two displacement gauges were located to measure the relative displacement between the steel tube and inner concrete. Displacement gauges were also arranged in middle and quarter section to measure displacement in vertical deflections. To observe the local ductility and deformation in middle section, displacement gauges were arranged in the top of steel tube and lateral face. Besides, in order to observe whether entire lateral bending or local ductility in web would occur, displacement gauges were arranged between supporting and quartered sections to observe the transverse displacement of web.

2.3 Test procedure and failure character

The distance of the two loading points was 1 m. Before the test, the specimen was elastically preloaded for several times to confirm the sensor could work properly and to eliminate the mechanical lag of strain gauges. The multi-stage loading method was used. Every grade of load was 50 kN in elastic range, and then reduced to 10 kN per grade after yielding of bottom flange. In the test process, the loading should be slow and continuous, and every grade of loading

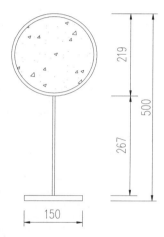

Figure 1. Cross-section of composite beam with concrete-filled tubular up-flange (Unit: mm).

Figure 2. Specimen of composite beam with concrete-filled tubular up-flange.

Figure 3. Failure mode of specimen.

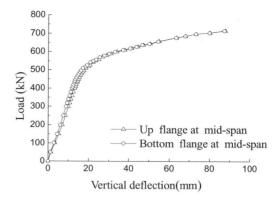

Figure 4. Load-displacement curves at mid-span.

should maintain 2 minutes approximately. From initial loading to the yielding of bottom flange, the specimen presented favorable working behavior as a whole and the deformation present linear increase. When the load attached to 300 kN, the bottom flange was yielded and the displacement in middle span was less than 20 mm. With the further increase of load, the displacement in the middle span increased aggregately and then a sound was heard, which signified failure of the natural connection between tube and inner concrete at the load of 640 kN (Wang 1996). The specimen attained ultimate capacity when load reaching 720 kN. In the entire testing process, there was neither apparent lateral displacement, nor obviously relative displacement between steel tube and inner concrete. There was no buckling found on the surface of concrete filled steel tube when specimen attained failure. The failure mode of the specimen is shown in Figure 3.

3 TEST RESULTS AND ANALYSIS

3.1 *The deformation of specimen*

3.1.1 *The load-displacement curve of specimen*

The relationships of load and displacement of bottom flange in middle span are showed in Figure 4. According to the tendency of load-displacement curves, the working process of specimen can be divided into four stages. Firstly, below the load of 300 kN, there is a linear relation between load and displacement, showing that the specimen is in elastic stage. Secondly, when the load reaching to the range of 500 kN to 600 kN, there is a non-linear relationship between load and displacement. With the expansion of yielding region from the bottom flange to interior section, the rigidity reduces and the stress redistributes, then reaching elastic-plastic stage. Thirdly, when loading beyond 600 kN, the displacement increases dramatically and there is still a linear relationship between load and displacement. But the rigidity of section is much smaller than that in elastic stage. As the section in middle span stepping into plastic stage, the capacity of specimen increases further and displacement in middle span accumulates at a faster speed. Then the specimen comes into hardening stage. Fourthly, when load exceeding 716 kN, the load is generally constant with the increase of displacement, and the load-displacement curve is nearly horizontal. The decline part of load-displacement curve is not observed in experiment, and this is used to testify that SCCGCFTF has better bending capacity and ductility (Guo et al. 2002). Acquired from Figure 4, with the increment of loading, the vertical displacement on the top of tube agrees with displacement of the bottom flange in middle span. This shows that there is no change in girder depth, no local convex in tube at middle span and no distortion in specimen.

3.1.2 *Transverse displacement at web in longitudinal direction*

The specimen was located in the east and west direction. The survey points were located at web in longitudinal direction to observe transverse displacement of web and determine whether entirely lateral buckling or local buckling in web would occur. In Figure 5, the negative value means transverse displacement to the south. Point DW5 was located at the web of middle span. The distances from middle span to point DW1, point DW2 and point DW3 increased gradually.

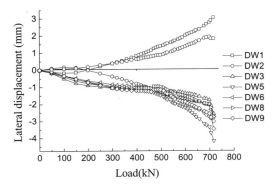

Figure 5. Lateral displacement-load curves at web.

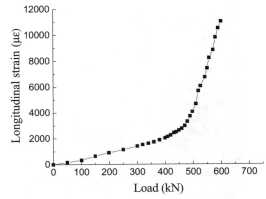

Figure 6. Load-longitudinal strain curves at mid-span.

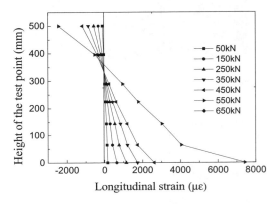

Figure 7. Strain distribution of mid-span section.

Point DW6, point DW8 and point DW9 were respectively symmetrical with point DW1, point DW2 and point DW3 about point DW5. It can be seen that the transverse displacement of the web decreases with the distance to middle span increased. The transverse displacement of web is generally symmetrical about middle span. The maximum transverse displacement is less than 5 mm at failure, showing that neither entirely lateral ductility nor local buckling in web occurred during experiment and that there is no instability before strength failure achieved.

3.1.3 *The relative displacement between steel tube and inner concrete*

Survey points were arranged at both ends of filled concrete to observe relative displacement between concrete and tube during entire test. It has been observed that the maximum relative displacement was 0.02 mm, which signified relative displacement could be ignored. As a result, it was rational to couple the longitudinal displacement by connecting the node of shell element of steel tube and concrete element to simulate the connection of steel tube and concrete.

3.2 *The strain of specimen*

3.2.1 *The load-longitudinal strain relationship of bottom flange in middle span*

The load-longitudinal strain curve of bottom flange in middle span can be seen in Figure 6. The strain at bottom flange reaches 1400 µε under the load of 300 kN, initially entering into yielding stage. Entire bottom flange is yielded when load exceeding 500 kN and specimen come into elastic-plastic stage as the yielding of exterior steel tube. In this situation, the strain increases at a larger speed. When middle span section entering into plastic stage entirely, the strain increases dramatically while the load increasing at a slower rate. It can be concluded from this experiment that advised section can make an effective use of both the compression of concrete and ductility of steel. The capacity of specimen can be further increased if high performance steel and concrete composite girder are formed by using high performance steel (HPS) and high performance concrete (HPC), which can show the superiorly advantages of composition between HPS and HPC (Sause et al. 2001).

3.2.2 *The distribution of strain along depth of girder at middle span*

Through allocating survey points along depth of girder at middle span to observe strain, it could be acquired the distribution of strain under increasing load and the change of location of neutral axis. The distributions of strain along depth of girder at each load grade are showed in Figure 7, in which positive values denote strain in tension and negative values reflect strain in compression. The longitudinal coordinate indicates the relevant location of survey points in the height; the point of zero in longitudinal coordinate denotes the bottom flange of specimen.

It can be learned from Figure 7 that deformation of specimen is in agreement with plane cross-section assumption generally. At the early stage of loading, the specimen is in elastic stage. The strain from web to the top of tube is linear distributed with the increment of load. Under the same load, the tension strain decreases gradually from bottom to upper section until attaining neutral axis. Under each grade of load, curves concerning the distribution of strain along the depth of girder have a common node and the location of neutral axis is unchanged. The increased speed of strain in compressive region is larger than that in compression region. Thus, increased load is taken by inner concrete, and bottom flange enters into elastic-plastic stage earlier than up-flange (Ji et al. 2007). With the further increment of load, the plastic region enlarges continuously along web. Then, the strain of web and tube increase significantly and the location of zero point of curves about the strain distribution along depth is rising, which indicates that the neutral axis offsets to compressive region.

Figure 8. FE mode of composite beam with concrete-filled tubular up-flange.

4 THE NUMERICAL ANALYSIS OF BENDING CAPACITY

The numerically analytical research was carried out by using the finite element software ANSYS. The dimension of specimen was determined initially based on linear static analysis and then was checked by calculation using eigenvalues and considering material nonlinearities, in order to ensure that critical load of stable failure was great than ultimate strength capacity. Then, there was a comparison between results from nonlinear static calculation and observed dates acquired in experiment.

4.1 The establish of specimen based on nonlinear analysis

To better simulate the specimen, the shell 143 in ANSYS was chosen for modeling steel. The finite element (FE) mode is shown in Figure 8.

The shell 143 is a three-dimensional shell element. The elastic modulus is 2.06×10^5 MPa and the Poisson's ratio is 0.3. The solid 65 was chosen for modeling concrete. The solid 65 is a three dimensional solid element, whose constitutive mode is the strain-stress mode considering "restriction effect coefficient ξ". The elastic modulus is 3.47×10^4 MPa and the Poisson's ratio is 0.2 as tested. The finite element mode and specimen have the same dimension.

4.2 The comparison between results of finite element mode and experiment

4.2.1 Bending capacity
From Table 1, it can be found that the results of nonlinear calculation in static are very close to dates observed in experiment, when considering material nonlinearity. It can be used to testify that finite element mode can stimulate the specimen in practice as expected and can caculate the bending capacity exactly, which serves as a basic mode for the calculation of bending capacity.

4.2.2 Load-displacement relationship in middle span
It is obvious to notice the concordance of numberical calculations and experiment for load-displacement relationship in middle span (Fig. 9).

Table 1. Comparison of FE analysis and experimental results.

Result	P_y kN	P_u kN	δ_y mm	δ_u mm	P_y/P_u	δ_y/δ_u**
Test	497	716	12.6	82.6	0.69	0.15
Calculation	504	762	12.4	84.47	0.66	0.15
Ratio*	101.4%	106.4%	98.4%	102.3%	–	–

* Ratio of calculated result to test result.
** P_y is yielding capacity when yielding of bottom flange; P_u is ultimate capacity at failure of girder; δ_y and δ_u is the displacement at mid-span under load of P_y and P_u respectively.

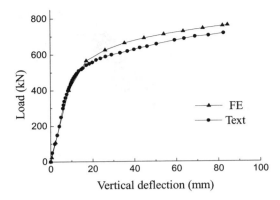

Figure 9. Comparison between FE analysis and experimental results for load-vertical deflection relationship.

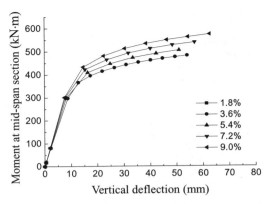

Figure 11. Vertical deflection at mid-span section of different steel ratio.

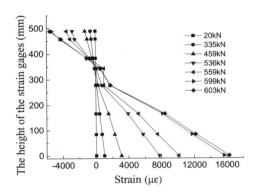

Figure 10. FE analysis strain distribution of mid-span section.

The load-displacement curves are basically consistent in elastic stage. In elastic-plastic stage, there are certain differences between these two curves. Under the same load, the calculated displacement is greater than date observed in experiment. This difference can be explained by various factors affecting deformation, such as the material nonlinearity of load-displacement relationship and initial imperfections in practical structures, which are beyond consideration since it is difficult to obtain the size and dimension of initial imperfection. Yet, the calculated load-deformation curves can generally reflect the regularity of displacement in middle span in practical specimens, thus supplying valuable reference to study the mechanical behavior of components.

4.2.3 *The distribution of strain along the depth of girder in middle span*

The distribution of strain along the depth of girder under different grades of load is shown in Figure 10, in which positive values represent tensile strain and negative values represent compressive strain. It can be observed that the distribution of strain in middle span presents triangular distribution along depth of girder before yielding, and the location of neutral axis is unchanged. When exceeding yielding load, the internal force is redistributated and the strains in both flanges no longer increase. Once attaining the ultimate capacity load, regions in both compression and tension attain completely plastic and strain along depth increases dramatically. It could be observed from the results of numberical calculation that the neutral axis was 300 mm away from bottom at initial stage of loading, which was approximate to the tested result of 341 mm. The numerical calculation of the strain distribution along depth is good accordance with the result observed in experiment. So the FE mode can reflect the mechanism of bending failure in practical structure.

4.3 *The analyze of parameters affecting bending behavior*

From the comparison of results obtained from experiment and numerical analyze, it can be concluded that finite element mode was able to reflect the behavior of practical component. Then, the finite element mode was used to discuss parameters further, which affecting moment-displacement curves of concrete filled steel tube up-flange composite girder.

4.3.1 *The affection of steel ratio in concrete filled steel tube*

From the moment-displacement curves of components with different steel ratios (Fig. 11), it can be found out that the ultimate bending strength in creases with the enhancement of steel ratio. The reason is that the greater proportion of capacity can be taken by greater proportion of steel, thereby enhancing entire rigidity of component. The bending capacity of component with steel ratio of 9.0% is 20% greater than that of component with steel ratio of 3.6%. The increment of displacement at initially loading stage is slower with the increment of steel ratio, since the elastic modulus of steel is greater than concrete and the initial rigidity of component is enhanced with increase in steel ratio.

4.3.2 *The affection of strength of inner concrete in steel tube*

The moment-displacement curves with various concrete strength were shown in Figure 12.

It can be found out that the bending capacity increase slightly when inner concrete is enhanced from C40 to C80, showing the slight contribution of inner concrete to the component's bending capacity. At the same time, there is neither obvious difference of rigidity in elastic stage, nor increase of displacement with the increase in strength of inner concrete, because there is no significant difference in elastic ratio among types of concrete with various strength. When concrete with higher strength is used, the larger displacement would be found in failure, which illustrates that high strength concrete can fully utilize the ductility of steel.

4.3.3 *The influence of yielding strength of steel*

Figure 13 reflects the moment-displacement curves with different yielding strength of steel. From Figure 13, it can be noticed that the ultimate bending capacity is proportional to the yielding strength of steel. The yielding strength of steel has almost no contribution to the rigidity of component in elastic stage. The deformation at failure is increased with increase in strength of steel. This is because the ultimate tensile stress would increase with increment in yielding strength of steel, while there is no significant difference in elastic modulus.

4.3.4 *The affection of shear span ratio*

The shear span ratio is an important factor affecting the behavior of bending component. Based on moment-displacement curves under different shear span ratios (Fig. 14), it can find that there is no big difference among macroscopically failure modes and moment-displacement relationships, when 2, 2.5, 3 and 4 are adopted as the shear span respectively. As a result, it can be considered that the bending behavior is unaffected by shear span ratio, if the shear span ratio of SCCGCFTF is not smaller than 2.

5 DESIGN RECOMMENDATION

By analyzing bending capacity of SCCGCFTF from experiment and numerical analysis, characteristics at bending failure mode can be acquired, thus supplying recommendations about dimensions in numerical analysis and design method. Shell element is chosen to simulate steel and three-dimensional solid element

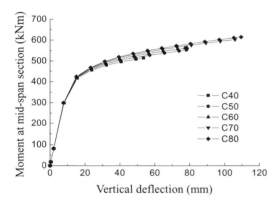

Figure 12. Vertical deflection at mid-span section with different concrete strength.

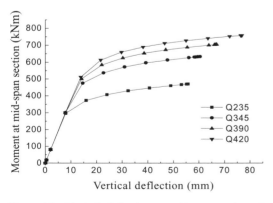

Figure 13. Vertical deflection at mid-span section of different steel yield strength.

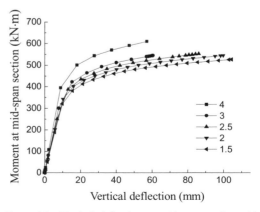

Figure 14. Vertical deflection at mid-span section with different shear-span ratio.

is used to mode concrete in the finite element mode of SCCGCFTF. At the same time, it is recognized that there is no relative displacement between steel and inner concrete. So their longitudinal displacement can be coupled and touching element is chosen to mode restriction passed from tube to inner concrete in transversal direction. The strain-stress mode of centric concrete considers "restriction effect coefficient ξ". These considerations ensure exact stimulation of both stress and deformation of SCCGCFTF.

When designing the dimension of SCCGCFTF, the results from linear elastic calculation can be used as rules to optimize dimensions and various parameters, but fail to reflect the behavior of girder in elastic-plastic stage. In order to obtain precise analysis and design in details, it is indispensable to take consideration of nonlinearity of material to get precise bending capacity.

The bending rigidity of SCCGCFTF is affected by contacted stress between tube and inner concrete. The compressive strength of concrete is enhanced due to constrain of steel tube, while the yielding stress of tube in longitudinal direction is decreased caused by the affection of ring stress. It is necessary to consider the affection of reduced yielding strength to bending capacity.

In the dimensional design of SCCGCFTF, the neutral axis should be put in web and the idealist location is the connection of tube and web. The design should ensure that the yielding of bottom flange is prior to the yielding of top of tube. Yet, yielding distance between top and bottom flange should be limited to avoid ineffective usage of material.

To prevent local buckling of up-flange of SCCGCFTF, the ratio of outer diameter to thickness of tube should meet the requirement in code for design of steel structures (2003). The stiffer can supply constrain to web and thus is contributable to the stability of web. The unconstrained flange in compression should be taken consideration according to code for design of steel structures (2003) when discussing the stability of web.

The steel ratio and its yielding strength are contributable to the bending capacity of SCCGCFTF. With enhancement in strength of inner concrete, the bending capacity would not increase obviously, but the ductility of steel can be utilized adequately. As a result, the relation between yielding strength of steel and strength of inner concrete should be adjusted according to requirements in design.

capacity, characteristics of deformation, and the alteration and distribution of stresses in middle span, thereby showing the mechanism of bending failure. The finite element program of ANSYS was used to conduct the static calculation considering material nonlinearity of SCCGCFTF. By comparison, results from calculation were in exactly agreement with the experiment. Finally, the recommendations about design of dimension were proposed.

ACKNOWLEDGMENT

The writers gratefully acknowledge the financial support provided by Fok Ying Tung Education Foundation (Grant No. 101078) and Program for New Century Excellent Talents in University of the Ministry of Education of the P.R. China (Grant No. NCET-07–0121).

REFERENCES

Guo, L.H., Zhang, S.M. & Wang, Y.Y. 2002. Experimental research and theoretical analysis on flexural behavior of high-strength concrete-filled steel tubes with circular sections. *Steel Construction* 17(6): 29–33.

Han, L.H. 2004. *Concrete filled steel tubular structures*. Beijing: China Science Press.

Ji, B.H., Hu, Z.Q, Chen, J.S. & Zhou, W.J. 2007. An experimental study on the behavior of lightweight aggregate concrete filled circular steel tubes under pure bending load. *China Civil Engineering Journal* 40(8):35–40.

Kim, B.G. 2005. *High Performance Steel Girders with Tubular Flange*. Ph.D. Thesis of Lehigh University: Bethlehem.

Nie, J.G. & Yu, Z.W. 1999. Research and practice of composite steel-concrete beams in china. *China Civil Engineering Journal* 32(2): 3–8.

Sause, R., Abbas, H., Kim, B.G., Driver, R. & Smith A. Innovative High Performance Steel Girders for Highway Bridges. In Azizinamini Atorod (ed.), *High performance materials in bridges; Proceedings of the international conference, July 29–August 3 2001*. Hawaii: Kona.

Smith, A. 2001. *Design of HPS Bridge Girders with Tubular Flange*. Ph.D. Thesis of Lehigh University: Bethlehem.

Wang, H.Q. 1996. *Experimental study on longitudinal splitting strength and ultimate flexural strength of composite steel-concrete beams with combined slab*. Master Thesis of Tsinghua University: Beijing.

National Standards of the People's Republic of China.2003. *Code for design of steel structures (GB 50017-2003)*. Beijing: China Planning Press.

6 CONCLUSION

Through experimental study and numerical analyze of SCCGCFTF, it could be acquired the bending

Parametric analysis of blind-bolted connections in a moment-resisting composite frame

H. Yao & H.M. Goldsworthy
The University of Melbourne, Melbourne, Australia

E.F. Gad
Swinburne University of Technology, Melbourne, Australia
The University of Melbourne, Melbourne, Australia

ABSTRACT: This paper presents a detailed parametric study on the behaviour of blind-bolted curved endplate connections to concrete-filled circular hollow sections. A detailed finite element model was developed and compared with the results from a full scale T-stub connection test. Comparisons between numerical and experimental data for tensile behaviour of the connection show satisfactory agreement. The models take into account material nonlinearities, geometrical curvature, and complex contact interactions between the blind bolt heads, circular tube wall, curved endplates and nuts. The various effects of tube wall thickness, endplate thickness, and anchorage extensions in the tubes on the connection behaviour have been explored using the developed finite element models. A moment-rotation curve has been derived based on the analysis. The favourable strength and stiffness of this innovative connection utilising one-sided fastening blind bolts with anchorage extensions showed that it could be an attractive solution to the conventional welded moment connections in a composite structural frame. The cogged extensions to the blind bolts are very effective in relieving the stress concentration on the thin tube wall.

1 INTRODUCTION

The use of structural hollow sections with or without concrete infill is continuing to increase in low to medium rise structural frames because of their natural aesthetic appeal and fast construction form. Concrete filled circular hollow sections are excellent load carrying structural members which provide a small footprint, enhance load carrying and fire resistant capacities, and reduce the potential for inward local buckling (Morino et al. 2001). However, their use is presently restricted by the difficulty of gaining access to the inside of the section to make conventional bolted connections to other structural members. Therefore, field welded beam-to-column connections remain the common practice for achieving moment resisting connections. As internal and through diaphragms tend to interfere with the placing of concrete within the tube, the external diaphragms are usually placed around the outside of the tube by welding in a moment-resisting frame (Nakada & Kawano 2003).

Damage observations of directly welded connections between steel beams and columns after the Northridge earthquake in 1994 and Kobe earthquake in 1995 raised concerns about the utilization of welds in steel moment resisting frames in regions of high seismicity (Swanson & Leon 2000). In a steel beam to concrete filled tube column connection, it is vital to prevent brittle fracture caused by local deformation.

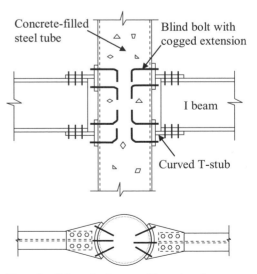

Figure 1. Schematic diagram of beam-to-column connection using blind bolts.

A novel blind-bolted connection has been developed to explore reliable alternatives to welded connections for connecting steel beams to concrete-filled circular tubular columns (Gardner & Goldsworthy 2005). Figure 1 shows a typical beam-to-circular column connection using blind bolts and double split Tees. Moment is resisted by tension in the top T-stub and compression in the bottom T-stub bearing on the tube wall. Shear action is assumed to be carried by both of top and bottom T-stubs. If additional shear capacity is required over and above what is provided by the T-stubs, web angle connections can be provided using blind bolts installed in a similar manner.

2 ONE SIDE FASTENING

The Ajax ONESIDE bolt is composed of a bolt with a circular head, a stepped washer, a split stepped washer, an optional sleeve (for shear) and a standard nut. The onesided blind bolts can be installed in a simple and effective way as indicated in Figure 2. This simple installation method reduces onsite labour requirements compared to welding.

The full structural strength of the high strength bolts as per AS4100 (SA 1998) can be achieved in tension by using ONESIDE blind bolts, provided that a pull-out mode is prevented. Figure 3 shows a comparison of components between an ONESIDE blind bolt and a standard bolt in joints. Two cases, namely, a) the same bolt size and b) the same hole-size are shown. The bolt

Figure 2. Typical blind bolt and installation tool.

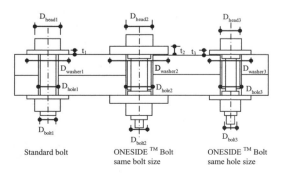

Figure 3. Comparison of standard bolts and blind bolts.

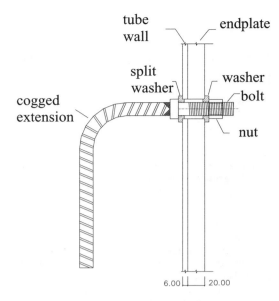

Figure 4. Blind bolt and cogged extension.

head diameter (D_{head}), the hole diameter (D_{hole}), the bolt diameter (D_{bolt}), the washer diameter (D_{washer}) and the washer thickness (t) are the fundamental design parameters of the blind bolts (Fernando 2005).

To improve the tensile behaviour of blind-bolted connections to the thin tube wall, an extension is provided to the blind bolt head by welding as shown in Figure 4. This extension is bent to form a cog in accordance with AS3600 (SA 2001) using N type reinforcing bar. Performance of the cogged extension anchored in concrete-filled steel tubes had been studied separately in an extensive experimental program (Yao et al. 2007).

3 EXPERIMENTAL PROGRAM

3.1 *T-stub connection test*

A full scale T-stub connection representing an interior beam-to-column joint has been tested (see Figure 5). The specimen consisted of a 323.9 × 6.0 mm circular hollow section of grade 350 with infilled concrete of 45 MPa characteristic compressive strength, two curved endplates (grade 300) of 20 mm thickness, and associated flared flange plates (grade 250) of 16 mm thickness. The endplates were fastened to the tube with 16 mm diameter Ajax blind bolts, which had a minimum tensile strength of 800 MPa and yield strength of 640 MPa. All the bolts were first tightened to a snug-tight condition and were then fully tensioned using the part-turn method. Cogged extensions were

Figure 5. Test setup.

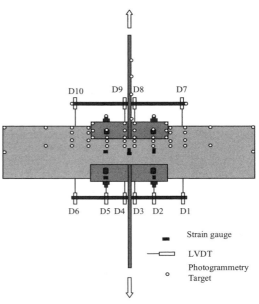

Figure 6. Instrumentation of T-stub specimen.

provided to the head of the blind bolts by using N type reinforcing bars of grade 500 MPa. Coupon testing of a welded bolt with cogged extension was performed as a check on the weld capacity between high strength blind bolts and reinforcing deformed bars. Failure occurred in the weld at a load of 115 kN, which is greater than the yield strength of the reinforcing bar.

The specimen was mounted horizontally into a universal testing machine with the two flared flanges clamped with friction grip jaws. Strain gauges were placed at various locations on tube wall, curved endplate, flared flange, and cogged extensions. The strains on the tube wall and curved endplate were monitored by gauges in the transverse and longitudinal directions. LVDTs were mounted on the specimen to record the deformation of the endplate, slip of the bolts, and displacement between the endplate and the tube. Retro-reflective photogrammetry targets were attached to the upper flange, endplate, bolt shank tips, tube wall, concrete core, and test rig to monitor the relative movements between those components during the loading process. Figure 6 shows the layout of strain gauges, photogrammetry targets, and transducers installed on the specimen. An initial photogrammetry survey was taken before the test to record the original conditions, then a series of surveys were performed at load levels of 100 kN, 200 kN, 300 kN, 400 kN, 500 kN, and 600 kN. Afterwards, the specimen was continuously loaded to failure in tension.

3.2 Test results

The specimen achieved a maximum tensile load of 690 kN. It failed due to the weld fracture between the blind bolt head and cogged extension at the middle bolt of the top T-stub. The load versus relative outward displacement, Δ_0, is provided in Figure 7. This displacement is the outward displacement at the centre of the endplate relative to the undeformed tube wall. It was obtained by taking the average of transducer values of (D7-D8) and (D10-D9) for the top plate and of (D1-D3) and (D6-D4) for the bottom plate.

The tensile load was shared between the membrane action of the tube wall activated by the internal washers bearing on the tube causing hoop stresses and anchorage of the cogged extensions within the infilled concrete. Nearly 45% of the tension load was carried by the cogged extensions and the remaining 55% was taken by hoop stress within the steel tube wall at load level at 600 kN. The overall tensile behaviour of the blind bolted T-stub connection was substantially better than it would have been without the extensions (Gardner & Goldsworthy 2005).

The hoop stress in the tube wall was a maximum at the centre of the tube and decreased in the manner shown in Figure 8 with distance away from the centre of the tube. The maximum transverse strain was

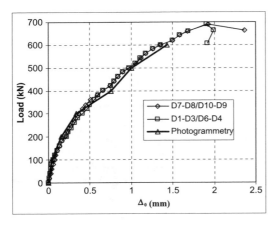

Figure 7. Load vs. outward displacement on T-stub test.

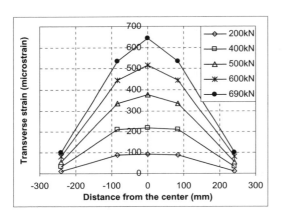

Figure 8. Transverse strain on tube of T-stub.

666 microstrain, which corresponds to a stress of 133 MPa at ultimate load.

4 NUMERICAL ANALYSIS

4.1 *Finite element modeling*

A three dimensional finite element (FE) model was specifically developed to simulate the full behaviour of blind-bolted T-stub connections using a general purpose finite element package ANSYS. An overall model of the T-stub connection to the CFT column is shown in Figure 9. Due to symmetry, only a quarter of the specimen was analysed as shown in Figure 10. Symmetrical boundary conditions were imposed on the two planes of symmetry, one longitudinal plane along the centre-line and one transverse plane at the middle height of the tube and the endplate.

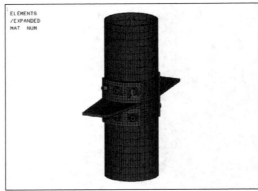

Figure 9. Overall model.

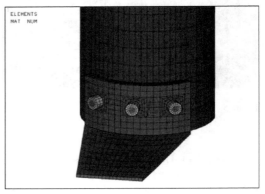

Figure 10. Quarter model.

In the three-dimensional model, eight-node isoparametric solid elements (SOLID45) were employed to model the circular hollow section, curved endplate, flange plate, round headed bolts, nuts and split washers. The mesh pattern of the curved endplate and flared flange plate is shown in Figure 10. The cogged bar extensions to the blind bolts were simulated by the non-linear spring elements (COMBIN39), which represented the interaction between the reinforcing bar and the surrounding concrete within a confined environment. The behaviour of the anchorage of springs was obtained through an analytical algorithm, which had been validated by an extensive experimental program (Yao et al. 2007). The springs were attached to the bolt head using a multi-point constraint approach with a pilot node. The configuration of solid elements representing the bolts, and spring elements representing the associated extensions, is shown in Figure 11. The bolt shank was split into two imaginary halves and pretension elements (PRETS179) were

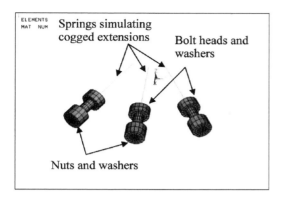

Figure 11. Bolts and extensions.

created to connect these two arbitrary parts. Pretension nodes are used to control and monitor the preloads in the bolts. When the pretension was applied on the node, the link element was in tension and pulled the two parts of the bolt shank towards each other to clamp the tube wall and the endplate together. By using the pretension element in the bolts, the physical bolt pretension was simulated. The level of pretension could be tracked during loading by pretension nodes on the pretension elements.

The solid bolt is the most realistic simulation approach for modelling a bolt as it can capture tensile, bending, and bearing loads with reasonable accuracy. It is considered a better approach compared to other alternative methods, such as line element or beam elements with coupled nodes. However intensive computation adds modelling time due to the number of solid elements required.

The concrete core was idealized as a rigid body so as to reduce the number of solid elements. The complex interaction between the surfaces of: (a) concrete core and steel tube; (b) curved endplate and steel tube; (c) bolt head/washer with steel tube; (d) nut/washer with endplate; and (e) bolt shank with endplate were modelled using surface-to-surface contact (CONTA174) and target (TARGE170) elements. These contact elements were overlaid on parts of the model that were analysed for interaction. The surface-to-surface contact elements have several advantages over node-to-node elements, as they support surface discontinuities, large sliding and deformations. Each bolt head and washer assembly was modeled as an integral unit. Each nut and washer assembly was modelled in the same way.

The material behaviour of the steel tube, endplate, flange, bolt and washer were described by bilinear stress-strain curves. The initial slope of the curve was taken as the elastic modulus, $E0$, of the material and the post yield stiffness identified as the tangent modulus, $E1$, was taken as 10% of the initial stiffness ($E1 = 0.1E0$). Nominal values were used to describe the material properties. The yield stress of the steel tube, curved endplate, flared flange, and high strength structural blind bolt and washer were 350 MPa, 310 MPa, 280 MPa, and 640 MPa respectively. The use of nominal properties rather than actual ones is expected to have very little influence on the overall load versus outward displacement behaviour of the connection (Sherbourne & Bahaari 1996, Swanson et al. 2002). Plasticity-based isotropic hardening using the von Mises yield criterion was employed to obtain the response of the connection in the inelastic region.

Nonlinearity in the behaviour of the blind-bolted T-stub connection was ascribed mainly to the nonlinear force vs. extension characteristics of the cog extensions into the concrete and to changes in the contact areas among endplate, tube wall, blind bolt head and nut assemblies. The modeling of these cogs using non-linear springs has been the subject of a rigorous experimental and theoretical study (Yao et al. 2007). There were also local nonlinear effects due to material plasticity. Therefore, this required that, in addition to multiple iterations per load step for convergence, the loads be applied in gradual increments, to characterize the actual load history. Two load stages were implemented to simulate the blind-bolted T-stub connection with full pretension. In the first stage, pretension loads were applied to the pretension elements. The effects of the preload on the bolt shank were preserved as an initial displacement. In the second stage, a normal tension load was applied to the flange.

4.2 Comparison of FE model with the test results

In the FE model, failure of the connection was determined by ultimate stress in the bolt assembly, maximum uniform elongation strains in the endplate and flare flange which indicate a mechanism in the plate, and also the weld capacity between bolt heads and cogged extensions. A limiting value of 0.1 for principal strains was specified in the model. The model could achieve an ultimate load at 800 kN if limited weld capacity was allowed to be exceeded. However, due to the limit of weld between bolt head and cogged extension, test on T-stub connection stopped at 690 kN. As observed in the experimental work, the FE model clearly demonstrates the benefit of the anchorage of the cogged extension within the concrete in reducing the pull-out displacement of the bolts. Furthermore, because of the load sharing between the cogged extension and tube wall, the connection achieves a high load capacity. The comparison between the experimental results and FE modelling results for the load versus relative outward displacement between the centre of the curved endplate and the concrete-filled steel

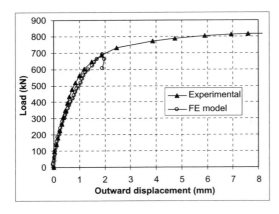

Figure 12. Comparison of FE model and experimental results.

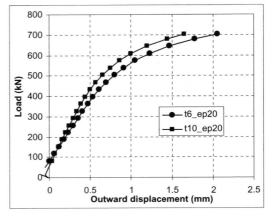

Figure 13. Load vs. outward displacement on model t10_ep20.

tube, Δ_0, is shown in Figure 12. The FE model represents the initial stiffness of the T-stub connection with good accuracy. The pretension effect on the bolts was released at the load step of 520 kN based on tracking of pretension nodes. The FE model was slightly stiffer compared to the test result after the load of 275 kN. In general, the FE simulation is in good agreement with the experimental results.

5 PARAMETRIC STUDY

5.1 Effect of tube wall thickness

The full range of thickness of the tube wall was designated to be 6 mm, 8 mm and 10 mm. In the model t10_ep20, the tube wall thickness was increased from 6 mm to 10 mm. All other variables, including the endplate thickness of 20 mm, were kept constant. The model was reconstructed to suit the change of tube wall. Figure 13 shows the load versus outward displacement for model tb10_ep20 compared with the initial model tb6_ep20. Both models have similar initial stiffness at the beginning stage.

After 200 kN, t10_ep20 tends to be displaced less due to its increased thickness. At the ultimate load of 700 kN, its outward displacement reaches only 1.65 mm as the thicker tube wall and associated cogged extensions greatly improve the bearing capacity and pullout resistance. A great proportion of load is carried by the membrane action in the tube wall and much smaller proportion goes to cogged extensions.

5.2 Effect of endplate thickness

The full range of endplate thickness for the T-stub connection was 12 mm, 16 mm, and 20 mm. In the

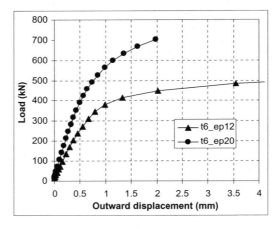

Figure 14. Load vs. outward displacement on model t6_ep12.

model of t6_ep12, the thickness of endplate was reduced from 20 mm to 12 mm with other variables kept constant. The whole model was rebuilt to accommodate the change in the endplate.

Figure 14 shows the load versus outward displacement for model tb6_ep12 compared with the initial model tb6_ep20. Model t6_ep12 is far more flexible than model t6_ep20. The thinner endplate deforms excessively beyond the load level of 300 kN. The great portion of the outward displacement is attributed to the deformation of the thinner endplate as a plastic hinge forms in the endplate. The capacity of the model is reduced to 500 kN at a displacement of 4 mm. The thinner endplate is prone to introducing prying action in the connection.

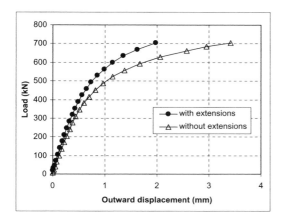

Figure 15. Effects of cogged extension on model t6_ep20.

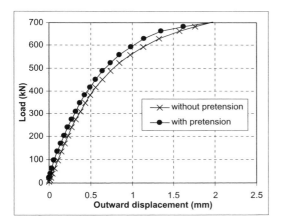

Figure 16. Effects of bolt pretension on model t6_ep20.

5.3 Effect of cogged extension

The cogged extensions were removed from the initial model of tb6_ep20. Without the cogged extensions attached to the blind bolts, the bolt heads directly bear against the internal tube wall and this results in a circular yield pattern occurring around the bolt holes at the tube wall. The localized deformation of the tube wall could lead to fracture and then bolt pullout. (Gardner & Goldsworthy 2005). The tensile load is carried exclusively by local bearing and the membrane action in the tube wall. The connection is quite flexible compared with the initial model with anchorage of extensions in the concrete-filled steel tubes. The effect of the cogged extension is shown in Figure 15.

5.4 Effect of bolt pretension

In this analysis, the pretension was removed from the initial model t6_ep20. The effect of blind bolt pretension on the connection behaviour is shown in Figure 16. Without the pretension load in the model, the connection is slightly more flexible. In the FE model, the surfaces of the curved endplate and tube wall are in a state of perfect contact. The preload on the individual bolt produces an initial clamped displacement of 0.017 mm. This helps to increase the initial stiffness of the T-stub connection.

6 MOMENT-ROTATION CURVE

A moment-rotation relationship (see Fig. 17) has been determined for a concrete-filled circular column with diameter of 323.9 mm and a thickness of 6 mm connected to a steel universal beam of 310UB 40.4 kg/m through the double split T connection. This represents a moment-resisting connection in a low-rise frame with a beam span of approximately 6 m. The design is based on capacity design principles so that the beam will reach its plastic moment capacity whilst the connection remains strong and stiff.

7 POTENTIAL COUPLED CONNECTION

The most essential characteristic of moment resisting frames in areas of high seismicity is the requirement that plastic hinges to form in the beams close to the connections. These plastic hinges provide ductility to dissipate energy hysteretically. As it is not feasible to

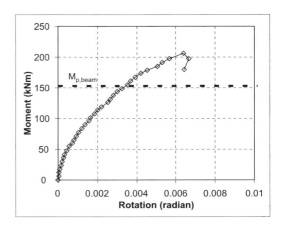

Figure 17. Moment-rotation curve.

develop large plastic deformations right at the blind-bolted T-stub connection, the ideal location for the plastic hinges to be formed in the beam section near the connection.

7.1 Reduced Beam Section (RBS)

Due to problems with welded connections encountered in the Northridge 1994 earthquake, extensive research has been conducted to explore ways of improving the performance of existing and new frames. One such approach is that of creating a "dog-bone" reduced section in the beam. In a reduced beam section moment connection, portions of the beam flanges are selectively trimmed in the region adjacent to the beam-to-column connection. Yielding and hinge formation are intended to occur primarily within the reduced section of the beam, therefore limiting the moment and inelastic deformation demands developed at the face of the circular column.

Figure 18 shows a beam-to-circular column connection with radius cuts on the beam flanges. Some concerns were raised that the presence of flange cuts might make the beam more prone to lateral torsional buckling and that supplemental lateral bracing should be provided at the RBS. Tests on RBS specimens with composite slabs (Engelhart et al. 2000) indicated that the presence of the slab provided a sufficient stabilizing effect that a supplementary brace at the RBS is not likely to provide significantly improved performance.

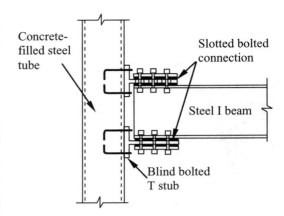

Figure 19. Blind bolted connection with SBC.

7.2 Slotted Bolted Connection (SBC)

An alternative method to RBS is to use energy dissipators as long as the added cost is small. There are several different types of passive energy dissipators available (Butterworth & Clifton 2000). The discussion here is limited to the slotted bolted connection developed at UC Berkeley (Grigorian & Popov 1994, Yang & Popov 1995). The slotted bolted connection is a modified bolted connection designed to dissipate energy through friction during rectilinear tension and compression loading cycles.

An SBC consists of five metal plates and a number of fastener bolts. Two brass shims are inserted between the stem plate and the splice plates in order to develop a constant friction. The stem plate is sandwiched directly between the brass shims in the assembly. Each plate, except the stem plate, has an equal number of circular bolt holes for bolting these five plates together. Elongated holes or slots are formed in the stem plate. Figure 19 shows an application of an SBC connection to connect the steel I beam to the blind-bolted T stubs attached to the concrete-filled steel tube.

8 CONCLUSIONS

A test on a large-scale blind-bolted steel T-stub connection to concrete-filled circular column has demonstrated the favourable strength and stiffness features that can be utilised in transferring moment between beam and circular column within structural frames for low-to-medium rise buildings. Adding a cogged extension to the blind bolts improves the behaviour of the T-stub connection as the load can be shared between the tube wall and cogged anchorage within the concrete. Thus, excessive deformation of

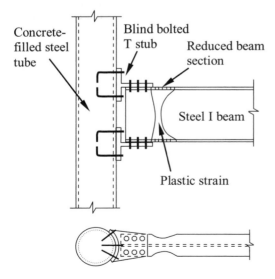

Figure 18. Blind bolted connection with RBS.

the tube wall can be avoided. A three-dimensional finite element model was developed to capture the full behaviour of the tested specimen. It has been used to simulate the connection performance and to explore the behavioural changes as a consequence of geometric variations. The feasibility of connecting a T-stub to the steel beam using a slotted bolted connection and reduced beam section to improve seismic behaviour has been explored.

ACKNOWLEDGEMENTS

The authors would like to express gratitude to Australian Tube Mills for donation of steel tubes for testing; GFC Industries for performing the welding between the cogged extensions to the bolts; AJAX Engineered Fasteners for their technical support.

REFERENCES

Ajax Engineered Fasteners 2002. ONESIDE brochure. *B-N012 data sheet*, Victoria.

ANSYS Inc. 2003. User's manual, ANSYS documentation, version 8.1, Southpointe, Canonsburg.

Butterworth, J.W. & Clifton, C.G.C. 2000. Performance of hierarchical friction dissipating joints in moment resisting steel frames. *Proc. 12th World Conf. on Earthquake Engineering*, Auckland, New Zealand, paper 0718.

Engelhart, M.D., Fry, G., Jones, S., Venti, M. & Holliday, S. 2000. Behavior and design of radius-cut reduced beam section connections. *Report No. SAC/BD-00/17*, SAC Joint Venture, Sacramento, CA.

Fernando, S. 2005. Joint design using ONESIDE™ structural fastener. *Technical note: AFI/03/012*, Ajax Fasteners, Victoria.

Gardner, A.P. & Goldsworthy, H.M. 2005. Experimental investigation of the stiffness of critical components in a moment-resisting composite connection. *Journal of Constructional Steel Research*, Vol. 61, No. 5, pp. 709–726.

Grigorian, C.E. & Popov, E.P. 1994. Energy dissipation with slotted bolted connections, *Report No. UCB/EERC-94/02*, Earthquake Engineering Research Center, University of California, Berkeley, California.

Morino, S. Uchikoshi, M. & Yamaguchi, I. 2001. Concrete-filled steel tube column system – its advantages. *International Journal of Steel Structures*, Vol. 1, No. 2001, pp. 33–34.

Nakada, K. & Kawano, A. 2003. Load-deformation relations of diaphragm-stiffened conncetions between H-shaped beams and circular CFT columns. *Proc. Intern. Conf. on Advances in Structures*, Hancock et al. (eds), Sydney, Australia.

Standards Australia. 1998. Steel Structures, AS4100, Standards Association of Australia, NSW, pp. 99–102.

Standards Australia. 2001. Concrete Structures, AS3600, Standards Association of Australia, NSW, pp. 130.

Sherbourne, A.N. & Bahaari, M.R. 1996. 3D simulation of bolted connections to unstiffened columns – I. T-stub connections. *Journal of Constructional Steel Research*, Vol. 40, No. 3, pp. 169–187.

Swanson, J.A. & Leon R.T. 2000. Bolted steel connections: tests on T-stub components. *Journal of Structural Engineering*, ASCE, Vol. 126, No. 1, pp. 50–56.

Swanson, J.A., Kokan, D.S., & Leon, R.T. 2002. Advanced finite element modeling of bolted T-stub connection components. *Journal of Constructional Steel Research*, Vol. 58, No. 5–8, pp. 1015–1031.

Yang, T-S., & Popov, E. P. 1995. Experimental and analytical studies of steel connections and energy dissipators. *Report No. UCB/ERRC-95/13*, University of California, Berkeley.

Yao, H., Goldsworthy, H.M., & Gad, E.F. (2007). Pullout behaviour of cogged deformed bars within concrete filled steel tubes, Proc. *23rd Biennial Conf. of the Concrete Institute of Australia*, Concrete Institute of Australia, Adelaide, Australia.

Steel-concrete composite full strength joints with concrete filled tubes: design and test results

O.S. Bursi & F. Ferrario
Department of Mechanical and Structural Engineering, University of Trento, Italy

R. Pucinotti
Department of Mechanics and Materials, Mediterranean University of Reggio Calabria, Italy

ABSTRACT: In this paper, a multi-objective advanced design methodology dealing with seismic actions followed by fire on steel-concrete composite full strength joints with concrete filled tubes is proposed instead of a traditional single-objective design where fire safety and seismic safety are achieved independently. Experimental tests together with numerical simulations of the fire behaviour were carried out to derive fundamental information about the performance of these joints. In detail a total of six specimens designed according to Eurocode 3, 4 and Eurocode 8 were subjected to monotonic and cyclic loadings up to collapse at the Laboratory for Materials and Structures Testing of the University of Trento, Italy. These specimens were detailed in order to exhibit a favourable fire behaviour after a severe earthquake. The major aspects of the cyclic behaviour of composite joints are presented and commented upon together with fire analysis results. Both the experimental activity and the numerical FE simulations demonstrate the adequacy of the proposed joint design.

1 INTRODUCTION

In the design of steel-concrete composite buildings the sequence of seismic and fire loadings are not taken into account. In fact, the seismic safety and the fire safety are considered separately. In reality, the risk of loss of lives increases if a fire occurs within the building after an earthquake. Where significant earthquakes can occur, fire after earthquake is a design scenario that should be properly addressed in any performance-based design.

This approach takes into account seismic safety and fire safety with regard to accidental actions as well as fire safety on a structure characterized by stiffness deterioration and strength degradation owing to seismic actions. The proposed design solution, were developed in an European research (Colombo & Bursi 2006). This research program is intended to develop fundamental data, design procedures and promotion of two types of ductile and fire-resistant composite beam-to-column joints with:

1. partially reinforced-concrete-encased column with I-section;
2. concrete filled tubular column with circular hollow steel section.

The project analyses the scenario in which a fire follows and earthquake, thus defining joint typologies for which, after being damaged by an earthquake, a residual load-bearing capacity is assured during a fire occurring after an earthquake. The design is performed in the modern context of performance-based engineering determining both the stiffness deterioration and the strength degradation of composite joints after seismic loading.

The present paper presents the results of experimental tests under monotonic and cyclic loading together with numerical simulations of the fire behaviour of steel-concrete composite full strength joints with concrete filled tubes.

2 COMPOSITE BUILDING DESIGN

The actions used in the design of the proposed joints were obtained by the analyses of two moment resisting frames having the same structural typology but different slab systems (160 mm composite steel-concrete slab high with structural profiled steel sheeting and 160 mm concrete slab high composed of electro-welded lattice girders).

The composite steel-concrete office-building was endowed with 5 floors with 3.5 m storey height. It was made up by three moment resisting frames placed at the distance of 7.5 m each in the longitudinal direction, while it was braced in the transverse direction. A different distance between the secondary beams was adopted for the two solutions that takes into account the different load bearing capacities of the two slab systems as well as the need to avoid propping systems during the construction phase. As a result, the main moment resisting frame is made up by two bays spanning 7.5 m and 10.0 m in the solution with steel sheeting with a distance between secondary beams equal to 2.5 m; and by two bays spanning 7.0 m and 10.5 m in the solution with lattice steel girders with a distance between secondary beams equal to 3.5 m (Figs 1, 2). All slabs were arranged in the direction parallel to the main frames. The main beams were IPE 400 while the secondary ones were IPE 300.

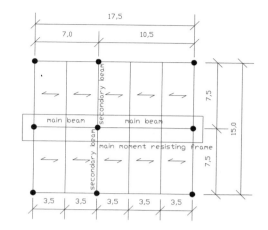

Figure 2. Plan and structure typology, slabs with profiled steel sheetings. Dimensions in metres.

2.1 Composite beams

Two different types of composite beams were checked according to point 6.1.1 of Eurocode 4 (UNI EN 1994-1-1. 2005). In the first one, the beam section was an IPE400 with steel grade S355 while the deck was a composite slab with a prefabricated lattice girder made by the Pittini Group (Fig. 3).

In this case, the slab reinforcement was performed by $3+3\phi 12$ longitudinal steel bars and by $5+5\phi 12@100$ mm plus $8+8\phi 16@200$ mm transversal steel bars. A mesh $6@200 \times 200$ mm completes the slab reinforcement. In the second type of beam as shown in Figure 4, the same beam section was adopted and a composite slab with profiled

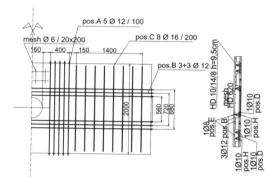

Figure 3. Specimens WJ-P -Slab with prefabricate lattice girders. Dimensions in mm.

steel sheeting was made following the rules given in Section 9 of Eurocode 4 (UNI EN 1994-1-1. 2005).

The slab reinforcement consisted of $3+3\phi 12$ longitudinal steel bars and of $4+4\phi 12@100$ mm and $7+7\phi 16@250$ mm transversal steel bars. Moreover, the same mesh $\phi 6@200 \times 200$ mm was adopted. The concrete class was C30/37 while the steel grade S450 was adopted for the reinforcing steel bars.

2.2 Composite columns

The columns of steel grade S355 were concrete-filled column with a CFT filled tubular column steel profile with a diameter of 457 mm and a thickness of 12 mm. Column reinforcements consisted of $8\phi 16$ longitudinal steel bars and stirrups $\phi 8@150$ mm.

The concrete class of composite columns was C30/37, while the steel grade S450 was adopted for the reinforcing steel bars as illustrated in Figure 5.

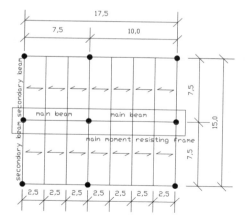

Figure 1. Plan and structure typology; slabs with prefabricate lattice girders. Dimensions in metres.

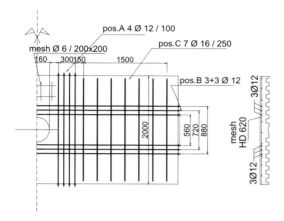

Figure 4. Specimens WJ-S—Slab with profiled Steel Sheetings. Dimensions in mm.

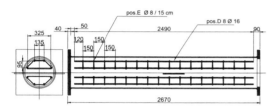

Figure 5. Column and column reinforcements.

2.3 Beam-to-column composite joints

The joint consisting of a steel concrete composite beam attached to a concrete-filled tubular column with circular hollow steel sections and is composed by two horizontal diaphragm plates and a vertical through-column plate (Figures 6 and 7).

The presence of a vertical through column plate shown in Figure 7 allows the transmission of shear forces owing to vertical loads from a beam to the other.

In the joint design, the transition zone between the column and the beam was assumed to be overstrength compared to the beam, thus forcing plastic hinges formation in adjacent beams (UNI EN 1998–1. 2005). As a result joints were detailed by using the component method as shown in Figure 8 (UNI EN 1993-1-8. 2005).

The following components were considered in the method: concrete slab in compression; upper horizontal plate in compression; vertical plate in bending and lower horizontal plate in tension, for sagging moment; reinforcing bars in tension, upper horizontal plate in tension; vertical plate in bending and lower horizontal plate in compression for hogging moment. The components concrete slab in compression and upper horizontal plate in tension were identified by means of FE models set with ABAQUS (Hibbitt et al., 2000). Figure 9 shows the FE model of a plate in tension to identify the effective width b.

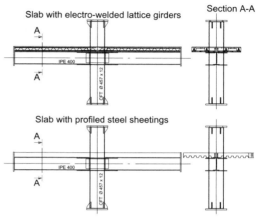

Figure 6. Beam-to-Column joint specimens.

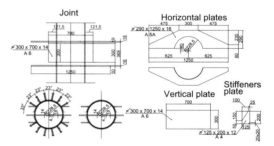

Figure 7. Steel-concrete composite beam-to-column joint. Dimensions in mm.

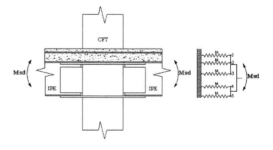

Figure 8. Mechanical model of an interior joint.

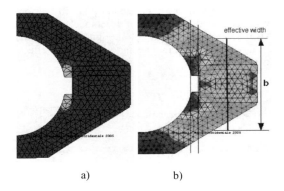

Figure 9. FE model of a plate in tension: (a) elastic stresses; (b) inelastic stresses and width.

3 EXPERIMENTAL TESTS

A total of 6 specimens was designed and fabricated according to Eurocode 4 (UNI EN 1994-1-1. 2005, UNI EN 1994-1-2. 2005) and Eurocode 8 (UNI EN 1998-1. 2005) provisions. The test specimens were interior subassemblage and the joints were made by two horizontal diaphragm plates split into two equal halves along the diagonal for fabrication convenience and easiness of assembling.

Once each side is properly placed on the pipe the two halves are attached with a full-joint penetration groove weld. In detail, the inferior plates are welded to column in the shop, while the superior plates are welded on site as should be understood from Figure 6. Flanges and web of each beam were connected to the horizontal plates and the vertical plate by welding, as depicted in Figure 7.

In all composite specimens the connections between steel beam and desk were made by Nelson 19 mm stud connectors with an ultimate tensile strength $f_u = 450$ MPa. The joints differ each other owing to the slab type and to additional Nelson 19 mm studs localized around the column in order to enforce a better force transmission between the column and the composite slabs as indicated in Figure 7. The joint specimens were subjected to monotonic and cyclic loadings up to collapse, according to the ECCS stepwise increasing amplitude loading protocol, modified with the SAC procedure (ECCS 1986,

SAC 1997): $e_y = 17.5$ mm. Hereafter the specimens are indicated as follows:

- WJ-P1—specimens with electro-welded lattice girders slab and no Nelson connectors around the column;
- WJ-P2—specimens with electro-welded lattice girders slab and Nelson connectors around the column;
- WJ-PM—specimens with electro-welded lattice girders slab and no Nelson connectors around the column;
- WJ-S1—specimens with profiled Steel Sheeting slab and no Nelson connectors around the column;
- WJ-S2—specimens with profiled Steel Sheeting slab and Nelson connectors around the column;
- WJ-SM—specimens with profiled Steel Sheeting slab and no Nelson connectors around the column.

3.1 Test set-up

The test set-up was designed in order to simulate the conditions of interior beam-to-column joints with concrete filled tubes within frame structures.

It consisted of a reaction wall, a hydraulic actuator (capacity 1000 kN, stroke ± 250 mm), a reinforced-concrete slab, a lateral frame designed to prevent specimen lateral displacements.

The main instrumentation employed is indicated in Figure 10 and detailed herein:

- 5 inclinometers were utilized in order to measure the inclinations of the zone adjacent to the joint and of the beams near the connection;
- 4 LVDTs detected the interface slip between the steel beam and the concrete slab and between the inferior horizontal plates and the beam flange;
- 2 LVDTs were employed in order to measure the connection deformation;
- 10 LVDTs were utilized in order to measure concrete slab deformations;
- 4 Omega strain gauges detected the deformations of the concrete slab;
- 8 strain gauges monitored axial deformations of the reinforcing bars in order to scrutinise the effective breadth of the reinforcing bars at each loading stage;
- 4 strain gauges monitored deformations of superior and inferior horizontal plates;
- 4 strain gauges recorded flange strains in order to estimate internal forces in steel beams;
- 2 load cell were set on the top of pendula and were utilized in order to measure horizontal and vertical components of axial forces;
- 1 digital transducer (DT500) was employed in order to measure the top column displacement;
- a FieldPoint acquisition system in order to acquire experimental data.

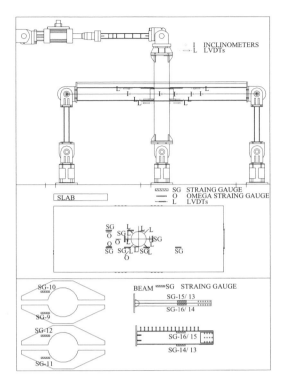

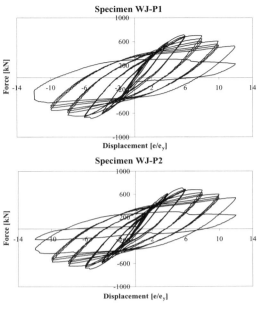

Figure 11. Specimens WJ-P1 and WJ-P2—Force-displacement relationships.

Figure 10. Main instrumentation on specimens.

3.2 Results

Experimental results are presented on the basis of hysteretic responses. In particular, Force-displacement (F-d) curves for all tested specimens are shown in Figures 11–14, while moment vs. rotation (M-ϕ) relationships are depicted in Figures 15–18.

All specimens exhibit a good performance in terms of resistance, stiffness, energy dissipation and ductility.

Both the overall force-displacement relationships and the moment-rotation relationships relevant to plastic hinges formed in the composite beams exhibit a hysteretic behaviour with large energy dissipation without evident loss of resistance and stiffness. In particular we can be observe that the hysteretic loops of moment-rotation relationship are unsymmetrical due to the different flexural resistance of the composite beam under hogging and sagging moments, respectively.

For all specimens, the experimental tests show a remarkable and progressive deterioration of strength, stiffness and energy absorption capacity as a consequence of the formation of a plastic hinge associated with local buckling of the beam flange. The collapse

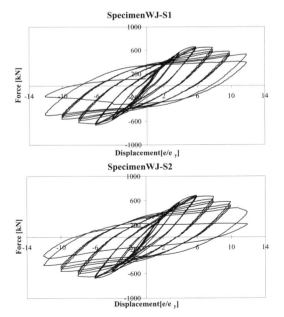

Figure 12. Specimens WJ-S1 and WJ-S2—Force-displacement relationships.

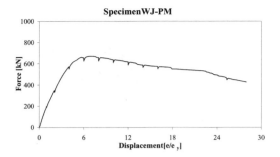

Figure 13. Specimen WJ-PM- Force-displacement relationship.

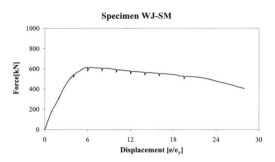

Figure 14. Specimen WJ-SM- Force-displacement relationship.

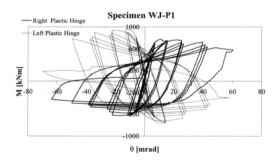

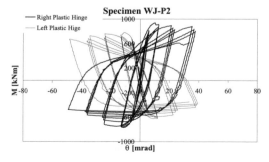

Figure 15. Specimens WJ-P1 and WJ-P2—Moment-rotation relationships of plastic hinges.

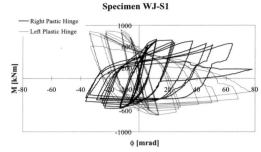

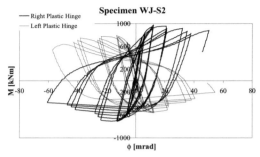

Figure 16. Specimens WJ-S1 and WJ-S2—Moment-rotation relationships of plastic hinges.

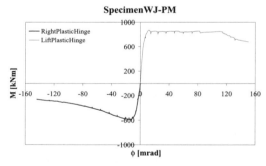

Figure 17. Specimen WJ-PM endowed with a slab with electro-welded lattice girders – Moment-rotation relationship.

of all specimens was associated with cracking of beam flanges.

Table 1 reports some major experimental results for each Test: in detail, the maximum applied Displacement (d), the maximum value of the Force (F), maximum values of both Sagging and Hogging Moment (M) and corresponding values of initial stiffness (k_θ), total number of Cycles performed during the tests (N_{tot}) are collected.

Monotonic tests were interrupted before failure of the specimens in order to reduce the risk of damaging the test equipment.

Table 1. Major experimental data.

Specimen	d mm	F kN	M kNm	k_θ kNm/mrad	N_{tot}
WJ-P1	210	+ 714.80 − 686.30	+ 962.46 − 767.72	+181.27 −141.00	21
WJ-P2	210	+ 708.90 − 695.40	+ 939.07 − 768.48	+183.43 − 158.00	21
WJ-S1	210	+ 661.70 − 662.00	+ 926.51 − 693.77	+220.32 − 113.50	21
WJ-S2	210	+ 686.83 − 673.46	+ 977.61 − 714.36	+208.25 −117.75	21
WJ-PM	490	+ 669.85	+ 874.15 − 591.14	+188.95 −105.19	Mon.
WJ-SM	490	+ 614.08	+ 814.74 − 528.03	+179.67 −81.79	Mon.

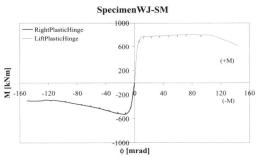

Figure 18. Specimen WJ-SM endowed with a slab with profiled steel sheetings – Moment-rotation relationship.

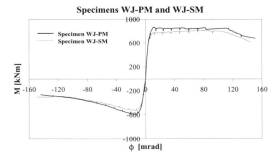

Figure 19. Comparison of (M-f) relationships of Specimens WJ-PM and Specimen WJ-SM.

Figure 19 shows moment vs. rotation relationships (M-ϕ) of specimens WJ-PM and WJ-SM, respectively. A better behaviour of the specimen-PM fabricated with the electro-welded lattice girders slab is evident in term of strength owing to the better composite action in the plastc hinge section. A numerical finite element (FE) model of the slab including friction was developed in a previous study by Ferrario et al. (Ferrario et al. 2007) with the objective to understand the activation of the transfer mechanisms between the slab and the beam proposed in the Eurocode 8 (UNI EN 1998-1. 2005).

The analyses demonstrated that the distribution of compression forces in the slab depend on the level of friction existing between the concrete slab and the composite column. Therefore, in order to activate some of the transfer mechanisms proposed in Eurocode 8, (i.e. Mechanism 1 and Mechanism 2), it was necessary to increase the level of friction between the concrete slab and the composite column of specimens.

As a result, Nelson stud connectors welded around the column were adopted in some specimens along the details indicated in Figure 7. Experimental results showed that the activation of the aforementioned mechanisms was evident and effective in the specimens fabricated with Nelson connectors welded around the column; while in the specimens fabricated without stud connectors around the columns Mechanisms 1 and 2 were less effective.

Figure 20 shows the total dissipated energy i.e. the Cumulative Energy for the specimens belonging to the WJ-P and WJ-S series. The dissipated energy in the plastic cycles is practically identical for all the specimens; after the 15th cycle which corresponds to a displacement equal to 10e_y, specimens with electro-welded lattice girders slab dissipate more energy than those with profiled steel sheetings owing to a better

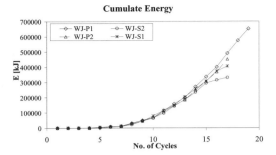

Figure 20. Dissipated energy of Specimens WJ-P and WJ-S.

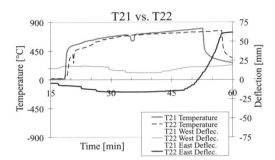

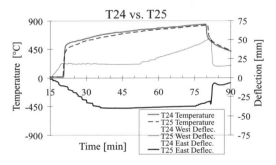

Figure 21. Performance of damaged (T21/T24) and undamaged (T22/T25) specimens.

influence of the force transfer between columns and slabs on plastic hinges.

Moreover, four fire tests on full-scale substructures were performed too. In particular, two specimens, labelled as T21 and T24, were pre-damaged to simulated damage owing to Type 1 spectrum compatible accelerograms at 0.4 g pga (UNI EN Eurocode 8-1, 2005), and two specimens (T22 and T25) were not, to clearly appreciate seismic damage effects on fire resistance. The Temperature vs. time curve imposed to the specimens T21-T22 and T24-T25 is shown in Figure 21(a) and (b), respectively.

Specimens T21 and T22 with profiled steel sheeting slabs exhibited failure owing to an excessive rate of deflection at approximately 40 minutes. The test on specimen T21 terminated after approximately 34 minutes owing to runaway deflection. Following the fire test, the profiled steel sheeting separated from the slab; then the slab cracked both along the surface and through the depth with extensive buckling at one hour both of the lower flange and the web of the adjacent east beam.

T24 and T25 specimens endowed with prefabricated slabs endured one hour of fire; however, in both cases specimens were very close to failure as indicated in Figure 21b, by an increasing rate of deflections towards the end of the test. However, at this stage, there was no permanent deformation and no sign of any significant damage from fire tests.

4 NUMERICAL SIMULATIONS

Different fire scenarios acting in the reference building were studied with the objective to evaluate the performance of composite beams, composite columns and beam-to-column joints under fire load for different times of exposure.

The Abaqus code (Hibbitt et al. 2000) was employed to perform thermal analyses of joints for different times of fire exposure, i.e. 15 min, 30 min and 60 min, respectively.

Moreover, thermal analyses with different fire scenarios were carried out by means of the SAFIR code (Franssen 2000) on a two-dimensional frame model.

Figure 22 shows the considerable reduction of the design capacity moment of the joint as a function of the time of fire exposure.

In detail, the hogging capacity moment becomes approximately the 20% of the initial value after 15 minutes of exposure, while it approaches about 5% of the initial value after 30 minutes.

Nevertheless, this time is enough to quit the building after a severe earthquake.

The joint endowed with prefabricated slab exhibits a lower concrete temperature compared to the joint endowed with the steel sheeting owing to the fact that the steel portion of the column heats the concrete slab near the joint but the slab has a uniform thickness around the column (Fig. 23).

Thermal analysis conducted by means of the Abaqus 6.4.1 code, for different times of fire exposure, demonstrated that the joint endowed with prefabricated slab exhibits a better fire behaviour compared to the joint endowed with steel sheeting as illustrated in Figure 24. Anew, this is due to the uniform thickness of the lattice girder with respect to the profiled steel sheeting.

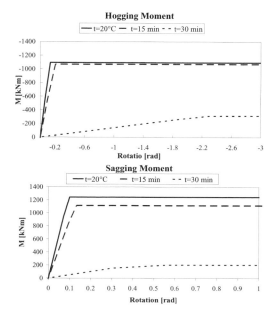

Figure 22. Moment-rotation relationships of the joint as a function of the time of fire exposure.

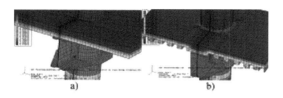

Figure 23. Temperature distribution of joints modelled by the FE Abaqus code: a) Joint with prefabricated slabs; b) Joint with profiled steel sheetings.

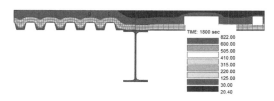

Figure 24. Temperature distribution in the composite beam with prefabricated slab and with steel sheeting slab.

The temperature of concrete inside the tube increases more quickly close to the joint in comparison with the portion of column distant from the joint owing to the influence of the vertical through-column web plate (Fig. 25).

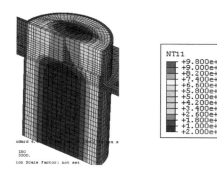

Figure 25. Temperature distribution in the composite concrete filled tube.

5 CONCLUSIONS

In this paper, a multi-objective advanced design methodology was proposed instead of a traditional single-objective design. Experimental cyclic tests and numerical simulations regarding the fire behavour of joints were carried out. Experimental results show how joint details influence beam-column subassemblage responses both in terms of seismic and fire performance.

In detail, test results exhibited rigid behaviour for the designed composite joints; as expected plastic hinges developed in beams owing to the capacity design.

It was evident that owing to local buckling effects, severe strength degradations appeared to exceed 20 per cent of maximum strength values with rotations of about 40 mrad. As a result, these joints are not suitable for high ductile structures to be used under seismic loading according to Eurocode 8 (UNI EN 1998-1. 2005). Nonetheless, the corresponding minimum plastic limit of 35 mrad remains rather high for typical European earthquakes.

Thermal analyses showed that steel elements, directly exposed to fire loading, had the same temperature of air, while for the concrete slab different behaviuors were observed depending on the adopted typology. In fact, specimens with profiled steel sheetings slab were characterized by temperatures higher than those endowed with electro-welded lattice girders. Moreover fire experiments showed the favourable behaviour of joints.

In sum we can affirm that both experimental and numerical results demonstrated the adequacy of the proposed joint design for concrete filled tubes to adequately face earthquake loading followed by fire.

ACKNOWLEDGMENTS

The writers are grateful for the financial support of UE under the project ECSC (RFS-CR-03034).

Nonetheless, conclusions of this paper are those of the author and not of the sponsor agency.

REFERENCES

Colombo A., Bursi O.S., 2006, Editors, "Mid-Term Report of Prefabricated Composite Beam-to-Concrete Filled Tube or Partially Reinforced-Concrete-Encased Column Connections for Severe Seismic and Fire Loadings, PRECIOUS Project, University of Trento.

ECCS 1986. Recommended Testing Procedure for Assessing the Behaviour of Structural Steel Elements under Cyclic Loads. ECCS Publication n° 45.

Ferrario F., Pucinotti R., Bursi O.S., Zandonini R. 2007, Steel-Concrete Composite Full Strength Joints with Concrete Filled Tubes. Part II: Experimental and Numerical Results, *Proceeding of 21st Conference of Steel Structures C.T.A., Catania, 1–3 October, pp. 397–404*, Dario Flaccovio Editor.

Franssen J.-M. 2000. "SAFIR; Non linear software for fire design". Univ. of Liege.

Hibbitt, Karlsson and Sorensen 2000. "ABAQUS User's manuals". 1080 Main Street, Pawtucket, R.I. 02860.

SAC 1997, Protocol for Fabrication, Inspection, Testing, and Documentation of Beam-Column Connection Tests and Other Experimental Specimens, report n. SAC /BD-97/02.

UNI EN1991-1-2. 2004. "Eurocode 1: Actions on Structures. Part 1-2 : General Actions – Actions on structures exposed to fire".

UNI EN 1992-1-2. 2005. "Eurocode 2: Design of concrete structures. Part 1-2: General rules – Structural fire design.

UNI EN 1993-1-1. 2005. "Eurocode 3: Design of steel structures. Part 1: General rules and rules for buildings".

UNI EN 1993-1-8. 2005. "Eurocode 3: Design of steel structures – Part 1-8: Design of joints".

UNI EN 1993-1-2. 2005. "Eurocode 3: Design of steel structures. Part 1-2: General rules – Structural fire design".

UNI EN 1994-1-1. 2005. "Eurocode 4: Design of composite steel and concrete structures – Part 1.1: General rules and rules for buildings".

UNI EN 1994-1-2. 2005. "Eurocode 4: Design of composite steel and concrete structures – Part 1.2: General rules – Structural fire design".

UNI EN 1998-1. 2005. "Eurocode 8: Design of structures for earthquake resistance – Part 1: General rules, seismic actions and rules for buildings".

Development of an effective shearhead system to eliminate punching shear failure between flat slabs and tubular columns

P.Y. Yan & Y.C. Wang
School of MACE, University of Manchester, UK

A. Orton
Corus Tubes, UK

ABSTRACT: This paper reports on the results of a test on a new shearhead system for tubular steel columns and concrete flat slabs. The shearhead eliminates punching shear failure between the flat slab and column while allowing use of thinner slabs. Furthermore the design allows service holes adjacent to the column to be made easier. To develop a practical system, the shearhead should not only possess sufficient punching shear resistance to eliminate punching shear failure, but should also be efficient for construction. This paper will present the results of a test and an assessment of three main design codes: Eurocode 2, British Standard BS 8110 and American Concrete Institute code ACI 318. The results of this assessment suggest that that British Standard BS 8110 provides the best prediction. Although the ACI 318 code is the only one that explicitly deals with shearheads, the ACI 318 prediction of the failure perimeter seems to be much lower than the test result. The reported test confirms that shearheads can be used with tubular columns.

1 INTRODUCTION

Flat slab structures lead to aesthetically pleasing designs, accelerated lead times and low construction costs. However a major flaw in such designs is the high shear stresses induced around the column head area which can cause brittle punching shear failure. To counteract this, most flat slab structures use heavy shear reinforcement across the column area. Flat slab structures are very often used in tall buildings, hospitals and laboratories and normally this kind of building will need good provision for the very large number of services specified. Very often this requires openings near the column and this can pose problems for reinforced concrete flat slab floors. Many attempts have been made in the past to enhance the punching shear capacity of the slab to column connection; generally these try to use some other kind of shear resistance system to replace normal shear reinforcement, such as the use of steel plates (Subedi & Baglin 2003), Carbon fiber reinforced polymers (CFRP) (Binici1 & Bayrak 2003), steel fiber (Choi et al. 2006, Tan & Paramasivam 1994) and shearheads (Corley & Hawkins 1968). This work is all predicated on the use of concrete columns but it is believed that the greatest advantages of all come from combining a shear system with a steel tubular column, filled with concrete or not.

Tubular columns have many advantages compared with concrete columns and have become popular in certain kinds of market. They allow easy and quick assembly and recycling. Because the steel tubular column may be erected at the same time as the concrete slab formwork and the tubular column may be poured at the same time as the slab, this type of construction gives major improvements in cycle time.

The new shearhead system described herewith was developed for tubular steel columns with concrete flat slabs so that it is quick to erect and allows service holes to be made adjacent to columns. The shearhead is different from existing systems; it does not require any shear reinforcement bars, only the normal flexural reinforcement, and can reduce slab depth, if this is not determined by deflection in the centre of the slab. It is also possible to design this system to achieve a safe, ductile failure mechanism which is not usually achieved in current practice using shear reinforcement or shear studs. To confirm this kind of structure's property, a serial of tests is at present being undertaken at the University of Manchester structural laboratory, supported by Corus Tubes.

The new shearhead system with tubular columns is presented in this paper and the first test result is compared with the provisions of EC2, BS 8110, ACI 318.

2 DESCRIPTION OF TEST SPECIMEN

2.1 Test step

The test rig is a self-supporting system as shown in Figure 1. The specimen was supported by four steel beams around the edge of the slab with 110 mm each side. The edge to edge distance between the support beams, i.e. clear span of the slab, is 1605 mm. A concentrated load was applied to the test specimen from the top of the tubular column by a 3000 KN hydraulic pumping jack combined with a load cell. This load was intended to produce the shear and bending moment combination which would be typical at an interior column of flat slab structure on a square grid of 5 meters with a total imposed load of 3.5 KN/m². For this test, care was taken to provide just sufficient flexural reinforcement so that failure was in shear. Load was added approximately in 10 kN load steps until the ultimate load was reached.

2.2 Material properties

Normal weight ready-mix concrete of nominal strength C40 (40 MPa) was used. The slab was cast in wooden film plate module and concrete was compacted by poker vibrators. Cubes and Cylinders of concrete were cast along with the slab for control purposes. Tensile tests were also conducted on the reinforcement and shear head steel. Table 1 gives the measured material properties.

2.3 Test specimen description

The test specimen was made up of an 1825 mm square, 200 mm thick slab with a centrally located 200 × 200 × 10 square tubular column through. Load was applied from the top so that the test slab was

Table 1. Material properties for test specimen.

		Concrete	Reinforcement	Shearhead
E	(MPa)		2.1×10^6	2.0×10^6
Yield stress	(MPa)		545.8	324.0
Tension stress	(MPa)		647.7	491.0
Elongation	(%)		18.3	37.0
Cube stress	(MPa)	45.0		
Split cylinder	(MPa)	3.36		

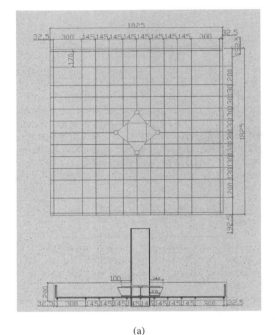

(a)

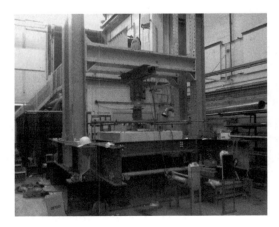

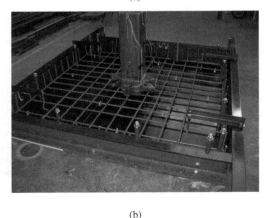

(b)

Figure 1. Test rig arrangement.

Figure 2. Test specimen and reinforcement arrangement.

reversed compared with that in a real case, with two layers of reinforcement on the bottom face but no reinforcement on the top face. T12 bars at 145 mm centres were used in the bottom layer and T12 bars at 130 mm centres used in the second to bottom layer over the central 1.25 meters to ensure the slab failed in shear and not bending. Over the rest of the area, T12 bars at 300 mm centres were used in the bottom layer and T12 bars at 200 mm centres in the second to bottom layer, as shown in Figure 2.

2.4 Shearhead construction

Four 102 × 44 × 7 joists with a top length of 140 mm and a bottom length of 100 mm were welded to fixing plates that, in turn, were welded to the column outside face. In order to make the shearhead structurally continuous, two 6 mm thick and 100 mm long fixing plates, on each column face, were made of equal width to the column face at one end and of equal width to the flange of the joist at the other end. Welds were made to each joist at both top and bottom flanges, as shown in Figure 3.

2.5 Instrumentation

Deflections at selected locations on the compression side of the slab were measured by linear variable displacement Transducers (LVDTs) shown in Figure 4.

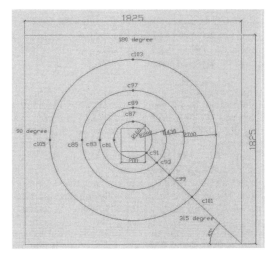

Figure 4. LVDT's position.

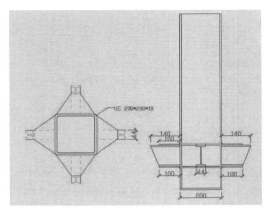

Figure 5. Bottom of slab at failure.

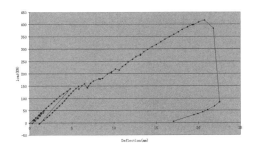

Figure 3. Shearhead.

Figure 6. Recorded load—column head deflection.

443

In addition, one more LVDT was put on the load cell bottom to measure the column center's displacement.

Electrical resistance strain gauges were used to measure the strain at selected locations both on reinforcement and shearhead arms. Some strain gauges were also put on the concrete compression face at corresponding positions to those on the reinforcement.

3 TEST RESULTS

Initial cracks appeared roughly 50 mm from the column face in the weak direction when the applied load was about 130 KN. The weak direction was the direction with less effective depth for the reinforcement. Afterwards, more cracks developed radially from the column face. Above 330 KN, a few new cracks developed, however the width of the cracks already near the column increased substantially. The test slab failed abruptly in punching shear at 416.77 KN. Figure 5 shows the bottom surface of the slab at failure.

The widths of the flexural cracks near the column were considerable, almost 1 mm. Furthermore the column penetrated about 22.5 mm into the slab. However, there was no sign of concrete compressive crushing near the column. The load versus column head deflection is shown in Figure 6, which clearly indicates brittle punching shear failure as the load reduced sharply once the maximum load was achieved.

4 ASSESSMENT OF CURRENT DESIGN METHODS

The main objective of this test is to confirm that a small shearhead to a tubular column is an effective structural system to resist punching shear. For this purpose, the test results will be compared against predictions using the three popular design methods (Albrecht 2002) for flat slab construction: Eurocode 2 Part 1.1 (CEN (2002), British Standard BS 8110 and American Concrete Institute code ACI 318-05. None of these codes deal with tubular columns. For this study, it is assumed that a tubular column may be considered as a reinforced concrete column. Of these three methods, only ACI318 includes shearheads. Therefore, in order to use EC2 and BS 8110, appropriate modifications (which will be described in detail later) will be made to enable the codes to be used to predict the test result.

All three methods employ the same basis of design calculation of punching shear resistance: the punching shear resistance is the critical shear area multiplied by shear stress per unit area at the critical shear perimeter.

The critical shear area is obtained by the critical shear perimeter length multiplied by the effective depth of the slab. In all three design codes, the effective depth of the slab is the same, being 168 mm.

Two generic methods may be used when analyzing the specimen performance: (1) treating the test specimen as a small column with shearhead; (2) treating the test specimen as a large column without shearhead, with the equivalent column cross-section size equal to the tube dimensions plus the total lengths of the shear heads (480 × 480 mm). Both approaches will be assessed.

4.1 Small column with shearhead

Since EC2 and BS 8110 do not deal with shearheads.

Therefore, for calculations using EC2 and BS 8110, the shearheads will be treated as equivalent shear reinforcements.

4.1.1 EC2

The value for punching shear resistance of concrete without shear reinforcement can be calculated by:

$$v_{Rd,c} = C_{Rd,c} k' (100 \rho_1 f_{ck}^*)^{\frac{1}{3}} \geq v_{min} \quad (1)$$

Giving $v_{Rd,c} = 0.581$ N/mm². In the equation above, $C_{Rd,c}$ is $0.18/\gamma_c$; v_{min} is calculated using equation 6.3 N in EC2;

$$k' = 1 + \sqrt{\frac{200}{d}} \leq 2.0 \ (d \text{ is in mm}) \quad (2)$$

f_{ck} is the cylinder strength of concrete, $f_{ck} = 35$ MPa.

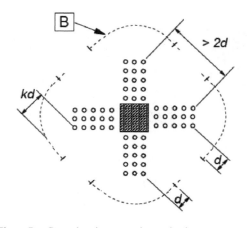

Figure 7. Control perimeters at internal columns.

To adapt EC2 to calculate the critical shear perimeter for the test specimen, it is assumed that the end of the shearhead is the last shear reinforcement. Since the direct distance between the ends of two adjacent perpendicular shearhead arms (340 mm) is greater than $2d$ (334 mm, d = effective depth of slab), the critical perimeter should be calculated according to Figure 7 which is extracted from EC2, giving

$$u_{2,ec2} = 2\pi \times 1.5d + 8d = 2927 \text{ mm} \quad (3)$$

Using the results above, the punching resistance capacity of the test specimen is:

$$v_{Rd,c}(u_{2,ec2}d) = 0.581 \times 2927 \times 168 = 285.70 \text{ KN} \quad (4)$$

4.1.2 BS 8110

The following equation of calculating the shear resistance of concrete without shear reinforcement at the critical shear perimeter is extracted from BS 8110:

$$v_c = 0.79 \times \left(\frac{100 A_s}{b_v d}\right)^{\frac{1}{3}} \times \left(\frac{400}{d}\right)^{\frac{1}{4}} \times \left(\frac{f_{cu}}{25}\right)^{\frac{1}{3}} \quad (5)$$

Where:

$$\frac{100 A_s}{b_v d} \leq 3, \quad \left(\frac{400}{d}\right)^{\frac{1}{4}} \geq 1 \quad (6)$$

Using the maximum value of $f_{cu} = 40$ N/mm² allowed in BS 8110, the above equation gives a value of $v = 0.720$ N/mm². If the actual concrete cube strength of 45 N/mm² is used, the shear stress $v'_c = 0.748$ N/mm².

Figure 8(a) is extracted from BS 8110, which shows how the critical shear perimeter may be calculated when shear reinforcement is used. It can be seen the critical punching shear perimeter is at a distance of $(0.25 + 0.75)d$ from the last shear reinforcement. For the test specimen, assuming the end of the shearhead is the location of the last shear reinforcement in BS 8110, and then the critical perimeter for the test specimen is as shown in Figure 8(b). From the scheme shown in figure 8(b), the critical perimeter (Figure 9) is:

$$u_{2,bs8110} = 8 \times (d + 240) = 3264 \text{ mm} \quad (7)$$

From the above calculated values, the punching shear capacity of the test specimen is:

$$v_c u_{2,bs8110} d = 0.720 \times 3264 \times 168 = 394.81 \text{ KN} \quad (8)$$

For $f_{cu} = 40$ N/mm², and

$$v'_c u_{2,bs8110} d = 0.748 \times 3264 \times 168 = 410.44 \text{ KN} \quad (9)$$

For $f_{cu} = 45$ N/mm².

4.1.3 ACI 318

ACI 318 is the only code that includes provisions for shearheads based on research by Corley and

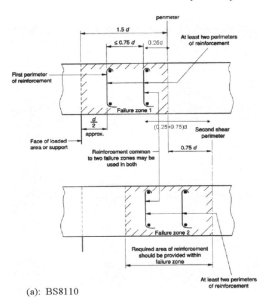

(a): BS8110

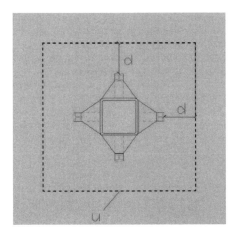

(b): Adapted for test specimen

Figure 8. Critical shear perimeter from BS 8110.

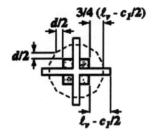

Figure 9. Critical area with shearhead.

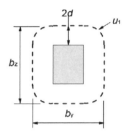

Figure 10. Basic control perimeter from EC2.

Hawkins. The new shearhead system has been checked according to the requirements for shearhead in ACI 318-11.12.4. Figure 10 shows how the critical perimeter should be calculated. Using the dimensions of the test specimen, the critical perimeter with shearhead is:

$$b_{o2} = 58.35 \text{ in} = 1482 \text{ mm} \qquad (10)$$

Using shearhead, the shear strength V_n should not be taken greater than $4\sqrt{f'_c}\, b_o d$ on the critical area defined in Figure 10.

From the above equation, the new shearhead system's shear resistance capacity is:

$$V_n = 4b_{o2}d\sqrt{f'_c} = 108.71 \text{ kips} = 492.57 \text{ KN} \qquad (11)$$

Where $f'_c = 4959.5$ psi $= 35$ MPa and $d = 6.6$ in $= 168$ mm.

4.2 Large column size without shearhead

The specimen may be considered as a plain column with larger cross-section dimensions of 480×480 mm, being made up of the dimensions of the tube (200×200 mm) and extended on both sides by the total length of the two shearheads in each direction ($2 \times 140 = 280$ mm).

4.2.1 EC 2

The basic control perimeter in EC 2 is normally taken to be at a distance 2d from the load area and should be constructed so as to minimize its length as shown in Figure 11. The control shear perimeter $u_{1,ec2}$ is:

$$u_{1,ec2} = 480 \times 4 + 2\pi \times (2 \times 168) = 4031 \text{ mm} \qquad (12)$$

The punching resistance capacity of the larger plain column head is:

$$V_{Rd}(u_1 d) = 0.581 \times 4031 \times 168 = 393.46 \text{ KN} \qquad (13)$$

4.2.2 BS 8110

The control perimeter is a rectangular shape with a distance of $1.5d$ from the boundaries of the load area as shown in fig. 11, giving:

$$u_{1,bs8110} = (480 + 3 \times 168) \times 4 = 3936 \text{ mm} \qquad (14)$$

Using this perimeter, the punching shear capacity of the test specimen is:

$$v_c u_{1,bs8110} d = 0.720 \times 3936 \times 168 = 476.10 \text{ KN} \qquad (15)$$

for $f_{cu} = 40$ N/mm², and

$$v'_c u_{1,bs8110} d = 0.748 \times 3936 \times 168 = 494.61 \text{ KN} \qquad (16)$$

for $f_{cu} = 45$ N/mm².

It is noted that the difference in the critical shear perimeter between a plain larger column head and a smaller column head with shear reinforcement is the extra distance of $0.5d$ on each side for the larger plain column head.

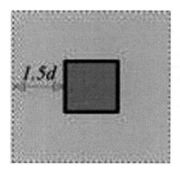

Figure 11. Control perimeter from BS 8110.

4.2.3 ACI 318

The control perimeter is a rectangular area with $0.5d$ from the edge of the load area without shearhead shown in Fig. 12, which gives:

$$b_{o1}' = 102.05 \text{ in} = 2592 \text{ mm} \qquad (17)$$

Then from ACI 318–11.12.2.1, the punching shear capacity is equal to:

$$V_c' = 4b_{o1}d\sqrt{f_c'} = 190.02 \text{ kips} = 860.98 \text{ KN} \qquad (18)$$

Figure 12. Critical area without shearhead from ACI.

4.3 Comparison

Table 2 compares the test result of ultimate punching shear load against results calculated using EC2, BS 8110 and ACI 318 for both a smaller column head with shearheads and a larger column head without shearhead. Furthermore, two results are given using BS 8110, for $f_{cu} = 40$ N/mm² (maximum value permitted in BS 8110) and $f_{cu} = 45$ N/mm² (actual measured value). Figure 13 compares the various calculated critical shear perimeters with the test observation, for both design assumptions.

The first impression is that although ACI 318 is the only code that explicitly considers shearheads, the ACI 318 values are the least accurate. Particularly, a comparison of the critical shear perimeter from ACI 318 calculation and the test observation indicates significant difference. This may be related to the difference between this test and the tests (Corley & Hawkins 1968) on which ACI318 was based. In this test, the shearhead was very short and only welded to the outside of the tubular column. In the tests for ACI318, the shearheads had long arms that went across the concrete column section. Nevertheless, since ACI318 predicts a very short critical perimeter and the final predicted punching shear resistance for the test specimen is much higher than the test

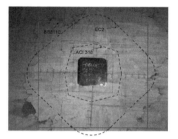

(a) Critical area for tube with shearheads

(b) Critical area for equivalent larger column without shearhead

Figure 13. Comparison of critical areas between test and code calculations.

Table 2. Comparison between calculation and test result.

Method	Perimeter (mm)	Stress (MPa)	Value (KN)	Difference (%)
Test	≈3500		416.77	
Smaller column head with shearheads				
EC 2	2927	0.581	285.70	−31.4
BS 8110 (40)	3264	0.720	394.81	−5.27
BS 8110 (45)	3264	0.748	410.44	−1.52
ACI 318	1482	1.978	492.57	18.2
Larger column head without shearhead				
EC 2	4031	0.581	393.46	−5.59
BS 8110 (40)	3936	0.720	476.10	14.2
BS 8110 (45)	3936	0.748	494.61	18.7
ACI 318	2592	1.977	860.98	106

result, this suggests that ACI318 may be inaccurate for predictions of both the critical perimeter and the concrete shear strength at the location of the critical shear perimeter.

Using BS 8110, the more accurate result is obtained by treating the test assembly as a column with shear reinforcement which has a zone length the same as the shearhead length. Furthermore, although the maximum permitted concrete stress in BS 8110 is 40 N/mm^2, using the higher measured concrete stress gives better prediction results. So the maximum permitted stress in BS 8110 may be relaxed.

Using EC2, treating the shearhead as a smaller column with equivalent shear reinforcement gives much lower strength than the test result. As shown in Figure 13(a), although EC2 predicted quiet well the parts of the critical shear perimeter at the ends of the shearheads, the parts of the critical shear perimeter linking them were wrongly predicted. If the test assembly was treated as a plain bigger column, the shape of the critical perimeter was slightly higher than observed as shown in figure 13(b), but the predicted punching shear resistance agrees quite well with the test result.

5 CONCLUSIONS

This paper has presented the test and design calculation results of a test on punching shear behaviour of a steel tube column with shearhead in flat slab construction. From the results, the following conclusions may be drawn:

1. Attaching shearheads to a tubular column is an effective method of enhancing punching shear resistance around the column head.
2. Although ACI 318 is the only code that explicitly deals with shearheads, the design calculation results using ACI318 are grossly inaccurate for the test.
3. Using BS 8110, the best correlation between code calculations and the test result was obtained by assuming the test specimen as a column with shear reinforcement that had a zone length equal to the length of the shearhead.
4. Using EC2, the best correlation between code calculations and test result was obtained by assuming the test specimen as a plain column with the overall dimension being equal to the tube dimension plus the total length of the two shearheads in the same direction.
5. Further tests and numerical simulations are being pursued to develop a better understanding so as to develop a more robust method for design.

ACKNOWLEDGEMENTS

The authors would like to thank technicians in the Structures Laboratory at the University of Manchester for their assistance in carrying out the test.

REFERENCES

N.K. Subedi & P.S. Baglin. 2003. Design of slab-column junction in flat slabs. *Proceedings of the Institution of Civil Engineers, Structures and Buildings*, 156(SB3): 319–331.

B. Binici1 & O. Bayrak. 2003. Punching Shear Strengthening of Reinforced Concrete Flat Plates Using Carbon Fiber Reinforced Polymers. *Journal of Structural Engineering, ASCE*, 129(9): 1173–1182.

K.K. Choi, M. M.R.T., H.G. Park & A.K. Maji. 2006. Punching shear strength of interior concrete slab–column connections reinforced with steel fibers. *Cement & Concrete Composites*, 29: 409–420.

K.H. Tan I & P. Paramasivam, 1994. Punching shear strength of steel fiber reinforced concrete slabs. *Journal of Materials in Civil Engineering*, 6(2): 240–253.

W.G. Corley & N.M. Hawkins. 1968. Shearhead reinforcement for slabs. *ACI Journal*, Vol. October 1968: 811–824.

U. Albrecht. 2002. Design of flat slabs for punching—European and North American Practices. *Cement & Concrete Composites*, 24: 531–538.

CEN 2002, Eurocode 2: Design of concrete structures—Part 1–1: General rules and rules for buildings. London: British Standards Institute.

BSI 1997, *British stand 8110: Structural use of concrete—Part 1: Code of practice for design and construction*. London: British Standards Institution.

American Concrete Institute 318 2005. *Building code requirements for structural concrete*. American Concrete Institute.

Special session: CIDECT President's Student Awards Research Award Competition

Cast steel connectors for seismic-resistant tubular bracing applications

J.C. de Oliveira
Cast ConneX Corporation, Toronto, Canada

ABSTRACT: While concentrically braced frames are among the most popular lateral load resisting systems in North America for medium- to low-rise steel structures, many such frames featuring Hollow Structural Section (HSS) brace members have been shown to be prone to unexpected connection failure in the event of an earthquake. This paper presents the use of a cast steel connector as an improvement over the reinforced, fabricated HSS bracing connections that are commonly used in seismic load resisting braced frames. The cast connector developed was shaped using solid modeling software, verified by finite element analysis, and cast to ASTM A958 standards. The connector concept was subsequently validated though cyclic inelastic and monotonic tensile testing of concentrically loaded HSS brace-connector assemblies. Since the completion of this research, this technology has been commercialized and may herald a new era of connection design for seismic-resistant HSS bracing.

1 INTRODUCTION

In North America, concentrically braced frames are amongst the most common lateral load resisting systems for medium- to low-rise steel structures. This is due to design and erection simplicity and the increased stiffness the system provides in comparison to other lateral systems. Hollow structural sections (HSS) are frequently specified as the bracing elements in braced frames due to their efficiency in carrying compressive loads, their improved aesthetic appearance, and because of the wide range of section sizes that are readily available.

In concentrically braced frames, the diagonal bracing element is subjected to predominantly axial loading. Statically loaded HSS braces are typically connected to the beam, column, or beam-column intersection via slotted HSS-to-gusset connections. Design of these connections for static loading accounting for shear lag effects has been studied by a number of researchers (British Steel 1992, Korol et al. 1994, Zhao and Hancock 1995, Cheng et al. 1996, Zhao et al. 1999, Wilkinson et al. 2002, Ling 2005, Willibald et al. 2006, Martinez-Saucedo 2006); these connections can be readily detailed according to prevailing design standards. However, in the event of an earthquake, the diagonal brace members of concentrically braced frames cyclically yield in tension and buckle in compression, thereby dissipating seismic energy. For this reason, brace connections in braced frames designed to be seismically resistant must be stronger than the probable tensile yield capacity of the connected brace member including any expected material overstrength.

Both in the laboratory and in the field as witnessed during post-earthquake reconnaissance (Tremblay et al. 1996, AIJ 1995, Bonneville and Bartoletti 1996), conventional slotted HSS-to-gusset connections have been shown to fail when subjected to cyclic inelastic loading. These premature failures, which may lead to structural collapse, occur due to a concentration of inelastic strain at the reduced section of the HSS connection. Thus, current North American seismic design provisions recommend the use of net-section reinforcement whenever slotted HSS-to-gusset connections are specified in seismic-resistant frames. As reinforcement is more readily fabricated for flat surfaces, the industry has tended toward the specification of square or rectangular HSS rather than circular HSS for seismic resistant braces. This is an unfortunate circumstance as the energy absorbing capability of North American cold-formed rectangular HSS under cyclic inelastic loading is now thought to be less than that implied by the current design provisions (Lee and Goel 1987, Shaback and Brown 2003). Further complicating seismic-resistant HSS connection design, there is wide variation and much dispute over the overstrength value for HSS steel grades in current American and Canadian steel design standards. To ensure HSS remain the industry choice for seismic resistant bracing, a connection detail for circular hollow section (CHS) braces that can withstand inelastic cyclic loading of the brace is required.

2 CAST STEEL SEISMIC BRACE CONNECTOR

In an effort to address the seismic brace connection dilemma, the author, under the supervision of Professors Jeffrey A. Packer and Constantin Christopoulos of the University of Toronto, developed a component to eliminate shear-lag in CHS-to-gusset connections that could also withstand cyclic inelastic loading of the CHS member. The component was designed to be shop welded to the ends of a CHS segment, with the connector-CHS assembly to be subsequently field-bolted to gusset plates in a steel braced frame as shown in Figure 1.

In this configuration, out-of-plane buckling of the brace during a compressive cycle of a seismic event is accommodated through the formation of plastic hinges in the gusset plates and at the midspan of the brace. As the brace is expected to yield in tension during an earthquake, a complete joint penetration (CJP) weld is required between the ends the CHS segment and the connectors.

Due to the geometric complexity of the connector, the components were designed to be produced using steel casting manufacturing. As casting manufacturing is predisposed to mass production, the research team elected to standardize the cast steel connectors, which was achieved as follows. The CHS connecting end of the cast connector was designed to accommodate any CHS of a given outer diameter, regardless of the section's wall thickness (Fig. 2). Since the CHS is welded to the connector using a CJP weld, any grade of CHS brace can be used, provided that the appropriate weld electrode is selected.

The gusset-receiving end of the connector was designed to accommodate a slip-critical connection for a load equal to the expected yield capacity of the heaviest-walled CHS section that could be accommodated by the connector, including material overstrength. This feature provides the option to design the connection using slip-critical or bearing type bolted connections for any size of CHS, with the number of bolts specified by the designer commensurate with the expected strength of the connected CHS member. This design flexibility also allows the designer to use whichever overstrength value is specified by the governing design code without having to completely re-design the connection (North American codified values for CHS material overstrength currently range from 1.1 to 1.6 depending on the grade and seismic design code). This presents a significant advantage over traditional reinforced HSS-to-gusset connections

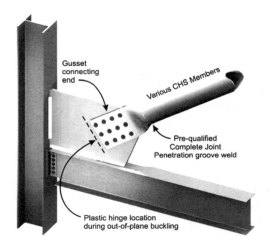

Figure 1. Seismic-resistant cast steel connector for circular hollow section bracing members shown in building frame.

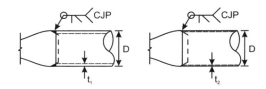

Figure 2. Connector accommodates various circular hollow section members of a given outer diameter, regardless of wall thickness and steel grade using the same complete join penetration (CJP) groove weld detail.

which entail detailing that is more sensitive to the grade and size of the connected HSS brace member. Further, the ability to shop-weld the CHS segment to the connector and field-bolt the connector-brace assembly to the frame eliminates the need for field welding of seismic-critical welds. Finally, the double-shear bolted connection on the gusset receiving end of the connector halves the number of bolts that would otherwise be required to bolt a shop-welded brace to the frame using splice plates. In sum, use of the standardized connector significantly simplifies connection design, detailing, fabrication, and erection of seismic-resistant HSS brace connections.

3 CAST CONNECTOR PROTOTYPE DESIGN

To validate the seismic-resistant connector concept through laboratory experimentation, a connector having the features presented above was required. A prototype connector was designed to accommodate

CHS having an outer diameter of 168 mm. This size was selected for several reasons. First, CHS of this diameter is readily available with a wide range of wall thicknesses from most North America steel tube manufacturers. Further, the nominal radius of gyration for most of the available 168 mm diameter tubes provides slenderness ratios that are below 200 at typical brace member lengths (a requirement for tension-compression braces). Finally, the yield capacity of 168 mm diameter CHS ranges from approximately 550 to 3,000 kN, depending on the wall thickness and steel grade selected. This wide range of yield forces provides the designer with the ability to select the appropriate level of lateral strength for each storey of a medium-rise structure while specifying the same cast connector throughout the building.

For the purpose of standardization, the gusset receiving end of the prototype connector was designed to resist the highest probable yield strength of the thickest walled 168 mm CHS brace members that are typically available in North America: HSS 168 × 13 CAN/CSA-G40.20/G40.21 (CSA 2004) Grade 350W and HSS 6.625 × 0.500 ASTM A500 (ASTM 2003) Grade C. This was achieved using 12 1-inch diameter ASTM A490 (ASTM 2006) bolts for connection to a 30 mm gusset plate. The 12 pretensioned high-strength bolts provide sufficient slip resistance (assuming a blast cleaned faying surface) to carry the probable yield strength of the largest available 168 mm CHS. Although the use of pretensioned bolts is required by North American seismic design codes, the use of slip-critical connections is not. However, increasing the number of bolts beyond the number that would be required for a bearing-type connection ensures that the connector will remain virtually fully elastic in the bolt region during a seismic event. This results in a more robust connection and ensures that all inelastic yielding is confined to the CHS segment.

Design of the prototype cast steel connector was carried out using 3-dimensional solid modeling software with consideration for the flow of force though the connector as well as the limitations of casting manufacturing. For sand casting, the steel casting process used to produce the prototype connectors, transitional geometry must be kept as smooth as possible to ensure quality casting. Further, it is beneficial for cast components to be shaped to promote directional solidification, thereby reducing the need for risering, chilling, or other costly casting considerations. Finally, components cast using the sand casting process must taper slightly from the parting line between the moulds into which the molten steel is poured. The foundry that was retained to manufacture the prototype connectors suggested minor design changes which improved the final component's

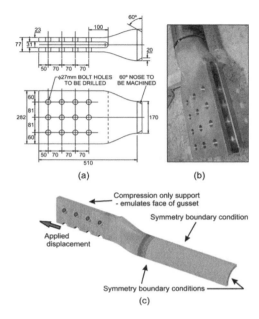

Figure 3. Prototype cast connector: (a) dimensional drawing, (b) connector tack-welded to CHS, (c) finite element model of prototype and CHS segment.

castability. Figure 3a shows the prototype's final dimensions while Figure 3b shows a photograph of a connector tack-welded to a CHS segment.

In typical concentrically braced frames, the brace member itself is the energy-absorbing element. Thus, according to the principles of capacity design, the cast connector must remain elastic during tensile yielding or compressive buckling of the brace. The elastic behaviour of the connector was established using finite element stress analysis prior to manufacturing.

For the purposes of finite element stress analysis, a 3-dimensional solid model of the connector was produced which also included a complete joint penetration groove weld between a 500 mm long (3 diameters) HSS 168 × 13 brace member of nominal diameter and thickness. Because of symmetry, finite element modeling of only a quarter of the assembly was required, as shown in Figure 3c. Finite element analysis was carried out in ANSYS (SAS 2004). Solid bodies were meshed using elements defined by 10 nodes and having three degrees of freedom at each node. As discussed by Moaveni (1999), higher-order elements like these are best suited for modeling solid bodies that are curved or have irregular boundaries.

Finite element stress analysis showed that when a tensile or compressive deformation was applied to the connector that caused a brace force corresponding to the design yield force, stresses in the casting were below the specified minimum yield stress of the cast material selected for the component. The stress distribution in the tubular brace was uniform a short distance from the welded connection, showing none of the shear lag effects associated with slotted tube-to-gusset connections. Validation of the finite element model used for the design of the connector was carried out and is presented below.

4 PROTOTYPE MANUFACTURING AND WELDING CONSIDERATIONS

The prototype cast steel connectors produced for this study were manufactured with steel produced to ASTM A958 Grade SC8620 Class 80/50 (ASTM 2000). This cast material has a chemical composition similar to that of a standard wrought steel grade and is considered a weldable base metal according to CSA W59 (2003). Material produced to this specification has a minimum yield stress of 345 MPa, a minimum ultimate tensile strength of 550 MPa, a minimum elongation of 22%, and a reduction in area of 35% in 50 mm. An additional Charpy V-Notch impact test value requirement of 27 Joules at −20 °C was specified to ensure the connection had a suitable toughness at the weld region between the connector and the brace.

To ensure that the castings were sound they were subjected to visual examination, non-destructive examination, and ultrasonic examination by the foundry. Coupons cut directly from a cast connector in two directions revealed a yield stress of 565 MPa, an ultimate tensile strength of 695 MPa, and an isotropic material response.

A professional welding engineer determined an appropriate welding procedure specification for the tube-to-casting complete joint penetration groove weld. Pre-qualification for this CJP welding procedure was authorized by the Canadian Welding Bureau after trial welds performed to the procedure by a certified welder were destructively examined by the Bureau. Figure 4 shows a chemically etched (10% Nital) section cut through the CJP groove weld between a cast connector and a HSS 168 × 9.5 member. This section clearly shows that the weld fully engages the entire cross-section of the CHS.

5 EXPERIMENTAL PROGRAM

Laboratory validation of the connector concept consisted of both cyclic inelastic and monotonic tensile testing of brace-connector assemblies. Cyclic testing was carried out on two brace-connector assemblies fabricated with different circular hollow sections that had been certified to both ASTM A500 Grade C and Grade B: HSS 168 × 6.4 and HSS 168 × 9.5. Testing was carried out in a 2750 kN MTS testing frame; size and stroke limitations of the machine required that both test specimens be made up of two cast steel connectors welded to the ends of a 1.4-metre long CHS segment. Welding between the CHS and cast connectors was carried out according to the pre-qualified CJP welding procedure previously described. The connectors were bolted with 12 1-inch diameter ASTM A490 high strength pre-tensioned bolts to 32 mm gusset plates that were shaped to fit the testing frame's hydraulic grips.

As the brace members tested were rather stocky (nominal slenderness, KL/r, in the range of 45) in comparison to typical brace lengths for concentrically braced frames, a cyclic loading protocol was developed which ensured that several inelastic cycles could be applied to the brace-connector assemblies prior to inducing fracture in the CHS segment. This was achieved by applying a non-symmetric protocol that cycled between tensile yielding of the brace-assembly and a compressive displacement associated with the onset of buckling. Once the compressive force being transmitted through the brace began to decrease, loading was reversed, thereby limiting the damage imparted to the CHS brace while maximizing the load transferred through the connections. After the pattern had been repeated over several cycles of inelastic loading, a large compressive displacement was applied to the assembly. This axial deformation caused a very large out-of-plane kink to form in the brace, proving that the brace connection was capable of transmitting large compressive forces when the brace was in its buckled configuration. This deformation also resulted in the onset of local buckling at the plastic hinge which

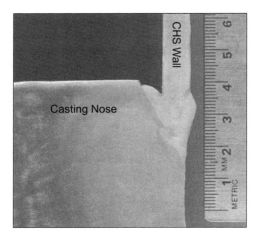

Figure 4. Section through a complete joint penetration groove weld between a prototype cast connector and a CHS member.

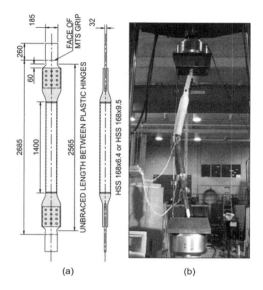

Figure 5. (a) Cyclic test specimen details, (b) specimen experiencing inelastic buckling during compression loading cycle.

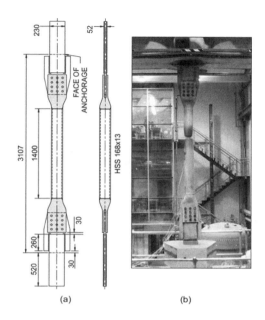

Figure 6. (a) Tension test specimen details, (b) speciment after fracture.

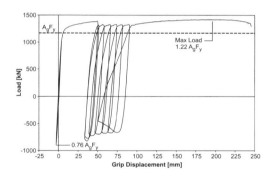

Figure 7. Cyclic load-displacement plot for HSS 168×6.4 assembly [$F_y = 382$ MPa as determined by the average of several tensile coupon tests and $A_g = 3063$ mm^2 determined by measurement of tube geometric properties].

formed at the midspan of the brace. The protocol was then completed by applying increasing tensile deformation until the brace member fractured. Because the braces tested in this manner were stocky, the compressive forces attained in the brace-connector assembly were much higher than those that would be expected in brace members of typical length. Thus the protocol that was applied was conservative with regard to the required compressive resistance of the connection. Figure 5 shows the brace-connector assembly details and a specimen experiencing inelastic buckling during a compressive cycle of the cyclic testing procedure.

After cyclic testing, two cast connectors were cut from the ends of one of the two brace-connector assemblies that had been tested. After machining the connector noses back to their original shape, the two connectors were welded to a 1.4-metre long HSS 168×13 tubular section, which had also been dual-certified to ASTM A500 Grades C and B, for tensile testing in a 5350 kN capacity testing machine. Figure 6 shows the specimen details and the fractured test specimen after completion of the monotonic tensile test.

6 TEST RESULTS

Figures 7 and 8 show the load-displacement plots produced during the cyclic testing of the HSS 168×6.4 and HSS 168×9.5 brace-connector assemblies, respectively. As is evident from the figures, the connectors were able to transmit tensile loads in excess of the measured CHS yield capacity in both of the tested specimens. Note that the peak tensile loads induced in the braces are significantly higher than those that would have been required by both the Canadian and American seismic design standards, particularly if the designer had specified ASTM A500 Grade B material but had received the dual-certified material used for these tests. A connection detailed using traditional slotted HSS-to-gusset connections and designed to carry the codified expected force for the ASTM A500 Grade B material would very likely have fractured if it were subject to the same test protocol.

Further, the tested connectors were capable of transmitting compressive loads in the presence of large out-of-plane brace deformations. Bolt slip was observed only once for each connector during the testing of the HSS 168 × 9.5 as the slip load was never exceeded in compression or in the testing of the HSS 168 × 6.4 brace assembly. After several high amplitude tension-compression cycles, the brace-assemblies exhibited plastic hinging at their gusset plates and local buckling at the tube section's midspan. The final failure mode in both tests was fracture at the brace's midspan after necking of the tube's cross-section had become clearly visible. Strain gauge readings and whitewash applied to the connectors showed that the connectors remained elastic during both tests.

Figure 9 shows the load-displacement plot produced during tensile testing of the HSS 168 × 13 brace-connector assembly. During this test, tensile displacement was monotonically increased beyond the peak load of 2970 kN up to rupture of the specimen, which occurred in the CHS segment a short distance above the brace midpoint after significant tube necking was observed. Readings from strain gauges showed that the connectors remained elastic in the gauged regions during the test. As with the cyclically tested specimens, whitewash applied to the connectors did not flake during any stage of the testing whereas whitewash applied to the CHS segment flaked, particularly in the necked region of the tube.

7 FINITE ELEMENT VALIDATION

Strains and deformations measured during laboratory testing were compared to the results of finite element analysis for the purpose of validating the numerical models used for the design of the connector. Unlike the initial analysis which used assumed material properties and nominal tube dimensions, finite element analysis performed for the purpose of validation was carried out using material properties determined from tensile testing of coupons cut from the brace members as well as the measured tube dimensions. Otherwise, all other finite element modeling assumptions were the same as those previously described. Figure 10 shows experimental surface strains in the connector compared to strains predicted by the finite element model at several load levels measured during testing of the HSS 168 × 9.5 assembly.

As seen in the figure below, the strain distribution predicted by finite element analysis across the width of the cast connector is consistent with the measured distribution. The correlation between laboratory

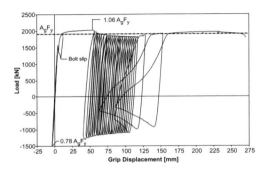

Figure 8. Cyclic load-displacement plot for HSS 168 × 9.5 assembly [F_y = 434 MPa as determined by the average of several tensile coupon tests and A_g = 4404 mm^2 determined by measurement of tube geometric properties].

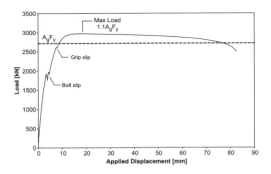

Figure 9. Static tensile load-displacement plot for HSS 168 × 13 assembly [F_y = 475 MPa as determined by the average of several tensile coupon tests and A_g = 5717 mm^2 determined by measurement of tube geometric properties].

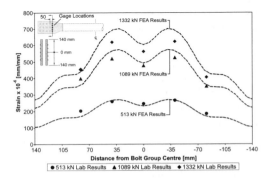

Figure 10. Measured surface strains across the cast steel connector and finite element predictions at given brace-assembly loads for the HSS 168 × 9.5 brace-assembly.

measurements and the finite element results serves as validation of the finite element modeling conducted for the purpose of designing the connector.

8 FULL-SCALE VALIDATION TESTING

Subsequent to the initial short-brace testing, a full-scale connector-brace assembly was tested in a 12,000 kN MTS testing frame. The brace assembly, fabricated using HSS 168 × 13 produced to ASTM

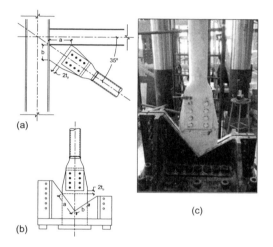

Figure 11. (a) Brace connection detail at beam-column intersection, (b) laboratory setup producing similar boundary conditions as those in the field, (c) laboratory test setup.

Figure 12. Cyclic inelastic testing of a full-scale HSS 168 × 13 brace assembly.

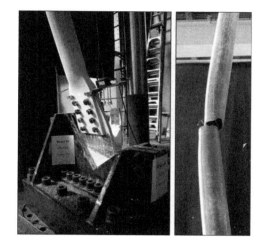

Figure 13. Connector remained elastic during cyclic inelastic loading of brace-assembly; plastic hinge forms in gusset plate (left), final brace-connector assembly failure occurs in the CHS brace after the onset of local buckling in the brace (right).

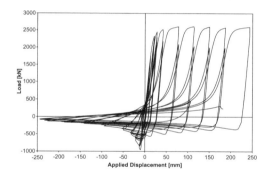

Figure 14. Cyclic load-displacement response of a full-scale (6.120 m) HSS 168 × 13 assembly.

A500 grade C, had an unbraced length of 6.120 metres, a nominal slenderness of approximately 110 and a nominal diameter-to-thickness ratio of 14.72. The setup for the full-scale test included boundary conditions that more closely simulated the gusset configuration and out-of-plane stiffness that would be present in the field, as shown in Figure 11.

Cyclic inelastic testing was carried out according to the symmetric testing protocol outlined in Appendix T of the AISC Seismic Provisions (AISC 2005). The brace performed well (Figs. 12–14) with the cast connectors easily outlasting the inelastic life of the CHS brace. The final fracture occurred at the midspan of the brace after the onset of local buckling.

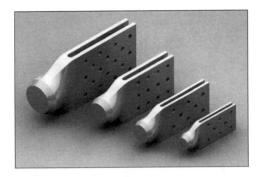

Figure 15. Cast ConneX™ High-Strength Connectors.

9 CONCLUSIONS

This paper presents a novel method of connecting HSS bracing members in seismic-resistant concentrically braced frames. Laboratory results showed that the brace assemblies fabricated using cast steel connectors performed very well when subjected to both cyclic and tensile inelastic loading. Further laboratory validation of the connector concept is currently underway and is focused on confirming that the excellent performance of the connectors is also achieved in braces of typical lengths, under a variety of loading protocols, and with testing criteria that more accurately reproduces the boundary conditions expected in the field with respect to member end rotations. A more detailed presentation of the development and testing of the connector concept is provided in de Oliveira et al. (2008).

Since the completion of this research, the seismic-resistant connector technology has been licensed to Cast ConneX Corporation, which distributes a line of seismic-resistant HSS connectors for use in braced frames across North America (Fig. 15). These innovative connectors may herald a new era of pre-engineered connection design and fabrication for arduously loaded HSS members.

ACKNOWLEDGEMENTS

The author gratefully acknowledges the guidance of Professors Jeffrey A. Packer and Constantin Christopoulos of the University of Toronto. Financial support for this research was provided by the Natural Science and Engineering Research Council of Canada (NSERC) and by the Ontario Centres of Excellence for Materials and Manufacturing. Atlas Tube Inc. provided steel tube and Walters Inc. (Ontario) generously donated their fabrication services.

REFERENCES

AISC. 2005. Seismic provisions for structural steel buildings, *ANSI/AISC 341-05 and ANSI/AISC 341s1-05*. Chicago: American Institute of Steel Construction.

Architectural Institute of Japan (AIJ). 1995. Reconnaissance report on damage to steel building structures observed from the 1995 Hyogoken-Nanbu (Hanshin/Awaji) earthquake. Tokyo: AIJ, Steel Committee of Kinki Branch.

ASTM. 2000. Standard Specification for Steel Castings, Carbon, and Alloy, with Tensile Requirements, Chemical Requirements Similar to Standard Wrought Grades. *ASTM A958-00*. West Conshohocken: ASTM International.

ASTM. 2003. Standard specification for cold-formed welded and seamless carbon steel structural tubing in rounds and shapes. *ASTM-A500-03a*. West Conshohocken: ASTM International.

ASTM. 2006. Standard specification for structural bolts, alloy steel, heat treated, 150 ksi minimum tensile strength. *ASTM-A490-06*. West Conshohocken: ASTM International.

Bonneville, D. & Bartoletti, S. 1996. Case Study 2.3: Concentric Braced Frame, Lankershim Boulevard, North Hollywood. 1994 Northridge Earthquake; Building Case Studies Project; Proposition 122: Product 3.2, SSC 94-06, Seismic Safety Commission State of California: 305–324.

British Steel. 1992. Slotted end plate connections. Report No. SL/HED/TN/22/-/92/D. Rotherham: Swinden Laboratories.

CSA. 2003. Welded Steel Construction (Metal Arc Welding). *CAN/CSA-W59-03*. Toronto: Canadian Standards Association.

CSA. 2004. General requirements for rolled or welded structural quality steel/structural quality steel. *CAN/CSA-G40.20-04/G40.21-04*. Toronto: Canadian Standards Association.

Cheng, J.J.R., Kulak, G.L., & Khoo, H. 1996. Shear lag effect in slotted tubular tension members. *Proc.1st CSCE Structural Specialty Conf.*: 1103–1114.

de Oliveira, J.C., Packer, J.A., & Christopoulos, C. 2008. Cast Steel Connectors for Circular Hollow Section Braces Under Inelastic Cyclic Loading. *Journal of Structural Engineering*. 134(3): 374–383.

Korol, R.M., Mirza, F.A., & Mirza, M.Y. 1994. Investigation of shear-lag in slotted HSS tension members. *Proc. 6th Intern. Symp. on Tubular Structures*: 473–482.

Lee, S., & Goel, S.C. 1987. Seismic Behavior of Hollow and Concrete Filled Square Tubular Bracing Members. UMCE87-11. Ann Arbor: University of Michigan.

Ling, T.W. 2005. The tensile behaviour of gusset-plate welded connections in very high strength (VHS) tubes. PhD Thesis, Dept. of Civil Engrg, Monash University, Melbourne.

Martinez-Saucedo, G. 2006. Slotted end connections to hollow sections. PhD Thesis, Dept. of Civil Engrg, University of Toronto, Toronto.

Moaveni, S. 1999. Finite Element Analysis: Theory and application with ANSYS. New Jersey: Prentice Hall.

Shaback, B., & Brown, T. 2003. Behavior of square hollow structural steel braces with end connections under reversed cyclic axial loading. *Canadian Journal of Civil Engineering*. 30: 745–753.

Swanson Analysis System (SAS). 2004. ANSYS Workbench release 9.0. Houston: Swanson Analysis System.

Tremblay, R., Bruneau, M., Nakashima, M., Prion, H.G.L., Filiatrault, A. & Devall, R. 1996. Seismic design of steel buildings: Lessons from the 1995 Hyogoken-Nanbu earthquake. *Canadian Journal of Civil Engineering*. 23: 727–756.

Wilkinson, T., Petrovski, T., Bechara, E., & Rubal, M. 2002. Experimental investigation of slot lengths in RHS bracing members. *Proc., 3rd Int. Conf. on Advances in Steel Structures*. New York: Elsevier Science and Technology: 205–212.

Willibald, S., Packer, J.A., & Martinez-Saucedo, G. 2006. Behaviour of gusset plate to round and elliptical hollow structural section end connections. *Canadian Journal of Civil Engineering*. 33: 373–383.

Zhao, X.L., Al-Mahaidi, R., & Kiew, K.P. 1999. Longitudinal fillet welds in thin-walled C450 RHS members. *Journal of Structural Engineering*. 125(8): 821–828.

Zhao, X.L., & Hancock, G.J. 1995. Longitudinal fillet welds in thin cold-formed RHS members. *Journal of Structural Engineering*. 121(11): 1683–1690.

High strength concrete-filled tubular steel columns in fire

O. Bahr
Institute for Steel Construction, Leibniz Universität Hannover, Hannover, Germany

ABSTRACT: Concrete-filling of hollow structural section steel columns is an effective means to increase the fire resistance significantly. In recent years, high strength concrete has often been favoured over normal strength concrete to achieve increased load-bearing capacity of the composite column. Nevertheless, high strength concrete has performance problems when exposed to fire due to faster degradation of properties. This paper addresses some important questions relating to constitutive laws of high strength concrete as well as its reasonable use. Investigations are based on both North American and European high temperature material properties. Effect of different types of concrete fillings on the fire resistance of the column was considered. Investigations included both plain and reinforced concrete-filling (bar or steel fiber reinforced). Moreover, design recommendations regarding its use are given.

1 INTRODUCTION

The ongoing urbanisation forces the construction of high-rise buildings for working and living. Columns used in high-rise buildings have to bear loads from many storeys. Moreover, high fire safety requirements have to be met. Thus, high-rise buildings are an ideal field of application for composite columns.

The use of hollow structural section (HSS) steel columns filled with high strength concrete (HSC) offers many advantages. The composite system can carry far higher loads than conventional composite columns. Apart from the mechanical advantages, the usable building space and therefore the cost-effectiveness of the project is increased. In addition, slender structures are architecturally appealing. In the aftermath of the World Trade Center disaster, the robustness of composite solutions with their inherent fire resistance is another advantage over traditional fire protection systems such as plaster board, intumescent coating or spray applied fire-proofing, which can quite easily be removed or damaged.

These advantages have led to the increased use of concrete-filled HSS steel columns in some of the recent tall buildings in Germany. The skyscraper "HighLight Munich Business Towers" (see Figure 1) is an example for the application of this type of technology. HSC is an innovative material. Hence two different descriptions for the material properties were compared to each other at first. These were the North American provisions according to the standard CAN/CSA-S16–01 and to the work of the researchers Kodur & Sultan (2003) as well as Cheng, Kodur & Wang (2004). In addition, the European material properties according to the European codes EN 1992-1-2 for concrete structures as well as EN 1994-1-2 for composite structures were considered. In the following, they are referred to as Eurocode 2 and Eurocode 4, respectively.

The regarded material models were implemented in the Finite Element program "BoFIRE". A detailed description of this program can be found in Schaumann, Bahr & Kodur (2006a). Using BoFIRE numerical studies were carried out concerning the effect of

Figure 1. HighLight Towers (Gooseman, 2008).

varying material properties on the heating of HSC-filled steel columns. The influence of different HSC-fillings was studied by comparing results of numerical studies with test results. These included plain HSC-fillings as well as HSC reinforced with bars or steel fibers.

Finally, parametric studies were performed to establish some design recommendations for the reasonable use of HSC.

2 MATERIAL PROPERTIES OF HSC

The numerical studies were carried out using both North American (NA) and European (EU) HSC material properties. The considered codes showed differences regarding the high temperature mechanical as well as the thermal properties. The following comparison includes the material properties of HSC that are relevant for structural analysis of fire-exposed members. These are the specific heat capacity, conductivity, expansion and mass loss (thermal properties) as well as the stress-strain relationship at elevated temperatures (mechanical property).

2.1 Thermal material properties

2.1.1 Specific heat capacity

Figure 2 shows the variation of the specific heat capacity as a function of temperature, as obtained from North American recommendation and European codes. According to Eurocode 2 thermal properties of normal weight concrete may be applied for HSC too. In contrast to this, the North American recommendations differ between concrete with calcareous and siliceous aggregates. Accordingly, the initially assumed specific heat capacity for siliceous aggregates is below the corresponding European values. However, the North American formula results in a peak value at 500 °C for concrete with siliceous aggregates. This is due to the presence of quartz that transforms at these temperatures Kodur & Sultan (2003). For concrete with calcareous aggregates two peaks occur. The small peak near 400 °C is caused by the removal of crystal water from the cement paste. The higher peak at temperatures of about 700 °C results from the dissociation of dolomite. At temperatures exceeding 800 °C similar values are obtained from the different formulas.

In general, the differences can be explained by the fact that the North American method takes physicochemical processes into account. In contrast to this, the European approaches neglect such phenomena. However, Eurocode 2 allows considering the beneficial effect of increased moisture content on the specific heat capacity in the temperature range from 115 °C to 200 °C, which is not presented here. According to Eurocode 4 a moisture content of 10% may occur for HSS filled with concrete.

2.1.2 Thermal conductivity

The different approaches for thermal conductivity can be seen in Figure 3. In Eurocode 2 thermal conductivity is defined by lower and upper limits. Higher values of thermal conductivity result from the North American provisions, whereas the minimum is given by the lower limit of Eurocode 2.

2.1.3 Variation of thermal elongation

The computed values for the variation of the thermal elongation $\varepsilon = \Delta l/l$ are summarized in Figure 4, where the influence of the type of aggregate becomes clear. Because of the transformation of quartz in the siliceous aggregates, thermal elongation increases for temperatures about 450 °C. For temperatures exceeding 650 °C thermal expansion remains constant. Thermal

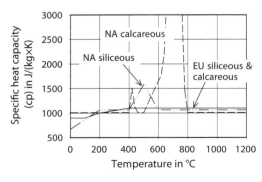

Figure 2. Comparison between North American (NA) and European (EU) material properties for specific heat capacity.

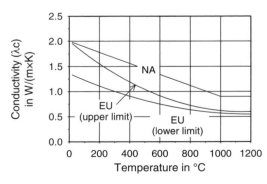

Figure 3. Comparison between North American (NA) and European (EU) material properties for thermal conductivity.

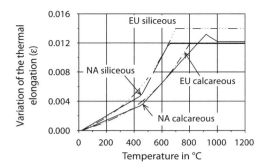

Figure 4. Comparison between North American (NA) and European (EU) material properties for the variation of the thermal elongation.

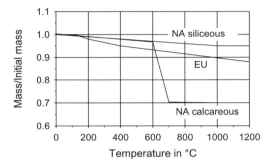

Figure 5. North American (NA) and European (EU) material properties for mass loss.

elongation for concrete with calcareous aggregates is considerably lower for temperatures less than 800 °C.

2.1.4 Mass loss

The results for the mass loss are presented in Figure 5. According to the North American material properties, the overall mass loss for concrete with siliceous aggregates only amounts to about 5%. Nevertheless, for temperatures exceeding 700 °C the mass of concrete with siliceous aggregates is sharply reduced to 70% of its initial mass. The loss is caused by the dissociation of dolomite in concrete. In contrast to this, no distinction with respect to mass loss is made in Eurocode 2 between concrete with siliceous and calcareous aggregates.

2.2 Mechanical material properties

The stress-strain relationship is of fundamental importance for accurate prediction of the load-bearing behaviour. Thus, resulting stress-strain curves according to EU and NA provisions were compared to each other for the case of HSC with siliceous aggregates and 60 MPa strength.

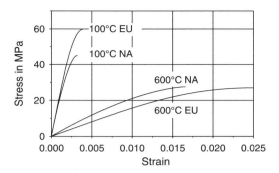

Figure 6. Comparison between North American (NA) and European (EU) stress-strain relationship at elevated temperatures.

To show the characteristics, the ascending branch of the stress-strain relationships is presented only for 100 °C and 600 °C in Figure 6. It is obvious that the North American stress-strain relations are more conservative since the peak stress is significantly reduced for temperature of 100 °C. In contrast to this, Eurocode 2 does not diminish the peak stress for the same temperature. This is also true for a temperature of 200 °C, which is not presented in Figure 6. For higher temperature of 600 °C, the peak stress according to both codes is almost equal. However, the North American code assumes a far more brittle HSC behavior since the strain at peak stress is less than the corresponding European value. The differences for other temperatures are less pronounced and thus not presented. Mathematical formulation of the North American material properties are presented in detail by Kodur & Sultan (2003) as well as by Schaumann, Bahr & Kodur (2006b).

3 EFFECT OF DIFFERENT MATERIAL PROPERTIES

3.1 Effect of different thermal properties

Diverging material properties were previously discussed at the material level. For illustrating the effect on the heating, a thermal analysis on a typical cross-section of a composite column was carried out. For this purpose both material properties were implemented in the computer program BoFIRE. The columns were then analyzed using first the North American and afterwards the European material properties.

A circular HSS steel column with outer diameter of 355.6 mm and wall thickness of 16 mm was selected for the analysis. HSC-filling with siliceous aggregates and 60 MPa strength was chosen (see Figure 7). The cross-section is exposed to ISO standard fire from all sides. Concrete humidity was not considered. For

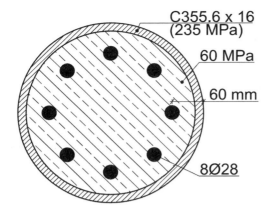

Figure 7. Cross-section of the examined composite column.

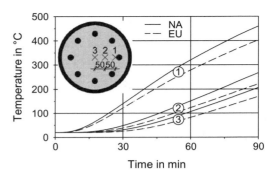

Figure 9. Comparison of cross-sectional temperatures.

The resulting temperature fields after 90 minutes exposure to ISO standard fire are presented in Figure 8. From the qualitative comparison it can be noticed that the North American material properties lead to faster temperature rise in the cross-section than the European material models.

Moreover, the development of temperatures was compared for distinct cross-sectional points, as indicated in Figure 9. It is obvious that the application of the North American thermal material properties leads to higher cross-sectional temperatures than using the European material formulations. This is valid for the whole heating range leading to differences in temperature between 37 °C (point 3) and 57 °C (point 1) after 90 minutes of heating. This difference of course influences the load-bearing behaviour of fire-exposed members, which is later-on analyzed in section 4.

3.2 *Effect of different types of HSC-filling*

A review of literature indicates that there is very limited fire test data on HSC-filled HSS columns exposed to fire (Kodur, 1998). The main objective of the reported tests was to determine the fire resistance of HSS steel columns filled with different types of HSC. The tests, which have been reported on HSS columns, include plain HSC-filling (columns C-46 and C-47), HSC-filling with steel fibers (columns C-36 and SQ-11) and HSC-filling with 4 × 15 mm reinforcement bars (column SQ-14). Investigated cross-section types are presented in Figure 10. Full details of the fire tests can be found in Kodur (1998).

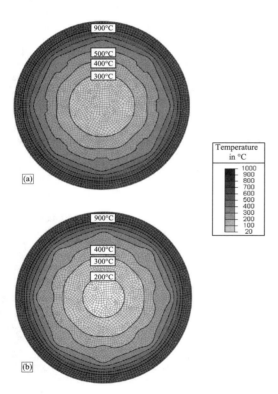

Figure 8. Heated cross-section after 90 minutes of ISO standard fire using the North American (a) and European material properties (b).

the European code provisions, the upper limit of the thermal conductivity was chosen, which is in accordance with Eurocode 4. For the density of concrete, a value of $\rho = 2{,}300$ kg/m³ was assumed.

The tested HSS columns were 3.81 m long with fixed end conditions (see Figure 10) and filled with calcareous aggregate (limestone) HSC. After the load was applied, the columns were exposed to heating according to standard fire ULC-S101-M89 until failure occurred. This fire exposure is equivalent to ASTM E-119 or ISO-834 standard fire. The fire resistance period of some of the tested columns (selected from literature) is calculated with the computer program BoFIRE.

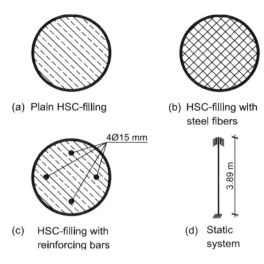

(a) Plain HSC-filling
(b) HSC-filling with steel fibers
(c) HSC-filling with reinforcing bars
(d) Static system

Figure 10. Cross-section and static system of circular HSC-filled HSS columns.

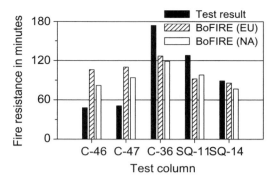

Figure 11. Comparison between tested and calculated fire resistance.

This allows verification of material properties given by the North American and European codes. Moreover, it facilitates to study the influence of concrete filling on fire resistance. The material properties of steel fiber reinforced HSC-filling are taken according to the published data by Kodur & Sultan (2003). The test data of the different HSS columns, taken from Kodur (1998), is summarized in Table 1, where 'C' in the column designation denotes circular cross-section and 'SQ' denotes square cross-section.

The tested and calculated fire resistance period of the different HSS columns is presented in Figure 11. A moisture content of 10% was considered in the computer program BoFIRE, which may occur for HSS filled with concrete according to Eurocode 4.

Regarding the numerical results for HSS columns filled with plain HSC it is obvious that the fire resistance is overestimated to a great extent. This is true for both the North American and European HSC material properties. With respect to the circular HSS columns C-46 and C-47 the gap between tested and calculated fire resistance increases for higher concrete compressive strength. However, the results of both numerical models are conservative for HSC-fillings with additional steel fibers (columns C-36 and SQ-11) and reinforcement bars (column SQ-14).

The divergence between computed and test fire resistance period for plain HSC-filling can be explained as follows. The steel HSS column expands faster than the concrete core under fire load, initiating cracks in the concrete. This effect can be considerable due to lack of any reinforcement.

In the worst case scenario, the cracks magnify and might separate the concrete cross-section into two halves as shown in Figure 12 (a). In the initial stage of fire exposure, the load is mainly carried by the steel section. However, after approximately 15 to 20 minutes the load has to be transferred from the steel section to the concrete core. This is because the steel section yields due to attaining critical temperature, which is linked with significant reduction of its strength.

In case of load eccentricities, which hardly can be avoided, the gap between the concrete cross-sections cannot be closed, which leads to local failure of the column (Kordina & Klingsch, 1983). An example of local buckling of a concrete-filled HSS column is shown in Figure 12 (b). As the program BoFIRE does not take such local effects into account, it might be reasonable to establish a three-dimensional model for

Table 1. Summary of fire resistance test data on HSC-filled HSS columns (Kodur, 1998).

Tube filling	Column	Cross-section	Test load in kN	Concrete strength in MPa	Fire resistance in minutes
Plain HSC	C-46	273.1 × 6.35	1,050	82.2	48
	C-47	273.1 × 6.35	1,050	107.0	51
Steel fibers	C-36	219.1 × 4.78	600	98.1	174
	SQ-11	203.2 × 6.35	900	99.5	128
Reinforcement bars	SQ-14	203.2 × 6.35	1,150	81.7	89

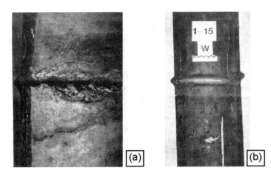

Figure 12. Separating concrete crack of an HSS column filled with plain concrete (a) and local buckling of a concrete-filled HSS column (b) (Kordina & Klingsch, 1983).

this problem in future. Nevertheless, by comparison to test data it is examined if the distribution of temperatures derived from this computer program is sufficient.

The comparison between calculated and measured temperatures is presented elsewhere (Schaumann, Bahr & Kodur, 2006b). It is shown by the comparisons that the calculated temperatures are in good agreement with the recorded temperatures in fire tests. Thus, these differences could not be seen as the reason for the high overestimation of the fire resistance of the HSS steel columns with plain HSC-filling.

4 EFFECT OF DIFFERENT CONSTITUTIVE LAWS ON LOAD-RESPONSE

Columns are mainly compression members and hence susceptible to bending with respect to column slenderness and load eccentricity. Thus, a numerical study was carried out to set a reasonable limitation for the use of HSC as HSS-filling. The parametric study was performed with a circular HSS cross-section C219.1 × 6 mm and hinged end conditions.

This corresponds to compressive strength of 50, 75, 95 and 115 MPa based on cube strength. The yield strength of the tube was 235 MPa.

The influence of slenderness was examined by varying the column length from 0.1 to 6 m with an assumed imperfection of L/1,000. Both the North American and European material properties for HSC were used. The cross-sections were heated for 30 minutes according to the ISO standard fire. After the heating process, the load was gradually increased until failure occurred.

The short heating period of 30 minutes already showed the unfavourable fire endurance of slender HSC-filled HSS columns. For longer fire exposure times, the reduction in ultimate load would be even more rapid.

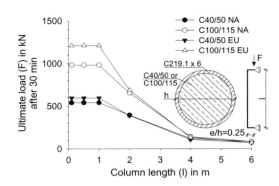

Figure 13. Ultimate load for HSC-filled HSS column with tube yield strength of 235 MPa and eccentric load of e/h = 0.25 using North American (NA) and European (EU) material properties at 30 minutes fire exposure.

The results for moderate eccentric loads with e/h = 0.25 are given in Figure 13. It can be seen that the North American provisions yield the more conservative results. Nevertheless, the difference between the approaches is insignificant for column lengths exceeding 2 m.

Due to its low tensile strength, the beneficial effect of HSC is sharply reduced for slender columns and load eccentricities causing bending moments. Further numerical studies can be found in Schaumann, Bahr & Kodur (2006a).

5 CONCLUSIONS

Based on the information presented, the following conclusions can be drawn:

- The computer code BoFIRE is capable of modelling the fire response of high strength concrete (HSC)—filled hollow structural steel (HSS) columns with different types of reinforcement (steel fibers and bar reinforcement).
- Local effects like gaping cracks and local plastic buckling, which might occur in concrete-filled HSS columns without reinforcement, are not covered by most conventional simulation tools. This might lead to unconservative results.
- Use of North American material property relations as compared to the European material properties leads to conservative fire resistance results.
- The use of plain HSC-filling in HSS columns is reasonable for non-slender columns with only moderate load eccentricity. High eccentricities might significantly reduce the fire resistance of HSC-filled HSS columns due to faster degradation of material properties.

REFERENCES

Canadian Standards Association 1994. Code for the Design of Concrete Structures for Buildings (CAN3-A23.3-M94). Mississauga, Canada.

Canadian Standards Association 2001. Limit States Design of Steel Structures (CAN/CSA-S16-01). Mississauga, Canada.

North American Standards Association 1989, Standard Methods of Fire Endurance Tests of Building Construction and Materials (ULC-S101-M89). Toronto, Canada.

European Committee for Standardization (CEN). EN 1991-1-2 (Eurocode 1) 2002. Actions on structures, Part 1-2: Actions on structures exposed to fire. Brussels, Belgium.

European Committee for Standardization (CEN). EN 1992-1-2 (Eurocode 2) 2002. Design of concrete structures, Part 1-2: General rules – Structural fire design. Brussels, Belgium.

European Committee for Standardization (CEN). EN 1994-1-2 (Eurocode 4) 2004. Design of composite steel and concrete structures, Part 1-2: General rules – Structural fire design. Brussels, Belgium.

Cheng, F.-P., Kodur, V.K.R. & Wang, T.-C. 2004. Stress-strain curves for high strength concrete at elevated temperatures. *Journal of Materials in Civil Engineering* 16(1): 84–94.

Gooseman, S. 2008. HighLight Munich Business Towers. Flickr.

Kodur, V.K.R. & Lie, T.T. 1995. Fire performance of concrete-filled hollow steel columns. *Journal of Fire Protection Engineering* 7(3): 89–98.

Kodur, V.K.R. 1998. Performance of high strength concrete-filled steel columns exposed to fire. *North American Journal of Civil Engineering* 25: 975–981.

Kodur, V.K.R. & MacKinnon, D.H. 2000. Fire endurance of concrete-filled hollow structural steel columns. *AISC Steel Construction Journal* 37(1): 13–24.

Kodur, V.K.R. & McGrath, R. 2003. Fire endurance of high strength concrete columns. *Fire Technology* 39(1): 73–87.

Kodur, V.K.R. & Sultan, M.A. 2003. Effect of temperature on thermal properties of high strength concrete. *Journal of Materials in Civil Engineering* 15(2): 101–107.

Kodur, V.K.R. 2003. Fire resistance design guidelines for high strength concrete columns. ASCE/SFPE Specialty Conference of Designing Structures for Fire and JFPE, Baltimore, MD, USA.

Kodur, V.K.R. 2006. Solutions for enhancing the fire resistance endurance of steel HSS columns filled with high strength concrete. *AISC Steel Construction Journal* 43(1): 1–7.

Kordina, K. & Klingsch, W. 1983. Fire resistance of composite columns and of solid steel columns – part I. Studiengesellschaft für Anwendungstechnik von Eisen und Stahl e.V., Project 35.

Lie, T.T. & Kodur, V.K.R. 1996. Fire resistance of steel columns filled with bar-reinforced concrete. *Journal of Structural Engineering* 122(1): 30–36.

Schaumann, P. 1984. Computation of steel members and frames exposed to fire (in German: 'Zur Berechnung stählerner Bauteile und Rahmentragwerke unter Brandbeanspruchung'). Technisch-wissenschaftliche Mitteilungen Nr. 84-4. Institut für konstruktiven Ingenieurbau, Ruhr-Universität Bochum, Germany.

Schaumann, P., Bahr, O. & Kodur, V.K.R. 2006a. Numerical studies on HSC-filled steel columns exposed to fire. Tubular Structures XI. Packer & Willibald, Taylor & Francis Group, London.

Schaumann, P., Bahr, O. & Kodur, V.K.R. 2006b. Fire resistance of high strength concrete-filled steel columns. Structures in Fire (SiF) 2006. Aveiro, Portugal.

Upmeyer, J. 2001. Fire design of partially encased composite columns by ultimate fire loads (in German: 'Nachweis der Brandsicherheit von kammerbetonierten Verbundbauteilen über Grenzbrandlasten'). Schriftenreihe des Instituts für Stahlbau der Universität Hannover, Heft 19, Germany.

Experimental study on SCFs of welded CHS-to-concrete filled CHS T-joints under axial loading and in-plane bending

W.Z. Shi
School of Civil Engineering, Tongji University, Shanghai, China

ABSTRACT: The present paper deals with an experimental investigation of the hot spot stress distributions and Stress Concentration Factors (SCFs) of welded T-joints composed of Circular Hollow Section (CHS) braces and Concrete Filled Circular Hollow Section (CFCHS) chords under tensile and compressive axial loadings and in-plane bending on the braces respectively, which is foundation for fatigue assessment of the joints. Static tests were performed on ten test specimens with different non-dimensional geometric parameters and concrete grades. The test results have been statistically interpreted, and indicate that the CHS-to-CFCHS T-joints exhibited evident improvement in rigidity and reduction in SCFs around the joints under the three loading conditions when compared with joints made of CHS. The difference in hot spot stresses between linear and quadratic extrapolation was slight for CFCHS chords and obvious for CHS braces. In addition, the CHS-to-CFCHS joints were found to have higher SCFs on chords under axial tension than under axial compression and SCFs at tension-side chord crowns were a little higher than those at compression-side chord crowns under in-plane bending.

1 INTRODUCTION

The application of concrete-filled circular hollow section (CFCHS) in large span arch bridges has been developed prosperously in China these years on account of their excellent structural properties and aesthetic appearance. The arch can be made of CFCHS chord members and circular hollow section (CHS) brace members (Fig. 1). The largest CFCHS arch bridge in China has come up to 460 meters (Zhou & Chen 2003). It is considered that such CFCHS trusses have better structural behavior than CHS trusses in both the whole structures and the connections between chords and braces.

As is known to all, fatigue behavior of welded CFCHS joints under daily vehicular loads is a significant problem that needs investigating, owing to high stress concentrations and weld flaws around the joints. In the last twenty years, a large amount of research on the fatigue behavior of welded CHS T-, Y-, X- and K-joints has been performed and the achievements have been integrated into IIW recommendations (Zhao et al. 2000) and CIDECT Report No.8 (Zhao et al. 2001), in which fatigue evaluation based on the advanced concept of hot spot stress is proposed. It is believed that the hot spot stress approach is also fit for the fatigue assessment of the welded CHS-to-CFCHS joints. The existence of concrete in the CHS chords certainly changes the original fatigue behavior of welded CHS joints to some extent. It is a new investigation aspect about how concrete exerts its effect on welded CHS-to-CFCHS joints.

The experimental study on hot spot stresses and the stress concentration factors (SCFs) of welded CHS-to-CFCHS T-joints under tensile and compressive axial loadings and in-plane bending on the braces is presented in this paper. Discussion and comparison of the extrapolation methods and the hot spot stress

Figure 1. Concrete filled tubular arch bridge in China.

distributions around the joints of all loading conditions are carried out to address the current lack of understanding of fatigue behavior of welded CHS-to-CFCHS T-joints.

2 EXPERIMENTAL SETUP

2.1 Specimen details

Ten test specimens of welded CHS-to-CFCHS T-joints were designed for the static tests, whose chord members were filled with concrete and brace members were kept empty. Different non-dimensional geometric parameters (β, γ, τ) and concrete grades were configured to reveal their effects on joint behavior (Table 1). Both the CHS braces and chords were made of steel Q345, whose mechanical test showed its yield stress was 324 MPa. Therefore there were four groups of test in ten specimens and each group had three specimens.

In order to avoid the stress distribution along the joint being influenced by the member end condition, the length of the chord and brace for each specimen was taken to be about $6d_0$ and $5d_1$ respectively

(Wingerde, 1992). The brace was connected to the chord by manual arc welding with full penetration welds according to AWS D1 (AWS 2000).

Each specimen was simply-supported at the two ends of the chord member. Three different loading conditions were considered respectively, which were

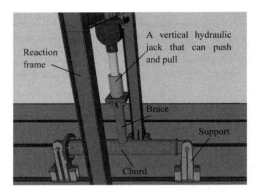

Figure 3. 3D schematic view of test specimen under axial loading.

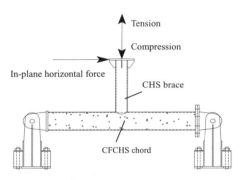

Figure 2. Plan schematic view of test specimen under axial loading and in-plane blending respectively.

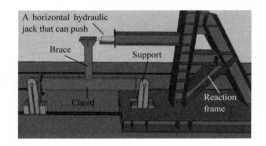

Figure 4. 3D schematic view of test specimen under in-plane bending.

Table 1. Details of test specimens.

Joint Nr.	$d_0 \times t_0$* (mm)	$d_1 \times t_1$* (mm)	$\beta = d_1/d_0$	$\gamma = d_0/2t_0$	$\tau = t_1/t_0$	Steel grade	Concrete grade
T1	245 × 8	133 × 8	0.54	15.31	1.00	Q345	C50
T2	180 × 6	133 × 6	0.74	15.00	1.00	Q345	C50
T3	133 × 4.5	133 × 4.5	1.00	14.78	1.00	Q345	C50
T4	245 × 8	133 × 6	0.54	15.31	0.75	Q345	C50
T5	245 × 8	133 × 4.5	0.54	15.31	0.56	Q345	C50
T6	245 × 8	133 × 8	0.54	15.31	1.00	Q345	C20
T7	245 × 8	133 × 8	0.54	15.31	1.00	Q345	C70
T8	203 × 8	140 × 8	0.69	12.69	1.00	Q345	C50
T9	203 × 10	140 × 10	0.69	10.15	1.00	Q345	C50
T10	203 × 12	140 × 12	0.69	8.46	1.00	Q345	C50

* d_0 and t_0 = diameter and thickness of chord respectively. d_1 and t_1 = diameter and thickness of brace respectively.

the tensile and compressive axial loadings and in-plane bending on the brace. Axial compressive and tensile loadings were applied to the end of the brace member through a vertical hydraulic jack that can push and pull, and horizontal force was applied to the end of the CHS brace through a hydraulic jack when the in-plane moment loading condition was implemented (Figs. 2–4). All loading conditions were performed in elastic stage.

2.2 Determination method of SCFs on CHS-to-CFCHS T-joints

Consulting the method used for a CHS joint (Zhao et al. 2000), four lines were arranged respectively on both the chord and the brace to measure the hot spot stresses at the weld toe of the CHS-to-CFCHS joint, namely two lines for crowns and the other for saddles (Fig. 5). Besides, three supplementary lines, 22.5 degrees apart each other in a quarter quadrant of intersection, were added respectively onto the chord and the brace for the sake of observing whether there would be any extreme points of hot spot stress between the crown and the saddle. A set of four strain gauges manufactured specially for the tests was used in each line of measurement, in which the four measuring points were disposed within the extrapolation region (Fig. 5). The SCFs were procured from converting the original strain concentration factors (SNCFs) measured using strain gauges, namely SCF = 1.2SNCF (Zhao et al. 2001). Both linear and quadratic extrapolation methods were adopted to examine their difference.

3 EXPERIMENTAL RESULTS AND DISCUSSION

It was found that the difference in hot spot stresses between linear and quadratic extrapolation was slight for CFCHS chords and evident for CHS braces under both axial loading and in-plane bending. Comparison between linear and quadratic extrapolation of ten test specimens is showed in Figure 6. No strong non-linear feature of hot spot stresses was found on the chords, and the difference in the hot

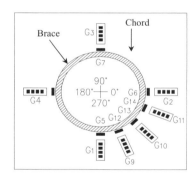

(a) Measurement lines

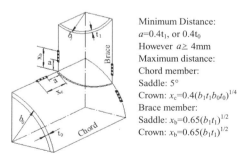

Minimum Distance:
$a=0.4t_1$, or $0.4t_0$
However $a \geq 4mm$
Maximum distance:
Chord member:
Saddle: 5°
Crown: $x_c = 0.4(b_1 t_1 b_0 t_0)^{1/4}$
Brace member:
Saddle: $x_b = 0.65(b_1 t_1)^{1/2}$
Crown: $x_b = 0.65(b_1 t_1)^{1/2}$

(b) Boundaries of extrapolation region

Figure 5. Measurement of hot spot stress for CHS-to-CFCHS joint.

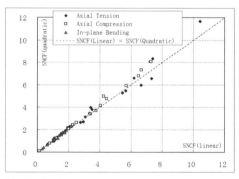

(a) On chords

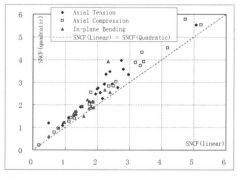

(b) On braces

Figure 6. Comparison between linear and quadratic extrapolation of ten test specimens.

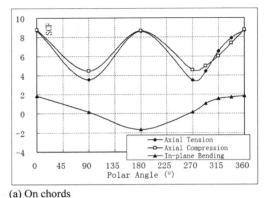
(a) On chords

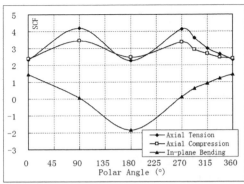
(b) On braces

Figure 7. Hot spot stress distributions around joint of T10 under three loading conditions.

Table 2. Comparison of the SCFs between experimental and calculational under axial loading.

Joint Nr.	Experimental value under compression (1)				Experimental value under tension (2)				Calculational values as CHS joints (3)				(1)max/ (2)max	(1)max/ (3)max	(2)max/ (3)max
	Chord		Brace		Chord		Brace		Chord		Brace				
	S*	C*	S	C	S	C	S	C	S	C	S	C			
T1	2.35	**4.86**	1.15	4.66	4.20	**6.79**	1.30	3.25	**15.28**	7.10	8.20	3.34	0.71	0.32	0.44
T2	0.94	**5.11**	0.26	5.02	1.50	**8.86**	1.43	4.30	**13.90**	8.22	7.07	3.76	0.58	0.37	0.64
T3	0.14	**8.00**	1.72	6.64	0.78	**12.50**	4.01	6.63	6.40	**11.66**	3.56	4.93	0.64	0.69	1.07
T4	1.52	**2.72**	1.69	1.51	1.99	**3.98**	2.56	1.85	**11.15**	5.32	7.24	3.07	0.68	0.24	0.36
T5	1.37	2.33	2.53	**3.63**	1.43	**4.08**	4.03	2.98	**8.14**	4.00	6.41	2.87	0.89	0.45	0.5
T6	2.99	**5.34**	3.40	5.22	3.30	**7.42**	3.51	3.32	**15.28**	7.10	8.20	3.34	0.72	0.35	0.49
T7	2.40	4.06	3.07	**4.70**	3.66	**6.54**	4.77	2.75	**15.28**	7.10	8.20	3.34	0.62	0.27	0.43
T8	2.08	**6.84**	2.29	4.48	2.16	**7.98**	3.29	3.51	**12.18**	7.00	5.41	3.72	0.85	0.68	0.80
T9	1.46	**7.78**	0.93	5.91	2.44	**8.77**	1.72	3.06	**10.01**	7.18	6.44	3.56	0.89	0.64	0.72
T10	4.52	**8.72**	3.41	2.43	3.52	**8.65**	4.18	2.29	**8.56**	6.84	4.73	3.82	1.01	1.02	1.01

* S-Saddle position; C-Crown position. The SCF highlighted is the maximum SCF among the four positions of the chord saddle, chord crown, brace saddle and brace crown.

spot strains between linear and quadratic extrapolation of chords was less than 15% (usually about 6%). At the same time, the difference in the hot spot strains between linear and quadratic extrapolation of braces was greater than 15% (usually about 25%). Consequently the linear extrapolation method was used on the chords and the quadratic extrapolation was used on the braces.

According to the test results of supplementary lines, no extreme point was found between the crown and the saddle. Figure 7 shows the stress distributions along the connection between CHS brace and CFCHS chord of the specimen T10 under axial compressive and tensile loadings and in-plane bending. The experimental SCFs of CHS-to-CFCHS T-joints under axial loading including both compression and tension and under in-plane bending on the braces are listed in Table 2 and 3 to compare with the calculated SCFs of CHS-to-CHS T-joints with the same sizes and under the same loadings as the CFCHS-to-CHS T-joints, which were obtained from the formula in the CIDECT Report No. 8 (Zhao et al. 2001).

It can be recognized that the maximum SCFs along the weld toes of the CHS-to-CFCHS joints were always lower than those of the CHS-to-CHS joints, whatever loading condition was applied on the brace. The maximum SCFs under axial loading usually occurred at the chord crown (except T5 and T7),

Table 3. Comparison of the SCFs between experimental and calculational under in-plane bending.

Joint NR.	Experimental values				Calculational values as CHS joints					
	Compressive side		Tensile side							
	(1)	(2)	(3)	(4)	(5)	(6)				
	Chord	Brace	Chord	Brace	Chord	Brace				
	C*						(1)/(5)	(2)/(6)	(3)/(5)	(4)/(6)
T1	1.82	2.06	2.21	2.32	4.40	3.21	0.41	0.64	0.50	0.72
T2	1.18	1.91	1.22	1.90	4.12	2.97	0.29	0.64	0.30	0.64
T3	0.49	2.32	0.37	3.07	3.43	2.54	0.14	0.91	0.11	1.21
T4	1.21	0.72	1.27	1.16	3.45	2.97	0.35	0.24	0.37	0.39
T5	0.75	3.19	1.35	2.60	2.70	2.75	0.28	1.16	0.50	0.95
T6	2.64	2.69	1.77	2.22	4.40	3.21	0.60	0.84	0.40	0.69
T7	1.33	2.25	2.11	1.80	4.40	3.21	0.30	0.70	0.48	0.56
T8	1.59	2.48	2.55	1.93	3.85	2.85	0.41	0.87	0.66	0.68
T9	1.74	2.58	1.95	1.46	3.42	2.64	0.51	0.98	0.57	0.55
T10	1.67	1.74	1.85	1.45	3.11	2.48	0.54	0.70	0.59	0.58

*C-Crown position. The SCF highlighted is the maximum SCF among the four positions of the chord saddle, Chord crown, brace saddle and brace crown.

whereas the maximum SCFs of CHS-to-CHS joints do at the chord saddle. The maximum SCFs under in-plane bending were located at the chord crown sometimes and at the brace crown ever and agah for CHS-to-CFCHS joints, whereas they do always at the chord crown for CHS-to-CHS joints. It indicate that the concrete filled in the chord exerts a distinct influence on the behavior of CHS-to-CFCHS joint. The concrete greatly increased the rigidity of the whole joint, restrained ovalisation of the chord and made the rigidity distribution around the intersection evener so that the hot spot stresses in the joint became lower and the positions of the maximum SCFs were changed.

It can be found that the SCFs of the CHS-to-CFCHS joints on chords under axial compression were quite different from that under axial tension (Table 2). The maximum SCFs under tension were usually about 30%~70% higher than those under compression, but their positions did not change, both being at the chord crown (except T5 and T7). It was also found that the SCFs of the CHS-to-CFCHS at the chord crown of tensile side were generally a little higher than those at the chord crown of compressive side under brace bending (Table 3). The phenomenon can be explained that the wall of the chord in the vicinity of intersection tended to be separated from the concrete inside the chord due to low cohesive ability between steel and concrete during tension, which weakened their cooperation and then increased the hot spot stress at the local area (Fig. 8).

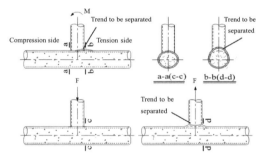

Figure 8. Abridged general view of the weld CHS-to-CFCHS T-joint deformation pattern under axial compressive and tensile loading and in-plane bending.

Within range of the parameters tested in the paper, the effect of the non-dimensional geometric parameters (β, γ, τ) of the tubular joint and concrete strength on SCFs are presented in Figure 9 for the axial loading and Figure 10 for the in-plane bending.

For axial loading condition, the effect of β on SCF_{max} of the CHS-to-CFCHS joint was significant. SCF_{max} increased with increasing β value. However, SCF_{max} decreases with increasing β value for the CHS-to-CHS joint. Both τ and γ exerted no important effect on SCF_{max} of the CFCHS-to-CHS joint under axial loading. This behavior is different from that of the CHS-to-CHS joint. SCF_{max} of the CHS-to-CHS joint increases obviously with increasing τ and γ respectively. Both τ and γ are the function of wall

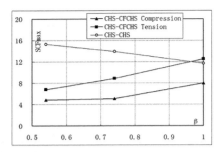

(a) SCF_{max} Vs. β

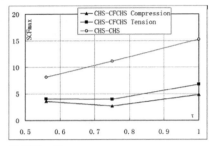

(b) SCF_{max} Vs. τ

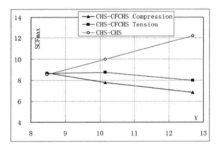

(c) SCF_{max} Vs. γ

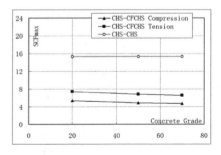

(d) SCF_{max} Vs. concrete grade

Figure 9. Effect of non-dimensional geometric parameters and concrete grade on SCF_{max} under axial force.

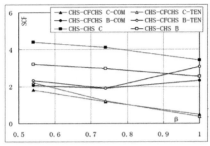

(a) SCF Vs. β

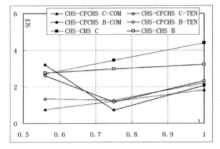

(b) SCF Vs. τ

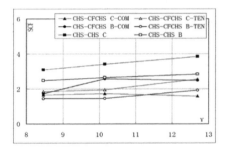

(c) SCF Vs. γ

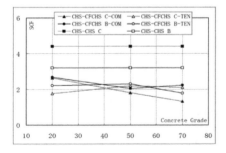

(d) SCF Vs. concrete grade

Figure 10. Effect of non-dimensional geometric parameters and concrete grade on SCF under in-plane bending.

474

Table 4. Comparison of SCFs between different loading conditions.

Comparative item	Axial force	In-plane bending
Locations of the maximum hot spot stresses	Chord crown	Chord crown or brace crown
$SCF_{CHS\text{-}to\text{-}CFCHS}$	0.14~12.50 or usually 0.3~0.9 $SCF_{CHS\text{-}to\text{-}CHS}$	0.37~3.07 or usually 0.2~0.9 $SCF_{CHS\text{-}to\text{-}CHS}$
$SCF_{tension}/SCF_{compression}$ on chord	Usually 1.0~2.0	Usually 0.8~2.0
Effect of β	Significant	Significant
Effect of τ	Slight	Significant
Effect of γ	Slight	Neglectable
Effect of concrete strength	Neglectable	Neglectable

thickness of the chord. The thickness of tubular chord seems to have no pivotal meaning after concrete is poured, because the tubular chord becomes solid and its diameter can be regarded as the whole thickness of the chord. Compared with the diameter of the tubular chord, the wall thickness of tube is so small that it can not play important role in the CFCHS chord.

For in-plane bending condition, the trend of SCF of CHS-to-CFCHS joints varying with β and τ was generally similar to that of CHS-to-CHS joints, but there was a little difference for the brace. The parameter γ almost did not influence SCF_{max}.

For both loading conditions, the concrete grade had not remarkable effect on SCF_{max} of the CHS-to-CFCHS joint within the period of elasticity, which indicates the main function of concrete is to enhance the stiffness of the joint, but the strength itself, no matter what grade is, has almost nothing to do with the SCFs of the joint.

Comparison in SCFs between the two different kinds of loading is given in Table 4, from which the same and different behavior can be clear at a glance.

4 CONCLUSIONS

Experimental study on the hot spot stresses and the stress concentration factors (SCFs) of welded CHS-to-CFCHS T-joints under axial loading and in-plane bending was carried out and presented in this paper. The experimental SCFs of the CHS-to-CFCHS joints were compared with the calculated SCFs of CHS-to-CHS joints based on CIDECT formula by using the same non-dimensional geometric parameters. The following conclusions can be drawn:

1. The linear extrapolation method used for measuring hot spot stresses of CHS-to-CHS joints was proved to be only suitable for the chords of CHS-to-CFCHS joints, and non-linear feature of hot spot stresses was found on the braces.
2. The hot spot stresses of CHS-to-CFCHS T-joints were found to decrease evidently and the locations of maximum SCF were changed after the CHS chord was filled with concrete. The SCFs of CHS-to-CFCHS joints were usually equal to 0.2~0.9 SCFs of CHS-to-CHS T-joints.
3. The CHS-to-CFCHS joints had different SCFs when they were loaded in compression and in tension on the braces. The maximum SCFs under compression on chords were lower than those under tension, although their positions were usually the same. SCFs at chord crown positions of compression side were lower than those on tension side under in-plane bending. The phenomena attributed to restraint provided by the concrete against the deformation of the chord in the condition of brace compression.
4. When subjected to axial loading, the maximum SCF around the CHS-to-CFCHS joint increased with increasing β within the range of 0.5 to 1.0, and the effects of τ and γ on the SCF around the CFCHS joint were not significant. When subjected to in-plane bending, effects of non-dimensional geometric parameters β and τ on SCFs were significant, whereas effect of γ was neglectable.
5. The concrete filled in the CHS chord improved the stiffness and strength of the joint significantly, but its grade, namely the strength of concrete itself, did not exert much influence on the SCFs around the joints for three loading conditions.

ACKNOWLEDGMENTS

The authors would like to thank Dr. K. Wang for his continuous support and help for the research of the paper. The authors also wish to thank the laboratory staff in the Department of Building Engineering at Tongji University for their assistance in the project. Natural Science Foundation of China supported the research in the paper financially through the grant No. 50478108.

REFERENCES

AWS. 2000. Structural Welding Code-Steel. *ANSI/AWS D1. 1–2000, 17th Edition*. Miami: American Welding Society.

Udomworarat P., Miki C., Ichikawa A., Sasaki E., Sakamoto T., Mitsuki K. & Hasaka T. 2000. Fatigue and ultimate strengths of concrete filled tubular K-joints on truss girder. *Journal of Structural Engineering, Japan Society of Civil Engineers, Vol. 46 A, March 2000*: 1627–1635.

Van Wingerde. 1992. *The Fatigue Behaviour of T and X Joints made of Square Hollow Sections*, The Netherlands: Herion.

Zhao X.L., Herion S., Packer J.A., Puthli R., Sedlacek G., Wardenier J., Weynand K., Winderde A. & Yeomans N. 2001. Design Guide for Circular and Rectangular Hollow Section Welded Joints under Fatigue Loading. *CIDECT and TUV-Verlag*, German.

Zhao X.L. & Packer J.A. 2000. Fatigue Design Procedures for Welded Hollow Section Joints. *IIW*, Cambridge: Abington Publishing.

Zhou S. & Chen S. 2003. Rapid Development of CFST arch bridges in China. *Proceedings of the International Conference on Advances in Structures: Steel, Concrete, Composite and Aluminium, ASSCCA'03*, June 2003: 915–920. Sydney.

Design Award Competition

Modular viewing tower

C. Joost & T.O. Mundle
University of Applied Sciences, Stuttgart, Baden-Württemberg, Germany

ABSTRACT: This paper describes the architectural concept and design elements of a modular viewing tower to be a new landmark in Stugart, Germany. The design of the viewing tower shows the creative potential of the support structure beyond pure functional architecture. Ease and transparency of the structure convince and excitingly contrast with the still existing stub of the destroyed first viewing tower made of red sandstone. The selection of a minimum amount of support structure elements and the intelligent employment of the outer shell intensifies the reduced character of the tower. Depending on the inner load the support structure elements are sophisticatedly built of hollow sections and steel bars and let experience the force paths.

1 ASSIGNMENT

The assignment is to develop a new viewing tower for Hasenberg that would be a new landmark in constructive engineering architecture.

2 LOCATION

The anticipated location for the planned viewing tower is a ridge on the edge of Stutgart. Stuttgart lies in a valley surrounded by elevated forrest areas which have long been a popular destination for excursions. At the weekend many residents use the opportunity to get away from the confines of the city and enjoy the magnificent views of Stuttgart from the surrounding heights.

3 HISTORY

Historically, a tower was erected on this site in 1879 in the style of a tower from the middle ages. With a height of 36 meters, it attracted up to 10 000 visitors a year. This tower at Hasenberg was the precursor to further tower constructions along the heights surrounding Stuttgart, towers which now act as orientation points for the cities public spaces. As a victim of the second world war, the original Hasenberg tower was destroyed in 1943. Today the visitors of the Hasenberg will only encounter the remaining stub of the original tower. The new Hasenberg Tower construction will reestablish the arrangement of Stuttgart's viewing towers. Erecting a second tower at this place has been a goal of Stuttgart's citizens in a long time.

4 BASIC APPROACH

Our basic approach is to develop a tower using a simple yet consistent system, which emphasizes the structural forces and only uses a limited number of modules. In order to provide the necessary presence of a landmark, we decided to use an external support structure.

The aesthetic of the tower therefore lies in the support structure and evolves out of the structural requirements and construction. In the approach the triangle provides the primary construct of a structurally stable vertical space framework. This dominating element will be the primary perception of the observer.

5 MODULAR SYSTEM

Based on the primary construct, we have developed a tower as a modular system which can be constructed from a minimum number of diverse building components. A high quantity of homogenous building components and a size suitable for transportation allows an economic, efficient prefabrication and swift, precise on site construction. This modular system would allow also for the construction of towers at similar places varying in size and proportion.

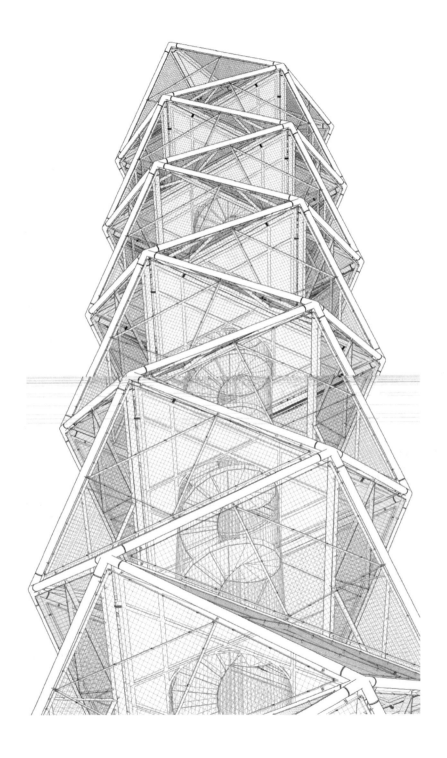

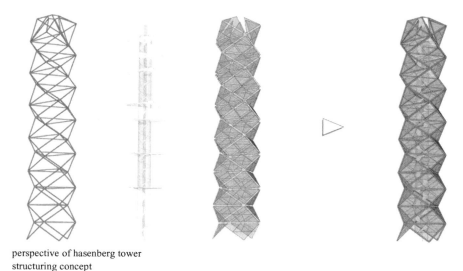

◄ perspective of hasenberg tower
▲ structuring concept
▶ drawings: groundplans, cross section, elevations

6 SUPPORT STRUCTURE

The guiding concept for the support structure is a constructive framework which, similar to a skeleton, absorbs and transfers all vertical and horizontal load, in addition to applied moment, to the foundation. All further building components are then fitted into this primary structure. The support structure of the tower consists of hollow sections welded to spacious cast joints. A total of 5 hollow section lengths in addition to 5 different moulded joints are employed, the apex load bearing framework having the most custom requirements.

7 SCULPTURAL QUALITY

Through the consequent use of the triangle, a tower with a sculptural quality emerges where the appearance shifts depending on the viewing point of the observer. Thereby a lifelike aspect is developed which involves the viewer, capturing attention and arousing interest.

The reduction of the number of elements used increases the impression of the sculptural character. Changes in light and shadows let the tower appear as a constantly changing object and this way merges with its environment.

8 STAIRCASE

A spiral staircase rises from the entrance walkway, leading the visitor to the upmost platform. The staircase is the second most important design element following the supporting framework. The individual steps are hung from the top section of the support structure via load bearing cables attached to a funnellike structure. Between the twelve load bearing cables further thinner cables will be introduced to act as a safegard against falling and to underline the vertical aspiration of the stair space. Stepping through a metal ring allows access to the individual platforms from the stairs.

9 PLATFORMS

The platforms consist of four components, three identical triangles, and the central section incorporating the well of staircase. This allows prefabrication of the individual components which only need to be joined together on site. These components are constructed from frames of U- and double-T-profiles on which corrugated sheet metal is welded as a floor. The number of platforms increases towards the top since plant cover severly limits viewing in the lower part of the tower.

The various viewing platforms unfold, due to their location, with starkly differing qualities-between the trees, amongst the treetops, and panorama above the treetops.

10 SIDE FACING

The side facing of the tower consists of a filigree metal frames spanned with a stainless steel net. These

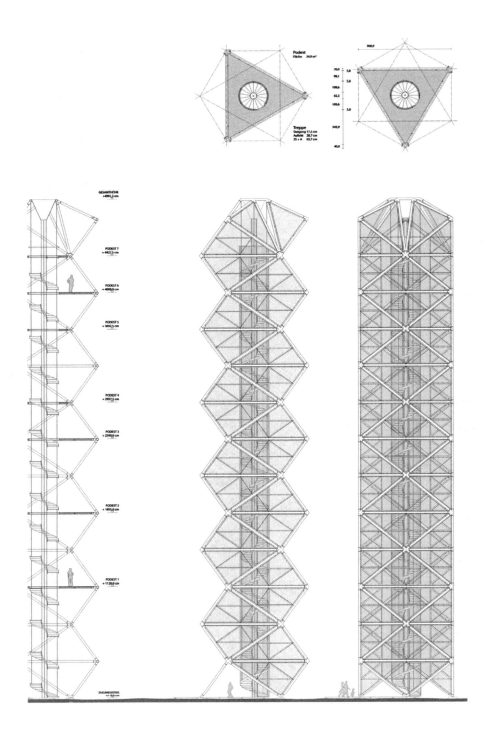

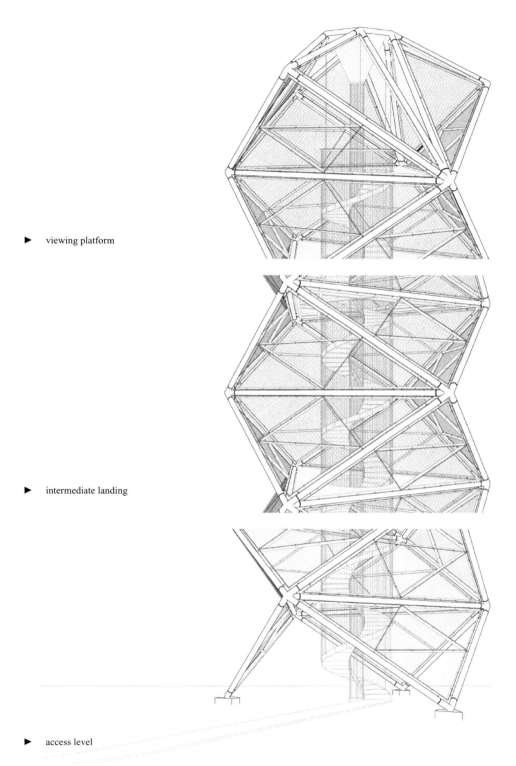

▶ viewing platform

▶ intermediate landing

▶ access level

act as a safeguard against falling and are bolted to the main load carrying structure via welded plates. In addition to the safety aspect they support the sculpture like appearance of the tower.

11 CONSTRUCTION

Simultaneous to the floor by floor construction of the load bearing frame, the intermediate platforms are inserted and attached. As soon as the tower and its platforms stand, the funnel and the twelve load bearing cables are attached to the apex of the load bearing structure. At this point the stair modules can be hoisted up and can be suspended piece by piece. Finally the remaining cables follow, along with the hand rail, and the safety net in the triangular framework.

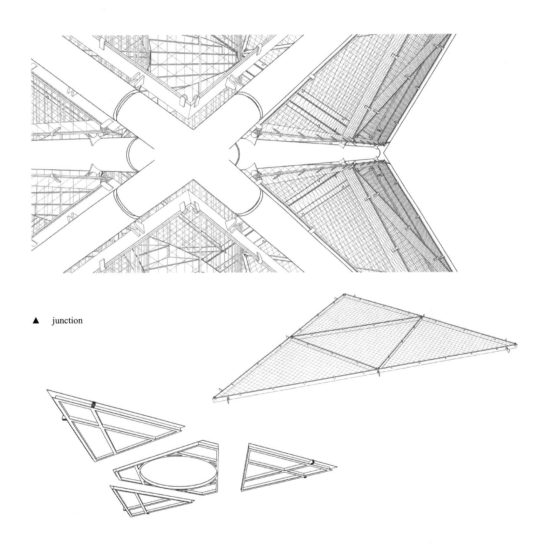

▲ junction

▲ components: side covering, platform modules
▶ details: support structure, stairs, platforms, connection
▶▶ model

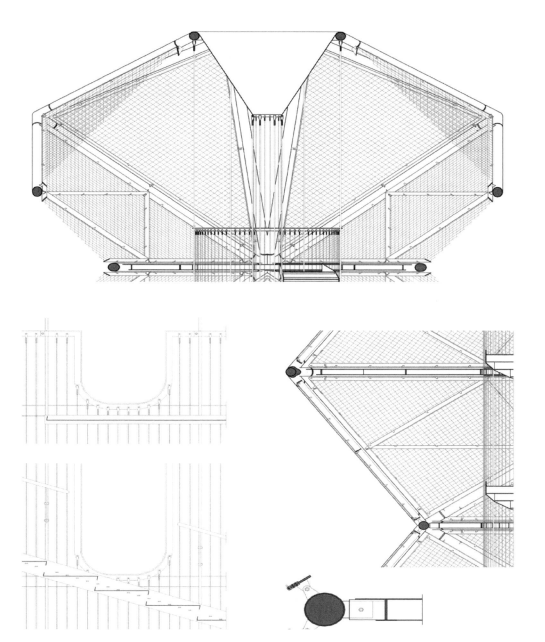

Design of a single large-span roof structure for Rotterdam Zoo 'Blijdorp'

S.J.C. Kieboom
Eindhoven University of Technology, Eindhoven, The Netherlands
Pieters Bouwtechniek Eindhoven B.V., Eindhoven, The Netherlands

ABSTRACT: My graduation project at the Department of Architecture, Building and Planning from Eindhoven University of Technology consists of a design project for the Rotterdam Zoo 'Blijdorp', The Netherlands. I created a design for a number of animals' accommodations underneath a single large-span roof structure. The project covered not only the aspects of design—both architectural and structural—but also the aspects of fabrication and erection of the building. Furthermore I found a solution to translate a blob design into a logical structure, using mathematical design principles. For the entire project it was the most obvious choice to use steel tubular profiles because of their aestetic and structural qualities.

1 THE DESIGN ASSIGNMENT

The Rotterdam Zoo (Diergaarde Blijdorp) in the Netherlands is currently modernising its park. Once this modernisation is completed, the animals' accommodations will resemble the animals' natural habitats as closely as possible. The park will be divided into different 'continents' and species native to a particular geographical location will be situated in the corresponding part of the park (Fig.1). Several aviaries will be constructed on the area where 'Europe' and 'Africa' meet, and the boundary will be symbolised by an aviary which will house European birds that normally migrate to Africa before winter sets in.

For these aviaries and the accompanying buildings—such as a restaurant, and a giraffe and gazelle accommodation—I have created an alternative design in which all these different functions are realised underneath a single large-span roof structure. To resemble the natural habitat as closely as possible, as well as maintaining a large flexibility for the interior layout, column-free areas was an important prerequisite for the design. Another important requirement was that large parts of the roofs could be opened. The building's site borders a curved railway, and I wanted to have the building's outline reflect this shape.

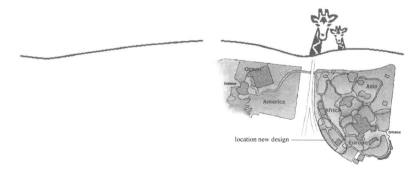

Figure 1. Plan Rotterdam Zoo 'Diergaarde Blijdorp'.

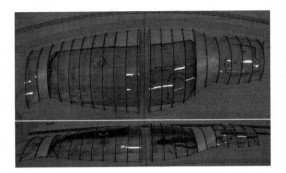

Figure 2. Design model, top view & front view.

2 THE DESIGN

Since the building is intended for a zoo, it has been designed to fit in with its natural surroundings (Fig. 2). This could be realised by using parallel arches for the roof structure. These arches consist of round steel tubular profiles that have been bent by a rolling mill and differ in height and span; these features give the structure an organic shape, which will fit in very well with its surroundings.

By using tubular profiles, it has been possible to create a strong design. The arches' closed profiling gives the construction a distinct look, but without interfering with the desired image of the animal accommodation. This is mainly due to the fact that these arch-shaped tubes enable the use of very slim profiling in this structural design. The tubular profiles can be bent into the shape of an arch through the process of rolling, and together these arches create the flowing lines of the main structure.

The arches' large spans create a column-free space, which is beneficial to the building's sense of openness. To maintain this openness, the parts of the structure that will be open to the public are only separated by netting or glass partitions.

The building consists of two parts, with the European birds on the one side of the structure and the African animals and birds on the other side. However, since the lines in the roof and the facade of both parts of the building merge into each other, it still looks like a coherent whole. Each part of the building is 80 metres in length. The largest arches span 63 metres and are 15.5 metres high. The distance between the arches is eight metres.

3 SHAPE AND CONFIGURATION OF THE ARCHES

The dimensions of the arches were first of all determined by the building's functional requirements. The arches' heights and spans have been adapted to meet the requirements of the different functions to be carried out under the roof. Therefore, the aviaries and the giraffe accommodation are covered by higher arches. Also, the site's curved boundary, which is caused by the railway running along the back of the site, and the need to realise (optically) flowing lines in order to maintain the building's organic shape have been influential.

Perhaps the most important factor was the need for a retractable roof. Subsequently, as a result of a model study, two design models were applied to find systematic solutions for the construction of an organic design, and the need to be able to open the roof. The two design models used were a model based on concentric circles and a model based on shifted circles.

In the model with concentric circles (Fig. 3), each strip between two arches is part of the surface of a cone, with the cone's axis as the centre of the arches. The strips between the arches can be filled in with plane quadrangular (trapezial) elements. Each particular strip can be filled in with the same elements, and the arches of that strip can be tied by the same secondary girders.

In the model with shifted circles (Fig. 4), each strip between two arches is part of the surface of an oblique cylinder in which both arches have the same radius. The strips between the arches can be filled in with plane quadrangular (parallel) elements. In this model, too, the same secondary girders can be used in a single strip.

In the final design both models have been applied, as this combination enables the construction of non-geometric shapes (Fig. 5). However, these models have

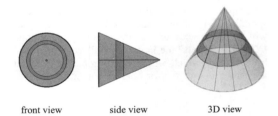

front view   side view   3D view

Figure 3. Concentric circle.

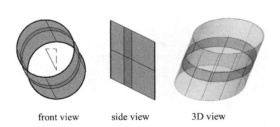

front view   side view   3D view

Figure 4. Shifted circles.

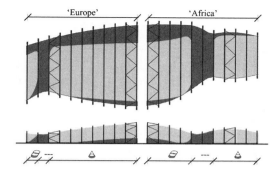

Figure 5. Design, top view & front view.

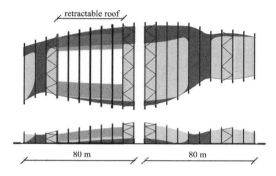

Figure 6. Design, retractable roof open.

not been applied to two transitional parts in order to meet the requirements mentioned above. The main decisive factor in this consideration was the aesthetic aspect of the structure's flowing lines.

The application of the model with concentric circles has made it possible to make part of the roofing covering the 'Europe' section retractable so that it can be opened (Fig. 6). This roof retracts under the arches and disappears over the fixed section of the roof.

4 THE STRUCTURAL DESIGN

In short, the building's structure consists of parallel interconnected arches that are made of round steel pipes. Hinges attach the ends of the arches to the structure's foundation. The ends of each single arch are interconnected through tie bars to absorb tension, and foundation blocks enable vertical load transfer. The arches are interconnected by secondary girders made of square steel tubular profiles (Fig. 7).

In-plane stability is realised by the arches themselves, and wind bracings realise out-of-plane stability (visible in Figs. 5 & 6). Two wind bracings have been applied to each of the two sections of the building. The façade's horizontal load is successively transferred to the secondary girders, wind bracings and the foundation. These wind bracings, too, consist of steel tubular profiles.

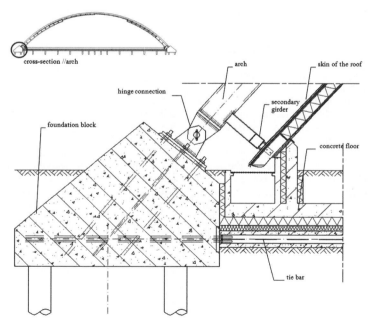

Figure 7. Detail of the connection between foundation, arch and roofing.

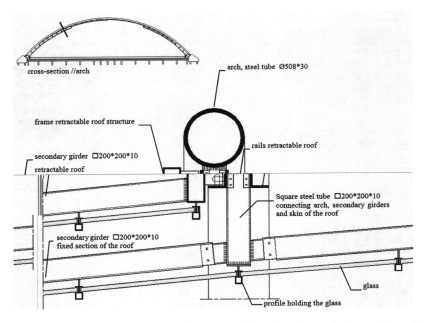

Figure 8. Detail (⊥arch) connection arch, secondary girder, skin of the roof and retractable roof.

5 CONNECTING THE STRUCTURE AND THE ROOFING

In order to give the building's interior a smooth look, the roofing has been attached underneath the secondary girders. The roofing consists mainly of sheets of glass. Part of the building has been closed with a wooden structure.

In order to connect the arches, the secondary girders, and the skin of the roof, square tubes have been welded perpendicular to the main arches. The secondary girders could then be attached to these tubes (Fig. 8). By using square tubular profiles, this method of connection could be applied to the entire structure. Each girder along an arch has been rotated slightly to enable the use of a single connection method to attach the girders to the main arch. The only way this could be realised was by using these square tubular profiles, since their strength characteristics remain unaltered when they are attached at an angle. Moreover, tubular profiles are best suited to absorb eccentric loads.

6 THE CONSTRUCTION PHASE

For the construction of this design, the arches will be delivered in two pieces. They will first be transported over the water, after which they will be transported by road for the final part of their journey. Once on site, the two pieces will be unloaded at the intended location of the arch, after which they will be assembled and erected. A single tower crane will be sufficient to carry out this work. A smaller, mobile crane can be used to place the secondary construction and the wind bracings.

Design and development of a public leisure center made up of tropical garden and an indoor swimming pool in the City of Valencia [Spain]

E. Fernandez Lacruz
Escuela Técnica Superior de Ingenieros Agrónomos, Universidad Politécnica de Valencia, Spain

ABSTRACT: This final year project proposes the building of a public leisure center, which consists of a tropical garden and an indoor swimming pool (see Figure 1), both within the same enclosure, so the installations can be enjoyed all through the year. In order to define a unique and compact volume, an organic form with its own identity was chosen during the enclosure's conceptual design phase. This volume needed to meet several requirements as the enclosure's configuration should adequately integrate the different uses for which the enclosure is projected (garden—hall—swimming pool), It should be structurally and economically feasible, there should be enough light in the garden and it should be singular and efficient in its structure.

1 INTRODUCTION

1.1 General

After several trials, a shape was found which met all of these requirements: a peanut shell. As will be explained the resulting form allows the different uses to be continuously independent

A zone for the garden, a different one for the swimming pool and an intermediate one which serves as exchanger between them

A large glass surface would allow the light into the garden, creating the proper microclimate; a relationship would be established between the external environment and the other spaces.

Building the structure is technically feasible. The solutions are known.

After actions identification, effort calculation and profile assignation optimization, the resulting structure turns out to be very light and with remarkably.

161 m tall, between 73,65 m and 44,47 m of light and a coronation height between 24,52 m and 13,25 m, the cover has a surface of 11.290 m². The structure leans discretely on a perimeter wall which also serves as foundations and container of the enclosure's different areas.

The inner space includes approximately 9230 m² in plant, in 3 levels: 3290 m² for the swimming pool zone, 1441 m$_2$ for the hall and access walkways and more than 4500 m² for the garden.

For the structural resolution of the surface, it is decided on a continuous reticular metallic structure, completely diaphanous, closed, glazed and made in its totality by CHS tubular steel profiles.

2 MODELLING

2.1 Profile-type obtaining

To obtain a parametric model, diverse simplifications were made in order to synthesize the peanut's shell surface. First, it was impossible to obtain a useful model. Second, in order to reduce the surface's complexity and achieve a clean form which nonetheless can be identified with its natural counterpart.

A series of profile-types which will serve as a basis for surface generation were obtained from several peanut shells (see Figure 2).

Several elevations, some of which are shown in the enclosed images, were obtained through longitudinal section (see Figure 3).

A vectorial curve interpolated using NURBS was obtained from these profiles

Two parallel methods for surface formation were tried. In the first one, surfaces were interpolated among different profiles which caused results with little coherence. However, the use of a single profile, and the generation of surfaces through revolution gave

Figure 1. Main view of the complex.

Figure 2. Peanut shells.

Figure 3. Peanut shape cuts.

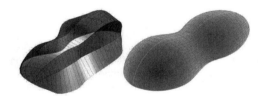

Figure 4. Preliminary models.

Figure 5. Final base model.

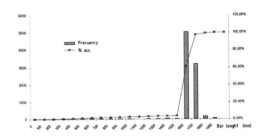

Figure 6. Bar length histogram.

rise to much more satisfactory results. Therefore, this last method was finally chosen (see Figure 4).

2.2 Discretization

O The next goal was to go from a non-dimensional surface to a discrete one, in order to obtain a bar model and thus to get a model for calculation. Therefore, a density for bars would have to be selected taking into account:

– Bar orientation.

– In order to reduce the number of bars and crystal cuts as much as possible, a quadrilaterals mesh was used, only resorting to triangle forms for the border close.
– In order to reduce costs, the surface cover should be generated through flat crystals. This forces the use of crystals with a defined maximum shape, which is established by the surface curvature.
– Length should be as regular as possible in order to simplify the implementation.

The last attainment was the most crucial as it brings the three together. The transnational surface method was used to obtain it. The surface's convergence in the directive curve surfaces forced the use of increasingly small bars and crystals, an aspect which had not been taken into account. Therefore, a cut line over the rotation baseline was set. In this way, a highly homogeneous reticule is obtained (see Figure 5 and Figure 6).

A guideline curve was divided by a hundredth of the original total length, which resulted in 1600 mm segments. The form which the surface then takes is obtained through small relative turns {8} between the grid squares. The surface produces then 3 types of modulations:

Quadrilaterals with parallel sides which are perpendicular to each other two to two, with all its vertexes in a same 1600 × 1600 mm plane.

Quadrilaterals with parallel sides, not perpendicular to each other, with all their vertexes in a same plan of variable dimensions.

Triangles of diverse sizes in the zones of covered encounter—laying of foundations.

Figure 8. Node detailing.

3 CALCULUS MODEL

3.1 Procedure

Calculations for the structure were carried out following a 3-dimensional bar model, through the use of non-lineal matrix models. The used regulations of reference have been the Eurocodes and its national transcriptions (CTE—action and structural design) along with internationally recognized design guides.

Taking into account the surface's singularity, this method will provide the most reliable results. A simplified flat model would result in conclusions far removed from reality

The use of a 3-dimensional calculation model allows making efforts compatible with the deformations produced by them.

The structural typology (from top to bottom) of the set is a reticular mesh like main lifting element of the closing, altogether with an orthogonal shared in common mesh wiring to each one of the reticules, acting like bracing element (see Figure 7). The global behavior can be assimilated to a laminar structure, that at local level works like a Vierendel beam (see Figure 9). Effective buckling lengths take values around 1.

4 TUBULAR STEEL PROFILES

4.1 Procedure

In order to build the enclosure, only hot formed tubular profiles (CHS) were used. The use of that kind of sections is the solution for several needs.

On one hand, the selected profile will have to face efforts in any direction of the plane. Therefore, profiles with remarkable mechanic properties with regard to one of its planes are discarded (I, T, H sections).

By avoiding the use of stiffeners, a tubular profile greatly simplifies manufacturing. Unions are made directly and continuously either in the workshop or in the construction site, substantially simplifying the building process. This kind of building allows for the manufacturing of big structure blocks which are situated in their final position once they are built in the ground.

In this case, CHS profiles allow for the covering of big lights in a completely transparent way and with great plastic art value.

This kind of sections offer greater resistance to fire, and a smaller quantity of products are needed for its protection, especially important in swimming pool zone, due to a smaller form factor.

Being a structure that works through compression, materials and section are optimally profited. This makes the use of steal with high elastic limit (S355) competitive.

Tubular steel sections makes possible simplify mechanisms of specific supports in the structure's

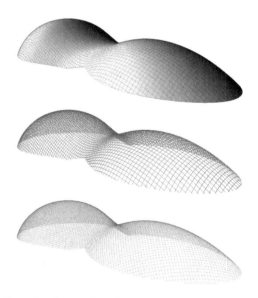

Figure 7. Structural typology.

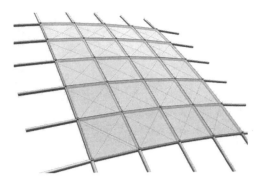

Figure 9. Modulus detailing. Bi—directional Vierendel beam..

Figure 10. Aerial view of Valencia.

nodes, to be used for load transmission. This allows the installation of low-emission glasses in the upper side of the profile. It also allows placing the cable web for bracing in the lower side (see Figure 8 and Figure 9).

5 CONCLUSSION

A tubular section, in particular circular profiles, allows obtaining an extremely slight structure, with an optimal advantage of profiles.

The resultant structure is made of (See Table 1):

– 8732 CHS tubes segments, with a total length of 13,719 meters.
– 19.416 meters of steel cable.
– 4576 glasses, with a total surface of 11290 m^2.

The resultant ratio steel/surface in plant results less than 10 Kg S355 / m^2.

Some additional figures can be found below.

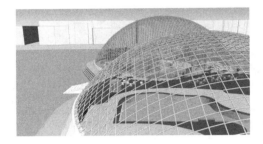

Figure 11. External view of swimming pool.

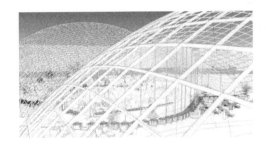

Figure 12. External view of swimming pool.

Table 1. Steel profile list.

Profile	Units	Length (m)	weight
CHS-50.4	5885	9,306.3	42,229.6
CHS-50.5	703	1,098.0	6,092.5
CHS-60′3.4	586	933.7	5,185.6
CHS-70.4	463	724.4	4,716.1
CHS-70.6	84	131.6	1,246.1
CHS-80.4	264	421.4	3,159.7
CHS-100.4	217	343.9	3,256.4
CHS-100.8	31	49.1	891.0
CHS-120.4	161	255.5	2,923.7
CHS-133.4	81	130.3	1,657.6
CHS-133.6	140	223.3	4,195.7
CHS-152′4.6	43	67.5	1,461.4
CHS-168′3.10	43	69.5	2,713.3
CHS-168′3.12.3	4.	8	224.3
CHS-193′7-.12′5	4	4.9	272.2
CHS-219′7.12′5	19	25.2	1,606.6

Figure 13. Entrance—hall.

Figure 14. Swimming pool view.

Figure 16. Aerial garden view.

Figure 15. Garden view.

*Parallel session 7: Welding & cast steel
(reviewed by R. Puthli & J. Ding)*

Selection of butt welding methods for joints between tubular steel and steel castings under fatigue loading

M. Veselcic, S. Herion & R. Puthli
Research Centre for Steel, Timber and Masonry, University of Karlsruhe, Germany

ABSTRACT: Research work at the University of Karlsruhe on fabrication of welded connections between hot-rolled tubes and steel castings is presented in this paper. Comparison is made between different parameter variations at the weld seams. This also includes the influence of different backing materials and influence of cast defects. Furthermore the cost effectiveness of the various measures is also taken into account to find a proper solution to the design of the butt weld. These investigations are intended to lead towards the development of a concept for improving the fatigue life for bridges using such connections.

1 INTRODUCTION

The use of joints between cast steel and structural hollow sections has increased in recent times. This is due to the favorable properties of cast steel now available. Cast steel in joints is particularly of interest in major prestigious projects because of the aesthetic design possibilities. Further advantages are the significant improvement of fatigue life and ease of fabricating complicated geometries.

The practical application of cast steel in bridges was presented in a previous paper at the 10th International Symposium on Tubular Structures in Madrid (Veselcic et al. 2003) and further, details concerning the cost effectiveness of different butt weld geometries were presented subsequently at the 11th International Symposium on Tubular Structures in Quebec (Veselcic et al. 2006).

Meanwhile, new product standards for delivery conditions, especially dealing with cast steel, have been established in Europe. The present research is aimed at providing data from evaluation based on fatigue test results for further standardization. The ongoing work from two research projects at the University of Karlsruhe on end-to-end CHS connections will be presented here. The projects are funded by CIDECT (Project 7 W—Fatigue of end-to-end connections) and FOSTA (Project P591- Economic use of structural hollow sections for highway and railway bridges), a German steel research association.

2 TESTING PROGRAMME

Detailed information on the manufacturing process is presented in former publications mentioned above. Details on quality levels and welding parameters described in this section are described in detail in Veselcic et al. (2006, 2007).

2.1 *Fabrication of the test specimens*

For the research project steels according to Table 1 are considered for the tests. With the use of high strength steel, a reduction of the member thickness can be realized in practice. To ensure good weldability, especially at the weld surface, the cast quality is chosen according Table 2. For the welding, pre-heating was omitted, which entails a significant cost reduction.

Altogether, three test series are performed. In the first series, only steel grade S355J2H with the appropriate cast steel quality is used. Following this, in the second series both steel grades are chosen for the specimens. For the last series, the steel quality will be S460, to be in line with future developments in steel structures.

The outer diameters of the hollow sections to be investigated are selected according to Table 3. They have been chosen on the basis of experience of the industrial participants in different working groups of the research project, in order to cover a wide range of applications. The cast steel wall thickness is chosen generally to be the same order as that of the hollow

Table 1. Steel grades for the hollow sections and corresponding cast steels.

Hollow sections	Standard	Cast steel	standard
S355J2H	EN 10210	G20Mn5(V)	EN 10293
S460NH	EN 10210	G10MnMo V 63	EN 10293

Table 2. Cast steel quality levels according to EN 126801.

Welding area	Quality level 1	V1 S1
Surface	Quality level 2	V2 S2
Within cast steel	Quality level 3	up to V3

Table 3. Testing program.

Test Series	Diameter [mm]	Wall thickness [mm]	
		Steel hollow section	Cast steel
1	193.7	20	20/30*
2	298.5	30	30/40*
3	508.0	50	50/60*

* depending on chosen variant.

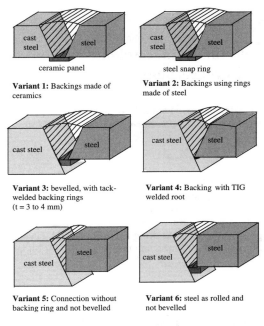

Figure 1. Tested variants of butt welds (schematic sketches).

section to be welded to it. For the variants with different wall-thicknesses, 10 mm is added to the cast steel components.

For the planned butt weld details, a total of six different design variants have been taken into account (Fig. 1).

For all variants, a weld preparation with grinding of the cast steel members can become necessary, to achieve the required surface quality. The detailed procedures should be clarified in advance with the foundry and the ultrasonic examiner.

Variants 1 and 2 only differ in the weld backing and have equal wall-thicknesses for the adjoining members. These variants are the optimum design in practice and should provide the best possible quality. However, to save costs, a grinding to achieve equal wall-thicknesses is omitted.

All other variants have different wall-thicknesses. To reduce costs, the tubular steel in variants 5 and 6 are not bevelled, but have a straight cut. Variants 3 and 6 are designed with a backing and variant 5 has none. In variant 4, a special TIG welded root, which is carefully executed, is provided.

The variants 5 and 6 are prone to welding defects due to the straight cut in the tubular steel members and should normally be avoided in critical areas. However, with lower stressed connections this method can be cost reducing if quality control is performed carefully.

After welding the test specimens, an ultrasonic inspection of the different variants is carried out to detect any defects. However, for practical applications, it should be pointed out that with a multitude of regulations for ultrasonic inspection, the necessary requirements should be specified at an early stage of a project to avoid subsequent disagreements between the client and contractor and to maintain quality standards.

2.2 *Welding parameters*

For the welding, the more practically oriented manual metal arc welding process was used. The results are therefore on the safe side. Dependent on the welder qualifications, the welding position and the component size, more defects may occur in comparison to metal active gas welding, for example. This should be considered in the assessment of test results.

The following welding positions (Figure 2) were selected in accordance to EN 287: PC (tube: fixed, axis: vertical, welding: transversly), PF (tube: fixed, axis: horizontal, welding: upwards) and PA (tube: rotating, axis: horizontal, welding: in bath) only used for the largest specimen. In this way, conditions similar to those on a building site were simulated.

In contrast to existing design rules, preheating is omitted, so as to reduce costs (approx. 15% of joint production costs). Nevertheless, humidity on the components must be avoided to eliminate hydrogen

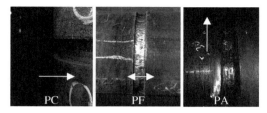

Figure 2. Welding positions.

Table 4. Shape of the weld according EN ISO 9692.

Wall thickness	Weld type	Geometry
20/30	V-weld	$\alpha = 2*20°$ gap b = 2 mm
55	single V-butt joint	$\beta = 2*15°$ gap b = 4–8 mm

Table 5. Testing program status.

Series	Diameter [mm]	Test set-up	Status
1	193.7	small tension tests (STT)	completed
		bending tests (BT)	completed
		tension tests (TT)	running
2	298.5	bending tests (BT)	completed
3	508.0	bending tests (BT)	completed

brittleness. This can be achieved e.g. with drying of the specimen parts before welding.

The weld preparation is based on EN ISO 9692 with the geometry provided in Table 4. For cost effective welds a reduction of the weld metal needed is essential.

2.3 Testing arrangement

Three different test rigs are used for the specimens, one for small test specimens under tension load (STT), another for the tension tests (TT) and the last one for bending tests (BT). An overview of the tests is presented in Table 5 and the testing parameters are presented in Table 6. The cycles to failure of a specimen is taken to be when the first crack occurs through the wall thickness (through crack).

All tests, except the small tension tests, are carried out with two butt welded joints (one at each end of a casting) tested simultaneously. For the small tension tests, specimens are cut out of butt welded joints between hollow sections and steel castings for testing.

For the tests with two butt welded joints either in variation cast steel—steel tube—cast steel or in the variation steel tube—cast steel—steel tube, there would always be only one set of test results biased towards the weaker joint, allowing a conservative safety margin for the evaluation. In a subsequent project, repair welding on the failed joint will be carried out to provide additional results if the second joint fails first, or instead to provide insights on the durability of repair welds if the repaired joint fails first. To date, all planned tests except the tension tests are completed, therefore providing a partial insight into the research work.

Figure 3. Full scale bending test (diameter 508 mm and wall-thickness 55 mm).

With the tension tests, a uniform stress along and across the section is possible, revealing any weak spots in the weld. However, high loads would have been necessary for the larger tests. Lower test loads and lower testing costs are achieved in a 4-point bending test. Also, higher load cycles are possible for the bending tests, resulting in a shorter testing time. However, only a small part of the butt weld along the circumference of the joint will have the highest stress. If this part of the weld is error-free, it is obvious that the expected fatigue life would be higher than in the tension tests, where the weakest part will fail first.

A special arrangement was made for the largest tests with a diameter of 508 mm and a wall-thickness of 55 mm (Fig. 3). Here, the test specimen was hung vertically in the test rig along the system axis in two bearing points. For the tests, the specimens were loaded with an out-of-balance exciter to induce the resonance frequency of the specimen with both bearing points located in the zero points of the oscillation. With this test set-up, a stress ratio of $R = -1.0$ and a testing frequency between 20 to 22 Hz was realized.

Table 6. Testing parameters.

Series	Diameter [mm]	Test Set-up	Welding position	Variants	Specimen per variant
1	193.7	STT	PC	all	4
		BT	PF	all	2
		TT	PC	2,3,5,6	3
2	298.5	BT	PC	2,5,6	4
3	508.0	BT	PA	2,5	2

3 TEST RESULTS

3.1 Inspection results

For all test specimens, ultrasonic inspections have been carried out. All the welding for the first series was performed without any special instructions given to the welder. In fact, the welder received no information other than that specified in the welding procedure sheets and was not informed that the welding was to be tested ultrasonically later on. As a result, the welders were not induced to give special attention to the welds, so that lower welding quality was observed than for the other series, where the welder was instructed beforehand of the non destructive tests.

All failures in the welds for the tests carried out up to date started at the backings. The test results were not influenced by the outer surface of the weld, therefore any irregularities on the outside were irrelevant. The welding for all different variants showed that in the first and second series, the variants 1 and 2 were the best for verifiability and quality of the welding. The variants 3 and 4 provided moderate results concerning the weld quality. A special problem was the welding of the square edge in variants 5 and 6. A lack of fusion, especially in the first series, was observed and during ultrasonic inspection, every weld pass could be recognized, which leads to a lower quality.

Nevertheless, all specimens were tested with the given defects in the weld. Especially for the bending tests, the most severe defect detected in the weld was turned into the most stressed area, to provide a possible starting point for the crack. In this way, the test results are on the safe side.

The welding position also influences the error occurrence. Welding positions PA and PF were better than the welding position PC. They were easier to manufacture and Position PA in particular led to only small errors compared to the size of the welds.

3.2 Strain gauge measurements

The stresses applied during the tests had also been confirmed in strain gauge measurements, which have been taken in an undisturbed area of the test

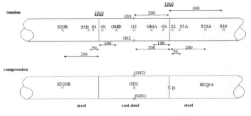

Figure 4. Strain gauge locations for test specimen.

specimens without stress peaks. Further strain gauge measurements had been carried out on the weld seam of the test specimens. A typical test specimen with strain gauge locations is shown in Figure 4.

The stresses from the finite element (FE) analysis of variant V3 in Figure 5 gave good agreement with those determined from the strain gauge measurements. In Figure 5, some measurements at various load levels are presented. To reduce the number of strain gauges, the measurements were performed in the following way: The test specimen was rotated into the different positions and then the stresses were recorded at each strain gauge. The rotation points were numbered 1 to 5 with the dedicated angle of 0° to 180°. Similar investigations were performed for different support distances "a" of 600, 700, 800 and 900 mm and, with adjusted load to get the same stresses in the specimen, only slight differences in the stresses were observed. Further strain gauge measurements on different test specimens showed the same behaviour.

3.3 S-N-curves

All fatigue tests have been stopped, when a crack through the wall-thickness (through crack) was visually observed. All failures for tests observed to date started at the weld root. After the through crack appeared, testing was continued to study crack growth. The crack propagated in the weld area or the heat affected zone. Within about 10,000 cycles, it propagated around about 33% of the circumference of the tube and the test was stopped.

In Figure 6, a test specimen with the measured crack growth is given. The crack initiation was determined with the displacement measurement of the testing machine during the test. With up to 0.2 mm displacement in the machine, no crack was visible

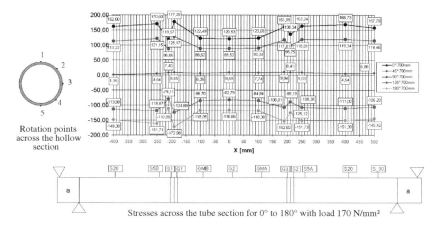

Figure 5. Stresses across the specimen section.

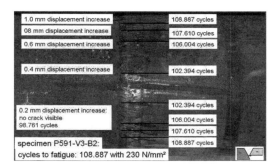

Figure 6. Crack propagation on test specimen P591-V3-B2.

Table 7. Fatigue parameters.

Series	Diameter [mm]	Test Set-up	Stress ratio R	Load levels N/mm²
1	193.7	STT	0.1	90/120
		BT	0.2	170/223
		TT	0.2	92/137/183
2	298.5	BT	0.2	120/130/170
3	508.0	BT	−1.0	see Table 8

Table 8. Large scale testing parameters.

Specimen/weld	Variant	Calculated stress in cross section [N/mm²]	Cycles to failure
P591-508-1-1	5	155	62,000
P591-508-1-2	2	135	72,000
P591-508-2-1	5	80	440,000
P591-508-2-2	2	72	470,000

during testing. A further rapid increase in displacement correlated with the crack propagation. With a 0.4 mm displacement and a total of 102,394 cycles of fatigue in the example in Figure 6, the first crack was observed. With a displacement increase up to 1.0 mm the testing was stopped at 108,887 cycles with a crack extending around 33% of the circumference.

It is intended to perform repair welding on the cracked joint and to continue testing with a subsequent research program.

The fatigue loading parameters for the tests are presented in Table 7. The maximum stress is converted to the load to be applied to the specimens on the basis of nominal dimensions of the structural hollow sections. The fatigue classes have been determined at 2 million load cycles. The fatigue curves in the following S_R-N_f-diagrams are plotted up to 5 million load cycles.

The method of testing for the large scale bending tests allowed the joints at each end of the steel casting to be tested with different stress ranges. To get a value for the cycles to failure, the tests were continued after the first joint showed a crack upto appearance of a through crack in the second joint. Furthermore the crack growth was investigated for the first crack.

Figure 7 gives the results of the test specimens under bending load. The different nominal stress ranges are calculated on the basis of the stress in the outer fibre of the steel tubes. The calculated stresses also corresponds to the strain measurements carried out. The line with the slope $m = 5$ is a lower bound of experimental results. End to end structural steel hollow section joints have a detail category of 71 (tension load) with $m = 3$ in EN 1993-1-9 (2005). The present work for joints between structural steel

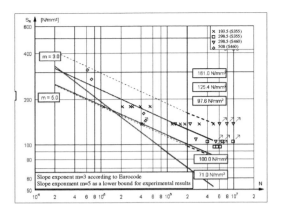

Figure 7. Test results for bending specimens up to through crack.

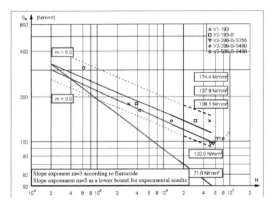

Figure 8. Test results for bending specimens variants 1 and 2.

hollow sections and cast steel show a tendency for a flatter slope of $m = 5$ and a detail category of 100.

In these figures both steel grades are evaluated together. The results are therefore applicable to the higher steel grade of S460, also. Some of the specimens in the second series with diameter of 298.5, especially in the lower stress ranges, did not exhibit any failure when they received more than 5 million cycles of failure (marked with arrows in Figs. 7–8). These experiments were stopped and will be continued in a subsequent project.

As observed by the ultrasonic inspection, the quality of the investigated welds of variants 1 and 2 in the first series is the best and variant 6 the worst. The results of variant 5 are better than for variants 3 and 4, attributed to the better weld quality. This is also the case for the second test series. Here, the results of variant 5 are almost as good as variant 2, what was not

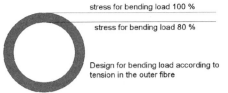

Figure 9. Stress at the inner side of the tube.

expected before. This shows that, when a good weld quality is maintained, a weld backing can be omitted. In Figure 8, a seperate evaluation of the variants 1 and 2 under bending load is given. With this detail, a category of 120 seems to be possible.

The tension tests are presently still being performed, so that only some test results are presented here. The individual variants cannot yet be clearly interpreted, but the fatigue behaviour in the variants is similar, namely, variants 1, 2 and 5 give better results than variants 3, 4 and 6. However, in the tension tests the whole joint is subjected to a uniform maximum stress and not only a small part. Therefore, the stress situation in the weld is different from that in a 4-point bending test, where only the outer fibre has the maximum stress. In the tests, the crack always started from the weld root. Here, under bending load, a reduced stress of only 80% of that in the outer fibres is applicable, in comparison to the tension tests (Fig. 9).

With this, the tension tests performed so far (1 for variant 2, 1 for variant 3 and 1 for variant 5) give a tendency towards a detail category 67 with $m = 5$, compared to 100 with $m = 5$ for the bending tests. One specimen of V6 (with arrow) was also tested, but not included in the analysis, since fatigue failure occurred at the support of the casting. The additional test for variant 2 shown in Figure 10 (cross with an arrow) is still being tested, and therefore not included in the above analysis.

The results for the small test specimens (see Fig. 11) under tension load are in accordance with the results for the variants in the bending tests. However, in the case of the small specimen tests, a reduced fatigue life is observed.

An evaluation of all small test specimens under tension load suggests a line with the slope of m = 5 through 57 N/mm² at 2 million load cycles as a lower bound. However, EN 1993-1-9 (2005) gives a detail category of 71 for curved plates. A separate analysis for the variants 1 and 2 also leads to this higher detail category. For flat plates with an eccentricity, a reduction factor k_s is to be used Equation 1:

$$k_s = \left(1 + \frac{6e}{t_1} \frac{t_1^{1.5}}{t_1^{1.5} + t_2^{1.5}}\right) \left(\frac{25}{t_1}\right)^{0.2} \quad (1)$$

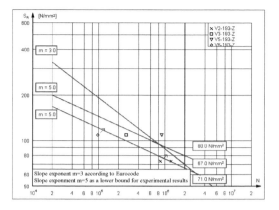

Figure 10. Test results for specimen under tension.

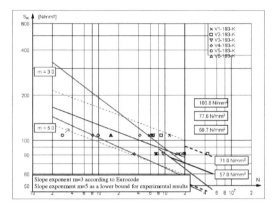

Figure 11. Test results for small test specimens under tension.

where e = eccentricity of the connection between both plates; t_1 and t_2 the wall thicknesses (with $t_2 \geq t_1$).

If this factor is applied to curved plates for the eccentricity as encountered in variants 3 to 6, a correction factor of 0.68 for the fatigue resistance is obtained. A reduction of 68% applied to detail category 71 leads to 48 N/mm² as a lower bound, with the test results exceeding 57 N/mm².

3.4 Further testing

The tests described above are not yet completed and still in progress. Only a part of the testing program has been evaluated, to provide a preliminary insight into the research work. After the tests are completed, repair of the failed test specimens under similar conditions to those at a building site is planned.

Fatigue tests will give information of the residual life of repaired end-to-end connections.

In a theoretical FE analysis, the basic hot spot stresses and detailed notch stresses for each variant are to be determined. In further investigations, effects concerning the practical design of the joints will be taken into account. The given geometrical tolerances and the influence of wall-thickness discrepancies will particularly be modeled, to ascertain this influence. Therefore, special importance is attached to the measurements of the test specimens. For the later assessment of results and further FE-analyses, a precise documentation will be made. In a second step, a fracture mechanics approach to the development of cracks in selected areas will be taken into account.

3.5 Cost effectiveness

Another point in the evaluation is the cost of producing each variant. The production costs necessary to fabricate the joints are related to the fatigue life of each joint, to determine which variant has the cheapest cost to fatigue life ratio. On this basis, a choice can be made, whether a cost effective production or a required fatigue life is preferred.

For the production costs, the following parameters are taken into account: The volume of the weld determines how much weld material is necessary. Furthermore, the welding of smaller volumes can be faster, so that the reduced labour costs are to be included. Another consideration is the accessibility and the verifiability of the weld defects in ultrasonic inspection. Here again, labour costs can be minimized with the right choice of variant. In addition, the production process itself has to be taken into account. The preparation of the parts to be welded on the flanks or the weld backing has an effect. Furthermore, for a cost effective design, the necessary cuts and chamfers on a tube should be minimized.

All these production processes lead to different production costs for each variant. Together with the fatigue life described in the previous section, Figure 12 gives some details. The research work is still continuing, so that the given values are only a first insight into the cost effectiveness.

In this diagram, variant 2 is used as a reference variant. The production costs and the fatigue life are set to 100% and the other variants are plotted in relation to this. The production costs are supplied by the companies involved in the research project or are gathered during the tests.

Variants 1 and 2 give identical results. The ceramic panel has no significant influence observed in the tests. Variants 3 and 6 have a low fatigue life, which is due to the defined crack initiation on the weld backing.

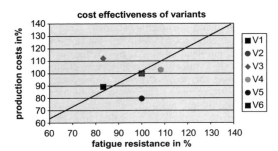

Figure 12. Fatigue life and production costs.

Variant 3 is the most expensive in production, with both the tubular steel and the casting requiring to have their ends appropriately processed. The best solution with regard to fatigue life is variant 4, where the weld root was given special treatment. Surprisingly, variant 5 had the same fatigue life as observed in variants 1 and 2 but with considerably reduced production costs. Further investigations will show whether these observations can be confirmed.

4 CONCLUSIONS

The properties of cast steel from reputable foundries are comparable nowadays to rolled steel, particularly with regard to weldability and toughness. Comparing only test specimens without weld defects, the failures in the joints between steel tubulars and steel castings detected during the investigations are comparable to those between tubular steel members. Also, casting defects within the tolerances of Table 2 have no observed influence on the fatigue life.

Experimental and numerical studies were initiated with the aim of establishing solutions to improve fatigue behavior of end-to-end connections with respect to costs and desired quality levels. The results are presented in this paper. The reported research work is aimed towards an economic design of butt welds in joints between hollow sections and cast steel nodes.

ACKNOWLDGEMENTS

The authors gratefully acknowledge the financial support of FOSTA (Forschungsvereinigung Stahlanwendung e.V.) and CIDECT for these investigations. The authors would also like to thank the members of the working group for the research project P591 "Wirtschaftliches Bauen von Straßen- und Eisenbahnbrücken aus Stahlhohlprofilen" and the research project CIDECT 7W "Fatigue of End-To-End CHS Connections".

REFERENCES

AD HP 5/3 and App.1. 2002. Manufacture and testing of joints–Non-destructive testing of welded joinst.

CIDECT Report – Project 7W. 2007. Fatigue of End-To-End CHS Connections, Sixth Interim Report, University of Karlsruhe.

DIN 1681. 1985. Steel castings for general purposes; technical delivery conditions.

EN 287: 2004. Qualification test of welders – Fusion welding–Part 1: Steels.

EN 1369. 1997. Founding, Magnetic particle Inspection.

EN 1370. 1997. Founding, Surface roughness inspection by visualtactile comparators.

EN 1371-1. 1997. Founding, Liquid penetrant inspection–Part 1: Sand, gravity die and low pressure castings.

EN 1559-1. 1997. Founding, Technical conditions of delivery.

EN 1559-2. 1997. Founding, Technical conditions of delivery, Part 2: Additional requirements for steel castings.

EN 1993-1-1. 1993. Design of steel structures – Part 1-1: General rules and rules for buildings.

EN 1993-1-9. 2005. Design of steel structures – Part 1-9: Fatigue.

EN 10210. 2006. Hot finished structural hollow sections of non-alloy and fine grain steels.

EN 10293. 2005. Steel castings for general engineering uses.

EN 12680. 2003. Founding–Ultrasonic inspection – Part 1: Steel castings for general purposes.

EN 12681. 2003. Founding–Radiographic inspection.

EN 17182: Steel castings with improved weldability and toughness for general purposes, DIN, Mai 1992.

EN 17205. 1992. Quenched and tempered steel castings for general purposes.

EN ISO 5817. 2003. Welding–Fusion-welded joints in steel, nickel, titanium and their alloys (beam welding excluded)–Quality levels for imperfections.

EN ISO 9692. 2004. Welding and allied processes–Recommendations for joint preparation.

FOSTA 4. Zwischenbericht. 2007. Wirtschaftliches Bauen von Straßen- und Eisenbahnbrücken aus Stahlhohlprofilen, University of Karlsruhe.

Herion, S., Veselcic, M & Puthli, R. 2007. Cast steel–new standards and advanced technologies, *5th International Conference on Advances in Steel Structures*, Singapore.

Herion, S. 2007. Guss im Bauwesen, Sonderdruck aus Stahlbau Kalender 2007, Ernst & Sohn.

Veselcic, M., Herion, S. & Puthli, R. 2003. Cast steel in tubular bridges–New applications and technologies, *Proceedings of the 10th International Symposium on Tubular Structures*, Swets & Zeitlinger ,Lisse, Netherlands.

Veselcic, M., Herion, S. & Puthli, R. 2006. Selection of butt-welded connections for joints between tubulars and cast steel nodes under fatigue loading, *Proceedings of the 11th International Symposium on Tubular Structures*, Taylor and Francis, London, United Kingdom.

Veselcic, M., Herion, S. & Puthli, R. 2007. Cast Steel and Hollow Sections–New Applications and Technologies, *Proceedings of the 17th International Offshore and Polar Engineering Conference (ISOPE-2007)*, Lisbon, July 1–7, 2007.

Welding recommendations for modern tubular steels

M. Liedtke & W. Scheller
Salzgitter Mannesmann Forschung, Duisburg, Germany

J. Krampen
Vallourec & Mannesmann Tubes, Düsseldorf, Germany

ABSTRACT: Good weldability is the necessary precondition for the use of modern tubular steels in structural components. Increasing requirements in terms of higher strength and toughness properties signify challenging tasks for fabricators. In particular small and medium-sized companies that are not so experienced in handling these grades require support from the tube supplier in the form of welding recommendations. In-house process development is, in the majority of cases, not economically feasible and this lack of knowledge could lead in the worst case to a misapplication or wrong handling of these steel tubes.
  In order to meet customers needs and avoid incorrect welding of these materials, Vallourec & Mannesmann Tubes has, in cooperation with the Salzgitter Mannesmann Forschung in Duisburg, Germany, evaluated an R&D program for supply of welding recommendations, custom-tailored to their modern steels in terms of chemical composition, wall thickness, heat treatment state and the appropriate welding processes.
  This presentation examines the technical and commercial implementation of this project and the success of the strategy for provision of welding recommendations to customers.

1 INTRODUCTION

Structural steel grades are used in a diverse range of industrial applications and steel construction projects. Figures 1 to 3 show examples of the use of these grades in the fabrication of legs for oil-rigs. Due to these mechanical properties and chemical composition, these steels have to be handled more or less carefully during fabrication by welding. Depending on the chemical composition of the steel grade and the hydrogen content in the filler material, the risk of high hardness values and cold-cracking in the deposited weld metal and heat affected zone(HAZ) can be reduced by pre-heating the weld area. Pre-heating, comprising at the very least the avoidance of condensation on the bevel edges, is necessary in most cases. The weld area must be kept free of moisture during welding. In general, the pre-heating temperature depends essentially on the carbon and alloying element contents of the base metal. Knowledge of the chemical composition is necessary for correct pre-heating. For the fabrication of structures and connections made of steel grades with high carbon equivalents and high strength levels, pre-heating is necessary before welding, since otherwise properties in the HAZ may deteriorate, as a result of high hardness values and/or cold-cracking. (EN 1011: 2000, EN ISO 3690: 2000)

The hardness of the microstructure adjacent to the fusion line is, as stated above, dependent on the chemical composition of the base metal and the cooling rate of the austenitized material. One of the most critical elements which influence the hardness of the base metal in this context is carbon. With the help of pre-heating, an increase in hardness in the areas

Figure 1. Application of structural steels for legs of oil-rigs.

Figure 2. Fabricated steel legs for oil-rigs.

Figure 3. Application of tubular steels in oil rigs.

close to the fusion line can be effectively avoided or restricted.

For a given base metal (steel grade), hardness can be decreased only by reducing the cooling rate of the areas of the joint heated to above Ac_3. The Ac_3 temperature is, in general, the temperature at which transformation of ferrite to austenite is completed during heating. This requirement is achieved by pre-heating. During this heat treatment, the components should preferably be heated at a constant rate and should be kept at this minimum temperature during welding. Arc-based welding processes are generally used in industrial applications. Therefore, the following section refers to these processes.

The factors influencing the properties in the weld area, the methods for pre-heating and the performance of the evaluated PRE-H-TEMP program are described below.

2 METHODS FOR PRE-HEATING JOINTS

Pre-heating prior to welding may take a number of forms. One of the simplest methods is pre-heating using a propane torch. This procedure is characterized by good mobility and autonomy in the field compared to other methods, such as heating blankets and induction coils, both of which require a power supply. The advantages of induction coils and heating blankets over propane torches are the shorter heating-up time to reach the working point, and more uniform heat distribution in the component. In addition, welders can work more efficiently during pre-heating because less heat is reflected from the work. Once welding has been completed, it is necessary to allow the weld area to cool to ambient temperature slowly and to shield this defined area against ambient conditions such as moisture and draught, in order to avoid excessively rapid cooling and thus the formation of an unfavourable microstructure in the weld zone. Slow cooling can, for example, be achieved by wrapping the weld area in special heat-resistant blankets.

2.1 Essential pre-heat temperature

The basic formula for calculation of the necessary minimum pre-heat temperature $T_{P,\min}$ is given by Equation 1. This formula is specified in the relevant standards, such as the German Stahl Eisen Werkstoffblatt (SEW) 088 (1993) code and EN 1011: 2000. As shown in the formula, the calculated pre-heat temperature depends on the carbon equivalent CET, heat input Q during welding, wall-thickness wt, and the hydrogen content of the filler material HD. For determining the minimum pre-heat temperature $T_{p,\min}$ the influence of these individual factors can be summarized as follows:

$$T_{P,\min} = 697 \times CET + 160 \times \tanh\left(\frac{wt}{35}\right) + 62^{0.35} \times HD + (53 \times CET - 32) \times Q - 328 \quad (1)$$

The wall thicknesses and chemical composition of the base metals to be used are known or can be provided by the supplier. Heat input Q is defined by the welding parameters in general and is dependent on factors such as welding technique (string or weaved-bead), welding position and travel speed. The necessary information concerning hydrogen content in the weld metal is provided by the consumables manufacturer. Equation 1 applies for HD contents in the 1 ml/100 g to 20 ml/100 g range. The amount of diffusible hydrogen in the weld metal can be measured using the currently valid (mercury) method described in ISO standard EN ISO 3690: 2000, or by means of equivalent analytical procedures. The classification of the hydrogen contents in weld metal, such as that produced using basic-coated electrodes, for examples, is shown in Table 1.

Equation 1 applies to steel grades with yield strengths of up to 1000 MPa and

- CET of 0.2% to 0.5%,
- Wall thickness of 10 mm to 90 mm,
- HD content of 1 ml/100 g to 20 ml/100 g,
- Welding heat input Q of 0.5 kJ/mm to 4.0 kJ/mm.

Values outside the ranges mentioned above are not examined here. The various factors included in the formula for calculation of the pre-heat temperature are described in more detail in the following sub-sections.

2.2 Carbon equivalent CET

The influence of chemical composition on the cold-cracking behaviour of the base metal is expressed by carbon equivalent CET. Carbon equivalent is calculated in accordance with SEW088 (1993) or the EN 1011: 2000 standard shown in Equation 2. This equation describes the influence of the individual alloying elements on the above-mentioned properties, referred to that of carbon.

$$CET = C + \frac{Mn + Mo}{10} + \frac{Cr + Cu}{20} + \frac{Ni}{40} \quad \text{in \%} \quad (2)$$

The validity of the CET depends on alloying element content. The corresponding chemical element content ranges (in % by mass) are stated in Table 2. These element contents correspond to steel grades relevant for typical structural component applications.

2.3 Heat input Q

Heat input in accordance with the relevant standards, which describe the welding parameters in general, is defined in Equation 3, in which k is a factor that takes account of the arc-based welding processes shown in Table 3. Factor k itself is the degree of thermal efficiency of the most widely used welding processes. This factor expresses the thermal efficiency of the particular welding process referred to that of the submerged-arc process, which has $k = 1.0$ (i.e. the highest degree of thermal efficiency).

$$Q = k \times \frac{U \times I}{v} \quad \text{in kJ/mm} \quad (3)$$

where
U = welding voltage;
I = welding current;
v = welding speed and
k = k-factor in Table 3.

Heat input Q depends in general on a number of variables, such as the welding parameters shown in Equation 3. In addition, factors such as the welding performance and welding position also have a great influence on Q. Parameters such as voltage and current can be determined via selection of the welding parameters on the power source unit. This data can be measured, on the one hand, by source internal measuring units or, on the other hand, by

Table 1. Classification of HD content in weld metal.

Amount of diff. hydrogen in weld metal *[cm³/100 g]	Evaluation
HD > 15	high
10 > HD ≤ 15	medium
5 > HD ≤ 10	low
HD ≤ 5	very low

*According to Schulze, G. (2003) the HD value is the volume of diffusible hydrogen in cm³, relating to 100 g applied weld metal.

Table 2. Range of alloying element content for the validity of the CET.

Element	Mass %
carbon	0.05–0.32
silicon	max. 0.8
manganese	0.5–1.9
chromium	max. 1.5
copper	max. 0.7
molybdenum	max. 0.75
niobium	max. 0.06
nickel	max. 2.5
titanium	max. 0.12
vanadium	max. 0.18
boron	max. 0.005

Table 3. Degree of efficiency of arc related welding processes.

Process no. (Acc. to EN ISO 3690 2000)	Process	Factor k
111	Shielded Metal Arc Welding	0.8
121	Submerged Arc Welding	1.0
131/135	Gas Metal Arc Welding	0.8
141	Gas Tungsten Arc Welding	0.6

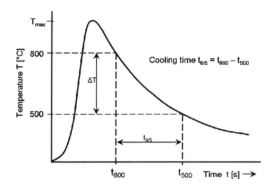

Figure 4. Typical temperature-time cycle and definition of cooling time $t_{8/5}$.

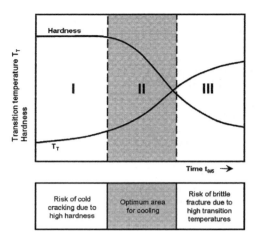

Figure 5. Influence of different $t_{8/5}$ values on hardness and transition temperature (EN ISO 3690: 2000).

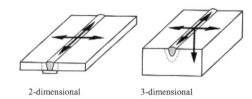

Figure 6. Variants of the heat flow (EN ISO 3690:2000).

means of external data acquisition systems. In addition, direct averaging of the individual indicated values can cause relatively high measurement errors in time-dependent process models (e.g. pulse processes). Welding speed, also being an important factor, is dependent on welding position (e. g. vertical up/down), the weld width or any oscillating movement of the torch which may be necessary.

2.4 Cooling time $t_{8/5}$

Another factor of major importance for the mechanical properties of the weld is the temperature-time cycle. This cycle is influenced in particular by weld geometry, heat input, pre-heating temperature and weld layer details. The temperature-time cycle during welding is generally expressed by cooling time $t_{8/5}$. During welding, time $t_{8/5}$ is used to characterize the temperature-time cycle of an individual weld bead and the time taken for a weld run and its HAZ to pass, during cooling, through a temperature range from 800 °C to 500 °C. This distinct cycle is shown in Figure 4. The corresponding influence of the cooling rate and effects are presented in Figure 5. For low $t_{8/5}$ values the transition temperature T_T is low and therefore the safety for brittle fracture is high, but the high hardness may increase the risk of cold cracking (area I). On the other hand, very high $t_{8/5}$ values lead to explicit lower hardness values, but the impact energy toughness is notably depleted due to the formation of soft ferrite microstructure components (area III).

Therefore, the welding parameters have to be selected in such a way, that the cooling time does not exceed or fall outside a particular range. This optimum range for cooling is within the area II in Figure 5. This graph is in general applicable for un- or low alloyed steel grades. The different "areas", where problems such as brittleness or cold cracking can occur with a wrong heat steering during welding, are considered in the software tool PRE-H-TEMP©, which is described in more detail in the next section.

Another output for fabricators is the heat flow characteristics in the joint segments during cooling which can be either 2- or 3-dimensional. In this context, cooling in the 2D direction means that the heat flow is only in the plain of the plate section and not in the wall thickness direction which is relevant for 3D heat flow. The wall thickness (wt) at the transition from 2- to 3- dimensional heat flow is called as the transition wall thickness ($wt_{transition}$). The transition wall thickness can be calculated by equating the formulas for calculating the 2- or 3-dimensinal heat flow and dissolve the equation system after the wall thickness wt. 2D is mainly valid for "thinner" wall thicknesses where $wt < wt_{transition}$, (EN 1011: 2000, EN ISO 4063: 2000). 3D ($wt > wt_{transition}$) means, that the heat can flow in all directions and is widely independent of the wall thickness. The possibilities of the heat flow characteristics are visualised in Figure 6.

written by: Dr.-Ing. Markus Weissenberg

Vallourec & Mannesmann Tubes

Material:	Material A
heat No.:	1234
order No.:	56789

Limits of applicability:
0,2% < CET < 0,5%
10 mm < wt < 90 mm
1 ml/100g < HD < 20 ml/100g
0,5 kJ/mm < Q < 4 kJ/mm

Dimensions [mm]:

Hollow section with circular profile			
D	250	T	20

Hollow section with rectangular/square profile			
B	H	T	

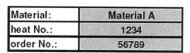

circular profile   rectangular profile   square Profile

All grey fields must be set by user!

Chemical composition (base material):

C	Si	Mn	Cr	Mo	Ni	Cu	wt	HD	Q	min. $t_{8/5}$	max. $t_{8/5}$
%	%	%	%	%	%	%	mm	ml/100g	kJ/mm	s	s
0,2	0,35	0,5	0,4	0,4	0,3	0,3	30	5	1,8	8	25

CET=	0,33	%	$t_{8/5}$ =	5,6	s, 3D
$T_{P,min}$=	98	°C	T_P=	149	°C
$T_{P,max}$=	200	°C			

CET=C + (Mn+Mo)/10 + (Cr+Cu)/20 + Ni/40

Q can be calculated if parameters are available, otherwise set Q directly in programme!

Heat input Q calculation:		1,8 kJ/mm
Current	[A]	100
voltage	[V]	31
welding speed	[mm/min]	105
k-factor		1

Figure 7. Screenshot of the software programme PRE-H-TEMP © with explanatory notes.

3 SOFTWARE PROGRAMME PRE-H-TEMP©

The PRE-H-TEMP© program shall be seen as a helpful tool for fabricators of structural components of different steel grades and in different industrial sectors by welding processes. The aim is to avoid any welding imperfections such as high hardness values, cold-cracking, etc. An example screenshot of the program's user interface is shown in Figure 7. Regarding to the above mentioned formulas, the relevant parameters have to be set by the user. The fields for user input of parameters have a grey background. These substantial parameters are as follows:

- Chemical composition of the base material—this data is submitted by the supplier of the steel components.
- Diffusible hydrogen content of the filler material (consumables)—this data can be received from the consumables manufacturer.
- Heat input (calculated by the welding parameters of current, voltage and weld speed).
- Wall thickness.

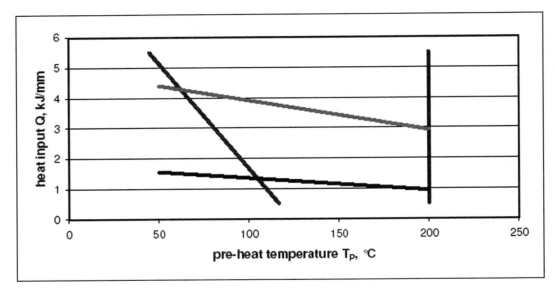

Figure 8. Process window depending on heat input and pre-heating temperature.

- Minimum and maximum cooling time ($t_{8/5}$ time) for bordering the optimum pre-heat area– these values are often empirical values.
- Maximum pre-heat temperature $T_{p,max}$ which depends on economical criteria.

One of the outputs supplied by the program is the carbon equivalent CET mentioned above, which is calculated automatically and displayed in the corresponding field.

The optimum process window for welding without any likelihood of problems such as cold-cracking, hardness-cracking, etc. occurring is within the area bounded by the four lines in Figure 8. The line on the left describes and illustrates the problems which can occur if a certain $t_{8/5}$ value or pre-heat temperature is not met during welding, i.e., the risk of cold-cracking. Below the lower line, high hardness as a result of low $t_{8/5}$ values (high cooling rates), and above the upper line, soft behavior due to excessively high $t_{8/5}$ values (low cooling rates), can occur. The right bar is used to delineate the maximum pre-heat temperature.

This software tool enables users to obtain important information for welding, in the form of recommended pre-heat temperatures and the applicable temperature range, and also the problems which can be expected if the temperature range is not met. Other information includes the cooling time ($t_{8/5}$ values) for the weld metal, the corresponding heat flow characteristics (2- or 3-dimensional) and carbon equivalent CET.

4 CONCLUSION

As discussed above, the risk of cold-cracking in the weld metal and heat affected zone is one of the primary problems in welding of structural steels and of steels in general. These problems are dependent on factors such as the chemical composition of the steel (base metal) and the hydrogen content of the filler material. One of the major methods of reducing these problems is the pre-heating of the weld area to eliminate moisture, and control of the cooling behaviour. Pre-heating can be performed in a number of ways, such as by using flame torches or heating blankets. In this context, Vallourec & Mannesmann Tubes has, in cooperation with Salzgitter Mannesmann Forschung, evaluated a software tool which provides welding recommendations (in this context, pre-heating temperatures) specifically for their modern steel grades, in order to meet the customers needs and avoid damage to these materials from incorrect welding. The critical parameters, such as chemical composition and the welding parameters (current, voltage, welding speed) can be entered in the software by the user. As an output, the user obtains the minimum pre-heat temperature for his application, and also the applicable heat flow characteristics. Minimum pre-heat temperature is displayed in the software mask, and also as a numerical value and in graphical form. The graphic also shows parameter dependencies and the possible process window in an easily comprehensible manner. It must be pointed out that this software is by no means exhaustive, but nonetheless provides useful aids for preheating strategies.

REFERENCES

EN 1011: 2000. *Welding – Recommendations for welding of metallic materials – Part 2: Arc Welding of ferritic steels, German version.*

EN ISO 4063: 2000. *Welding and related processes – List of processes and indenture numbers.*

EN ISO 3690: 2000. *Welding and allied processes – Determination of hydrogen content in ferritic steel arc weld metal.*

Kasuya, T., Yurioka N. 1995. Determination of necessary preheat temperature to avid cold cracking und varying ambient temperature. *ISIJ international,* Vol. 35, No. 10, pp. 1183–1189.

Schulze, G. 2003. *The metallurgy of welding, 3rd Edition.* Berlin: Springer.

Stahl-Eisen-Werkstoffblatt SEW088 Beiblatt 1, 4. Ausgabe (only available in german language) 1993. Schweißgeeignete Feinkornbaustähle – Richtlinien für die Verarbeitung, besonders für das Schmelzschweißen – Kaltrisssicherheit beim Schweißen; Ermittlung angemessener Mindestvorwärmtemperaturen.

Non-linear finite element simulation of non-local softening for high strength steel material

R.Y. Xiao, F.M. Tong, C.S. Chin & F. Wang
Civil and Computational Engineering Centre, Swansea University, Swansea, UK

ABSTRACT: The capability of numerical simulation for the stress—strain relation beyond the elastic-plastic region has been limited by softening modeling for brittle materials. An attempt has been made to develop a modeling technique based on the experimental behaviour of the high strength steel under normal and cyclic loading conditions. It is capable of capturing the complete stress distribution on the analysis. Different validation cases have been conducted to investigate the feasibility of this technique and results obtained have been satisfactory. The aim is to create a material model and numerical technique for high strength tubular structure analysis.

1 INTRODUCTION

The problem encountered in computational analysis in the softening region is the inability of conventional computational algorithms to handle matrix non-positivity (Tong 2007). Solution adopting Newton-Raphson (N-R) procedures for softening analysis will be aborted due to non-convergence, even if the softening stress-strain relation was defined. The arc-length option, readily available in ANSYS does provide an alternative for tension softening and in particular, compression buckling analysis. This option is capable of passing through the unstable region of the stress-strain curve and over into the strain softening region, preventing divergence by avoiding the numerical complexities that accompany the N-R solutions. However, it provides no indication of the stress states which corresponds to the failure modes and it indicates some further limitations (Crisfield 1991; Tong 2007). A few softening models have been developed and proposed (Mazars and Pijaudier-Cabot 1989; Feenstra et al. 1998; Komori 2002; Jefferson 2003; Celentano et al. 2004; Ling 2004; Rots and Invernizzi 2004). Not all of these have been widely accepted. Their implementation into finite element software for commercial applications is even more limited. This reflects the need for continuous extensive research in this field.

In this paper, a modelling technique capable of predicting the full stress distribution is proposed. The fundamental concept of this technique has adopted a Tension Softening Material (TSM) model proposed (Xiao and Chin 2004a, 2004b; Chin 2006; Xiao et al 2008) where it was applied on conventional and fibrous cementitious composites. A similar concept was used in the development of the proposed softening technique to investigate its adaptability for high strength steel materials. The aim of the research is to develop a material model and numerical technique to handle the high strength tubular steel structure computational analysis.

2 THE CONSTITUTIVE THEORY OF THE STRAIN SOFTENING MODELLING TECHNIQUE

The proposed softening technique was built based on credible laboratory observations on material behaviour in a series of normal and cyclic conditions. When the peak point of each cyclic loop is connected, it forms a similar hardening and softening path as a single monotonic test would follow. This principle of increasing (and decreasing) stresses with increasing load cycles was adopted in this model for the softening behaviour, as the hardening branch could be numerically solved in a straight-forward manner by selecting the available nonlinear options.

This approach is characterized by the Voce Equation 1 (Voce 1948) for the hardening and reloading stages whereby Equation 2 was proposed to define the monotonic softening behaviour;

$$\sigma = k_0 + R_0 \varepsilon^{pl} + R_\infty(1 - e^{-b\varepsilon\varepsilon^{pl}}); \text{ if } \varepsilon^{pl} \leq \varepsilon^{pl}_{peak} \quad (1)$$

$$\sigma = \sigma_{ult} + R_0^{soft}(\varepsilon^{pl} - \varepsilon^{pl}_{peak}) + R_\infty^{soft}\left[1 - e^{b^{soft}(\varepsilon^{pl} - \varepsilon^{pl}_{peak})}\right];$$

$$\text{if } \varepsilon^{pl} > \varepsilon^{pl}_{peak} \quad (2)$$

where k_0 = elastic limit (N/mm²); R_0 = plastic hardening modulus (N/mm²); R_∞ = asymptotic stress (N/mm²);

σ_{ult} = ultimate tensile strength (N/mm²); R_0^{soft} = plastic softening modulus (N/mm²); R_∞^{soft} = asymptotic softening modulus (N/mm²); constants b and b^{soft} (dimensionless) which controls the elastic-plastic hardening and softening transition fillets. σ is the current stress state (N/mm²), ε^{pl} the corresponding equivalent plastic strain (dimensionless) and ε^{pl}_{peak} is the equivalent plastic strain at peak stress.

When the ultimate tensile strength is reached, the reloading Voce equation takes the same expression but a slightly different definition than previously defined. Since we have little interest with the possible paths in which the reloading could follow, apart from its peak point, R0 can be considered as redundant. This is replaced by a negative constant R0* which no longer is the plastic hardening modulus to serve as a dummy value, giving rise to the bifurcation point in the Voce equation. For each subsequent reloading procedure, the pair of stress-strain states corresponds to a certain point along the monotonic softening branch. Hence, the softening region has been differentiated from the hardening region. Therefore, in the softening reloading region we have;

$$\sigma = k_0^* + R_0^* \varepsilon^{pl} + R_\infty^* (1 - e^{-b^* \varepsilon^{pl}}) \qquad (3)$$

At each reloading phase, the Voce equation is reassigned with a new set of parameters. By employing such an approach to simulate the material property change at each subsequent step, the solution process would avoid unconvergence due to negative stiffness.

Another Equation 4 was also investigated to provide an alternative to the exponential softening Equation 2 to define the descending branch of the stress strain curve. This equation was originally adopted for fibre-reinforced concrete (FRC) (Li et al. 1998). It consists of two coefficients α and β which defines the horizontal and vertical asymptotic stress values.

$$\sigma = \frac{\left(\dfrac{\varepsilon}{\varepsilon_{peak}}\right)}{\alpha\left[\left(\dfrac{\varepsilon}{\varepsilon_{peak}}\right) - 1\right]^\beta + \left(\dfrac{\varepsilon}{\varepsilon_{peak}}\right)} \sigma_{ult} \qquad (4)$$

where ε_{peak} = peak nominal strain; ε = current nominal strain; σ_{ult} = ultimate tensile strength (N/mm²); and coefficients α and β (dimensionless).

3 THE FINITE ELEMENT MODEL

Several validation cases have been carried out to determine the feasibility and efficiency of the proposed modelling technique (Tong 2007). The geometries were modelled by adopting solid elements for 3-dimensional modelling as shown in Figure 1, 4, 6. The initial material properties were assigned for the pre-peak analysis. As the peak stress is approached, a new set of updated material properties is re-assigned to the selected elements or element components. The complete analysis therefore consists of a series of material property update procedures at each subsequent softening loadstep. The linear elastic properties, i.e. Young's (elastic) modulus and Poisson's ratio, remain consistent and unchanged throughout the analysis.

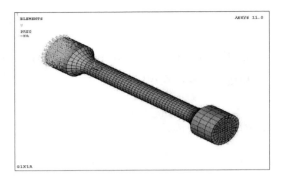

Figure 1. The element mesh of G1X1A with its applied boundary condition(s).

The selection of boundary conditions for the finite element models is rather straightforward. The one-end of the tension specimen was constrained in the x, y and z directions while the pressure (surface) loads were applied at the other end of the specimen.

4 VALIDATION OF EXPERIMENTAL TEST CASES

The following sections present a few cases which demonstrate the capability of the modelling technique. These will be presented in the sequence as material properties and gauge dimensions, element mesh and boundary condition, stress-deformation response and the stress distribution plots at peak and fracture.

The stress contours at the peak and fracture stress states with the conventions SMN and SMX represent the minimum and maximum stresses in the longitudinal direction, respectively. DMX is the maximum displacement between two ends of a specimen.

4.1 Circular solid steel specimen G1X1A

Case one of a circular solid steel rod has been modeled as detailed in Table 1. A comparison of the numerical results with the test data is shown in Figure 2. The stress distribution is indicated in Figure 3.

Table 1. Material properties and gauge dimensions of specimen G1X1A.

Source	Material properties			Gauge area	
	Young's modulus	Yield stress	Ultimate tensile strength	Length	Diameter
Barret 1999	205,000 MPa	125 MPa	350 MPa	25.4 mm	5.0 mm

Table 2. Material properties and gauge dimensions of specimen DP800.

Source	Material properties			Gauge area	
	Young's modulus	Yield stress	Ultimate tensile strength	Length	Width
Xin 2005	205,000 MPa	500 MPa	780 MPa	40.0 mm	20.0 mm

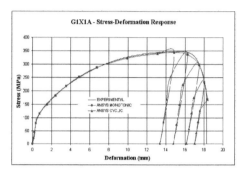

Figure 2. Monotonic and cyclic stress-deformation response of G1X1A.

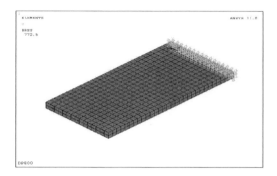

Figure 4. The element mesh of DP800 with the applied boundary condition(s).

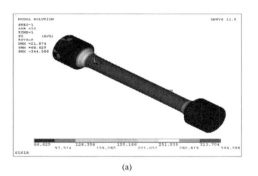
(a)

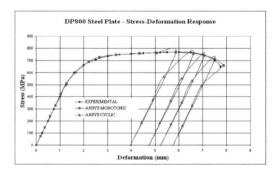

Figure 5. Comparison of numerical and test result for DP800.

4.2 Dual phase steel strip; specimen DP800

Case two of a steel strip has been used. Table 2 lists its details. The FE mesh is plotted in Figure 4. Figure 5 shows that comparison of the numerical result and test data matches well.

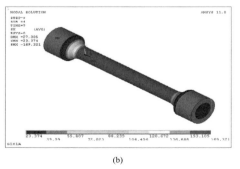
(b)

Figure 3. The stress distribution of specimen G1X1A at a) peak and b) fracture stress.

4.3 Non-heat-treated (NHT) circular hollow section (CHS); specimen FTS1A

Case three of a circular hollow section has been used as in Table 3. The FE mesh is shown in Figure 6.

Table 3. Material properties and gauge dimensions of specimen FTS1A.

Source	Material properties			Gauge Area		
	Young's modulus	Yield stress	Ultimate tensile strength	Length	Outer diameter	Thickness
Jiao and Zhao 2001	198,600 MPa	200 MPa	486 MPa	320 mm	38.1 mm	1.58 mm

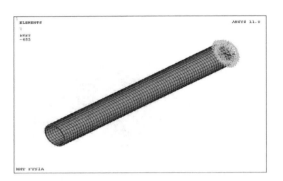

Figure 6. The element mesh of FTS1A with its applied boundary condition(s).

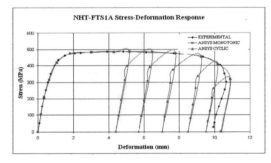

Figure 7. Comparison of numerical and test results for FTS1A.

The numerical and test results have been compared in Figure 7.

From Figures 2, 5, 7, the complete loading and unloading stress paths which were captured throughout the solution can be observed as to correspond to the monotonic response.

5 DISCUSSIONS

The effect of large deformation in the softening FEA has been achieved. Solution runs with large deformation have resulted in non-convergence in come conditions. This is due to the excessive stress and deformation localization occurring in the boundary region which results in this premature non-convergence. Although this could be overcome by increasing the stiffness in the boundary region, the large deformation option is kept deactivated in this preliminary study for simplicity.

For the future development, large deformation and geometrical instability should be implemented, especially if a modelled material exhibits considerable ductility. A ductile failure mode is characterized by the necking phenomenon whereby strain will be localized in this region when the yield stress is reached.

Three cases presented in the paper have demonstrated the capability of the proposed modelling technique to capture the complete stress distribution response in the analyses. Especially, it has demonstrated that it is able to handle with high strength material analysis. It has proved to be adequate with regard to the overall structural material behaviour. Therefore, it promises good prospects for further investigation of this modelling technique.

6 CONCLUDING REMARKS

Numerical simulation up to the peak stress has been proved only adequate for simple analysis. However, it provides no indication for the possible failure mode that is to follow with its corresponding stress distribution. The computational finite element simulation of the softening response is not possible in a direct manner by applying the softening theory of plasticity along with the conventional computational algorithm. Hence, an attempt was made to overcome this problem which has resulted in the development of the proposed modelling technique. This technique utilizes the loading-unloading-reloading characteristics which are obtained in experimental results along with a series of material properties updates. To demonstrate its capability and accuracy several validation cases have been used. It has been adopted in uniaxial tension test material specimens, whereby the results have been satisfactory but have yet to be implemented in more complicated multi-axial structural problems. This will be conducted in the further course of this research. The model used for the research on high strength tubular joint analysis will be reported separately.

REFERENCES

Barret, Z. (1999). The tensile test and the stress strain curves. Swansea, Swansea University.

Celentano, D.J., E. E. Cabezas, C. M. Garcia and A. E. Monsalve (2004). "Characterization of the mechanical behaviour of materials in the tensile test: experiments and simulation." Modelling and Simulation in Materials Science and Engineering 12.

Chin, C. S. (2006). Experimental and computational analysis of fibre reinforced cementitious composites. Civil & Computational Engineering Centre. Swansea, University of Wales Swansea. Ph.D.

Crisfield, M. A. (1991). Nonlinear finite element analysis of solids and structures, John Wileys & Sons.

Feenstra, P. H., J. G. Rots, A. Arnesen, J. G. Teigen, K. V. Hoiseth. (1998). A 3D constitutive model for concrete based on a co-rotational concept. Proceedings of EURO-C 1998.

Jefferson, A. D. (2003). "Craft—a plastic-damage-contact model for concrete II. Model implementation with implicit return-mapping algorithm and consistent tangent matrix." International Journal of Solids and Structures (40).

Jiao, H. and X. L. Zhao (2001). "Material ductility of very high strength (VHS) circular steel tubes in tension." Thin-Walled Structures 39.

Komori, K. (2002). "Simulation of tensile test by node separation method." Journal of material processing technology.

Li, Z. J., F. M. Li, T. Y. Chang, Y. W. Mai (1998). "Uniaxial tensile behaviour of concrete reinforced with randomly distributed short fibres." ACI Materials Journal.

Ling, Y. (2004). "Uniaxial true stress-strain after necking." AMP Journal of Technology 5.

Mazars, J. and G. Pijaudier-Cabot (1989). "Continuum damage theory—application to concrete." Journal of Engineering Mechanics of cohesive-frictional materials (ASCE) (115).

Rots, J. G. and S. Invernizzi (2004). "Regularized sequentially linear saw-tooth softening model." International Journal of Numerical And Analytical Methods in Geomechanics.

Tong, F. M. (2007). Nonlinear Finite Element Simulation of Non-local Tension Softening for High Strength Steel Material. Civil & Computational Engineering Centre. Swansea, Swansea University. MPhil.

Voce, E. (1948). "The relationship between stress and strain for homogeneous deformation." Metallurgist.

Xiao, R. Y. and C. S. Chin (2004a). Fracture and Tension Softening of High Performance Fibrous Concrete. Proceedings of the Seventh International Conference on Computational Structures Technology, Stirling Scotland, Civil-Comp Press.

Xiao, R. Y. and C. S. Chin (2004b). Nonlinear finite element modelling of the tension softening of conventional and fibrous cementitious composites. ACME Association of Computational Mechanics in Engineering, United Kingdom.

Xin, Y. (2005). Optimisation of the microstructure and mechanical properties of DP800 strip steel. Material Research Centre. Swansea, Swansea University. MPhil.

Xiao R. Y., Chin, C. S and Taufik, S, (2008) Nonlinear finite element modelling of the tension softening behaviour of brittle materials, Journal of Computer Methods in Applied Mechanics and Engineering (in press).

Experimental research and design for regular and irregular cast steel joints in tubular structures

Y.Y. Chen, X.Z. Zhao & L.W. Tong
Tongji University, Shanghai, China

ABSTRACT: The geometrical configuration of cast steel joints can be basically classified as regular and irregular ones. The former is manufactured in batch, the behavior of which can be studied systematically and designed following theoretical or empirical formula. The latter, however, has to be analyzed numerically or tested individually in most cases. In this paper, design methodology for both geometrical regular and irregular cast steel joints is studied. The behavior and design rules for two types of regular X-shaped cast steel joints, one with solid joint zone and the other with a feature of four-way joint zone, are studied and reported. General design criteria and important issues for irregular cast steel joints are also discussed. It is indicated that numerical analysis and physical test supplement each other due to their own advantages, especially to the complicated irregular joints.

1 INTRODUCTION

Cast steel joints are widely used in structural steel buildings in China in recent years. One reason of this increasing application is the need to connect multiple members in compact or narrow space, where welding is very difficult to perform with reliable quality and bolt connection is neither a satisfying option. A recent case is the 500-meter-high building in China, the Shanghai World Financial Tower, in which cast steel part is used to connect huge columns stretching in tree type. The requirement for aesthetic outlook of the joint, especially for those exposed structural connections in public buildings, also promotes the usage of cast steel joints.

On the other hand, the advanced development of high-strength and weldable casting steel makes the application possible. Although the present design code for steel structures in China specifies the design strength of cast steel being mere 200 MPa, a new specification has recently been drafted which raises the strength up to near 300 MPa (CECS235: 2008), based on the properties of newly produced cast steel. Furthermore, it is expected that cast steel with even higher strength can be put into practice (Herrion et al. 2007). Since the high quality cast steel has met the specification requirement of strength, elongation, toughness and weldability, the methodology adopted for common low carbon steel and low alloy steel is also suitable in studying the mechanical properties and designing of cast steel components.

The geometrical configuration of cast steel joints can be basically classified as regular and irregular ones. The former is manufactured in batch, the behavior of which can be studied systematically and be designed following theoretical or empirical formula. The latter, however, has to be analyzed numerically or tested individually in most cases.

2 REGULAR CAST STEEL JOINTS

Two typical regular cast steel joints are studied. One is with solid core in the joint zone designated as SC joint, the other with a feature of cross-junction joint zone and called CJ joint.

2.1 SC joint

SC joint was first used in the roof truss structures of Shanghai Pudong International Exhibition Center in China (Fig. 1). Figure 2 shows its geometric details.

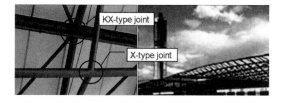

Figure 1. Shanghai Pudong International Exhibition Center.

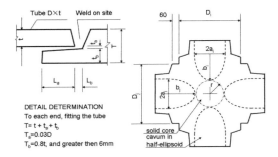

Figure 2. Details of SC joint.

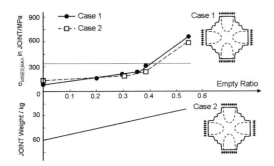

Figure 3. Maximum stress of SC joint with variation of 'empty ratio'.

In extreme condition SC joint can be a completly solid chunk of steel. The strength of the joint should be fully ensured, but at a cost of the joint weight being large and thus increasing the forces of members. So it is necessary to lighten the joint weight for the sake of safety and cost saving.

Cavum in half-ellipsoid are practically opened in SC joint as shown in Figure 2. With the increase of cavum depth, the stress in the joint induced by member force becomes great. It is clear that there is a limitation of the cavum space where the SC joint reaches its full strength. An example is illustrated in Figure 3, where an X-type SC joint connecting four CHS tubes is to be designed assuming the sections of the CHS tubes are identical. The thickness of the edge rim of the joint, T, is not less than two times of the tube wall as shown in Figure 2. In order to investigate the effect of empty ratio of SC joint on joint resistance, the cavum depth is taken as the only variable of geometrical parameters. Here, 'empty ratio' is defined as the cavum volume to the gross volume of whole joint. Take axial stress of CHS tubes exerted on joint rim is equal to yield strength of 345 MPa, it can be seen that when empty ratio is around 0.4 the maximum stress in joint reaches the material yield strength. When the half-ellipsoid cavum penetrate the solid core, the 'empty ratio' is about 0.53.

It is shown from the above example that a minimum core should be kept for safety. The size of the solid core depends on the design criteria to be adopted. The geometry of the SC joint can be designed based on the criteria and construction condition illustrated in the flowchart (Fig. 4).

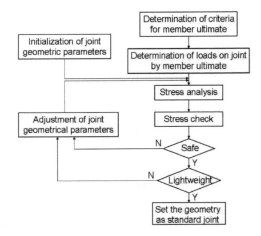

Figure 4. SC joint design flow chart.

As for X-type SC joint connecting CHS members, assuming the loads applied to the joint, axial force P and moment M, satisfy the correlation Equation 1, and the radius of solid core, r, is not less then 2T (Fig. 2), the stress in joint should meet the requirement of Equation 2a or Equation 2b, in condition that yield strength fy of cast steel is the same as that of steel CHS tubes.

$$\frac{M}{M_p} = 1 - 1.18 \left(\frac{P}{P_y}\right)^2, \quad \text{if } 0 \le \frac{P}{P_y} \le 0.65 \quad (1a)$$

$$\frac{M}{M_p} = 1.43 \left(\frac{P}{P_y}\right)^2, \quad \text{if } 0.65 < \frac{P}{P_y} \le 1 \quad (1b)$$

where P_y and M_p are yield axial force and plastic moment of the CHS member section, respectively.

$$\sqrt{0.5[(\sigma_1 - \sigma_2)^2 + (\sigma_2 - \sigma_3)^2 + (\sigma_3 - \sigma_1)^2]} \le f_y \quad (2a)$$

$$\sigma_1 \le (1.1 \sim 1.2) f_y, \quad \text{if } \sigma_3 > 0 \quad (2b)$$

(a) for CHS members (b) for RHS members

Figure 5. CJ joints.

where σ_1, σ_2 and σ_3 are principal stresses at a point, and tensile stress is taken as positive. If the principal stresses are all tensile, to prevent the brittle fracture, the major principal stress must be limited as shown in Equation 2b.

The same design procedure can be followed for joint with other geometrical configuration or loading combination, so that 'standard' joint is able to be brought forth.

2.2 CJ joint

Figure 5 shows the CJ joints both for CHS and RHS members. Figure 6 describes the elements of X-type CJ joint connecting RHS members. Plate I is a continuous element over the joint surface, while Plate II curves at the edge of the joint. Thus the stress flow path in Plate II turns at the edge, and most inter force will be transferred through Plate I, which is the essential difference of CJ joint from SC joint.

The moment in the tube axes plane is called in-plane moment, and that in the perpendicular plane out-of-plane moment. When the tube member is bent in plane, the stress distribution in the member far away from the joint is similar to that in a normal beam; while in the joint zone, there is magnificent stress concentration due to the poor ability of Plate II which can not transfer stress effectively. Experimental research and numerical study reveal that the concentration of stress and excessive plastic deformation in local part of the joint result in the low joint rigidity and poor joint capacity resisting both axial force and moment if the joint wall does not thicken noticeably compared with that of the tube wall it connects (Chen et al. 2006).

For simplicity, X-type CJ joint connecting RHS members with same section is studied.

2.2.1 Joint rigidity against bending moment

Based on the tests and FEM analysis, the initial bending rigidity is determined as follows, with reference to Figure 7 for the parameters.

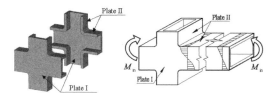

Figure 6. Geometry and stress flow character of CJ type joint connecting RHS joint.

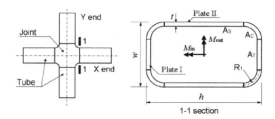

Figure 7. Typical section of CJ joint and the parameter.

$$K_{in} = 0.0142 E w^2 t \left[\left(\frac{h_0}{h}\right)^2 + 1.08\right]\left(\frac{w}{w_0}\right)^{0.65}\left(\frac{t_0}{t} + 11\right) \quad (3)$$

$$K_{out,XX} = 1.007 E h^2 t \exp\left(1.2 \frac{w}{w_0}\right) \quad (4a)$$

$$K_{out,YX} = -1.674 E h^2 t \left(\frac{h}{h_0}\right)\left[\frac{w}{(w - 0.548 w_0)}\right] \quad (4b)$$

In those equations, subscripts 'in' and 'out' refer to in-plane and out-of-plane bending rigidity, respectively; subscript 'XX' means the in-plane moment due to unit rotation angle of the same end, while subscript 'YX' means the in-plane moment in the given end but induced by unit rotation at the end of neighboring perpendicular member. All the equations are validated in the range of $w \in [100, 350]$, $h \in [100, 350]$ and $t \in [5,12]$, in mm, while standard RHS section ($w_0 \times h_0 \times t_0$) is $120 \times 280 \times 10$ mm. In all cases, R_1 is taken as $3t$. By numerical analysis according to orthogonal test, the maximum error of the above equations is less than 6%.

2.2.2 Joint resistance to in-plane moments

Non-linear FEM model was generated and validated using load-displacement curves obtained from test (Fig. 8). The moment capacity is defined as the one corresponding to the point at the curve where tangent stiffness becomes one tenth of its initially elastic stiffness.

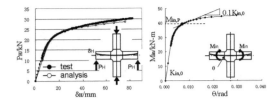

Figure 8. Load-displacement under in-plane moment.

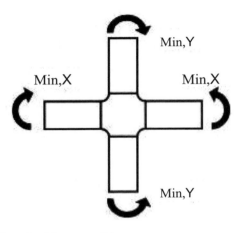

Figure 9. Moment condition.

Table 1. Comparison of in-plane moment capacity.

Case	$w \times h \times t$ mm	Mp by FEM kN-m	Mp by Eq.(6)	error %
1	120 × 160 × 10	38.0	37.4	−1.6
2	140 × 160 × 8	38.5	38.8	0.8
3	120 × 220 × 8	31.0	30.5	−1.6
4	140 × 220 × 10	50.0	50.1	0.2
5	150 × 280 × 15	90.0	89.3	−0.8
6	200 × 280 × 12	110.0	114.2	3.8
7	150 × 360 × 12	72.5	71.4	−1.5
8	200 × 360 × 15	145.0	147.8	1.9

The safety of CJ joint under in-plane moments should satisfy Equation 5, referring to Figure 9.

$$M_{in,X}^2 + M_{in,Y}^2 \leq M_{in,P}^2 \quad (5)$$

Moment capacity $M_{in,P}^2$ is defined as Equation 6.

$$M_{in,p} = M_T + M_C + M_S + M_S' \quad (6)$$

The components M_T, M_C and M_S represent the contribution of different portions in joint section, Plate I (A_T), curved corner (A_C) and Plate II (A_S) as shown in Figure 7, to the whole moment capacity, while M_S' is a special local component in Plate II produced by constraint effect of Plate I. According to FEM results, this component contributes about 5% of total moment resistance, varying with the wall thickness of the joint. Those components are calculated by Equation 7,

$$M_T = W_T(1.05 f_y), M_c = W_c(0.95 f_y), M_S = W_S \cdot \sigma_S$$
$$M_S' = W_{SE} \cdot \sigma_{SS} \quad (7)$$

The section moduli of different portions are given by Equation 8a and equivalent strength σ_S and σ_{SS} by Equation 8b.

$$W_T = 2(w - 2R_1)^2 \frac{t}{4},$$

$$W_C = \pi \left(R_1 - \frac{t}{2}\right) t \left[(w - 2R_1) + 2\frac{76t}{15\pi}\right],$$

$$W_S = (h - 2R_1) \cdot t \cdot (w - t) \text{ and } W_{SE} = 2(h - 2R_1)\frac{t^2}{6}$$
$$(8a)$$

$$\sigma_S = 4.13 \times 10^{-4} E \frac{h_0}{h} \left(\frac{t}{h}\right)^{0.2} \frac{f_y}{345}, \quad \sigma_{SS} = 1.05 f_y - \sigma_S$$
$$(8b)$$

Predictions by Equations 6–8 show good agreement with that from FEM analysis, as shown in Table 1.

2.2.3 *Joint resistance to axial forces*

Since the resistance is mainly provided by Plate I, the joint behaves soft soon after Plate I yields. It can be observed from the axial load versus deformation curves as shown in Figure 10. It is different from what shown in Figure 8 where the in-plane moment increases stably with the development of inelastic deformation. Hence, the axial force corresponding to the noticeable turn point in load-deformation curve is defined as yield axial force.

The safety requirement under axial load is determined by Equation 9.

$$\left(\frac{N_X}{N_u} - 0.35\right)^2 + \left(\frac{N_Y}{N_u}\right)^2 \leq 1, \text{ if } |N_Y| \geq |N_X| \quad (9a)$$

$$\left(\frac{N_X}{N_u}\right)^2 + \left(\frac{N_Y}{N_u} - 0.35\right)^2 \leq 1, \text{ if } |N_Y| < |N_X| \quad (9b)$$

$$N_u = \left(\frac{2A_T}{\sqrt{3}} + A_C\right) f_y, \quad (9c)$$

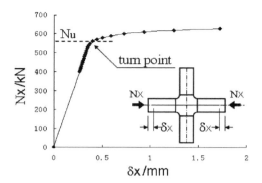

Figure 10. Load-deformation curve under axial loads.

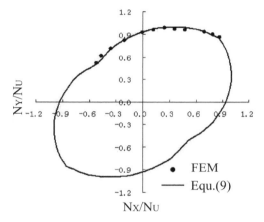

Figure 11. Ultimate state of axial loads in two directions.

where, $A_T = 2(w-2R_1)t$ and, $A_C = 2\pi(R_1-0.5t)t$.

In Equation 9c the contribution of Plate I to the capacity is neglected. Tensile force is taken as positive and compressive force negative. Figure 11 describes the inter-relation of the four end loads under ultimate by Equations 9a–9b.

2.2.4 Joint resistance to out-of-plane moment

The characteristic of stress distribution in a joint subjected to out-of-plane moments is similar to that under axial force, except from that one of Plate I is tensioned and the other one compressed. By comparison with numerical analysis results, the strength condition should satisfy Equation 10.

$$\left(\frac{M_{out,X}}{M_{out,p}} - 0.35\right)^2 + \left(\frac{M_{out,Y}}{M_{out,p}}\right)^2 \le 1, \quad \text{if } |M_{out,Y}| \ge |M_{out,X}| \tag{10a}$$

$$\left(\frac{M_{out,X}}{M_{out,p}}\right)^2 + \left(\frac{M_{out,Y}}{M_{out,p}} - 0.35\right)^2 \le 1, \quad \text{if } |M_{out,Y}| < |M_{out,X}| \tag{10b}$$

$$M_{out,p} = \left(\frac{2W_T}{\sqrt{3}} + W_C\right) f_y, \tag{10c}$$

3 IRREGULAR CAST STEEL JOINTS

3.1 Numerical analysis on irregular casting joints

To meet various requirements of aesthetics, casting procedure and construction condition, cast steel joints in general are irregular in geometric configuration. It is thus impossible to establish mathematical formulae for all types of irregular joints. FEM is a powerful alternate method in analyzing stress development for joint safety verification.

Principle design rules are proposed in recent draft *Technical Specification for Application of Connections of Structural Steel Casting* (CECS235: 2008). A few key points are described as below.

3.1.1 Criteria of strength

The joint itself need to remain in elastic in use, but while stress concentration can not be completely avoided, certain plastic deformation should be allowed. As shown in Equation 11, factor α shall range between 1.0~1.2, and β 1.1~1.2, depending on the engineer's judgement: complete elastic design or partial plastic one.

$$\sqrt{0.5[(\sigma_1 - \sigma_2)^2 + (\sigma_2 - \sigma_3)^2 + (\sigma_3 - \sigma_1)^2]} \le \alpha f_y \tag{11a}$$

or $\quad \sigma_1 \le \beta f_y, \quad \text{if } \sigma_3 > 0 \tag{11b}$

The components of stress in Equation 11 are defined same as that in Equation 2.

3.1.2 Utilization of fully plastic capacity of joint

Use of fully plastic capacity of joint is not recommended by recent research. On reason is that in most cases the joint wall is thick due to the casting procedure, thus the joint is strong enough on the whole compared with member capacity, except possible peak value of stress in spot of stress concentration. The other reason is that the joint behavior in fully plastic deformation has not been thoroughly studied yet, especially lack of the test validation and evaluation index.

3.2 Loading tests on irregular cast steel joint

Though FEM analysis is able to give detailed stress distribution in joints under various loads, there are still some limitations: uncertainty of the stress response which closely depends on element refining technique, especially at the location of hot stress, as well as the variation of output relating to the mechanical assumption behind the element stiffness matrix. There are cases where tests on cast steel joints are required. Since 2000, more than ten types of irregular cast steel joint have been tested in the laboratory of Tongji University. These experiences gave us lessons which should be paid attention during the tests.

3.2.1 Size of specimen

The defects caused in casting process are unavoidable and relates directly to the casting procedure which is decided according to the size of casting objective. The full-scale or large scale specimen is used as long as the loading capacity of the test facility permits as such defects are expected to have effect on the actual strength of the joint.

3.2.2 Realization of loading condition

As a connection part of a structural system, the cast steel joint is usually subjected to complicated loads by external load and through surrounding members. Since to achieve an accurate simulation of all the actual loading conditions in a test is not always easy, the main loading components and the resulted stress effects are required to be similar to those in actual structure.

Figure 12 shows an example of a cast steel joint in a mast linking six cables in a suspend roof structure. The cables shall apply tension up to 5000 kN each and the cantilever portion above the joint will exert moment on the joint produced by wind load. Analysis results indicated that most stress in the joint is induced by tension of the cables, so the relative small moment effect can be ignored in test. Figure 12 also shows the test facility.

3.2.3 Realization of restraint condition

Joint boundary condition including both loads and restraints should be properly simulated in a test. Cast steel joints were used as pin-connections to link the basement and the V-shape columns arranged in out-

Figure 12. Loading test example 1.

Figure 13. Loading test example 2.

ring of the Olympic Laoshan Velodrome as shown in Figure 13. When the joint with two columns are isolated from the whole structure and tested in laboratory, this part could not keep balance. For a proper simulation, horizontal actuator is used to limit the lateral displacement of the test specimen. However the end restricted horizontally may move freely in vertical direction owing to a special rail along which the horizontal actuator can slide smoothly.

3.2.4 Integration of testing and numerical simulation

It is necessary to combine the test data and numerical analysis results to understand the behavior of cast steel joint. Test data provide important but not only partial information of the specimen, while a calibrated FE model may reveal more detailed results. For instance, strain gages on the surface of a specimen can only measure three strain components of six ones. In the above mentioned loading test example 1, FE results agree well with test results if only three surface strain components are compared, while the entire effect of six strain components shows 10 percentage difference from the surface strains. Numerical analysis is needed as a supplement in obtaining the whole picture of stress development inside the joint. Needless to say, validity of FE model shall be checked by test results.

4 CONCLUSIONS

In this paper, the focus is given to how to design cast steel joint in the view of structural engineering, using high strength and high quality cast steel material.

Two typical X-type joints, with solid core and empty core respectively are investigated. It is shown that even though these joints are quite different from traditional joints, it is possible to establish strength equations for safety check, given the geometric configuration is regular.

In General, cast joints are designed individually to meet the requirements of aesthetical appearance, construction and rational stress flow. Numerical analysis and experiments are both important and supplement to each other.

ACKNOWLEDGEMENT

Postgraduates at Tongji University, Bian, R.N., Peng, L., Wang, G.N. & Gu, M. contributed to the research work. The research was supported by the National Science Foundation of China (50578117).

REFERENCES

CECS. 2007. *Technical Specification for Connections of Structural Steel Casting*. (In draft, in Chinese).

Chen, Y.Y., Peng, L., Zhao, X.Z. and Wang, Y. 2006. Rigidity and ultimate capacity of cross-junction type joint for tubular structures. In W. Kanok-Nukulchai, S. Munasinghe & N. Anwar (eds), *Real Structures: Bridge and Tall Buildings, Proceedings of The Tenth EASEC*, Volume 4: 13-418, Location: Press.

Herrion, S., Veselcic, M. & Puthli, R. 2007, Cast steel–new standard and advanced technologies. In Y.S. Choo & J.Y.R. Liew (eds), *Proceedings of 5th Advanced in Steel Structures*, Volumn II: 149–154. Singapore: Research Publishing.

Non-linear analysis of ultimate loading capacity of cast tubular Y-joints under axial loading

Y.Q. Wang, Y. Jiang, Y.J. Shi & P. Sun
Department of Civil Engineering, Key Laboratory of Structural Engineering and Vibration of China Education Ministry, Tsinghua University, Beijing, China

ABSTRACT: In this paper, a number of parametric analyses on the ultimate load capacity of cast Y-joints are performed. The geometry model is established using the CAD software Solidworks. Non-linear analysis is carried out using the commercial finite element program ANSYS. Parametric equations of the ultimate loading capacity derived from the results of the finite element analysis are presented for the normal range of basic shapes of Y-joints under axial loading. The ultimate loading capacity sensitivity of the cast tubular joints to variations in the geometric parameters has been assessed. As well as the parameters which governing the stress in welded joints, an additional parameter ρ defined by Edwards (1985) has been introduced to characterize the size of the fillet. Results and interpretation of the sensitivity of ultimate loading capacity of cast tubular joints to the parameter ρ are presented, and it provides references for the construction design and the revision of Chinese code.

1 INTRODUCTION

Cast tubular joints are widely applied in bridges, high-rise and industrial constructions, because from a structural and aesthetic point of view they are superior to directly welded joints. Cast tubular joints have lower stress concentrations and higher ultimate loading capacity than welded joints because they have better shapes of fillet and no welds in the critical areas. Figure 1 shows the comparison between the cost of cast tubular joints and welded joints, and Figure 2 shows stress concentration factor of the two joint types (Marstion 1990). The two figures indicate the obvious advantages in application of cast tubular joints.

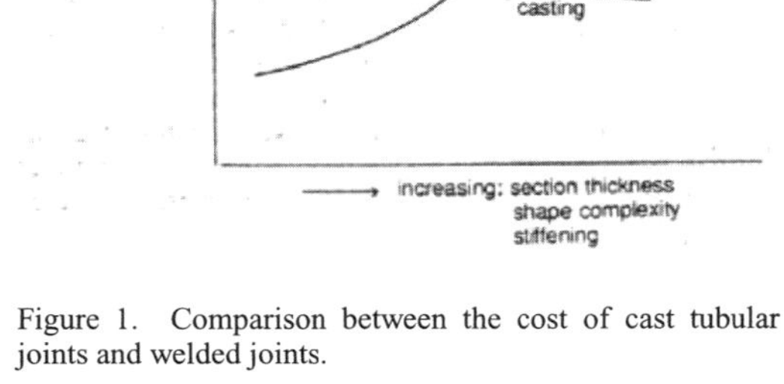

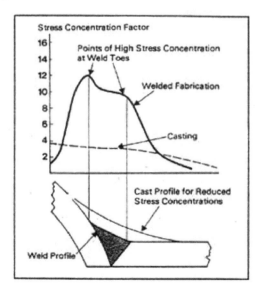

Figure 1. Comparison between the cost of cast tubular joints and welded joints.

Figure 2. Comparison between the stress concentration factor of cast tubular joints and welded joints.

For structural engineers, it is important to understand the behavior of cast tubular joints which have already been widely used in many projects. Sets of parametric equations for welded joints have been published in many design codes, in which the joints' ultimate loading capacity are related to its geometric parameters. These equations are widely used in the welded tubular joints design and it seems appropriate that a similar approach should be available for designers intending to use cast tubular joints. The equations presented in this paper are intended to enable the designers to predict ultimate loading capacity of Y-joints with a variety of different geometric parameters. The analysis method introduced in this paper will be the base of further research on ultimate loading capacity of complex cast tubular joints.

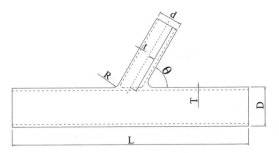

Figure 3. Dimensions of Y-joints.

2 THEORETICAL MODEL

Dimensions of cast tubular Y-joints are shown in Figure 3, and key Parameters are defined as follows:

F = Axial load in brace;
θ = Angel between brace and chord;
$a = 2L/D$;
$\beta = d/D$;
$\gamma = D/2T$;
$\tau = t/T$;
$\rho = R/(DT)^{0.5}$.

Based on large amount of experimental data and FEM results, Edwards, C.D. & Fessler, H. (1985) established theoretical model for computing stress concentration factor of cast tubular Y-joints (Equation 1). The equation can be used in conditions of either axial load or moment load in or out of plane.

$$S = C \cdot f_1(\alpha) \cdot f_2(\beta) \cdot f_3(\theta) \cdot \gamma^x \cdot \rho^y \cdot \tau^z \quad (1)$$

where $f_1(\alpha), f_2(\beta)$ and $f_3(\theta)$ are functions of α, β and θ.

In this paper, a new theoretical model of cast tubular Y-joints is built partially according to the T-joints model (Sun 2005), which was established based on Equation 1 by the linear regression methods.

Equation 2 is the ultimate strength formula of cast tubular T-Joints under axial loading ($F_{t,u}$) (Sun 2005). And Equation 3 is the rough relationship between ultimate load capacity of cast Y-joints ($F_{y,u}$) and T-joints ($F_{t,u}$). Then, $F_{y,u}$ can be obtained by Equation 4. With necessary amendment, Equation 4 can be used to estimate the ultimate load capacity of cast Y-joints.

$$\frac{F_{t,u}}{f_y \cdot T^2} = \frac{7.22 \cdot \gamma^{0.41\beta - 0.28\beta^2} \cdot (1+C_1) \cdot (1+\frac{\alpha}{\gamma})^{-1.24} \rho^{0.132}}{\left(1 - \frac{0.8\psi_2}{\pi}\right)\sin(0.8\psi_2)(1+C_1) - \left[1 - \frac{\arcsin(0.8\beta)}{\pi}\right]0.8\beta[1+\cos(0.8\psi_2)] + \frac{0.7}{\gamma^2}} \quad (2)$$

where $C_1 = \sqrt{1-(R_4\beta)^2}$; $\psi_2 = 1.2 + 0.8\beta^2$

$$F_{y,u} = \frac{F_{t,u}}{\sin\theta} \quad (3)$$

$$\frac{F_{y,u}}{f_y \cdot T^2} = \frac{7.22 \cdot \gamma^{0.41\beta - 0.28\beta^2} \cdot (1+C_1) \cdot \left(1+\frac{\alpha}{\gamma}\right)^{-1.24} \rho^{0.132}}{\sin\theta \cdot \left\{\left(1 - \frac{0.8\psi_2}{\pi}\right)\sin(0.8\psi_2)(1+C_1) - \left[1 - \frac{\arcsin(0.8\beta)}{\pi}\right]0.8\beta[1+\cos(0.8\psi_2)] + \frac{0.7}{\gamma^2}\right\}} \quad (4)$$

Table 1. Nominal dimensions and numerical results of the axially loaded cast Y-joints.

Joint	Nominal dimensions			Numerical results		
	T mm	d mm	θ	$F_{y,u}$ kN	$F_{t,u}$ kN	κ^*
YA5M45	28	165.1	45°	2341	1994	0.83
YA5M60	28	165.1	60°	2026	1994	0.88
YA6M45	16	165.1	45°	1106	942	0.83
YA6M60	16	165.1	60°	935	942	0.86
YA7M45	11	165.1	45°	601	500	0.85
YA7M60	11	165.1	60°	502	500	0.87
YA9M45	28	244.5	45°	2619	2231	0.83
YA9M60	28	244.5	60°	2319	2231	0.90
YA10M45	16	244.5	45°	1386	1153	0.85
YA10M60	16	244.5	60°	1225	1153	0.92
YA11M45	11	244.5	45°	872	709	0.87
YA11M60	11	244.5	60°	753	709	0.92
YA13M45	28	323.9	45°	2824	2349	0.85
YA13M60	28	323.9	60°	2468	2349	0.91
YA14M45	16	323.9	45°	1619	1363	0.84
YA14M60	16	323.9	60°	1401	1363	0.89
YA15M45	11	323.9	45°	1081	899	0.85
YA15M60	11	323.9	60°	934	899	0.90

$^*\kappa = F_{y,u}/(F_{t,u}/\sin\theta)$, which is the ratio of numerical results to results obtained from Equation 4.

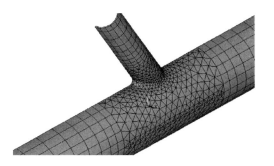

Figure 4. FEM model of Y-joints.

Table 2. FEM boundary of Y-joints.

Degrees of freedom	Nodes on plane $X = 0$	Nodes on plane $Z = 0$	Nodes with $Z = 0.5L$ and $Y = 0$
u_x	0	Free	Free
u_y	Free	Free	0
u_z	Free	0	Free
$\varnothing_x$	Free	0	Free
$\varnothing_y$	0	0	Free
$\varnothing_z$	0	Free	Free

3 FEM ANALYSIS

In order to verify the accuracy of Equation 4, 18 joints whose geometry model were established using the CAD software Solidworks, have been analyzed numerically by the commercial finite element program ANSYS. For all joints, the wall thickness ratio τ was set to 0.63, α was 12, the fillet radius ratio ρ was 0.40, and the diameter of chord was set to 406.4 mm. The material for the cast Y-joints was taken to be cast steel with yield stress $F_y = 300$ N/mm², modulus of elasticity $E = 206$ GN/m² and Poisson's ratio $\upsilon = 0.3$. The material was assumed to be elastic/perfect plastic.

The general characteristics of the FEM analyses were as follows:

– The nominal dimensions of the joints, summarized in table 1, were used to model the joints.
– Half of each joint was modeled.
– 10-node solid elements and 20-node solid elements were used to mesh the joints.
– The number of elements used to model the joints was about 2000. The finite element mesh and boundary conditions used for the joints are separately shown in figure 4 and table 2.
– The load has been applied to the brace tip by displacement control. In this way, it is easy to obtain the descending branch of load-displacement curve.
– The Von Mises yield criterion has been used.

From the values of κ in Table 1, it is easy to find that all of the 18 joints' numerical results are less than that obtained from Equation 4, as the axial load in chord weakens the ultimate load capacity of joints. Another result is that, all κ of the joints of $\theta = 45°$ are generally less than that of $\theta = 60°$. It should also be explained by the weakening effect of axial load in chord. Obviously, axial load in chord with $\theta = 45°$ is greater than that of $\theta = 60°$.

In Chinese code (Code for design of steel structure, GB50017–2003), reduction factor shown in Equation 5 is advised to counteract weakening effect of axial load in chord.

$$f(n') = 1.0 - \frac{0.25n'}{\beta} \leq 1 \quad (5)$$

where n' is the ratio of compression stress to yield stress of chord steel.

After computing $f(n')$ for all joints, new differences are shown in Table 3. Obviously, for all joints, reduction factor $f(n')$ is greater than κ, which is obtained in Table 1. Thus, the results amended by $f(n')$ are still not conservative. And the difference between numerical results and theoretical results should be explained by non-uniform stiffening effect at the chord-brace intersection. Different from T-joints, the chord-brace intersection of Y-joints has two typical areas, intersection near acute angle (θ) and

Table 3. Compensation factor for ultimate load capacity of Y-joints.

Joints	κ	f(n')	ø' = f(n')/κ
YA5M45	0.83	0.90	1.08
YA5M60	0.88	0.94	1.07
YA6M45	0.83	0.92	1.11
YA6M60	0.86	0.95	1.11
YA7M45	0.85	0.94	1.10
YA7M60	0.87	0.96	1.11
YA9M45	0.83	0.92	1.11
YA9M60	0.90	0.95	1.06
YA10M45	0.85	0.93	1.10
YA10M60	0.92	0.96	1.04
YA11M45	0.87	0.94	1.08
YA11M60	0.92	0.96	1.05
YA13M45	0.85	0.94	1.10
YA13M60	0.91	0.96	1.06
YA14M45	0.84	0.94	1.12
YA14M60	0.89	0.96	1.08
YA15M45	0.85	0.94	1.11
YA15M60	0.90	0.96	1.07

intersection near blunt angle ($180°-\theta$). The stiffening effect at intersection near acute angle is greater than that near blunt angle. And in process of FEM analysis, steel near the area of blunt angle yield earlier than that near acute angle. This is the reason that numerical results are lower than theoretical results.

New appropriate factor have to be defined to make sure the theoretical formula for ultimate load capacity of Y-joints is more accurate, and results are conservative enough. To get this new factor, in this paper, factor ø' in Table 3 is analyzed and improved by statistical method. The mean value μ of ø' in Table 3 is 1.09, and standard deviation σ is 0.025. With confidence level of 95%, the new compensation factor ø is defined in Equation 6.

$$\varphi = \mu + 1.645\sigma = 1.13 \qquad (6)$$

Thus, unified procedures and formulas for calculating ultimate load capacity of Y-joints can be proposed to designers as follows.

1. Calculate estimated value for ultimate load capacity $F'_{y,u} = F_{t,u} / \sin\theta$ by Equation 4.
2. Calculate reduction factor $f(n')$ and ø, and get the final ultimate load capacity of Y-joints $F_{y,u}$ by Equation 7.

$$F_{y,u} = f(n') \frac{F'_{y,u}}{\varphi} \qquad (7)$$

4 CONCLUSION

In this paper, formulas including factor α, β, γ, τ, ρ for ultimate load capacity of Y-joints are advised. To make sure that the theoretical model for ultimate load capacity of Y-joints is more accurate, new factor ø is defined to amend old theoretical formulas. New factor, unified formulas and simple procedures can help designers make sure that the calculated ultimate load capacity of Y-joints is conservative.

REFERENCES

Edwards, C. D. & Fessler, H. 1985. Stress Concentrations in Cast Corner Joints of Tubular Structures. *Developments in Marine Technology* 2: 465–473.

GB50017-2003, Code for design of steel structure, Beijing.

Hyde, T. H., Fessler, H. & Khalid, Y. A. 1997. Effect of cracks at one saddle on the static strength of die cast tin-lead alloy T, Y, and YT tubular joints. *Proceedings of the Institution of Civil Engineers-Structures and Buildings* 122(4): 379–389.

Marston, G. J. 1990. Better cast than fabricated. *Foundryman* 83(3): 108–113.

Sun, P., Wang, Y. & Shi, Y. 2005. Non-linear Analysis for Ultimate Loading Capacity of Cast Tubular T-joints. *Proceedings of the International Conference on Offshore Mechanics and Arctic Engineering- OMAE*, v1 B: 629–636.

Van der vegte, G. J. 1995. The Static Strength of Uniplanar and Multiplanar Tubular T- and X-joints. *International Journal of Offshore and Polar Engineering* 5(4): 308–316.

*Parallel session 8: Static strength of members/frames
(reviewed by J.A. Packer & B. Uy)*

Structural performance of stainless steel oval hollow sections

T.M. Chan
University of Warwick, UK

L. Gardner
Imperial College London, UK

ABSTRACT: Stainless steel is gaining increasing usage in the construction industry and tubular members represent the most widely used structural component. Cold-formed square, rectangular and circular hollow sections are readily available in stainless steel, and more recently, oval hollow sections have been introduced. Oval hollow sections are not currently included in structural design codes, neither for carbon steel nor stainless steel. This paper describes recent tests and finite element modelling performed on stainless steel oval hollow sections in compression and bending. A total of six stub column tests, six in-plane bending tests and eight column buckling tests are reported. Tests have been performed about both the major and minor axes. The results of the tests have been analysed and comparisons have been made with the structural performance of hot-rolled carbon steel elliptical hollow sections. Design rules including cross-section slenderness limits and a column buckling curve have been proposed, and progress towards the development of comprehensive design rules for stainless steel oval hollow sections is reported.

1 INTRODUCTION

The physical characteristics of stainless steel make it well suited to use in construction; it possesses high strength and stiffness (comparable with carbon steel), very high ductility (approximately two times that of carbon steel), and excellent corrosion resistance, which means, suitably specified that it requires no protective coatings (Gardner 2005). Stainless steel also offers better retention of strength and stiffness than carbon steel at elevated temperatures. The principal disincentive for the application of stainless steel in construction is the initial material cost, though considered on a whole life basis, cost comparisons with carbon steel become more favourable. Structural design guidance is available for stainless steel in Eurocode 3: Part 1.4 (EN 1993-1-4 2006); the design rules are harmonised, where possible, with those developed for structural carbon steel in Eurocode 3: Part 1.1 (EN 1993-1-1 2005), thus enabling relatively straightforward transition between the two materials. Stainless steel exhibits a rounded stress-strain curve (Fig. 1), which necessitates the use of a proof stress (generally the 0.2% proof stress) in place of a yield stress in the calculation of structural resistance, and use of the secant modulus in serviceability calculations. Strain hardening under load is also significant though the resulting enhancement in structural resistance is not currently incorpo-

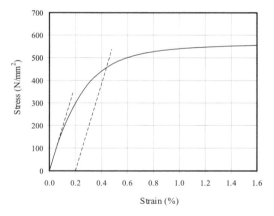

Figure 1. Rounded stainless steel stress-strain response and use of 0.2% proof stress.

rated in design codes. A method for harnessing this increase in resistance has been developed (Gardner & Nethercot 2004a, Gardner & Ashraf 2006).

Exploiting both its corrosion resistance and aesthetic appeal, stainless steel is often employed for exposed structural elements, a number of examples of which are given by Baddoo et al. (1997). Tubular construction is also becoming increasingly popular for such applications. Until recently, the primary tubular cross-section shapes available to structural engineers and architects have been square, rectangular and circular hollow sections. Now, hot-rolled carbon steel elliptical hollow sections (EHS) and cold-rolled stainless steel oval hollow sections (OHS) are available for structural applications. These sections offer the architectural attributes of circular hollow sections, together with the structural advantages associated with sections of differing properties about the two principal axes. This paper focuses on the compressive and flexural responses of cold-rolled stainless steel oval hollow sections, though comparisons are made with the results from previous studies on carbon steel elliptical hollow sections (Chan & Gardner 2006, Gardner & Chan 2007).

2 EXPERIMENTAL STUDY

2.1 Introduction

A series of tensile material tests, compressive stub column tests, in-plane bending tests and column buckling tests were carried out to investigate the structural behaviour of stainless steel oval hollow sections. All tests were performed in the Structures laboratory of the Department of Civil and Environmental Engineering, Imperial College London. A total of six tensile coupons, six stub columns, six beams and eight columns were tested. Three section sizes were employed—OHS 121 × 76×2, OHS 121 × 76 × 3 and OHS 86 × 58 × 3. All tested material was austenitic stainless steel, grade 1.4401 (316), which contains approximately 18% chromium and 10% nickel (EN 10088-1 2005). All specimens were cold-rolled and seam welded and supplied by the stainless steel tube manufacturers, Oval 316. The minimum specified yield strength (0.2% proof strength) for grade 1.4401 stainless steel is 240 N/mm² for cold-rolled sheet material (up to 6 mm in thickness) and 220 N/mm² for hot-rolled sheet material (up to 12 mm in thickness) (EN 10088-2 2005). However, the material strength of stainless is considerably enhanced during the section forming process due to cold-working, as indicated by the results of the tensile coupon tests given in Table 1 (where the coupons were machined from the completed oval sections), and from the results of previous studies. This section summarises the testing apparatus, the experimental procedures and the test results obtained.

2.2 Tensile coupon tests

Tensile coupon tests were performed to establish the basic material stress-strain response; this was subsequently utilised during the analysis of the member test results and in the development of numerical models. The tests were carried out in accordance with EN 10002-1 (2001).

Parallel coupons were machined longitudinally from the two flattest portions of the cross-sections (i.e. along the centrelines of the minor axis) using a tipped slot-drill. Longitudinal curving of the coupons was observed as machining progressed, indicating the presence of through-thickness residual stresses. However, no attempt was made to straighten the coupons (by plastic bending) prior to tensile testing. Two coupon tests, designated TC1 and TC2, were performed for each section size.

The key results from the six coupon tests are summarised in Tables 1 and 2, where E is the initial tangent modulus, $\sigma_{0.2}$ and $\sigma_{1.0}$ are the 0.2% and 1% proof strengths respectively, σ_u is the ultimate tensile strength, ε_f is the strain at fracture and n and $n'_{0.2,1.0}$ are strain hardening exponents for the compound Ramberg-Osgood model described by Gardner & Ashraf (2006). The adopted compound model was developed on the basis of a two-stage version (Mirambell & Real 2000, Rasmussen 2003) of the original Ramberg-Osgood expression.

2.3 Stub column tests

A total of six stainless steel oval hollow section (OHS) stub columns were tested in pure axial compression to assess load carrying capacity and deformation capacity. Full load-end shortening curves were recorded, including into the post-ultimate range. The nominal

Table 1. Measured tensile material properties.

Coupon designation	E (N/mm²)	$\sigma_{0.2}$ (N/mm²)	$\sigma_{1.0}$ (N/mm²)
OHS 121 × 76 × 2-TC1	193900	380	426
OHS 121 × 76 × 2-TC2	193300	377	419
OHS 121 × 76 × 3-TC1	194100	420	460
OHS 121 × 76 × 3-TC2	190400	428	467
OHS 86 × 58 × 3-TC1	194500	339	368
OHS 86 × 58 × 3-TC2	194500	331	349

Table 2. Measured tensile material properties (continued).

Coupon designation	σ_u (N/mm²)	ε_f	R-O coefficients	
			n	$n'_{0.2,1.0}$
OHS 121 × 76 × 2-TC1	676	0.61	7.8	2.9
OHS 121 × 76 × 2-TC2	672	0.60	8.9	2.9
OHS 121 × 76 × 3-TC1	578	0.58	9.7	4.0
OHS 121 × 76 × 3-TC2	583	0.58	8.2	4.0
OHS 86 × 58 × 3-TC1	586	0.62	14.0	1.8
OHS 86 × 58 × 3-TC2	597	0.62	13.5	1.3

lengths of the stub columns were chosen such that they were sufficiently short not to fail by overall buckling, yet still long enough to contain a representative residual stress pattern. The stub column lengths were taken as two times the larger cross-sectional dimension. The stub column tests were carried out in a self-contained 300 T Amsler hydraulic testing machine. The tests were load-controlled through an Amsler control cabinet. The end platens of the testing arrangement were fixed flat and parallel. Four linear variable displacement transducers (LVDTs) were used to determine the end shortening of the stub columns between the end platens of the testing machine. Four linear electrical resistance strain gauges were affixed to each specimen at mid-height, and at a distance of four times the material thickness from the major axis. The strain gauges were initially used for alignment purposes, and later to modify the end shortening data from the LVDTs to eliminate the elastic deformation of the end platens.

Measurements of major and minor axis diameters (2a and 2b, respectively), material thickness t and stub column length were taken at four different points for each specimen. The mean measured dimensions for the six stub column specimens are presented in Table 3; cross-section geometry and notation is defined in Figure 2. Two stub column tests, designated SC1 and SC2, were performed for each section size. The circumferences of the oval specimens were traced to determine their exact cross-sectional profiles. These were subsequently fitted with the equation of an ellipse, which was found to provide a suitably accurate representation of the geometry. The cross-sectional area A of the test specimens was calculated as the circumference (determined along the centreline of the thickness) multiplied by the material thickness.

Compression tests on stub columns reveal the average compressive response of the cross-sections. Ultimate failure is due to local buckling of the cross-section. For cross-sections comprising slender elements local buckling may occur in the elastic range. For more stocky cross-sections, local buckling may occur following significant inelastic deformation.

Measured end shortening readings from the LVDTs were modified on the basis of the strain gauge readings to account for the elastic deformation of the end platens (that are present in the LVDT measurements). Thus, true end shortening values were derived and are utilised in the remainder of this study. Typical load-end shortening curves from the 86 × 58 × 3 stub column tests are shown in Figure 3, whilst a summary of the key test results including ultimate load F_u and end shortening at ultimate load δ_u is given in Table 4. In general, the test results indicate good correlation between the repeated stub column tests (SC1 and SC2).

Table 3. Measured geometric properties of stub columns.

Stub column designation	2a (mm)	2b (mm)	t (mm)	L (mm)
OHS 121 × 76 × 2-SC1	123.70	76.94	1.84	242.0
OHS 121 × 76 × 2-SC2	124.14	76.60	1.84	242.0
OHS 121 × 76 × 3-SC1	121.53	77.14	2.94	241.7
OHS 121 × 76 × 3-SC2	121.56	77.09	2.95	241.2
OHS 86 × 58 × 3-SC1	85.63	57.22	3.11	171.9
OHS 86 × 58 × 3-SC2	84.67	58.98	3.12	171.6

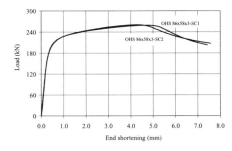

Figure 3. Load-end shortening curves for OHS 86 × 58 × 3 stub columns.

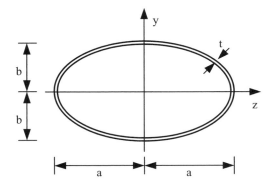

Figure 2. Geometry and notation of oval (elliptical) hollow sections.

Table 4. Key results from stub column tests.

Stub column designation	A (mm²)	F_u (kN)	δ_u (mm)
OHS 121 × 76 × 2-SC1	577.3	234	1.55
OHS 121 × 76 × 2-SC2	577.1	235	1.63
OHS 121 × 76 × 3-SC1	902.1	444	2.77
OHS 121 × 76 × 3-SC2	904.0	442	2.71
OHS 86 × 58 × 3-SC1	674.4	259	4.69
OHS 86 × 58 × 3-SC2	678.4	260	4.25

All stub columns exhibited a similar failure mode whereby the two larger faces of the specimen buckled locally. Photographs of the deformed 86 × 58 × 3 stub column specimens are shown in Figure 4.

The results of the stub column tests are analysed and discussed in Section 4, whereupon comparisons with existing compression tests on stainless steel circular hollow sections and carbon steel elliptical hollow sections are made. Design recommendations (slenderness parameters and limits) are set out thereafter.

2.4 In-plane bending tests

Bending tests were carried out about the major and minor axes in order to study the moment-rotation behaviour of stainless steel OHS beams. The specimen dimensions were chosen to cover a range of cross-section slenderness values. The beams were simply supported on rollers, as shown in Figure 5, and were loaded at mid-span by a 50 T hydraulic actuator. Displacement transducers were employed to measure mid-span deflection and end-rotation. All beams failed by local buckling of the compressed portion of the cross-section in the region of the maximum bending moment. The results are examined in Section 4 of this paper.

Mean measured properties of the beams are given in Table 5; beams designated B1 were subjected to minor axis bending, while those designated B2 were under major axis bending. The symbol L refers to the lengths of the beams between simple supports.

Key results from the beam tests are presented in Table 6. The calculated elastic and plastic moment capacities are designated M_{el} and M_{pl} respectively, M_u refers to the maximum moment achieved during testing, and R is the calculated rotation capacity (defined as the observed plastic rotation beyond M_{pl}, normalised by the elastic rotation at M_{pl}). These results are analysed and used for the assessment of slenderness limits in Section 4.

2.5 Pin-ended column tests

Flexural buckling tests about the major and minor axes were carried out in order to obtain ultimate load carrying capacity data to enable the proposition of a suitable buckling curve for stainless steel oval hollow section columns. The tests were conducted on pin-ended columns of lengths ranging from 700 mm to 3100 mm

Figure 4. Failure mode for OHS 86 × 58 × 3 stub columns.

Figure 5. Simple roller supports employed in beam tests.

Table 5. Measured geometric properties of beams.

Beam designation	2a (mm)	2b (mm)	t (mm)	L (mm)
OHS 121 × 76 × 2-B1	77.27	123.82	1.92	1106.1
OHS 121 × 76 × 2-B2	78.44	121.79	1.91	1102.9
OHS 121 × 76 × 3-B1	77.08	121.79	3.01	1116.3
OHS 121 × 76 × 3-B2	78.74	121.35	3.03	1108.1
OHS 86 × 58 × 3-B1	57.21	85.68	3.18	801.5
OHS 86 × 58 × 3-B2	57.17	85.47	3.17	801.9

Table 6. Key results from beam tests.

Beam designation	M_{el} (kNm)	M_{pl} (kNm)	Test M_u (kNm)	R
OHS 121 × 76 × 2-B1	4.68	5.90	6.51	2.3
OHS 121 × 76 × 2-B2	5.94	8.00	9.00	4.1
OHS 121 × 76 × 3-B1	7.76	9.95	11.78	4.6
OHS 121 × 76 × 3-B2	10.20	13.86	16.32	6.6
OHS 86 × 58 × 3-B1	3.25	4.25	5.13	11.8
OHS 86 × 58 × 3-B2	4.09	5.61	7.84	16.5

in order to cover a spectrum of member slenderness. Member slenderness was defined though Equation 1, with values ranging between 0.34 and 2.07.

$$\bar{\lambda} = \sqrt{A\sigma_{0.2}/N_{cr}} \quad (1)$$

where A is the cross-sectional area, $\sigma_{0.2}$ is the material 0.2% proof stress and N_{cr} is the elastic buckling load of the column. The nominal section dimensions were 86 × 58 × 3 mm and the tests were carried out in a 400T capacity rig. The pin-ended boundary conditions were attained using knife-edges.

Two displacement transducers were placed at the mid-height of the columns to measure the lateral deflection. Two further displacement transducers were placed on each of the end plates to measure the end rotation, while an additional two were employed to measure the end shortening. The mean measured dimensions for the eight column specimens are presented in Table 7. It should be noted that the column lengths L_{cr} are measured between the pin ends, and the designations 'CMI' and 'CMA' refer to minor and major axis buckling, respectively.

The key results from the column tests are summarised in Table 8.

The results from the column tests are combined with existing test data on stainless steel hollow section columns and analysed in Section 4 of this paper and a suitable buckling curve is proposed.

3 NUMERICAL MODELLING

A numerical study was performed to generate ultimate capacity data to supplement the test results for pin-ended columns. This study was performed using the finite element (FE) package ABAQUS V6.6 (2006), and is briefly described in this paper. Eight FE models incorporating material and geometric nonlinearities were initially produced to replicate the experimental column results. The nonlinear material stress-strain behaviour of stainless steel was incorporated into the FE models by fitting a compound material model (Gardner & Ashraf 2006, Mirambell & Real 2000, Rasmussen 2003) to data obtained from the tensile coupon tests.

Having validated the numerical models against the tests, a series of parametric studies were performed to investigate variation in cross-section and member slenderness. Global geometric imperfections were taken as the lowest global buckling mode. Typical failure modes from the FE models are shown in Figure 6. The FE models were generally able to accurately reflect the observed physical response of the test specimens in terms of general load-lateral deflection behaviour, ultimate load carrying capacity and failure mode. For the more slender columns, the FE models exhibited higher failure loads than the test specimens; this is believed to be due to unintentional eccentricity induced during testing of the longer samples.

Table 7. Measured geometric properties of columns.

Column designation	2a (mm)	2b (mm)	t (mm)	L_{cr} (mm)
86 × 58 × 3-CMI1	85.41	57.16	3.11	699.5
86 × 58 × 3-CMA1	85.48	56.84	3.09	700.6
86 × 58 × 3-CMI2	86.05	56.21	3.11	1499.6
86 × 58 × 3-CMA2	85.91	56.70	3.15	1500.5
86 × 58 × 3-CMI3	86.18	56.22	3.11	2299.3
86 × 58 × 3-CMA3	85.97	56.37	3.10	2300.3
86 × 58 × 3-CMI4	86.02	56.33	3.12	3100.3
86 × 58 × 3-CMA4	85.91	56.06	3.13	3100.3

Table 8. Key results from column tests.

Column designation	A (mm²)	$\bar{\lambda}$	N_u (kN)
86 × 58 × 3-CMI1	672.7	0.46	181.8
86 × 58 × 3-CMA1	668.5	0.34	196.9
86 × 58 × 3-CMI2	672.9	1.00	116.1
86 × 58 × 3-CMA2	681.4	0.72	150.8
86 × 58 × 3-CMI3	674.2	1.54	72.3
86 × 58 × 3-CMA3	671.5	1.10	92.8
86 × 58 × 3-CMI4	675.0	2.07	40.6
86 × 58 × 3-CMA4	676.0	1.48	51.1

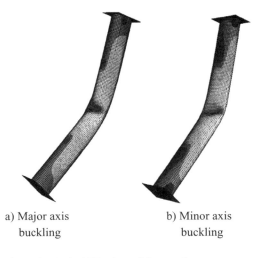

a) Major axis buckling   b) Minor axis buckling

Figure 6. Typical FE column failure modes.

4 ANALYSIS OF RESULTS AND PROPOSALS

4.1 Cross-sections in compression

The stub column tests described in Section 2.3 were used to determine a suitable Class 3 slenderness limit for stainless steel OHS in compression. The ultimate load carrying capacities from the stub column tests have been normalised by the yield load (defined as the measured tensile 0.2% proof strength multiplied by the measured gross cross-sectional area), and plotted in Figure 7. Normalised resistances of greater than unity indicate a Class 1–3 cross-section (i.e. fully effective in compression), whilst values less than unity indicate failure by local buckling prior to the attainment of the yield load, and therefore a Class 4 section. The adopted slenderness parameter $D_e/t\varepsilon^2$ (Equation 2), which is based on an equivalent circular section, was proposed by Gardner & Chan (2007) and Chan & Gardner (2008).

$$\frac{D_e}{t\varepsilon^2} = 2\frac{(a^2/b)}{t\varepsilon^2} \qquad (2)$$

where D_e is the equivalent diameter $= 2a^2/b$ (in compression), which is two times the maximum local radius of curvature of the section, assumed to be an ellipse, and ε is a modification factor for material strength, defined as $\varepsilon = (235/f_y)^{0.5}$ for carbon steel. To reflect the difference in stiffness between carbon steel and stainless steel, the definition of ε is modified to that given by Equation 3, and this is adopted throughout this study. In EN 1993-1-4 (2006), Young's modulus of stainless steel is given as 200000 N/mm².

$$\varepsilon = \left(\frac{235}{\sigma_{0.2}} \frac{E}{210000}\right)^{0.5} \qquad (3)$$

The normalised compression test results are plotted with comparative stainless steel circular hollow section (CHS) data (Kuwamura 2001, Gardner & Nethercot 2004b, Bardi & Kyriakides 2006, Lam & Roach 2006) and carbon steel EHS data (Chan & Gardner 2008) in Figure 7. From the results, it may be observed that the current slenderness limit of 90 (as given in both EN 1993-1-1 and EN 1993-1-4) may be safely adopted as the Class 3 limit for stainless steel OHS in compression.

4.2 Cross-sections in bending

For OHS in bending, different regions of the section (of varying local curvature) are subjected to varying stress levels, and hence the point of initiation of buckling is less clear than in the pure compression case. In minor axis bending, the maximum local radius of curvature ($=a^2/b$) coincides with the maximum compressive stress, and hence, the slenderness parameter of Equation 2 remains appropriate. For major axis bending, this is not the case, and a suitable slenderness parameter has been proposed by Chan & Gardner (2008) on the basis of theoretical and experimental studies, as given by Equations 4 and 5.

$$\frac{D_e}{t\varepsilon^2} = 0.8\frac{(a^2/b)}{t\varepsilon^2} \qquad \text{for } a/b > 1.357 \qquad (4)$$

$$\frac{D_e}{t\varepsilon^2} = \frac{(b^2/a)}{t\varepsilon^2} \qquad \text{for } a/b \leq 1.357 \qquad (5)$$

These measures of slenderness have also been adopted in this study for stainless steel OHS.

In bending, cross-sections are generally placed into one of four behavioural classes. Class 1 sections are able to reach the full plastic moment M_{pl} and have a minimum rotation capacity (generally $R > 3$), Class 2 are also able to reach M_{pl} but do not meet the minimum rotation capacity requirements, Class 3 sections can reach the elastic moment M_{el}, but not M_{pl}, and Class 4 sections contain very slender elements which fail by local buckling prior to the attainment of M_{el}. The results from the bending tests have been assessed against these requirements in order to propose suitable slenderness limits for stainless steel OHS.

The Class 3, 2 and 1 slenderness limits are assessed by making comparison of the test results with the elastic and plastic bending moment resistances and rotation capacity requirements in Figures 8, 9 and 10,

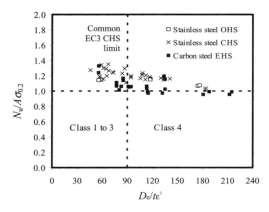

Figure 7. Analysis of stub column test data.

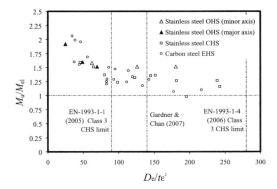

Figure 8. Assessment of Class 3 slenderness limit in bending.

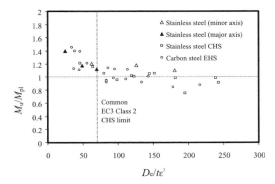

Figure 9. Assessment of Class 2 slenderness limit in bending.

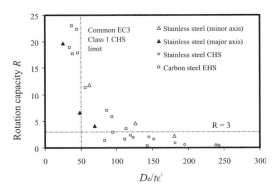

Figure 10. Assessment of Class 1 slenderness limit in bending.

respectively. Comparative stainless steel circular hollow section (CHS) data (Rasmussen & Hancock 1993a, ECSC 2000) and carbon steel EHS data (Chan & Gardner 2008) has also been added to the figures.

From the comparisons, it may be broadly concluded that, by employing the slenderness parameters proposed for carbon steel EHS (Gardner & Chan 2007, Chan & Gardner 2008), the current cross-section classification limits for stainless steel CHS given in EN 1993-1-4 (2006) can safely be adopted for stainless steel OHS. An exception to this may be the Class 3 slenderness limit given in EN 1993-1-4 of 280—this seems overly relaxed given the scarcity of data in this region, and the intermediate value proposed by Gardner & Chan (2007) of 140 seems more appropriate.

4.3 Column buckling

The experimentally derived column buckling data have been added to existing test data (Rasmussen & Hancock 1993b, Talja & Salmi 1995, Ala-Outinen & Oksanen 1997, Burgan et al. 2000, Young & Hartono 2002, Young & Liu 2003, Liu & Young 2003, Gardner & Nethercot 2004c, Young & Lui 2006, Gardner et al. 2006) on stainless steel square, rectangular and circular hollow section columns (SHS, RHS and CHS, respectively), and compared with the current design curve provided in EN 1993-1-4 (2006) for stainless steel hollow sections—see Figure 11. This curve has a limiting slenderness (plateau length) λ_0 of 0.4 and an imperfection factor α of 0.49.

The comparison (Fig. 11) reveals distinctly different behaviour between the SHS and RHS columns and the CHS and OHS columns. The current buckling curve for stainless steel hollow sections shown in Figure 11 may be seen to be broadly representative of the SHS and RHS test data, but the CHS and OHS data falls someway below this curve. The superior buckling response of the SHS and RHS columns may be partly explained by the enhanced strength due to cold-working in the corner regions of the sections. Ashraf et al. (in press) and Ellobody & Young (2007) recognised the deficiency in the current provisions for CHS columns. Ashraf et al. (in press) proposed a

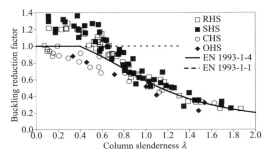

Figure 11. Column buckling test results for stainless steel hollow sections (SHS, RHS, CHS and OHS).

new buckling curve defined by a limiting slenderness value λ_0 of 0.05 and an imperfection factor α of 0.49, which is found to be in good agreement with the majority of the test data. Adoption of this curve for both CHS and OHS stainless steel column buckling was therefore proposed by Gardner et al. (2008), a recommendation supported in the present paper. Of the five existing buckling curves in Eurocode 3 (EN 1993-1-1 2005), curve 'd', which has a limiting slenderness λ_0 of 0.2 and an imperfection factor α of 0.76, is the most appropriate and is plotted on Figure 11.

5 CONCLUSIONS

A series of tests on stainless steel oval hollow sections (OHS) has been described in this paper. The experimental programme comprised tensile material tests, six stub column tests, six in-plane bending tests and eight column buckling tests. The test results were supplemented by some numerically generated results. On the basis of the generated structural performance data, design rules for stainless steel OHS, including slenderness limits in compression and bending, and a column buckling curve, have been proposed. It should be noted that although some of the proposed slenderness limits appear conservative, this reflects the limited test data and the desire to maintain consistency with existing proposals for carbon steel elliptical hollow sections. Further research into stainless steel oval hollow sections is on-going.

ACKNOWLEDGEMENTS

The authors are grateful to Oval 316 for the supply of test specimens and technical data, and would like to thank Gemma Aubeeluck, Namrata Ghelani and Marios Theofanous for their contributions to this research.

REFERENCES

ABAQUS. 2006. *ABAQUS, Version 6.6*. Hibbitt, Karlsson & Sorensen, Inc. Pawtucket, USA.

Ala-Outinen, T. & Oksanen, T. 1997. *Stainless steel compression members exposed to fire*, Research Note 1684, VTT Building Technology, Finland.

Ashraf, M., Gardner, L. & Nethercot, D.A. (in press). Resistance of stainless steel CHS columns based on cross-section deformation capacity. *Journal of Constructional Steel Research*.

Baddoo, N.R., Burgan, R. & Ogden, R. 1997. *Architects' Guide to Stainless Steel. SCI-P-179*. The Steel Construction Institute, UK.

Bardi, F.C. & Kyriakides, S. 2006. Plastic buckling of circular tubes under axial compression-part I: Experiments. *International Journal of Mechanical Sciences*, Vol. 48, No. 8, pp. 830–841.

Burgan, B.A., Baddoo, N.R. & Gilsenan, K.A. 2000. Structural design of stainless steel members-comparison between Eurocode 3, Part 1.4 and test results. *Journal of Constructional Steel Research*, Vol. 54, No. 1, pp. 51–73.

Chan, T.M. & Gardner, L. 2006. Experimental and numerical studies of elliptical hollow sections under axial compression and bending. *Proc. 11th Intern. Symp. on Tubular Structures*. Québec City, Canada: 163–170.

Chan, T.M. & Gardner, L. 2008. Compressive resistance of hot-rolled elliptical hollow sections. *Engineering Structures*, Vol. 30, No. 2: 522–532.

Chan, T.M. & Gardner, L. 2008. Bending strength of hot-rolled elliptical hollow sections. *Journal of Constructional Steel Research*, Vol. 64, No. 9: 971–986.

ECSC 2000. *Final Report. ECSC project—Development of the use of stainless steel in construction. Document RT810, Contract No. 7210 SA/ 842*, The Steel Construction Institute, UK.

Ellobody, E. & Young, B. 2007. Investigation of cold-formed stainless steel non-slender circular hollow section columns. *Steel and Composite Structures*, Vol. 7, No. 4: 321–337.

EN 10002-1. 2001. *Metallic materials—Tensile testing, Part 1: Method of test at ambient temperature*. European Standard, CEN.

EN 10088-1. 2005. *Stainless steels—Part 1: List of stainless steels*. CEN.

EN 10088-2. 2005. *Stainless steels—Part 2: Technical delivery conditions for sheet/plate and strip for general purposes*. CEN.

EN 1993-1-1.2005. *Eurocode 3: Design of steel structures—Part 1.1: General rules–General rules and rules for buildings*. CEN.

EN 1993-1-4.2006. *Eurocode 3: Design of steel structures—Part 1.4: General rules–Supplementary rules for stainless steel*. CEN.

Gardner, L. 2005. The use of stainless steel in structures. *Progress in Structural Engineering and Materials*, Vol. 7, No. 2: 45–55.

Gardner, L. & Ashraf, M. 2006. Structural design for non-linear metallic materials. *Engineering Structures*, Vol. 28, No. 6: 926–934.

Gardner, L. & Chan, T. M. 2007. Cross-section classification of elliptical hollow sections. *Journal of Steel and Composite Structures*, Vol. 7, No. 3: 185–200.

Gardner, L. & Nethercot, D.A. 2004a. Stainless steel structural design: A new approach. *The Structural Engineer*, Vol. 82, No. 21: 21–28.

Gardner, L. & Nethercot, D.A. 2004b. Experiments on stainless steel hollow sections–Part 1: Material and cross-sectional behaviour. *Journal of Constructional Steel Research*, Vol. 60, No. 9: 1291–1318.

Gardner L. & Nethercot D.A. 2004c. Experiments on stainless steel hollow sections–Part 2: Member behaviour of columns and beams. *Journal of Constructional Steel Research*, Vol. 60, No. 9: 1319–1332.

Gardner, L., Talja, A. & Baddoo, N.R. 2006. Structural design of high-strength austenitic stainless steel. *Thin-Walled Structures*, Vol. 44, No. 5: 517–528.

Gardner, L., Theofanous, M. & Chan, T.M. 2008. Buckling of tubular stainless steel columns. *Proc. 5th Intern. Conf.*

on *Coupled Instabilities in Metal Structures*. Sydney, Australia: 199–207.

Kuwamura, H. 2001. *Local buckling consideration in design of thin-walled stainless steel members*. Lecture at Pusan National University.

Lam, D. & Roach, C. 2006. Axial capacity of concrete filled stainless steel circular columns. *Proc. 11th Intern. Symp. and IIW Int. Conf. on Tubular Structures*, Quebec City, Canada: 495–501.

Liu, Y. & Young, B. 2003. Buckling of stainless steel square hollow section compression members. *Journal of Constructional Steel Research*, Vol. 59, No. 2: 165–177.

Mirambell, E. & Real, E. 2000. On the calculation of deflections in structural stainless steel beams: an experimental and numerical investigation. *Journal of Constructional Steel Research*, Vol. 54, No. 1: 109–133.

Rasmussen, K.J.R. 2003. Full range stress-strain curves for stainless steel alloys. *Journal of Constructional Steel Research*, Vol. 59, No.1: 47–61.

Rasmussen, K.J.R. & Hancock, G.J. 1993a. Design of cold-formed stainless steel tubular members II: Beams. *Journal of Structural Engineering*, ASCE, Vol. 119, No. 8: 2368–2386.

Rasmussen, K.J.R. & Hancock, G.J. 1993b. Design of cold-formed stainless steel tubular members. I: Columns. *Journal of Structural Engineering*, ASCE, Vol. 119, No. 8: 2349–2367.

Talja, A. & Salmi, P. 1995. *Design of stainless steel RHS beams, columns and beam-columns*. Research Note 1619, VTT Building Technology, Finland.

Young, B. & Hartono, W. 2002. Compression tests of stainless steel tubular members. *Journal of Structural Engineering*, ASCE, Vol. 128, No. 6: 754–761.

Young, B. & Liu, Y. 2003. Experimental investigation of cold-formed stainless steel columns. *Journal of Structural Engineering*, ASCE, Vol. 129, No. 2: 169–176.

Young, B. & Lui, W.M. 2006. Tests on cold formed high strength stainless steel compression members. *Thin-Walled Structures*, Vol. 44, No. 2: 224–234.

Tubular Structures XII – Shen, Chen & Zhao (eds)
© 2009 Taylor & Francis Group, London, ISBN 978-0-415-46853-4

Study on columns with rounded triangular section in Henan TV tower

M.J. He, F. Liang & R.L. Ma
Department of Building Engineering, Tongji University, Shanghai, China

ABSTRACT: Structural members with Circular Hollow Section (CHS) are commonly used in high steel TV towers due to their low wind pressure coefficient and high overall buckling capacity under axial compressive loads. However, the outer columns of the 388 m high Henan TV tower, currently a landmark in China's Henan province, adopted the Rounded Triangular Sections (RTS) in order to show a unique architectural appearance of the tower. This makes it very difficult for structural designers to determine the structural form and load-carrying capacity of the columns as well as the connection method. Considering the size of the structural members, feasibility of the connections, construction difficulty, construction time, and project cost, designers decided to use the RTS columns with specially designed flange-plates extruded in both sides of the column wall. A series of experimental studies on the scaled structural models and finite element analyses were carried out to investigate the mechanical behavior of the RTS columns and the associated flange connections.

1 INTRODUCTION

Circular hollow sections (CHS) members are commonly used as the main structural elements in high steel towers, particularly in the super high towers such as the 385 m high Kiev TV Tower and the 336 m high Heilongjiang Dragon Tower, etc. (Wang et al. 2004, He et al. 2001). The main reasons of adopting CHS members in high steel towers are: (1) wind load is the dominated load for high towers and the wind pressure coefficient for CHS is lower than any other types of cross sections; and (2) axial compression force is the dominated force in the main structural members. CHS is central symmetric and has the same radius of gyration about any axis of the cross section. However, the outer columns of the 388 m high Henan TV tower, currently a landmark in China's Henan province, adopted the rounded triangular section (RTS) in order to present a unique architectural appearance of the tower. This makes it very difficult for structural designers to determine the structural form and load-carrying capacity of the columns as well as the connection method. Considering the size of the structural members, feasibility of the connections, construction difficulty, construction time, and project cost, designers decided to use the RTS columns with special flange-plates extruded in both sides of the column wall. Usually, during manufacturing, the length of structural member segments should be no longer than 15 m due to the limitation of the shipping and the equipment for hot-dip galvanization.

In order to assure the construction efficiency and the quality of assembling, bolted connections were used in this structure. This paper presents a study on the connection method and the mechanical behavior of the connections. Meanwhile, static tests and finite element analyses were conducted to study the behavior of the RTS columns under eccentric compression loads.

2 DESCRIPTION OF THE HENAN TV TOWER

The Henan tower consists of five major parts: a bottom building, an outer column-frame, an inner core truss, observation decks, and a top truss for supporting antennas, as shown in Figure 1. The bottom building is a six-story steel frame structure with a total construction area of 36000 m², functioning as the entertainment, shopping, and administration center of the tower. The 12-storey observation decks with construction area of 7000 m², located at a height of 200 m, is used for sightseeing and installing emission equipment etc. One platform in the observation decks can offer a 360 degree view of Zhengzhou city. The 110 m high top truss is designed to support the broadcasting antennas. The outer column-frame and the inner core truss work together as the main structure of the tower to carry all the vertical and horizontal loads. The inner core is a regular decagon space truss with circumcircle diameter of 14 m. The outer column-frame consists of 10 RTC columns.

Five columns are positioned clockwise and the other fives are positioned counterclockwise from bottom to top. The ten columns surround the inner core truss and form "X" joints at the intersections. The outer column-frame and the inner core truss are connected together with the braces at the heights of 35 m, 70 m, 105 m, 148 m, 188 m, and 218.8 m respectively to ensure the whole structural rigidity and stability.

3 STRUCTURAL STYLE OF OUTER COLUMN

To meet the requirement of the architectural appearance of the tower, two different methods are available to design the outer columns: (1) the architectural method which encloses columns into required shape with sheathing boards such as steel boards or aluminum boards; and (2) the structural method which directly adopts RTS columns.

For the architectural method, the lattice truss columns could be a choice. As shown in Figure 2, three CHS pipes are linked together with additional webs. However, through comparisons and analyses, some adverse concerns were identified:

1. Lattice truss column style will result in quite large column size. Since the outer column-frame carries most of the vertical dead and live loads as well as the horizontal loads due to wind and earthquake actions (Wang et al. 2004), the design forces in the columns are very large. In the lattice truss column, a part of cross section is close to its centroid. Therefore, its radius of gyration will be smaller than a RTS with the same cross section area. And the width of the lattice truss column at the bottom of the tower would be about 6 m, which seems to be large compared with the radius of the inner core truss which is 7 m.
2. A lot of problems in design and fabrication of the "X" joints for the lattice truss column need to be solved. A total of 25 "X" joints exist in the outer column-frame. If the lattice truss column is adopted, four lattice truss columns will be connected at each joint. The "X" joints would be too complex to design, fabricate, transport and install.
3. The fabrication cost will increase since a large number of enclosing boards are needed. Based on the calculation, the total area of the enclosing boards will exceed 30000 m². However, these enclosing boards are only for decoration and have no contributions to structural load-carrying capacity.
4. It's hard to control the construction time on the project schedule since the lattice truss columns requires a large amount of welding, especially the on-site welding. Besides, the quality of on-site welding is difficult to control.

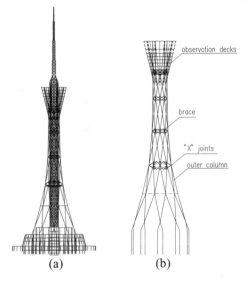

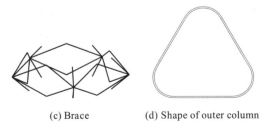

(a) The elevation of the tower (b) Outer column-frame

(c) Brace (d) Shape of outer column

Figure 1. Henan TV tower.

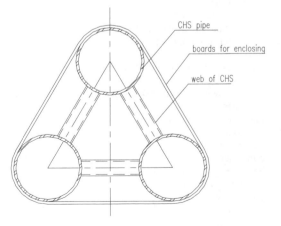

Figure 2. Lattice truss column.

5) For lattice truss columns, all structural members will be concealed by the enclosing boards. Thus, these members and joints will be difficult for inspection and maintenance during the service life of the structure.

Concerning about the above issues, solutions with RTS columns have been proposed. To prevent the local buckling of column wall, a few stiffening plates are added, as shown in Figure 3.

Compared with the lattice truss column, the RTS column has the following advantages:

1. It has a lager section area and section modulus, so that the width of the section at the bottom of the tower is only 2 m~2.7 m, which satisfies the requirement of the architectural appearance, i.e. 2 m~3 m.
2. It is much easier to design and fabricate the "X" joints where four columns intersect.
3. It does not need the enclosing boards while satisfying the requirement of the architectural appearance. The project cost is reduced.
4. Most of the welding can be done in the workshop. On the construction site, most of the structural members are connected by the bolted connections. The construction quality can be controlled well.
5. The outer column-frame is exposed. It is convenient to check the health condition of the structure during its life time (Ma et al. 2006).

4 CONNECTION OF THE OUTER COLUMNS

The wall thickness of the outer columns is 35 mm and the wall thickness of those in the "X" joint is 40 mm.

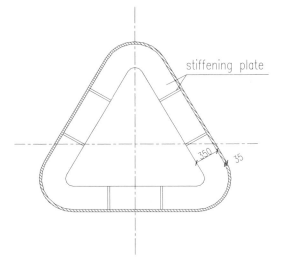

Figure 3. RTS column.

Therefore, it is very hard to do on-site welding to connect these columns at the height of 170 m. It is well known that the hot-dip galvanization is to provide a long-term corrosion resistance for steel members. With the fluorocarbon coatings, the silvery white color of the surface can retain for about 20 years. However, the on-site welding will cause damages to the hot-dip galvanizing coatings. And it is difficult to repair. In order to ensure the construction quality and protect the coating, it is desirable to finish the welding work in the workshop and only assemble the bolted connections on site.

The length of the straight passage of the outer columns exceeds 40 m, but the longest hot-dip galvanizing bath in China is 18 m long and the transportation vehicle is 12 m long. Also considering the capacity of the hoisting cranes, the columns must be divided into several segments with maximum length of 12 m, which can be spliced on site.

Flange-plate connection is a common splicing method in space truss structures. In steel tower design, most of the structural members are regarded as axially loaded members, and the bending stiffness at the both end of members are released in model analysis. Flange-plate connection is a kind of semi-rigid joint, which is required to transfer the loads instead of the stiffness. However, in the Henan TV tower, the outer column-frame is designed as a space rigid frame structure. Every part of the column should have the same bending strength and stiffness. And very little rotation is allowed between each side of the flange splice joint.

In order to satisfy the requirements above, a kind of special flange extruded in two sides of the column wall has been designed, as shown in Figure 4. The contact surfaces of the flange are required to be milled, and connected with frictional high-strength bolts. The pre-tightening force must strictly follow the requirements in the building code of China to ensure that no rotation would occur between each side of the flange splice joint under the design loads.

Since the dominated design force in the flange splice joint is tension, the mechanic behavior of the flange splice joint can be studied with the maximum design tensile force. Take the column with side length of 2.2 m as an example. The maximum tensile force in the column is 13229 kN, coupled with the moment of 1478 $kN \cdot m$ around the axis 2 and the moment of $-4327\ kN \cdot m$ around the axis 3 (Fig. 4a).

The thickness of the flange plate is 50 mm. The Φ30 bolts with tensile capacity of 224 kN are spaced at 100 mm o.c. Calculation results show that the maximum moment in the flange is 12.2 $kN \cdot m$ and the maximum normal stress is 258 N/mm^2, less than the design strength.

As all the bolts of the flange are under tension, the contact surfaces will probably rotate around the axis

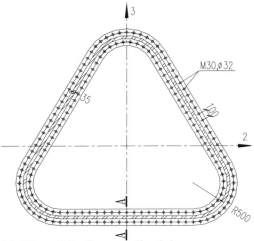

(a) Plan of the flange splice joint

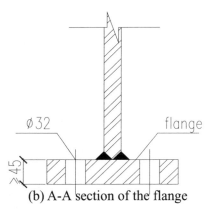

(b) A-A section of the flange

Figure 4. The special flange splice joint.

2 and 3 (CMC 2006, Shi et al. & Wang et al. 2005). Maximum tensile force N^b_{max} in the bolt can be calculated by equation (1) recommended in the building code of China (GB 50135-2006):

$$N^b_{max} = \frac{My_n}{\sum(y'_i)^2} + \frac{N}{n_0} \quad (1)$$

where M = the moment on the section; N = the tension on the section; y'_i = the distance from the bolt to the axis; and n_0 = the number of the bolts. The calculated maximum tensile force in the bolt is 168 kN, less than the capacity of 224 kN. Therefore, the contact surfaces will not separate, and the bending stiffness of the outer column is continuous at the location of the splice joint.

5 MODEL TEST FOR THE COLUMN

The overall structural behavior of the tower was pre-assessed by the finite element software Sap2000. The outer columns were considered as rigid frame members, and the sections were chosen based on the strength and stability capacity. In order to evaluate the mechanical behavior of the outer column, load-carrying capacity tests and FE modeling of the RTS columns with different section sizes were carried out to study the stress distribution in the columns under the design load. For the columns, since the eccentric compression is the governing loading scenario for design, only compression test was conducted.

5.1 Design and construction of the scaled model

The objective of the tests was to assess the mechanical behavior of the RTS columns with flange splice joint. Three types of columns were considered: 1) side length of 2.2 m with column length 19.7 m; 2) side length of 2.6 m with column length of 21 m; and 3) side length of 2.8 m with column length of 34 m. The scaling factor of the test specimens must be calculated according to the similitude theory. The capacity of the actuator (jack) and the capacity of the crane in the workshop as well as the space in the laboratory should also be taken into account when designing a scaled model test (Fan et al. 2006).

Because the forces applied on the test specimens were large, a self-equilibrium loading system was designed and fabricated for the tests (Fig. 5). Two piers were designed. One pier was fixed and the other could slide. Pre-stressed steel cables were used to connect the piers together and provide tension during the tests. Table 1 showed the scale factors considering the similitude relationships. Because the test programs of three types of columns were similar, only the test of the third type of column with side length of 2.8 m is presented herein.

5.2 Test program

A total of 21 load steps were used in the test and the applied peak load was 30% higher than the design load, as shown in Table 2, in which P is the applied load. Eccentric loading was adopted to impose the moment effect on the column. There were 27 strain gauges and 25 strain gauge rosettes installed on the test specimen, especially on the locations of the weld seams and the rounded vertices. The compression load was applied step by step. For each load step, test data were collected. After the data was verified, the load for the next step was applied.

5.3 Test results and data analysis

During the test, no cracks were observed in the weld seams and the surface of the column experienced no

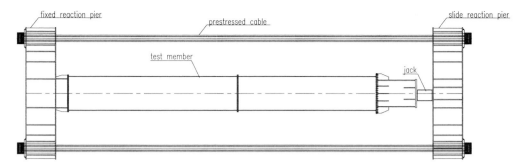

Figure 5. The plan of the test.

Table 1. Similitude relationships.

Physical	Unit	Scale
Dimension	Length	1/3
Material	Strain	1
	Young's modulus	1
	Stress	1
Load	Force	1/9
	moment	1/27

Table 2. Load program.

Load step	P(kN)	Load step	P(kN)
1	0	12	3780
2	440	13	3880
3	880	14	3980
4	1320	15	4080
5	1760	16	4180
6	2200	17	4280
7	2640	18	4380
8	3080	19	4480
9	3280	20	4580
10	3480	21	4680
11	3680		

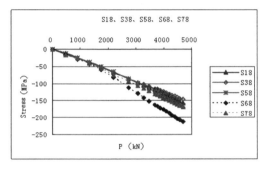

Figure 6. The stress on the vertexes.

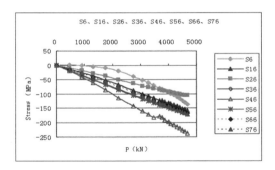

Figure 7. The stress on the plate.

local buckling. The test data showed that the strain at different locations of the column remained almost in a linear elastic range with the increase of the applied load. Figures 6 and 7 show the relationship between the stresses and the applied load, in which the horizontal axis is the load (P), and the vertical axis is the stress. Since P was an eccentric load, the maximum strain was the compression stain in the compressive zone. The maximum compression strain of the column was $1151\mu\varepsilon$ at the spot S46 with the corresponding equivalent stress 237 MPa, less than the design strength of steel Q345. It indicated that the column should be safe under the design load. Because of the humid weather, some data at the spot S8, S9, S10, S28, S41, S42, S45, S47, S48, and S61 were lost and not listed herein.

6 FINITE ELEMENT ANALYSIS

The mechanical behavior of the RTS columns with flange splice joint was also investigated using the finite element software ANSYS (Fig. 8). The shell element SHELL181 with four nodes was used to model the column walls (Tong et al. 2007).

The maximum load in the tests wall applied on the model. The stresses of the spots where the strain gauges were installed in the test were extracted. Table 3 gives the comparisons between the FE model calculated stresses and the test results, which shows a

Table 3. Comparison of the stresses in test and FEA (N/mm²).

Spot	S1	S2	S3	S4	S5	S6	S7	S8	S9	S10
Test	−99.1	−159.2	−57.1	−156.8	−150.2	−135.8	−123.4			
FEM	−175.1	−170.6	−163.3	−151.4	−148.6	−146.5	142.8	−142.8	−157.9	−167.3
Spot	S11	S12	S13	S14	S15	S16	S17	S18	S19	S20
Test	−84.7	−96.6	−104.0	−136.0	−157.6	−157.6	−164.0	−156.6	−133.9	−112.7
FEM	−153.1	−152.5	−147.6	−138.3	137.0	−135.6	−133.4	−138.8	−143.8	−150.2
Spot	S21	S22	S23	S24	S25	S26	S27	S28	S29	S30
Test	−54.8	−65.3	−80.3	−96.0	−144.8	−106.5	−148.5		−130.8	−97.9
FEM	−150.8	−128.7	−127.3	−145.8	−124.4	−124.0	−143.6	−144.2	−126.2	−128.1
Spot	S31	S32	S33	S34	S35	S36	S37	S38	S39	S40
Test	−103.6	−73.7	−95.0	−139.1	−159.9	−167.3	−160.3	−146.3	−121.5	−126.1
FEM	−138.5	−142.7	−144.5	−143.6	−148.7	−149.2	−145.7	−145.3	−146.0	−143.8
Spot	S41	S42	S43	S44	S45	S46	S47	S48	S49	S50
Test			−63.2	−77.0		−237.3			−121.7	−286.8
FEM	−180.3	−195.6	−200.4	−194.9	−210.9	−212.4	−201.0	−199.8	−204.3	−197.9
Spot	S51	S52	S53	S54	S55	S56	S57	S58	S59	S60
Test	−97.2	−77.9	−96.6	−147.5	−172.4	−171.2	−177.2	−163.8	−137.4	−139.9
FEM	−124.7	−133.7	−141.0	−147.3	−157.0	−159.1	−155.6	−154.2	−146.7	−137.2
Spot	S61	S62	S63	S64	S65	S66	S67	S68	S69	S70
Test		−99.3	−139.3	−115.2	−160.7	−166.4	−125.0	−211.8	−169.1	−134.3
FEM	−118.7	−110.8	−121.0	−154.0	−140.2	−143.1	−167.0	−164.7	−123.8	−115.2
Spot	S71	S72	S73	S74	S75	S76	S77	S78	S79	S80
Test	−73.1	−96.0	−112.5	−127.7	−161.7	−170.4	−179.2	−167.5	−129.8	−101.8
FEM	−108.3	−121.2	−135.3	−142.6	−165.9	−170.0	−165.8	−163.0	−146.2	−128.0

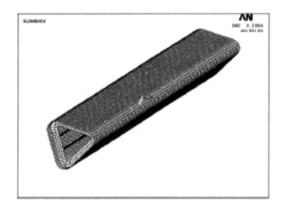

Figure 8. Finite element model.

reasonable agreement between the test results and the model simulation. Only at a small number of spots such as S1, S11, and S21, the model overpredicted the stress response. This is because in the test, these spots were close to the weld seams of the column plate and stiffening plate, and the weld seams also carried part of the load. However, the FE model did not consider these weld seams. Overall, the model simulation indicated that the test method was rational and the test data were reliable. It is believed that the RTS columns should be safe under the design loads.

7 CONCLUSIONS

Using finite element modeling and structural testing, this paper presented a study on a type of specially designed columns with rounded triangular section (RTS) in the Henan TV tower. Based on the experimental results and the modeling results, some conclusions can be drawn:

1. The RTS columns provide a unique architectural appearance, and the construction of the columns and "X" joints is feasible.
2. Under the design tensile force coupled with bending moments, the contact surfaces of the flange plates will not separate; bending stiffness of section at the splice joint is continuous as well as the load-carrying capacity.
3. Under the compressive force and bending moment with 30% higher than the design load, the RTS column spliced with flange-plate connections remained still in elastic range. The column was believed to have adequate safety margin. The section of the column class can be considered as class B in the stability analysis (CMC 2003).
4. With adoption of RTS columns and flange-plate splice connections, the length of a segment of the column can be limited to 12 m. Thus, the hot-dip galvanizing, transportation, and hoisting are feasible.

REFERENCES

He Minjuan, Ma Renle & Zhu Liang. 2001. Some Key Technologies about Heilongjiang Broadcasting and TV Steel Tower. *Special structure* 18(1): 13–16.

China Ministry of Construction. 2003. *Code for design of steel structures (GB 50017-2003)*. Beijing: China planning Press.

Ma Renle & Wang Zhaomin. 2004. *Tower structures*. Beijing: Science Press.

Chen Hong, Li Shaofu, Shi Gang, Shi Yongjiu, & Wang Yuanqing. 2005. Experimental study on bolt resistance for various structural steel end-plate connections. *journal of harbin institute of technology* 37(1): 66–69.

Chen Youquan, Chen Yiyi, Wei Chaowen & Wang Sufang. 2005. Research on neutral axis location of high-strength bolts of end-plate connections. *Journal of architecture and civil engineering* 22(3): 45–54.

China Ministry of Construction. 2006. *Code for design of high-rising structures (GB 50135-2006)*. Beijing: China planning Press.

Chen Yiyi, Fan Zhong, Hu Tianbing, Li Ming, Peng Yi, Zhao Lihua, & Zhao Xianzhong. Design and research of double-chord KK-connections of welded thin-wall box-section truss of the National Stadium. *Journal of building Structures* 28(3): 41–48.

Liang Feng & Ma Renle. 2006. Structural Design and Test Research of Henan Broadcast and TV Tower. *Special Structures* 23(3): 1–4.

Chen Yiyi, Wang Wei & Zhao Xianzhong. 2007. State of the art and key issues on performance-based design of steel tubular joints. *China civil engineering journal* 40(11): 1–8.

Chen Yangji, Chen Yiyi, Chen Zuo, Lin Gao, Lin Yinru, Tong Lewei & Wang Bin. 2007. Static behavior of circular hollow section joints with curved chords. *Journal of building structures* 28(1): 28–34.

Numerical modeling and design approach of aluminum alloy tubular columns

J.H. Zhu
College of Civil Engineering, ShenZhen University, ShenZhen & PR China

B. Young
Department of Civil Engineering, The University of Hong Kong, Hong Kong

ABSTRACT: A numerical investigation and design of aluminum alloy circular hollow section non-welded and welded columns are presented in this paper. A non-linear finite element model was developed and verified against fixed-ended column tests. The column specimens were fabricated using aluminum extrusions of heat-treated aluminum alloys. Both columns with and without transverse welds at the ends of the specimens were investigated. The welded columns were modeled by dividing the column into different portions along the column length, so that the heat-affected zone softening at both ends of the welded columns was included in the simulation. The geometric imperfections and material non-linearities obtained from a previous test program were incorporated in this study. The verified finite element model was used for a parametric study of fixed-ended aluminum alloy circular hollow section columns. The column strengths predicted by the finite element analysis are compared with the design strengths calculated using the current American, Australian/New Zealand and European specifications for aluminum structures. The column strengths are also compared with the design strengths predicted by the direct strength method, which was developed for cold-formed carbon steel members. Design rules are proposed for aluminum alloy circular hollow section columns with transverse welds at the ends of the columns. Reliability analysis was performed to justify the design rules.

1 INTRODUCTION

Aluminum members are being used increasingly in structural applications. The current American Aluminum Design Manual (AA 2005), Australian/New Zealand Standard (AS/NZS 1997) and European Code (EC9 2000) for aluminum structures provide design rules for compression members. Schafer & Peköz (1998) developed a new design method called the direct strength method for cold-formed steel structures. The test data used in the development of column design for the direct strength method were based on concentrically loaded pin-ended cold-formed steel columns for certain cross sections and geometric limits (Schafer 2000, 2002). The direct strength method has been adopted by the North American Specification (AISI 2001, 2004) for cold-formed steel structures. Zhu & Young (2006a, 2006b) showed that the direct strength method, with some modification, can be used in the design of aluminum alloy square hollow section (SHS) and rectangular hollow section (RHS) columns. For the purpose of obtaining accordant design rules of different cross-sections, the direct strength method was used in this study for the design of aluminum alloy circular hollow section (CHS) columns.

One disadvantage of using aluminum as a structural material is that heat-treated aluminum alloys could suffer loss of strength in a localized region when welding is involved, and this is known as heat-affected zone (HAZ) softening. Previous research (Mazzolani 1995 & Sharp 1993) indicated that welds have significant effect on column strength. The test program presented by Zhu & Young (2006c) showed that transverse welds at the ends of the CHS columns reduce the column strength by nearly 46%. In addition, it was also shown that the design rules in the current American, Australian/New Zealand and European specifications are generally quite conservative for aluminum alloy welded columns of circular hollow sections (Zhu & Young 2006c). Hence, it is necessary to obtain more accurate design rules for aluminum alloy columns containing transverse welds.

Finite element analysis (FEA) has been widely used in structural design. Compared with physical experiments, FEA is relatively inexpensive and time efficient, especially when a parametric study of cross-section geometry is involved. In addition, FEA is more convenient for investigation involving geometric imperfections of structural members, whereas this could be difficult to investigate through physical tests. Although FEA is a useful and powerful tool for structural analysis and design, it is important to obtain an accurate and reliable finite element model (FEM) prior to a parametric study of FEA being carried out. A non-linear FEM for aluminum columns of SHS and RHS, with and without transverse welds, has been developed by Zhu & Young (2006a). In this study, a non-linear FEM for aluminum columns of CHS has been developed and verified against experimental results.

The purpose of this paper is firstly to investigate the behaviour and design of aluminum CHS columns using non-linear FEA. The verified FEM is used for a parametric study of cross-section geometries. Secondly, the current direct strength method is used for the design of aluminum non-welded and welded columns of CHS. Thirdly, design rules for aluminum welded columns of CHS are proposed based on the current direct strength method. The column strengths predicted by the FEA are compared with the design strengths calculated using the American Aluminum Design Manual, Australian/New Zealand Standard and European Code for aluminum structures, as well as the direct strength method (DSM) and proposed design rules. Lastly, reliability analysis is performed to assess the reliability of these design rules.

2 SUMMARY OF TEST PROGRAM

2.1 Column tests and material properties

Experimental results of aluminum alloy circular hollow sections compressed between fixed ends have been reported by Zhu & Young (2006c). The test specimens were fabricated by extrusion using 6063-T5 and 6061-T6 heat-treated aluminum alloys. The test program included 21 fixed-ended CHS columns with both ends welded to aluminum end plates, and 8 fixed-ended CHS columns without the welding of end plates. In this paper, the term "welded column" refers to a specimen with transverse welds at the ends of the column, whereas the term "non-welded column" refers to a specimen without transverse welds. The testing conditions of the non-welded and welded columns are identical, other than the absence of welding in the non-welded columns. The experimental program included four test series with different cross-section geometry and type of aluminum alloy, as shown in Table 1, using the symbols illustrated in Figure 1. The measured cross-section dimensions of each specimen are detailed in Zhu & Young (2006c). The specimens were tested between fixed ends at various column lengths ranging from 300 to 3000 mm. The test rig and operation are also detailed in Zhu & Young (2006c). The experimental ultimate loads (P_{Exp}) and failure modes observed at ultimate loads obtained from the non-welded and welded column tests are detailed in Zhu & Young (2006c). The non-welded and welded material properties for each series of specimens were determined by longitudinal tensile coupon tests as detailed in Zhu & Young (2006c). The measured material properties obtained from the coupon tests are summarized in Table 2.

Table 1. Nominal specimen dimension of test series.

Test series	Type of material	Dimension $D \times t$ (mm)
N-C1	6063-T5	50×1.6
N-C2	6063-T5	50×3.0
H-C1	6061-T6	50×1.6
H-C2	6061-T6	50×3.0

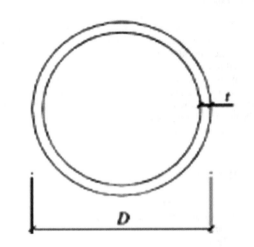

Figure 1. Definition of symbols.

Table 2. Measured non-welded and welded material properties of tensile coupons.

Specimen	E (GPa)	$\sigma_{0.2}$ (MPa)	σ_u (MPa)	ε_f (%)
N-C1-W	73.0	71.3	120.9	9.9
N-C1-NW	66.7	194.6	214.4	10.0
N-C2-W	71.6	75.3	109.7	6.9
N-C2-NW	67.1	185.9	207.7	10.4
H-C1-W	72.6	92.5	148.3	10.7
H-C1-NW	67.1	286.7	310.1	10.7
H-C2-W	71.7	94.3	161.2	10.9
H-C2-NW	70.2	278.9	284.3	11.7

Note: 1 ksi = 6.89 MPa; NW = non-welded tensile coupon; W = welded tensile coupon.

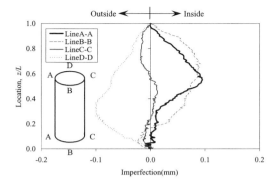

Figure 2. Measured initial local geometric imperfection profiles for Series N-C1 specimen of 300 mm in length.

2.2 Measured local and overall geometric imperfections

In this study, the initial local geometric imperfections were measured on four CHS specimens with a length of 300 mm. The specimens were cut from specimens belonging to the same batch of specimens as the column tests. Hence, the measured local geometric imperfections are considered nearly the same order of imperfections as the column specimens. A Mitutoyo co-ordinate Measuring Machine (CMM) with an accuracy of 0.001 mm was used to measure the initial local geometric imperfections. The measurements were taken at the longitudinal quarter lines A-A, B-B, C-C and D-D of each specimen as shown in Figure 2. Readings were taken at regular intervals of 2 mm along the specimen length. The measured local imperfection profiles for Series N-C1 are shown in Figure 2. The vertical axis is plotted against the normalized location along specimen length and the horizontal axis is plotted against the measured local geometric imperfections. The maximum measured local geometric imperfections were 6.3%, 2.0%, 10.4% and 16.0% of the section thickness for Series N-C1, N-C2, H-C1 and H-C2, respectively.

Initial overall geometric imperfections were measured on all specimens prior to testing, except for the short specimens of 300 mm in length, as detailed in Zhu & Young (2006c). The maximum measured overall geometric imperfections at mid-length were 1/1732, 1/1432, 1/562 and 1/854 of the specimen length for Series N-C1, N-C2, H-C1 and H-C2, respectively.

3 FINITE ELEMENT MODELING

An accurate and reliable non-linear finite element model (FEM) for aluminum non-welded and welded columns of square and rectangular hollow sections has been developed using the finite element program ABAQUS (2004), as presented by Zhu & Young (2006a). In this study, the FEM was used for the simulation of aluminum alloy circular hollow section columns tested by Zhu & Young (2006c). The development of the FEM is detailed in Zhu & Young (2006a). In the FEM, the measured cross-section dimensions, material properties and initial geometric imperfections of the test specimens were modeled. The fixed-ended boundary condition was modeled by restraining all the degrees of freedom of the nodes at both ends of the column, except for the translational degree of freedom in the axial direction at one end of the column. The nodes other than the two ends were free to translate and rotate in any directions. Material non-linearity was included in the FEM by specifying the true values of stresses and strains. The plasticity of the material was simulated by a mathematical model, known as the incremental plasticity model, in which the true stresses and true plastic strains were calculated in accordance with ABAQUS (2004). The geometric imperfections were included in the FEM by using the Eigenvalue analyses. The displacement control loading method was simulated in the FEA that was identical to the loading method used in the column tests. The S4R general-purpose shell elements were used in the FEM. A finite element mesh size of 5 × 5 mm (length by width) was used in the modeling of the non-welded and welded columns of CHS. The welded columns were modeled by dividing the columns into different portions along the column length. Therefore, the heat-affected zone (HAZ) softening at both ends of the columns was simulated. The welded columns were separated into three parts, the HAZ regions at both ends of the columns, and the main body of the columns that are not affected by welding. In this study, different lengths of HAZ extension at both ends of the welded columns were considered, that are equal to 20 and 15 mm. The material properties obtained from the non-welded and welded tensile coupon tests were used for the main body and the HAZ regions of the welded columns, respectively.

4 TEST VERIFICATION

The developed finite element model was verified against the experimental results. For the non-welded columns, the ultimate loads and failure modes predicted by the FEA are compared with the experimental results as shown in Table 3. It is shown that the ultimate loads (P_{FEA}) obtained from the FEA are generally in good agreement with the experimental ultimate loads (P_{Exp}). The ultimate loads predicted by the FEA are slightly lower than the experimental ultimate

Table 3. Comparison of test and FEA results for non-welded columns.

Specimen	Experimental		FEA		Comparison
	P_{Exp} (kN)	Failure mode	P_{FEA} (kN)	Failure mode	P_{Exp}/P_{FEA}
N-C1-NW-L300	48.5	Y	46.3	Y	1.05
N-C1-NW-L1000	45.9	F	43.6	F	1.05
N-C2-NW-L300	102.4	Y	87.1	Y	1.18
N-C2-NW-L1000	86.1	Y	76.0	F	1.13
H-C1-NW-L300	75.9	Y	69.6	Y	1.09
H-C1-NW-L1000	71.7	Y	65.6	F	1.09
H-C2-NW-L300	129.6	Y	124.3	Y	1.04
H-C2-NW-L1000	119.6	F	117.1	F	1.02
				Mean	1.08
				COV	0.048

Note: F = flexural buckling; Y = yielding.

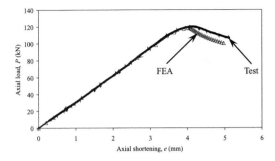

Figure 3. Comparison of experimental and FEA axial load-shortening curves for Specimen H-C2-NW-L1000.

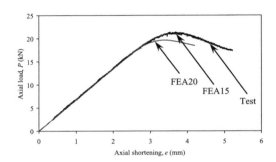

Figure 4. Comparison of experimental and FEA axial load-shortening curves for Specimen N-C1-W-L2350.

loads. The mean value of the experimental-to-FEA ultimate load ratio is 1.08 with a corresponding coefficient of variation (COV) of 0.048 for the non-welded columns, as shown in Table 3.

For the welded columns, both the ultimate loads predicted by the FEA using the HAZ extension of 20 mm (P_{FEA20}) and 15 mm (P_{FEA15}) are compared with the experimental results as shown in Zhu (2006). It is shown that the P_{FEA15} are in better agreement with the experimental ultimate loads compared with the P_{FEA20}.

The failure modes at ultimate load obtained from the tests and FEA for each specimen are also shown in Zhu (2006). The observed failure modes included yielding, flexural buckling, and failure in the heat-affected zone. The failure modes predicted by the FEA are identical to those observed in the tests, except for the specimens N-C2-NW-L1000 and H-C1-NW-L1000. Figure 3 shows the comparison of the load-shortening curves obtained from the test and predicted by the FEA for the non-welded specimen H-C2-NW-L1000. It is shown that the FEA curve follows the experimental curve closely, except that the loads predicted by the FEA are slightly lower than the experimental loads in the post-ultimate range. Figure 4 also shows the load-shortening curves for the welded specimen N-C1-W-L2350. The load-shortening curves predicted by the FEA using the HAZ extension of 20 and 15 mm are shown in Figure 4.

5 PARAMETRIC STUDY

The finite element model closely predicted the experimental ultimate loads and failure modes of the tested aluminum circular hollow section columns. Hence, the model was used for an extensive parametric study. The parametric study included 80 specimens that consisted of 16 series, as shown in Table 4. Each series contained 5 specimens with column lengths of 500, 1200, 2000, 2700 and 3500 mm. The specimen series were labelled such that the type of aluminum alloy, section thickness and welding condition could be identified, as shown in Table 4. For example, the label "T5-t0.4-NW" defines the following specimen series:

- The first letter indicates the type of material of the specimen, where "T5" refers to the aluminum alloy 6063-T5, and "T6" refers to the aluminum alloy 6061-T6;
- The second part of the label "t0.4" indicates the section thickness of the specimen, where the letter "t" refers to the section thickness and the following digits are the thickness of the CHS in millimetres (0.4 mm). Table 4 shows the cross-section

Table 4. Cross-section dimensions of the series for parametric study.

Series	Type of material	Diameter D (mm)	Thickness t (mm)	Area A (mm^2)
T5-t0.4-NW	6063-T5	75.4	0.4	94.2
T5-t0.6-NW	6063-T5	75.6	0.6	141.4
T5-t1.0-NW	6063-T5	76.0	1.0	235.6
T5-t2.0-NW	6063-T5	77.0	2.0	471.2
T6-t0.4-NW	6061-T6	75.4	0.4	94.2
T6-t0.6-NW	6061-T6	75.6	0.6	141.4
T6-t1.0-NW	6061-T6	76.0	1.0	235.6
T6-t2.0-NW	6061-T6	77.0	2.0	471.2
T5-t0.4-W	6063-T5	75.4	0.4	94.2
T5-t0.6-W	6063-T5	75.6	0.6	141.4
T5-t1.0-W	6063-T5	76.0	1.0	235.6
T5-t2.0-W	6063-T5	77.0	2.0	471.2
T6-t0.4-W	6061-T6	75.4	0.4	94.2
T6-t0.6-W	6061-T6	75.6	0.6	141.4
T6-t1.0-W	6061-T6	76.0	1.0	235.6
T6-t2.0-W	6061-T6	77.0	2.0	471.2

Note: NW = non-welded column series, W = welded column series.

dimensions of each series using the nomenclature defined in Fig. 1; and
• The following part of the label "NW" indicates the welding condition of the specimen, where the letter "NW" refers to the non-welded column, and the letter "W" refers to the welded column.

The material properties of the specimens of 6063-T5 alloy investigated in the parametric study are identical to the material properties of Series N-C1 in the experimental program for the non-welded and welded material, whereas the material properties of the specimens of 6061-T6 alloy are identical to the material properties of Series H-C1 in the experimental program for the non-welded and welded material, as detailed in Zhu & Young (2006c). The local imperfection magnitude was 10% of the section thickness, and the overall imperfection magnitude was 1/2000 of the column length used in the parametric study. The size of the finite element mesh was kept at 5 × 5 mm (length by width) for the non-welded and welded columns. The welded columns were modeled with 15 mm HAZ extension at both ends of the columns. The column strengths (P_{FEA}) obtained from the parametric study are detailed in Zhu (2006).

6 DESIGN APPROACHES

6.1 Current design rules for aluminum structures

The American Aluminum Design Manual (AA 2005), Australian/New Zealand Standard (AS/NZS 1997) and European Code (EC9 2000) for aluminum structures provide design rules for aluminum columns with and without transverse welds. The design rules in the AA Specification for calculating the design strengths of non-welded aluminum columns are based on the Euler column strength. The inelastic column curve, based on the tangent modulus, is well approximated by a straight line using buckling constants (Sharp 1993). The buckling constants were obtained from Tables 3.3–3 and 3.3–4 of Part I-B of the AA Specification. Local buckling strength can be calculated using an empirical formula which was first developed by Clark & Rolf (1964). The design rules in the AS/NZS Standard for calculating the design strengths of non-welded aluminum columns are generally identical to those in the AA Specification, except that the AS/NZS Standard reduces the yield load of the column using a parameter k_c which is not included in the AA Specification. The EC9 Code adopts the Perry curve for column design, and values of the imperfection factors are listed in Table 5.6 of the Code. The effects of local buckling on column strength are considered by replacing the true section with an effective section. The effective cross-section is obtained by employing a local buckling coefficient ρ_c to reduce the thickness of the element in the section.

The strength of aluminum columns with transverse welds (welded column) depends on the location and number of welds (AA 2005). For CHS columns with transverse welds at the ends only, the design equations given by the AA and AS/NZS specifications are identical to the design equations of non-welded columns. However, the design strength of CHS welded columns is calculated using the welded mechanical properties and the buckling constants obtained from Table 3.3–3 of Part I-B of the AA Specification regardless of temper before welding. The EC9 Code uses a factor ρ_{haz} to consider the weakening effects of welding on column strength, and ρ_{haz} is equal to 0.60 and 0.50 for the 6000 Series alloys of T5 and T6 conditions, respectively, as shown in Table 5.2 of the Code.

6.2 Direct strength method for aluminum alloy non-welded columns

The direct strength method has been proposed by Schafer & Peköz (1998) for laterally braced flexural members undergoing local or distortional buckling. Subsequently, the method has been developed for concentrically loaded pin-ended cold-formed steel columns undergoing local, distortional, or overall buckling (Schafer 2000, 2002), which allows for interaction of local and overall buckling as well as interaction of distortional and overall buckling. Zhu & Young (2006b) reported that the modified direct strength method can be used for the design of aluminum alloy columns of SHS and RHS. CHS are investigated in this study. As summarized in the North American Specification (AISI 2001, 2004) for cold-formed steel structures, the column design rules of the direct strength method that considered local and overall flexural buckling are shown in Equations. 1–3. The values of 0.15 and 0.4 are the coefficient and exponent of the direct strength equation, respectively, that were calibrated against test data of concentrically loaded pin-ended cold-formed steel columns for certain cross sections and geometric limits.

$$P_{DSM} = \min(P_{ne}, P_{nl}) \quad (1)$$

$$P_{ne} = \begin{cases} \left(0.658^{\lambda_c^2}\right) P_y & \text{for } \lambda_c \leq 1.5 \\ \left(\dfrac{0.877}{\lambda_c^2}\right) P_y & \text{for } \lambda_c > 1.5 \end{cases} \quad (2)$$

$$P_{nl} = \begin{cases} P_{ne} & \text{for } \lambda_l \leq 0.776 \\ \left[1 - 0.15\left(\dfrac{P_{crl}}{P_{ne}}\right)^{0.4}\right]\left(\dfrac{P_{crl}}{P_{ne}}\right)^{0.4} P_{ne} & \text{for } \lambda_l > 0.776 \end{cases} \quad (3)$$

where $P_y = f_y A$; $\lambda_c = \sqrt{P_y/P_{cre}}$; $\lambda_l = \sqrt{P_{ne}/P_{crl}}$.

A = Gross cross-section area;
f_y = Material yield strength which is the static 0.2% proof stress ($\sigma_{0.2}$) using the non-welded material properties in this paper;
$P_{cre} = \pi^2 EA/(l_e/r)^2$, critical elastic buckling load in flexural buckling for CHS columns;
P_{crl} = Critical elastic local column buckling load;
E = Young's modulus;
l_e = Column effective length; and
r = Radius of gyration of gross cross-section.

The nominal axial strengths (P_{DSM}) are calculated for the two cases, as shown in Equations. 2 and 3, respectively, where P_{ne} refers to the nominal axial strength for flexural buckling, and P_{nl} refers to the nominal axial strength for local buckling as well as interaction of local and overall buckling. The nominal axial strength, P_{DSM}, is the minimum of P_{ne} and P_{nl}, as shown in Equation. 1. In calculating the axial strengths, the critical elastic local buckling load (P_{crl}) of the cross section was obtained from a rational elastic finite strip buckling analysis (Papangelis & Hancock 1995). Figure 5 shows the local buckling (cross-section distortion) of a CHS generated from the finite strip analysis. Figure 6(a) and Figure 6(b) show comparisons of FEA and experimental results against the direct strength curves of flexural buckling and interaction of local and flexural buckling, respectively, for the non-welded columns. The unfactored design strengths (P_{DSM}) calculated using the direct strength method are compared with the numerical and test results of aluminum non-welded columns of CHS, as detailed in Zhu (2006).

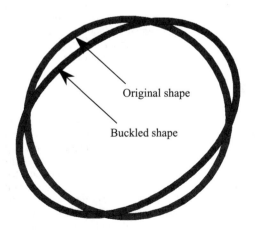

Figure 5. Local buckling (cross-section distortion) of CHS generated from finite strip analysis.

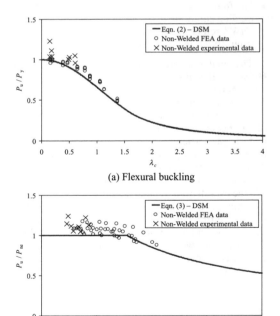

Figure 6. Comparison of FEA and experimental data with direct strength method (P_{DSM}) for non-welded columns.

6.3 Proposed design equation based on direct strength method for aluminum alloy welded columns

A design equation was proposed based on the current direct strength method for aluminum alloy CHS columns with transverse welds at both ends of the columns (welded columns). The research reported by Zhu & Young (2007) indicated that the effects of transverse welds on aluminum stub column strengths are varied with the overall diameter-to-thickness ratio of the CHS. The heat-affected zone softening factor proposed by Zhu & Young (2007) is used in this study. The unfactored design strengths (P_{DSM-W}) is obtained based on the column strengths (P_{DSM}) calculated using Equations. 1–3 and multiplied by the factors α_1 and α_2 due to welding, as shown in Equation. 4. The proposed design equation was calibrated with the welded column strengths obtained from the parametric study presented in this study, as well as the test results reported by Zhu & Young (2006c). The unfactored design strengths (P_{DSM-W}) are compared with the numerical and test results of aluminum welded columns of CHS, as detailed in Zhu (2006).

$$P_{DSM-W} = \alpha_1 \alpha_2 P_{DSM} \qquad (4)$$

where $\alpha_1 = 1.3(D/t)^{-0.19}$ and $\alpha_2 = 0.6(30/L)^{-0.12}$
P_{DSM-W} = Welded column strength;
P_{DSM} = Column strength calculated using Equations 1–3;
D = Overall diameter of CHS;
L = Column length in mm, but $L \geq 1200$; and
t = Thickness.

7 COMPARISON OF NUMERICAL AND EXPERIMENTAL RESULTS WITH DESIGN PREDICTIONS

The nominal axial strengths (unfactored design strengths) predicted by the AA Specification (P_{AA}), AS/NZS Standard ($P_{AS/NZS}$), and European Code (P_{EC9}) for aluminum structures, as well as the current direct strength method (P_{DSM}) and proposed design equation (P_{DSM-W}) are compared with the column strengths obtained from the parametric study (P_{FEA}) and experimental program (P_{Exp}) (Zhu & Young 2006c. The FEA results are also compared with the column design curves obtained from the design rules, as detailed in Zhu (2006). The design strengths are calculated using the material properties for each series of specimens, as shown in Table 2, and the 0.2% proof stress ($\sigma_{0.2}$) was used as the corresponding yield stress. The design strengths of non-welded column specimens are calculated using the non-welded material properties. In calculating the design strengths of welded columns, P_{AA} and $P_{AS/NZS}$ are calculated using the welded material properties, as specified in the AA and AS/NZS specifications, whereas P_{EC9} are calculated using the non-welded material properties as required by the EC9 Code. The design strengths (P_{DSM-W}) of the proposed design Equation. 4 are calculated using the non-welded material properties. The fixed-ended column specimens were designed as concentrically loaded compression members, and the effective length (l_e) was taken as one-half of the column length (L), as recommended by Young & Rasmussen (1998).

The reliability of the design rules for aluminum columns has also been evaluated using reliability analysis. Reliability analysis is detailed in the AA Specification, and the ratio of dead (DL) to live (LL) load of 0.2 was used in the analysis. In general, a target reliability index of 2.5 for aluminum alloy columns as a lower limit is recommended by the AA Specification. If the reliability index is greater than or equal to 2.5 ($\beta \geq 2.5$), then the design is considered to be reliable. The AA and AS/NZS specifications provide different resistance factors (ϕ) for compression members with different failure modes. The resistance factor varies with the slenderness parameter for the flexural buckling failure mode. The resistance factor is a constant and equal to 0.85 for local buckling or interaction of local and overall buckling failure mode. The observed failure modes of the columns in this study included local buckling, flexural buckling, interaction of local and overall buckling, and failure in the heat-affected zone (HAZ). Hence, the resistance factor of the columns given by the AA and AS/NZS specifications ranged from 0.76 to 0.95. In calculating the reliability indices of the AA and AS/NZS design rules, the resistance factor for the columns is chosen to be equal to 0.85 for all failure modes in this study. The EC9 Code provides a constant resistance factor of $1/1.1 = 0.91$ for compression members, which is used in the reliability analysis. The reliability of the direct strength method and proposed design rule for aluminum columns is also evaluated, and the resistance factor is equal to 0.85. The load combination of 1.2DL + 1.6 LL is used in the analysis for the AA Specification, the direct strength method and the proposed design rule. The load combinations of 1.25DL + 1.5LL and 1.35DL + 1.5 LL are used in the analysis for AS/NZS and EC9 specifications, respectively. The statistical parameters M_m, F_m, V_M, and V_F are the mean values and coefficients of variation (COV) of material and fabrication factors. These values are obtained from Section 9 of Part I-B of the AA Specification, where $M_m = 1.10$, $F_m = 1.00$, $V_M = 0.06$, and $V_F = 0.05$. The statistical parameters P_m and V_p are the mean value and the coefficient of variation of test-to-predicted load ratios, respectively.

The comparisons are detailed in Zhu (2006). For the non-welded columns, it is shown that the column strengths predicted by the AA and AS/NZS specifications are quite close with the numerical and experimental results, except for the long columns that are slightly unconservative. The design strengths calculated using the EC9 Code are generally more conservative compared with the predictions given by the AA and AS/NZS specifications. The design strengths (P_{DSM}) predicted by the current direct strength method are generally conservative and reliable for all the non-welded column series, as shown in Zhu (2006). It is shown that the current direct strength method could be successfully used in the design of aluminum non-welded columns of circular hollow sections. The non-welded column design curves predicted by the AA, AS/NZS and EC9 specifications, as well as the direct strength method are shown in Figure 7 for non-welded column Series T6-t0.6-NW.

For the welded columns, it is shown that the design strengths calculated using the AA, AS/NZS and EC9 specifications are quite conservative. It is also shown that the current direct strength method is not suitable for the design of aluminum CHS welded columns.

The design equation based on the direct strength method for aluminum alloy CHS columns with transverse welds at both ends of the columns (welded columns) is shown in Equation. 4 of the paper. The unfactored design strengths (P_{DSM-W}) calculated using the proposed design rules are generally conservative for the welded columns. The welded column design curves predicted by the AA, AS/NZS and EC9 specifications, as well as the current direct strength method and the proposed design rules are shown in Figure 8 for welded column Series T6-t0.4-W. It is shown that the proposed design equation based on the direct strength method can be used for the design of aluminum columns of circular hollow sections with transverse welds at the ends of the columns.

8 SUMMARY AND CONCLUSIONS

Numerical investigation and design of aluminum alloy circular hollow section columns have been presented in this paper. A non-linear finite element model incorporating geometric imperfections and material non-linearities was developed and verified against experimental results. The column specimens were fabricated by extrusion using heat-treated aluminum alloy of 6061-T6 and 6063-T5. The ultimate loads and failure modes predicted by the finite element model are in good agreement with the experimental results. A parametric study was performed using the verified finite element model that included 80 specimens with column lengths ranging from 500 to 3500 mm. The column strengths obtained from the experimental and numerical investigations were compared with the design strengths calculated using the current American, Australian/New Zealand and European specifications for aluminum structures, as well as the direct strength method that was developed for cold-formed steel members. It is shown that the design strengths predicted by the current direct strength method are generally conservative for aluminum non-welded columns of circular hollow sections. A design equation was proposed based on the current direct strength method for aluminum alloy circular hollow section columns with transverse welds at the ends of the columns. It is also shown that the design strengths calculated using the proposed design rules are generally conservative for the aluminum welded columns. The reliability of the current and proposed design rules was evaluated using reliability analysis and it is shown that the proposed design rules are reliable. It is thus recommended to use the proposed design rules in the design of aluminum columns of circular hollow section containing transverse welds at the ends of the columns.

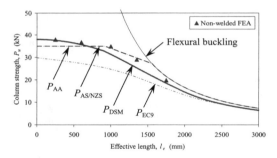

Figure 7. Comparison of FEA and design column strengths for Series T6-t0.6-NW.

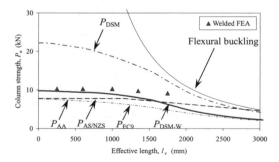

Figure 8. Comparison of FEA and design column strengths for Series T6-t0.4-W.

REFERENCES

AA. 2005. *Aluminum Design Manual*. The Aluminum Association, Washington, D.C.

ABAQUS. 2004. *ABAQUS analysis user's manual, Version 6.5*. ABAQUS, Inc.

AISI. 2001. *North American Specification for the design of cold-formed steel structural members*. American Iron and Steel Institute, Washington, D.C.

AISI. 2004. *Supplement to the North American Specification for the design of cold-formed steel structural member*. American Iron and Steel Institute, Washington, D.C.

AS/NZS. 1997. *Aluminum structures Part 1: Limit state design, Australian/New Zealand Standard AS/NZS 1664.1:1997*. Standards Australia, Sydney, Australia.

Clark, J.W. & Rolf, R.L. 1964. Design of aluminum tubular columns. *Journal of structural division*, Vol. 90:259.

EC9. 2000. *Eurocode 9: Design of aluminum structures—Part 1–1: General rules – General rules and rules for buildings*, DD ENV 1999-1-1:2000, Final draft Oct 2000. European Committee for Standardization.

Mazzolani, F.M. 1995. *Aluminum alloy structures*, 2nd Edition. E & FN Spon, London.

Papangelis, J.P. & Hancock, G.J. 1995. Computer analysis of thin-walled structural members. *Computer & Structures*, Vol. 56, No. 1: 157–176.

Schafer, B.W. & Peköz, T. 1998. Direct strength prediction of cold-formed steel members using numerical elastic buckling solutions. *Proc. 14th Intern. Spec. Conf. on cold-formed steel structures*, University of Missouri-Rolla, Rolla, Mo: 69–76.

Schafer, B.W. 2000. *Distortional buckling of cold-formed steel columns*. August Final Report to the American Iron and Steel Institute, Washington, D.C.

Schafer, B.W. 2002. Local, distortional, and Euler buckling of thin-walled columns. *Journal of Structural Engineering*, Vol. 128, No. 3: 289–299.

Sharp, M.L. 1993. *Behaviour and design of aluminum structures*. McGraw-Hill, New York.

Young, B. & Rasmussen, K.J.R. 1998. Tests of fixed-ended plain channel columns. *Journal of Structural Engineering*, Vol. 124, No. 2: 131–139.

Zhu, J.H. 2006. *Behavior and design of aluminum alloy structural members*. PhD thesis, Department of Civil Engineering, The University of Hong Kong, Hong Kong.

Zhu, J.H. & Young, B. 2006a. Aluminum alloy tubular columns – part I: finite element modeling and test verification. *Thin-Walled Structures*, Vol. 44, No. 9: 961–968.

Zhu, J.H. & Young, B. 2006b. Aluminum alloy tubular columns – part II: parametric study and design using direct strength method. *Thin-Walled Structures*, Vol. 44, No. 9: 969–985.

Zhu, J.H. & Young, B. 2006c. Experimental investigation of aluminum alloy circular hollow section columns. *Engineering Structures*, Vol. 28, No. 2: 207–215.

Zhu, J.H. & Young, B. 2007. Effects of transverse welds on aluminum alloy columns. *Thin-Walled Structures*, Vol. 45, No. 3: 321–329.

Experimental research on beams with tubular chords and corrugated steel webs

J. Gao & B.C. Chen
College of Civil Engineering, Fuzhou University, Fuzhou, China

ABSTRACT: Experiments on beams with tubular chords and corrugated steel webs were carried out. The differences among the three specimens lie in the chords of hollow steel tube or Concrete Filled Steel Tube (CFST). In specimen A, all chords are hollow steel tubes; in specimen C, all chords are CFST; while in specimen B, the upper chords are CFST but the lower chords are hollow steel tubes. Stresses, displacements, failure modes and ultimate load carrying capacity of the beams are analyzed and compared with each other. Experimental results show that the flexural rigidity and ultimate carrying capacity is improved considerably if the chords are CFST compare to hollow steel tubes. Joint failure in tubular trusses can be avoided in such a beam with tubular chords and corrugated webs. Based on the Quasi-Plane-Section assumption, the formula to calculate the ultimate load of this new composite structure is presented and the predicted results agree well with the test results.

1 INTRODUCTION

Tubular structures are widely used in civil structures. Experiments on tubular truss with steel tubular web members show that they collapsed due to joint failure whether concrete is filling into the tubular chords (Huang & Chen 2006). In order to prevent the joint failure and improve the behavior, we can use corrugated steel web into the tubular truss instead of steel tubular web members. Besides good behaviors expected, it is obvious that no tubular joints in such a beam will make its manufacture easier and cost reduced.

This paper describes the experiment carried out on three beam specimens with tubular chords and corrugated steel webs. The aim of this study is to determine the mechanical behaviour of this new structure and the effects of corrugated web.

2 SPECIMENS AND SET-UP

The experimental program involved three specimens with cross sections of four tubular chords (two in upper and two in lower) connected vertically by corrugated webs and transverse by tubular bracings throughout the whole length of 3008 mm (Fig. 1). The tubular chord is 89 mm in diameter with a 1.8 mm thick well. The web, with 2 mm thick, was welded continuously to the tubular chords from its two sides. Steel tubes and plate stiffeners were adopted in the loading position and support sections to prevent web from local buckling, as shown in Figure 1(a). The corrugation amplitude was 30 mm and the wavelength was 140 mm, as shown in Figure 1(c). The average yield stress of the steel material was found to be 428 MPa, and the average ultimate strength is 533 MPa. The tested Young's elastic modulus of steel is 2.09×10^5 MPa. With regard to concrete, the average ultimate strength was found to be 55Mpa with the Yong's elastic modulus of 0.347×10^5 MPa.

The three specimens had same in profiles and steel materials. Their difference in the chord was hollow steel tube or concrete filled steel tube (CFST). In Specimen A, all chords were hollow steel tubes; In Specimen C, all chords were CFST; however, in Specimen B, the upper chords are CFST while the lower chords were hollow steel tubes. The differences between the specimen A, B and C in this paper and the specimen B0, B1 and B2 in Huang & Chen (2006) were the web members. They were corrugated steel plates in the formers (beams) while steel tubular bars in the latter (trusses).

The test specimens were simply supported at both ends with a distance of 2880 mm. and subjected to four-point loading. The supports allowed the girder to rotate and the lateral supports prevent the girders from displacing and rotating laterally. Details of the test setup are shown in Figure 2. Rosettes and transducers were installed to measure strains of steel tubes and displacement of the specimens as shown in Figure 3.

The specimens were loaded in a 5,000 kN capacity testing machine. The load was applied to the specimens as two equal line loads across the top chords.

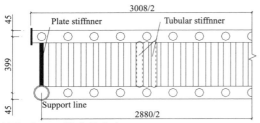

(a) Specimen profile (half-view)

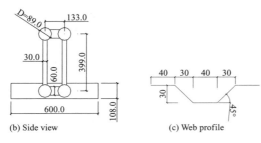

(b) Side view (c) Web profile

Figure 1. Dimensions of test specimen.

Figure 2. Test set-up.

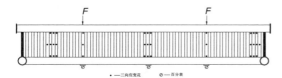

Figure 3. Distribution of strain gauges.

3 TEST RESULTS

3.1 Test procedure

The measured load-lateral deflection curves at the mid-span positions are shown in Figure 4. From these curves, three distinct phases were observed as elastic (linear), elastic-plastic, and plastic phase. Filling concrete into steel tubes will increase the beam's rigidity and strength, specimen C has the largest initial slope, longest linear phase in its load vs. deflection curve and largest ultimate loads, followed by specimen B, then specimen A.

The plastic phases and the failure modes of the three specimens are quite different due to the section difference. The plastic phase in specimen A is very short because it failed by local buckling of the upper hollow steel tube subjected to compression. Both specimen B and C have good ductility for their upper chords are CFST members.

As shown in Figure 5, the upper tubes in specimen A collapsed suddenly at about 3 L/8 distances away from the left end when the load reached 190 kN. The corrugated webs were still in elastic range at this time. The maximum longitudinal stain in the upper tube is -1985×10^{-6}, which is smaller than yield strain 2048×10^{-6}, and the maximum longitudinal strain in the lower tube is 2786×10^{-6}, which has been in plastic range. However, the maximum shear strain in corrugated web is 1804×10^{-6}. Therefore, the failure of specimen A is caused by the local folding the upper tubes, which is similar to the failure mode in Mohamed et al. (1997), which failed due to the buckling of the upper plate.

For specimen B with upper chords of CFST while the lower chords of hollow steel tubes, the loading was

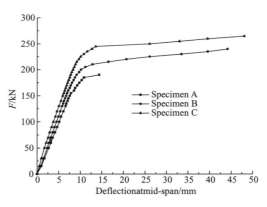

Figure 4. Load-deflection curve.

Figure 5. Failure mode of Beam A.

Figure 6. Failure mode of Beam C.

stopped at 240 kN because the load can't be increased at that time but the deflection increased rapidly. Its maximum longitudinal strain in the upper and lower flange are -6184×10^{-6} and 19632×10^{-6} respectively under failure load, which are far beyond the yield strain. The corresponding shear strain in the corrugated steel web is 2568×10^{-6}. So there is no obvious strength failure mode for specimen B.

As far as the specimen C is concerned, the upper and lower chords were both CFST members. When the applied load was 230 kN, the longitudinal strain in the lower chords exceeded the yield strain. After that, the shear strain in the corrugated steel web at L/8 section increased suddenly, and the web near the end plate budged slowly. So the failure of specimen C was due to the yield of the lower chords, which resulted in the local buckling of the corrugated steel webs. When the specimen failed, the corresponding maximum longitudinal strain in the upper and lower chords are -4045×10^{-6} and 4579×10^{-6} respectively, while the maximum shear strain in the corrugated steel web was 31138×10^{-6}. The local buckling of the web is shown in Figure 6.

Comparing the tested ultimate loads, 240 kN in specimen B is 1.263 times of 190 kN in specimen A of, and 265 kN in specimen C is 1.104 times of that in specimen B. This indicates that the load carrying capacity can be enhanced by filling concrete in both upper and lower hollow steel tubular chords. Furthermore, filling concrete into upper chord will avoid its local buckling and improved the beam ultimate load greatly, while filling concrete into lower chord will not change the failure mode but only enhance the steel tube strength about 1.1 times.

The tested ultimate loads in specimens B0, B1 and B2 by Huang & Chen (2006), were 92.5kN, 107.5 kN, and 147.5 kN respectively. It is obvious that the ultimate loads of the specimens in this paper are much larger.

3.2 Deflection curve analysis

Figure 7 shows the deflection curves of the three beams. It should be noted that they are nearly symmetric under each load increment. The shape of each deflection curve agrees very well with half-sinusoid curve, it means that this structure is close to solid beam. Comparing the curves in Figure 5, it is evident that all the curves have almost the same shapes, i.e. filling concrete into the tube chords or not will not affect the deflection shapes. However, with the chords changing from tubes to CFST members, the flexural rigidity improves greatly, and the deflection under the same load decrease.

3.3 Longitudinal strain in chords

Figure 8 shows the longitudinal strain in both upper and lower chords varying with load. Figure 6(a) indicates that the upper chord of specimen A is always in elastic range during the test, the maximum strain is only -1985×10^{-6}, smaller than the yield stain 2048×10^{-6}. After local buckling, its strain decrease as shown in the short line turned in opposite direction.

For specimen B and C, the upper chords are both CFST members, in which the steel tube and concrete carry the compression simultaneously, so the strain of the steel tube is much smaller than that in specimen A. Moreover, there are obvious nonlinear stage for specimen B and C.

Comparing with the upper chords, the load-strain curve of the lower chord doesn't have much difference among the three specimens, shown in Figure 6(b).

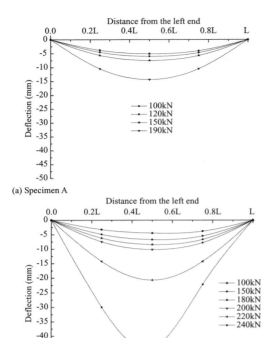

(a) Specimen A

(b) Specimen B

(c) Specimen C

Figure 7. Deflection curve of beam.

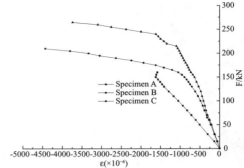

(a) Load-strain curve of upper chord at mid-span

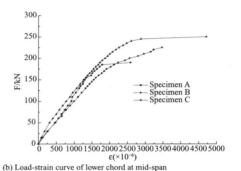

(b) Load-strain curve of lower chord at mid-span

Figure 8. Load-strain curve of chord.

Due to the small tension strength of concrete, filling concrete has small influence on the behavior of lower chords which are under tension. For specimen A, when the upper chords buckled, the tension in the lower chords increased suddenly. As can be noted from the figure, the load-strain curve of specimen C is higher than that of specimen B and developed horizontally at the last stage, which imply that the core concrete is helpful for the mechanical behavior of CFST members at late stage. When a CFST subjected to tension, the core concrete will restrain the radial contraction of the steel tube and improve its tensile strength about 1.1 times of that subjected to uniaxial tension (Zhang 2003).

3.4 Strain distribution in corrugated steel webs

3.4.1 Normal strain

The strain distribution will be shown at two vertical sections for each specimen, i.e. mid-span section and L/8 section. The normal strain in the direction of the beam's horizontal axis is plotted along the depth of the beam for the two sections, as shown in Figures 9 and 10. As can be noted from the figures, the strains are equal to zero except for much localized areas of the web adjacent to the chords. The strain in the flange increased linearly to the yield strain of the flange material at failure. As discussed earlier, the contribution of the web to the moment carrying capacity of the specimens is negligible. This is because the corrugated web has no stiffness perpendicular to the

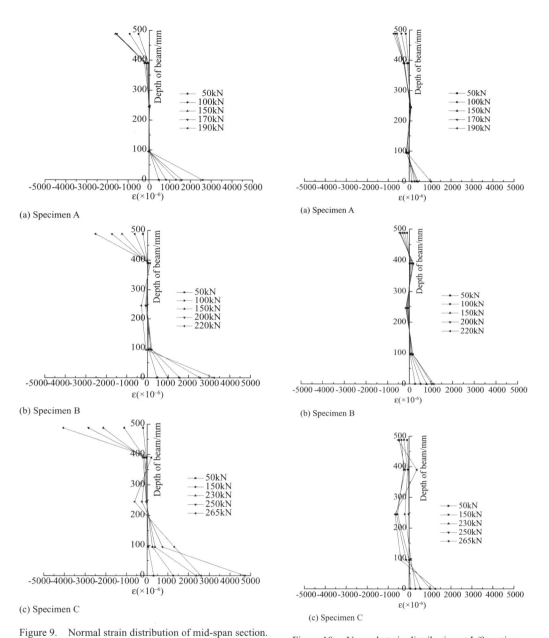

Figure 9. Normal strain distribution of mid-span section.

Figure 10. Normal strain distribution at L/8 section.

direction of the corrugation, except for a very small distance that is adjacent to and restrained by the flanges.

Comparing with Figure 9, the difference between normal strain of chords and webs are smaller in Figure 10, especially for specimen C, the normal strain in web is even higher than that in chords at late stage.

3.4.2 *Shear strain*

The shear strain of the corrugated steel web is tested by strain rosette. According to the strain gauge readings from the specimen, the shear strain, v, can be obtained as follows:

$$v = 2\varepsilon_{45°} - (\varepsilon_{0°} + \varepsilon_{90°}) \quad (1)$$

where: $\varepsilon_{45°}$, $\varepsilon_{0°}$ and $\varepsilon_{90°}$ denote the strain in the angle which the subscript describe.

The calculated shear strain by Equation (1) is located at the L/8 section. The load-shear strain curves at L/8 section of the three specimens are plotted in Figure 9. It is shown that the shear strain in specimen A and B is always in elastic phase, while for specimen C, when the applied load reach 230 kN, the shear strain developed rapidly. All the three specimens have approximately equal slope at the early stage in Figure 11, which imply that filling concrete into the steel tube has small influence on the shear stiffness of the section. In specimen A and B, the failure is due to buckling of the compression hollow tubes or yielding of tensile tubes, so the shear strain in the webs is in elastic, but for specimen C, after filling concrete into both upper and lower tubes, the load carrying capacity is improved and the failure is due to lower chords yielding followed by web buckling near L/8 section. So at the ultimate load, the shear strain in the web reaches very large value and the material becomes perfectly plastic.

Figure 12 shows the shear strain distribution along the web depth. It is shown that the distribution vary with the load increasing, that is to say, when the applied load is in elastic range, the shear strain increase at a higher rate, so the section can't remain uniform. Comparing with the specimen A and B, the shear strain in specimen C is much higher under the same load. This is also true in Figure 11 that the normal stain in web is very large at the late stage.

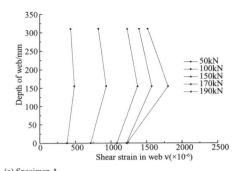

(a) Specimen A

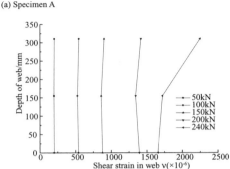

(b) Specimen B

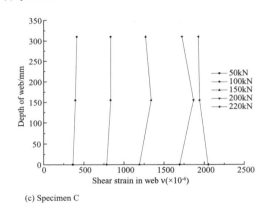

(c) Specimen C

Figure 12. Shear strain distribution at L/8 section.

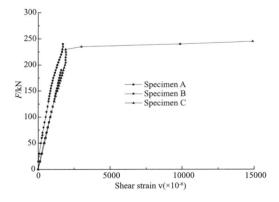

Figure 11. Load-shear strain curves of L/8 section.

4 ULTIMATE LOAD-CARRYING CAPACITY

Research on girders with corrugated steel web shows that the moment capacity of a panel with corrugated web is almost equal to the moment capacity based on flange yielding with no contribution from the web (Huang et al. 2006, Li 2004, Song 2002, Wan 2005). In other words, the contribution of the corrugated steel web to the beam bending strength can be ignored.

It was indicated that although the longitudinal strain of the whole cross section can't satisfy the Plane-section assumption, however, the upper and lower chords still obey the Quasi-plane section, that is to say the cooperation between upper and lower chords can be ensured due to the high shear stiffness of corrugated steel web (Wu 2005). Therefore, the ultimate load carrying capacity of beam with corrugated steel webs can be calculated as the way of truss. With regard to the beam proposed in this paper, the ultimate load carrying capacity and deflection can be calculated according to the section as Figure 13.

According to the force equilibrium,

$$N_1 - N_2 = 0 \\ N_1 \times \frac{h}{2} + N_2 \times \frac{h}{2} + M_1 + M_2 = M \quad (2)$$

where all the symbols are denoted in Figure 13.

Based on the test results, the upper and lower chords can keep in plane at every stage. So the moment carrying by the upper and lower chords can be calculated based on their own bending rigidity as follows:

$$M_1 = \frac{(EI)_1}{(EI)}M = \alpha_1 M \quad M_2 = \frac{(EI)_2}{(EI)}M = \alpha_2 M \quad (3)$$

Where: $\alpha_1 = (EI)_1/(EI)$, $\alpha_2 = (EI)_2/(EI)$, $(EI)_1 = E_s I_{s1} + E_c I_{c1}$, $(EI)_2 = E_s I_{s2} + E_c I_{c2}$

E_s, E_c is Young's elastic modulus of steel and concrete respectively; A_{s1}, A_{s2} is the area of single steel tube in upper and lower chord respectively; I_{s1}, I_{s2} is bending moment of inertia of single steel tube in upper and lower chord respectively; I_{c1}, I_{c2} is bending moment of inertia of single steel tube in upper and lower chord respectively; h is distance between the center of upper and lower chords.

From Equation (2) and (3), the internal force, N_i, of upper and lower chords can be obtained as follows:

$$N_i = \frac{(1 - \alpha_1 - \alpha_2)M}{h} \quad (4)$$

Thus, both N_i and M_i can be got by Equation (3) and (4). Then the ultimate load carrying capacity of the beam can be obtained by $P_s = M/0.72$. Considering the different section type of the three specimens, the ultimate capacity of the chords can be calculated as follows:

1. For specimen A, both upper and lower chord are hollow steel tube, so $\alpha_1 = \alpha_2$. Due to the lower chord is in tension, $N_2 = 2f_y A_s$ (two tubes in lower chord), P_{s2} can be obtained. The upper chord is in compression but restrained by the corrugated steel web, so the failure in upper chord is either local buckling or material yield. The critical strength for local buckling, σ_{cr}, can be calculated by Equation (5). Taking the minimum value of σ_{cr} and yield stress of steel σ_s, the ultimate strength N_s and M_s can be obtained, and then substituting them into Equation (6) to get P_{s1}. In this case, $P_{s1} < P_{s2}$.

$$\sigma_{cr} = \frac{\pi^2 E(t^2/12)}{(1-\mu^2)L^2} \quad (5)$$

$$\frac{N}{N_s} + \frac{M}{M_s} \leq 1 \quad (6)$$

2. For specimen B and C, the upper chord are CFST members, so the load carrying capacity is calculated as $N_1 = 2\varphi f_{sc} A_{sc}$, where f_{sc} and A_{sc} are axial strength and area of CFST members, introducing in detail in DL/T 5085-1999. With regard to the lower chord, the load carrying capacity of specimen B, with hollow steel tube, is $N_2 = 2f_y A_s$, while due to the lower chord of specimen C is CFST members, so $N_2 = 2*1.1f_y A_s$. It is noted that $N_1 > N_2$.

The calculated ultimate load carrying capacity of the three specimens is given in Table 1, accompanied with the test results.

Table 1. Ultimate load carrying capacity (kN).

Specimen	Experimental results	Calculated results	Error (%)
A	190.0	181.5	4.7
B	240.0	231.3	3.7
C	265.0	262.2	1.2

5 SUMMARY AND CONCLUSIONS

1. Experiments indicate that beams comprised of tubular chords and corrugated steel webs have much greater ultimate loads and rigidities than

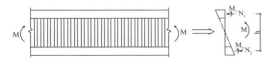

Figure 13. Sketch for load-carrying capacity.

circular truss with tubular chords and tubular web members, because the joint failure is avoided in the new beam by substituting the tubular web members for corrugated steel webs.
2. Load versus deflection curves of specimens in this paper are half-sinusoid curves, similar with that of prestressed concrete beams with corrugated steel webs.
3. Failure of specimen A, without filling concrete into the steel tube, is due to the local buckling of upper chord. For specimen B, the lower chord reaches yield stress at the ultimate load. Failure of specimen C is the crippling of steel web resulted from the yield of the lower tube chords. The ultimate load-carrying capacity of the three specimens is listed descending as specimen C, specimen B and specimen A. It is indicated that filling concrete into the upper chords improves the ultimate load greatly, the effect is smaller when filling concrete into the lower chord.
4. Equations based on Quasi-plane Section Assumption are developed to calculate the ultimate capacity of beam with tubular chords and corrugated steel webs subjected to patch loading. These equations give very good results comparing with the experimental results.
5. Circular section truss with corrugated steel web has high bending resistance and ductility. Even when the deflection at mid-span reaches L/50, the beam can still carry load.

REFERENCES

Chen B.C. 2007. *Concrete filled steel tubular arch bridge.* Beijing: China Communications Press.

DL/T 5085-1999, *Design specification for steel-concrete composite structure.*

Huang L., Hikosaka H. &Komine K. 2002. Modelling of accordion effect in corrugated steel web with concrete flanges. *Proceedings of the Sixth International Conference on Computational Structures Technology.* Civil-Comp Press, Paper 113.

Huang W.J. & Chen B.C. 2006. Experimental Research on Concrete-Filled Steel Tube Truss Girder Under Bending. *Journal of Building Structures*, 23(1): 29–33.

Li H.J., Ye J.S., Wan S., Qian P.S. & Jiang Z.G. 2004. Experimental research on prestressed concrete box girderwith corrugated steel webs. *China Journal of Highway and Transport*, 17(4): 31–36.

Mohamed Elgaaly, Anand Seshadri & Robert W. Hamilton. 1997. Bending strength of steel beams with corrugated Webs[J]. *Journal of Structural Engineering*, 123(6): 772–782.

Nakamura Shun-ichi. 2000. New structural forms for steel/concrete composite bridges. *Structural Engineering International*, 10(1): 45–50.

Song J.Y., Zhang S.R, Wang T. & Lv J.M. 2002. A theorectical analysis and experimental study on the flexural behavior of externally prestressed composite beam with corrugated steel webs. *China Civil Engineering Journal.* 37(11): 50–55.

Wu W.Q., Ye J.S., Wan S. & Hu C. 2005. Quasi plane assumption and its application in steel-concrete composite box girders with corrugated steel webs. *Engineering Mechanics.* 22(5): 177–180.

Wan S., Chen J.B., Yuan A.H. & Yu W.B. 2005. Experimental Study and Predigestion Calculate of Prestressed Concrete Box-Girder With Corrugated Steel Webs. *Journal of East China Jiaotong University.* 22(1): 11–14.

Xu Y., Zhu W.Y. & Yang Y. 2005. Calculation of ultimate moment capacity of prestressed concrete box-girder bridge with corrugated steel webs. *Journal of Chang'an University(Natural science edition)*, 25(2): 60–64.

Zhong Shantong. 2003. The concrete-filled steel tubular structures (Third Edition). Beijing: Press of Tsinghua University.

Simulation of CFRP strengthened butt-welded Very High Strength (VHS) circular steel tubes in tension

H. Jiao
Maintenance Technology Institute, Department of Mechanical Engineering, Monash University, Australia

X.L. Zhao
Department of Civil Engineering, Monash University, Australia

ABSTRACT: The debonding failure of CFRP strengthened butt-welded VHS tubes under static tension loading was simulated using finite element method. Material properties of the parent metal, the heat-affected-zone (HAZ), the carbon fibre reinforced polymer (CFRP) and epoxy were taken into account. The traction separation law of the cohesive zone model was used to simulate the debonding between CFRP and the VHS tube during the non-linear analysis process. The influence of CFRP bond length on the load carrying capacity was studied. The ultimate loads from the FE model were compared to those obtained from the experiments.

1 INTRODUCTION

Extensive studies on FRP repairing and strengthening of steel structures have been conducted in recent years by Zhao & Zhang (2007). A CFRP-epoxy strengthening system was used by Jiao & Zhao (2004) to strengthen butt-welded Very High Strength circular steel tubes. From the experimental results, it was found that the connection strength was increased significantly. The load carrying capacity was doubled for CFRP strengthened butt-welded VHS tube, compared to that without CFRP strengthening. The major failure mode was a mixture of debonding of CFRP along the tube interface and CFRP fibre fracture. This failure mode indicated that the strength of epoxy resin and CFRP tensile strength played an import role. It was also found from the experimental tests that the load carrying capacity increased with the bond length of the CFRP until CFRP length reached an effective bond length. For Araldite 420 used in the tests, an effective bond length of 75 mm was obtained.

Theoretical studies were conducted by some researchers to understand the strengthening effect of CFRP on steel structures. Different methodologies were adopted in these studies, such as the fracture mechanics approach, and finite element method by Al-Emrani & Kliger (2006), Buyukozturk et al. (2004), Colombi (2006), Colombi & Poggi (2006), Hollaway et al. (2006), Lenwari et al. (2006) and Liu et al. (2007).

In this study, a FE model was established using ABAQUS®. The cohesive zone element type with the traction separation law was used to simulate the debonding between the CFRP and VHS steel tube under static loading. Different CFRP bond lengths were set as input in the model to understand the debonding process and to verify the effective CFRP bond length.

2 ESTABLISHMENT OF FE MODEL

2.1 Model geometry and element types

In order to compare the simulation results with the experiments, seven specimens in the study of Jiao & Zhao (2004) with bond lengths of 15, 30, 50, 75, 100, 125 and 150 mm were selected in the model. The diameter and wall thickness of the VHS tube in the model were the same as the specimens in Jiao & Zhao (2004). Detailed tube diameter and wall thickness are listed in Table 2. With the geometric feature of axial symmetry, only a quarter of the tube was included in the model as shown in Figure 1.

To simulate the butt-weld, based on the measurements by Jiao (2003), an 8 mm length of heat affected zone (HAZ) was added in the middle of the tube model. The reduced integration 3D element type, C3D8R, was used for VHS tube and HAZ. The element size was 2 mm for VHS tube and 1 mm for the HAZ. The butt-weld in Figure 1 illustrates the location of the

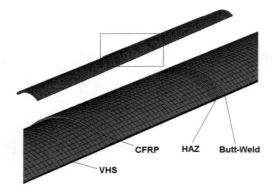

Figure 1. Model geometry of CFRP strengthened butt-welded VHS tune (a quarter).

Table 1. Material properties.

Material	Young's modulus (GPa)	Tensile strength (MPa)
VHS	198.4	
HAZ	197	737
CFRP	230	
Araldite 420	1.901	28.6

weld. For simplicity, different properties of the weld metal were not included in the model.

A zero thickness cohesive element, COH3D8, was used to simulate the interface between CFRP and VHS tube.

Four layers of CFRP were applied in the model, with a thickness of 0.13 mm for each layer. The layup direction of CFRP layers was in the longitudinal direction along the tube. The CFRP bond length in this paper refers to the distance from one end of the CFRP to the centre of the HAZ. The length of CFRP on the other side of the HAZ was made longer in order that failure happened in the designated area. The element type for CFRP was C3D8R (8 nodes 3D element).

2.2 *Material properties*

The non-linear true stress-strain data for VHS and HAZ obtained from the experiments by Jiao & Zhao (2004) and Jiao & Zhao (2001) were put in the model. The Young's Modulus and the ultimate strength are listed in Table 1. The tensile strength of the HAZ is 737 MPa.

From the material data sheet and tests conducted by Boyd, Winkle et al. (2004), Fawzia (2007) and Fawzia et al. (2006), the Young's modulus of Araldite 420 adopted in the model was 1.901 GPa with a tensile strength of 28.6 MPa. The maximum nominal stress criterion was used in the model to initialize the debonding between CFRP and steel interface.

From the material data sheet, the Young's modules of CFRP was 230 GPa.

2.3 *Boundary and load conditions*

All nodes at one end of the tube were fixed in the longitudinal direction. As one quarter of the VHS tube was included in the model, symmetry boundary conditions were applied on the two edges in the tangential direction. Based on the displacement measurements at failure of tests by Jiao (2003), a 3.5 mm displacement was applied on the other end of the tube as a loading condition.

3 SIMULATION RESULTS

3.1 *Failure mode*

For the butt-weld VHS tube without CFRP strengthening, necking happened in the HAZ as shown in Figure 2. For CFRP strengthened butt-welded VHS tubes, debonding initially happened at one end of the CFRP as shown in Figure 3(a). As the load in creased, debonding developed towards the HAZ, until the ultimate load was reached, followed by necking in the HAZ as shown in Figure 3(b). This failure mode is different from the mixed failure mode, i.e. debonding of CFRP plus fibre break. The CFRP was modeled as an elastic material. Therefore, no failure of the CFRP could be shown in the model.

3.2 *Load carrying capacity*

In order to compare the FE results with the experimental results of full section tubes, the ultimate load obtained from the model was multiplied by four, since only a quarter of a full section tube was included in

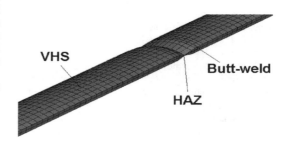

Figure 2. Necking of butt-welded VHS tube without CFRP strengthening.

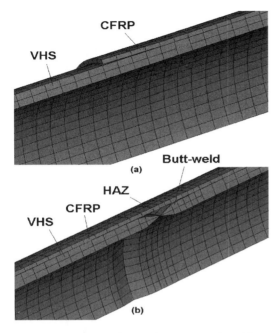

Figure 3. Debonding failure of CFRP-strengthened butt-welded VHS tube.

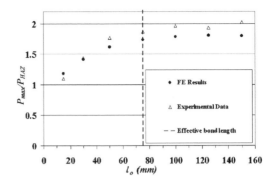

Figure 4. Comparison of P_{max}/P_{HAZ} versus CFRP bond length l_o.

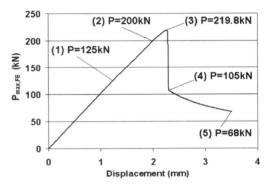

Figure 5. Typical load-displacement curve debonding process during loading (Specimen with bond length of 50 mm).

the model. The tube diameters, wall thickness, the ultimate load from the FE model $P_{max,FE}$, the ultimate load from experimental tests $P_{max,test}$, the ultimate load of HAZ P_{HAZ} and their ratios are listed in Table 2. P_{HAZ} was obtained from $f_{HAZ}A_s$, where f_{HAZ} is the HAZ tensile strength (737 MPa) and A_s is the tube cross-sectional area. In order to verify the effective bond length of 75 mm in Jiao & Zhao (2004), specimens with CFRP bond length from 15 mm to 150 mm were selected. It can be seen from Table 2 that the longer the bond length, the higher the load carrying capacity.

The ratios of $P_{max,FE}/P_{HAZ}$ and $P_{max,test}/P_{HAZ}$ versus the CFRP bond length (l_o) were shown in Figure 4. It can be seen from Figure 4 that the results from the FE model have satisfactory agreement with the experimental results in the effective bond length. From the FE results, it can be verified that the effective bond length of 75 mm was appropriate for this CFRP strengthening system. It is noticed that the ultimate loads are slightly (within 5%) lower that the experimental results. This could be due to the mixed fibre break was not included in the FE model.

A typical load-displacement curve (from model with bond length of 50 mm) is shown in Figure 5. It can be seen from Figure 5 that there is a sudden load drop when CFRP debonding and necking happened in the HAZ. The numbers (1) to (5) in Figure 5 correspond to the stress state described in the next section.

3.3 Debonding process during loading

The debonding progress and stress distribution in a CFRP-bonded butt-welded VHS tube, with a CFRP

Table 2. Comparison of simulation and experimental results.

D	t	l_o	$P_{max,FE}$	$P_{max,test}$	P_{HAZ}	$P_{max,FE}/P_{HAZ}$	$P_{max,test}/P_{HAZ}$
38	1.6	15	158.3	147.8	134.4	1.178	1.100
38	1.6	30	188.6	191.8	133.7	1.441	1.435
38	1.6	50	219.8	240.4	136.3	1.613	1.764
38	1.6	75	237.1	255.3	136.6	1.736	1.869
38	1.6	100	239.6	263.5	134.3	1.784	1.962
38	1.6	125	241.2	257.9	133.6	1.805	1.930
38	1.6	150	241.7	272.7	134.6	1.795	2.026

bond length of 50 mm, are shown from Figure 6 to Figure 10. It can be seen that the stress in VHS tubes is higher in areas outside the CFRP bonding than that in the bonding area. Debonding happened initially at the end of the CFRP away from the HAZ. As loading increases, debonding develops from the end towards the HAZ until it reaches the HAZ.

Figure 11 and Figure 12 show the Von Mises stress distribution of the cohesive elements at (1) $P = 125$ kN and point (2) $P = 200$ kN in Figure 5. It can be seen that before debonding happens at the end of CFRP, the stress concentration is at the end of the CFRP. This was also observed by Fawzia et al. (2006). As the load increases, the maximum stress moves in the HAZ direction.

Since element removal was used in the model, when debonding happened, cohesive elements within the debonding area would be removed. Therefore, the debonding process can be illustrated by showing the cohesive elements. Figure 13 to Figure 15 shows the cohesive elements together with the VHS elements (CFRP elements are not displayed) for a butt-welded VHS tube with a CFRP bond length of 50 mm. It can be seen from Figure 13 that debonding happened at the end of the CFRP bonding area away from the HAZ when the load approached 217 kN. As the load increased, debonding developed towards the HAZ as shown in Figure 14. However, no debonding happened in the HAZ area until necking developed in the HAZ as shown in Figure 15.

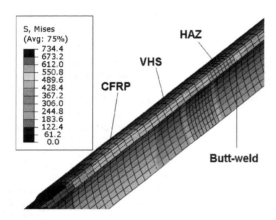

Figure 6. Stress distribution of butt-welded VHS tube with load = 125 kN (Point (1) in Figure 5).

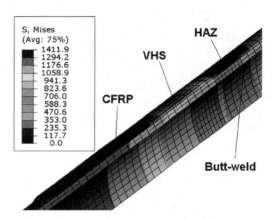

Figure 8. Stress distribution of butt-welded VHS tube with load = 219.8 kN (Point (3) in Figure 5).

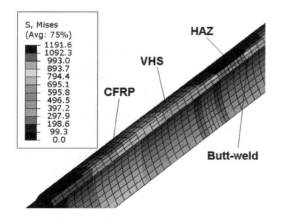

Figure 7. Stress distribution of butt-welded VHS tube with load = 200 kN (Point (2) in Figure 5).

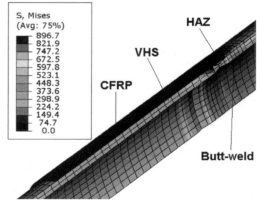

Figure 9. Stress distribution of butt-welded VHS tube with load = 105 kN (Point (4) in Figure 5).

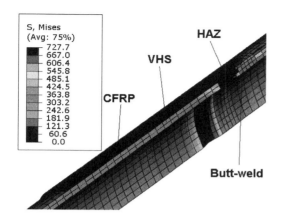

Figure 10. Stress distribution of butt-welded VHS tube with load = 68 kN (Point (5) in Figure 5).

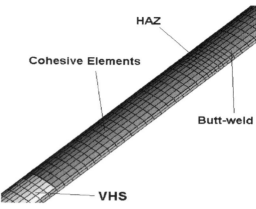

Figure 13. Debonding of Cohesive elements (with load = 217 kN).

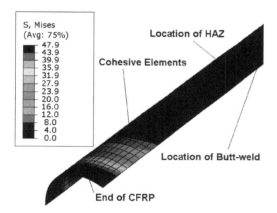

Figure 11. Von Mises stress distribution of cohesive elements with load = 125 kN (Point (1) in Figure 5).

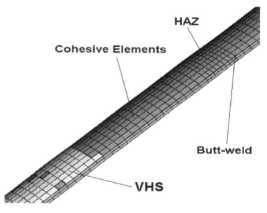

Figure 14. Debonding of Cohesive elements (with load = 219.8 kN (Point (3) in Figure 5).

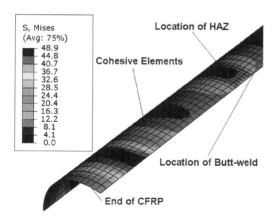

Figure 12. Von Mises stress distribution of cohesive elements with load = 200 kN (Point (2) in Figure 5).

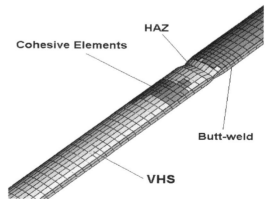

Figure 15. Debonding of Cohesive elements (with load = 105 kN Point (4) in Figure 5).

4 SUMMARY AND CONCLUSIONS

CFRP strengthened butt-welded VHS steel tubes under static loading were simulated using ABAQUS®. Cohesive elements were used to simulate the debonding between CFRP and VHS steel tube. Different CFRP bond lengths were used.

The failure mode of CFRP debonding and necking in the HAZ were simulated in the FE model. The ultimate loads from the FE model were in satisfactory agreement with the experimental results. The effective bond length of 75 mm obtained from the test results was verified by the FE model. It should be noted that this effective bond length may be suitable for the material properties used in this study only, including tube dimensions, material properties of steel, epoxy and CFRP.

REFERENCES

Al-Emrani, M. & Kliger, R. 2006. Experimental and numerical investigation of the behaviour and strength of composite steel-CFRP members. *Advances in Structural Engineering* 9(6): 819–831.

Boyd, S. W., Winkle, I. E. & Day, A. H. 2004. Bonded butt joints in pultruded GRP panels—An experimental study. *International Journal of Adhesion and Adhesives* 24(3): 263–275.

Buyukozturk, O., Gunes, O. & Karaca, E. 2004. Progress on understanding debonding problems in reinforced concrete and steel members strengthened using FRP composites. *Construction and Building Materials* 18(1): 9–19.

Colombi, P. 2006. Reinforcement delamination of metallic beams strengthened by FRP strips: Fracture mechanics based approach. *Engineering Fracture Mechanics* 73(14): 1980–1995.

Colombi, P. & Poggi, C. 2006. An experimental, analytical and numerical study of the static behavior of steel beams reinforced by pultruded CFRP strips. *Composites Part B (Engineering)* 37(1): 64–73.

Fawzia, S. 2007. *Bond Characteristics Between Steel Members and Carbon Fibre Reinforced Polymer Composites*. Ph.D Thesis, Department of Civil Engineering, Monash University, Clayton, Australia.

Fawzia, S., Al-Mahaidi, R. & Zhao, X.-L. 2006. Experimental and finite element analysis of a double strap joint between steel plates and normal modulus CFRP. *Composite Structures* 75(1–4): 156–162.

Hollaway, L. C., Photiou, N. K. & Chryssanthopoulos, M. K. 2006. Strengthening of an artificially degraded steel beam utilising a carbon/glass composite system. *Construction & Building Materials* 20(1–2): 11–21.

Jiao, H. 2003. *The Behaviour of Very High Strength (VHS) Members and Welded Connections*. Ph.D Thesis, Department of Civil Engineering, Monash University, Clayton, Australia.

Jiao, H. & Zhao, X. L. 2001. Material Ductility of Very High Strength (VHS) Circular Steel Tubes In Tension. *Thin-Walled Structures* 39(11): 887–906.

Jiao, H. & Zhao, X. L. 2004. CFRP strengthened butt-welded very high strength (VHS) circular steel tubes. *Thin-Walled Structures* 42(7): 963–978.

Lenwari, A., Thepchatri, T. & Albrecht, P. 2006. Debonding strength of steel beams strengthened with CFRP plates. *Journal of Composites for Construction* 10(1): 69–78.

Liu, H. B., Xiao, Z. G., Zhao, X. L. & Al-Mahaidi, R. 2007. Fracture Mechanics Analysis of Cracked Steel Plates Repaired with Composite Sheets. *1 st Asia-Pacific Conf. on FRP in Structures (APFIS07)*, Hong Kong, Dec.

Zhao, X.-L. & Zhang, L. 2007. State-of-the-art review on FRP strengthened steel structures. *Engineering Structures* 29(8): 1808–1823.

*Parallel session 9: Static strength of joints
(review by J. Wardenier & Q.L. Zhang)*

Reanalysis of the moment capacity of CHS joints

G.J. van der Vegte
Delft University of Technology, Delft, The Netherlands

J. Wardenier
Delft University of Technology, Delft, The Netherlands
National University of Singapore, Singapore

X.D. Qian & Y.S. Choo
Center for Offshore Research and Engineering, National University of Singapore, Singapore

ABSTRACT: Current design formulae for Circular Hollow Section (CHS) T- and X-joints loaded by brace in-plane or out-of-plane bending moments are based on a limited number of tests available at the early eighties. In the last two decades, many experimental studies as well as numerical research programmes were conducted in this area. In this paper, a reanalysis is made of the brace moment capacity of T- and X-joints. The available data are checked against possible member failures, while the experimental data are carefully screened with respect to outliers. Detailed comparisons are made between FE results and "reliable" test data obtained from different research programmes. Based on the selected data, the existing strength equations are updated.

1 INTRODUCTION

Recently, a new set of recommendations has been established by IIW Subcommission XV-E for the design of circular hollow section (CHS) joints under predominantly static loading. Extensive reanalyses of both existing data, additional experimental evidence that became available after the previous recommendations (IIW 1989) and well-calibrated finite element (FE) data has formed the basis for the new design formulations. Companion papers e.g. van der Vegte et al. (2008) give an overview of the joint capacity equations derived for axially loaded K-gap, T- and X-joints, while Qian et al. (2008) confirm the validity of these equations for thick-walled joints. Zhao et al. (2008) summarize the revised edition of the IIW recommendations and briefly describe the major changes made to the previous edition.

In the previous IIW recommendations (1989), the strength equations adopted for joints under bending moments, are based on the equations presented by Wardenier (1982). For joints loaded by either brace in-plane or out-of plane bending moments, the number of available tests is significantly less than that for axially loaded joints. For joints under in-plane bending moments, the strength description was based on the experiments carried out by Akiyama et al. (1974), Gibstein (1976), Sparrow (1979), and Yura et al. (1980), while the modified function of Gibstein (1976) was used to describe the in-plane bending moment capacity.

Experimental research on joints loaded by out-of-plane bending was even less than that for in-plane bending moments (Akiyama et al. 1974, Yura et al. 1980). For this type of loading, the strength formula was related to axially loaded X-joints.

Table 1 summarizes the design strength equations for both types of moment loading in a format similar to that in the API (2007), i.e. for joints subjected to brace moment loading:

$$\frac{M_1^* \sin \theta_1}{f_{y0} t_0^2 d_1} = Q_u Q_f \quad (1)$$

where Q_u represents the basic joint strength or reference strength function, and Q_f accounts for the influence of the chord stress.

Table 1. Current design rules (Wardenier et al. 1991) for T-, Y- and X-joints subjected to brace bending.

$$M_1^* = Q_u Q_f \frac{f_{y0} t_0^2}{\sin \theta_1} d_1$$

Brace in-plane bending	Brace out-of-plane bending
$Q_u = 4.85 \beta \gamma^{0.5}$	$Q_u = \dfrac{2.7}{1 - 0.81\beta}$

In the past 25 years, new data became available, both experimental and numerical. This paper gives an overview of the references presenting the new data. Experimental programmes often consist of series of experiments whereby one or two parameters are varied. However, unlike numerical data, experimental data inevitably include a certain amount of scatter, even if duplicate tests are carried out under seemingly identical conditions. Furthermore, a series of experiments conducted by one laboratory may not necessarily give the same results as a comparable series carried out somewhere else. The current study not only identifies and highlights trends within certain series of experiments but also makes comparisons between various series of experiments. This approach, combined with comparison with numerical data, provides an excellent tool for identifying possible outliers in experimental programmes. The screened database is then used to update the existing strength equations.

Table 2. Overview of research conducted for T- and X-joints subjected to in-plane bending moments.

In-plane bending

T-joints

	β	2γ	d_0 [mm]	Test FE
Akiyama et al. (1974)	0.19–0.46	35.1–95.3	165–457	Test
Gibstein (1976)	0.33–0.81	17.5–41.5	219–299	Test
Stamenkovic & Sparrow (1983)	0.42–1.0	18.8–33.4	114	Test
Stamenkovic (1984)	0.42–0.80	21.6–45.9	168–273	Test
Puthli (Stol et al. 1985)	0.35–1.0	16.0–49.3	168	Test
Healy & Zettlemoyer (1993)	0.30–0.95	16.0–80.0	305–508	FE
van der Vegte (1995)	0.25–1.0	14.5–50.8	406.4	FE

X-joints

	β	2γ	d_0 [mm]	Test FE
Stamenkovic (1984)	0.42–1.0	18.6–44.2	114–272	Test
van der Vegte et al. (1990)	0.60	40.9	409	Test
van der Vegte (1995)	0.25–1.0	14.5–50.8	406.4	FE
Choo et al. (2007)	0.80–1.0	19.6–35.3	407–508	Test & FE

Table 3. Overview of research conducted for T- and X-joints subjected to out-of-plane bending moments.

Out-of-plane bending

T-joints

	β	2γ	d_0 [mm]	Test FE
Akiyama et al. (1974)	0.19–0.46	35.2–95.3	165–457	Test
Yura et al. (1980)	0.34–0.90	45.7	507	Test
Puthli (Stol et al. 1985)	0.35–1.0	16.0–49.3	169	Test
Kurobane et al. (1991)	0.28–1.0	26.7–48.1	216–218	Test
Dexter et al. (1993)	0.37	35.6	1067	Test
Lee et al. (1994)	0.37–1.0	20.0–80.0	1000	FE
van der Vegte (1995)	0.25–1.0	14.5–50.8	406.4	FE

X-joints

	β	2γ	d_0 [mm]	Test FE
van der Vegte et al. (1990)	0.60	40.0	408	Test
van der Vegte (1995)	0.25–1.0	14.5–50.8	406.4	FE

2 NEW DATA

Table 2 summarizes the various references for T- and X-joints under in-plane bending moments. For each reference, the ranges of the chord diameter d_0, the brace-to-chord diameter ratio β and the chord diameter-to-thickness ratio $2\gamma = d_0/t_0$ are given as well. Table 3 gives a similar list for the joints under out-of-plane bending moments.

3 EXAMINATION OF DATABASE

In the following sections, various test series will be discussed separately in order to identify deviating data within a particular series. Furthermore, comparisons between the various research results are presented.

Table 2 lists two publications made by Stamenkovic & Sparrow (1983) and Stamenkovic (1984) with respect to the in-plane bending capacity of T- and X-joints. Detailed analyses of these results showed that the strength values reported in both references can be significantly different from (i.e. either higher or lower than) the results found by other researchers. It should be noted that Stamenkovic & Sparrow (1983) do not provide material properties for the brace, while Stamenkovic (1984) does. Hence, some of the differences can be attributed to brace failures, while other differences remain inexplicable. Due to page limitations, the analyses of these data are not presented in the current study. For completeness, it should be mentioned that the data of both references which are in line with the findings of other researchers, are included in the database.

3.1 T-joints loaded by brace in-plane bending

Figure 1 shows selected results obtained by Gibstein (1976). The presented data are in line with the expected trend i.e. joints with increasing 2γ values exhibit increasing non-dimensional strengths $M_{1,u,ipb} \sin\theta_1 / f_{y0} t_0^2 d_1$.

Because of the limited number of geometric parameters that determine the strength of T-joints under in-plane bending moments, it is possible to make comparisons between series of results obtained from different references. Of major interest are the comparisons between experimental data and numerical results. For example, Figure 2 illustrates selected experiments conducted by Puthli (Stol et al. 1984) and Gibstein (1976) and the FE results from van der Vegte (1995). The two upper lines refer to test and FE data with 2γ values ranging from 48.8 to 50.8. The lower three curves represent the data for thicker walled joints with 2γ varying from 25 to 29. For both cases, an excellent agreement is observed between the data from the various sources. Nevertheless, two remarks can be made:

i. Puthli's data point with $\beta = 0.68$ and $2\gamma = 29.0$ seems somewhat low if compared with the trend found by Gibstein's and van der Vegte's data, especially when looking at the slope of the line connecting this data point with the smaller β point. Gibstein's data for $\beta = 0.64$ and $\beta = 0.84$ which almost perfectly match van der Vegte's FE curve for this 2γ value, confirm this.

ii. van der Vegte's FE data for $\beta = 1.0$ and $2\gamma = 50.8$ gives slightly higher predictions than Puthli's tests, even though no welds are modeled for $\beta = 1.0$.

Other researchers who conducted numerical analyses on T-joints loaded by brace in-plane bending moments with geometric parameters ($2\gamma = 48$) close to those of Puthli (Stol et al. 1984) and van der Vegte (1995) are Healy & Zettlemoyer (1993). Figure 3 illustrates data with $48 < 2\gamma < 51$ for each of the three studies. Healy did not model the weld geometry in any of his FE analyses. Further, the chord length parameter α is taken as 16, which should cause a reduction of strength as compared to van der Vegte's FE analyses who used $\alpha = 12$. As shown in Figure 3, for $\beta = 0.95$ and $2\gamma = 48.0$, Healy indeed finds strength values in between Puthli's and van der Vegte's results.

In addition, Figure 3 shows that for the configurations studied, Healy's data seem to confirm that the effect of the brace angle θ is in line with $\sin \theta$.

Another comparison of interest is that between the experiments of Akiyama et al. (1974) and Healy & Zettlemoyer's FE data (1993) for relatively thin walled joints. Both research programmes include thin walled joints, with 2γ values up to 95 for Akiyama, and 80 for Healy. Figure 4 plots both sets of data. Healy's results (with $2\gamma = 80$) are well in line with Akiyama's data for $2\gamma = 72.4$ and 95.3. Figure 4 further plots a comparison between Akiyama's data for $2\gamma = 36.0$ and van der Vegte's FE data (1995) for $2\gamma = 25.4$ and 36.9. The data are in good agreement.

3.2 X-joints loaded by brace in-plane bending

Figure 5 illustrates FE data analysed by van der Vegte (1995) and experimental data obtained from recent tests by Choo et al. (2007). It becomes clear

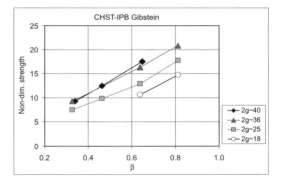

Figure 1. Tests by Gibstein (1976).

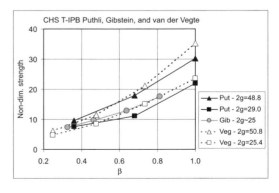

Figure 2. Selected results of Gibstein (1976), Puthli (Stol et al. 1984), and van der Vegte (1995).

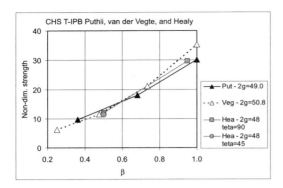

Figure 3. Selected results of Puthli (Stol et al. 1984), Healy & Zettlemoyer (1993) and van der Vegte (1995).

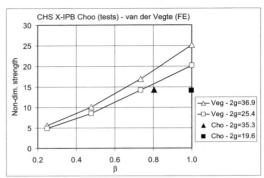

Figure 5. Test results of Choo et al. (2007) and numerical results of van der Vegte (1995).

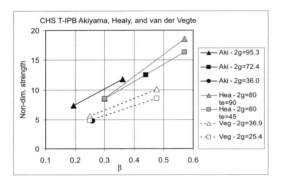

Figure 4. Selected results of Akiyama et al. (1974), Healy & Zettlemoyer (1993) and van der Vegte (1995).

Table 4. Details of tests by van der Vegte et al. (1990) and Choo et al. (2007).

	d_0				f_{y0}	Non-dim.
	mm	α	β	2γ	N/mm²	strength test
XM-18	408.5	11.9	0.60	40.9	318	14.44
X1	508.0	10.2	0.80	35.3	295	14.25
X2	407.0	12.8	1.0	19.6	312	14.11

that the strength values found in the tests by Choo et al. (2007) are relatively low. Especially the data point with β = 0.8 (specimen X1) seems low for the following reasons:

i. van der Vegte et al. (1990) conducted a single test on an X-joint (XM-18) subjected to brace in-plane bending with β = 0.6 and 2γ = 40.9. Although this data point is not shown in Figure 5, it agrees well with (i.e. is slightly above) van der Vegte's curve marked by 2γ = 36.9. Joint X1 tested by Choo has a β value = 0.8 which is considerably larger and a 2γ value = 35.3 which is only slightly smaller than the corresponding values in van der Vegte's test XM-18. Hence, it would be expected that the non-dimensional strength of test X1 by Choo et al. (2007) would be higher than van der Vegte's result. As shown in Table 4, this is not confirmed by the test data: Choo's test gives a non dimensional strength which is below the value found by van der Vegte (1990).

ii. FE simulations conducted by Choo et al. (2007) on the two joints tested show a good match between the FE prediction and the actual behaviour for joint X2. However, for joint X1, the in-plane bending strength found in the FE analysis is 13% higher than the experimental strength.

On the other hand, Choo et al. (2004) reanalysed four of van der Vegte's FE analyses on un-reinforced X joints (with β = 0.25 and 0.73 for 2γ = 36.9 and 50.8) loaded by brace in-plane bending. Instead of the quadratic, thick shell elements employed in van der Vegte's study, Choo et al. (2004) used quadratic, solid elements to model joints. A comparison between the moment-rotation curves obtained from both studies reveals a very good match between the predictions. In view of the more accurate representation of the welds in the solid element model, Choo's computed strength values are slightly higher than those predicted by van der Vegte. The maximum difference in ultimate strength between both studies is only 6%, confirming the reliability of both FE models used.

3.3 *T-joints loaded by brace out-of-plane bending*

For T-joints loaded by brace out-of-plane bending, the following test series are discussed individually in order to identify deviating data within each programme:

- Puthli (Stol et al. 1984) conducted eight tests, summarized in Figure 6. The three joints with $\beta = 0.35$–0.36 follow the trend that for a thinner chord, the non-dimensional strength increases. However, for the three joints with $\beta = 0.68$, this trend is not observed. Here, the strength of the joint with $2\gamma = 16$ seems too high, since the strength of this joint is well above the thinner walled joint with $2\gamma = 49$. On the other hand, the strength of the joint with $2\gamma = 29$ seems much too low. This joint has an ultimate strength, which is of the same order as that of the joint with a much smaller β value of 0.35.
- Kurobane et al. (1991) conducted ten experiments on T-joints loaded by out-of plane bending moment, illustrated in Figure 7. At first sight, peculiar behavior is observed for the joints with $\beta = 0.53$. The non-dimensional strength of the three data with $32 \leq 2\gamma \leq 35$ is expected to be within the ranges for the joints with $2\gamma = 27$ and 48, which is not found in the tests. However, for the particular joints, the yield strength of the chord f_{y0} (and the f_{u0}/f_{y0} ratio) varies significantly ($298 \leq f_{y0} \leq 832$ N/mm^2). A similar observation is made for tests by Kurobane et al. (1991) on T-joints under axial loading (van der Vegte et al. 2008).

Because two of the 2γ values considered by Kurobane et al. (1991) and Puthli (Stol et al. 1984) are almost similar, a comparison is possible between the various test series. Figure 8 shows the non-dimensional strength data for $28 \leq 2\gamma \leq 29$ and $48 \leq 2\gamma \leq 49$. In addition, the test results with a comparable 2γ value ($2\gamma = 46$) obtained from Yura et al. (1980) are displayed. Figure 8 reveals that the tests of Kurobane and Puthli give similar results, except for $\beta = 1.0$ where Kurobane's data are higher (for $2\gamma = 48$) and lower (for $2\gamma = 27$) than Puthli's corresponding points. For all β values, the data points found by Yura et al. (1980) are below the thin-walled series ($48 \leq 2\gamma \leq 49$) of both Kurobane and Puthli. Of interest to note is that Yura et al. (1980) were the first researchers who used a deformation/rotation limit to define the "end" of a test.

Another comparison of interest is that between the numerical results obtained from Lee & Dexter (1994) and van der Vegte (1995). Figure 9 presents selected results of both studies. In general, for β values ≤ 0.8, the results of Lee and van der Vegte show a good match, where the data by van der Vegte are enclosed by Lee's data, i.e. for the relatively thick-walled joints with $2\gamma = 20$, Lee's data are below van der Vegte's curve for $2\gamma = 25.4$, while for the thin-walled joints, Lee's data for $2\gamma = 60$ are higher than van der Vegte's data for $2\gamma = 50.8$. However, for β values > 0.8, Lee's strength predictions for thin walled joints are higher than van der Vegte's results. A possible explanation could be the inclusion of the weld geometry into the FE model. Unlike Lee & Dexter (1994), van der Vegte (1995) did not model the geometry of the welds for joints with $\beta = 1.0$.

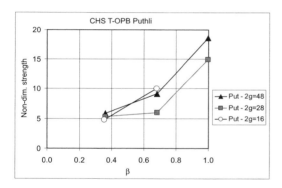

Figure 6. Tests by Puthli (Stol et al. 1984).

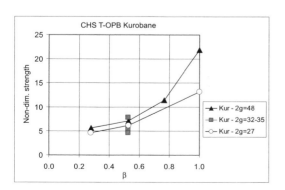

Figure 7. Tests by Kurobane et al. (1991).

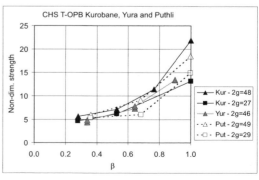

Figure 8. Selected tests by Kurobane et al. (1991), Puthli (Stol et al. 1984) and Yura et al. (1980).

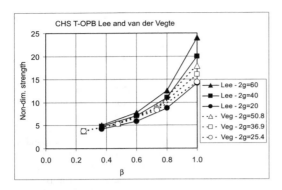

Figure 9. FE results of Lee & Dexter (1994) and van der Vegte (1995).

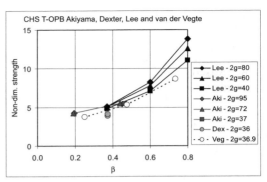

Figure 11. Test results of Akiyama et al. (1974), Dexter et al. (1993) and numerical results of Lee & Dexter (1994) and van der Vegte (1995).

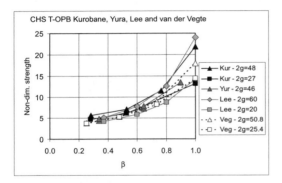

Figure 10. Selected test results of Kurobane et al. (1991) and Yura et al. (1980) and FE results of Lee & Dexter (1994) and van der Vegte (1995).

Figure 10 shows the combined results of selected experimental research (Kurobane et al. 1991, Yura et al. 1980) and numerical analyses (Lee & Dexter 1994, van der Vegte 1995). It is observed that Yura's data ($2\gamma = 46$) and van der Vegte's results (for $2\gamma = 50.8$) are almost identical for all β values considered. Lee's data for $2\gamma = 60$ is well in line with van der Vegte's and Yura's results for small β values, but underpredicts Kurobane's data for small β values. For $\beta = 1.0$, Lee's predictions for $2\gamma = 60$ are higher than all other results, which is in accordance with the trend that a higher 2γ value should result in a higher non-dimensional strength.

A comparison between results of Akiyama et al. (1974), Dexter et al. (1993), Lee & Dexter (1994), and van der Vegte (1995) illustrated in Figure 11 reveals that for low β values, the series of Akiyama, Lee and van der Vegte match well. For example, van der Vegte's data for $2\gamma = 36.9$ and Akiyama data point for $2\gamma = 37$ are close. Dexter's double tests on two identical $\beta = 0.37$ T-joints give almost similar strength values, although both results are lower than Lee's and van der Vegte's predictions. For thin walled joints with $60 \leq 2\gamma \leq 95$ and a small β value, Akiyama and Lee give strength predictions which are well in line.

3.4 X-joints loaded by brace out-of-plane bending

For X-joints under out-of-plane bending moments, only the results of van der Vegte (1995) are available. No comparisons can be made with the results of other studies.

3.5 N-joints loaded by brace out-of-plane bending

For N-joints under out-of-plane bending moments, only four tests by Yura et al. (1980) are found in the literature, whereby two series of two identical tests can be distinguished. However, for the series with $\beta = 0.64$, the ultimate out-of-plane bending moments inexplicably show a significant difference between the two tests ($M_{1,u,opb} = 82.5$ and 117.8 kNm).

4 NEW STRENGTH FORMULAE

4.1 In-plane bending moment

Similar to the approach adopted in the CIDECT recommendations (Wardenier et al. 1991), the strength for in-plane bending moments is related to the format derived by Gibstein (1976), whose equation is based on a modified punching shear model (Wardenier, 1982).

$$M_{1,u,ipb} = c\, \beta\, \gamma^{0.5}\, \frac{f_{y0}\, t_0^2}{\sin \theta_1}\, d_1 \qquad (2)$$

Table 5 summarizes the statistical data for the coefficient c in Equation 2 for the various series of experiments and FE analyses, whereby only series with three or more data are included. Figure 12 presents the comparison between Equation 2 and all data for T-joints under brace in-plane bending (except brace failures), while Figure 13 shows a similar comparison for X-joints.

In these figures, deviating data can be easily spotted. For example, the relatively high data points marked by XM-21, XM-22 and XM-23 in Figure 13 are some of the peculiar tests by Stamenkovic (1984).

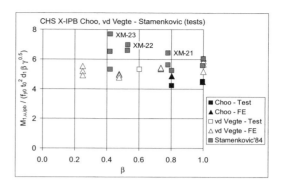

Figure 13. Comparison between the in-plane bending strength of X-joints and Equation 2.

Table 5. Comparison between the brace in-plane bending strength and Equation 2.

	No. of data	Mean	CoV	Min.	Max.
T-joints—tests					
Akiyama	4	4.83	0.096	4.38	5.48
Gibstein	13	6.05	0.050	5.59	6.78
Puthli	6	5.43	0.113	4.35	6.20
Stamenkovic'83	17	5.25	0.211	3.63	6.73
Stamenkovic'84	6	5.23	0.191	3.48	6.24
T-joints—FE					
Healy	13	4.82	0.111	4.18	6.38
van der Vegte	14	5.69	0.128	4.80	7.01
X-joints—tests					
Stamenkovic'84	10	6.19	0.127	5.23	7.69
X-joints—FE					
van der Vegte	14	5.33	0.07	4.76	6.08

Based on a statistical analysis on selected data, in line with the procedure described by van der Vegte et al. (2008), the design value adopted for c = 4.3.

4.2 Out-of-plane bending moment

Similar to the approach adopted in the CIDECT recommendations (Wardenier et al. 1991), the strength for out-of-plane bending moments $M_{1,u,opb}$ is related to the strength of axially loaded X-joints $N_{1,u}$ through a factor $c.d_1$ (Wardenier 1982):

$$M_{1,u,opb} = c\, d_1\, N_{1,u} \quad (3)$$

For the mean ultimate capacity $N_{1,u}$ of X-joints, the equation derived by van der Vegte et al. (2008) is used:

$$N_{1,u} = 3.16 \left(\frac{1+\beta}{1-0.7\beta} \right) \gamma^{0.15} \frac{f_{y0}\, t_0^2}{\sin\theta_1} Q_f \quad (4)$$

Table 6 summarizes the statistical data for the coefficient c in Equation 3 for the various series of experiments and FE analyses, whereby only series with three or more data are included. Figures 14–15 illustrate the results obtained for T-, N- and Y-joints from the various references.

Similar as for T- and X-joints under brace in-plane bending, the deviating data described in the previous sections can be simply traced, especially for series which exhibit a relatively large CoV such as the data of Kurobane et al. (1991) and Puthli (Stol et al. 1984).

Of interest is the relation between the out-of-plane bending and axial strength for the FE series of

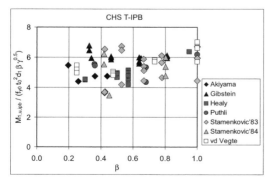

Figure 12. Comparison between the in-plane bending strength of T-joints and Equation 2.

X-joints under brace out-of-plane bending analysed by van der Vegte (1995), illustrated in Figure 16. This graph compares the c values obtained from Equation 3, either using (i) $N_{1,u}$ calculated from Equation 4 or (ii) actual $N_{1,u}$ values found in the FE analyses of X-joints under axial brace load. Because the strength equation for axially loaded X-joints (Eq. 4) is intentionally assumed to be conservative for $\beta = 1.0$, due to the lower deformation capacity of these joints, the values of the coefficient c are rather high for $\beta = 1.0$, causing a sharp increase of the CoV. Using the actual strength of the X-joints instead of the numbers derived with Equation 4, the values of the coefficient c shown in Table 7 and illustrated in Figure 16 give a much better match between the results for $\beta < 1.0$ and $\beta = 1.0$.

Based on a statistical analysis in line with the procedure described by van der Vegte et al. (2008), a

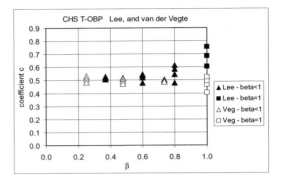

Figure 15. Comparison between the out-of-plane bending strength of T-joints and Equation 3—FE data.

Table 6. Comparison between the brace out-of-plane bending strength and Equation 3.

	No. of data	Mean	CoV	Min.	Max.
T-joints—tests					
Akiyama	5	0.511	0.113	0.42	0.57
Kurobane	10	0.566	0.184	0.40	0.70
Puthli	8	0.540	0.166	0.40	0.72
Yura	6	0.514	0.043	0.48	0.55
N-joints—tests					
Yura	4	0.495	0.167	0.39	0.57
T-joints—FE					
Lee	16	0.555	0.138	0.48	0.75
vd Vegte	15	0.493	0.061	0.41	0.53
X-joints—FE					
vd Vegte $\beta < 1.0$	11	0.505	0.035	0.48	0.54
vd Vegte $\beta = 1.0$	4	0.721	0.133	0.60	0.83

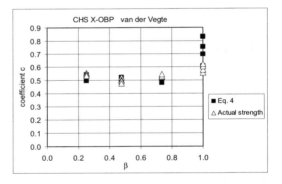

Figure 16. Comparison between out-of-plane bending strength of X-joints and Equation 3 for van der Vegte's FE data (1995).

Table 7. Comparison between brace out-of-plane bending strength and axial strength for X-joints analysed by van der Vegte (1995).

	No. of data	Mean	CoV	Min.	Max.
Axial strength based on Equation 4					
$\beta < 1.0$	11	0.505	0.035	0.48	0.54
$\beta = 1.0$	4	0.721	0.133	0.60	0.83
Axial strength based on actual strength data					
$\beta < 1.0$	11	0.522	0.046	0.47	0.55
$\beta = 1.0$	4	0.588	0.048	0.55	0.62

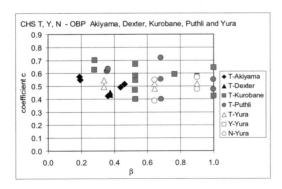

Figure 14. Comparison between the in-plane bending strength of T-joints and Equation 3—experiments.

design value of 0.41 is adopted for the coefficient c in Equation 3.

5 SUMMARY AND CONCLUSIONS

A reanalysis is made of the brace moment capacity of T- and X-joints. The available (experimental and numerical) data are checked against possible member failures and are carefully screened with respect to outliers. Comparisons are made between the various results, not only within series of individual references, but also between results obtained from different sources. Based on the selected data, the existing strength equations for joints under brace in-plane or out-of-plane bending moments are updated. The following conclusions can be drawn:

- Care should be taken with the results of Stamenkovic & Sparrow (1983) and Stamenkovic (1984). In both references, strength values are reported which can be significantly different from (i.e. either lower or higher than) the results of other researchers.
- The available experimental data exhibit a large scatter due to a variety of reasons. Even within series of experiments conducted by the same researcher, considerable but inexplicable differences are observed.
- It can be concluded that after careful screening, examination of, and comparisons between the experimental and numerical data, an excellent agreement can be found between the experimental results and the predictions obtained with FE analyses.
- Table 8 lists the equations recommended for the design of brace in-plane and out-of-plane bending strength for all joints (i.e. K-, T-, Y-, and X-joints).

REFERENCES

Akiyama, N., Yajima, M., Akiyama, H. & Ohtake, A. 1974. Experimental study on strength of joints in steel tubular

Table 8. Recommended design rules (IIW 2008) for all joints subjected to brace in-plane bending or out-of plane bending.

$$M_1^* = Q_u \, Q_f \, \frac{f_{y0} \, t_0^2}{\sin \theta_1} \, d_1$$

Brace in-plane bending   Brace out-of-plane bending

$Q_u = 4.85 \beta \gamma^{0.5}$ $Q_u = 1.3 \left(\dfrac{1+\beta}{1-0.81\beta} \right) \gamma^{0.15}$

$Q_f = (1-|n|)^{(0.45-0.25\beta)}$

where n = maximum chord stress ratio.

structures. *Journal of Society of Steel Construction*, Vol. 10, No. 102, pp. 37–68, (in Japanese).

American Petroleum Institute 2007. Recommended practice for planning, designing and constructing fixed offshore platforms—Working stress design. API RP 2 A, 21st Ed, Supp. 3, American Petroleum Institute, USA.

Choo, Y.S., Liang, J.X., Vegte, G.J. van der & Liew, J.Y.R. 2004. Static strength of doubler plate reinforced CHS X-joints loaded by in-plane bending. *Journal of Constructional Steel Research*, Vol. 60, No. 12, pp. 1725–1744.

Choo, Y.S., Chen, Z., Wardenier, J. & Gronbech, J. 2007. Static strength of simple and grouted tubular X-joints subjected to in-plane bending. *Proc. 5th Intern. Conf. on Advances in Steel Structures*, Singapore, Vol. II, pp. 183–194.

Dexter, E.M., Haswell, J.V. & Lee, M.M.K. 1993. A comparative study on out-of-plane moment capacity of tubular T/Y joints. *Proc. 5th Intern. Symp. on Tubular Structures*, Nottingham, UK, pp. 675–682.

Gibstein, M.B. 1976. The static strength of T-joints subjected to in-plane bending moments. Det Norske Veritas Report No. 76–137, Oslo, Norway.

Healy, B.E. & Zettlemoyer, N. 1993. In-plane bending strength of circular tubular joints. *Proc. 5th Intern. Symp. on Tubular Structures*, Nottingham, UK, pp. 325–344.

IIW 1989. Design recommendations for hollow section joints—Predominantly statically loaded. 2nd Edition, Intern. Institute of Welding, Subcommission XV-E, Annual Assembly, Helsinki, Finland, IIW Doc. XV-701-89.

IIW 2008. Static design procedure for welded hollow section joints—Recommendations. 3rd Edition, Intern. Institute of Welding, Subcommission XV-E, IIW Doc. XV-1281-08.

Kurobane, Y., Makino, Y., Ogawa, K. & Maruyama, Y. 1991. Capacity of CHS T-joints under combined OPB and axial loads and its interactions with frame behavior. *Proc. 4th Intern. Symp. on Tubular Structures*, Delft, The Netherlands, pp. 412–423.

Lee, M.M.K. & Dexter, E.M. 1994. A parametric study on the out-of-plane bending strength of T/Y-joints. *Proc. 6th Intern. Symp. on Tubular Structures*, Melbourne, Australia, pp. 433–440.

Qian, X.D., Choo, Y.S., Vegte, G.J. van der & Wardenier, J. 2008. Evaluation of the new IIW CHS strength formulae for thick-walled joints. *Proc. 12th Intern. Symp. on Tubular Structures*, Shanghai, China.

Sparrow, K.D. 1979. Ultimate strengths of welded joints in tubular steel structures. *PhD Thesis*, Kingston Polytechnic, UK.

Stamenkovic, A. & Sparrow, K.D. 1983. Load interaction in T-joints of steel circular hollow sections. *Journal of Structural Engineering*, ASCE, Vol. 109, No. 9, pp. 2192–2204.

Stamenkovic, A. & Sparrow, K.D. 1984. In-plane bending and interaction of CHS T- and X-joints. *Proc. 1st Intern. Symp. on Tubular Structures*, Boston, USA, pp. 543–554.

Stol, H.G.A., Puthli, R.S. & Bijlaard, F.S.K. 1984. Static strength of welded tubular T-joints under combined loading. TNO-IBBC Report B-84-561/63.6.0829, The Netherlands.

Vegte, G.J. van der, Koning, C.H.M. de, Puthli, R.S. & Wardenier, J. 1990. Static behaviour of multiplanar welded joints in circular hollow sections. Stevin Report

25.6.90.13/A1/11.03, TNO-IBBC Report BI-90-106/63.5.3860, Delft, The Netherlands.

Vegte, G.J. van der 1995. The static strength of uniplanar and multiplanar tubular T- and X-joints. *PhD Thesis*, Delft University of Technology, Delft, The Netherlands, Delft University Press, The Netherlands, ISBN 90-407-1081-3.

Vegte, G.J. van der, Wardenier, J., Zhao, X.-L. & Packer, J.A. 2008 Evaluation of new CHS strength formulae to design strengths, *Proc. 12th Intern. Symp. on Tubular Structures,* Shanghai, China.

Yura, J.A., Zettlemoyer, N. & Edwards, I.F. 1980. Ultimate capacity equations for tubular joints. *Proc. Offshore Technology Conf.*, OTC 3690, Houston, USA.

Wardenier, J. 1982. *Hollow section joints*. Delft University Press, The Netherlands, ISBN 90-6275-084-2.

Wardenier, J., Kurobane, Y., Packer, J.A., Dutta, D. & Yeomans, N. 1991. *Design guide for circular hollow section (CHS) joints under predominantly static loading.* Verlag TUV Rheinland GmbH, Köln, Germany, ISBN 3-88585-975-0.

Zhao, X.-L., Wardenier, J., Packer, J.A. & Vegte, G.J. van der 2008. New IIW (2008) static design recommendations for hollow section joints. *Proc. 12th Intern. Symp. on Tubular Structures,* Shanghai, China.

Experimental research and parameter analysis on rigidity of unstiffened tubular X-joints

G.Z. Qiu & J.C. Zhao
Department of Civil Engineering, Shanghai Jiaotong University, Shanghai, China

ABSTRACT: In order to check the influence of rigidity of unstiffened tubular joints on structural behavior and stability, experimental research is carried out on axial rigidity, bending rigidity both in-plane and out-of-plane of X-joints on the background of a single-layer tubular structure gymnasium. Detailed experimental scheme and results are presented in the paper, and the fracture mechanism of specimens is analyzed. Systematic parameter analysis of tubular X-joints is performed using Ansys program. The influence of joint parameters, including ratio of chord diameter to its thickness, ratio of branch diameter to that of chord, ratio of branch thickness to that of chord, the angle between branch and chord, and chord stress etc, on rigidity of tubular X-joints is analyzed. Test results and analysis show that when the difference between branch diameter and that of chord is bigger, the axial and bending rigidity of joints are weaker, the influence of semirigid of tubular joints on structural behavior and especially stability behaviour is not negligible. While the branch diameter approaches to that of main tube, both axial and bending rigidity of joints are strong, and thus they can meet the requirements of rigid joint basically.

Keywords: tubular structure; unstiffened tubular joint; experiment on joint rigidity; parameter analysis.

1 INTRODUCTION

Tubular truss structures are applied widely in China in recent years, in which the unstiffened tubular joints are used mostly. At present, the structures with tubular joints are usually taken as planar or spatial hinge-connected systems in design, so the method to calculate the load bearing capacity of joint in the present "Code for design of steel structures" (GB50017-2003) (the shortened form "Steel Code" is used in following text) only takes the axial force of members into account. While within the range of usual structure dimension, the juncture of two connective members may distorts locally under load, which can cause relative deformation and rotation of the joint, and the rigidity behavior appears whether in elastic or in elastic-plastic stage. However, if the joints are taken as semi-rigid, their stiffness formula is not given in the "steel code", it's impossible to build the computational model according to practical stiffness of tubular joints, so the joints have to be taken as hinge or rigid connection in tube structures design, and the load bearing capacity of the structure can not be judged rationally, which usually causes either wasting or structural unsafety. Furthermore, the stiffness of tubular joints is closely relative to member stability and structural dynamical performance.

Based on a practical engineering project, the experimental research on stiffness of tubular joint is carried out by taking X-joint as an example, the parameters influencing joint stiffness are analyzed, and some meaningful conclusions are drawn by finite element analysis, which can supply reference for practical engineering design.

2 EXPERIMENTAL RESEARCH ON RIGIDITY OF UNSTIFFENED TUBULAR X-JOINTS

2.1 Project background

The gymnasium is located in Minhang District of Shanghai. The shape of steel roof is elliptic. Membrane structure with framework is adopted as roof structure as shown in Figure 1. The longitudinal span of the gymnasium is 89.624 m, and the transverse one is 78.214 m, the highest point is 26.826 m, and the projection area is about 5500 sq.m. The large-span one-layer tubular structure is adopted and the computational schematic of initial design scheme is shown as Figure 1. The shape of the structure is very complicated, which is the combination of multi-direction curves. There are many kinds of structural members which are different each other in size. Tubular joints are used in this structure. The steel tubes running through joints continuously are called main tubes, while the tubes intersecting with main tubes are branch tubes, which are welded with main tubes directly. Because of the small curvature of members, the joints are approximate to be in the same plane, which can be called planar tubular X-joints.

In the present "Steel Code", for the stiffness of tubular joints, the checking formula is given only under the condition that the branch tube is axially loaded, the moment is not taken into account. While in fact, the members may bear great moment, whether the structural safety can be ensured with the joints intensity? Whether the tubular joints of single-layer tube structure can be taken as joints with elastic bending stiffness? How about the influence of tubular joints stiffness on mechanical behavior and stability of the whole structure? Based on these questions, the experimental research is carried out to study the axial stiffness, in-plane and out-of-plane bending stiffness, aiming at the typical joint marked with a circle in Figure 1.

2.2 Experiment scheme

2.2.1 Specimen

In actual structure, the section of main tube is $\Phi 710 \times 16$ and that of branch is $\Phi 406 \times 12$, the specimens are manufactured with the ratio of 1:2, and for the reason of market supply, $\Phi 351 \times 8$ and $\Phi 219 \times 6$ are used as main tube and branch respectively in the experiment. Steel grade is Q345B. The main parameters of real joint and specimen are shown in Table 1.

2.2.2 Load setting

1. Axial rigidity test. Apply axial force on both branches synchronously, the intensity and rigidity of the counterforce setting must be enough to avoid load offsetting caused by biggish joint distortion, otherwise, the accuracy of displacement measure can not be ensured. Meanwhile, simple convenient

Table 1. The parameters of real joint and specimen.

Parameters	D/T	D/t	d/D	t/T
Real joint	44.375	33.833	0.572	0.75
Specimen	43.875	36.5	0.624	0.75

* D—main tube diameter, T—main tube thickness, d—branch diameter, t—branch thickness.

manufacturing and setting should be taken into account.

Based on the above requirements, the load setting is designed as shown in Figure 2(a), Figure 2(b) is the photo of the test.

2. Out-of-plane bending stiffness test. In the plane being vertical to the specimen, both jacks apply vertical load at the ends of branches synchronously, so the moment is generated at the joint. Because of the joint rotation caused by moment, the branches and jacks will turn. The moment is generated at the joint by the friction between jacks and counterforce beam. For eliminating the minus moment, rollers are placed between jacks and counterforce beam.

The load setting is shown as Figure 3(a), and Figure 3(b) shows the photo of the test.

3. In-plane bending stiffness test. In the specimen plane, jacks apply vertical load at the end of branches, so the moment is generated at the joint. The load setting is shown as Figure 4(a), and Figure 4(b) shows the photo of the test.

2.2.3 Test method

The method to study axial rigidity is simple, seen as Figure 2, axial force is applied at the end of the branch, which can be read directly or be controlled, the axial displacement at the end of member can be measured, and the joint deformation can be computed by deleting the member shrinkage, so the curve of axial force-displacement can be gotten directly, then axial rigidity of joint can be obtained.

There are two methods to study bending rigidity of joint, which are direct method and indirect method respectively. For the direct method, joint moment and relative angle of adjacent members are required, the advantage of this method is that the bending rigidity can be denoted by the result directly, while it's difficult to measure the joint moment directly, then calculation is needed. Because the relative angle measured in the test includes member shrinkage, it should be subtracted. The indirect method measures the load and corresponding displacement at the end of member which suffers from transverse force, the moment can be calculated, the curve of moment-displacement then can be drawn, and the bending rigidity be analyzed according to theoretical computation and

Figure 1. The computational schematic of the steel roof.

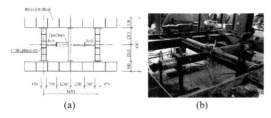

Figure 2. Test on axial rigidity of the joint.

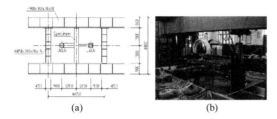

Figure 3. Test on out-of-plane rigidity of the joint.

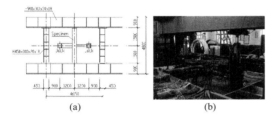

Figure 4. Test on in-plane rigidity of the X-joint.

comparison. The experiment scheme is established based on the later method in this paper, the expectant moment can be supplied on the joint by controlling the transverse force, the deformation of control point or relative angle of adjacent members are measured respectively to study the in-plane and out-of-plane bending rigidity, then the bending rigidity of the joint can be obtained.

2.3 Analysis of the results

2.3.1 Axial rigidity test

The failure mode of the joint in axial rigidity test is shown as Figure 5. From the figure it can be seen that there is great deformation while the joint is destroyed, under the axial force of the branch, the failure modality of the joint is buckling of the main tube wall, which is aroused by the local bending caused by the vertical force. The deformation of the main tube wall connected with the branch $\Phi 219 \times 6$ is large, while that of the end of branch $\Phi 351 \times 8$ is very small. The axial rigidity is strong when the ratio of branch diameter to that of main tube is equal to 1. The axial rigidity of the joint at the end of branch $\Phi 219 \times 6$ is mainly analyzed in this paper for its influence on the structural mechanical behavior is obvious.

Figure 7 shows the curve of axial loads on branch $\Phi 219 \times 6$ and its displacements, of which, the displacements of point 32 and 33 are measured respectively by two displacement gauges fixed at the end of branch $\Phi 219 \times 6$, the location of these two points are shown in Figure 6. In Figure 7, the curve of analysis is the result of finite element analysis on the model joint by program ANSYS. Figure 7 shows that axial rigidity of the joint decreases gradually with the increasing axial force, the relationship between axial rigidity and axial force is nonlinear. However, there hasn't been a valid method to simulate the influence of the nonlinear relationship for analyzing real structures. To simplify the analysis of real structures, the single rigidity is used in stead of the nonlinear one. Calculating the intensity of the joint according to the present "Steel Code", the maximum axial force of branch $\Phi 219 \times 6$ can be obtained as 218.6 kN. According to the curves in Figure 8, the slope of adjacent two points under the axial force 218.6 kN is taken as axial rigidity of the joint, the branch deformation is ignored for the short length and great axial rigidity, the average value of axial rigidity of the two points is 172.0 kN/mm, seen as Table 2. If a branch $\Phi 219 \times 6$ is 5 m long in real structure, its axial rigidity is $EA/l = 165.4$ kN/mm, when taking the semi-rigid behavior of the tubular joint at one end of the branch into account, the tube and joint can be taken as being in series, and the axial rigidity being calculated is 84.3 kN/mm. From the calculation it is seen that the axial rigidity of a 5 m

Figure 5. The failure mode of the joint in axial rigidity test.

Figure 6. Location of the two points.

long branch Φ219 × 6 decreases 49.0% while taking one joint axial rigidity into account, so the axial rigidity of tubular joint can not be ignored.

2.3.2 *Out-of-plane bending rigidity test*

When a tubular joint bears out-of-plane bending, there are two possible modes of the joint deformation under the uneven stress transferred by the branch tubes. If the main tube wall generates enough relative deformation, the relative rotation at the root of branch appears obviously; if the main tube wall has great rigidity, not only the above relative rotation will not appear, but also the bending length of the branch is shorten relatively, presenting the character of rigid region at the joint.

The main tube wall of the specimen is relatively thin in this test, the first deformation mode mentioned above appears under the out-of-plane moment. At the connection of the small branch and the main tube, local sunken-in is obvious on the main tube wall on the compressed side, seen as Figure 8, it is judged as elastic-plastic local instability of the tube without breakage of the welds. While at the connection of the big branch and main tube, there is no visible deformation on the tube wall. That means that when the ratio of branch diameter to that of main tube is equal to 1, the out-of-plane bending rigidity of the joint is biggish. The failure mode of the whole specimen belongs to the elastic-plastic local instability of the main tube wall.

For the small branch Φ219 × 6, according to the out-of-plane load and the rotation angle measured in the test, the curve of branch's rotation angle changing with out-of-plane moment is seen as Figure 9, of which, the curve of theoretical analysis is the result of finite element analysis on the specimen joint by program ANSYS. With the increasing of the out-of-plane moment, the out-of-plane bending rigidity of the joint decreases gradually, the relationship between them is nonlinear. To simplify the calculation, the single out-of-plane bending rigidity is used to displace the nonlinear one. The out-of-plane bending rigidity corresponding to the maximum out-of-plane moment in real structure is taken as the reference value, the maximum moment converting from the real joint to the experimental one is 18 kN.m, the corresponding out-of-plane bending rigidity is about 932.6 kN.m/rad, while that of the 5 meters-length branch Φ219 × 6 is $4EI/l$ = 3755 kN.m/rad, the former is much smaller than the later. So the influence of the

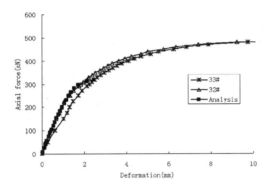

Figure 7. The curve of axial force—deformation at the end of small branch.

Figure 8. Failure mode of the joint in out-of-plane bending rigidity test.

Table 2. Axial rigidity of the joint.

	point32	point33	average value	analysis value	suggest value
axial rigidity (kN/mm)	185.2	158.7	172.0	160.0	172.0

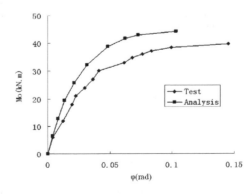

Figure 9. Curve of out-of-plane moment-rotation of branch Φ219 × 6.

out-of-plane bending rigidity of the joint on the structural mechanical behavior and global stability can not be neglected.

2.3.3 In-plane bending rigidity test

The failure modality of the joint in in-plane bending rigidity test is shown as Figure 10, being similar to the out-of-plane bending rigidity test, at the connection of the small branch and main tube, obviously local sunken-in appears on the main tube wall on the compressed side, that is elastic-plastic local instability of the tube without breakage of the welds, while at the connection of the big branch and chord, there is no visible deformation on the chord wall. That means that when the ratio of branch diameter to that of chord is equal to 1, the in-plane bending rigidity of the joint is biggish. For the branch Φ219 × 6, according to the load and the deformation of the control point, the curve of branch rotation angle changing with in-plane moment is shown as Figure 11. From the Figure. it is seen that the in-plane bending rigidity of the joint decreases gradually with the increasing of the in-plane moment, the relationship between them is also nonlinear, while the branch rotation changes with the in-plane moment linearly when the moment is small enough, it is less than 40 kN.m in this test, and the in-plane bending rigidity keeps constant basically, because the in-plane moment is small in the real structure, the rigidity of the linear phase can be taken as the reference value.

3 PARAMETER ANALYSIS ON THE RIGIDITY OF TUBULAR X-JOINT

3.1 Definition of the parameters

The parameters influencing the rigidity of tubular X-joint mainly include geometrical parameters without dimension of the main tube and branch and the internal forces of the members. The geometrical parameters without dimension of the joint are the ratio of branch diameter to that of main tube $\beta = d/D$, ratio of main tube diameter to its thickness $\gamma = D/(2T)$, ratio of branch thickness to that of main tube $\tau = t/T$, the angle between branch and main tube θ, and the ratio of two branch diameters while they are different d_1/d_2. The meaning of each parameter is seen as Figure 12. The relative relationship of the main tube and branch is established by these parameters, while the absolute value of the joint dimension can not be determined. The parameter D is selected as a factor to determine the absolute dimension of the joint, which can be established uniquely as long as the above parameters and "D" are given.

To learn about the influence of parameters on the axial rigidity of tubular X-joint, they are analyzed respectively. A series of models are used in analysis, keeping the main tube diameter D = 500 mm constantly, changing a parameter while keeping others unchanged, so the difference of joint rigidity is caused just by the changing

Figure 10. Failure mode of the joint in in-plane bending rigidity test.

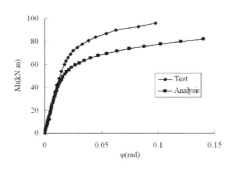

Figure 11. Curve of in-plane moment-rotation of branch Φ219 × 6.

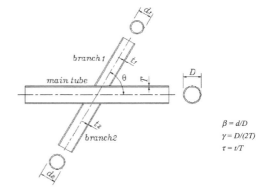

Figure 12. Definition of the parameters.

parameter, and the variety regulation of the joint rigidity changing with every parameter can be obtained. Two branch diameters of the model are supposed to be equal in this paper, the axial stress of the main tube is not taken into account. For the limit of the paper length, only the influence of the parameters β, γ, τ, and θ on out-of-plane bending rigidity is analyzed in this paper.

3.2 The finite element model

The finite element program ANSYS is used to analyze the tubular joints, 4-node elastic-plastic shell element SHELL181 is selected to establish the finite element model, quadrangular free mesh is adopted to create mesh. To decrease the calculation error, the mesh at the connection area of main tube and branch is densified. The typical steel material Q235 is used, and the influence of welds is not taken into account in the finite element analysis.

3.3 The influence of parameter β

The applicable range of the calculation formula about the bearing capacity of tubular joints is presented by the present "Steel Code", $0.2 \leq \beta \leq 1.0$, $d_i / t_i \leq 60$, $d / t \leq 100$, $\theta \geq 30°$, so the range of β is 0.2~1.0 in this paper. Other parameters of the models are $\gamma = 15.625$, $\tau = 0.5$, $\theta = 90°$ respectively.

The deformation characteristic of the joints is summarized as follows while branch bears the out-of-plane moment. When β is small, mainly local joint deformation occurs, the main tube wall connecting with the pressed side of branch is concaves inward, while that at the tensional side is convex outward, and the global deformation is more obvious with the increasing β, under the pressure of the pressed side of branch, the main tube wall becomes convex outward gradually, while sunken-in generated at the tensional side.

The internal force of the joint under out-of-plane moment also changes with β. When β is small, the joint suffers from load locally, plastic zone is mainly distributed in the connection of branch and main tube. With the increasing β, the joint suffers from load from locally to globally, besides the connection of branch and main tube, the convex and concave zone on the main tube wall also turns into plastic zone.

The curves of joint rotation $\varnothing$ changing with out-of-plane moment Mo under different β are shown as Figure 13.

From Figure 13, it can be seen that the joint rotation $\varnothing$ increases rapidly with the increasing out-of-plane moment Mo when Mo amounts to a certain value, which means that the out-of-plane bending rigidity decreases quickly and the obviously nonlinear behavior appears. Furthermore, the joint capability of bearing out-of-plane moment increases with the

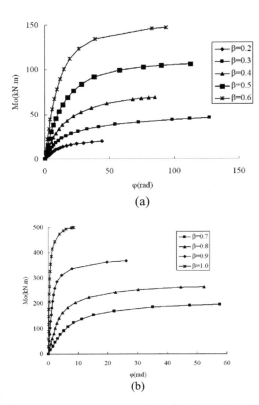

Figure 13. The curve of joint moment-rotation ($0.2 \leq \beta \leq 1.0$).

increasing β. It is concluded by analyzing the rigidity of the models that the out-of-plane bending rigidity is biggish when $\beta \geq 0.6$, the joint can be taken as rigid connection approximately in static analysis.

3.4 The influence of parameter γ

The ratio of main tube diameter to its thickness in tubular structures should not exceed 100 ($235/f_y$) according to the active "Steel Code", considering the possible range of real structures, the value of γ varies within 10~50 in this paper. Other parameters of the computational models are $\beta = 0.5$, $\tau = 0.5$, $\theta = 90°$.

The curves of joint rotation $\varnothing$ changing with out-of-plane moment Mo under different γ are shown as Figure 14.

Figure 14 shows that the joint capability of bearing out-of-plane moment decreases rapidly with the increasing γ when parameter γ changes within 10~50, and the out-of-plane bending rigidity increases rapidly with the increasing γ. When $\gamma < 20$, before most area of the joint comes into plasticity at the initial stage of out-of-plane moment effect, the out-of-plane bending rigidity Ko keeps constant basically, and the rigidity Ko decreased rapidly when local area of the joint comes into plasticity. While the joint comes into plasticity under the moment of about 10% branch bending intensity when $\gamma > 30$, the rigidity Ko decreases with the increasing out-of-plane moment. And when $\gamma \geq 25$, the joint capability of bearing out-of-plane moment is less than the branch bending intensity, under the effect of out-of-plane moment, the model joints breaks prior to the branch, the out-of-plane bending intensity of this kinds of joints should be paid enough attention in structural design.

3.5 *The influence of parameter τ*

The branch thickness is less than that of main tube normally, so the value of parameter τ is in 0.2~1.0. Other parameters of the model joints are $\beta = 0.6$, $\gamma = 12.5$, $\theta = 90°$.

The curves of joint rotation φ changing with out-of-plane moment Mo under different τ are shown as Figure 15.

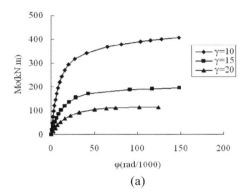

(a)

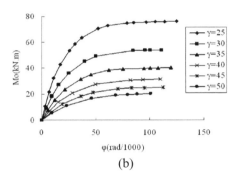

(b)

Figure 14. The curve of joint moment-rotation ($\gamma = 10~50$).

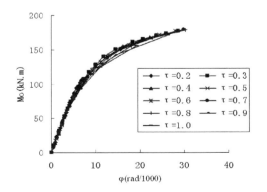

Figure 15 The Curve of joint moment-rotation ($\tau = 0.2~1.0$).

It is seen from Figure 15 that for $\tau = 0.2~1.0$, the curves of moment-rotation coincide basically, changing the parameter τ while keeping other parameters constant, that is just changing the branch thickness, the influence on internal force of the tubular X-joint is less enough to be neglected. From the above analysis of parameter γ, it is concluded that the chord thickness T has great effect on out-of-plane bending rigidity of the joint, and the influence is embodied by parameter γ.

It is obtained from the analysis of joint rigidity that when $\tau \leq 0.5$ and before the joint moment amounts to the maximum design value of branch moment, the rigidity of joint is far more than that of branch with moderate length, the joint can be taken as rigid connection in static analysis for real structures.

3.6 *The influence of parameter θ*

According to the active "Steel Code", it is not advised that the angle between branch and main tube or between two branches is less than 30°, so the value of parameter θ is set within the range 30°~90°, other parameters of the model joints are $\beta = 0.6$, $\tau = 0.6$, $\gamma = 12.5$.

When θ increases gradually within 30°~90°, the deformation of the joint under out-of-plane moment increases rapidly, and the deformation zone is centralized so much more. With the increasing parameter θ, the moment borne directly by main tube wall increases while the effect zone is reduced relatively, this consequentially induces the maximum deformation of the joint to increase gradually. When θ approaches to 30°, the stress concentrated zone is much larger than that of $\theta = 90°$ because of the bigger connection zone of branch and main tube, and the long stress concentrated strap is obvious at the side of the connection joining branch with main tube.

The curves of joint rotation φ changing with out-of-plane moment Mo under different θ are shown as Figure 16.

Figure 16 shows that the joint capability of bearing out-of-plane moment decreases with the increasing θ, and its bearing capability is in the inverse ratio

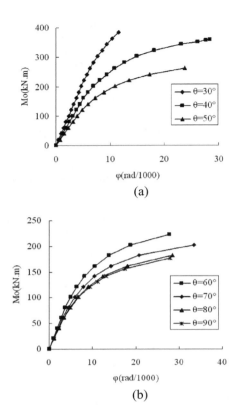

Figure 16. The curve of joint moment-rotation ($\theta = 30°{\sim}90°$).

of sin(θ) approximately. The out-of-plane bending rigidity of the joint decreases with the increasing parameter θ. The initial out-of-plane bending rigidity of the joint is biggish, when the out-of-plane bending moment suffered by branch is sub-moment, in the structure whose internal force is mainly axial force, the influence of out-of-plane bending rigidity of the joint can be ignored and the end of branch can be taken as rigid connection in static analysis. However, when $\theta \geq 70°$ and the joint moment amounts to the maximum design value of branch moment, the out-of-plane bending rigidity of joint is almost equal to that of branch with moderate length, the influence of joint rigidity can not be ignored in structure design.

4 CONCLUSIONS

1. The failure modes of all the specimens are similar. All the modes are elastic-plastic stability failure of the main tube wall connected with the small branch. It indicates that the rigidity and intensity of the joint are weak when the value of d/D is small under the same other parameters, the joint needs to be strengthened in structural design. When the diameters of branch and main tube are similar or same, the joint rigidity is biggish and satisfies the requirement of rigid connection approximately.
2. Parameters β, γ, θ have great influence on the rigidity of tubular X-joint, while the influence of parameter τ-or the branch thickness-on internal force of the joint is less enough to be neglected. The main tube thickness T has great influence on out-of-plane bending rigidity, which is embodied by parameter γ.
3. To obtain the critical value of the geometrical parameters between rigid and semi-rigid joint and the rigidity value of semi-rigid joint, more experimental results and numerical analysis data are needed.

REFERENCES

CHEN Yiyi, WANG Wei, etc. Experiments on bending rigidity and resistance of unstiffened tubular joints, *Journal of Building Structures*, Vol. 22, No. 6, Dec., 2001.

SHU Xingping, ZHU Shaoning, etc. Full-scale experiment research on CHS joints of steel roof of He Long, *Journal of Building Structure*, Vol. 25, No. 3, June, 2004.

National Standard of PRC. Code for design of steel structures (GB50017-2003), Beijing, 2003.

Tabuchi, M., Kanatani, H., Kamba, T., "The local strength of welded RHS T-joints subjected to bending moment," IIW Doc. XV-563-84, 1984.

Dexter, E.M., Lee, M.M. K., Kirkwood, M.G., Overlapped K joints in circular hollow section under axial loading, *Journal of Offshore, Mechanics and Arctic Engineering*, Vol.118, February, 1996.

Numerical investigations on the static behaviour of CHS X-joints made of high strength steels

O. Fleischer, S. Herion & R. Puthli
Karlsruhe University, Research Centre for Steel, Timber and Masonry, Karlsruhe, Germany

ABSTRACT: The use of high strength steels with yield strengths up to $f_y = 690$ N/mm² (S690) for welded hollow section joints is hampered due to the lack of adequate provisions in design standards. The design recommendations of EN1993-1-8 (2005) are based on experimental and numerical investigations using steel grades S235 or S355 and the detailed investigations are under way to verify how these recommendations affect higher strength steels. Based upon the few detailed experimental investigations available dealing with the load carrying behaviour of connections made of high strength steels (Noordhoek et al. 1998), EN1993-1-8 (2005) gives a reduction factor of $\alpha = 0.9$ to the design formulae for connections made of steels with yield strengths $355 < f_y \leq 460$ N/mm². In EN1993-1-12 (2007), this reduction factor is further reduced to $\alpha = 0.8$ for steels with yield strengths $460 < f_y \leq 690$ N/mm². Both factors are independent of the geometry of the joints. This paper reports observations from the first numerical investigations.

1 INTRODUCTION

Based on previous investigations mainly carried out in the scope of different CIDECT research programmes, design formulae for X-joints have been developed which are now the basis of a number of design recommendations and standards. The design resistance $N_{i,Rd}$ in EN 1993-1-8 (2005) is also based on this research. However, for steels with yield strengths $355 < f_y \leq 460$ N/mm², a reduction factor of $\alpha = 0.9$ is specified in this design standard. For higher strength steels with $460 < f_y \leq 690$ N/mm² $\alpha = 0.8$ is specified in EN1993–1–12 (2007).

These reduction factors are based upon larger deformations than those considered here, if chord plastification is the governing failure mode and the lower ductility of high strength steels, which influences the punching shear and effective width strengths. (Liu et al. 1993).

In the framework of a CIDECT (Puthli et al. 2007) and a FOSTA research project (FOSTA Project P715 2008), numerical and experimental investigations are being carried out, to provide an insight into the differences in the load carrying behaviour of CHS X-joints using steel grade S355 and high strength steel grades S460 and S690.

In this paper, details of the numerical model validated with experimental results (Noordhoek et al. 1998) are presented.

Based on the numerical model, results of parameter studies considering CHS X-joints under compression and tensile load using a deformation crite-rion (Lu et al. 1994) are given and evaluated with reference to the design equation in EN1993-1-8 for chord plastification. The results of this numerical work will give more evidence regarding the reduc-tion factors by varying the geometrical and material parameters of the joints.

2 NUMERICAL INVESTIGATIONS

2.1 *General*

For the numerical investigations, the finite element package ABAQUS 6.5 had been used.

To be able to cover a wide parameter range in the finite element analyses, a large number of models have to be analysed. Since it is impracticable to model every joint manually, a script based on the embedded Python interface (program code) had been developed, which enables an automatic numerical model generation by defining the load situation, the geometrical and the material properties and the mesh refinement.

2.2 *Element type*

For X joints high strain gradients can be expected at the connection. Since solid elements offer only 3 integration points over the thickness, a high mesh refinement through the thickness of the hollow section is required to obtain sufficient accuracy for non-linear numerical analyses. This results in lengthy calculation times (CPU-times) and large disc space. Therefore shell elements are chosen for modeling the connections.

Based upon experience from previous work, thick shell elements (S8R) are selected. The mesh refinement is determined by convergence analyses and comparing with experimental test results (Noord-hoek et al. 1998).

For the complete model only quadratic 8-noded elements are used. Simpson's integration method is used, with 9 integration points over the thickness.

2.3 Model discretisation

The joints are modeled using the nominal dimensions of the sections. The elements are arranged at the mid-surfaces of the sections.

Since only symmetrical X-joints with brace inclinations of $\Theta_I = 90°$ are investigated, symmetry conditions are used and only 1/8 of the joint is considered for the numerical investigations.

To avoid the influence of the supports on the joint behaviour, a brace length of 5 times the brace diameter d_i (on the shorter length) and a distance of 5 times the chord diameter d_0 at both ends of the chord was chosen (Figure 1).

The degrees of freedom at the support are coupled in order to give uniform displacements and rotations.

For the analysis, axial compressive and tensile loads N_i are prescribed at the reference point of the brace. The force is applied in increments. The modified Newton-Raphson method with Riks arc length technique is used in ABAQUS for the iterative-incremental solution procedure.

2.4 Welds

In this paper, only wall thicknesses of the braces with $t_i > 8$ mm are investigated and therefore only butt welds are considered (Figure 2).

Although fillet welds are considered for joints with brace wall thicknesses $t_i \leq 8$ mm in the research study (Puthli et al. 2007) they are not presented in this paper.

2.5 Material behaviour

A nonlinear material behaviour (modulus of elasticity $E = 2.1 \times 10^6$ N/mm², Poisson's ratio $\mu = 0.3$) with isotropic hardening is used for the model. The uniform strains are considered to be $A_g = 10\%$ for all steel grades considered.

The yield f_y and ultimate strengths f_u are nominal values taken from standards.

Because ABAQUS needs true stress σ_{true} and true plastic strain e_{true} material data (Figure 3) the engineering stresses σ and strains ε (Table 1) are converted by using Cauchy's law.

True stress: $\qquad \sigma_{true} = \sigma (1 + \varepsilon)$ (1)

True plastic strain: $\qquad \varepsilon_{true} = \ln(1 + \varepsilon) - \sigma/E$ (2)

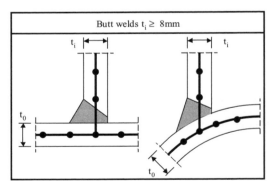

Figure 2. Modeling of butt welds.

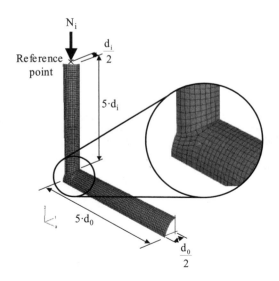

Figure 1. Typical mesh of a joint.

Table 1. Nominal material data used for the finite element investigations.

Material	f_y [N/mm²]	f_u [N/mm²]	A [%]
S355 *	355	470	22
S460 *	460	540	17
S690 **	690	770	14

* EN10210-1 (2006) and EN10219-1 (2006).
** EN10025-6 (2005).

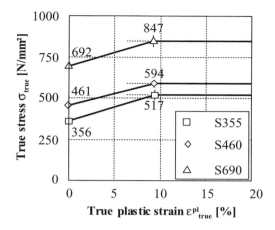

Figure 3. True stress-strain relationship for the numerical investigations of steel grades S355, S460 and S690.

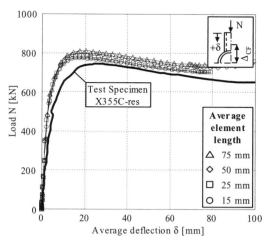

Figure 5. Convergence analyses for thick shell elements S8R by comparison of load deflections.

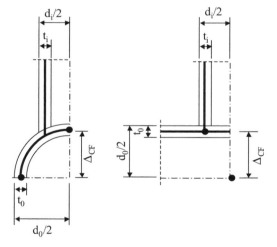

Figure 4. Determination of the chord face indentation Δ_{CF}.

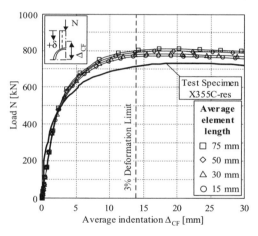

Figure 6. Convergence analyses for thick shell elements S8R by comparison of loads for chord indentations Δ_{CF}.

For the chords, braces and the welds (butt welds) the same material properties have been assumed.

2.6 Determination of ultimate loads

For the determination of ultimate loads $N_{i.u}$, the deformation limit from previous work (Lu et al. 1994) is taken, which limits the load at indentations of 3% of the chords diameters $\Delta_{CF} \leq 3\% \cdot d_0$ (for Δ_{CF} see Figure 4). If the load is limited by the deformation criterion, chord failure is assumed to be the governing failure mode of the connection.

Since cracking cannot be simulated with the numerical analyses, the evaluation of the numerical results is only based on ultimate loads $N_{i.u}$ (def. limit). For these loads, tests have shown that the connections have enough deformation capacity and that an indentation of $\Delta_{CF} = 3\% \cdot b_0$ can develop even for high strength steels (Noordhoek et al. 1998).

3 VALIDATION OF THE NUMERICAL MODEL

3.1 General

Since the experimental investigations have not been completed, experimental results of tests carried out at Delft University (Noordhoek et al. 1998) are used for the numerical validation.

Table 2. Measured dimensions yield and ultimate strengths of specimen S355C-res (Noordhoek et al. 1998).

Specimen	d_0 [mm]	t_0 [mm]	d_i [mm]	t_i [mm]	f_{y0} [N/mm²]	f_{yi} [N/mm²]
S355C-res	460,0	10,5	370,0	9,4	382	563

Table 3. Results of convergence analyses.

Average element length [mm]	No. of elements	$N_{i,u}$ [kN]	$N_{i,max}$ [kN]	$\dfrac{N_{i,u}}{N_{i,u,Test}}$	$\dfrac{N_{i,max}}{N_{i,max,Test}}$
75	161	801	816	1.14	1.09
50	304	790	798	1.12	1.07
30	871	778	784	1.10	1.05
15	3590	767	772	1.09	1.03

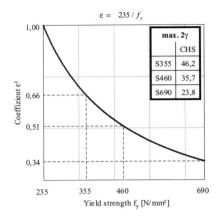

Figure 7. Maximum diameter to wall thickness ratio d/t in relation to the yield strength f_y.

For the validation, the measured dimensions of the sections and material properties (Table 2) of the CHS X joint with a brace inclination of $\Theta_i = 90°$ under compression load (specimen S355C-res, Noordhoek et al. 1998) has been used. The lengths of chord $l_0 = 2700$ mm and braces $l_i = 1230$ mm in the numerical model were the same as in the experiments.

This specimen has the lowest geometrical imperfections of the 8 specimens in the test series and is therefore considered most suitable for the present validation. Imperfections have a high influence on the joint deformations and thus on the overall load displacement curves as well as on the relative deformation Δ_{CF} of the connection, which are both used for the validation. Also, modeling imperfections is a very time consuming and cumbersome process that is not considered necessary for the present work.

In the convergence analyses, only negligible difference of the ultimate $N_{i,u}$ (def. limit, Figure 6) and the maximum loads $N_{i,max}$ (Figure 5) for average element lengths smaller than 30 mm is observed (Table 3).

Minor differences of the numerical determined ultimate $N_{i,u}$ (def. limit) and maximum loads $N_{i,max}$ to the experimental loads of the test specimen X355C-res ($N_{i,u,Test}$ = 705 kN and $N_{i,max,Test}$ = 747 kN; Noordhoek et al. 1998) are caused by the misalignment of the braces and the out of plane brace angles, which deviate marginally from 180° as reported by Noordhoek et al. (1998).

For the following parameter studies, an average element length of 30 mm is therefore used as mesh refinement.

4 PARAMETER STUDIES

4.1 Parameter range

To investigate the load carrying behaviour of joints with different steel grades in detail, extensive FE parameter studies are carried out for X-joints subjected to compression and tension loading of the braces. The parameter study is based upon nominal dimensions and material properties (Table 1).

In the parameter studies, the slenderness of the chord and brace sections are limited according to the regulations given in EN1993-1-8:

Chord: $\quad 10 \leq 2\gamma = d_0/t_0 \leq 40$

Braces: $\quad 10 \leq 2\gamma = d_0/t_0 \leq 50$

The limitation of the sections to cross section class 2 ($d/t \leq 70 \cdot \varepsilon^2$), for which the yield strength f_y has to be considered in the coefficient ε (Figure 7) is not taken into account.

The research project only considers $0.6 \leq \beta = d_i/d_0 \leq 0.8$, because this range was taken to be the most important for both, chord plastification and punching shear. The numerical investigations of CHS X-joints with brace angles of $\Theta_i = 90°$ considered here therefore only cover width ratios $0.6 \leq \beta = d_i/d_0 \leq 0.8$ (with increment size $\Delta\beta = 0.05$), chord slenderness $15 \leq 2\gamma = d_0/t_0 \leq 40$ (increment size $\Delta 2\gamma = 5$) and wall thickness ratios $0.4 \leq \tau = t_i/t_0 \leq 1.0$ (increment size $\Delta\tau = 0.2$). All parameters given in Table 4 are based on a basic chord diameter $d_0 = 323.9$ mm. Each combination is calculated for the steel grades S355,

Table 4. Investigated parameter range for axially loaded CHS X-joints (compression and tensile load)—332 FE-Models.

2γ	τ	Width ratio β = d_i/d_0														
		0.60			0.65			0.70			0.75			0.80		
		①	②	③	①	②	③	①	②	③	①	②	③	①	②	③
15	1.0	1)	1)	1)	1)	1)	1)	✓	✓	✓	✓	✓	✓	✓	✓	✓
	0.8	1)	1)	1)	1)	1)	1)	✓	✓	✓	✓	✓	✓	✓	✓	✓
	0.6	1)	1)	1)	1)	1)	1)	✓	✓	✓	✓	✓	✓	✓	✓	✓
	0.4	1)	1)	1)	1)	1)	1)	✓	✓	✓	✓	✓	✓	✓	✓	✓
20	1.0	✓	✓	✓	✓	✓	✓	✓	✓	✓	✓	✓	✓	✓	✓	✓
	0.8	✓	✓	✓	✓	✓	✓	✓	✓	✓	✓	✓	✓	✓	✓	✓
	0.6	✓	✓	2)	✓	✓	2)	✓	✓	2)	✓	✓	2)	✓	✓	2)
	0.4	2)	2)	2)	2)	2)	2)	2)	2)	2)	2)	2)	2)	2)	2)	2)
25	1.0	✓	✓	✓	✓	✓	✓	✓	✓	✓	✓	✓	✓	✓	✓	✓
	0.8	✓	✓	✓	✓	✓	✓	✓	✓	✓	✓	✓	✓	✓	✓	✓
	0.6	2)	2)	2)	2)	2)	2)	2)	2)	2)	2)	2)	2)	2)	2)	2)
	0.4	2)	2)	2)	2)	2)	2)	2)	2)	2)	2)	2)	2)	2)	2)	2)
30	1.0	✓	✓	✓	✓	✓	✓	✓	✓	✓	✓	✓	✓	✓	✓	✓
	0.8	✓	✓	✓	✓	✓	✓	✓	✓	✓	✓	✓	✓	✓	✓	✓
	0.6	2)	2)	2)	2)	2)	2)	2)	2)	2)	2)	2)	2)	2)	2)	2)
	0.4	2)	2)	2)	2)	2)	2)	2)	2)	2)	2)	2)	2)	2)	2)	2)
35	1.0	✓	✓	✓	✓	✓	✓	✓	✓	✓	✓	✓	✓	✓	✓	✓
	0.8	2)	2)	2)	2)	2)	2)	2)	2)	2)	2)	2)	2)	2)	2)	2)
	0.6	2)	2)	2)	2)	2)	2)	2)	2)	2)	2)	2)	2)	2)	2)	2)
	0.4	2)	2)	2)	2)	2)	2)	2)	2)	2)	2)	2)	2)	2)	2)	2)
40	1.0	✓	✓	✓	✓	✓	✓	✓	✓	✓	✓	✓	✓	✓	✓	✓
	0.8	2)	2)	2)	2)	2)	2)	2)	2)	2)	2)	2)	2)	2)	2)	2)
	0.6	2)	2)	2)	2)	2)	2)	2)	2)	2)	2)	2)	2)	2)	2)	2)
	0.4	3)	3)	3)	3)	3)	3)	3)	3)	3)	3)	3)	3)	3)	3)	3)

① S355 acc. EN1993-1-9 (see Table 1).
② S460 acc. EN1993-1-8 (see Table 1).
③ S690 acc. EN1993-1-12 (see Table 1).
1) slenderness of braces $d_i/t_i < 10$.
2) wall thickness of braces $t_i < 8$mm, not included at present.
3) wall thickness of brace $t_i < 2$mm.

S460 and S690 for compression and tension loading of the brace (Table 4).

4.2 Evaluation of numerical results

Joints which achieve the joint resistance $N_{i,u}$ at smaller indentations $\Delta_{CF} < 3\% \cdot d_0$ are not used in the evaluations, since any possible cracking cannot be identified by the numerical investigations and failure modes such as punching shear, brace failure or chord plastification cannot be distinguished from each other.

For all connections of the FE parameter studies (Table 4), EN1993-1-8 gives chord plastification as the predominant failure mode.

$$N_{i,Rd} = \sigma \cdot \frac{k_p \cdot f_{y0} \cdot t_0^2}{\sin\Theta_i} \cdot \frac{5.2}{(1 - 0.81 \cdot \beta)} \quad (3)$$

where α = steel grade reduction factor; k_p = reduction factor to consider chord pre-stress, f_{y0} = yield strength of the chord; t_0 = wall thickness of the chord, Θ_i = brace inclination; $\beta = d_i/d_0$ = width ratio.

In addition to chord plastification and punching shear failure, brace failure had been checked by:

$$N_{i,Rd} = A_i \cdot \frac{f_{yi}}{1.1} = \pi \cdot t_i \cdot (d_i - t_i) \cdot \frac{f_{yi}}{1.1} \quad (4)$$

where A_i = cross section area; t_i, d_i, = wall thickness and diameter of the braces; f_{yi} = yield strength of braces. For braces in cross-section class 4 ($d_i/t_i > 90 \cdot \varepsilon^2$) the critical axial buckling stress $\sigma_{x,Rcr}$ (EN1993-1-6, 2007) has to be used.

To obtain the reduction factor α for the investigated yield strengths S460 and S690, the ratio ($N_{i,u}/N_{i,Rd}$)$_{S355}$ of the ultimate loads and the design resistances of chord plastification for joints made of steel grade S355 is normalized by the corresponding ratio ($N_{i,u}/N_{i,Rd}$) for the steel grades S460 and S690.

$$\frac{\left(\dfrac{N_{i,u}}{N_{i,Rd}}\right)_{S355}}{\left(\dfrac{N_{i,u}}{N_{i,Rd}}\right)} = \frac{(N_{i,u})_{S355}}{(N_{i,u})} \cdot \frac{\sigma}{1.0} \cdot \frac{f_{y0}}{355} = 1 \quad (5)$$

Assuming that the reduction factor α for steel grade S355 ($f_y = 355$ N/mm²) is equal to $\alpha = 1.0$ and that the connections should offer the same reliability for chord plastification, the reduction factors α for the steel grades S460 and S690 is as follows:

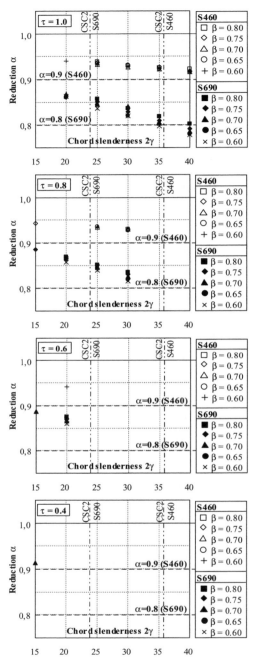

CSC: Limits of cross section class 2 ($d/t \leq 70 \cdot \varepsilon^2$) (S460 and S690)

Figure 8. Reduction factors for joints under compression load.

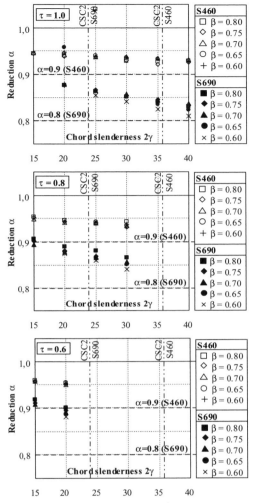

CSC : Limits of cross section class 2 ($d/t \leq 70 \cdot \varepsilon^2$) (S460 and S690)

Figure 9. Reduction factors for joints under tensile load.

$$\sigma = \frac{(N_{i,u})}{(N_{i,u})_{S355}} \cdot \frac{355}{f_{y0}} \quad (6)$$

The reduction factors α obtained on this basis are given in Figure 8 for joints under compression and in Figure 9 under tensile load.

5 RESULTS

Since only those ultimate loads that are based upon the deformation criterion of Lu et al. (1994) are considered in the evaluations, the reduction factors presented here are only related to chord plastification.

The results of the numerical analyses confirm the assumption of EN1993-1-8 (2005) and EN1993-1-12 (2007), that the reduction factor α of the design resistances $N_{i,Rd}$ for joints made of steel grade S690 are lower than for joints made of steel grade S460.

For the compression loaded CHS X joints investigated (Figure 8), it can be observed that the reduction factor α for joints made of steel grade S460 is always higher than the factor of 0.9 given in EN1993-1-8. For a chord slenderness $2\gamma = 15$, an average reduction factor of $\alpha = 0.94$ is obtained. With increase in chord slenderness, a small decrease of the reduction factor can be observed down to $\alpha = 0.92$ for a chord slenderness of $2\gamma = 40$. However, cross section class 2 for S460 is limited to $2\gamma \leq 35.7$ (Figure 7), where the smallest reduction factor of $\alpha = 0.93$ is obtained.

For joints under compression load using steel grade S690 (Figure 8), a reduction factor of $\alpha = 0.88$ is obtained for a chord slenderness of $2\gamma = 15$. With an increase of chord slenderness, a more distinct decrease of the reduction factor down to $\alpha = 0.78$ for chord slenderness $2\gamma = 40$ is obtained, which is even smaller than the reduction factor $\alpha = 0.8$ of EN1993-1-12 (2007). However, cross section class 2 for S690 is limited to $2\gamma \leq 23.8$ (Figure 7), where the smallest reduction factor of $\alpha = 0.85$ is obtained.

For the tension loaded CHS X-joints (Figure 9), the load carrying behaviour is similar to the compression loaded joints, but the reduction factors obtained for steel grades S460 and S690 are higher than for joints under compression loads and therefore ignored at present. Should a separate (more favourable) design equation be found necessary for tension loading, the improved reduction factors can be included at some later stage.

From the evaluation of the numerical analyses considering chord plastification (Figure 8 & Figure 9), the chord slenderness 2γ and yield strength f_y are identified as the main influencing parameters for the magnitude of the reduction factor α. The width ratio β and the thickness ratio τ can be observed to have only minor influence for the investigated steel grades and load cases.

6 CONCLUSIONS

The results presented in this paper form a limited preliminary investigation on the current research to establish reduction factors for chord plastification to be applied to the design resistance when high strength steels are used in hollow section connections. For chord plastification as governing failure mode, the influence of yield strength f_y (S355, S460 and S690), and the geometrical parameters on the reduction factors are investigated by finite element analyses.

At this stage, the investigations cover symmetrical, butt welded CHS X-joints with brace inclinations of $\Theta_i = 90°$ and wall thicknesses larger than 8 mm, under compression and tensile loading, for $0.6 \leq \beta = d_i/d_0 \leq 0.8$. Comparison is made between the behaviour of joints using steel grade S355 and identical joints using the higher grade steels, to obtain the reduction factors.

For the joints and the load conditions (compression and tension) considered, the reduction factors of EN1993-1-8 (2005) and EN1993-1-12 (2007) are found to be conservative in comparison to those obtained from numerical investigations using nominal geometrical and material values, if the geometric parameters of the connections are in the permitted range of ($d/t \leq 70 \cdot \varepsilon^2$), as given in EN1993-1-8 (2005).

For slender chord members with $d/t \geq 70 \cdot \varepsilon^2$ (cross section classes 3 and 4 which are not permitted at present due to the above limitation), decreasing reduction factors with increasing slenderness ratios are observed. The restriction of only permitting design of joints with member slenderness $d/t \geq 70 \cdot \varepsilon^2$ particularly restricts the parameter range for sections with higher yield strengths, so that including slender members in future standards would increase the use of higher strength steels in the future.

However, only joints with $\beta \leq 0.8$ are included in the numerical investigations at the present stage. Connections with a higher width ratio $\beta > 0.8$ offer a lower deformation capacity, so that punching shear or brace failure may be the predominant failure modes. Since these failures cannot be determined by numerical investigations but may influence the level of the reduction factors, the results of the experimental work have to be evaluated before final conclusions can be drawn.

7 FUTURE WORK

At present, only connections with a brace wall thickness of $t_i \geq 8$ mm are considered. The numerical work for smaller wall thicknesses ($t_i < 8$ mm) is already carried out but has not yet been included in the evaluation on CHS X-joints and will be included in the final evaluations.

For these CHS X-joints, in-plane and out-of-plane bending have also to be included in the investigations.

The effect of decreasing the uniform strain A_g, which might be lower for high strength steels, has to be considered on the load carrying behaviour and the reduction factor α. Decreasing uniform strains could result in punching shear or brace failure as the governing failure mode, since the stresses are not able to redistribute.

Statistical evaluation of test results for failure modes which cannot be determined by numerical investigations will be carried out at the Munich University of Applied Science to obtain the reduction factor α for punching shear and brace failure modes.

The influence of prestressed chords will be taken into account according to the work Liu et al. (2006).

Based on the results of the ongoing experimental and numerical investigations, the necessary reduction of the design resistances in relation of the yield strength will be determined and changes will be proposed where necessary.

ACKNOWLEDGEMENTS

This paper was written in commemoration Prof. Dr.-Ing. F. Mang, who initiated first research efforts in the scope of the use of high strength steels in hollow section constructions at Karlsruhe University (Mang, 1978).

The authors would like to thank the FOSTA research group P715 and CIDECT for financial support, V & M Tubes and Voest Alpine Krems for the supply of materials to carry out the investigations.

REFERENCES

EN 10025-6. 2005. Hot rolled product of structural steels– Part 6: Technical delivery conditions for flat products of high yield strength structural steel in the quenched and tempered condition: German version.

EN 10210-1. 2006. Hot finished structural hollow sections on non-alloy and fine grain steels – Part 1: Technical delivery conditions: German version.

EN 10210-2. 2006. Hot finished structural hollow sections on non-alloy and fine grain steels – Part 2: Tolerances dimensions and sectional properties: German version.

EN 10219-1. 2006. Cold formed structural hollow sections of non-alloy and fine grain steels – Part 1: Technical delivery conditions: German version.

EN 10219-2. 2006. Cold formed structural hollow sections of non-alloy and fine grain steels – Part 2: Tolerances, dimensions and sectional properties: German version.

EN 1993-1-1. 2005. Eurocode 3: Design of steel structures – Part 1.1: General rules and rules for buildings: German version.

EN 1993-1-12. 2007. Eurocode 3 –Design of steel structures – Part 1.12: Additional rules for the extension of EN 1993 up to steel grades S700: German version.

EN 1993-1-6. 2007. Eurocode 3: Design of steel structures – Part 1.6: Strength and stability of shell structures: German Version.

EN 1993-1-8. 2005. Eurocode 3: Design of steel structures – Part 1.8: Design of joints: German version.

FOSTA Project P715. 2008. Überprüfung und Korrektur der Abminderungsbeiwerte für Hohlprofilverbindungen aus Werkstoffen mit Streckgrenzen zwischen 355 N/mm^2 und 690 N/mm^2 (Control and correction of reduction factors for structural hollow sections with yield stresses between 355 N/mm^2 and 690 N/mm^2). Forschungsvereinigung Stahlanwendung e.V., *Final Report*: to be published in 2009. Düsseldorf, Germany.

Kurobane Y. 1981. New developments and practices in tubular joint design. *IIW Doc. XV-488-81* and *IIW Doc. XIII-1004-81*.

Liu, D.K. & Wardenier, J. 1993. Effect of the yield strength on the static strength of uniplanar K joints in RHS. *IIW Doc. XVE-04-293*.

Lu, L.H., de Winkel, G.D., Yu. Y., & Wardenier, J. 1994. Ultimate deformation limit for tubular joints, *Proc. 6th Intern. Symp. on Tubular Structures*. Melbourne, Australia.

Mang, F. 1978. Untersuchungen an Verbindungen von geschlossenen und offenen Profilen aus hochfesten Stählen. *AIF—Nr. 3347*: Karlsruhe University, Germany.

Noordhoek, C. & Verheul, A. 1998. Static strength of high strength steel tubular joints, *CIDECT Report No. 5 BD-9/98*. Delft University of Technology, Delft, Netherlands.

Packer, J.A., Wardenier, J., Kurobane, Y., Dutta, D. & Yeomans, N. 1995. Design Guide for circular hollow sections (CHS) joints under predominantly static loading. *CIDECT Construction with hollow steel sections*. TÜV Verlag: Köln.

Puthli, R.S., Fleischer, O. & Herion, S. 2007. Adaption end extension of the valid design formulae for joints made of high-strength steels up to S690 for cold-formed and hot-rolled sections, *CIDECT Report No. 5 BS-8/07*: Karlsruhe University, Germany.

Vegte, G. J. van der, Wardenier, J. & Makino, J. 2007. Effect of chord load on ultimate strength of CHS X joints. *International Journal of Offshore and Polar Engineering*, 2007, Vol. 17, No. 4:301–308.

Vegte, G. J. van der. 1995. *The static strength of uniplanar and multiplanar tubular T- and X-joints*, PhD thesis. Delft: Delft University Press.

Wardenier J., Vegte, G.J. van der & Liu, D.K. 2007. Chord Stress Functions for K Gap Joints of Rectangular Hollow Sections", *ISOPE 2007*. Lisbon, Portugal.

Local joint flexibility of tubular circular hollow section joints with complete overlap of braces

W.M. Gho
Marine Engineering Services Pte Ltd, Singapore
(Formerly, School of Civil and Environmental Engineering, Nanyang Technological University, Singapore)

ABSTRACT: This paper discusses the design concern of Local Joint Flexibility (LJF) of tubular circular hollow section joints with complete overlap of braces under basic loading. The results of the finite element analysis showed that the gap element is a very important geometrical parameter that significantly affects the LJF. Under lap brace loading, there exist through brace axial and bending load that resulting in axial and rotational deformation on the chord wall. These loadings are in general transferred from the lap brace to the chord via the shell bending, beam action or combined shell and beam behavior of the gap element. In view of these load-transfer mechanisms that affecting the joint LJF, use of existing Y-joint formulae to predict LJF of the joints with complete overlap of braces may not necessarily be appropriate.

1 INTRODUCTION

The analysis of tubular structures of circular hollow section (CHS) commonly considered members as one-dimensional beam elements with rigid joint connections. The member end forces at joints are assumed no effect on chord wall deformations. However, in reality, the chord wall at the joint locally deforms under brace loads. This localized deformation at joints showed to significantly affect the deflection, buckling and stress distribution of structures in both elastic and elastic-plastic load conditions. (Bouwkamp et al. 1980; Elnashai & Gho 1992).

One of the earliest recommendations for predicting local joint flexibility (LJF) due to brace in-plane (IPB) and out-of-plane bending (OPB) was stated in DnV (1977). UEG (1985) subsequently published a set of equations for computing LJF of single brace welded joints under basic loadings for design. In recent years, LJF of tubular joints had been recommended for reduction of secondary bending moments and for inelastic relaxation in ultimate strength analysis of structures (API RP2A 2000). The importance of LJF for fatigue life estimates at secondary connections as well as for pushover analyses with joint failures in system collapse mechanism was also emphasized.

There exist several techniques to determine LJF of welded tubular joints. Fessler et al. (1986a, 1986b) investigated LJF of single brace, cross and simple gapped K-joints based on small-scaled acrylic models. Chen et al. (1993) and Kohoutek & Hoshyari (1993) used semi-analytical and -empirical approach to quantify LJF of simple gapped K- and T/Y-joints respectively. Buitrago et al. (1993) proposed equations as well as methodology for predicting LJF of gapped and overlapped joints based on finite element method. All the abovementioned joint configurations are subject to basic loadings.

For CHS joints with complete overlap of braces that consist of two separate T/Y-joints, the existing equations for predicting LJF of simple joints such as T/Y-joints may not necessarily be appropriate (Figure 1). The reason is that the stress interaction of the two T/Y-joints has not been considered. Gao & Gho (2008) and Gho & Yang (2008) recently showed that the joints with complete overlap of braces at small gap size could have higher ultimate capacity and lower stress concentration than simple T/Y-joints. Despite research on the ultimate capacity and stress concentration has been examined, the LJF of the joint with complete overlap of braces is yet to be addressed.

The current paper presents limited numerical results of the LJF of CHS joints with complete overlap of braces under basic loadings based on finite element analysis. The main objective is to highlight the impact of gap sizes on the behavior as well as to address the applicability of existing formulae for predicting the joint LJF. The results of the study will be useful to assess the need for specific parametric equations or to incorporate adjusting factor considering the effect of gap sizes in existing equations to better predict the LJF of joints with complete overlap of braces.

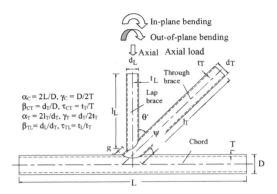

Figure 1. Geometrical parameters of CHS joint with complete overlap of braces.

2 DETERMINATION OF LOCAL JOINT FLEXIBILITY

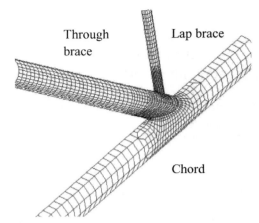

Figure 2. FE model of the joint with complete overlap of braces.

To determine LJF due to local chord wall deformation at the joint, the beam bending displacement of members must be eliminated from the total measured displacement. In the finite element (FE) model (Fig. 2), the local deformation at the joint could directly be measured in the direction normal to the chord (or through brace) axis from the chord (or through brace) ovality without considering the beam-bending movement. This measured the distortion of circular cross section into an oval shape under axial load, IPB and OPB. The deformation at the crown toe, saddle and crown heel of the chord and the through brace that subject to lap brace axial load, IPB and OPB is measured to determine the joint flexibility coefficient (f_{coeff}) through following formulation. In Figure 3, δ_1 and δ_3, δ_2 and δ_4, are the respective axial deformations at the crown and saddle measured perpendicular with respect to chord face.

Axial flexibility coefficient:

$$f_{coeff} = \frac{\delta}{P}\sin\theta\,(ED) \quad (1)$$

IPB flexibility coefficient:

$$f_{coeff} = \frac{\Phi_{IPB}}{M_{IPB}}(ED^3) \quad (2)$$

OPB flexibility coefficient:

$$f_{coeff} = \frac{\Phi_{OPB}}{M_{OPB}}(ED^3) \quad (3)$$

Where,
E Young's modulus (GPa)

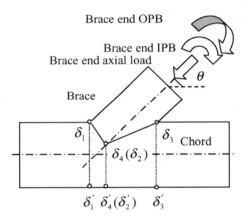

Figure 3. Points of measured deformation at joint.

D Chord diameter (mm)
d Brace diameter (mm)
t Brace wall thickness (mm)
θ Brace angle
P Brace end axial load (kN)
M_{IPB} Brace end in-plane Bending (kNm)
M_{OPB} Brace end out-of-plane Bending (kNm)

$$\delta = \frac{(\delta_1 - \delta_1') + (\delta_2 - \delta_2') + (\delta_3 - \delta_3') + (\delta_4 - \delta_4')}{4}$$

$$\Phi_{IPB} = \frac{\delta_3 - \delta_1}{(d-t)}(\sin\theta)$$

$$\Phi_{OPB} = \frac{\delta_4 - \delta_2}{(d-t)}(\sin\theta)$$

3 FINITE ELEMENT MODELING

A commercial FE package MARC (2005) was adopted to determine the LJF of CHS joints with complete overlap of braces for investigation. A 4-node thick shell element (Type 75) was used for modeling the mid face thickness of all members, which the effect of transverse shear deformation was considered. The weld size was not included as the difference of LJF for joints with and without the weld was negligible. In the analysis, in view of the symmetrical properties about the member longitudinal axis, a half model was considered reasonable for the joints subject to lap brace axial load and IPB (Fig. 2). However, a full model was used to simulate the joint subject to lap brace OPB. The mesh of the model had been optimized by varying the mesh density at the joint intersections and compared with the test results obtained by Fung et al. (2001).

In the joint model, the end condition of the chord and the through brace was assumed pinned. The inclination angles between the chord and the through brace as well as the through and the lap brace were 45^0. The load condition was unbalanced as the load was distributed according to the stiffness of members at the joint. The chord and the through brace length parameter was 6 times while the lap brace was 3 times their respective nominal diameter to avoid any short member effects.

3.1 Verification of FE model

The joint model had been verified against the experimental results of the ultimate capacity of completely overlapped K(N)-joints conducted by Fung et al. (2001). With the removal of chord, the LJF of Y-joint model that consists of through and lap brace subject to brace end axial load could also compare with that computed from the selected existing parametric equations derived by Fessler et al. (1986a), Chen et al. (1993) and Buitrago et al. (1993) with results as shown in Figure 4.

It was noted in Figure 4 that f_{coeff} increased with decreasing β_{TL}. The Chen et al. equation gave an incorrectly high f_{coeff} at $\beta_{TL} = 0.331$ as it was outside the range of validity. The LJF of the Y-joint model agreed reasonably well with that of Fessler et al. equation, with maximum difference of 17.5%. Buitrago et al. equation gave higher f_{coeff} with maximum difference about 19%. However, the maximum difference of f_{coeff} between Fessler and Buitrago et al. equation was 21.4%.

The joint model under lap brace IPB was compared with the existing parametric equations based on two separate Y-joints of the chord and through brace and the through brace and lap brace. In both the Y-joint models, the length of the chord and the through brace was adjusted to 6 times their respective nominal diameter. The chord IPB f_{coeff} in Figure 5 showed that the Y-joint model compares very well with Chen et al. equation, with maximum difference of 13.4%. On the other hand, the comparison with Fessler and Buitrago et al. equations showed that the difference of chord IPB f_{coeff} were 26 and 23% respectively. The comparison of chord IPB f_{coeff} between Chen and Fessler et al. equation was 14.6%.

The through brace IPB f_{coeff} in Figure 6 showed that the Y-joint model gives highest values with maximum difference of 13, 23 and 30% in comparison to Chen, Buitrago and Fessler et al. equations respectively.

In view of the FE models created in idealised conditions, as well as the extent of differences of f_{coeff} among existing parametric equations, the comparison of f_{coeff} of the assumed Y-joint models of joint with complete overlap of braces with respect to the existing parametric equations is considered acceptable. However, the impact on the LJF due to the difference of results generated among the experimental, numerical and analytical approach is yet to be examined in greater details.

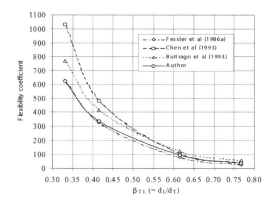

Figure 4. Verification of FE model (Axial flexibility).

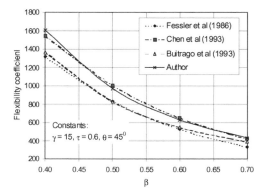

Figure 5. Verification of FE model (Chord IPB flexibility).

4 INFLUENCE OF GEOMETRIC PARAMETERS

4.1 Axial flexibility

In Figures 7 and 8, the joint LJF is measured on the through brace interfaces with the lap brace. The geometrical properties of these two members were constant with $\tau_{CT} = 0.75$. Figure 7 showed that the joint f_{coeff} increases with gap sizes. It was not significantly affected by β_{CT}. The joint f_{coeff} at $2d_T$ was about 5 times of that at $0.19d_T$.

Figure 8 shows the comparison of joint f_{coeff} based on two extreme chord wall thickness of $\tau_{CT} = 0.375$ and 1.00 with corresponding γ_C of 15 and 40 respectively. The difference of f_{coeff} for the two curves was small, about 25%, but the difference of f_{coeff} with respect to gap sizes was substantial. The joint f_{coeff} at $3d_T$ was 6.7 times of that at $0.19d_T$. The change of curvature of the two curves at small gap size could be attributed to the high bending stiffness of gap element due to restraining effect of chord wall.

The joint LJF due to the effect of through brace diameter in Figure 9 showed that the joint f_{coeff} increases with gap sizes. The large increment of f_{coeff} at small β_{TL} indicated the significance of γ_T effect. At $2.5d_T$, the joint f_{coeff} at small β_{TL} (=0.331) almost 13 times higher than that at large β_{TL} (=0.768). In all cases, the difference of joint f_{coeff} increased with decreasing β_{TL}.

A similar behaviour of joint f_{coeff} due to the effect of through brace wall thickness is observed in Figure 10. Two through brace wall thickness of $\tau_{TL} = 0.747$ and 0.28 (or $\tau_{CT} = 0.375$ and 1.00) were considered. The joint f_{coeff} decreased with increasing wall thickness and the reduction at large gap size was drastic, about 6.2 times at $2.5d_T$. The results revealed the less significant of LJF for joints with thicker wall through brace.

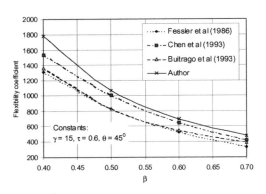

Figure 6. Verification of FE model (Through brace IPB flexibility).

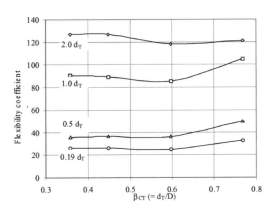

Figure 7. Effect of chord diameter (Axial flexibility).

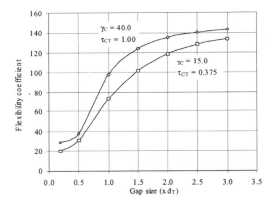

Figure 8. Effect of chord wall thickness (Axial flexibility).

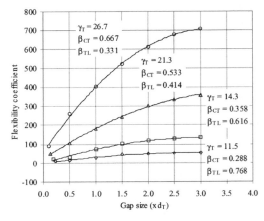

Figure 9. Effect of through brace diameter (Axial flexibility).

The load component at the joint was distributed according to the member stiffness. At small gap size, part of the lap brace axial load was directly transferred onto the chord wall via the axial and bending action of gap element. However, at large gap size, there exist short segment of through brace joining the chord to transfer the load as beam member. The through brace end forces resulted in axial and rotational deformations on the chord wall affected the through brace LJF interfaced with the lap brace.

The geometrical properties of the chord were found less impact on joint LJF compared to that of the through brace. Since the through brace length parameter was 6 times its nominal diameter, a gap size of $3d_T$ was approximately equivalent to the mid length of the through brace. Thus, if the existing Y-joint parametric equations were used, the LJF of the joints with complete overlap of braces at small gap size could be well overestimated.

4.2 In-plane bending flexibility

For the joint under lap brace IPB, the study was limited to examine the joint f_{coeff} on the chord and the through brace due to the effect of γ_c and β_{CT}.

The most noticeable behavior of the curves in Figure 11 is the change of curvatures with increasing gap sizes, particularly at large γ_c. The change of joint f_{coeff} at gap size of 0.3 to $0.5d_T$ was small, about 3.5%. The chord joint f_{coeff} decreased with increasing gap size and reached minimum at $1.0d_T$ and increasing thereafter. This abnormal behavior could be attributed to the load coupling effect of IPB at the crown toe and heel of the through brace that resulted in uneven deformations of the chord wall.

In the current study, the deformation was measured at the chord crown toe and heel. At small gap size, the deformation at these two locations was not identical. The lap brace IPB was mainly transferred through the shell bending of gap element joining the chord. Owing to the restraining effect of chord wall, this effect was found more profound. However, with the beam action of short segment at large gap size, the joint became flexible and the chord f_{coeff} increased again after the minimum.

Figure 12 shows the LJF of the through brace interfaces with the lap brace. The maximum difference of joint f_{coeff} at $0.5d_T$ was small, about 4.2%. However, γ_c showed less impact on the through brace LJF compared to that on the chord LJF.

The quadratic behavior of the curves in Figure 13 showed that the minimum f_{coeff} shifted with decreasing β_{TL} to larger gap size. This behavior was quite similar to that observed in Figure 11 except the curve curvature at small gap size. The reason could be due to the chord wall stiffness as medium τ_{TL} was considered in Figure 13. The study showed the importance of geometrical properties of the through and lap brace that controlled the magnitude of axial and bending load on the chord wall. The results also reflected that

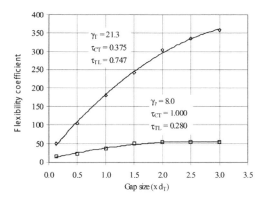

Figure 10. Effect of through brace wall thickness (Axial flexibility).

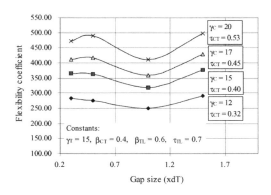

Figure 11. Effect of Chord radius-to-thickness ratio (Chord IPB flexibility).

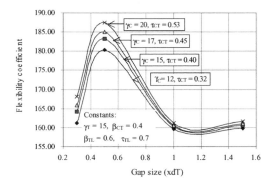

Figure 12. Effect of Chord radius-to-thickness ratio (Through brace IPB flexibility).

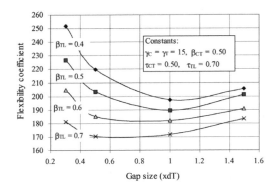

Figure 13. Effect of through-to-lap brace diameter ratio (Chord IPB flexibility).

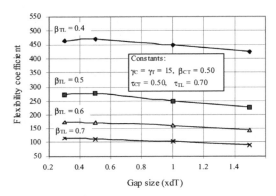

Figure 14. Effect of through-to-lap brace diameter ratio (Through brace IPB flexibility).

the minimum chord f_{coeff} was not necessarily occurred at small or large gap size.

However, the through brace f_{coeff} with increasing gap size is not significantly affected by β_{TL} (Figure 14). Unlike the chord subject to brace axial and rotational deformation, the through brace f_{coeff} was due directly to the rotational effect of lap brace IPB.

4.3 Out-of-plane bending flexibility

In Figures 15 to 18, the joint LJF is measured on the through brace interfaces with the lap brace. The joint LJF under lap brace OPB was expected higher f_{coeff} than that under lap brace axial load and IPB, as the joint was locally behaved less flexible, particularly on the chord wall, with members subjected to torsion. Owing to the load coupling effect of OPB at the saddles of the through brace joining the chord, there exist axial and bending loads resulted from the shell bending and membrane action of gap element on the chord wall that affecting the joint LJF. It can be seen in Figure 15 that the joint f_{coeff} decreases with increasing gap size and this effect is found more profound at small β_{CT}. The difference of joint f_{coeff} at gap size of 0.19 and $2d_T$ was 40%. A similar behaviour of f_{coeff} decreases with increasing gap size as in Figure 15 could also be observed in Figure 16.

However, the geometrical properties of the chord showed less impact on the joint LJF compared to

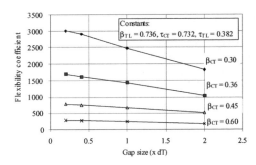

Figure 15. Effect of chord diameter (OPB flexibility).

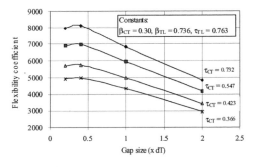

Figure 16. Effect of chord wall thickness (OPB flexibility).

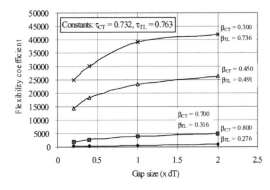

Figure 17. Effect of through brace diameter (OPB flexibility).

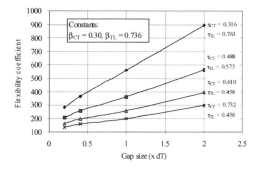

Figure 18. Effect of through brace wall thickness (OPB flexibility).

that of the though brace. In Figure 17, the joint f_{coeff} increases with β_{TL} but with decreasing β_{CT}. This behavior was not unexpected. As the through brace (or the lap brace) diameter was smaller, the bending and membrane action of gap element became more effective to yield higher joint LJF. On the other hand, the joint became more rigid with less LJF as the diameter of the two members about identical. This can be seen in Figure 17 that the difference of joint f_{coeff} at gap size of 0.19 and $2d_T$, with $\beta_{TL} = 0.738$ (or $\beta_{CT} = 0.30$), is about 40%.

In Figure 18, the joint f_{coeff} increases almost linearly with gap size. The maximum difference of joint f_{coeff} at small β_{CT} (= 0.30) and large β_{TL} (= 0.736) was about 68%. This large difference of joint f_{coeff} illustrated the significant impact of through brace wall thickness on joint LJF. It is interesting to note from these figures that the joint f_{coeff} decreases but increases with gap sizes, under varying chord and through brace geometrical parameters respectively. The results also showed that the joint became less flexible at smaller gap size.

5 DISCUSSIONS AND CONCLUDING REMARKS

The numerical results presented in this paper showed that the LJF of the joint with complete overlap of braces is significantly affected by through brace geometrical parameters. The gap size of the short segment of the through brace joining the chord is an important parameter that affecting the joint LJF. Generally, the joint LJF increases with the gap size, particularly at large γ_T and small β_{TL}. However, there are cases for the joint under lap brace bending that the joint LJF decreases with the specific gap sizes.

In the load transfer mechanism, there exists a portion of loads transmitted to the chord via the shell bending of gap element and the beam behavior of short segment at small and large gap size respectively. The magnitude of the load transfer is dependent upon the member stiffness at the joint. Under lap brace loading, the through brace end forces result in axial and rotational deformations of the chord wall. At smaller gap size, the high bending stiffness of gap element due to the restraining effect of chord wall causes the joint to become locally less flexible.

In view the existing Y-joint formulae are developed without considering the effect of flexibility at boundary, as well as the effect of load transfer through the short segment of through brace joining the chord not being considered, it can be concluded that the Y-joint formulae are not suitable for predicting the LJF of the joint with complete overlap of braces, despite the joint having two separate Y-joint configurations.

REFERENCES

American Petroleum Institute (API) (2000), *Recommended practice for planning, designing and constructing fixed offshore platforms*, API RP2 A, Washington: D.C, USA.

Bouwkamp, J.G., Hollings, J.P., Maison, B.F. & Row, D.G. (1980), Effects of Joint Flexibility on the Response of Offshore Towers, *Offshore Technology Conference*, OTC 3901, Houston, Texas: 455–464.

Buitrago, J, Healy, B.E. & Chang, T.Y (1993), Local Joint Flexibility of Tubular Joints, *Offshore Mechanics and Arctic Engineering*, OMAE, Glasgow, UK: 405–417.

Chen, B., Hu, Y. & Xu, H. (1993), Theoretical and Experimental Study on the Local Flexibility of Tubular Joints and its Effect on the Structural Analysis of Offshore Platforms, *Proceedings of the 5th International Symposium on Tubular Structures*, Nottingham, UK: 543–550.

Det Norske Veritas (DnV) (1977), *Rules for the Design, Construction and Inspection of Fixed Offshore Structures*, Norway.

Elnashai, A.S. & Gho, W. (1992), Effect of Joint Flexibility on Seismic Response Parameters of Steel Jackets, *Proceedings of the 2nd International Offshore and Polar Engineering, ISOPE*, San Francisco, USA: 475–480.

Fessler, H., Mockford, P.B. & Webster, J.J. (1986a), Parametric Equations for the Flexibility Matrices of Single Brace Tubular Joints in Offshore Structures, *Proceedings Institution of Civil Engineers*, Part 2 (81): 659–673.

Fessler, H., Mockford, P.B. & Webster, J.J. (1986b), Parametric Equations for the Flexibility Matrices of Multi-Brace Tubular Joints in Offshore Structures, *Proceedings Institution of Civil Engineers*, Part 2 (81): 675–696.

Fung, T.C., Soh, C.K., Gho, W.M & Qin, F. (2001), "Ultimate capacity of completely overlapped tubular joints I: An experimental investigation." *Journal of Construction Steel Research*, 57(8): 855–880.

Gao, F. & Gho, W.M. (2008), Parametric equations to predict SCF of axially loaded completely overlapped tubular CHS joints, *Journal of Structural Engineering*, ASCE (In press).

Gho, W.M. & Yang, Y. (2008), Parametric equation for static strength of tubular CHS joints with complete overlap of braces, *Journal of Structural Engineering*, ASCE (In press).

Hoshyari, I. & Kohoutek, R. (1993), Rotational and Axial Flexibility of Tubular T-Joints, *Proceedings of the 3rd International Offshore and Polar Engineering, ISOPE*, Singapore: 192–198.

MARC. (2005), *User's Guide*, MARC analysis and Research Corporation, California, USA.

Underwater Engineering Group (UEG) (1985), *Design guidance on tubular joints in steel offshore structures, Report UR33*, London, UK.

Experimental research on large-scale flange joints of steel transmission poles

H.Z. Deng & Y. Huang
Department of Building Engineering, Tongji University, Shanghai

X.H. Jin
Guang Dong Electric Power Design Institute, Guangzhou, China

ABSTRACT: Steel poles are more and more used in transmission line systems especially in urban areas in recent years. This paper is concerned with the experimental study on ultimate strength of flange joints of steel transmission poles for Baihuadong transmission lines being constructed in Guangdong province in China. The tubes are made of steel Q420B to China National Code. The biggest diameter of the tube is 2100 mm. To verify the reliability of the flange joint and study the rotational axis location of bolts, three specimens were tested under monotonic loading. Through the experiment, the ultimate capacity and failure mode of flange joints were obtained. By comparing with the results between experiment and finite element analysis, a new rotational axis for calculating bolt force is proposed, which can be used as a reference in engineering practice.

1 INTRODUCTION

With its advantages in construction convenience, less land space occupation and aesthetic merit, steel poles are widely used in transmission lines in urban and suburban areas (Fig. 1). Bolted flange joints subjected to axial forces and bending moments are one most common connection pattern in steel tubular construction. A typical bolted flange joint with stiffening ribs used in the steel poles is shown in Figure 2, which is mainly subjected to bending moments. The design of flange joints requires determination of the flange dimensions and bolt arrangements. Calculation of bolt forces is important for the design of both bolt and flange. Conventional design is usually carried out under the assumption that the bolts are subjected to axial tensile force and the rotational axis of bolt group is tangent to the tube surface. The allowable bolt axial force N^b_{tmax} should be (DL/T-5130, 2001 and DL/T5154, 2002) (Fig. 3)

$$N^b_{tmax} = \frac{M \cdot y_1}{\sum y_i^2} + \frac{N}{n} \leq N^b_t \quad (1)$$

where M = bending moment acting at the flange joint; N = axial force acting at the flange joint; n = total number of bolts in the flange joint; y_i = distance from the first bolt centerline to the rotational axis; y_1 = distance from the first bolt centerline to the rotational axis; N^b_t = design value of bolt tensile capacity.

For the large-scale flange joints in the Baihuadong transmission lines being constructed in Guangdong Province, China, the largest tube diameter is 2100 mm, the largest flange plate diameter is 2950 mm. As the reliability of Equation 1 for use in the design of bolts is uncertain, so is the safety and reliability of the flange joints. Although much research work has been

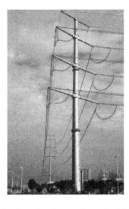

Figure 1. Typical steel poles in the suburbia.

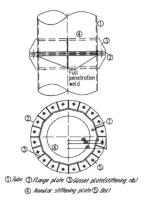

① Tube ② Flange plate ③ Gusset plate(stiffening rib)
④ Annular stiffening plate ⑤ Bolt

Figure 2. Typical circular flange joint in a tubular structure.

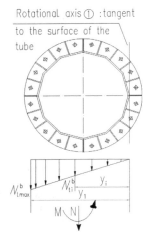

Figure 3. Model for calculating bolt force in a flange joint.

carried out for flange joints without stiffening ribs (Cao,1993 and Cao, 1996), there is little experimental study for flange joints with stiffening ribs.

This paper presents an experimental study on flange joints. Test results indicate that the conventional design for bolts is unsafe. By comparing with the results between experiment and finite element analysis, a new rotational axis for calculating bolt force is presented.

2 OUTLINE OF EXPERIMENTS

2.1 The principle of loading

As flange joints of steel transmission poles are mainly subjected to bending moment (M) and axial force (N), the moment was carried out by applying ec-centric force with hydraulic oil jacks (Fig. 4) in the test. M and N are calculated as

$$M = N_1 \cdot L_1 + N_2 \cdot L_2 \qquad (2)$$

$$N = N_1 - N_2 \qquad (3)$$

where N_1, N_2 = the load applied by the hydraulic oil jacks; L_1, L_2 = the arm of loading.

2.2 Properties of specimens

Three specimens (WFLH1-A, WFLH1-B and WFLH2) of bolted flange joints with a scale of 1:3.5 were

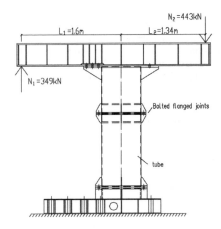

Figure 4. Loading sketch (Design load).

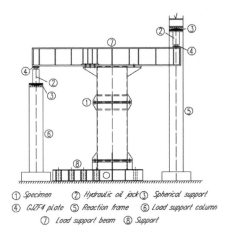

① Specimen ② Hydraulic oil jack ③ Spherical support
④ GJZF4 plate ⑤ Reaction frame ⑥ Load support column
⑦ Load support beam ⑧ Support

Figure 5. Test specimen and loading arrangement.

Figure 6. Testing set-up.

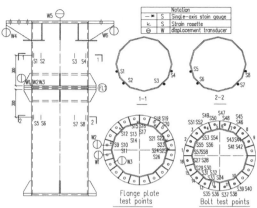

Figure 7. Test points arrangement.

Table 1. Details of specimens

D_t* mm	δ_t* mm	D_b* mm	D_f* mm	δ_f* mm	δ_g* mm	H_g* mm
660	10	33	840	22	10	150

* D_t = the tube diameter; δ_t = the tube thickness;
D_b = the bolt diameter; D_f = the flange plate diameter;
δ_f = the flange plate thickness;
δ_g = the gusset plate thickness;
H_g = the gusset plate height.

Figure 8. Test points

Table 2. Material properties of steel.

Material	Measured yield strength MPa	Measured tensile strength MPa	Measured elastic modulus MPa
Steel (Grade 420B)	433	584	212907
Steel (Grade 345B)	364	527	204183

Table 3. Material properties of bolt (M33 grade 8.8).

Material	Yield strength MPa	Tensile strength MPa	Elastic modulus MPa	Pre-toque N.m
Bolts	640	800	206000	445

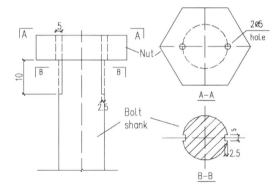

Figure 9. Bolts for strain gauge.

tested under monotonic loads. A sketch of a typical connection specimen is shown in Figures 5–6. In the loading process the specimens were free to slip at the top due to the GJZF4 Plate. The details of these specimens are shown in Table 1. Three specimens were identical except that WFLH1-A and WFLH1-B had an annular stiffening plate inside the tube in order to study stress concentration in the tube areas connected to gusset plates (stiffening ribs).

Full penetration welds are applied between the tubes and flange plates. The other welds, including the welds between the gusset plates and flange plates, the gusset plates and tubes, are fillet welds.

The tube is made of steel Q420B (nominal yielding stress f_y = 420MPa) to China National Code. Others

Figure 10. Gauged bolts.

are made of steel Q345B (nominal yielding stress f_y = 345 MPa), and high-strength bolts (Grade 8.8) were used. The material properties of the steel and bolts were obtained from tensile tests on coupons and from the bolts' certificate of quality, as shown in Table 2 and 3. The elastic modulus of the bolts is taken as 206 000 N/mm^2.

All the bolts were tightened by the calibrated wrench method; the pretorque value of the bolts was 445 N.m. The two flange plates were in contact with each other.

2.3 Test points

All the test points are shown in Figures 7–8. Displacement transducers (Nos.W1-W3) were positioned to measure the relative displacement of flange plates, Nos.W4-W6 were used to monitor the displacement at the loading points. Strain gauges (Nos.S1-S8) measure the strain of the tube. Strain rosettes (Nos. S9-S26) measure the strain of flange plates.

In order to obtain the bolt stress, two shallow slots were grooved symmetrically on the unthreaded portion of the bolt shank, and in each slot a strain gauge was fixed (Nos.S27-S58) and covered with resin for protection. Two little holes were pierced on the nut as outlets for data wire (Figs. 9–10). The bolt tension strains are calculated by averaging the values from the two strain gauges.

3 TEST RESULTS AND DISCUSSION

3.1 Outline of results

The test results of specimens are summarized in Table 4, and the ultimate capacity of flange joints is calculated by Equations 2–3. The failure mode of specimens is shown in Figure 11. As the applied load increased, the tube connected to top of the gusset plates was found to yield first, then the compressive zone and the tensile zone of the tube started to yield one after the other. The plastic area in the tube expanded as the load increased. Once local buckling happened in the compressive zone of the tube, the test was stopped. In the whole loading process, the flange plates, bolts and welds did not fail and showed no obvious deformation. It was also found that an annular plate inside the tube can effectively reduce stress concentration in the tube connected to the top of gusset plates (Fig. 12).

Table 4. Test results.

Specimen number	Ultimate capacity		Failure mode
	M (MPa)	N (kN)	
WFLH1-A	1727.1	−140.2	Buckling of tube
WFLH1-B	1846.5	−151.1	Buckling of tube
WFLH2	1844.5	−150.3	Buckling of tube

Figure 11. Failure mode of the specimens.

 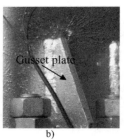

a) b)

Figure 12. Comparison of stress concentration in the tube connected to the top of gusset plate.
a) An annular stiffening plate inside the tube (WFLH1-B);
b) Without an annular stiffening plate (WFLH2)

3.2 Flange plates

During the loading process, the flange plates separated from the tensile zone of the tube. The finite element analysis and experimental results are similar to each other (Figs. 13–14). The relative displacement of two flange plates is presented in Figure 15, which indicates that deformation of the flange plate is very small. It is therefore concluded that flange plates have enough stiffness and there is no prying force in the tension region of the plates. The Von Mises stress (σ_s) of the flange plate is shown in Figures 16–17. It is noticed that the maximal σ_s appears on the tensile zone.

3.3 Bolts

The axial stress of a bolt can be calculated by averaging the values from the two strain gauges. The bolt tension stress generated by the applied moment (including pre-tension force) is shown in Figures 18–22. Figures 18–20 show stress variation of

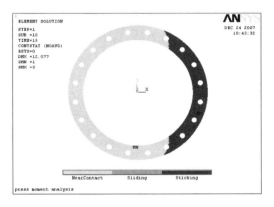

Figure 13. Contact state of flange plates.

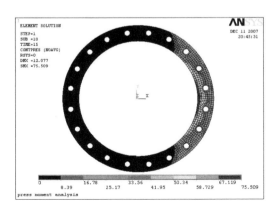

Figure 14. Contact stress of flange plates.

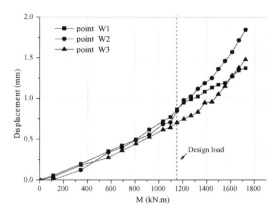

Figure 15. Two flange plates' relative displacement (WFLH1-A).

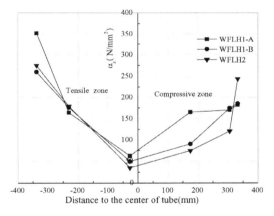

Figure 16. Flange plate stress (Design load).

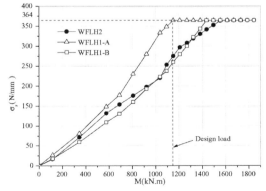

Figure 17. Flange plate stress (Test point S9, S10, S11).

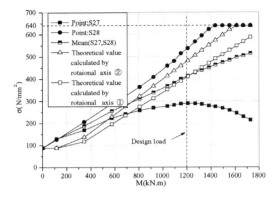

Figure 18. Bolt-16 stress (WFLH1-A).

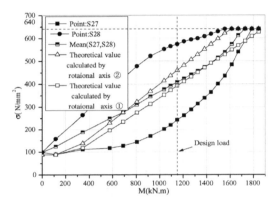

Figure 19. Bolt-16 stress (WFLH2).

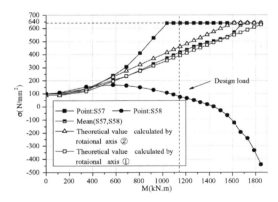

Figure 20. Bolt-1 stress (WFLH2).

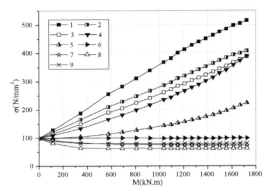

Figure 21. Bolt stress (mean) (WFLH1-A).

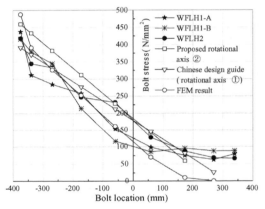

Figure 22. Comparison of bolt stress (mean).

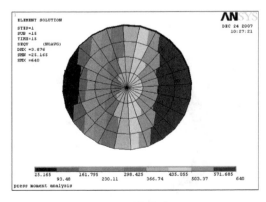

Figure 23. Stress diagraph of the first row bolt.

the first row bolts (bolt-1 and bolt-16) with the load increasing. It is obvious that the stress of every bolt is uneven, which suggests flexural behavior of every bolt during the loading process. This conclusion is also confirmed by FEM analysis (Fig. 23). Figures 21–22 show the mean stress of each bolt. As expected,

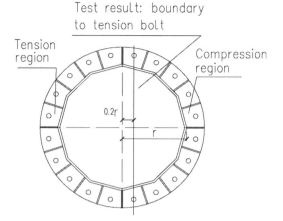

Figure 24. Location of boundary to tension bolts.

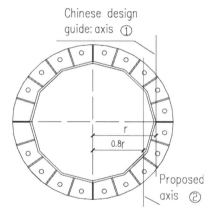

Figure 25. Location of rotational axis for calculation of bolt force.

Table 5. Comparison of the first-row bolt stress (N/mm²).

Specimen number	WFLH1-A		WFLH1-B		WFLH2	
Bolt number	1	16	1	16	1	16
(1): tests	438	402	417	387	418	408
(2): Axis ①		392				
(3)*	12	2.5	6.5	−1.2	6.7	4.1
FEM		488				
Axis ②		460				

*(3) denotes the comparison (%): {(1)−(2)}/(2)

the maximum mean tensile stress appears on the first row bolts. According to Figures 21–22 and the FEM an alysis (Figures 13–14), location of the boundary to tension bolts in the flange joint can be concluded as shown in Figure 24. It is found from Figure 22 that the maximum tensile stress of bolt in the test and FEM analysis exceeds the value calculated by eqn (1). A new rotational axis ② for calculating bolt force is therefore proposed, and the distance to the centerline of the tube is $0.8\,r$ (r = the radius of the tube) (Fig.25). Comparison of different scenario with the first-row bolt stress is shown in Figure 22 and Table 5.

4 CONCLUSION

Static loading experiments with three specimens have been conducted to analyze the ultimate strengthand failure mode of large-scale flange joint. It is found that the flange plates have enough strength and prying force does not exist in the flange plates connected to the tensile zone of the tube. The boundary to tension bolts in the flange joint is also obtained. A new rotational axis ② for calculating bolt force is recommended for large-scale flange joints, and maximal bolt force will be increased by 17.3% according to the axis ②. The joints designed by the new axis are more reliable in the application of projects.

REFERENCES

Cao, J.J. & Bell, A.J., (1993) Elastic analysis of a circular flat flange joint subjected to axial force, Int. J. Press. Ves. & Piping, 55(3) (1993): 435–449.

Cao, J.J. & Bell, A.J., (1996) Determination of bolt forces in a circular flange joint under tension force, Int. J. Press. Ves. & Piping, 68(3) (1996): 63–71.

DL/T5130 (2001).Technical regulation for design of steel tran-smission pole. China Electric Power Press, Beijing, China. (in Chinese)

DL/T5154 (2002). Technical Regulation of Design for Tower and Pole Structures of Overhead Transmission Line. China Electric Power Press, Beijing, China. (in Chinese)

Parallel session 10: Fire

Fire resistance design of externally protected concrete filled tubular columns

Y.C. Wang
University of Manchester, Manchester, UK

A.H. Orton
Corus Tubes, UK

ABSTRACT: This paper presents the development of a method to evaluate fire resistance of externally protected Concrete Filled Tubular (CFT) columns. A design program for CFT columns, based on the composite design method in EN 1994-1-1 but modified to take into account mechanical property changes of steel and concrete at elevated temperatures, has recently been developed to calculate fire resistance of CFT column at elevated temperatures. This paper will first present a summary of this method and validation of the design software, by comparison of prediction results against a large number of fire tests previously carried out by others. The main part of this study will assess the current Corus Tubes design method for CFT columns with external fire protection and propose modifications where appropriate. Finally, this paper presents a comparison of fire resistance between CFT columns and reinforced concrete columns to demonstrate a few anomalies that designers should be aware.

1 INTRODUCTION

Concrete filled tubular (CFT) steel columns have a number of advantages compared to bare steel or reinforced concrete columns and they are now being more widely used in the construction of building structures. One important advantage is their high fire resistance. It is often possible to develop fire design solutions using CFT columns in which external fire protection is not required. Where fire protection is necessary, a thin layer of intumescent coating is the best solution. Compared to reinforced concrete columns, the steel tube of a CFT column prevents the infill concrete from spalling in fire.

In the UK, a number of fire resistant design guides may be used for CFT columns, the three main ones being EN 1994-1-2 (CEN 2005b), Corus Tubes guide for CFT columns (Hicks & Newman 2002) and the CIDECT guide (Twilt et al. 1994).

In EN 1994-1-2, no distinction is made between unprotected and externally protected CFT columns. In the main text of EN 1994-1-2, only axially loaded columns are dealt with and the calculation method is the same as the ambient temperature calculation method in EN 1994-1-1 (CEN 2004), except that calculations of the cross-sectional properties (plastic resistance and flexural stiffness) of a CFT column should use the reduced mechanical strength and stiffness of steel and concrete at elevated temperatures.

In Annex H of EN 1994-1-2 (CEN 2004a), an alternative method is given. This method was originally developed by Guyaux and Janss (Guyaux and Janss 1970, Twilt 1988) for composite column design at ambient temperature.

To deal with bending moments, Annex H of EN 1994-1-2 proposes the following simple equation:

$$N_{equ} = \frac{N_{fi,Sd}}{\varphi_s \varphi_\delta} \quad (1)$$

where $N_{fi,Sd}$ is the design axial load of the column in fire. φ_s and φ_δ are modification factors depending on the amount of reinforcement, and column dimensions and eccentricity. N_{equ} is the axial load capacity of the column without eccentricity. In other words, the compressive resistance of the column with eccentricity is reduced from that without eccentricity by the combined modification factor $\varphi_s \varphi_\delta$. The values of φ_s and φ_δ may be obtained from graphs H.1 and H.2 in Annex H of EN 1994-1-2. As will be shown later in this paper, this method of dealing with eccentricity is grossly inaccurate.

The Annex H method for axial compression is antiquated and has not been adopted as the basis of composite column design at ambient temperature in EN 1994-1-1. Also the Annex H method for eccentricity is not accurate. Therefore, the Annex H method is not recommended for fire resistant design of CFT columns.

The CIDECT guide (Twist et al. 1994) for fire resistance of tubular columns refers to the computer program POTFIRE, which is based on the method in Annex H of EN 1994-1-2.

In the current version of the Corus Tubes design guide for CFT columns (Hicks & Newman 2002), a distinction is made between unprotected and externally protected CFT columns. For unprotected CFT columns, the calculation method broadly follows the ambient temperature design method for composite columns in EN 1994-1-1 (CEN 2004b), but uses a different bending moment—axial force interaction equation. For externally protected CFT columns, the approach is based on that developed by Edwards (1998, 2000). In this method, the same load ratio—limiting temperature relationship for steel columns in BS 5950 Part 8 (BSI 1990) is used to find the limiting temperature of the steel tube. The same limiting temperature of the CFT tube is then used to calculate the required fire protection thickness as for a bare steel tube, but by using a reduced section factor for the steel tube to take into account the heat sink effect of the concrete infill.

Based on the above brief review, it is clear that design methods for CFT columns are segmented. Therefore, despite extensive collective experiences of developing design methods for fire resistant design of CFT columns, there is still a need to find a better method, which should be consistent with the ambient temperature design method for composite columns and should be applicable to externally unprotected or protected columns under either axial load or combined axial load and bending moments. In fact, the basis of this method already exists and it is the method in EN 1994-1-1. Therefore, this paper will: (1) validate the method based on EN 1994-1-1 for fire resistant design; (2) assess the accuracy of the methods in EN 1994-1-2 Annex H and the current Corus Tubes design guide for CFT columns.

2 PROPOSED DESIGN METHOD

The method in EN 1994-1-1 is a general method for calculations of load carrying capacity of composite columns at ambient temperature. Under fire condition, the mechanical properties in the column cross-section will be different due to different materials used and different temperatures in different parts of the composite cross-section. The generic definition of a composite column is that it is a column made of materials with different mechanical properties. Therefore, the composite column at the fire limit state is merely another composite column that has a different set of mechanical properties. Hence, the general method for composite columns at ambient temperature should still be applicable to fire resistant design. In fact, the fire resistant design method in Annex H of EN 1994-1-2 was originally developed for composite column design at ambient temperature. However, since this method predates the currently adopted method in EN 1994-1-1 for composite columns, and the ambient temperature design method in EN 1994-1-1 has been thoroughly calibrated and also well-accepted, it is logical that the fire resistant calculation method should be aligned to the ambient temperature design method in EN 1994-1-1 for composite columns.

For axially loaded CFT columns, this is the situation in the main text of EN 1994-1-2 except for two modifications: (1) the column buckling curve "c" is used in EN 1994-1-2 to account for possible magnification of second order effects of initial imperfections in fire; (2) the flexural stiffness of the composite cross-section is modified by introducing a set of modification factors to different materials of the cross-section to take into consideration differential thermal strains.

For CFT columns with combined compression and bending moment, it is proposed to again adopt the ambient temperature design method for composite columns in EN 1994-1-2 and discard the method in Annex H of EN 1994-1-2.

3 VALIDATION OF THE PROPOSED METHOD

The proposed method has been developed into a design package (Firesoft). In addition, the Firesoft package also includes finite element modelling of heat transfer to calculate temperatures in the composite cross-section exposed to fire on all sides.

The validation for temperature calculation is carried out by comparison between Firesoft calculations and calculations using the general finite element package ABAQUS (HKS 2003). Firesoft and ABAQUS results are virtually identical for the following case: fire protection density 1200 kg/m^3, specific heat 1000 J/kg. K, thermal conductivity 0.1 W/(m.K) and fire protection thickness 1.909 mm.

For validation of strength calculations, extensive comparisons between Firesoft predictions and fire resistance tests have been carried out for the 48 fire tests on unprotected concrete filled tubular columns by Kordina & Klingsch (1983, CIDECT research project 15C) and Renaud (2004, CIDECT research project 15R) and by Lie and his co-workers (Lie & Chabot 1990 & 1992, Chabot & Lie 1992) at the National Research Council of Canada. These comparisons demonstrate the validity of the proposed method. Figure 1 shows a summary of comparison between all the aforementioned test

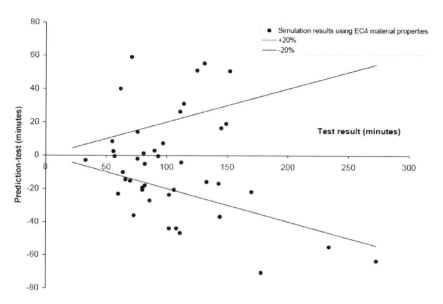

Figure 1. Comparison between firesoft predictions and test results of fire resistance times.

results and calculation results. Most of the results are within ±20% or on the safe side. Further more detailed information may be obtained from Wang & Orton (2008).

4 ASSESSMENT OF CURRENT DESIGN METHODS

The introductory section of this paper has already hinted at the inappropriateness of using the few main existing design methods. This section will use the results of Firesoft calculations to demonstrate that the existing design methods have severe limitations in their applications. This assessment will focus on the method in Annex H of EN 1994-1-2 for CFT columns with bending moments and on the current version of the Corus Tubes design guide for CFT columns with external fire protection.

4.1 Annex H of EN 1994-1-2 for CFT columns with combined compression and bending moment

To check the accuracy of equation (1), column strengths have been calculated for the following cases:

Sections: SHS 300 × 300×6.3 and SHS 300 × 300 × 16 (only for 8T20 bars with axial distance = 30 mm).

Axial load eccentricity: 0, ¼D (75 mm), ½ D(150 mm)

Reinforcement: 4T20 bars (bar axial distance = 30 mm), 8T20 bars (bar axial distance = 30 mm or 50 mm), 12T20 bars (bar axial distance = 30 mm)

Column effective length: 9 m and 3 m, giving effective length to column dimension ratio of 30 and 10 respectively.

For cross-section SHS 300 × 300 × 6.3, the reinforcement percentage is 0, 1.52%, 3.04% and 4.56% respectively. For cross-section SHS 300 × 300 × 16, the reinforcement percentage from 8T20 bars is 3.5%. According to Annex H of EN 1994-1-2), the values of the reduction coefficient $\varphi_s \varphi_\delta$ for the different combinations of column cross-section size, column length and reinforcement ratios are given in Table 1.

Table 2 presents the results of the Firesoft calculations. The results are presented as the ratio of the column axial strength with eccentricity to that without eccentricity to enable direct comparison with the two modification factors in equation 1. From these tables, it can be observed that:

1. The strength ratio is a variable at different fire ratings. In contrast, the value of $\varphi_s \varphi_\delta$ in EN 1994-1-2 does not change for different fire ratings.
2. The concrete cover to reinforcement has some influence on the strength ratio. A greater concrete cover to the reinforcement gives a higher strength ratio at higher fire ratings and a slightly lower ratio

Table 1. Values of ratio of column strength with eccentricity to column strength without eccentricity ($\varphi_s \varphi_\delta$) from Annex H of pr EN 1994-1-2.

Eccentricity	Column height 9 m for reinforcement ratio of					Column height 9 m for reinforcement ratio of				
	SHS300 × 300 × 6.3				SHS300 × 300 × 6.3	SHS300 × 300 × 6.3				SHS300 × 300 × 6.3
	0 (0%)	4T20 (10.5%)	8T20 (3.04%)	12T20 (4.56%)	8T20 (3.5%)	0 (0%)	4T20 (10.5%)	8T20 (3.04%)	12T20 (4.56%)	8T20 (3.5%)
¼D (75 mm)	0.228	0.445	0.517	0.539	0.526	0.203	0.396	0.461	0.481	0.469
½D (150 mm)	0.171	0.334	0.389	0.406	0.396	0.141	0.276	0.321	0.345	0.326

Table 2. Ratio of column axial strength with eccentricity to column axial strength without eccentricity, calculated using Firesoft.

L(m)	Rebar	Eccentricity	SHS300 × 300 × 6.3 at fire resistance rating of				SHS300 × 300 × 6.3 at fire resistance rating of			
			R30	R60	R90	R120	R30	R60	R90	R120
3	0	¼D (75 mm)	0.418	0.214	0.154	0.134				
		½D (150 mm)	0.242	0.089	0.07	0.066				
	4T20	¼D (75 mm)	0.499	0.339	0.190	0.182				
		½D (150 mm)	0.349	0.200	0.108	0.109				
	8T20	¼D (75 mm)	0.498	0.414	0.349	0.328	0.578	0.467	0.385	0.358
		½D (150 mm)	0.309	0.239	0.202	0.191	0.438	0.285	0.223	0.206
	8T20	¼D (75 mm)	0.479	0.439	0.407	0.390				
		½D (150 mm)	0.348	0.252	0.247	0.242				
	12T20	¼D (75 mm)	0.486	0.460	0.460	0.439				
		½D (150 mm)	0.308	0.285	0.289	0.270				
9	0	¼D (75 mm)	0.459	0.312	0.304	0.324				
		½D (150 mm)	0.329	0.202	0.198	0.314				
	4T20	¼D (75 mm)	0.520	0.456	0.393	0.404				
		½D (150 mm)	0.387	0.328	0.257	0.282				
	8T20	¼D (75 mm)	0.557	0.559	0.553	0.555	0.601	0.585	0.558	0.552
		½D (150 mm)	0.425	0.430	0.423	0.427	0.462	0.456	0.428	0.423
	8T20	¼D (75 mm)	0.550	0.571	0.602	0.617				
		½D (150 mm)	0.417	0.443	0.475	0.494				
	12T20	¼D (75 mm)	0.569	0.602	0.631	0.629				
		½D (150 mm)	0.437	0.474	0.508	0.506				

*Axis distance of rebar to inside of steel tube = 50 mm.

at lower fire ratings. Again, EN 1994-1-2 does not take this into account.

3. Comparing the values of strength ratio for the 9 m columns, it appears that the ratios according to EN 1994-1-2 are on the safe side. But for the 3 m columns, the ratios from BS EN 1994-1-2 can be either safe or unsafe.

To summarise, although the method in Annex H of EN 1994-1-2 is simple, it gives grossly inaccurate results and should not be used to deal with CFT columns under eccentricity.

4.2 Corus tubes design guide for CFT columns with external fire protection

For CFT columns with external fire protection, the Corus Tubes design guide is based on the work of Edwards (1998). This method makes the following two important assumptions:

1. The limiting temperature of steel of a CFT column is the same as the limiting temperature of a bare steel column, which may be calculated using the limiting temperature—load ratio relationship given in BS 5950 Part 8 (BSI 1990);

2. The steel temperature of a CFT column can be assessed on the basis of a bare steel tube with increased steel thickness to take into account heat sink effect of the infill concrete.

These two assumptions will be examined separately.

4.2.1 Steel limiting temperature of CFT columns

Using the Firesoft program, steel limiting temperatures of CFT columns have been calculated for different tubular sizes, fire protection thickness, level of bending moment and column effective length. The thermal properties of the insulation material were assumed to be: density 1200 kg/m^3, specific heat 1000 J/(kg.K) and thermal conductivity 0.06 W/(m.K).

Figures 2(a) and 2(b) compare the BS 5950 Part 8 limiting temperatures with Firesoft predictions for CFT columns with CHS 323.9 × 5 and CHS 323.9 × 16 sections. The calculations considered CFT columns with axial load only ($e = 0$, Figure 2(b)) and with combined axial load and eccentricity of the diameter of the tube ($e = D$, Figure 2(a)). For each column, the simulations were performed for two fire protection thicknesses to investigate the effects of different heating rates and for two column lengths (4 m and 10 m) to investigate the effect of column slenderness. For reference, the non-dimensional slenderness of the 10 m long columns was 1.38 and 1.22 respectively for the CHS 323.9 × 5 and CHS 323.9 × 16 columns. The steel grade was assumed to be S355 and concrete cylinder strength was 40 N/mm^2.

Figure 2(a) indicates that for CFT columns with the assumed very large eccentricities, the BS 5950 Part 8 limiting temperatures appear to be appropriate. This is not surprising because these columns are almost under pure bending wherein the steel tube contributes the most to the resistance of the column.

In contrast, for CFT columns under pure axial compression, Figure 2(b) shows two clear tendencies: (1) because of slenderness effects, the steel tube limiting temperatures are different, particularly for the higher load ratios. BS 5950 Part 8 does not take this into account and the BS 5950 Part 8 limiting temperature for long columns can be rather unsafe. (2) The steel tube limiting temperature also depends to some extent on the fire protection thickness. This is related to the different rates of increase in concrete temperatures. For the same CFT column but with two different fire protection thicknesses, if the steel tube temperature is the same, the CFT column with a thicker fire protection will be heated longer so the concrete temperatures will be higher than those of the CFT column with a thinner fire protection. Therefore, if the steel temperature is the same, the CFT column with a higher fire protection thickness will have a lower strength, or if the applied load is the same, then the steel tube limiting temperature will be lower. This is shown in Figure 2(b). Nevertheless, the effect of different fire protection thicknesses on steel tube limiting temperature is relatively small (compare the solid symbols for the 4 m long columns and the empty symbols for the 10 m long columns), with the maximum being 42 °C for a load ratio of 0.7.

Therefore, for realistic applications when the fire protection thickness and column length are likely to be within the ranges of these parameters used in the above study, it may be assumed that the limiting steel tube temperature is only related to the load ratio of the CFT column. This limiting temperature may be calculated using the Firesoft package. For short CFT columns (non-dimensional slenderness ≤0.8, corresponding to $\lambda \leq 70$ in BS 5950 Part 8 for short steel columns), the load ratio—limiting temperature relationship in BS 5950 Part 8 for steel columns may be used.

4.2.2 Steel temperature in protected CFT columns

Having obtained the steel tube limiting temperature, the current Corus Tubes guidance then treats a CFT column as an empty tube for the purpose of calculating the steel tube temperature. The only modification is to use an increased tube thickness to take into consideration the heat sink effect of the infill concrete. The increased equivalent steel tube thickness is:

$$t_{se} = t_s + t_{ce} \qquad (2)$$

with $t_{ce} = 0.15 b_i$ for $b_i < 12\sqrt{T}$

$t_{ce} = 1.8\sqrt{T}$ for $b_i \leq 12\sqrt{T}$

where:
t_s is the original steel tube thickness (mm).
t_{ce} is the increase in steel tube thickness for temperature calculation due to heat sink effect of the concrete core (mm).
b_i is the minimum dimension of the concrete core (mm).
T is the fire resistance time (minutes).

The above recommendation was based on a limited study by Edwards (1998, 2000). Its validity for general application is doubtful. Use CHS 323.9 × 5 as example. For T = 60 minutes, $t_{ce} = 1.8\sqrt{T}$, giving $t_{se} = t_s + t_{ce} = 18.94$ mm, which gives a section factor (defined as the ratio of the exposed surface area to the heated volume of steel or for a 2-D tubular cross-section, the exposed perimeter length divided by the cross-sectional area) = 56.07 m^{-1}. Figure 3 compares the empty steel tube temperatures calculated using the increased steel tube thickness of 18.94 mm and the well validated temperature calculation equation for protected steel in the fire design part of Eurocode 3

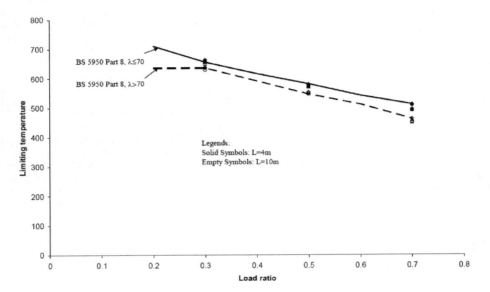

(a) $e = 0$

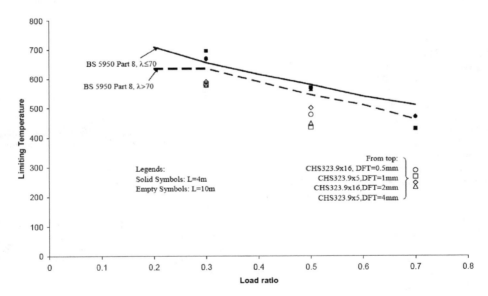

(b) $e = D$

Figure 2. Comparison between Firesoft calculations and BS 5950 Part 8 steel limiting temperatures, different fire protection thicknesses and two different tube sizes (CHS 323.9 × 5 & CHS 323.9 × 16).

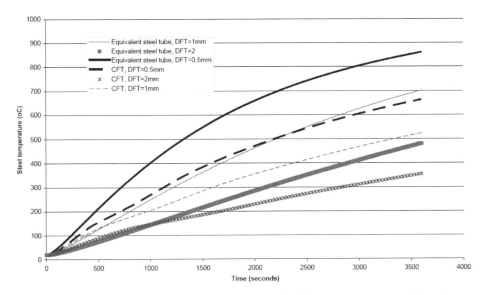

Figure 3. Comparison of steel temperatures between firesoft and EN 1993-1-2 calculations for different fire protection thicknesss.

for steel structures (EN 1993-1-2, CEN 2005a) and the CFT steel tube temperatures using Firesoft for three different nominal intumescent coating fire protection thicknesses (DFT = 0.5, 1, 2 mm). The solid curves represent the calculation results using the increased steel tube thickness and the dotted curves represent the Firesoft calculation results. Clearly, the differences between the two sets of results are too high.

From the above limited assessment of current practice, it is clear that if intumescent coating fire protection to CFT columns is needed but reliable thermal properties of the intumescent coating for CFT applications are not available, a new method of specifying the fire protection thickness is necessary. This should be carried out in two steps. First, the Firesoft package may be used to calculate the steel tube limiting temperature. Alternatively for short columns (non-dimensional slenderness ≤0.8), the load ratio—limiting temperature relationship in the British Standard BS 5950 Part 8 for fire resistant design of short columns ($\lambda \leq 70$) may be used to obtain the steel tube limiting temperature. Afterwards, the engineer should then consult with the intumescent coating manufacturer to obtain an appropriate fire protection thickness for the steel tube limiting temperature.

5 FURTHER COMMENTS

Although the main aim of this paper is to present a different fire resistant design method for CFT columns, it is worthwhile commenting on two issues that are relevant to fire resistance of CFT columns when engineers decide whether and how CFT columns may be used. These two issues are: (1) use of high strength concrete (HSC, cylinder strength > 60 N/mm^2) in CFT columns; (2) comparison of fire resistance between CFT columns and reinforced concrete (RC) columns.

5.1 Using HSC in CFT columns

One main advantage of using CFT columns is to be able to reduce the column dimensions. Therefore, it is understandable that engineers would consider using HSC filling to further reduce column dimensions. However, the use of HSC may or may not be beneficial to fire resistance of a CFT column. By definition, HSC has higher strength and stiffness than normal strength concrete (NSC) at ambient temperature, therefore HSC filled CFT columns have higher load carrying capacity than NSC filled CFT columns at ambient temperature. However, at elevated temperatures in fire, the strength and stiffness of HSC deteriorate much more rapidly than NSC. Therefore, if the load ratio (applied load for fire resistant design to load carrying capacity at ambient temperature) of a HSC filled CFT column and a NSC filled CFT column is the same, the HSC filled CFT columns may have lower fire resistance. In fact, according to Schaumann et al (2006), the deterioration in stiffness of HSC at high temperatures is such that for relatively slender CFT columns (e.g. 4 m high CHS 219.1 × 6), a HSC filled CFT column and a NSC filled CFT column would have nearly the same load carrying capacity in fire, indicating no advantage at all in using HSC filling.

Therefore, HSC filling is best considered when the CFT column is short. If HSC filling is used to reduce the dimensions of a CFT column (which by definition would mean that the CFT column would be fully loaded), it is important to understand that the fire resistance of the smaller CFT column with HSC filling may be lower than that of a larger CFT column with NSC filling.

5.2 Comparison between CFT and RC columns

If two columns, one using CFT and one using RC, have identical dimensions and identical loading and support conditions, one would expect that the fire resistance of the CFT column to not lower than that of the RC column. However, because of an invalid assumption in BS 8110 Part 2 (BSI 1985), the RC column may have considerably higher fire resistance (or load carrying capacity at the same fire rating) than the CFT column.

In BS 8110 Part 2, to achieve adequate fire resistance for an RC column, the RC column has been specified minimum dimensions and minimum cover to reinforcement. These minimum values are intended to limit the temperature rise in the reinforcement below 550 °C so that the reinforcement would retain at least 50% of its strength at ambient temperature. In these specifications, there is no consideration of deterioration in stiffness of the materials at high temperatures. Therefore, one can infer that the BS 8110 Part 2 rules should only be applied to short columns. However, since BS 8110 Part 2 does not have any restriction on column length, the rules on RC column dimensions and cover to reinforcement may be applied to any column. Assuming the engineer applies the rules of BS 8110 Part 2 within the unstated limit to short RC columns, since BS 8110 Part 2 does not define short columns, all he or she has to do is to follow BS 8110 Part 1 (BSI 1997) and limit the column length to minimum cross-section dimension (L/b) to 15. Whilst an RC column of $L/b = 15$ may be classified as a short column at ambient temperature, because of more rapid deterioration in concrete stiffness than its strength at high temperatures, the same RC column will become more slender (by a factor of about 2) for fire design and buckling of the column should be included in the design calculations. Whilst an engineer can not be prevented from using BS 8110 Part 2 beyond its unstated limit of applicability, it is recommended (for safety) to follow the more strict approach in EN 1992-1-2 (CEN 2004a). In EN 1992-1-2, the tabulated method (equivalent to the BS 8110 Part 2 specifications for minimum dimensions and cover to reinforcement) can only be used if the RC column fulfils definition of a short column ($L/l = 30$, corresponding to $L/b = 8.7$).

6 CONCLUSIONS

This paper has presented an alternative method for calculating fire resistance of concrete filled tubular columns and an assessment of the accuracy of the existing design methods in the three main design guides for CFT columns: EN 1994-1-2, CIDECT guide and the current Corus Tubes guide for CFT columns. The following conclusions may be drawn:

1. The alternative method is entirely based on the ambient temperature design method in EN 1994-1-1 for composite columns. This ambient temperature design method has been thoroughly calibrated for composite columns at ambient temperature. After modification of material properties to include the effects of elevated temperatures, this method is considered suitable for fire resistant design of CFT columns. Comparisons between predictions using this method and an extensive range of fire tests on CFT columns with and without fire protection, with and without eccentricity demonstrate validity of this method.

2. The proposed method has been implemented into a design package (Firesoft). The temperature calculation function of this software has also been verified by comparing temperature predictions of the software with predictions using an established finite element package ABAQUS.

3. The current method in Annex H of EN 1994-1-2 (which is also adopted in the CIDECT design guide) on CFT columns with eccentricity is found to be inaccurate. The method in the same appendix for CFT columns under axial compression is considered antiquated and is extremely time consuming to implement in practice. Therefore, this paper does not recommend using the method in Annex H of EN 1994-1-2.

4. For short CFT columns with external fire protection, the current Corus Tubes recommendation to use the same load ratio-steel limiting temperature as in BS 5950 Part 8 may be considered acceptable. However, for long columns, the steel limiting temperatures can be much lower. The thickness of fire protection may also have some effect on the steel limiting temperatures of CFT columns.

5. It is not recommended to use the method in the current Corus Tubes design guide to assess steel temperatures in CFT columns with external fire protection. This method has been derived based on a very narrow range of fire test results. To enable CFT columns with external fire protection to be

used, manufacturers of the fire protection materials should be requested to provide thermal properties of their fire protection materials.
6. It is not recommended to use HSC for fire resistance purpose, particularly if the column slenderness is high.

ACKNOWLEDGEMENT

The authors would like to like to acknowledge the contribution of Dr. J Ding in developing Firesoft.

REFERENCES

BSI (British Standards Institution) 1985. BS 8110: Structural use of concrete, Part 2: Code of practice for special circumstances, British Standards Institution.
BSI (British Standards Institution) 1990. BS 5950: Structural use of the steelwork in building, Part 8: Code of practice for fire resistant design, British Standards Institution.
BSI (British Standards Institution) 1997. BS 8110: Structural use of concrete, Part 1: Code of practice for design and construction, British Standards Institution.
CEN (European Committee for Standardisation) 2004a. EN 1992-1-2, Eurocode 2: Design of concrete structures, Part 1.2: General rules—Structural fire design, British Standards Institution.
CEN (European Committee for Standardisation) 2004b. EN 1994-1-1, Eurocode 4: Design of composite steel and concrete structures, Part 1.1: General rules and rules for buildings, British Standards Institution.
CEN (European Committee for Standardisation) 2005a. EN 1993-1-2, Eurocode 3: Design of Steel Structures, Part 1.2: General rules – Structural Fire Design, British Standards Institution.
CEN (European Committee for Standardisation) 2005b. EN 1994-1-2, Eurocode 4: Design of Composite Steel and Concrete Structures, Part 1.2: General rules – Structural Fire Design, British Standards Institution.
Chabot, M. and Lie, T.T. 1992. Experimental studies on the fire resistance of hollow steel columns filled with bar-reinforced concrete. IRC Internal Report No. 628. Ottawa: National Research Council of Canada.
Edwards, M. 1998. The performance in fire of concrete filled SHS columns protected by intumescent paint. Proceedings of the 8th International Symposium on Tubular Structures. Rotterdam: Balkema.
Edwards, M. 2000. The performance in fire of fully utilised concrete filled SHS columns with external fire protection. Proceedings of the 9th International Symposium on Tubular Structures. Rotterdam: Balkema.
Guyaux, P. & Janss, J. 1970. Comportement au flambement des tubes en acier remplies de beton. CRIF, MT 65.
Hicks, S.J. & Newman, G.M. 2002. Design guide for concrete filled columns. Corus Tubes.
HKS 2003. ABAQUS Standard User's Manual. Hibbit, Karlsson & Sorenson Inc.
Kordina, K. & Klingsch, W. 1983. Fire resistance of composite columns of concrete filled hollow sections. CIDECT Report No. 15C1/C2 – 83/27, Comite International pour la Developpement et l'Etude de la Construction Tubulaire (CIDECT).
Lie, T.T. & Chabot, M. 1990. A method to predict the fire resistance of circular concrete filled hollow steel columns. Journal of Fire Protection Engineering, 2(4): 111–126.
Lie, T.T. and Chabot, M. 1992. Experimental studies on the fire resistance of hollow steel columns filled with plain concrete, NRC-CNRC Report No. 611. Ottawa: National Research Council of Canada.
Renaud, C. 2004a. Improvement and Extension of the Simple Calculation Method for Fire Resistance of Unprotected Concrete Filled Hollow Columns, Research report, CIDECT 15Q, Comite International pour la Developpement et l'Etude de la Construction Tubulaire (CIDECT).
Renaud, C. (2004b). Unprotected Concrete Filled Columns Fire Tests—Verification of 15Q, Research report, CIDECT 15R, Final report, Comite International pour la Developement et l'Etude de la Construction Tubulaire (CIDECT).
Schaumann, P., Bahr, O. & Kodur, V.K.R. 2006. Numerical studies of HSC-filled steel columns exposed to fire. Packer, J.A. & Willibald, S. (ed.), Proceedings of Tubular Structures XI Conference. Taylor & Francis.
Twilt, L. 1988. Design charts for the fire resistance of concrete filled HSS columns under centric loading, TNO report BI-88-134, Delft, Holland.
Twilt, L., Hass, R., Klingsch, W., Edwards, M. & Dutta, D. 1994. CIDECT design guide for structural hollow section columns exposed to fire, Comite International pour la Developpement et l'Etude de la Construction Tubulaire (CIDECT).
Wang, Y.C. & Orton, A.H. 2008. Fire resistant design of concrete filled tubular columns. The Structural Engineer (in press).

Interaction diagrams for concrete-filled tubular sections under fire

J.B.M. Sousa Jr.
DECIV, Escola de Minas, Universidade Federal de Ouro Preto, Ouro Preto, MG, Brazil

R.B. Caldas & R.H. Fakury
DEES, Universidade Federal de Minas Gerais, Belo Horizonte, MG, Brazil

ABSTRACT: Strength interaction diagrams are useful tools for composite steel and concrete cross section analysis and design. They relate ultimate values of bending moments and axial force, allowing quick assessment of a cross section strength. Under fire action, the behavior of the cross section changes considerably, and these diagrams become dependent on the temperature field. The fire action introduces thermal strains and modifies the stress-strain relationships. The purpose of this paper is to present an algorithm for the construction of strength interaction diagrams for concrete-filled hollow sections subjected to fire. The diagrams are obtained by a stepwise variation of the deformed configuration, under assumptions of conventional ultimate strain values for concrete and steel. The procedure is illustrated by the construction of interaction diagrams for circular and square concrete-filled hollow sections.

1 INTRODUCTION

Concrete-filled hollow sections present an excellent load-carrying capacity under ambient temperature. The concrete helps prevent local buckling of the steel, and the confinement effect improves the structural behavior. Moreover, spalling is prevented by the steel tube. The behavior of such columns under fire action has received considerable attention in the last years, and a lot of numerical and experimental research for concrete-filled hollow sections and members have been carried out.

Three design approaches for structures under fire action are possible, according to Franssen & Dotreppe (2004): experimental tests, numerical modeling and simplified methods.

Experimental (furnace) tests provide useful information, but are expensive, time consuming, restricted to small specimen sizes and may not take into account member continuity and restraint.

Numerical modeling is the most powerful tool for predicting the behavior of structures under fire. Developments in numerical methods and computer hardware, associated to enhanced knowledge of the properties of concrete and steel at elevated temperatures, enabled the development of general-purpose as well as specialized programs capable of structural analysis under fire. However, these sophisticated programs are not always available to the practicing engineer and have therefore been employed mostly for research. Simplified methods, on the other hand, are readily available from design codes. These codes prescribe experimental, empirical based correlations and minimum dimensions (tabulated data), as well as simple and advanced calculation methods for determining the fire resistance of concrete structures.

The failure mechanism of composite columns is more difficult to predict than that of beams and slabs. Not only the crushing of concrete or yielding of steel and reinforcement may occur, by also column overall buckling.

Whichever method is employed for column design, knowledge on the behavior of the cross section under high temperature is highly desirable. Interaction bending moment-axial force curves and surfaces, for uniaxial and biaxial bending respectively, are useful tools for the design of cross sections and columns at ambient temperature, but few works have been developed for fire situations.

The purpose of this work is to present a procedure to obtain interaction diagrams of concrete-filled tubular sections subjected to fire action, similar to the ones commonly employed at ambient temperature design. Axial force-bending moment envelopes are obtained taking into account material property degradation due to temperature increase and the additional thermal strains. To illustrate the procedure, interaction diagrams for circular (CHS) as well as square (SHS) are presented and compared with the ambient temperature situation.

2 THERMAL ANALYSIS

The first step on the analysis of the cross section under fire is the evaluation of the temperature field. Several different numerical procedures may be employed with this objective, such as finite differences or finite elements.

In the present work the temperature fields were obtained using an explicit finite element scheme, with four-noded elements. The thermal analysis provides as a result the temperatures at every node of the mesh and also gives an approximate distribution for every point of the section, enabling the evaluation of the resistant forces and bending moments with a fiber approach.

3 MATERIAL PROPERTIES

For the concrete, Figure 1 depicts the stress-strain relationship accounting for temperature effects (Eurocode 2, 2004). For design purposes and development of the interaction diagrams, the ultimate compressive strain at temperature T is conservatively adopted as the value correspondent to the peak stress. It may be noted from Figure 1, however, that under higher temperatures larger strains may be reached on the descending branch. Tensile stresses in concrete are not taken into account.

For steel, a bilinear stress-strain relation with maximum absolute strain of 0.10 is adopted, considering the strength reduction at 0.2% for class N reinforcement, indicated by Eurocode 2 for use with simplified cross-section calculation methods. The expressions for the steel yield stress are:

Stiffness reduction factors and thermal strains for concrete and steel are once again taken from Eurocode 2.

4 INTERACTION DIAGRAMS

Under fire conditions, the presence of thermal strains and the material property degradation introduce considerable difficulties for the cross section limit state analysis. This work assumes the ultimate mechanical strain as the value which corresponds to the peak stress, which also depends on the temperature of the point, turning it impossible to know at which point of the section a limit state will be reached first. The mechanical (stress-inducing) strains are given by:

$$\varepsilon(x,y) = \varepsilon_0 + k_x y - k_y x - \varepsilon_{th} \quad (1)$$

or

$$\varepsilon(\xi,\eta) = \varepsilon_0 + k_0 \eta - \varepsilon_{th} \quad (2)$$

where ε_{th} is the thermal strain, function of the temperature, and the generalized strain variables may be expressed relative to systems xyz or $\xi\eta\zeta$, as shown in Figure 2.

A limit state is conventionally attained whenever $\varepsilon(x,y)$ in *any* fiber of the cross section or reinforcement reaches its temperature-dependent limit value. If both the thermal strain ε_{th} and the concrete compressive strain limit ε_{cu} depend on the temperature field $T(x,y)$, or $T(\xi,\eta)$, it is possible to evaluate these terms for each fiber (finite element) and obtain points of the surface $\varepsilon_{cu}(\xi,\eta)$, which approximately represents the bounds on the total compressive strain. If any point of the plane deformed section touches this limit surface, an ultimate state is characterized.

For a fixed neutral axis orientation angle α, it is possible to define an auxiliary variable that traces the evolution of the deformed configuration as the various

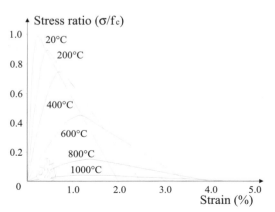

Figure 1. Stress strain relations for concrete.

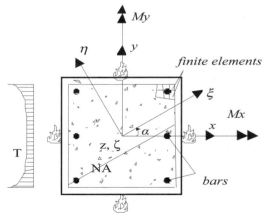

Figure 2. Cross section coordinate systems.

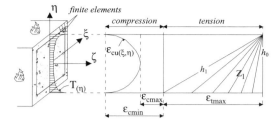

Figure 3. Zone Z_1, steel at limit state in tension.

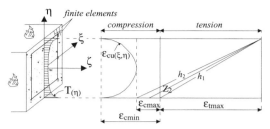

Figure 4. Zone Z_2, flexural limit state.

possibilities of limit state design are considered. This choice is not unique, and one of the possibilities is adopted here as a continuous and dimensionless variable, named β, whose numerical values are chosen to give a description as simple as possible, and which was introduced in a previous work by Caldas et al. (2007).

In a section subjected to a generic temperature field due to fire action, and considering compressive strains negative, let $\varepsilon_{c\max}$ and $\varepsilon_{c\min}$ be the maximum and minimum strictly admissible values for the total compressive strain, obtained from the $\varepsilon_{cu}(\xi,\eta)$ field (Fig. 3), and let $\varepsilon_{t\max}$ be the maximum admissible tensile strain, which in the present work was conventionally adopted as 1% or 0.01.

Starting from a limit state characterized by excessive plastic deformation on the steel tube, zone Z_1 is defined (Fig. 3), where the top fiber is $\varepsilon_{t\max}$ and the bottom fiber takes values from $\varepsilon_{t\max}$ to zero. Lines h_0 and h_1 enclose this region, in which concrete has no resistance, and setting $0 \leq \beta \leq 1$, one gets

$$\varepsilon^{\eta\max} = \varepsilon_{t\max} \quad \text{and} \quad \varepsilon^{\eta\min} = (1-\beta)\varepsilon_{t\max} \quad (3)$$

where $\varepsilon^{\eta\max}$ and $\varepsilon^{\eta\min}$ are the strains on the fibers with maximum and minimum η coordinate respectively.

The next zone is defined by keeping the positive strain at the maximum value and entering the compression zone for the concrete, characterizing a flexural state. Lines h_1 and h_2 enclose zone Z_2 (Fig. 4). Taking $1 < \beta \leq 8$, one gets

$$\varepsilon^{\eta\max} = \varepsilon_{t\max} \quad \text{and} \quad \varepsilon^{\eta\min} = (\beta-1)(\varepsilon_{c\max}/7) \quad (4)$$

The next zone considers deformed configurations with the concrete further into the compressive range, and with reduction on the maximum tensile strain of the top fiber. It starts on line h_3, where the top fiber has maximum admissible tensile strain and finishes on line h_4, with the top fiber just entering the compression range. The bottom fiber strain $\varepsilon^{\eta\min}$ lies between $\varepsilon_{c\max}$ and $\varepsilon_{c\min}$, and has to be evaluated by a trial process for each value of β. The solution is

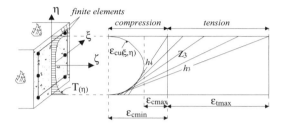

Figure 5. Zone Z_3 flexural limit state.

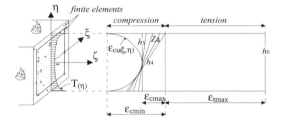

Figure 6. Zone Z_4, full compression limit state.

found when the deformed configuration touches the limit surface. Setting $8 < \beta \leq 18$, zone Z_3 (Fig. 5) is defined by:

$$\varepsilon^{\eta\max} = (18-\beta)(\varepsilon_{t\max}/10), \quad \varepsilon^{\eta\min} = f\left(\varepsilon^{\eta\max}(\beta)\right) \quad (5)$$

Proceeding further, zone Z_4 (Fig. 6) lies between lines h_4 and h_5. The cross section is fully compressed, and h_5 corresponds to a uniform compressive strain distribution (but not to pure compression, unless in the case of symmetric geometries and temperature distributions). This zone may be represented, for $18 < \beta \leq 26$, by

$$\varepsilon^{\eta\max}(\beta-18)(\varepsilon_{c\max}/8), \quad \varepsilon^{\eta\min} = f(\varepsilon^{\eta\max}(\beta)) \quad (6)$$

where once again $\varepsilon^{\eta\min}$ should be found iteratively for a fixed β.

For β > 26, there is a completely analogous set of zones, but now with compression mostly on the upper part of the section. The equations for these new zones are given by

$$\varepsilon^{\eta min}(34-\beta)(\varepsilon_{cmax}/8) \text{ and}$$
$$\varepsilon^{\eta max} = f(\varepsilon^{\eta min}(\beta)), \quad \text{for} \quad 26 < \beta \leq 34 \quad (7)$$

$$\varepsilon^{\eta min}(\beta-34)(\varepsilon_{tmax}/10) \text{ and}$$
$$\varepsilon^{\eta max} = f(\varepsilon^{\eta min}(\beta)), \quad \text{for} \quad 34 < \beta \leq 44 \quad (8)$$

$$\varepsilon^{\eta min} = \varepsilon_{tmax} \text{ and}$$
$$\varepsilon^{\eta max} = (51-\beta)(\varepsilon_{cmax}/7), \quad \text{for} \quad 44 < \beta \leq 51 \quad (9)$$

$$\varepsilon^{\eta min} = \varepsilon_{tmax} \text{ and}$$
$$\varepsilon^{\eta max} = (\beta-51)\varepsilon_{tmax}, \quad \text{for } 51 < \beta \leq 52 \quad (10)$$

With $\varepsilon^{\eta max}$ and $\varepsilon^{\eta min}$ defined, evaluated at the fibers with coordinates η^{max} and η^{min}, the generalized strain measures may be obtained:

$$k_o = (\varepsilon^{\eta max} - \varepsilon^{\eta min})/(\eta_{max} - \eta_{min}) \quad (11)$$

$$\varepsilon_o = \varepsilon^{\eta max} - k_o \eta_{max} \quad (12)$$

$$k_x = k_o \cos(\alpha) \quad (13)$$

$$k_y = k_o \sin(\alpha) \quad (14)$$

With the generalized strains it is possible to obtain the axial force and bending moments performing the integration of the stresses in the cross section, using a fiber approach which may be based on the existing finite element mesh:

$$N_z = \sum_{i=1}^{n}(\sigma(\varepsilon)A)_i \quad (15)$$

$$M_x = \sum_{i=1}^{n}(\sigma(\varepsilon)yA)_i \quad (16)$$

$$M_y = -\sum_{i=1}^{n}(\sigma(\varepsilon)xA)_i \quad (17)$$

With the aid of the β parameter, with $0 \leq \beta \leq 52$, and for discrete values of α (neutral axis angle), one may

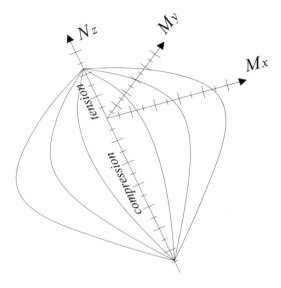

Figure 7. Interaction surface for a cross section.

evaluate as many (N_z, M_x, M_y) points as necessary to get a description of the interaction surface by its approximate meridian lines (Fig. 7). The intersection of this surface with one of the zero moment planes gives N-M interaction diagrams for uniaxial bending.

5 EXAMPLES

In this section, sets of interaction diagrams for different exposure times to the standard fire ASTM E119 are built for concrete-filled hollow sections. Although the procedure may be applied to a generic temperature field distribution, the examples are for a heating regime with equal temperature on all the exterior faces of the sections. This allows the construction of interaction curves which involve uniaxial bending only. The generic situation was analyzed by Caldas et al. (2007) for RC sections.

5.1 *Concrete-filled CHS under fire*

A circular tubular section with 350 MPa yield strength, external diameter 273.1 mm and 6.35 mm thickness is filled with concrete with 47 MPa compressive strength. Four reinforcement bars of 400 MPa yield strength and 19.5 mm diameter are embedded in the concrete, with 23 mm of concrete cover. This column was analyzed by Lie (1994).

In the thermal analysis the concrete was considered as having calcareous aggregate with 10% moisture and thermal conductivity given by the Eurocode 2 (2004). The section was exposed to

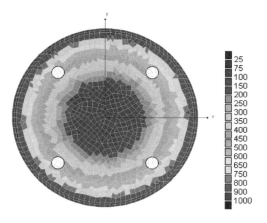

Figure 8. Temperature for 90 min exposure.

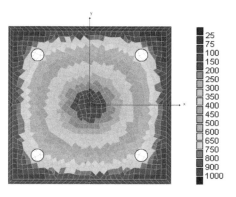

Figure 10. Temperature distribution.

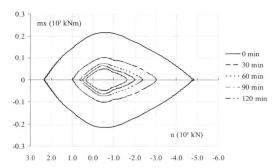

Figure 9. Interaction diagrams for CHS.

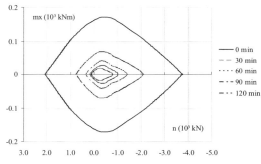

Figure 11. Interaction diagrams for SQS.

the standard ASTM E119 fire and the interaction diagrams are built for 30, 60, 90 and 120 minutes of fire exposure. The cross section is discretized with 1200 four-node elements.

Figure 8 shows the temperature distribution for a 90 min fire exposure, and Figure 9 depicts the interaction diagrams for 0 (ambient temperature), 30, 60, 90 and 120 minutes. Due to the symmetry of the cross section and the heating regime, there is a contraction of the surface as temperature rises. It may be noted that in the first 30 minutes the section experiences the largest reductions. This is the phase when the steel of the tube loses most of its strength and gradually the internal reinforced concrete becomes responsable for the cross section resistance.

5.2 Concrete-filled SHS under fire

A square tubular cross section with external dimensions 203.2 mm, thickness 6.35 mm and filled with concrete has four 16 mm bars with 23 mm concrete cover. Similar considerations about the concrete are made. This section was analyzed by Lie and Irwin (1995).

Figure 10 shows the temperature distribution for a 90 min fire exposure, and Figure 11 depicts the interaction diagrams for 0 (ambient temperature), 30, 60, 90 and 120 minutes.

6 SUMMARY AND CONCLUSIONS

The purpose of this work was to apply a procedure previously developed for reinforced concrete sections for the evaluation of axial force-bending moment interaction curves and surfaces for concrete-filled hollow sections subjected to fire.

The interaction diagrams obtained from the formulation provide an improved understanding of the cross section behavior when subjected to high temperature.

Two sets of interaction diagrams, for a concrete-filled circular hollow section and for a concrete-filled square hollow section were obtained with the procedure.

Some assumptions made for the construction of the diagrams, however, need to be addressed by further numerical and/or experimental research, such as the choice for the limit strain for a concrete fiber, which was taken here in order to produce results on the safe side.

The proposed procedure will be a step on the development of a method for design of concrete-filled hollow section columns under fire conditions.

ACKNOWLEDGEMENTS

The authors wish to thank V&M Tubes and CNPq for the financial support.

REFERENCES

Caldas, R., Sousa Jr., J.B.M. & Fakury, R.H. 2007. Interaction diagrams for reinforced concrete sections subjected to fire, submitted to Engineering Structures.

Eurocode 2. 2004. Design of Concrete Structures, Part 1.2: General Rules, Structural Fire Design. BS EN 1992-1-2:2004.

Eurocode 4. 2005. Design of Composite Steel and Concrete Structures, Part 1–2: General Rules, Structural Fire Design. BS EN 1994-1-2: 2005.

Franssen, J.M. & Dotreppe, J.C. 2004. Fire tests and calculation methods for circular concrete columns. *Fire Technology* 39: 89–97.

Lie, T.T. 1994. Fire Resistance of Circular Steel Columns Filled with Bar-Reinforced Concrete. *Journal of Structural Engineering* 120: 1489–1509.

Lie, T.T. & Irwin, R.J. 1995. Fire Resistance of Retangular Steel Columns Filled with Bar-Reinforced Concrete. *Journal of Structural Engineering* 121: 797–805.

Experimental study of the behaviour of joints between steel beams and concrete filled tubular (CFT) columns in fire

J. Ding
Bodycote Warringtonfire, UK

Y.C. Wang
School of Mechanical, Aerospace and Civil Engineering (MACE), The University of Manchester, UK

A. Orton
Corus Tubes, CIDECT, UK

ABSTRACT: This paper presents the experimental study of the performance of joints between steel beams and Concrete Filled Tubular (CFT) columns in simple construction under fire conditions. Different types of joints were investigated, including fin plate, T-stub, reverse channel and endplate. The fire tests of this research were carried out in a structural assembly, consisting of a steel beam and two CFT columns in the form of a "rugby goalpost". In total, 10 tests were carried out in the Fire Testing Laboratory of the University of Manchester. In each test, loads were applied to the beam and then the structural assembly was exposed to the standard fire condition in a furnace while maintaining the applied loads. Eight of the 10 test specimens were tested to failure during heating. All the joints behaved well before the steel beams had reached their limiting temperatures for flexural bending. The failure modes of the test specimens were always in the joint regions when the steel beams were in considerable catenary action. By appropriate design and protection of the joints, it is possible for the steel beams to develop catenary action and survive very high temperatures. In the other two tests, the test assembly was heated to temperatures close to the limiting temperature of the steel beam and then cooled down while still maintaining the applied loads on the beam. During cooling, the beams developed high tension forces, however there was no structural failure in the assemblies.

1 INTRODUCTION

Joints are critical elements of any building structure and fire resistance is an important safety requirement in building design. Recent observations from real fires (Newman et al. 2000) show that, on some occasions, the accumulative effects of a number of factors (including hogging bending moment, tension field action in shear and high cooling strain or pulling in effect at large deflections of the connected beam) could make the tension components of the joints fracture. The greatest danger of joint fracture may be progressive collapse of the structure. Despite extensive research studies of joint behaviour at ambient temperature, very little information is available on joints between steel beam and CFT column under fire conditions.

Observations from fire tests and real fire accidents in other types of steel structure indicate that there is strong interaction between joints and the connected beams and columns, and that the behaviour of the joints in complete structures is different from that in isolation. In general, steel beams restrained by the adjacent structure would go through a number of phases in behaviour when exposed to fire (Wang 2002), including combined flexural bending and compression during the early stage of fire exposure (as a result of restrained thermal expansion of the beam), through pure flexural bending at temperatures around the beam's limiting temperature under pure bending, and finally to catenary action at the late stage of fire exposure when the beam deflection is very large. As a result, the forces in the joints will change during the course of a fire exposure. In particular, catenary action in the steel beam can allow the beam to survive very high temperatures and it is possible to utilize catenary action in the steel beam to remove fire protection (Wang and Yin 2006). However, it is important that the joints have sufficient resistance to the catenary forces. Isolated joint tests under fixed loads, either in bending, tension, compression or their combinations, cannot capture the changing nature of joint performance and its impact on the adjacent structural behaviour. Therefore, it is essential to carry out fire tests on structural assemblies consisting of joints and other structural members.

In total, 10 fire tests were carried out on structural assemblies each consisting of two CFT columns and a steel beam in a "rugby goalpost" arrangement (Ding 2007). Temperature results from the fire tests are presented in another paper (Ding & Wang 2007). The aim of this paper is to provide the experimental results of structural behaviour.

2 TEST SET-UP AND SPECIMENS

2.1 Test set-up

The fire tests were carried out in the Fire Testing Laboratory of the University of Manchester. In each test, loads were applied to the beam and then the structural assembly was exposed to the standard fire condition in a furnace while maintaining the applied loads. The furnace is a rectangular box having internal dimensions 3000 mm × 1600 mm × 900 mm. The interior faces of the furnace are lined with ceramic fibre materials of thickness 200 mm that efficiently transfer heat to the specimen. An interior full-height honeycomb ceramic wall, as shown in Figure 1, was constructed to ensure uniform heating.

The average temperature inside the furnace was programmed to rise according to the ISO834 standard time-temperature curve. However, due to high thermal mass of the test assembly, especially that of the wet concrete, in relation to the furnace size, the furnace temperatures could not follow the standard fire curve. The furnace temperatures were measured with the aid of 6 control thermocouples, which are placed in two lines opposite each other at 430 mm apart (Figure 1). These thermocouples are embedded in ceramic tubes and their measuring tips project out-side the ceramic tubes and are located in the mid height of the furnace.

The arrangement of the members to be tested is in the form of a complete "rugby goalpost" frame as shown in Figure 2. The beam was mainly unprotected. In order to simulate the heat-sink effect of the concrete slab to the top flange in realistic structures, the top flange of the beam was wrapped with 15 mm thick ceramic fibre blanket. The columns were unprotected. They were restrained from lateral movement (so as to provide axial restraint to the beam) at the ends and were free to move in the longitudinal direction at the top. As shown in Figure 3, the lateral restraint at each column end was provided by a bracket bolted to the strong reaction frame around the furnace. To allow free axial movement of the columns, the bracket bolted to the top end of the column was slotted and there was a 20 mm gap between the column top and the top member of the strong reaction frame. Two transverse point loads were applied to the beam using two independent hydraulic jacks connected to the top member of the strong reaction frame surrounding the furnace to form a self-equilibrating system. Since the test beam had no lateral restraint, it could experience lateral-torsional buckling. In order to ensure that the loading jacks remained attached to the beam, each loading jack was inserted into a steel bracket which was then clamped on the top flange of the beam, as shown in Figure 4. Also in order for the two loading jacks to avoid sliding towards each other during the beam's large deflection and rotation, a steel bar (wrapped in insulation) was used to separate the two loading jacks, as shown in Figure 4.

The nominal load ratio applied on the steel beam was 0.5 in nine tests and was 0.25 in test 7. The nominal load ratio is defined as the ratio of the applied load during the fire test to the nominal load-carrying

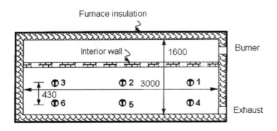

Figure 1. Plan view of schematic arrangement of the furnace.

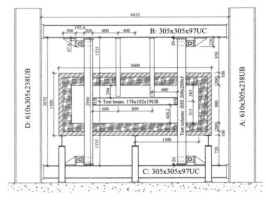

Figure 2. Elevation view of schematic arrangement of test.

(a) Top end.

(b) Bottom end.

Figure 3. Restraint device between column end and reaction frame.

Figure 4. Steel brackets between beam top flange and the loading jacks.

capacity of the beam with S275 grade steel at room temperature. Although axial load in the column would influence joint performance, the columns in these tests were not loaded so as to give scope for the fire tests to investigate the development of catenary action in the beams should the joints be adequate.

2.2 Structural instrumentation

Due to high temperatures in the fire test, the means of measuring strains and displacements were very limited. No strain gauge was placed on the specimen, as they would be quickly destroyed during the fire exposure. Therefore, only the temperature distributions and some displacements were measured during the tests. In particular, a large number of thermocouples were installed on each test specimen in the beam, the joints and the columns to record detailed temperature distributions.

In order to resolve the forces in the test assembly, the horizontal reaction forces (shear) in the columns were measured by four pin load cells attached at the four ends of the two columns as shown in Figure 3. Displacement transducers were placed near the ends of the beam and on the columns to measure the deformations, as shown in Figure 5. A displacement transducer with 400 mm stroke was located at the mid-span of the beam to measure the maximum expected beam deflection. All displacement transducers were inserted into ceramic tubes to minimize high-temperature effects.

Loads were applied on the beam before fire exposure. These loads were continuously monitored by using load cells, and attempts were made to maintain the applied loads throughout the fire test. However, as the rate of increase in beam deflection became very high during the transition from flexural bending to catenary action in the beam, there were short periods when the applied loading failed to catch up with the deflection.

2.3 Test specimens

In total, ten tests were carried out. Table 1 summarizes the test details. Each test assembly consists of two concrete filled tubular (CFT) columns and a steel beam. The cross-section size of the beams in all tests was the same, being grade S275 universal beam section $178 \times 102 \times 19$UB. Four types of joints were

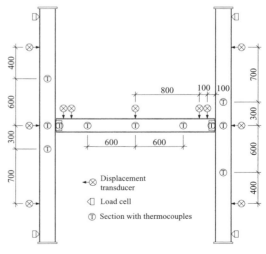

Figure 5. Locations of measurement devices on test specimen.

Table 1. Summary of test parameters.

Test load No.	Tube size	Beam length (mm)	Joint type	Steel tube	Material strength (MPa)			Applied per jack (kN)
					Concrete (cube)	Steel beam	Joint	
1	SHS 200 × 5	1980	fin plate	342	40.7	337	–	30
2	SHS 200 × 5	1953	bolted T	–	40.5	384	427	30
3	SHS 200 × 12.5	1953	bolted T	492	52.8	384	427	30
4	SHS 200 × 12.5	1832	reverse channel	492	39.1	384	–	30
5	SHS 200 × 5	1832	reverse channel	–	44.4	384	–	30
6	SHS 200 × 5	1984	extended endplate	–	51.0	323	311	60
7	SHS 200 × 5	1984	extended endplate	–	54.8	323	311	30
8	CHS 193.7 × 5	1812	reverse channel	406	45.7	323	365	45
9	CHS 193.7 × 5	1926	fin plate	406	45.3	323	311	30
10	CHS 193.7 × 5	1812	reverse channel	406	46.2	323	365	30

investigated in the fire tests: (1) fin plate welded to the tubular wall; (2) T-section bolted to the tubular wall using Molabolts (a type of blind bolt, see www.molabolt.co.uk for details); (3) reverse channel welded to the tubular wall and (4) endplate bolted to the tubular wall using Molabolts.

On each concrete filled tube, a 20 mm thick endplate was welded to each end. The top endplate had a circular hole of 150 mm diameter to enable concrete casting. The columns were held in a vertical position when filling the self-compact concrete. Drying tests were conducted to measure the moisture content of the concrete. The average water percentage was 4.92%.

3 OBSERVATIONS AND FAILURE MODES

3.1 *Test 1 (fin plate)*

Before the furnace was ignited, the loads were applied to the beam gradually. No visible deformation was found in the specimen at this stage. After 23.2 min of fire test, the beam deflection increased rapidly. As a result, the applied loads failed to catch up with the deflection. At 24.5 min of fire test, the displacement transducer at the mid-span of the beam (Figure 6) was detached from the beam, and did not give a meaningful reading thereafter. The sub-frame failed at 30.4 min with fracture at the weld in the left hand joint, as shown in Figure 6(a). The beam experienced large twist and vertical deflections as shown in Figure 6(b) and (c).

3.2 *Test 2 (bolted T-stub)*

The loads were applied to the beam gradually at room temperature. After 23.7 min of fire exposure, the beam deflection started to increase very quickly and the applied loads failed to catch up with the deflection. The sub-frame failed at 30 min with pull-out of the beam web at the left end, as shown in Figure 7(a). After fracture, the beam web formed a sharp edge, indicating tension failure. Due to the large twist in the beam, parts of the T-stubs bolted to the beam were also twisted to one side, as shown in Figure 7(a). Consequently, the bolts between the left T-stub and the beam experienced large rotations as shown in Figure 7(b). Parts of the T-sections bolted to the steel tubes were pulled out slightly. The steel tube walls were also pulled out slightly around the joint area but there was no indication of failure in the tubes, see Figure 7(a). There was no noticeable damage to the blind bolts connected to the steel tubes. Therefore, the large rotation of the joint was mainly a result of bearing deformations of the web of the steel beam and T-stubs around the bolt holes as shown in Figure 7(c).

3.3 *Test 3 (bolted T-stub)*

Test 3 was the same as Test 2 except for using thicker steel tubes and applying 100 mm long 15 mm thick mineral wool fire protection to the joints at the ends of the beam. The fire protection was applied to eliminate the beam web failure mode observed in Test 2. After applying the target loads at room temperature, no clear deformation was found in the specimen. After 25.5 min of fire test, the beam deflection increased rapidly and the applied loads could not be maintained. After 30 min of fire exposure, the beam deflection started to increase at a lower rate and the applied loads were increased to the original level. The applied loads were then maintained until the sub-frame failed at 38.2 min after fire exposure. As shown in Figure 8(a), failure of this sub-frame was due to fracture of the stem of the T-section on the left hand side. Figure 8(a) also shows a detailed image of the failure surface. No noticeable deformation could be seen in the steel tubes and there was no sign of

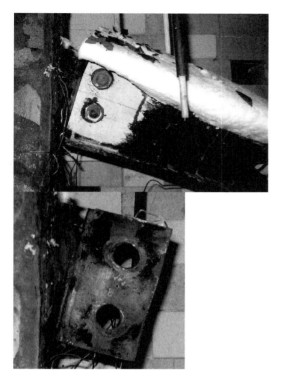

(a) Fracture of weld in left hand side joint.

(b) Overall deformed shape.

(c) Large twist of the beam.

Figure 6. Behaviour and failure mode of Test 1.

tube failure. As in Test 2, the large joint rotation on the right hand side (Figure 8(b)) was mainly from bearing deformations of the bolt holes on the beam web, see Figure 8(c).

3.4 Test 4 (reverse channel)

The beam was loaded gradually at room temperature to the target loads. After 23.3 min of fire test, the beam deflection increased rapidly and the applied loads could not catch up with the deflection for a short period. After 28.3 min of the fire test, the beam deflection started to increase at a lower rate and the applied loads were increased to the original level. The applied loads were then maintained until the sub-frame failed at 33 min after fire exposure. As shown in Figure 9(a), failure of this sub-frame was due to fracture in the beam web on the left hand side. The fracture length was the same as the weld length and the fractured web formed a sharp edge, indicating tension failure. Due to large deflections in the beam and the flexibility of the joint, the endplates and the front faces of the reverse channels deformed significantly to give the very large rotation on the right hand side of the test specimen where there was no observation of fracture as shown in Figure 9(b).

3.5 Test 5 (reverse channel)

Test 5 was the same as Test 4 except for using a different tube thickness and protecting the beam web at both ends by 100 mm to prevent beam web failure observed in Test 4. Before fire ignition, the beam was loaded gradually to the target loads and no obvious deformation was found in the specimen. After 25.2 min of the fire test, the beam deflection increased rapidly and the applied loads dropped and could not be maintained for a short period. After 28.7 min of the fire test, the beam deflection started to increase at a lower rate. After 29.5 min of the fire test, the applied loads were successfully increased to the original level and were then maintained until the end of the test. After 41.4 min of fire exposure, the hydraulic jack on the left hand side ran out of travel and was unable to maintain the load on the beam. After 46.2 min of fire exposure, the hydraulic jack on the right hand side also ran out of travel. When the test was finally terminated at 51.5 min, no fracture failure in the subframe could be observed from the view aperture on the furnace door. However, after the specimen was taken out of the furnace, fracture was found in the steel tube on the left hand side where the reverse channel was welded, as shown in Figure 10(a). There was also fracture in the weld between the endplate and the beam web on the left hand side, as shown in Figure 10(b). The crack in the steel tube revealed the concrete core, which did not show any damage at all. Due to the large catenary force in the beam, the endplates and the reverse channels distorted significantly, as shown in Figure 10(c), again illustrating the extraordinary deformation capability of the reverse channel joint.

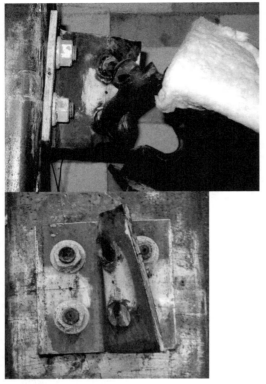

(a) Pull-out of beam web and deformation of T-stub in left hand side joint.

Left bottom, left top, right bottom, right top
(b) Deformed bolts between T-stubs and beam web.

(c) Deformed bolt holes of T-stub webs

Figure 7. Behaviour and failure mode of Test 2.

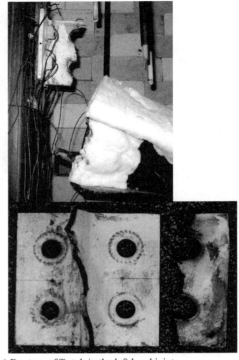

(a) Fracture of T-stub in the left hand joint.

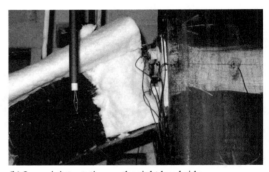

(b) Large joint rotation on the right hand side.

(c) Deformed bolt holes in steel beam web.

Figure 8. Behaviour and failure mode of Test 3.

(a) Fracture of beam web on the left hand side joint.

(a) Fractured weld on tube on the left hand joint (viewed from back of furnace).

(b) Large joint rotation on the right hand side.

Figure 9. Behaviour and failure mode of Test 4.

(b) Fractured weld on left hand beam.

(c) Deformed reverse channel on left hand side.

Figure 10. Behaviour and failure mode of Test 5.

3.6 Tests 6 and 7 (extended endplate)

The two tests behaved similarly, so only observations of Test 6 are described. The applied load in this test was twice the load applied in Tests 1–5, based on the assumption that the joint in this test would develop the full bending moment capacity of the steel beam whilst the joints in the previous tests would behave as simply supported. The loads were applied to the beam gradually at room temperature and no visible deformation was found in the specimen before the furnace was ignited. After 22.3 min of fire test, the beam deflection increased rapidly and the applied loads could not be maintained. The sub-frame failed at 25.2 min after fire exposure with fracture and pull-out of the Molabolts on both sides, as shown in Figure 11(a). Due to the large catenary force in the beam, the endplates distorted significantly. After the specimen was taken out of the furnace, a cut-out was made on the steel tube of the left hand side CFT column to reveal the concrete inside the steel tube where the Molabolts were connected, as shown in Figure 11(b). It can be seen that apart from the surface, the concrete core was not damaged. After the Molabolts had been taken out, it could be seen that although the gaps of the Molabolts were filled with concrete, they did not expand much, as shown in Figure 11(c). Clearly, the prototype Molabolt used in this test would not be usable for practical applications. The steel tube walls were pulled out slightly around the bolts, but there was no indication of failure in the tubes.

3.7 Test 8 (reverse channel)

The applied load in this test was 1.5 times that of Tests 1–5 on the assumption that the joint would be able to develop about 50% of the beam bending moment capacity, so maintaining the same nominal load ratio as in Tests 1–5. The beam was loaded gradually at room temperature to the target loads. No clear deformation was found in the specimen before fire ignition. After 22.2 min of the fire test, the beam deflection increased rapidly and the applied loads could not catch up with the deflection for a short period. After 25.3 min of the fire test, the beam deflection started to increase at a lower rate and the

Left   Right

(a) Pull-out of bolts on the left hand side.

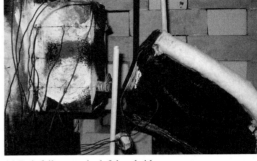

(a) Bolt failure on the left hand side.

(b) Concrete inside the left hand side column.

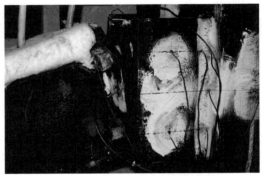

(b) Fracture of reverse channel on the right hand side.

Figure 12. Behaviour and failure mode of Test 8.

Left top (2) right bottom (1), right middle (2), right top (2)
(c) Fractured Molabolts between CFT columns and endplates.

Figure 11. Behaviour and failure mode of Test 6.

reverse channel connection be used in practice, the brittle shear failure of the reverse channel should be avoided so as to enable the reverse channel connection to develop substantial ductility.

3.8 Test 9 (fin plate)

This test arrangement was similar to test 1. The main aim of this test was to provide some experimental information on joint behaviour in cooling. Before the furnace was ignited, the loads were applied to the beam gradually. After 26 min of fire ignition, the beam deflection increased rapidly and the applied loads failed to catch up with the deflection. The furnace was stopped after 26.6 min of fire ignition and a fan in the exhaust was turned on to provide forced cooling. However, this did not stop the beam deflection from increasing. After 28.7 min of the test (2.1 min after cooling started), the beam deflection started to increase at a lower rate and the applied loads were increased to the original level. The loads were then maintained until the end of the test. After 36.3 min of the test (9.7 min after cooling started), the beam deflection started to decrease slowly. The furnace door was opened after 106.7 min of the test

applied loads were increased to the original level. However, the beam deflection started to increase faster at 28.3 min of the fire test and the hydraulic jack on the left hand side failed to maintain the load. The sub-frame failed at 29.7 min after fire exposure due to thread-stripping of the nuts on the left hand side, see Figure 12(a). Due to the large catenary force, bending moment and shear force in the beam, the front face of the right hand side reverse channel fractured, as shown in Figure 12(b). Comparing this with the extremely ductile performance of the reverse channels in tests 4 and 5, it suggests that should

(80.1 min after cooling started). The test was terminated after 331 min with no fracture and failure of the sub-frame. Due to expansion and bending of the steel beam, the bottom flange of the beam was bearing against the CFT columns during the heating stage and marks were left on the tubes due to this bearing action, as shown in Figure 13. Due to the large twist in the beam, the fin plates were also twisted to one side. Little deformation was found in the bolts between fin plates and beam web.

3.9 Test 10 (reverse channel)

This test was also a cooling test. The loads were applied to the beam gradually at room temperature. To prevent lateral-torsional buckling of the beam, the furnace was stopped after 23.1 min of fire ignition and a fan in the exhaust was turned on. After 29.7 min of the test (6.6 min after cooling started), the beam deflection started to decrease slowly. However, after 56 min of the test (32.9 min after cooling started), the beam deflection started to increase slowly again. The furnace door was opened after 158 min of the test (134.9 min after cooling started). The test was terminated after 320 min with no fracture and failure of the sub-frame. The loads were successfully maintained throughout the test. The beam was bent slightly, as shown in Figure 14. However, there was no visible deformation in the steel tubes and connection regions.

Figure 13. Behaviour of Test 9, left hand side.

Figure 14. Behaviour of Test 10, overall deformed shape.

Table 2. Main test results.

Test No.	Appox. load ratio[a]	Pure bending of steel beam		Failure or end of fire test			Failure mode
		Max beam Time (min)	Max beam temperature (°C)	Time (min)	Max beam temperature (°C)	Max beam deflection (mm)	
1	0.5	24.0	726	30.4	769	–	Crack of the weld
2	0.5	25.3	746	30.0	754	265	Fracture in beam web
3	0.5	27.2	736	38.2	790	263	Fracture in T-section
4	0.5	24.5	741	33.0	744	288	Fracture in beam web
5	0.5	26.6	763	51.6	873	Over 302	–
6	0.5[b]	23.2	723	25.2	738	175	Fracture and pull-out of Molabolts
7	0.25[b]	31.0	772	36.8	797	150	Fracture of Molabolts
8	0.5	23.5	732	29.7	766	229	Thread-stripping of the nuts
9	0.5	–	–	331	37.7 (end of test)	202	–
10	0.5	–	–	320	34.2 (end of test)	56	–

*Calculated according to lateral-torsional buckling resistance of simply supported steel beam with nominal S275 grade steel.
**Calculated according to rotationally restrained steel beam with equal connection bending moment capacity as the lateral-torsional buckling resistance of simply supported steel beam with nominal S275 grade steel.

4 SUMMARY OF TEST RESULTS

The steel beams in tests 1–8 went through three phases. During the early stage of fire exposure, the beam started to bend and the deflection was increasing slowly. When the beam temperature reached the limiting temperature of the axially unrestrained beam for lateral-torsional buckling resistance (as indicated by zero axial force), the beam started to twist accompanied by a drastic increase in the beam vertical deflection. During this phase, it was difficult to maintain the applied loads. Finally when the beam deflection was very large, the beam went into catenary action with a much reduced rate of increase in deflection. During this phase, the applied loads were successfully increased to the original level in some of the tests, which enabled long periods of successful observation and measurement of catenary action in the beams until fracture of some of the joint components. Because the left hand side of the furnace experienced higher temperatures than the right hand side, structural failure normally started from the left hand side of the structural assembly.

Table 2 summarizes the main test results. It can be seen from the results that the final beam temperatures and failure times are considerable higher than the limiting temperatures of the axially unrestrained beams and the corresponding times.

5 CONCLUSIONS

This paper has presented a description and experimental results of 10 fire tests on beam-concrete filled tubular column assemblies. The following conclusions may be drawn:

1. It is possible for moderately complex joints to CFT columns (T-stub, reverse channel) to develop substantial catenary action, whereby the steel beams could survive much higher temperatures than the limiting temperatures for pure bending. The fin plate joint had little stiffness to develop catenary action in the beam and the resistance of the joint to catenary action in the beam is low. The reverse channel joint appeared to have higher stiffness and strength than the T-stub joint to resist catenary action in the steel beam.
2. Among the few connections tested in this study, the reverse channel connection appears to have the best combination of desirable features: moderate construction cost, ability to develop catenary action, and extremely high ductility (rotational capacity) through deformation of the web of the channel. It appears that thinner reverse channels (Tests 4 and 5) would be preferable to thicker ones (Test 8).
3. The steel beams were able to accommodate very large deflections. Therefore, by appropriate design and protection of the joint components, it is possible for the steel beams to develop catenary action and survive very high temperatures.
4. During cooling, considerable tension forces can be developed in the steel beam and the joints. Although there was no fracture of connections during cooling in the tests (Tests 9 and 10) due to short span of the beams, it is important to consider this effect in design.

REFERENCES

Ding, J. 2007. *Behaviour of restrained concrete filled tubular (CFT) columns and their joints in fire*. Ph.D thesis: School of MACE, University of Manchester.

Ding, J. & Wang, Y.C. 2007. Temperature in unprotected joints between steel beam to concrete filled tubular columns in fire. *Fire Safety Journal* [under review].

Newman, G.M. & Robinson, J.T. & Bailey, C.G. 2000. *Fire safety design: A new approach to multi-storey steel-framed buildings*. London: SCI Publication P288.

Wang, Y.C. 2002. *Steel and composite structures, behaviour and design for fire safety*. London: Spon Press.

Wang, Y.C. & Yin, Y.Z. 2006. A simplified analysis of catenary action in steel beams in fire and implications on fire resistant design. *Steel and Composite Structures* 6(5): 367–386.

A numerical investigation into the temperature distribution of steel hollow dual tubes

M.B. Wong
Department of Civil Engineering, Monash University, Melbourne, Australia

ABSTRACT: A feasibility study on the use of steel hollow dual tubes is presented. The main advantage of using the dual tubes as columns is that both tubes are used as structural members when there is no fire and the outer tube is used as fire protection to the inner tube when there is fire. Numerical simulation of a fire in a compartment containing this type of column shows that the inner tube can maintain a substantial level of capacity during the course of fire. Various combinations of size of the outer and inner tubes are used to determine the optimum ratio to be used for fire engineering design.

1 INTRODUCTION

The use of steel tubular members in buildings always requires the installation of fire protection to the members due to the rapid deterioration in strength of steel when exposed to fire. Other methods to maintain strength of steel tubes during the course of fire include the use of concrete filled steel tubes (Han 2001), double-skin steel tubes (Zhao 2006), and steel tubes protected by an exterior layer of concrete or insulation material. The double-skin steel tubes with concrete in-fill between the tubes seem to be a promising alternative in terms of its efficiency in performance. However, concrete is a poor thermal insulator and in many cases the exterior steel tube has still to be fire-protected in order to achieve adequate fire rating.

An earlier preliminary experimental investigation into the use of steel columns similar to the double-skin steel tubes without concrete in-fill has been carried out. The idea is based on the fact that air is an excellent thermal insulator and the heat transfer between the two steel tubes is only through radiation. The results show that the inner steel tube has a temperature range of 100 °C to 200 °C lower than that of the outer tube in a compartment with maximum air temperature of about 600 °C. This means that the inner one of the hollow steel dual tubes is able to maintain substantial strength during a fire under the protection of the outer tube. This paper describes a numerical investigation into the temperature distribution of this type of hollow steel dual tubes system under exposure to fire. An analytical heat transfer analysis is conducted for a combination of tubes with different thickness. The strength deterioration rates of the inner and outer tubes are then evaluated. The investigation shows that the combined strength of the tubes during a fire may be able to sustain the loading at fire limit state. The results may be used to determine the optimum ratio of tube thicknesses for structural fire engineering design.

2 PRELIMINARY EXPERIMENT

A simple experiment using the steel hollow dual tubes shown in Figure 1 was carried out some time earlier as a preliminary investigation into the temperature difference between the two tubes. Due to the limited capability of the furnace, only the temperatures of the steel tubes were measured in this preliminary study. The two tubes were separated by a gap of constant thickness maintained by a steel rod through both tubes. The tubes were heated to a maximum of about 600 °C within 90 minutes and then the furnace was switched off. The results are shown in Figure 2.

Figure 1. Dual tubes in furnace.

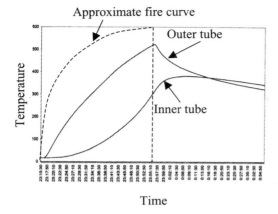

Figure 2. Dual tubes of 114.3 × 4.8 and 88.9 × 3.2.

It can be seen from Figure 2 that there is a substantial difference of more than 200°C in temperature between the outer and inner tubes for most of the heating period. The delay for the inner tube in temperature rise gives the possibility that the inner tube may survive and sustain the loading at fire limit state during a fire, particularly when the fire has limited burning fuel. Indeed, most fires have limited burning fuel and the temperature rise of such natural fires in a compartment can be obtained using the method given in EC3 (CEN 2002).

3 STEEL TUBE TEMPERATURES

Heat is transferred from the external fire to the outer tube by convection and radiation. In the numerical study, the computation of the radiation component of the heat transfer is based on the gas combustion theory proposed by Ghojel (1998), subsequently modified by Wong and Ghojel (2003) to enable analytical procedure to be performed on spreadsheet. Alternative formulation for this heat transfer model presented in a form similar to EC3 has also been proposed recently by Ghojel and Wong (2005). In the current study, the radiation component of heat transfer is calculated according to formulation given in Wong and Ghojel (2003).

Since the gap between the outer and inner tubes is filled with air only, the heat transferred from the outer tube to the inner one is mainly by radiation. In this case, the heat transfer formulation is based on the assumption that the two tubes are considered as two parallel plates with grey surfaces so that the quantity of heat is influenced by the mutual projections of surface areas of the two tubes.

3.1 Heat transfer formulations

A typical cross-section of dual tube column is shown in Figure 3.

Heat transfer between the external fire and the outer tube within a period of time Δt causes an increase in steel temperature by ΔT_{s1} given by

$$\Delta T_{s1} = \frac{\left(\frac{H_p}{A}\right)_1}{c_s \rho_s}[q_c + q_r]\Delta t \tag{1}$$

where c_s = specific heat of steel in J/kg °C = 472 + $0.00038 T_s^2 + 0.2 T_s$, ρ_s = density of steel = 7850 kg/m³, H_p/A = section factor = perimeter/cross-sectional area in m⁻¹. The subscript 1 refers to the outer tube whereas 2 to the inner tube.

The heat fluxes q_c due to convection and q_r due to radiation emitting from the fire to the outer tube are given by

$$q_c = h_c(T_g - T_{s1}) \tag{2}$$

where h_c is the convective heat transfer coefficient, assumed to be 25 W/m²K.

$$q_r = \sigma[\varepsilon_g(T_g + 273)^4 - \alpha_g(T_{s1} + 273)^4] \tag{3}$$

where $\sigma = 5.67 \times 10^{-8}$ W/m² K⁴.

The emissivity of gas ε_g in a burning compartment has been derived (Wong and Ghojel 2003) as a function of the gas temperature T_g:

$$\varepsilon_g = 0.458 - 1.29 \times 10^{-4} T_g \tag{4}$$

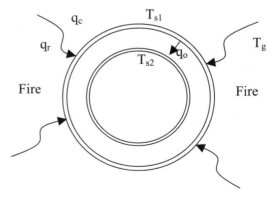

Figure 3. Heat transfer to dual tubes.

The absorptivity of gas α_g has been derived (Wong & Ghojel 2003) as

$$\alpha_g = 1.64148 \times \frac{(T_g + 273)^{0.41547}}{(T_{s1} + 273)^{0.63605}} \quad (5)$$

The heat flux q_o by radiation emitting from the outer tube to the inner one is given by

$$q_o = \varepsilon_r \sigma [(T_{s1} + 273)^4 - (T_{s2} + 273)^4] \quad (6)$$

It should be noted that Equation 6 is formulated generally for radiation between two parallel plates.

The heat transfer between the tubes is related to the relative areas of the tubes A_1 (outer tube) and A_2 (inner tube). For the present case, the resultant emissivity ε_r is given by

$$\varepsilon_r = \frac{1}{\frac{1}{\varepsilon_2} + \frac{A_2}{A_1}\left(\frac{1}{\varepsilon_1} - 1\right)} \quad (7)$$

where ε_1 and ε_2 are the respective emissivities of the steel surfaces of the outer and inner tubes. Since the interior surfaces are not affected by the fire, it can be assumed that $\varepsilon_1 = \varepsilon_2$. For the interior surfaces of steel tubes where the surface texture would not be affected by the fire, ε_1 and ε_2 can be taken as 0.32 (Drysdale 1998), a maximum value generally taken for mild steel surfaces. For example, a value of $\varepsilon_r = 0.2$ can be adopted for an inner to outer diameter ratio of 0.875.

As a result of receiving the heat flux q_o, the temperature of the inner tube is increased within Δt by an amount given by

$$\Delta T = \left(\frac{H_p}{A}\right)_2 [\ldots] \quad (8)$$

For a thin-walled steel tube, the section factor H_p/A is approximately equal to $1/t$ where t is the thickness of the wall in metres. Therefore the size effect on the steel temperature is independent of the diameter of the steel tube.

4 RELATIVE STRENGTHS OF STEEL TUBES

A steel column may fail by two modes: yielding and buckling. For short columns where yielding is the predominant failure mode, the strength capacity depends mainly on the cross-sectional dimensions. For slender columns where flexural buckling is the predominant failure mode, the strength depends mainly on the slenderness of the column. In this study, both the outer and inner tubes, at fire limit state, are assumed to fail independently and the total column strength capacity can be calculated as the sum of the two tubes.

At fire limit state, the loading is calculated according to the appropriate load combinations. To compare the load levels between fire limit state at temperature T and ultimate limit state at room temperature, a loading ratio λ can be defined as

$$\lambda = \frac{\gamma_a G + \psi_a Q}{\gamma_b G + \psi_b Q} \quad (9)$$

where γ_a, ψ_a, γ_b and ψ_b are values of the partial factors for loading. The subscript 'a' refers to fire limit state and 'b' to ultimate limit state. In Australia, typical values are $\gamma_a = 1.0$ and $\psi_a = 0.4 \sim 0.6$ depending on the type of building and occupancy for fire situation; $\gamma_b = 1.2$ and $\psi_b = 1.5$. In EC0 (2001), the corresponding values are $\gamma_a = 1.0$ and $\psi_a = 0.5 \sim 0.9$; $\gamma_b = 1.35$ and $\psi_b = 1.5$. A relative strength ratio η of the tubes in fire is defined as

$$\eta = \frac{Strength\ (in\ fire)}{Strength\ (no\ fire)} \quad (10)$$

When $\eta > \lambda$, the column will survive in fire. Usually, Q/G lies between 0.5–0.8. With this range of Q/G, λ lies between 0.62 and 0.55 for common office buildings according to Australian design standard. When using EC0, the corresponding range of λ lies between 0.59 and 0.54. Basically, the range of λ is very similar in both codes.

4.1 Strength for short columns

The capacity of a short steel column is calculated by its squash load which varies with temperature due to the varying yield stress. According to the Australian steel design code (Standards Australia 1998), the variation of yield stress with temperature is given by

$$k_{yT} = \frac{f_{yT}}{f_{y20}} = \frac{905 - T}{690} \quad (11)$$

where f_{yT} = yield stress at temperature T.

For an outer tube with thickness t_1, diameter d_1, and yield stress f_{yT1} and an inner tube with thickness t_2, diameter d_2, and yield stress f_{yT2}, the ratio η_s of the squash load for a completely yielded section at elevated temperature to that at room temperature can be calculated as

$$\eta_s = \frac{\pi d_1 t_1 f_{yT1} + \pi d_2 t_2 f_{yT2}}{(\pi d_1 t_1 + \pi d_2 t_2) f_{y20}}$$

$$= \frac{(k_{yT})_1 + \left(\frac{d_2}{d_1}\right)\left(\frac{t_2}{t_1}\right)(k_{yT})_2}{1 + \left(\frac{d_2}{d_1}\right)\left(\frac{t_2}{t_1}\right)} \quad (12)$$

4.2 Strength for slender columns

The capacity of a slender column is calculated by its buckling strength which varies with both yield stress and modulus of elasticity at elevated temperatures. If the capacity of the column at elevated temperature is expressed as

$$P_T = K_{T1} \pi d_1 t_1 f_{yT1} + K_{T2} \pi d_2 t_2 f_{yT2} \quad (13)$$

where K_{T1} and K_{T2} are coefficients related to the buckling of the tubes at elevated temperatures. At room temperature, the buckling coefficient of a tube of length L is

$$K \propto \frac{1}{\left(\frac{L}{r}\right)^2} \quad (14)$$

where L/r is the slenderness of the tube. The radius of gyration for a thin-walled tube can be approximated as $r \approx 0.353 d$.

At elevated temperatures, the slenderness of the column is dependent on $\sqrt{k_{yT}/k_{ET}}$ (EC3 2002) so that the buckling coefficient at elevated temperatures varies with the slenderness as

$$K_T \propto \frac{1}{\left(\left(\frac{L}{r}\right)\sqrt{\frac{k_{yT}}{k_{ET}}}\right)^2} \quad (15)$$

where k_{ET} is the variation of the modulus of elasticity with temperature.

The value of $\sqrt{k_{yT}/k_{ET}}$ varies, depending on the adopted values of k_{yT} and k_{ET} from the codes. Wong and Tan (1999) compared the values of $\sqrt{k_{yT}/k_{ET}}$ from the American, Australian and British codes, and found that all three varied in different ways although they fluctuated around the value of 1.0. It has been argued (Wong 2006) that for calculating the buckling strength of columns the value of $\sqrt{k_{yT}/k_{ET}}$ can

be conservatively taken as 1.0. This is the value of $\sqrt{k_{yT}/k_{ET}}$ adopted in the current study.

The load ratio of the buckling strength of the column at elevated to room temperatures can be expressed as

$$\eta_c = \frac{K_{T1} \pi d_1 t_1 f_{yT1} + K_{T2} \pi d_2 t_2 f_{yT2}}{K_1 \pi d_1 t_1 f_{y20} + K_2 \pi d_2 t_2 f_{y20}}$$

$$= \left(\frac{K_{T1}}{K_1}\right) \left(\frac{(k_{yT})_1 + \left(\frac{K_{T2}}{K_{T1}}\right)\left(\frac{d_2}{d_1}\right)\left(\frac{t_2}{t_1}\right)(k_{yT})_2}{1 + \left(\frac{K_2}{K_1}\right)\left(\frac{d_2}{d_1}\right)\left(\frac{t_2}{t_1}\right)}\right) \quad (16)$$

where K_1 and K_2 are the buckling coefficients for the outer and inner tubes respectively at room temperature. From Equations 14 and 15,

$$\frac{K_{T1}}{K_1} = \frac{1}{\left(\frac{k_{yT}}{k_{ET}}\right)} = 1 \quad (17)$$

$$\frac{K_{T2}}{K_{T1}} = \left(\frac{d_2}{d_1}\right)^2 \frac{\left(\frac{k_{yT}}{k_{ET}}\right)}{\left(\frac{k_{yT}}{k_{ET}}\right)} = \left(\frac{d_2}{d_1}\right)^2 \quad (18)$$

$$\frac{K_2}{K_1} = \left(\frac{d_2}{d_1}\right)^2 \quad (19)$$

Hence,

$$\eta_c = \frac{(k_{yT})_1 + \left(\frac{d_2}{d_1}\right)^3 \left(\frac{t_2}{t_1}\right)(k_{yT})_2}{1 + \left(\frac{d_2}{d_1}\right)^3 \left(\frac{t_2}{t_1}\right)} \quad (20)$$

Equations 9, 12 and 20 are used to check if the dual tube column will survive in a fire for which $\eta > \lambda$ is satisfied.

5 CASE STUDY

To demonstrate that a steel dual tube column system without passive fire protection may survive in a real fire, a case study is carried out for such a system under a relatively severe natural fire. The natural fire curve to be used for predicting the temperatures of the tubes is constructed according to EC1 (2002). Details of the case study are given below.

5.1 A laboratory engulfed in fire

A chemical laboratory in a compartment of a building has the following characteristics:
Compartment: height = 2.5 m; length = 6 m; width = 5 m
1st Opening height = 2 m
1st Opening width = 1 m
2nd Opening height = 1 m
2nd Opening width = 3 m
Boundary enclosure properties:
  Concrete density = 2400 kg/m³
  Specific heat = 1125 J/kgK
  Thermal conductivity = 1.2 W/mK
Active fire protection measures:
  Sprinklers = no
  Water supply = none
  Smoke alarm = no
  Transmission to fire brigade = no
  Safe access route = no
  Fire fighting device = yes
  Smoke exhaust system = no

The properties of the dual tube column in the compartment are given as:
Outer tube—Diameter = d_1, thickness, t_1, $H_p/A = 1/t_1$.
Inner tube: Diameter = d_2, Thickness, t_2, $H_p/A = 1/t_2$.

There are five combinations of dimensions to be studied as shown in Table 1.

The five cases given in Table 1 represent various possible combinations of three steel tube sizes to form a dual tube column system. The thicknesses of the tubes range from amongst the smallest to amongst the largest being used in practice. The above information is used to construct a natural fire curve using the method given in EC1 (2002). In this particular example, the gas temperature in the compartment rises rapidly to a peak of above 800°C at about 40th minute and then drops off to below 300°C within a period of 120 minutes. The load ratios for both short and slender columns are calculated for all five cases. The temperature development of the tubes for the five cases and the load ratios are shown in Figures 4–8. The numerical results are shown in Table 2.

In all five cases, there are significant differences in temperature between the outer and inner tubes during the temperature rising part of the two hour period. It can be seen that for all cases the temperature of the inner tube always decreases when the wall thickness of the outer tube increases. However, the trend is that the larger the wall thickness of the inner tube relative to the outer one, the higher is the load ratio. In other words, the wall thickness of the inner tube should be made as large as possible in order to ensure a stronger column during fire. It should be noted from Table 2 that the time at which the maximum temperature difference between the tubes occurs does not necessarily coincide with the time at which the lowest

Table 1. Properties of dual tubes.

Case	Outer tube		Inner tube	
	$(H_p/A)_1$ (m⁻¹)	t_1 (mm)	$(H_p/A)_2$ (m⁻¹)	t_2 (mm)
1	156	6.4	78.7	12.7
2	105.3	9.5	78.7	12.7
3	78.7	12.7	78.7	12.7
4	78.7	12.7	156	6.4
5	105.3	9.5	105.3	9.5

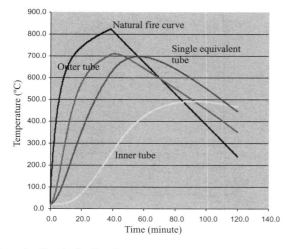
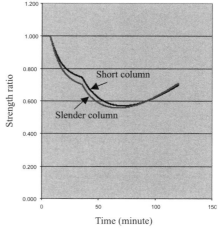

Figure 4. Results for Case 1.

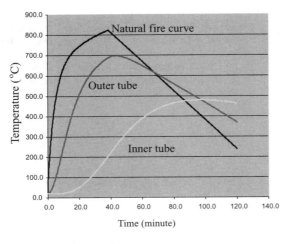
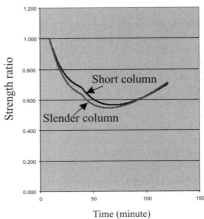

Figure 5. Results for Case 2.

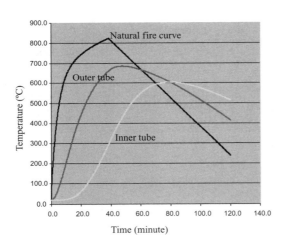
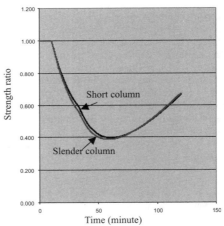

Figure 6. Results for Case 3.

Table 2. Results for dual tubes.

Case	t_2/t_1	d_2/d_1	Maximum temperature difference (°C)	At time (minute)	Minimum load ratio (short column)	At time (minute)	Minimum load ratio (slender column)	At time (minute)
1	1.984	0.875	523	24	0.570	71	0.557	67
2	1.337	0.875	510	29	0.565	69	0.546	64
3	1	0.875	497	33	0.562	69	0.540	63
4	0.504	0.875	425	27	0.396	61	0.387	58
5	1	0.875	482	27	0.480	64	0.465	60

656

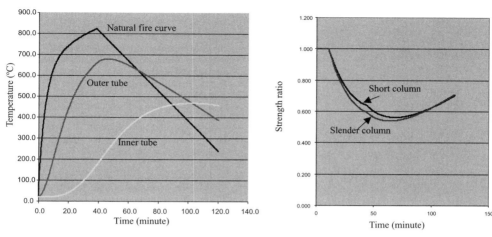

Figure 7. Results for Case 4.

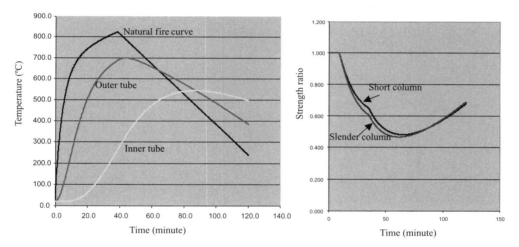

Figure 8. Results for Case 5.

load ratio occurs. The load ratio for slender column is lower in all cases.

Case 1 gives the highest load ratio of 0.557 at which it has a good chance to survive in this fire as this ratio is comparable with the applied load ratio. When using an equivalent single steel tube with wall thickness equal to the sum of the two tubes for Case 1, the maximum temperature in the single tube is about 700 °C as shown in Figure 4. At this temperature, the equivalent single tube column would have failed. Finally, it is noted that the pattern of the temperature developments of the dual tubes in all cases are similar to that obtained from the experiment.

6 CONCLUSIONS

A novel approach to the use of steel tubular columns is presented. The main feature of the column is that it is made of two concentric steel tubes. The advantage of this arrangement is that the outer tube can be used as fire protection for the inner tube during fire, delaying the temperature rise of the latter. With such arrangement, the strength of the inner tube would not be subjected to rapid deterioration by the effect of high temperature so that the column has a good chance of survival in a real fire. Numerical case study confirms that if the wall thickness of the inner tube is made relatively larger than the outer one, the strength of the inner tube will be most effective and maximised in fire. When there is no fire, both tubes are used as normal structural columns.

As fire protection constitutes a large proportion of the total construction cost, the dual tube column system is cost effective by eliminating the need for installing fire protection to the columns. Even under extreme fire conditions with very high temperature, installing a thin layer of fire protection material between the tubes would be adequate. As this layer of fire protection is out of sight and does not need to satisfy aesthetic requirements, installation procedure should be easy and the cost would be low.

REFERENCES

EC0. 2001. *prEN 1990. Basis of structural design, CEN.*
EC1. 2002. *prEN 1991 – Actions on structures. Part 1–2: General Actions – Actions on structures exposed to fire.* CEN.
EC3, 2002. *prEN 1993-1-2. Design of Steel Structures. Part 1.2: General Rules – Structural Fire Design.* CEN.
Drysdale, D. 1998. *An introduction to fire dynamics.* 2nd Ed. Wiley.
Ghojel, J. 1998. A new approach to modelling heat transfer in compartment fires. *Fire Safety Journal* 31: 227–237.
Ghojel, J.I., Wong, M.B. 2005. Heat transfer model for unprotected steel members in a standard compartment fire with participating medium. *Journal of Constructional Steel Research* 61: 825–833.
Han, L.H. 2001. Fire performance of concrete filled steel tubular beam-columns. *J. Const. St. Res* 57(6): 697–711.
Standards Australia. 1998. *AS4100 – Steel Structures.* SA.
Wong, M.B. 2006. Effect of torsion on limiting temperature of steel structures in fire. *ASCE, J. of Structural Engineering* 132(5): 726–732.
Ghojel, J.I. & Wong, M.B. 2003. Spreadsheet method for temperature calculation of unprotected steelwork subject to fire. *The Structural Design of Tall and pecial Buildings* 12(2): 83–92.
Tan, K.H. & Wong, M.B. 1999. Local buckling of steel plates at elevated temperatures. *Proceedings of the 16th Australasian Conference on the Mechanics of Structures and Materials.* Sydney: 467–471.
Han, L.H. & Zhao, X.L. 2006. Double skin composite construction. *Progress in Structural Engineering and Materials* 8(3): 93–102.

Thermoelastic behaviour of elastically restrained tubular steel arches

Y.L. Pi & M.A. Bradford
Centre for Infrastructure Engineering and Safety, School of Civil and Environmental Engineering, The University of New South Wales, NSW, Australia

ABSTRACT: Because arches resist external loads mainly by axial compression, tubular steel arches are an efficient structural form that is finding extensive contemporary use, in particular in China. The behaviour of tubular steel arches under external mechanical loading has been studied previously, but studies of the behaviour of tubular steel arches under thermal loading do not appear to be reported in the open literature. In practice, tubular steel arches are often supported by other structural members, which provide elastic types of restraint at the arch ends, which participate in the structural response of the arch and which may influence significantly the in-plane thermoelastic behaviour and possible buckling of the arch. This paper presents a thermoelastic analysis for the behaviour of a rotationally restrained tubular steel circular arch that is uniformly heated, and it also studies the possible thermoelastic buckling of rotationally restrained shallow arch.

1 INTRODUCTION

The thermoelastic analyses of the straight members have been presented in a number of books and papers and the thermoelastic behaviour of restrained straight members is well understood (Bradford 2006a, Bažant & Cedolin 2003, Gere & Timoshenko 1991, Gere 2002, Nowinski 1978). When a straight member is subjected to a uniform temperature field and the axial displacement of its ends is prevented, the uniform thermal expansion tends to produce uniform thermal axial stresses in the member without axial mechanical strains as shown by Gere & Timoshenko (1991) and Gere (2002).

An unrestrained slender arch that is uniformly heated expands freely and its ends tend to move outwards and to rotate. When the ends of the arch are prevented from motion in the horizontal and vertical directions, the thermal expansion will produce reactive forces at both ends of the arch. For equilibrium, the horizontal reactive forces at both ends need to be equal and opposite to each other. If the ends of the arch are also restrained from rotation, the thermal expansion may also produce reactive moments at the ends. To satisfy the static equilibrium, the reactive moments at both ends also need to be equal and opposite to each other. These end restraining reactions will produce axial compressive and bending actions in the arch, which vary along the arch axis and in turn produce compressive stresses in the arch in many cases. On the other hand, with an increase of the uniform temperature, an end-restrained arch tends to displace upwards and the length of the arch tends to increase. As a result, tensile strains may be produced in the arch. This forms a special case that compressive stresses in the arch produced by a uniform temperature field may correspond to tensile strains. The thermal deformations, strains and stresses produced by a uniform temperature field play important roles in comprehensive serviceability limit state, strength limit state, and stability limit state designs of tubular steel arches under a uniform temperature field. Hence, an exact thermoelastic analysis is much needed in order to predict accurate thermal deformations, strains and stresses. However, very few studies of the thermoelastic analysis of arches appear to have been reported.

Although the axial compressive force produced by the uniform heating varies along the arch axis, an assumption that the axial compressive force in the arch produced by the uniform heating is uniform along the arch axis was sometimes used in the thermoelastic analysis for arches (Bradford 2006b, 2006c). This assumption greatly simplifies the thermoelastic analysis of arches, but needs to be justified and the accuracy of the results based on this assumption is also need to be investigated.

It is known that the thermal axial compressive action in a straight member, which is subjected to a uniform temperature field, increases with an increase of the temperature and when the temperature reaches a certain critical value, the corresponding thermal

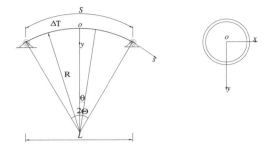

(a) Rotationally restrained arch under a uniform temperature field ΔT
(b) Typical tubular section

Figure 1. Rotationally restrained arch.

compressive action may cause the straight member to buckle suddenly out of the direction of the expansion in a bifurcation mode (Bažant & Cedolin 2003, Nowinski 1978). The arch under a uniform temperature field may be considered to have similar buckling behaviour. However, there are differences between the deformation behaviour between an arch and a straight member under a uniform heating. An arch under a uniform heating will primarily deform in both the radial (transverse) and axial directions, while primary deformations of a straight member are only in the axial direction. In addition, the primary radial displacements of an arch due to uniform heating are symmetrical. As the temperature increases, the symmetric radial displacements continue to increase. At a certain critical temperature, the symmetric radial displacements of an arch may bifurcate into antisymmetric displacements which lead to antisymmetric buckling of the arch. Bradford (2006b, 2006c) used a nonlinear analytical method to investigate the nonlinear buckling of an axial-elastically restrained shallow arch under a uniform temperature field and found that the nonlinear thermoelastic buckling of the shallow arch cannot occur under the uniform temperature field.

In practice, tubular steel arches are often supported by other structural members which provide elastic types of restraint at the arch ends, which participate in the structural response of the arch and which may influence significantly the in-plane thermoelastic behaviour and possible buckling of the arch.

The purpose of this paper is to present a systematic treatment of the thermoelastic analysis for rotationally restrained tubular steel circular arches that are subjected to a uniform temperature field (Figure 1). A virtual work method is used to derive differential equations of equilibrium for an exact thermoelastic analysis of arches under a uniform temperature field. An approximate thermoelastic analysis is also presented. Furthermore, classical buckling theory is used to investigate the in-plane bifurcation buckling of the arches under a uniform temperature field.

2 BASIC ASSUMPTIONS

The following assumptions are used in this investigation: 1. Deformations of the arch are elastic and satisfy the Euler-Bernoulli hypothesis, i.e. the cross-section remains plane and perpendicular to the arch axis during deformation. 2. The states of deformation and temperature are treated as time-independent, i.e. the derivative of the temperature T with respect to the time t vanishes $dT/dt = 0$ and so this separates the analysis of the temperature field from that of the displacement field and makes the problem uncoupled. 3. The temperature field $\Delta T = (T - 20\ °C)$ is uniform over the entire arch, i.e. the temperature gradient also vanishes $\nabla(\Delta T) = 0$, where T is the uniform temperature and the ambient temperature is assumed to be 20 °C. 4. Because the thermoelastic analysis of slender arches is carried out with the same degree of rigour as that accepted in the theory of elasticity, expansions of the cross-sections are assumed to be so small that they can be disregarded in the analysis (Gere 2002, Gere & Timoshenko 1991, Nowinski, 1978, Timoshenko & Goodire 1970).

3 EXACT ANALYSIS

3.1 Strains and stresses

The strain at an arbitrary point y of the cross-section of an arch is given by (Pi et al. 2002, 2007).

$$\epsilon = \tilde{w}' - \tilde{v} - y(\tilde{v}'' + \tilde{w}')/R \qquad (1)$$

where $()' \equiv d()/d\theta, ()'' \equiv d^2()/d\theta^2$, μ is the angular coordinate (Figure 1), the dimensionless displacements $\tilde{v}$ and $\tilde{w}$ are defined by $\tilde{v} \equiv v/R$ and $\tilde{w} \equiv w/R$, v and w are the radial and axial displacements, y is the coordinate of the point in the axis oy, and R is the radius of the arch.

From the Duhamel-Neumann equation (Nowinski 1978), the stress produced by the uniform temperature field ΔT can be expressed as

$$\sigma = E(\epsilon - \alpha \Delta T) \qquad (2)$$

where E is the Young's modulus at the temperature T.

3.2 Differential equations of equilibrium

The differential equations of equilibrium for the thermoelastic analysis of an arch that is subjected to a uniform temperature field ΔT can be derived by using the principle of virtual work, which requires

$$\delta \Pi = \int_V \sigma \delta \epsilon dV + \sum_{i=\pm\Theta} k_\phi \tilde{v}'_i \, \delta \tilde{v}'_i = 0, \quad (3)$$

for all kinematically admissible infinitesimal variations of deformations $\delta \tilde{v}, \delta \tilde{v}', \delta \tilde{v}'', \delta \tilde{w}, \delta \tilde{w}'$, where V indicates the volume of the arch, $\delta()$ denotes the Lagrange operator of simultaneous variation, k_ϕ is the stiffness of rotational elastic restraints.

By substituting Equations 1 and 2, the statement of the principle of virtual work given by Equation 3 can be rewritten as

$$\delta \Pi = \int_{-\Theta}^{\Theta} [NR(\delta\tilde{w}' - \delta\tilde{v}) - M(\delta\tilde{v}'' + \delta\tilde{w}')]d\theta \\ + \sum_{i=\pm\Theta} k_\phi \tilde{v}'_i \, \delta \tilde{v}'_i = 0, \quad (4)$$

where the axial force N is given by

$$N = \int_A \sigma dA = AE(\tilde{w}' - \tilde{v} - \alpha\Delta T), \quad (5)$$

and the bending moment M is given by

$$M = \int_A \sigma y dA = -EI_x(\tilde{v}'' + \tilde{w}')/R, \quad (6)$$

in which Θ is the half included angle of the arch (Figure 1), A is the area of the cross-section and I_x is the second moment of area of the cross-section about its major principal axis.

Integrating Equation 4 by parts and substituting Equations 5 and 6 leads to the differential equations of equilibrium as

$$(\tilde{v}^{iv} + \tilde{w}''') - R^2(\tilde{w}' - \tilde{v} - \alpha\Delta T)/r_x^2 = 0 \quad (7)$$

for the radial direction, and

$$(\tilde{v}''' + \tilde{w}'') + R^2(\tilde{w}'' - \tilde{v}')/r_x^2 = 0 \quad (8)$$

for the axial direction; and to the static boundary conditions as

$$(\tilde{v}'' + \tilde{w}') \pm \gamma_\phi \tilde{v}' = 0 \text{ at } \theta = \pm\Theta, \quad (9)$$

where $r_x \equiv \sqrt{I_x/A}$ and the dimensionless elastic stiffness γ_ϕ of the rotational restraints is defined by $\gamma_\phi \equiv k_\phi R/EI_x$.

The geometric boundary condition is given by

$$\tilde{v} = 0, \text{ at } \theta = \pm\Theta. \quad (10)$$

3.3 Solutions

Solutions for $\tilde{v}$ and $\tilde{w}$ can be obtained by solving Equations 7 and 8 simultaneously and using the boundary conditions given by Equations 9 and 10 as

$$\tilde{v} = \alpha\Delta T(R^2 + r_x^2)(1 + \gamma_\phi\Theta)\Xi \times \\ \{[\theta\sin\theta S + (\Theta C + S)\cos\theta - (CS + \Theta)]\} \quad (11) \\ + 2\alpha\Delta T R^2(\cos\theta - C)(\Theta C - S)\Xi,$$

and

$$\tilde{w} = \alpha\Delta T(R^2 + r_x^2)(1 + \gamma_\phi\Theta)(\Theta\sin\theta C - \theta\cos\theta S)\Xi \\ + 2\alpha\Delta T R^2(\Theta\sin\theta - \theta S)(C + \gamma_\phi S)\Xi, \quad (12)$$

where $C \equiv \cos\Theta$, $S \equiv \sin\Theta$, and the constant $\Xi \equiv [(1 + \gamma_\phi\Theta)\Gamma + \Phi]^{-1}$ with $\Gamma \equiv (R^2 + r_x^2)(CS + \Theta)$ and $\Phi \equiv 2R^2(\Theta C^2 - 2CS - \gamma_\phi S^2)$. The solutions given by Equations 11 and 12 reduce to those for pin-ended arches when $\gamma_\phi = 0$, and for fixed arches when $\gamma_\phi = \infty$.

Substituting the solutions for $\tilde{v}$ and $\tilde{w}$ given by Equations 11 and 12 into the strain expression given by Equation 1 leads to the strains ϵ at a point y of the cross-section μ as

$$\epsilon = \alpha\Delta T - 2\alpha\Delta T\Xi S \\ \times [(yR + r_x^2)(1 + \gamma_\phi S)\cos\theta - yR(\gamma_\phi\Theta + C)]. \quad (13)$$

Variations of the strains ϵ at the centroid ($y = 0$), bottom fibre ($y = h/2$), and top fibre ($y = h/2$) of the arch crown ($\theta = 0°$) with the included angle 2Θ given by Equation 13 are shown in Figures 2–4 for pin-ended ($\gamma_\phi = 0$), rotationally restrained ($\gamma_\phi = 15$) and fixed arches ($\gamma_\phi = \infty$). It can be seen that the centroid and top fibre have tensile strains, and that in most cases, the bottom fibre also has tensile strains. It is only for shallow arches with a small included angle that the bottom fibre has compressive strains. It can also be seen that effects of the stiffness of the rotational restraints on the strains are significant.

From the Duhamel-Neumann equation given by Equation 2, the stresses at the point y of the cross-section θ can be obtained as

$$\sigma = E(\epsilon - \alpha T) = -2E\alpha\Delta T\Xi S \times \\ [(yR + r_x^2)(1 + \gamma_\phi\Theta)\cos\theta - yR(\gamma_\phi S + C)]. \quad (14)$$

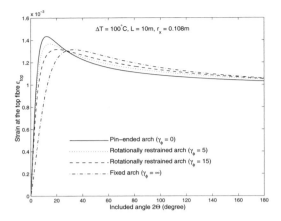

Figure 2. Strains at top fibres of the arch crown.

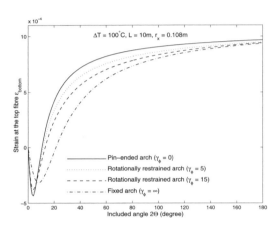

Figure 3. Strains at bottom fibres of the arch crown.

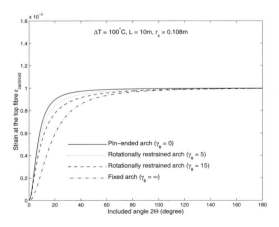

Figure 4. Strains at centroids of the arch crown.

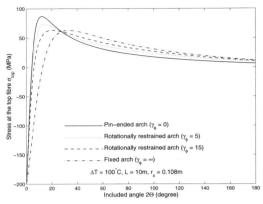

Figure 5. Stresses at top fibres of the arch crown.

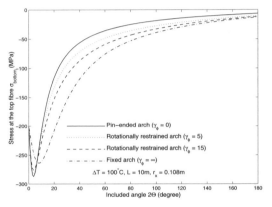

Figure 6. Stresses at bottom fibres of the arch crown.

Variations of the stresses σ at the centroid, bottom fibre, and top fibre of the arch crown ($\theta = 0°$) with the included angle 2Θ given by Equation 14 are shown in Figures 5–7 for pin-ended, rotationally restrained and fixed arches. Generally, it can be seen from Figures 5–7 that the stress at the centroid is compressive for shallow arches and is nearly equal to zero for deep arches. However, the corresponding strain at the centroid (membrane strain) is tensile as shown in Figures 2–4. This forms a special case for which thermal tensile membrane strains are associated with compressive or near-zero membrane stresses. The stresses at the bottom fibres (Figure 6) of very shallow arches are very high under the temperature $\Delta T = 100$ °C and may reach the yield stress of the steel already. It can be seen that effects of the stiffness of the rotational restraints on the stresses are significant. It can also be seen from Figures 2–7 that the strains and stresses produced by the uniform temperature field in shallow arches are much higher that those in deep arches.

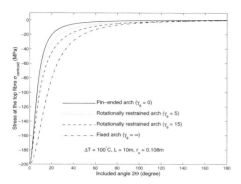

Figure 7. Stresses at centroids of the arch crown.

Substituting Equation 14 into Equations 5 and 6 leads to the axial force N and the bending moment M along the arch axis as

$$N = -2EI_x \alpha \Delta T (1 + \gamma_\phi \Theta) S \cos\theta \, \Xi. \quad (15)$$

and

$$M = EI_x \alpha \Delta TL[\gamma_\phi S + C = (1 + \gamma_\phi \Theta)\cos\theta]\Xi. \quad (16)$$

Equations 15 and 16 show that both the bending moment M and the axial compressive force N vary along the arch length.

The reaction Q_s in the axial direction at both ends of the arch ($\theta = \pm\Theta$) can be obtained directly from Equation 15 as

$$Q_s = \mp 2EI_x \alpha \Delta T (1 + \gamma_\phi \Theta) CS\Xi, \quad (17)$$

and the radial reaction Q_y at $\theta = \mp\Theta$ can be obtained from the bending moment M given by Equation 16 as

$$Q_y = M'/R = \mp 2EI_x \alpha \Delta T (1 + \gamma_\phi \Theta) S^2 \Xi. \quad (18)$$

The horizontal and vertical reactive resultants are then calculated from the axial and radial reactions Q_s and Q_y as

$$H = Q_s C + Q_y S$$
$$= \pm 2EI_x \alpha \Delta T (1 + \gamma_\phi \Theta) S\Xi \text{ at } \theta = \mp\Theta, \quad (19)$$

and

$$V = Q_s S - Q_y C = 0 \text{ at } \theta = \mp\Theta, \quad (20)$$

which indicates that the horizontal reactions are equal and opposite to each other and that there are no vertical reactions at both ends of an arch under a uniform temperature field.

4 APPROXIMATE ANALYSIS

The effects of the axial deformations on the curvature changes are very small and so can be ignored in the analysis. In this case, the strain given by Equation 1 is simplified to

$$\epsilon = \tilde{w}' - \tilde{v} - y\tilde{v}''/R. \quad (21)$$

An approximate analysis can be developed by substituting the simplified strain given by Equation 21 into the virtual work statement given by Equation 3. In this case, the principle of virtual work can be restated as

$$\int_{-\Theta}^{\Theta} [NR(\delta\tilde{w}' - \delta\tilde{v}) - M\delta\tilde{v}''] d\theta + \sum_{i=\pm\Theta} k_\phi \tilde{v}_i' \delta\tilde{v}_i' = 0. \quad (22)$$

Integrating Equation 22 by parts leads to the differential equations of equilibrium as

$$-\tilde{v}^{iv} + R^2(\tilde{w}' - \tilde{v} - \alpha\Delta T)/r_x^2 = 0 \quad (23)$$

for the radial direction, and

$$\tilde{w}'' - \tilde{v}' = 0 \quad (24)$$

for the axial direction; and to the static rotational boundary conditions as

$$\tilde{v}'' \pm \gamma_\phi \tilde{v}' = 0 \text{ at } \theta = \pm\Theta. \quad (25)$$

The geometric boundary conditions are the same as those given by Equation 10.

The solutions for $\tilde{v}$ and $\tilde{w}$ can be obtained by solving Equations 23 and 24 simultaneously and using the boundary conditions given by Equations 10 and 25 as

$$\tilde{v} = 15R^2 \alpha \Delta T \Theta (\Theta^2 - \theta^2) \Gamma_s [(\theta^2 - \Theta^2) \times (1 + \gamma_\phi \Theta) - 4\Theta^2]/8, \quad (26)$$

and

$$\tilde{w} = -\alpha\Delta T \theta R^2 \Gamma_s [(1 + \gamma_\phi \Theta)\Theta(\theta^2 - \Theta^2) \times (3\theta^2 - 7\Theta^2) + 20\Theta^3(\Theta^2 - \theta^2)]/8, \quad (27)$$

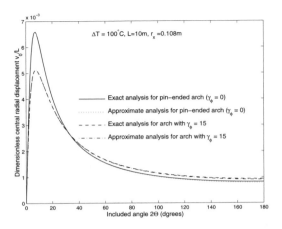

Figure 8.   Comparison for central radial displacements.

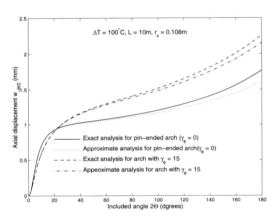

Figure 9. Comparison for axial displacements at a quarter arch length.

with $\Gamma_s = [45\Theta r_x^2(1+\gamma_\phi\Theta) + R^2\Theta^5(6+\gamma_\phi\Theta)]^{-1}$.

Approximate solutions for the radial displacement $\tilde{v}$ and axial displacement $\tilde{w}$ given by Equations 26 and 27 are compared with the exact solutions given Equations 11 and 12 in Figures 8–9. It can be seen that the approximate solution for $\tilde{v}$ almost coincides with the exact solution while the approximate solution for $\tilde{w}$ agrees well with the exact solution for shallow arches. There are small differences between the approximate and exact displacements $\tilde{w}$ for deep arches.

Substituting the approximate solutions for $\tilde{v}$ and $\tilde{w}$ into Equation 21 leads to the approximate strains as

$$\epsilon = \alpha\Delta T - 15\alpha\Delta T\Theta\{6(1+\gamma_\phi\Theta)r_x^2 \\ + [\Theta^2(3+\gamma_\phi\Theta) - 3\theta^2(1+\gamma_\phi\Theta)]yR\}\Gamma_s/2, \quad (28)$$

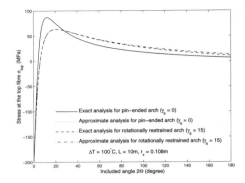

Figure 10. Comparison of approximate and exact stresses at top fibres of arch crown.

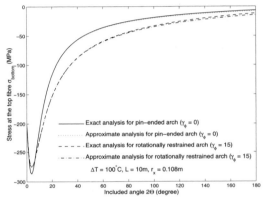

Figure 11. Comparison of approximate and exact stresses at bottom fibres of the arch crown.

and then the approximate stresses can be obtained by substituting Equation 28 into the Duhamel-Neumann equation given by Equation 2 as

$$\sigma = -15E\alpha\Delta T\Theta\{6(1+\gamma_\phi\Theta)r_x^2 \\ + [\Theta^2(3+\gamma_\phi\Theta) - 3\theta^2(1+\gamma_\phi\Theta)]yR\}\Gamma_s/2. \quad (29)$$

Approximate stresses given by Equation 29 are compared in Figures 10–11 with the exact solutions given by Equation 14. It can be seen that the approximate stresses are almost the same as the exact solutions for shallow arches, and very close to the exact solutions for deep arches.

The approximate axial compressive force N and the bending moment M are then obtained by substituting the stresses given by Equation 29 into Equations 5 and 6 as

$$N = -45EI_x\alpha\Delta T\Theta(1+\gamma_\phi\Theta)\Gamma_s, \quad (30)$$

and

$$M = 15EI_x \alpha \Delta T \Theta R \times [(3\theta^2 - \Theta^2) \times (1 + \gamma_\phi \Theta) - 2\Theta^2] \Gamma_s / 2. \quad (31)$$

It can be seen from Equations 30 and 31 that in the approximate analysis, the axial compressive force N is a constant while the bending moments vary along the arch axis.

The axial reaction Q_s and the radial reaction Q_y at $\theta = \pm \Theta$ can be obtained from Equations 30 and 31 as

$$Q_s = \mp 45 EI_x \alpha \Delta T \Theta (1 + \gamma_\phi \Theta) \Gamma_s, \quad (32)$$

and

$$Q_y = -M'/R = \mp 45 EI_x \alpha \Delta T \Theta^2 (1 + \gamma_\phi \Theta) \Gamma_s. \quad (33)$$

In the same way as for the exact analysis, the horizontal and vertical reactions H and V at $\theta = \pm \Theta$ can be obtained as

$$H = \mp 45 EI_x \alpha \Delta T (1 + \gamma_\phi \Theta)(C + \Theta S) \Gamma_s, \quad (34)$$

and

$$V = \pm 45 EI_x \alpha \Delta T (1 + \gamma_\phi \Theta)(\Theta C - S) \Gamma_s, \quad (35)$$

which indicates that the approximate analysis produces incorrect vertical reactions, which should vanish according to the exact analysis.

5 BIFURCATION BUCKLING OF SHALLOW ARCHES

From the previous analysis, the thermal internal actions N and M increase with an increase of the uniform temperature field ΔT. When ΔT reaches a certain critical value, the corresponding thermal internal actions may cause the arch to buckle suddenly in a bifurcation mode. Bifurcation buckling is an adjacent deformation which is characterized by the fact that, as the thermal internal actions reach their critical values, the arch passes from its prebuckling equilibrium configuration $(N, M, \Delta T, \tilde{v}, \tilde{w})$ to an infinitesimally close buckled equilibrium configuration defined by $(N, M, \Delta T, \tilde{v} + \tilde{v}_b, \tilde{w} + \tilde{w}b)$ under the constant thermal actions N and M, and constant temperature ΔT, where $\tilde{w}_b$ and $\tilde{v}b$ are the buckling displacements. Because both the thermal expansion $\alpha \Delta T$ and internal stress resultant N are constant during the bifurcation buckling, $N = AE[\tilde{w}' + \tilde{w}'_b - (\tilde{v} + \tilde{v}_b) - \alpha \Delta T]$ in the buckled configuration remains the same as $N = AE[\tilde{w}' - \tilde{v} - \alpha \Delta T]$ in the prebuckling configuration. This leads to

$$\tilde{w}'_b - \tilde{v}_b = 0 \quad (36)$$

which indicates that the axial buckling deformation vanishes and this is the widely used axial inextensibility condition for bifurcation buckling of an arch (Pi et al. 2002, Timoshenko and Gere 1961).

To perform a buckling analysis, second order terms $(\tilde{v}' + \tilde{w})^2 / 2$ need to be included in the strain given by Equation 1 as

$$\epsilon = \tilde{w}' - \tilde{v} + (\tilde{v}' + \tilde{w})^2 / 2 - y(\tilde{v}'' + \tilde{w}')/R \quad (37)$$

Applying the principle of virtual work to the equilibrium at both the prebuckling configuration and the buckled configuration, substituting the nonlinear strain given by Equation 37 and the bending moment M given by Equation 6, and using the axial inextensibility condition given by Equation 36 lead to the virtual work statement for buckling equilibrium as

$$\sum_{\theta = \pm \Theta} k_\phi \tilde{v}'_b \delta \tilde{v}'_b + \int_{-\Theta}^{\Theta} [NR(\tilde{v}'_b + \tilde{w}_b) \delta \tilde{v}'_b + EI_x (\tilde{v}''_b + \tilde{v}_b)(\delta \tilde{v}''_b + \delta \tilde{v}_b)/R] d\theta = 0. \quad (38)$$

Integrating Equation 38 by parts leads to the differential equation of equilibrium for the buckling deformation as

$$EI_x (\tilde{v}_b^{iv} + 2\tilde{v}''_b + \tilde{v}_b)/R - [NR(\tilde{v}'_b + \tilde{w}_b)]' = 0 \quad (39)$$

and to the static boundary condition as

$$(\tilde{v}''_b + \tilde{v}_b) \pm \gamma_\phi \tilde{v}'_b = 0, \quad \text{at } \theta = \pm \Theta. \quad (40)$$

The geometric boundary condition is given by

$$\tilde{v}_b = 0, \quad \text{at } \theta = \pm \Theta. \quad (41)$$

Because the exact axial compressive forces N vary along the arch axis, numerical methods need to be used to solve Equation 39 for bifurcation buckling. However, from the approximate analysis, the axial compressive forces N in shallow arches can be assumed to be uniform along the arch axis, and then the differential

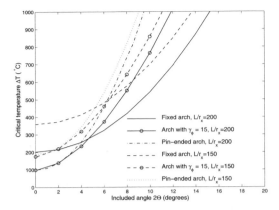

Figure 12. Bifurcation buckling of shallow arches.

equation of equilibrium for buckling deformation given by Equation 39 can be simplified to

$$\tilde{v}_b^{iv} + (1+\mu^2)\tilde{v}_b'' + \mu^2 \tilde{v}_b = 0, \quad (42)$$

where the parameter μ is defined as

$$\mu^2 = 1 - NR^2/(EI_x) \quad (43)$$

The solution of Equation 42 is given by

$$\tilde{v}_b = c_1 \sin\theta + c_2 \cos\theta + c_3 \sin(\mu\theta) + c_4 \cos(\mu\theta) \quad (44)$$

Substituting this solution into the boundary conditions given by Equations 40 and 41 leads to a group of four homogeneous algebraic equations for c_1–c_4. The existence of nontrivial solutions for c_1–c_4 requires that the determinant of the coefficient matrix of the group of equations vanishes, which leads to the characteristic equations as

$$\frac{\tan(\mu\Theta)}{\mu\Theta}[(\mu^2-1)\tan\Theta + \gamma_\phi] - \gamma_\phi \frac{\tan\Theta}{\Theta} = 0 \quad (45)$$

and

$$\frac{\cot(\mu\Theta)}{\mu\Theta}[\gamma_\phi - (\mu^2-1)\cot\Theta] - \gamma_\phi \frac{\cot\Theta}{\Theta} = 0 \quad (46)$$

The solution of Equation 45 corresponds to antisymmetric bifurcation buckling, in which case the buckling deflections $\tilde{v}_b$ upward tend to extend the arch while those downward tend to shorten the arch, and one compensates for the other and so the axial inextensibility condition for bifurcation buckling given by Equation 36 is satisfied. The lowest solution of Equation 45 can be obtained as

$$\mu\Theta = \eta\pi. \quad (47)$$

The antisymmetric bifurcation buckling load for shallow arches can be obtained by substituting this solution into Equation 43 as

$$N_{cr} = -EI_x[(\eta\pi)^2 - \Theta^2]/(S/2)^2. \quad (48)$$

For shallow arches, the critical temperature ΔT_{cr} can be obtained by substituting the axial compressive force given by Equation 30 into Equation 48 as

$$\Delta T_{cr} = \beta[(\eta\pi)^2 - \Theta^2], \quad (49)$$

where the factor β is given by

$$\beta = \frac{1}{\alpha}\{[\Theta^2(6+\gamma_\phi\Theta)]/[45(1+\gamma_\phi\Theta)] + 1/\lambda^2\} \quad (50)$$

and $\lambda = S/(2r_x)$ is the slenderness of an arch.

Typical variations of the critical temperature ΔT_{cr} for shallow arches given by Equation 49 with the included angle 2Θ are shown in Figure 12. It can be seen that as the included angle 2Θ increases, the critical temperature ΔT_{cr} increases and becomes very high rapidly. However, from Figures 5–7, the stresses in an arch may reach the yield stress of the steel at a temperature much lower than the critical temperature for elastic bifurcation buckling. And from Figure 8, the central radial deflection of the arch may also be excessive at a temperature much lower than the critical temperature. Hence, in practice, thermoelastic bifurcation buckling is possible only for very shallow arches, which is consistent with that found by Bradford (2006b). Nevertheless, the critical temperature for thermoelastic buckling of an arch is an important reference temperature and so is useful for the comprehensive analysis and design of an arch. It can also be seen from Figure 12 that effects of the stiffness of the rotational restraints on the bifurcation buckling are significant.

When Equation 46 is satisfied, mathematically, the symmetric solutions $\mu\Theta$ and subsequently the critical temperature ΔT_{cr} can be obtained. However, the symmetric solutions do not satisfy the axial inextensibility condition Equation 36 for bifurcation buckling. Consequently, the symmetric solutions obtained from Equation 46 do not represent bifurcation buckling modes of an arch.

6 CONCLUSIONS

A systematic treatment of the thermoelastic analysis for rotationally restrained tubular steel circular arches that are subjected to a uniform temperature field was presented. It was found that the strains and stresses in the arch under a uniform temperature field form a special case that tensile membrane strains correspond to compressive or near-zero membrane stresses. It was also found that the strains and stresses produced by the uniform temperature field in shallow arches are much higher than those in deep arches. Under a uniform temperature field, the stresses in shallow arches may reach the yield stress of the steel and the central radial deflection of shallow arches may be excessive. Hence, the thermal effect of the uniform temperature field needs to be allowed for in a comprehensive arch design, and ignoring the thermal effects on the structural behaviour of an arch may lead to a premature and unexpected failure caused by thermal expansion.

An approximate analysis for the thermoelastic behaviour of arches was also presented and compared with the results of the exact analysis. It was found that the approximate analysis provides simple and accurate solutions. The approximate analysis implies an assumption that the axial compressive forces can be treated as uniform along the arch axis, and this assumption can facilitate classical closed-form solutions for thermoelastic bifurcation buckling of shallow arches under a uniform temperature field. It was shown by the classical bifurcation buckling analysis that in practice, thermoelastic bifurcation buckling is possible only for very shallow arches.

It has been found that effects of the stiffness of the rotational restraints on the deformations, stresses, internal actions, and on the bifurcation buckling of arches under a uniform temperature field are significant.

7 ACKNOWLEDGEMENTS

This work has been supported by the Australian Research Council through a Discovery Project and a Federation Fellowship awarded to the second author.

8 REFERENCES

Bažant, Z.P. & Cedolin L. 2003. *Stability of structures*, Mine-ola, New York, Dover Publication.

Bradford, M.A. 2006a. Elastic analysis of straight members at elevated temperature. Advances in Structural Engineering, 9(5):611–618.

Bradford, M.A. 2006b. In-plane nonlinear behaviour of circular pinned arches with elastic restraints under thermal loading, *International Journal of Structural Stability and Dynamics* 6(2): 163–177.

Bradford, M.A. 2006c. Buckling of tubular steel arches subjected to fire loading. *11th International Symposium on Tubular Structures*, Quebec City, Canada, 433–438.

Gere, J.M. & Timoshenko, S.P. 1991. Mechanics of materials, 3rd SI Edition. Chapman and Hall, London.

Gere, J.M. 2002. Mechanics of materials, 5th Edition. Cheltenham Eng., Nelson Thorns, New York.

Nowinski, J.L. 1978. Theory of thermoelasticity with applications. Sijthoff & Noordhoff International Publishers, Alphen aan den Rijn, The Netherlands.

Pi, Y.-L., Bradford, M.A. & Uy, B. 2002. In-plane stability of arches. International Journal of Solids and Structures 39(1): 105–125.

Pi, Y.-L., Bradford, M.A. & Tin-Loi, F. 2007. Nonlinear analysis and buckling of elastically supported circular shallow arches. International Journal of Solids and Structures 44(7–8): 2401–2423.

Timoshenko, S.P. & Gere, J.M. 1961. Theory of elastic stability. McGraw-Hill, New York.

Timoshenko, S.P. & Goodier, J.N. 1970. Theory of elasticity, 3rd Edition. McGraw-Hill, London.

Author index

Anđelić, M. 67
Azuma, K. 127, 145

Bahr, O. 461
Baig, M.N. 407
Benčat, J. 243
Boissonnade, N. 331
Bonet, J.L. 103
Borges, L. 135, 351
Bradford, M.A. 659
Bursi, O.S. 431

Caldas, R.B. 635
Cao, G.F. 183
Cao, Y. 209
Chan, T.M. 323, 535
Chen, B.C. 43, 563
Chen, Y.Y. 193, 235, 305, 389, 521
Chiew, S.P. 119, 375
Chin, C.S. 515
Choo, Y.S. 271, 579
Cibulka, M. 243

D'Alambert, F.C. 37
de Oliveira, J.C. 451
Deng, H.Z. 615
Ding, J.M. 61
Ding, J. 641
Dong, P. 153
Duan, L. 413

Elghazouli, A.Y. 219
Euler, M. 381

Fakury, R.H. 635
Falah, N. 75
Fan, Z. 21
Fan, X.W. 21
Fan, J.S. 407
Fernandez Lacruz, E. 491
Ferrario, F. 431
Filippou, F.C. 103
Fleischer, O. 293, 597

Gad, E.F. 421
Gao, J. 563

Gardner, L. 219, 323, 535
Gho, W.M. 607
Goldsworthy, H.M. 421
Gu, W. 111
Gu, M. 359

Han, L.H. 85
He, Z.J. 61
He, M.J. 545
Hemerich, E. 67
Herion, S. 499, 597
Hernández-Figueirido, D. 103
Hrvol, M. 243
Huang, Y. 615

Iwashita, T. 127, 145

Jaspart, J.P. 331
Jiang, Y. 529
Jiao, H. 571
Jin, X.H. 615
Joost, C. 479

Kaneko, T. 249
Kieboom, S.J.C. 487
Krampen, J. 507
Kuhlmann, U. 381
Kurobane, Y. 127, 145

Lazarević, D. 67
Lee, M.J. 93
Lee, E.T. 93
Lee, C.K. 119
Li, B.R. 413
Liang, F. 545
Lie, S.T. 119, 367, 375
Liedtke, M. 507
Lin, X.G. 359
Liu, W. 85
Liu, C.B. 389
Lv, X.D. 235

Ma, R.L. 545
Maddox, S.J. 341
Makino, Y. 281
Marshall, P.W. 281
Martinez-Saucedo, G. 227
Mashiri, F.R. 153, 163

Mundle, T.O. 479
Murphy, D. 51

Nguyen, T.B.N. 119
Nie, J.G. 407
Nip, K.H. 219
Nussbaumer, A. 135, 351

Oh, Y.S. 93
Orton, A.H. 201, 625
Orton, A. 441, 641

Packer, J.A. 173, 227, 261, 313
Peng, Y. 21
Pi, Y.L. 659
Portoles, J.M. 103
Pucinotti, R. 431
Puthli, R. 3, 293, 499, 597

Qian, X. 271
Qian, X.D. 579
Qiu, G.Z. 589

Romero, M.L. 103

Sasaki, S. 249
Scheller, W. 507
Shao, Y.B. 375
Shen, Z.Y. 305
Shen, B. 389
Shi, W.Z. 469
Shi, Y.J. 529
Šošić, I. 67
Sousa Jr, J.B.M. 635
Stacey, A. 341
Starossek, U. 75
Sun, H.L. 21
Sun, G.S. 111
Sun, C.Q. 389
Sun, P. 529

Takenaka, H. 249
Tanaka, N. 249
Tao, Z. 399
Tong, L.W. 359, 389, 521
Tong, F.M. 515
Tremblay, R. 227

Uy, B. 399

van der Vegte, G.J. 261, 271, 281, 313, 579
Venuto, M. 51
Veselcic, M. 499
Voth, A.P. 173

Wang, Z. 21
Wang, G.N. 193
Wang, W. 235, 305
Wang, Y. 235
Wang, C.S. 413
Wang, Y.C. 441, 625, 641

Wang, F. 515
Wang, Y.Q. 529
Wardenier, J. 261, 271, 281, 313, 579
Weynand, K. 331
Wilkinson, T. 209
Wong, M.B. 651
Wu, H.L. 61

Xiao, R.Y. 515
Xu, Y.J. 235

Yan, P.Y. 441
Yang, Y.F. 85
Yao, N.L. 183

Yao, H. 421
Ye, Q. 51
Young, B. 553

Zhai, X.L. 413
Zhang, Y.H. 341
Zhang, B.F. 367
Zhao, Y.H. 111
Zhao, X.L. 153, 163, 261, 359, 389, 571
Zhao, X.Z. 193, 521
Zhao, X.-L. 313
Zhao, J.C. 589
Zhu, J.H. 553